Lecture Notes in Computer Science 1563

Edited by G. Goos, J. Hartmanis and J. van Leeuwen

Springer

Berlin
Heidelberg
New York
Barcelona
Hong Kong
London
Milan
Paris
Singapore
Tokyo

Christoph Meinel   Sophie Tison (Eds.)

STACS 99

16th Annual Symposium
on Theoretical Aspects of Computer Science
Trier, Germany, March 4-6, 1999
Proceedings

Springer

Series Editors

Gerhard Goos, Karlsruhe University, Germany
Juris Hartmanis, Cornell University, NY, USA
Jan van Leeuwen, Utrecht University, The Netherlands

Volume Editors

Christoph Meinel
FB IV – Informatik, Universität Trier
D-54286 Trier, Germany
E-mail: meinel@ti.uni-trier.de

Sophie Tison
LIFL, Université de Lille I, Bâtiment 3
F-59655 Villeneuve d'Ascq Cedex, France
E-mail: tison@lifl.fr

Cataloging-in-Publication data applied for

Die Deutsche Bibliothek - CIP-Einheitsaufnahme

STACS <16, 1999, Trier>:
Proceedings / STACS 99 / 16th Annual Symposium on Theoretical Aspects of
Computer Science, Trier, Germany, March 4 - 6, 1999. Christoph Meinel ; Sophie
Tison (ed.). - Berlin ; Heidelberg ; New York ; Barcelona ; Hong Kong ; London
; Milan ; Paris ; Singapore ; Tokyo : Springer, 1999
(Lecture notes in computer science ; Vol. 1563)
ISBN 3-540-65691-X

CR Subject Classification (1998): F, D.1, D.3, G.1-2, I.3.5, E.1

ISSN 0302-9743
ISBN 3-540-65691-X Springer-Verlag Berlin Heidelberg New York

© Springer-Verlag Berlin Heidelberg 1999
Printed in Germany

Typesetting: Camera-ready by author
SPIN: 10702947   06/3142 – 5 4 3 2 1 0   Printed on acid-free paper

Preface

The Symposium on Theoretical Aspects of Computer Science (STACS) is held annually, alternating between France and Germany. The current volume constitutes the proceedings of the 16th STACS conference, organized jointly by the Special Interest Group for Theoretical Computer Science of the Gesellschaft für Informatik (GI) in Germany, and Maison de l'Informatique et des Mathématiques Discrètes (MIMD) in France.

The conference took place in Trier – the oldest town in Germany, with more than 2 millennia of history. Previous symposia of the series were held in Paris (1984), Saarbrücken (1985), Orsay (1986), Passau (1987), Bordeaux (1988), Paderborn (1989), Rouen (1990), Hamburg (1991), Cachan (1992), Würzburg (1993), Caen (1994), München (1995), Grenoble (1996), Lübeck (1997), and Paris (1998). All proceedings of the series have been published in the Lecture Notes of Computer Science series of Springer-Verlag.

STACS has become one of the most important annual meetings in Europe for the theoretical computer science community. This time, altogether 300 authors from 36 countries on five continents submitted their papers. Each submission was sent to five members of the program committee for review. During the program committee session 51 out of the 146 submissions were accepted for presentation. In two of the selected papers the same result was proved independently. The program committee decided to include the final versions of both papers in the proceedings (Icking, Klein, Langetepe: *An Optimal Competitive Strategy for Walking in Streets* and Semrau, Schuierer: *An Optimal Strategy for Searching in Unknown Streets*) and to have the result presented in a joint talk during the conference. The program committee was impressed by the high scientific quality of the submissions as well as the broad spectrum of topics they covered within the area of theoretical computer science. In spite of this a number of good papers submitted had to be rejected due to general limitations of the conference.

The program committee consisted of S. Albers (Saarbrücken), R. Amadio (Marseille), R. Cori (Bordeaux), J. Esparza (München), J. Hromkovič (Aachen), C. Kenyon (Paris), J. Köbler (Ulm), D. Krizanc (Ottawa), Ch. Meinel (Trier, chair), A. Petit (Cachan), S. Rudich (Pittsburgh), J. Sgall (Praha), R. Silvestri (Roma), S. Tison (Lille, co-chair), and P. Widmayer (Zürich). We wish to thank all the members for their work in evaluating the significance and scientific merits of the submitted papers. The program committee was assisted by numerous reviewers from all over the world whose names are listed on the next pages. We would like to thank all these scientists for their efforts. Without their voluntary work it would have been impossible to organize such a high quality conference.

We also thank the three invited speakers of this meeting for accepting our invitation and sharing with us their insights on interesting developments in our area.

We acknowledge the support of the following institutions for this conference: Deutsche Forschungsgemeinschaft, European Commission, Institut für Telematik (Trier), and Universität Trier.

Finally, we would like to thank the members of the organizing committee consisting of J. Bern, C. Damm, Ch. Meinel, and M. Mundhenk.

Trier, February 1999   Sophie Tison and Christoph Meinel

List of Reviewers

Ahamd, M.
Allender, E.
Alt, H.
d'Amore, F.
Annexstein, F.
Arbab, F.
Arnold, A.
Arvind, V.
Asarin, E.
Augros, X.

Babel, L.
Badouel, E.
Bailleux, O.
Balcázar, J. L.
Bampis, E.
Baudron, O.
Beaudry, M.
Beauquier, D.
Beauquier, J.
Becker, B.
Beigel, R.
Bellare, M.
Bellegarde, F.
Benhamou, B.
Bérard, B.
Bernot, G.
Berstel, J.
Betrema, J.
Biehl, I.
Blum, A.
Blum, L.
Boasson, L.
Böckenhauer, H.-J.
Bodlaender, H. L.
Boissonnat, J. D.
Bond, J.
Bose, P.
Bouajjani, A.
Boufkhad, Y.

Bougé, L.
Boulet, P.
Bradfield, J. C.
Brandstädt, A.
Brieden, A.
Brodal, G. S.
Bucciarelli, A.
Buchsbaum, A. L.
Buhrman, H.
Burlet, M.

Cadoli, M.
Cai, J. Y.
Carton, O.
Cassaigne, J.
Caucal, D.
Cazals, F.
Cece, G.
Chang, R.
Cheong, O.
Choffrut, C.
Chretienne, P.
Clarkson, K.
Clementi, A.
Clerbout, M.
Comon, H.
Contejean, E.
Cosmo, R. D.
Courcelle, B.
Couvreur, J. M.
Creignou, N.
Crochemore, M.
Cusick, T.

Dam, M.
Damaschke, P.
Damm, C.
Darte, A.
DasGupta, B.
Dassow, J.

Dauchet, M.
De Agostino, S.
De Rougemont, M.
Delahaye, J. P.
Delmas, O.
Delzanno, G.
Denis, F.
Denise, A.
Di Cosmo, R.
Diekert, V.
Dietzfelbinger, M.
Donini, F. M.
Dowek, G.
Dubois, O.
Durand, B.
Duris, P.
Dutertre, B.

Eidenbenz, S.
Eilam, T.
Eiter, Th.
Engebretsen, L.
Engelen, B.
Engelfriet, J.
Epstein, D.
Esik, Z.

Fages, F.
Fellows, M.
Fernau, H.
Ferrand, G.
Ferreira, A.
Finkel, A.
Fleischer, R.
Formenti, E.
Fortnow, L.
Franciosa, P. G.
Fraysseix de, H.
Fribourg, L.
Frieze, A.

Furbach, U.

Gabow, H.
Galli, N.
Galmiche, D.
Ganzinger, H.
Gasarch, W.
Gavoille, C.
Gilleron, R.
Gilman, R.
Glasnak, V.
Glenn, J.
Goemans, M.
Goerdt, A.
Gore, R.
Gottlob, G.
Goubault-Larrecq, J.
Gröpl, C.
Grädel, E.
Grandjean, E.
Grigorieff, S.
Gross, T.
Guerrini, S.

Habib, M.
Habsieger, L.
Handke, D.
Hardin, T.
Harju, T.
Håstad, J.
Havel, I.
Heckmann, R.
Hertrampf, U.
Heun, V.
Hjaltason, G. R.
Hofbauer, D.
Hoffmann, F.
Hofmeister, Th.
Honkala, J.
Hutchinson, D.

Jacob, G.
Jakubiec, L.
Janin, D.
Jarecki, S.
Jerrum, M.
Juban, L.
Jukna, S.
Julia, S.

Kapidakis, S.
Karhumäki, J.
Karpinski, M.
Kaufmann, M.
Kemp, R.
Kessler, Ch.
Kilian, J.
King, V.
Kirchner, H.
Kirousis, L.
Klasing, R.
Kloks, T.
Knudsen, L. R.
Koiran, P.
Konig, J.-C.
Kozen, D.
Krajíček, J.
Kranakis, E.
Krandick, W.
Kreowski, H. J.
Kriegel, K.
Krogh, B.
Kucherov, G.
Kühnemann, A.
Kündig, A.
Kummer, O.
Kunde, M.
Kupferman, O.
Kutrib, M.
Kutylowski, M.

Ladner, R.
Laforest, C.
Lakhnech, Y.
Lange, S.
Laplante, S.
Lapoire, D.
Laroussinie, F.
Latteux, M.
Lautemann, C.
Lavault, C.
Lazard, S.
Lefmann, H.
Levaire, J. L.
Li, M.
Lickteig, T.
Liskiewicz, M.
Litovski, I.
Loescher, B.
Lohrey, M.
Lopez-Ortiz, A.
Lorys, K.
Lucks, S.
Lugiez, D.

Maffray, F.
Maheshwari, A.
Majerech, V.
Makowsky, J.
Maler, O.
Marcinkowski, J.
Margara, L.
Margenstern, M.
Marion, J. Y.
Marquis, P.
Martin, B.
Masini, A.
Mathieu, P.
Maurras, J. F.
Mayordomo, E.
Mayr, E. W.

Mayr, R.
Mazoyer, J.
Meer, K.
Menzel, W.
Meo, M. C.
Merz, S.
Messeguer, X.
Messner, J.
Metivier, Y.
Meyer auf der Heide, F.
Meyer, U.
Miltersen, P. B.
Monien, B.
Monti, A.
Moran, S.
Morain, F.
Morin, P.
Morvan, C.
Morvan, M.
Muller, J. M.
Mundhenk, M.
Muscholl, A.

Näher, S.
Narayanan, L.
Narbel, P.
Neumann, A.
Neyer, G.
Niedermeier, R.
Niehren, J.
Nipkow, T.
Nisan, N.
Niwinski, D.
Nonnengart, A.
Nourine, L.

Ogihara, M.
Olive, F.
Ostrovsky, R.
Ottmann, Th.

Otto, F.

Pacholski, L.
Pagnucco, M.
Parigot, M.
Paschos, V. Th.
Paul, C.
Paulin-Mohring, C.
Paun, Gh.
Persiano, P.
Picaronny, C.
Pino-Perez, R.
Pistore, M.
Pitassi, T.
Pitt, L.
Plateau, G.
Pointcheval, D.
Postow, B.
Pottier, F.
Poupard, G.
Prade, H.
Prömel, H. J.
Pudlák, P.

Rampon, J.
Rauzy, A.
Razborov, A.
Recker, F.
Reinhardt, K.
Reischuk, R.
Remmele, P.
Richomme, G.
Risch, V.
Robson, M.
Röckl, C.
Roman, J.
Roos, Th.
Roos, Y.
Rossmanith, P.
Rothe, J.

Rougemont de, M.
Rusu, I.

Sadowski, Z.
Safer, T.
Saheb, N.
Sais, L.
Sakarovitch, J.
Salomaa, K.
Sanders, P.
Sangiorgi, D.
Santosh, A.
Sauerhoff, M.
Savicky, P.
Schaeffer, G.
Schaerf, M.
Schirra, S.
Schlechta, K.
Schlude, K.
Schnoebelen, Ph.
Schnorr, C.
Schöning, U.
Schott, R.
Schuler, R.
Schulman, L.
Schulz, K. U.
Schulze, K.
Séébold, P.
Seese, D.
Seibert, S.
Seidl, H.
Seifert, J. P.
Senizergues, G.
Serna, M. J.
Shankar, S.
Shende, S.
Sibeyn, J. F.
Sinclair, S.
Skillicorn, D.
Smid, M.

Spinrad, J.

Staiger, L.

Stamatiou, Y.

Stamm, Ch.

Standh, R.

Steger, A.

Stephan, F.

Steyaert, J. M.

Stolzenburg, F.

Straubing, H.

Sykora, O.

Szekely, G.

Talbot, J. M.

Tamassia, R.

Tate, S.

Temam, O.

Terlutte, A.

Tesson, P.

Theobald, T.

Thibon, J. Y.

Thiele, L.

Thierauf, T.

Thigarajan, S. P.

Thomasset, F.

Tiga, S.

Toda, S.

Todinca, I.

Torán, J.

Trevisan, L.

Tripakis, S.

Tromp, J.

Truszczynski, M.

Trystram, D.

Ulber, R.

Unger, W.

Valk, R.

Vallée, B.

Vauquelin, B.

Verbitsky, O.

Vergne, M.

Vidal-Naquet, G.

Vigneron, L.

Vitanyi, P.

Vitter, J. S.

Vogler, H.

Vogler, W.

Vrain, C.

Wätjen, D.

Waack, S.

Wagner, K. W.

Walicki, M.

Walukiewicz, I.

Warnow, T.

Watanabe, O.

Wattenhofer, R.

Weil, P.

Weiskircher, R.

Wilf, H.

Wilke, T.

Wimmel, H.

Wing, J.

Woeginger, G. J.

Wolf de, R.

Wolper, P.

Worsch, T.

Young, N.

Yu, S.

Yvinec, M.

Žák, S.

Zaroliagis, Ch.

Zatopianski, J.

Zhu, Y.

Zielonka, W.

Zissimopoulos, V.

Table of Contents

Algorithms for Selfish Agents
Mechanism Design for Distributed Computation

Noam Nisan*

Institute of Computer Science, Hebrew U., Jerusalem and IDC, Herzliya.

Abstract. This paper considers algorithmic problems in a distributed setting where the participants cannot be assumed to follow the algorithm but rather their own self-interest. Such scenarios arise, in particular, when computers or users aim to cooperate or trade over the Internet. As such participants, termed agents, are capable of manipulating the algorithm, the algorithm designer should ensure in advance that the agents' interests are best served by behaving correctly.

This exposition presents a model to formally study such algorithms. This model, based on the field of mechanism design, is taken from the author's joint work with Amir Ronen, and is similar to approaches taken in the distributed AI community in recent years. Using this model, we demonstrate how some of the techniques of mechanism design can be applied towards distributed computation problems. We then exhibit some issues that arise in distributed computation which require going beyond the existing theory of mechanism design.

1 Introduction

A large part of research in computer science is concerned with protocols and algorithms for inter-connected collections of computers. The designer of such an algorithm or protocol always makes an implicit assumption that the participating computers will act as instructed – except, perhaps, for the faulty or malicious ones.

With the emergence of the Internet as *the* platform of computation, this assumption can no longer be taken for granted. Computers on the Internet belong to different persons or organizations, and will likely do what is most beneficial to their owners – act "selfishly". We cannot simply expect each computer on the Internet to faithfully follow the designed protocols or algorithms. It is more reasonable to expect that each selfish computer will try to manipulate it for its owners' benefit. An algorithm or protocol intended for selfish computers must therefore be designed in advance for this kind of behavior!

Such protocols and algorithms will likely involve payments (or other trade) between the selfish participants. One can view this challenge (of designing protocols and algorithms for selfish computers) as that of designing automated trade

---

* This research was supported by grants from the Israeli ministry of Science and the Israeli academy of sciences.

rules for the Internet environment. The normal practices of human trade, while clearly relevant, cannot be directly applied due to the much greater complexity involved and due to the automated nature of the trade.

The view taken in this paper is that of a systems' engineer that has certain technical goals for the global behavior of the Internet. We view the selfishness of the participants as an obstacle to our goals, and we view the trade and payments involved as a way to overcome this obstacle. In economic terms, we desire a virtual "managed economy" of all Internet resources, but due to the selfishness of the participants we are forced to obtain it using the "invisible hand" of "free markets". Our goal is to design the market rules as to ensure the desired global behavior.

We first present a formal model that allows studying these types of issues. The model relies on the rationality of the participants and is game-theoretic in nature. Specifically, it is based upon the theory of mechanism design. The model is directly taken from the author's joint work with Amir Ronen [18], and is similar in spirit to some models studied in the distributed AI community. After presenting the model we present some of the basic notions and results from mechanism design in our distributed computation setting. We do not intend here to give a balanced or exhaustive survey of mechanism design, but rather to pick and choose the notions that we feel are most applicable to our applications in distributed computation. Finally, we present some scenarios that arise in distributed computation that require going beyond the existing theory of mechanism design.

Before getting into the model, we will mention some of the application areas we have in mind, and shortly mention some of the existing work in computer science along this and similar tracks.

2   Sample Scenarios

We shortly sketch below three (somewhat related) application areas that we feel require these types "selfish algorithms". These application areas are each quite wide in their scope, involve complicated optimizations of resources, and directly involve differing goals of the participants. Most of the works cited below lie in one of these areas.

2.1   Resource Allocation

The aggregate power of all computers on the Internet is huge. In a "dream world" this aggregate power will be optimally allocated online among all connected processors. One could imagine CPU-intensive jobs automatically migrating to CPU-servers, caching automatically done by computers with free disk space, etc. Access to data, communication lines, and even physical attachments (such as printers) could all be allocated across the Internet. This is clearly a difficult optimization problem even within tightly linked systems, and is addressed, in various forms and with varying degrees of success, by all distributed operating systems.

The same type of allocation over the Internet requires handling an additional problem: the resources belong to different parties who may not allow others to freely use them. The algorithms and protocols may, thus, need to provide some motivation for these owners to "play along".

2.2 Routing

When one computer wishes to send information to another, the data usually gets routed through various intermediate routers. So far this has been done voluntarily, probably due to the low marginal cost of forwarding a packet. However, when communication of larger amounts of data becomes common (e.g. video), and bandwidth needs to be reserved under various quality of service (QoS) protocols, this altruistic behavior of the routers may no longer hold. If so, we will have to design protocols specifically taking the routers' self-interest into account.

2.3 Electronic Trade

Much trade is taking place on the Internet and much more is likely to take place on it. Such trade may include various financial goods (stocks, currency exchange, options), various information goods (video-on-demand, database access, music), many services (help desk, flower delivery, data storage), as well as real goods (books, groceries, computers) . This trade will likely involve sophisticated programs communicating with each other trying to find "the best deal". In addition, this will also raise the possibility of various brokerage services such as information providers, aggregators, and other types of agents. Clearly any system that enables such programs to efficiently trade with each other needs to offer general economic efficiency while very strongly taking into account the fact that all participants have totally differing goals.

3 Existing Work

Game theory, Economics, and Computer Science

In recent years there have been many works that tried to introduce economic or game-theoretic aspects into computational questions. The approach presented here is part of this trend, but is much narrower, taking specifically the direction of mechanism design. The reader interested in the wider view may start his exploration e.g. with the surveys [8, 13], the book [21], the web sites [3, 1, 2], or the papers in the conference [4].

Mechanism design

The field of mechanism design (also known as implementation theory) aims to study how privately known preferences of many people can be aggregated towards a "social choice". The main motivation of this field is micro-economic, and the tools are game-theoretic. Emphasis is put on the implementation of various types of auctions.

In the last few years this field has received much interest, especially due to its influence on large privatizations and spectrum allocations [16]. An introduction to this field can be found in [15, chapter 23] [20, chapter 10], and an influential web site in [17].

Mechanism design in Computer Science

One may identify three motivations for combining mechanism design with computational questions.

> **Auction implementation:** As auctions become more popular as well as more complicated, they are often implemented using computers and computer networks. Many computational implementation questions result. These range from purely combinatorial ones regarding optimization in complex combinatorial auctions to systems questions regarding communication and performance issues in wide-scale auctions.
>
> **Leveraging Market Power:** In the "real world" the invisible hand of free markets seems to yield surprisingly good results for complex optimization problems. This occurs despite the many underlying difficulties: decentralized control, uncertainties, information gaps, limited computational power, etc. One is tempted to apply similar market-based ideas in computational scenarios with similar complications, in the hope of achieving similarly good results.
>
> **Handling Selfishness:** This is the approach taken here, and it views mechanism design introduced into computational problems as a necessary evil, required to deal with the differing goals of the participants.

Even though these motivations are different philosophically, research often combines aspects from all approaches. Below we shortly sketch some of previous work done introducing mechanism design into different branches of computer science, without attempting to further classify them.

Distributed AI

In the last decade or so, researchers in AI have studied cooperation and competition among "software agents". The meaning of agents here is very broad, incorporating attributes of code-mobility, artificial-intelligence, user-customization, and self-interest.

A subfield of this general direction of research takes a game theoretic analysis of agents' goals, and in particular uses notions from mechanism design [21] [22] [7]. A related subfield of Distributed AI, sometimes termed market-based computation [26] [8] [25], aims to leverage the notions of free markets in order to solve distributed problems. These subfields of DAI are related to our work.

Communication Networks

In recent years researchers in the field of network design adopted a game theoretic approach (See e.g. [11]). In particular mechanism design was applied to various problems including resource allocation [12], cost sharing, and pricing [23].

4   The Model

In this section we formally present the model. It is taken from the author's joint work with Amir Ronen [18].

The model is concerned with computing functions that depend on inputs that are distributed among n different agents. A problem in this model has, in addition to the specification of the function to be computed, a specification of the goals of each of the agents. The solution, termed a mechanism, includes, in addition to an algorithm computing the function, payments to be handed out to the agents. These payments are intended to motivate the agents to behave "correctly".

Subsection 4.1 describes what a mechanism design problem is. In subsection 4.2 we define what a good solution is: an implementation with dominant strategies. Subsection 4.3 defines a special class of good solutions: truthful implementations, and states the well-known fact that restricting ourselves to such solutions loses no generality.

4.1   Mechanism Design Problem Description

Intuitively, a mechanism design problem has two components: the usual algorithmic output specification, and descriptions of what the participating agents want, formally given as utility functions over the set of possible outputs (outcomes).

Definition 1 (Mechanism Design Problem) *A mechanism design problem is given by an output specification and by a set of agent's utilities. Specifically:*

1. *There are n agents, each agent i has available to it some **private** input $t^i \in T^i$ (termed its type). Everything else in this scenario is public knowledge.*

2. *The output specification maps to each type vector $t = t^1...t^n$ a set of allowed outcomes o.*

3. *Each agent i's preferences are given by a real valued function: $v^i(o, t^i)$, called its valuation. This is a quantification of its value from the outcome o, when its type is t^i, in terms of some common currency. I.e. if the mechanism's outcome is o and in addition the mechanism hands this agent p^i units of this currency, then its utility will be $u^i = p^i + v^i(o, t^i)$[1]. This utility is what the agent aims to optimize.*

In this paper we only discuss optimization problems. In these problems the outcome specification is to optimize a given objective function. We present the definition for minimization problems.

---

[1] This is termed "semi-linear utility". In this paper we limit ourselves to this type of utilities.

Definition 2 (Mechanism Design Optimization problem) *This is a mechanism design problem where the outcome specification is given by a positive real valued objective function $g(o, t)$ and a set of feasible outcomes F. The required output is the outcome $o \in F$ that minimizes g.*

4.2   The Mechanism

Intuitively, a mechanism solves a given problem by assuring that the required outcome occurs, when agents choose their strategies as to *maximize their own selfish utilities*. A mechanism needs thus to ensure that players' utilities (which it can influence by handing out payments) are compatible with the algorithm.

Notation: We will denote $(a^1, ... a^{i-1}, a^{i+1}, ... a^n)$ by a^{-i}. (a^i, a^{-i}) will denote the tuple $(a^1, ... a^n)$

Definition 3 (A Mechanism) *A mechanism $m = (o, p)$ is composed of two elements: An outcome $o = o(a)$, and an n-tuple of payments $p^1(a)...p^n(a)$. Specifically:*

1. *The mechanism defines for each agent i a family of strategies A^i. Agent i can choose to perform any $a^i \in A^i$.*
2. *The first thing a mechanism must provide is an outcome function $o = o(a^1...a^n)$.*
3. *The second thing a mechanism provides is a payment $p^i = p^i(a^1...a^n)$ to each of the agents.*
4. *We say that a mechanism is an* implementation with dominant strategies *(or in short just an implementation) if*
   - *For each agent i and each t^i there exists a strategy $a^i \in A^i$, termed dominant, such that for all possible strategies of the other agents a^{-i}, a^i maximizes agent i's utility. I.e. for every $a'^i \in A^i$, if we define $o = o(a^i, a^{-i})$, $o' = o(a'^i, a^{-i})$, $p^i = p^i(a^i, a^{-i})$, $p'^i = p^i(a'^i, a^{-i})$, then $v^i(t^i, o) + p^i \geq v^i(t^i, o') + p'^i$*
   - *For each tuple of dominant strategies $a = (a^1...a^n)$ the outcome $o(a)$ satisfies the specification.*

4.3   The Revelation Principle

The simplest types of mechanisms are those in which the agents' strategies are to simply report their types.

Definition 4 (Truthful Implementation) *We say that a mechanism is* truthful *if*

1. *For all i, and all t^i, $A^i = T^i$, i.e. the agents' strategies are to report their type. (This is called a direct revelation mechanism.)*
2. *Truth-telling is a dominant strategy, i.e. $a^i = t^i$ satisfies the definition of a dominant strategy above.*

A simple observation, known as the *revelation principle*, states that without loss of generality one can concentrate on truthful implementations.

Proposition 4.1 *([15], page 871) If there exists a mechanism that implements a given problem with dominant strategies then there exists a truthful implementation as well.*

Proof: (sketch) We let the truthful implementation simulate the agents' strategies. I.e. given a mechanism $(o, p^1, ...p^n)$, with dominant strategies $a^i(t^i)$, we can define a new one by $o^*(t^1...t^n) = o(a^1(t^1)...a^n(t^n))$ and $(p^*)^i(t^1...t^n) = p^i(a^1(t^1)...a^n(t^n))$. □

5 Applying Existing Mechanism Design Theory

In this section we present several well known mechanisms. While these mechanisms are the usual ones one would find in a standard text on mechanism design, we present them in a distributed-computation setting. The implementations provided are all truthful ones, i.e. they follow this pattern:

1. Each agent reports its input to the mechanism.
2. The mechanism computes the desired outcome based on the reported types.

3. The mechanism computes payments for each agent.

The challenge in these examples is to determine these payments as to ensure that the truth is indeed a dominating strategy for all agents.

5.1 Maximum

Story:[2]
A single server is serving many clients. At a certain time, the server can serve exactly one request. Each client has a private valuation t^i for his request being served. (The valuation is 0 if the request is not served.) We want the most valuble request to be served.

Failed attempts:
One might first attempt to simply ignore all payments (i.e. set $p^i = 0$ for all i). This however is clearly insufficient since it motivates each agent to exaggerate his valuation, as to get his request executed. The second attempt would be to let the winning agent pay his declaration. I.e. set $p^i = -t'^i$ for the agent i that declared the highest t'^i (and $p^i = 0$ for all others). This also fails since the agent with highest t^i is motivated to reduce his declaration to slightly above the second highest valuation offered. This will result in his request still being served, and his payment reduced. In case agent i has imperfect information about the others

---

[2] This is an auction and the solution presented is Vickrey's well-known second price auction [24].

this strategic behavior may lead him to accidently declare a lower value than the second valuation, which will result in a sub-optimal allocation.

Solution:
The agent that offers the highest valuation for his request pays the second highest price offered. I.e. $p^i = -t^j$, where i offers the highest price and j the second highest. All other agents have $p^k = 0$.

Analysis:
To see why this is a truthful implementation, consider agent i and consider a lie $t'^i \neq t^i$. If this lie does not change the allocation, then nothing is gained or lost by agent i since his payment is also unaffected by his own declaration. If this lie gets his request served, then $t'^i > t^j > t^i$ and he gains t^i of utility from his valuation of the served request, but he loses t^j on payments, thus his total utility would be $t^i - t^j < 0$, as opposed to 0 in the case of the truth. On the other hand, if his lie makes him lose the service, then his utility is now 0, as opposed to a positive number which it was in the truthful case.

5.2   Threshold

Story:[3]
A single cache is shared by many processors. When an item is entered into the cache, all processors gain faster access to this item. Each processor i will save t^i in communication costs if a certain item X is brought into the cache. (I.e its valuation of loading X is $t^i > 0$, and of not loading it, 0.) The cost of loading X is a publicly known constant C. We want to load X iff $\sum_i t^i > C$.

Failed attempts:
We may first attempt to just divide the total cost between the n participating agents, i.e. set $p^i = -C/n$ for all i. This however motivates any agent with $t^i > C/n$ to announce his valuation as greater than C, and thus assure that X is loaded. We may, as a second attempt, let each agent pay the amount declared (or perhaps something proportional to it.) In this case, however, we will be faced with a free-rider problem, where agents will tend to report lower valuation than the true ones so as to reduce their payments. This, when done by several agents, may result in the wrong decision of not loading X.

Solution:
In case X is loaded, each agent pays a sum equal to the minimum declaration required from him in order to load X, given the other's declarations. I.e. the only case where $p^i \neq 0$, is when $\sum_{j \neq i} t^j \leq C < \sum_j t^j$, in which case $p^i = \sum_{j \neq i} t^j - C$ (a negative number).

The analysis is left to the reader. Alternatively, this example may be seen to be a special case of the example below.

This example can be generalized to the case where t^i can be negative as well.

---

[3] This is known as the "public project" problem, and the solution is known as the Clarke tax [5].

5.3   Shortest Path

Story:

We have a communication network modeled by a directed graph G, and two special nodes in it x and y. Each edge e of the graph is an agent. Each agent e has private information (its type) $t^e \geq 0$ which is the agent's cost for sending a single message along this edge. The goal is to find the cheapest path from x to y (as to send a single message from x to y). I.e the set of feasible outcomes are all paths from x to y, and the objective function is the path's total cost. Agent e's valuation is 0 if his edge is not part of the chosen path, and $-t^e$ if it is. We will assume for simplicity that the graph is bi-connected.

Solution:

The following mechanism ensures that the dominant strategy for each agent is to report his true type t^e to the mechanism. When all agents honestly report their costs, the cheapest path is chosen: The outcome is obtained by a simple shortest path calculation. The payment p^e given to agent e is 0 if e is not in the shortest path and $p^e = d_{G-e} - (d_G - t'^e)$ if it is. Here t'^e is the agents' reported input (which may be different from its actual one), d_G is the length of the shortest path (according to the inputs reported), and d_{G-e} is the length of the shortest path that does not contain e (again according to the reported types).

Analysis:

First notice that if the same shortest path is chosen with t'^e as with t^e then the payment and thus utility of the agent does not change. A lie $t'^e > t^e$ will cause the algorithm to choose the shortest path that does not contain e as opposed to the (correct one) which does contain it iff $d_{G-e} - d_G < t'^e - t^e$. This directly implies that e's utility would have been positive had e been chosen in the path (as opposed to 0 when its not chosen), thus the truth is better. A similar argument works to show that $t'^e < t^e$ is worse than the truth.

Many other graph problems, where agents are edges, and their valuations proportional to the edges' weights, can be implemented by a VCG mechanism. In particular minimum spanning tree and max-weight matching seem natural problems in this setting. A similar solution applies to the more general case where each agent holds some subset of the edges.

Algorithmic Problem: How fast can the payment functions be computed? Can it be done faster than computing n versions of the original problem? For the shortest paths problem we get the following equivalent problem: given a directed graph G with non-negative weights, and two vertices in it x, y. Find, for each edge e in the graph, the shortest path from x to y that does not use e. Using Disjktra's algorithm for each edge on the shortest path gives an $O(nm \log n)$ algorithm. Is anything better possible? Maybe $O(m \log n)$?

5.4   Utilitarian Functions

Arguably the most important positive result in mechanism design is what is usually called the generalized Vickrey-Groves-Clark (VCG) mechanism [24] [10] [5]. All previous examples are, in fact, VCG mechanisms. In this section we present the general case.

The VCG mechanism applies to mechanism design optimization problems where the objective function is simply the sum of all agents' valuations.

Definition 5 *An optimization mechanism design problem is called* utilitarian *if its objective function satisfies* $g(o, t) = \sum_i v^i(o, t^i)$.

Definition 6 *We say that a direct revelation mechanism* $m = (o(t), p(t))$ *belongs to the VCG family if*

1. $o(t) \in \arg\max_o(\sum_{i=1}^n v^i(t^i, o))$.
2. $p^i(t) = \sum_{j \neq i} v^i(o(t), t^i) + h^i(t^{-i})$ *where* $h^i()$ *is an arbitrary function of* t^{-i}.

Theorem 5.1 *(Groves [10])* A VCG mechanism is truthful.

Proof: (sketch) Let $d^1, \ldots, d^n$ denote the declaration of the agents and $t^1, \ldots, t^n$ denote their real types. Suppose that truth telling is not a dominant strategy, then there exists d, i, t, d'^i such that

$$v^i(t^i, o(d^{-i}, t^i)) + p^i(t^i, o(d^{-i}, t^i)) + h^i(d^{-i}) <$$

$$v^i(t^i, o(d^{-i}, d'^i)) + p^i(t^i, o(d^{-i}, d'^i)) + h^i(d^{-i})$$

But then

$$\sum_{i=1}^n v^i(o(d^{-i}, t^i), t^i) < \sum_{i=1}^n v^i(o(d^{-i}, d'^i), t^i)$$

In contradiction for the definition of $o()$. $\square$

Thus a VCG mechanism essentially provides a solution for any utilitarian problem (except for the possible problem that there might be dominant strategies other than truth-telling). It is known that (under mild assumptions) VCG are the only truthful implementation for utilitarian problems ([9]).

5.5   More Issues in Mechanism Design

The examples presented here demonstrate only the most basic notions from the field of mechanism design. Many more issues addressed by the theory of mechanism design are applicable to the distributed computation setting. We briefly mention just some of the issues commonly studied by mechanism design (and other branches of game theory) that we feel may find applications in distributed computation.

Bayesian-Nash equilibrium: Our notion of a solution was very strong, requiring dominant strategies. Weaker notions of equilibrium are also often considered, in particular Bayesian-Nash equilibrium.

Non semi-linear utilities: We assumed that the utility of each agent is additive in the money. More general types of utilities may be considered, where money influences the utility in an arbitrary manner.

Budgets: We did not put any requirements on the sums of money involved in a mechanism. At least two types of constraints are widely studied: constraining the total money spent by the mechanism (either to as large a negative amount as possible, or to 0 – budget balance), and considering budget limitations of the agents.

Common value models: We assumed that each agent has a known valuation function that is independent from the others. One may alternatively assume a valuation that is common to all agents but is not fully known by them.

Repeated Games: We only considered a single instance of a problem. One may clearly consider repeated instances.

Coalitions: We only considered manipulation by a single agent. Clearly one may study coalitions of agents.

6   Beyond Existing Mechanism Design

We feel that the application of existing mechanism design in distributed computation, as demonstrated above, is just a first step. Many of the considerations of distributed computation are quite different from the ones usually considered in mechanism design. Addressing these considerations will thus require new research. In this section we exhibit several scenarios in distributed computation that raise questions that indeed go beyond the current scope of mechanism design.

6.1   Task Scheduling

Story:

A computer has k tasks it wishes to execute, and can execute each of them on any one of n servers. Each server i knows, for every task j, the time t_j^i it requires to execute this task. Each server's cost is proportional to the time it spends on executing the tasks assigned to it. Our goal is to have all tasks completed as soon as possible (i.e. to minimize the completion time of the last task.)

This problem was considered in [18]. Here are some of the issues raised by this problem and addressed there. Similar issues arise in many other problems in distributed computation.

Issues:

Non-utilitarian Problem: The goal in this example is non-utilitarian. Thus, the VCG mechanism cannot be applied and new mechanisms need to be invented.

Impossibility: It is possible to prove that no mechanism perfectly solves this problem. As is common in Computer Science, one should try to overcome this impossibility. In particular, the following approaches may be considered (and were all studied in [18]):

Approximation: Find a mechanism that approximates the optimal solution as well as possible.

Randomization: In Computer Science as well as in game theory randomization often helps. In turns out that for this problem, randomized mechanisms can provably do better than deterministic ones.

Model Extensions: Every model is an imperfect abstraction of reality. One may incorporate useful attributes of reality into the model as to make an impossible result possible. In [18] the model was extended by assuming that the mechanism need only compute the payments *after* the tasks were actually executed, giving it additional information.

Computational Intractability: Even from a purely algorithmic point of view, the task scheduling problem is intractable (NP-complete). When adding the requirements of a mechanism things only get worse. In particular, standard ways of overcoming the computational intractability (such as tractable approximations) have complicated interactions with the requirements of mechanism design.

6.2   Maximum Independent Set

Story:

There are n processors connected in a linear array (i.e. each processor i is connected to $i-1$ and to $i+1$). Each processor wants to execute a single job, and values it at $t^i \geq 0$. The problem is that executing the job requires exclusive access to the common link with each of its neighbors. Thus no two consecutive processors can execute their job. Our goal is to execute the set of tasks with maximal valuation, i.e. to find an independent set S of processors that maximizes $\sum_{i \in S} t^i$.

Model Restriction:

In this story we want to find a decentralized solution. I.e. we want to design a protocol, that runs on *these computers*, using only the *available communication links*, and without assuming any central trusted computer, or any other communication links.

Solution:

Our protocol has two phases a left-to-right phase and a right-to-left phase. In the left-to-right phase, each processor places a bid R^i for link on its right. These offers are computed by each processor in turn as follows: $R^1 = t^1$, and for $1 < i < n$, $R^i = max(t^i - R^{i-1}, 0)$. In the right-to-left phase each processor places a bid L^i on the link to its left as follows: $L^n = t^n$, and for $1 < i < n$, $L^i = max(t^i - L^{i+1}, 0)$. Processor i wins the left link iff $L^i > R^{i-1}$ and wins the right link iff $R^i \geq L^{i+1}$. It can execute its task (i.e. is chosen to be in S) if it has won both links. In this case its payment is $-p^i = R^{i-1} + L^{i+1}$ (i.e. the second price on each of links it has won).

Analysis:

There are many issues to consider here:

Algorithmic correctness: One may verify that R^i is the difference between the weight of the maximum weight independent set in $1...i-1$ and the weight of the maximum weight independent set in $1...i$. Similarly, L^i is the difference between of the weights of the maximum weight independent sets in $i+1...n$ and $i...n$. Clearly i should be chosen to be in S if $t^i > L^{i+1} + R^{i-1}$ (ties can be broken arbitrarily), which is exactly what this protocol does. This protocol can be viewed as a dynamic programming solution of this problem.

Domination of the Truth: Assume that the players' strategies are limited to acting according to some fixed valuation t'^i. Such a model may be called the "honest but selfish" case. In this case one may observe that the protocol achieves the VCG mechanism that is a solution since the problem is indeed utilitarian.

Dishonesty: A more general model would allow all strategies made possible by the protocol. In this case the processors could act according to a different t'^i in each phase. One may verify that in this model the truth is no longer dominant. Yet, truth is still a Nash equilibrium.

Ensuring Honesty: There are various ways to augment the model as to force the processors to be consistent in both phases, and thus essentially force the "honest but selfish" situation. In particular, if processors $i-1$ and $i+1$ can communicate with each other then they can catch i's dishonesty. Such communication may alternatively be implicitly achieved by using cryptographic signatures.

Decentralized Payments: The payments in this solution were to be given to some party outside of the n involved processors. It would have been nice to have a mechanism where the payments are only transferred between connected processors.

6.3   Decentralized Auction

Story:

A single item is to be auctioned over the Internet among n humans (each with his own computer).

Restriction:

There is no trusted entity. In particular we do not trust the auctioneer to faithfully execute the auction rules or to keep any secrets. In the absence of such a trusted entity we would like to ensure two goals:

— The auction is executed according to the published auction rules (e.g. second price).

– No information about bids is leaked to any participant, beyond the results of the auction which become public knowledge. I.e. only the identity of the winner (but not his bid), and the amount of the second highest bid (but not the identity of the bidder) become known.

Solution:
The celebrated "oblivious circuit evaluation" cryptographic protocols [19, 14, 6] exactly achieve this goal (as long as not too many of the participants collude to lie). These cryptographic protocols can faithfully carry out any distributed computation without leaking any information to the participants. What cannot, in principle, be ensured by cryptography is that the participants reveal their inputs. This, however, is ensured by the mechanism. We should note that these cryptographic protocols, while theoretically tractable, are quite impractical.

7   Acknowledgments

The notions expressed in this paper are derived from my joint work with Amir Ronen, who has also helped with the writing of this paper. I thank Dov Monderer, Motty Perry, and Moshe Tennenholtz for helpful discussions.

References

[1] Comet group technical reports.
Web Page: http://comet.ctr.columbia.edu/publications/techreports.html.

[2] The information economy.
Web Page: http://www.sims.berkeley.edu/resources/infoecon/.

[3] Market-oriented programming.
Web Page: http://ai.eecs.umich.edu/people/wellman/MOP.html.

[4] First international conference on information and computation economies ice-98.
Web Page: http://www.cs.columbia.edu/ICE-98/, October 1998.

[5] E. H. Clarke. Multipart pricing of public goods. *Public Choice*, pages 17–33, 1971.

[6] C. Crepeau D. Chaum and I. Damgard. Multiparty unconditionally secure protocols. In *20th STOC*, 1988.

[7] Eithan Ephrati and Jeffrey S. Rosenschein. The clarke tax as a concensus mechanism among automated agents. In *Proceedings of the national Conference on Artificial Intelligence*, pages 173–178, July 1991.

[8] Donald F. Ferguson, Christos Nikolaou, and Yechiam Yemini. Economic models for allocating resources in computer systems. In Scott Clearwater, editor, *Market-Based Control: A Paradigm for Distributed Resource Allocation*. World Scientific, 1995.

[9] J. Green and J.J. Laffont. Characterization of satisfactory mechanism for the revelation of preferences for public goods. *Econometrica*, pages 427–438, 1977.

[10] T. Groves. Incentives in teams. *Econometrica*, pages 617–631, 1973.

[11] Y.A Korilis, A. A. Lazar, and A. Orda. Architecting noncooperative networks. *IEEE Journal on Selected Areas in Communication (Special Issue on Advances in the Fundamentals of Networking)*, 13(7):1241–1251, September 1991.

[12] A.A. Lazar and N. Semret. The progressive second price auction mechanism for network resource sharing. In *8th International Symposium on Dynamic Games*, Maastricht, The Netherlands, July 1998.

[13] Nathan Lineal. Game theoretic aspects of computing. In *Handbook of Game Theory*, volume 2, pages 1339–1395. Elsevier Science Publishers B.V, 1994.

[14] S. Golwasser M. Ben-Or and A. Wigderson. Completeness theorems for fault-taulerent distributed computing. In *20th STOC*, 1988.

[15] A. Mas-Collel, W. Whinston, and J. Green. *Microeconomic Theory*. Oxford university press, 1995.

[16] J. McMillan. Selling spectrum rights. *Journal of Economic Perspectives*, pages 145–162, 1994.

[17] Market design inc. Web Page: http://www.market-design.com.

[18] Noam Nisan and Amir Ronen. Algorithmic mechanism design. Avilable at http://www.cs.huji.ac.il/~amiry.

[19] S. Micali O. Goldreich and A. Wigderson. Proofs that yield nothing but their validity and a methodology of cryptographic protocol design. In *27th FOCS*, 1986.

[20] M. J. Osborne and A. Rubistein. *A Course in Game Theory*. MIT press, 1994.

[21] Jeffrey S. Rosenschein and Gilad Zlotkin. *Rules of Encounter: Designing Conventions for Automated Negotiation Among Computers*. MIT Press, 1994.

[22] Tuomas W. Sandholm. Limitations of the vickrey auction in computational multiagent systems. In *Proceedings of the Second International Conference on Multiagent Systems (ICMAS-96)*, pages 299–306, Keihanna Plaza, Kyoto, Japan, December 1996.

[23] S. Shenkar, Clark D. E., and Hertzog S. Pricing in computer networks: Reshaping the research agenda. *ACM Computational Comm. Review*, pages 19–43, 1996.

[24] W. Vickrey. Counterspeculation, auctions and competitive sealed tenders. *Journal of Finance*, pages 8–37, 1961.

[25] W.E. Walsh and M.P. Wellman. A market protocol for decentralized task allocation: Extended version. In *The Proceedings of the Third International Conference on Multi-Agent Systems (ICMAS-98)*, 1998.

[26] W.E. Walsh, M.P. Wellman, P.R. Wurman, and J.K. MacKie-Mason. Auction protocols for decentralized scheduling. In *Proceedings of The Eighteenth International Conference on Distributed Computing Systems (ICDCS-98)*, Amsterdam, The Netherlands, 1998.

The Reduced Genus of a Multigraph*

Patrice Ossona de Mendez

CNRS UMR 0017, E.H.E.S.S.
54 Bd Raspail, 75006 Paris, France
pom@ehess.fr

Abstract. We define here the reduced genus of a multigraph as the minimum genus of a hypergraph having the same adjacencies with the same multiplicities. Through a study of embedded hypergraphs, we obtain new bounds on the coloring number, clique number and point arboricity of simple graphs of a given reduced genus. We present some new related problems on graph coloring and graph representation.

1 Introduction

Graph Coloring is a central topic in Graph Theory and numerous studies relate coloring properties with the genus of a graph. The maximum chromatic number among all graphs which can be embedded in a surface of genus g is given by Heawood's formula [14], as established by Ringel and Youngs for $g > 0$ [22]; the case $g = 0$, which is the Four Color Theorem, has been established by Appel and Haken [1][2] (see [24] for a simpler proof). Other approaches have related the chromatic number with other graph invariants: For instance, Szekeres and Wilf [28] gave the simple upper bound $\chi(G) \leq \mathrm{sw}(G) + 1$ where $\mathrm{sw}(G)$ is the maximum among all induced subgraphs H of G of the minimum degree of the vertices of H (the value $\mathrm{col}(G) = \mathrm{sw}(G) + 1$, the coloring number of G, is actually an upper bound for the choice number of G). On the other hand, it has been established that, for any positive integer n, there exists an n-chromatic graph G containing no triangles (see [10][18][19] and [21] for constructions).

As a natural generalization of the graphs, Berge introduced the concept of hypergraph [3]: a *hypergraph* is a pair $\mathcal{H} = (X, \mathcal{E})$, where X is a finite set and $\mathcal{E}$ is a family $(E_i, i \in I)$ of subsets of X, such that: $E_i \neq \emptyset$ ($\forall i \in I$) and $\bigcup_{i \in I} E_i = X$. The elements of X and $\mathcal{E}$ are respectively the *vertices* and the *edges* (or *hyperedges*) of the hypergraph. Different generalizations of the concept of planarity to hypergraphs have been proposed (See [16], for instance). The first generalization is due to Zykov [36]. He proposed to represent the edges of a hypergraph by a subset of the faces of a planar map. Walsh [34] has shown that Zykov's definition (as well as another definition due to Cori [5]) is equivalent to the following: A hypergraph is *planar* if and only if its vertex-edge incidence graph is planar [34] (see also [17][35]). This planarity concept is easily extended

---

* This work was partially supported by the **Esprit LTR Project no 20244-ALCOM IT**.

C. Meinel and S. Tison (Eds.): STACS'99, LNCS 1563, pp. 16–31, 1999.
© Springer-Verlag Berlin Heidelberg 1999

to oriented surfaces: the *genus* of a hypergraph is defined as the genus of its incidence graph. The *adjacency multigraph* of a hypergraph $\mathcal{H}$ is the multigraph on the same vertex set in which the number of edges incident to two vertices x and y is the same as the number of edges of $\mathcal{H}$ containing x and y (the edges of the hypergraphs become edge-disjoint cliques of the adjacency multigraph). The reduced genus of a multigraph G is then defined has the minimum genus of a hypergraph whose adjacency multigraph is G.

In Section 2, we recall some basic definitions on hypergraphs and introduce the corresponding notations used through this paper. In Section 3 we introduce the charge of a vertex that allows to prove the existence of a vertex having relatively few neighbors and, also, to bound the maximal cardinality of a set of mutually adjacent vertices. In Section 4, new charge functions lead to improved bounds in the cases of linear hypergraphs and triangle free linear hypergraphs. The results obtained in this section lead in Section 5 to new bounds on the coloring number, clique number and point arboricity of simples graphs of a given reduced genus. Section 6 introduces some new problems on graph representation.

2   Preliminaries

A *hypergraph* is a pair $\mathcal{H} = (X, \mathcal{E})$, where X is a finite set and $\mathcal{E}$ is a family $(E_i, i \in I)$ of subsets of X, such that: $E_i \neq \emptyset$ $(\forall i \in I)$ and $\bigcup_{i \in I} E_i = X$. The elements of X and $\mathcal{E}$ are respectively the *vertices* and the *edges* (or *hyperedges*) of the hypergraph. Two vertices $x, y \in X$ are *adjacent* if they both belong to some edge of $\mathcal{H}$; two edges E_i, E_j are *adjacent* if their intersection is not empty. A vertex $x \in X$ is *incident* to an edge $E_i \in \mathcal{E}$ if x belongs to E_i (see [3]). In the following, we do not allow a hypergraph to have a loop, that is a single element edge. The *neighbor set* $N(v)$ of a vertex v is the set of the vertices to which v is adjacent. the *degree* $d(v)$ of a vertex v is the number of edges including v. The maximum cardinality of an edge of the hypergraph is denoted $\max |E|$. The *clique number* $\omega(\mathcal{H})$ of the hypergraph $\mathcal{H}$ is the maximum cardinality of a set of pairwise adjacent vertices.

A hypergraph is *k-uniform* if all its edges have cardinality k; a 2-uniform hypergraph is nothing but a multigraph. A hypergraph is *linear* if any two edges have at most one common element:

$$\forall i \neq j, \quad |E_i \cap E_j| \leq 1$$

Linearity somehow extends the notion of simple graph (a 2-uniform linear hypergraph is nothing but a simple graph).

The *sub-hypergraph* of $\mathcal{H}$ *induced* by a subset $Y \subseteq X$ is the hypergraph $\mathcal{H}_Y = (Y, \mathcal{E}_Y)$, where

$$\mathcal{E}_Y = \{E_i \cap Y, \quad E_i \in \mathcal{E}; E_i \cap Y \neq \emptyset\}$$

Definition 1. *The* incidence graph Incid($\mathcal{H}$) *of $\mathcal{H}$ is the colored bipartite graph on $(X, \mathcal{E})$ defined by the vertex-edge incidence, a vertex being colored white (resp. black) if it belongs to X (resp. $\mathcal{E}$).*

As a special case, if G is a graph, $\mathrm{Incid}(G)$ is the bicolored vertex-edge incidence graph of G.

Remark 1. A hypergraph $\mathcal{H}$ is linear if and only if $\mathrm{Incid}(\mathcal{H})$ is C_4-free.

Definition 2. *The* adjacency multigraph $\mathrm{Adj}(\mathcal{H})$ *of a hypergraph $\mathcal{H}$ is the multigraph on the same vertex set in which the number of edges incident to two vertices x and y is the same as the number of edges of $\mathcal{H}$ containing x and y (the edges of the hypergraphs become edge-disjoint cliques of the adjacency multigraph).*

Remark that $\mathrm{Adj}(\mathcal{H})$ is a simple graph if and only if $\mathcal{H}$ is linear.

Definition 3. *The* genus *of a hypergraph is the genus of its incidence graph.*

Remark that the genus of a multigraph G is the same when considered as a multigraph and when considered as a 2-uniform hypergraph.

3 Embedded Hypergraphs

3.1 Charges

In order to prove the existence of vertices with a "small" number of neighbors, we introduce the charge of a vertex of an embedded hypergraph. This function is closely related Euler's formula and could allow the adaptation of powerful technics to embedded hypergraph, such as the discharging introduced by Heesch [15] and which is a crucial tool for the unavoidability part of the proofs of the Four Color Theorem.

Definition 4. *The* charge $\xi(v)$ *of a vertex v is defined by:*

$$\xi(v) = \sum_{E \ni v} \left(3 - \frac{6}{|E|} \right) \tag{1}$$

Lemma 1. *The sum of the charges of the vertices of a hypergraph $\mathcal{H}$ on n vertices and genus g is at most $6n + 12(g-1)$.*

Proof. The number of vertices of $\mathrm{Incid}(\mathcal{H})$ is $n + |\mathcal{E}|$, where $\mathcal{E}$ is the edge set of $\mathcal{H}$. Hence, the number of faces of the embedding is $|E(\mathrm{Incid}(\mathcal{H}))| - n - |\mathcal{E}| - 2(g-1)$. As $\mathrm{Incid}(\mathcal{H})$ is a simple vertex-bipartite map, each face of an embedding has length at least 4. Therefore, $2|E(\mathrm{Incid}(\mathcal{H}))|$ is greater or equal to 4 times the number of faces of the embedding. Altogether, we get: $|E(\mathrm{Incid}(\mathcal{H}))| \leq 2n + 2|\mathcal{E}| + 4(g-1)$.

Thus, we have:

$$\sum_v \xi(v) = \sum_v \sum_{E \ni v} \left(3 - \frac{6}{|E|}\right)$$

$$= 3 \sum_v d(v) - 6 \sum_E \sum_{v \in E} \frac{1}{|E|}$$

$$= 3|E(\mathrm{Incid}(\mathcal{H}))| - 6|\mathcal{E}|$$

$$\leq 6n + 12(g-1)$$

□

Lemma 2. *For any hypergraph $\mathcal{H}$ of genus g with at least an edge of cardinality at least 3, there exists a map of genus g, which vertex-face hypergraph $\mathcal{H}^+$ satisfies the following conditions:*

- *any edge of $\mathcal{H}$ of size at least 3 is an edge of $\mathcal{H}^+$,*
- *any edge of $\mathcal{H}$ of size at most 2 is included into an edge of $\mathcal{H}^+$,*
- *the maximal cardinality of an edge of $\mathcal{H}^+$ is equal to the maximal cardinality of an edge of $\mathcal{H}$,*
- *the number of neighbors of a vertex v of $\mathcal{H}^+$ (which is an upper bound for the number of neighbors of v in $\mathcal{H}$) is bounded by:*

$$|N^+(v)| \leq \sum_{v \ni E} (|E| - 2) \tag{2}$$

(where the sum is taken over all the edges of $\mathcal{H}^+$ containing v)

Proof. Quadrangulate the embedding of Incid($\mathcal{H}$) by adding new black vertices of degree 3. Then, remove all the black vertices of degree 2. This is the incidence graph of a hypergraph $\mathcal{H}^+$. By construction, all but the last properties are obviously satisfied. As no vertex of $\mathcal{H}^+$ may have degree 1 and as two edges of $\mathcal{H}$ incident to a vertex v which are consecutive when considering the neighbors of v in the embedding of Incid($\mathcal{H}$) have a non empty intersection, the last property follows. □

3.2 Vertices with Few Neighbors

Theorem 1. *For any integer $g \geq 0$, there exists a constant $M(g)$, such that any hypergraph of genus g with at least an edge of cardinality $M(g)$ has a vertex v with at most $2 \max |E| - 4$ neighbors.*

Proof. If a hypergraph $\mathcal{H}$ on n vertices has genus g, there exists a vertex v of $\mathcal{H}^+$ with charge $\xi(v) \leq 6 + \frac{12}{n}(g-1)$. As $n \geq \max |E|$ and as $3 - \frac{6}{|E|}$ tends to 3 as $|E|$ goes to infinity, if $M(g)$ is large enough, the maximum number of neighbors of v is achieved for v incident to two edges of cardinality $\max |E|$. the number of neighbors of v in $\mathcal{H}^+$) is hence bounded by $2 \max |E| - 4$. □

Theorem 2. *For any integer $g \geq 0$, there exists a constant $M'(g)$, such that any hypergraph of genus g with at least an edge of cardinality $M'(g)$ and such that $\mathcal{H}^+$ has no vertex of degree less than 3 has a vertex v with at most $\max |E| + 4$ neighbors.*

Proof. As previously, if $M'(g)$ is large enough, the maximum number of neighbors of v is achieved for v incident to one edge of cardinality $\max |E|$, one edge of cardinality 6 and one edge of cardinality 2. the number of neighbors of v in $\mathcal{H}^+$) is hence bounded by $\max |E| + 4$.   □

3.3   Maximal Cliques

Lemma 3. *For any integer $g \geq 0$, there exists a constant $C(g)$, such that, for any hypergraph of genus g on $n \geq C(g)$ vertices, if all the vertices are pairwise adjacent, then there is an edge of cardinality at least $\frac{2n}{3}$.*

Proof. As the complete graphs of genus at most g have a size bounded by a function of g, we may assume (by choosing $C(g)$ large enough) that $\mathcal{H}$ has at least an edge of size at least 3. Furthermore, as the adjacency multigraph of $\mathcal{H}$ is a partial graph of the one of $\mathcal{H}^+$ and has the maximal size of the edges of $\mathcal{H}$ and $\mathcal{H}^+$ are equal, we may consider $\mathcal{H}^+$ in place of $\mathcal{H}$.

Then, we have to consider the following cases:

- There exists a vertex v of degree 1.
  This vertex is then incident to an edge of cardinality n.
- All the vertices have degree at least 3.
  Then, there exists a vertex having at most $\max |E| + 4$ neighbors. Thus, $\max |E|$ is greater or equal to $n - 5$.
- All the vertices have degree at least 2 and there exists a vertex v of degree 2 (incident to edges E_α and E_β)
  Let $A = E_\alpha \setminus E_\beta$ and $B = E_\beta \setminus E_\alpha$. Then, $A, B, E_\alpha \cap E_\beta$ forms a partition of the vertex set of $\mathcal{H}$. Assume that E_α and E_β have cardinality strictly smaller than $\frac{2n}{3}$. It follows that A and B have cardinality strictly greater than $\frac{n}{3}$.
  - No vertex of $A \cup B$ has degree 2.
    Let $\mathcal{H}'$ be the hypergraph obtained from $\mathcal{H}$ by deleting all the vertices of $E_\alpha \cap E_\beta$ except v and adding an edge E including v, a vertex of A and a vertex of B (it can be easily done in such a way that the genus of $\mathcal{H}'$ is at most g). Then, as the adjacency multigraph of $\mathcal{H}'$ is a complete graph on $|A| + |B| + 1$ vertices (plus some parallel edges), the number of neighbors of any vertex $x \in E_\alpha$ is given by $|N(x)| = |A| + |B|$. According to this value and the fact that all the vertices of $\mathcal{H}'$ have degree at least 3, the minimal value of the charge of x is achieved if x belongs to E_α, an edge of cardinality 3 and an edge of cardinality $|B| - 1$. Then $\xi(x)$ is at least:

$$\xi(x) \geq \left(3 - \frac{6}{|A| + 1}\right) + 1 + \left(3 - \frac{6}{|B| - 1}\right)$$

$$\geq 7 - \frac{6}{\frac{n}{3} + 1} - \frac{6}{\frac{n}{3} - 1}$$

If n is large enough, this value is strictly greater than $6 + \frac{12}{n}(g-1)$. As the same holds for the vertices in E_β, a contradiction follows.

- The exists a vertex $w \in A \cup B$ of degree 2.

  Without loss of generality, we may assume that w belongs to A and is incident to edges E_α and E_γ.

  * there exists vertex $z \in A$ which does not belong to E_γ.

    By contracting the edges of $\mathrm{Incid}(\mathcal{H})$ incident to z, the vertices corresponding to E_α, E_β, z and those corresponding to the vertices in B form a complete bipartite graph homeomorphic to $K_{3,|B|}$. Hence, $|B|$ is bound by a constant depending on g, what contradicts $|B| > \frac{n}{3}$.

  * all the vertices of E_α belong to $E_\beta \cup E_\gamma$.

    Then, every vertex belongs to two edges among $E_\alpha, E_\beta, E_\gamma$. Hence, $|E_\alpha| + |E_\beta| + |E_\gamma| \geq 2n$ and one of these edges has cardinality at least $\frac{2n}{3}$.

$\square$

As a direct consequence of this lemma, we have:

Theorem 3. *For any integer $g \geq 0$, there exists a constant $C(g)$, such that, for any hypergraph $\mathcal{H}$ of genus g:*

$$\omega(\mathcal{H}) \leq \max(C(g), \frac{3}{2} \max |E|) \tag{3}$$

Remark 2. By an easy technical proof, one checks that $C(0) = 4$, that is: Any planar hypergraph $\mathcal{H}$ having a clique number at least 5 has an edge of size at least $\frac{2}{3}\omega(\mathcal{H})$.

4 Embedded Linear Hypergraphs

4.1 Linear Charges

We shall give an alternate definition of the charge of a vertex taking advantage of the linearity assumption, the linear charge.

Definition 5. *The* linear charge $\xi_l(v)$ *of a vertex v is defined by:*

$$\xi_l(v) = \sum_{E \ni v} \left(4 - \frac{6}{|E|}\right) \tag{4}$$

Lemma 4. *The sum of the linear charges of the vertices of a linear hypergraph $\mathcal{H}$ of genus g with n vertices is bounded by $6n + 12(g-1)$.*

Proof. The number of vertices of $\mathrm{Incid}(\mathcal{H})$ is $n + |\mathcal{E}|$, where $\mathcal{E}$ is the edge set of $\mathcal{H}$. Hence, the number of faces of the embedding is $|E(\mathrm{Incid}(\mathcal{H}))| - n - |\mathcal{E}| - 2(g-1)$. As $\mathrm{Incid}(\mathcal{H})$ is a C_4-free simple vertex-bipartite map, each face of an embedding has length at least 6. Therefore, $2|E(\mathrm{Incid}(\mathcal{H}))|$ is greater or equal to 6 times the number of faces of the embedding. Altogether, we get: $2|E(\mathrm{Incid}(\mathcal{H}))| \le 3n + 3|\mathcal{E}| + 6(g-1)$.

Thus, we have:

$$\sum_v \xi_1(v) = \sum_v \sum_{E \ni v} \left(4 - \frac{6}{|E|}\right)$$

$$= 4\sum_v d(v) - 6\sum_E \sum_{v \in E} \frac{1}{|E|}$$

$$= 4|E(\mathrm{Incid}(\mathcal{H}))| - 6|\mathcal{E}|$$

$$\le 6n + 12(g-1)$$

$\square$

4.2   Vertices with Few Neighbors

Theorem 4. *For any integer $g \ge 0$, there exists a constant $M_l(g)$, such that any linear hypergraph of genus g with at least an edge of cardinality $M_l(g)$ has a vertex v with at most $\max|E| + 1$ neighbors.*

Proof. If a hypergraph $\mathcal{H}$ on n vertices has genus g, there exists a vertex v of $\mathcal{H}$ with charge $\xi_1(v) \le 6 + \frac{12}{n}(g-1)$. As $n \ge \max|E|$ and as $4 - \frac{6}{|E|}$ tends to 4 as $|E|$ goes to infinity, if $M_l(g)$ is large enough, the maximum number of neighbors of v is achieved for v incident to one edges of cardinality $\max|E|$ and two edges of cardinality 2 or one edge of cardinality $\max|E|$ and one edge of cardinality 3. the number of neighbors of v in $\mathcal{H}$ is hence bounded by $\max|E| + 1$. $\square$

4.3   Maximal Cliques

Lemma 5. *For any integer $g \ge 0$, there exists a constant $C_l(g)$, such that, for any linear hypergraph of genus g on $n \ge C_l(g)$ vertices, if all the vertices are pairwise adjacent, then there is an edge of cardinality at least $n - 1$.*

Proof. As $\mathcal{H}$ is linear, there exists a vertex v having at most $\max|E| + 1$ neighbors. Hence, $\max|E| \ge n - 2$. Assume that the maximal cardinality of an edge of $\mathcal{H}$ is $n - 2$ and that this value is achieved by an edge E. Let a and b be the two vertices of $\mathcal{H}$ that do not belong to E. As $\mathcal{H}$ is linear, no edge different from E may include (at least) two vertices of E. The same way, at most one edge include both a and b and at most one vertex of E. Therefore, by deleting E and at most one vertex, we get a 2-uniform linear hypergraph of genus at most g which adjacency multigraph is homeomorphic to $K_{3,n-3}$ (plus some parallel edges). If $C_l(g)$ is large enough, the complete bipartite graph $K_{3,n-3}$ has genus strictly greater than g and we are lead to a contradiction. $\square$

As a direct consequence, we get:

Theorem 5. *For any integer $g \geq 0$, there exists a constant $C_l(g)$, such that any linear hypergraph of genus g has a clique size bounded by:*

$$\omega(\mathcal{H}) \leq \max(C_l(g), \max|E| + 1) \tag{5}$$

Remark 3. A technical easy analysis shows that $C_l(0) = 4$, that is: any planar linear hypergraph $\mathcal{H}$ which has a clique of size at least 5 has an edge of cardinality at least $\omega(\mathcal{H}) - 1$.

4.4   Triangle-Free Linear Hypergraphs

We shall say that a linear hypergraph is *triangle-free* if any triplet of mutually adjacent vertices is included into an edge. Obviously, a triangle free linear hypergraph $\mathcal{H}$ satisfies:

$$\omega(\mathcal{H}) = \max|E| \tag{6}$$

For such hypergraphs, we may also define a new charge function $\xi_\triangle$ by:

$$\xi_\triangle(v) = \sum_{E \ni v} \left(3 - \frac{4}{|E|} \right) \tag{7}$$

As a linear hypergraph $\mathcal{H}$ is obviously triangle-free if and only if $\mathrm{Incid}(\mathcal{H})$ is C_6-free, we get, with similar arguments as those used in Lemma 1 and Lemma 4, that the sum of the charges of the vertices of a triangle-free linear hypergraph $\mathcal{H}$ with n vertices is bounded by $4n + 8(g-1)$. With similar arguments as those used in Theorem 1 and Theorem 1, we get:

Theorem 6. *For any integer $g \geq 0$, there exist a constant $M'(g)$, such that any triangle-free linear hypergraph of genus g with at least an edge of cardinality $M'(g)$ has a vertex v with at most $\max|E|$ neighbors.*

5   Graphs with Given Reduced Genus

As any multigraph is its own adjacency multigraph, we may define:

Definition 6. *The reduced genus $\widetilde{\gamma}(G)$ of a multigraph G is the minimum genus of a hypergraph which adjacency multigraph is G.*

Remark that, if $\gamma(G)$ denotes the genus of the graph G,

$$0 \leq \widetilde{\gamma}(G) \leq \gamma(G) \tag{8}$$

For instance, the reduced genus of a triangle-free multigraph is equal to its genus and the reduced genus of a complete graph is equal to 0.

A generalization of the triangle-free simple graphs is the clique partitioned graphs:

Definition 7. *A simple graph* G *is a* clique partitioned graph *if no two maximal cliques of* G *have more than one vertex in common.*

Examples of clique partitioned graphs are triangle free graphs and line graphs of triangle-free graphs. It is not difficult to check that any clique partitioned graph of reduced genus g is the adjacency graph of a triangle-free linear hypergraph of genus g.

Problem 1. For a fixed integer $g \geq 0$, what is the complexity of the following decision problem: "Given a graph G, decide whether the reduced genus of G is at most g"?

5.1   Coloring

We first recall some basic definitions on graph coloring. The concept of *list coloring* was introduced by Vizing [30] and independently by Erdös, Rubin and Taylor [11]: Assuming that a list $L(v)$ is associated to each vertex v of a graph G, a mapping $f : V \to \bigcup_{v \in V} L(v)$ is a *list coloring* if f is a proper coloring and $f(v) \in L(v)$ holds for all $v \in V$. If $|L(v)| = k$ holds for all $v \in V$, L is a *k-assignment*. The *choice number* $\mathrm{ch}(G)$ of G is the smallest k such that every k-assignment of G admits a strict list coloring. For $\mathrm{ch}(\mathcal{H}) \leq k$, G is said to be *k-choosable*. The *coloring number* $\mathrm{col}(G)$ of a graph G is the largest integer k such that G has a subgraph with minimum degree $k - 1$. If $\omega(G)$ and $\chi(G)$ denote, as usual, the *clique number* and the *chromatic number* of the graph G, we have:

$$\omega(G) \leq \chi(G) \leq \mathrm{ch}(G) \leq \mathrm{col}(G) \tag{9}$$

According to the definition of the coloring number, we have, as a direct consequence of Theorem 4:

Theorem 7. *For any integer* $g \geq 0$, *there exists a constant* $H_l(g)$, *such that any simple graph of reduced genus* g *has a coloring number bounded by:*

$$\mathrm{col}(G) \leq \max(H_l(g), \omega(G) + 2) \tag{10}$$

And, according to Theorem 6:

Theorem 8. *For any integer* $g \geq 0$, *there exists a constant* $H'(g)$, *such that any clique partitioned graph of reduced genus* g *has a coloring number bounded by:*

$$\mathrm{col}(G) \leq \max(H'(g), \omega(G) + 1) \tag{11}$$

On the other hand, the following results are know on planar graphs:

- Every planar graph is 4-colorable (Appel and Haken [1][2]; Robertson, Sanders Seymour and Thomas [23][24])
- Every planar graph is 5-choosable (Thomassen [29])
- There exists a non-4-choosable planar graph (A 3-colorable non-4-choosable graph has been independently obtained by H. Ben Meeki and H.A. Kierstead and by M. Voigt and B. Wirth [32])

- Every triangle-free planar graph is 3-colorable (Grötzsch [13])
- Every triangle-free planar graph is 4-choosable (obvious, as $\mathrm{col}(G) \leq 4$)
- There exists a non-3-choosable triangle-free planar graph (Voigt [31])

These results suggest the following problems:

Problem 2. For any integer $g \geq 0$, what is the smallest value $C(g) \geq 0$, such that the choice number of any simple graph G with reduced genus g is bounded by:

$$\mathrm{ch}(G) \leq \max(C(g), \omega(G) + 2) \tag{12}$$

In particular, if a simple graph G has reduced genus 0, is G $(\omega(G)+2)$-choosable?

Problem 3. Does any simple graph G with reduced genus g have a chromatic number bounded by:

$$\chi(G) \leq \max(H(g), \omega(G) + 1) \tag{13}$$

where $H(g) = \frac{1}{2}\left(7 + \sqrt{1 + 48g}\right)$ is Heawood's function?
   If G is clique partitioned, can the bound be improved as follows?

$$\chi(G) \leq \max(H(g) - 1, \omega(G)) \tag{14}$$

Remark that the bound $\omega(G) + 1$ would be optimal:

Proposition 1. *For all $k \geq 2$, there exists a simple graph G with reduced genus 0, such that $\omega(G) = k$ and $\chi(G) = k + 1$.*

Proof. If $k = 2$, consider C_5.
   Let $\mathcal{H}$ be the linear planar hypergraph with vertex set

$$V = \{a, b, c, \alpha_2, \ldots, \alpha_k, \beta_2, \ldots, \beta_{k-1}\}$$

and edge set $\mathcal{E} = \{E_1, \ldots, E_{2k+1}\}$, where

$$\begin{aligned}
E_1 &= \{a, \alpha_2, \ldots, \alpha_k\} \\
E_i &= \{b, \alpha_i\} \quad (2 \leq i \leq k) \\
E_{k+1} &= \{b, \beta_2, \ldots, \beta_{k-1}, \alpha_k\} \\
E_{k+i} &= \{c, \beta_i\} \quad (2 \leq i \leq k - 1) \\
E_{2k} &= \{c, \alpha_k\} \\
E_{2k+1} &= \{a, c\}
\end{aligned}$$

and let $G = \mathrm{Adj}(\mathcal{H})$ be the adjacency multigraph of $\mathcal{H}$. Obviously, $k \leq \chi(G) \leq k + 1$. Assume $\chi(G) = k$ and let $\mathrm{Color} : V \to [1, \ldots, k]$ be a proper coloring of G. Then, considering the cliques induced by E_1 and E_i for $2 \leq i \leq k$, we get $\mathrm{Color}(a) = \mathrm{Color}(b)$. Similarly, considering E_{k+1} and E_{k+i} for $2 \leq i \leq k$, we get $\mathrm{Color}(b) = \mathrm{Color}(b)$. Thus, $\mathrm{Color}(a) = \mathrm{Color}(c)$ and Color is not a proper coloring, according to the edge E_{2k+1}. Hence, $\chi(G) = k + 1$. $\square$

Also, the condition that G is a simple graph (and not a multigraph) is crucial, as adding parallel edges allows to drastically decrease the reduced genus. Actually, we have:

Proposition 2. *For all $k \geq 1$, there exists a multigraph G which is the adjacency graph of a $2k$-uniform planar hypergraph $\mathcal{H}$, and such that $\chi(G) \geq \frac{5}{4}\omega(G)$.*

Proof. Consider the hypergraph $\mathcal{H}$ with vertex set

$$V = \{a_1, \ldots, a_k, b_1, \ldots, b_k, c_1, \ldots, c_k, d_1, \ldots, d_k, e_1, \ldots, e_k\}$$

and edge set $\mathcal{E} = \{E_1, \ldots, E_5\}$, where

$$E_1 = \{a_1, \ldots, a_k, b_1, \ldots, b_k\}$$
$$E_2 = \{b_1, \ldots, b_k, c_1, \ldots, c_k\}$$
$$E_3 = \{c_1, \ldots, c_k, d_1, \ldots, d_k\}$$
$$E_4 = \{d_1, \ldots, d_k, e_1, \ldots, e_k\}$$
$$E_5 = \{e_1, \ldots, e_k, a_1, \ldots, a_k\}$$

Then, $\omega(G) = 2k$. Consider any optimal proper strict coloring of G and let $C_1, \ldots, C_5$ be the color sets used by vertices $a_1, \ldots, a_k$, resp. $b_1, \ldots b_k$, resp. $c_1, \ldots, c_k$, resp. $d_1, \ldots, d_k$, resp. $e_1, \ldots, e_k$. Then, any C_i has cardinality k and three of the C_i have an empty intersection. Thus,

$$\chi(G) = \left| \bigcup_i C_i \right|$$
$$= \sum_i |C_i| - \sum_{i \neq j} |C_i \cap C_j|$$
$$= \sum_i |C_i| - \frac{1}{2} \sum_i \left| L_i \cap \left(\bigcup_{j \neq i} L_j \right) \right|$$
$$\geq \frac{1}{2} \sum_i |C_i|$$
$$\geq \frac{5k}{2}$$

□

5.2   Maximal Cliques

Computing the clique number of a graph is an NP-complete problem. However, we shall prove that, for any fixed $g \geq 0$, this problem is polynomial. More precisely:

Theorem 9. *For any fixed $g \geq 0$, there exists a polynomial algorithm that enumerates all the maximal cliques of any simple graph G with reduced genus g.*

Proof. Consider a linear hypergraph $\mathcal{H}$ of genus g, which adjacency multigraph is G. According to Theorem 5, the maximal cliques of G have a simple structure: either they are of size at most $C_l(g)$, or all the vertices of the clique (but maybe one) belong to a same edge of $\mathcal{H}$. Hence, the number of maximal cliques is bounded by $P_g(n) = \sum_{i=1}^{C_l(g)} n^i + |\mathcal{E}| + n|\mathcal{E}|$. As the number of edges $|\mathcal{E}|$ of $\mathcal{H}$ is at most n^2, $P(n)$ is a polynomial function of n.

Consider the following recursive algorithm computing the set $\mathcal{F}(G)$ of the maximal cliques of G:

- If G is empty, $\mathcal{F}(G) = \{\emptyset\}$;
- Otherwise, $\mathcal{F}(G)$ is the set of the maximal sets in $\mathcal{F}(G-v) \cup \{K \cup \{v\}, K \in \mathcal{F}(G-v) \text{ and } K \cap N(v) \neq \emptyset\}$, where v is any vertex of G.

As the number of maximal cliques of a graph of reduced genus is bounded by a polynomial function of n and as the deletion of a vertex may not increase the reduced genus, the preceding algorithm computes the set of the maximal cliques of G in polynomial time.   □

5.3   Point Arboricity

Definition 8. *The* point arboricity $\rho(G)$ *of a simple graph G is the minimum number of subsets into which the vertex set of G may be partitioned so that each subset induces an acyclic subgraph.*

From [4], it follows that:

$$\left\lceil \frac{\chi(G)}{2} \right\rceil \leq \rho(G) \leq \left\lceil \frac{\mathrm{col}(G)}{2} \right\rceil \tag{15}$$

So that, according to Theorem 7, for any integer $g \geq 0$, if $\omega(G)$ is large enough, then:

$$\left\lceil \frac{\omega(\mathcal{H})}{2} \right\rceil \leq \rho(\mathcal{H}) \leq 2 + \left\lceil \frac{\omega(\mathcal{H})}{2} \right\rceil \tag{16}$$

For a planar graph G, it is proved in [4] that the bound $\rho(G) \leq 3$ is sharp. For a triangle-free planar graph, $\mathrm{col}(G) = 4$ so that $\rho(G) \leq 2$ and the bound is obviously sharp. Moreover, according to Proposition 1, for any integer $k \geq 2$, there exists a simple graph G of reduced genus 0, such that $\omega(G) = k$ and $\chi(G) = k + 1$. For such a simple graph, we get :

$$\rho(G) \geq \left\lceil \frac{\omega(G) + 1}{2} \right\rceil = 1 + \left\lfloor \frac{\omega(G)}{2} \right\rfloor \tag{17}$$

Thus, if true, the following bound would be sharp.

Problem 4. For any integer $g \geq 0$, does there exist a constant $P(g)$ such that any simple graph of reduced genus g and clique number $\omega(G) \geq P(g)$ has a point arboricity bounded by:

$$\rho(G) \leq 1 + \left\lfloor \frac{\omega(G)}{2} \right\rfloor \tag{18}$$

6   Representations of Hypergraphs

It appears that a planar graph representation which may be easily extended to planar linear hypergraphs is the visibility representation.

6.1   Visibility Representation

A useful way of representing planar graphs is the *visibility representation*:

Theorem (Rosenstiehl, Tarjan [25]). *A graph is planar if and only if it has a visibility representation, that is a representation in the plane in which the vertices are represented by horizontal line segments, the edges are represented by vertical line segments, and such that:*

- *no two horizontal (resp. vertical) segments intersect,*
- *no two segment cross,*
- *an edge is incident to a vertex if and only if the corresponding segments touch each other.*

This theorem has been extended to toroidal visibility representations (see [20]).

This representation may be extended to planar hypergraphs and (up to a reformulation of the original theorem) we get:

Theorem (de Fraysseix, Ossona de Mendez, Pach [8]). *A hypergraph is planar if and only if it has a "visibility representation", that is a representation in which the vertices are represented by horizontal line segments, the edges are represented by vertical line segments, and such that:*

- *no two horizontal (resp. vertical) segments intersect,*
- *no two segment cross,*
- *an edge is incident to a vertex if and only if the corresponding segments touch each other.*

6.2   Straight Line Representation

It is a classical result independently established by Wagner [33], Fáry [12] and Stein [26] (which is also a consequence of the Steinitz's theorem on convex polytopes [27]) that any simple planar graph has a straight line representation in the plane.

Problem 5. Has any planar linear hypergraph a straight line representation, that is a representation in which:

- vertices are represented by pairwise distinct points,
- edges are represented by pairwise non-overlapping segments,
- a vertex v is incident to an edge E if and only if the point representing v belongs to the segment representing E.

If we relax the "straight line" condition is the following true?

Problem 6. Has any hypergraph of genus g a representation on a surface of genus g in which:

- vertices are represented by pairwise distinct points,
- edges are represented by pairwise non-overlapping arcs,
- a vertex v is incident to an edge E if and only if the point representing v belongs to the arc representing E.

In a "dual" form (dual in the sense of dual hypergraphs, not in the sense of algebraic or planar duality), Problem 5 may be restated as follows:

Problem 7. Is any simple graph of reduced genus 0 the intersection graph of a family of straight line segments in the plane?

This problem extends the following conjecture on planar graphs:

Conjecture 1. Any simple planar graph is the intersection graph of a family of straight line segments in the plane.

A weaker conjecture may be stated on planar multigraphs (it has been partially solved in [6] and some advancement on the "stretching" problem of the arcs may be found in [7]):

Conjecture 2. Any planar multigraph is the intersection multigraph of a family of arcs.

Actually, Problem 6 is an extension of this problem, as it can be restated as follows:

Problem 8. Is any multigraph of reduced genus g the intersection multigraph of a family of arcs on a surface of genus g?

6.3   Representation by Contact of Triangles

Other representation may probably be extended to planar linear hypergraphs. For instance, the representation by contact of triangles:

Theorem ([9]). *Any planar graph is a contact graph of triangles.*

Problem 9. Is any planar linear hypergraph the contact hypergraph of triangles in the plane?

References

1. K. Appel and W. Haken. Every planar map is four colorable. part I. discharging. *Illinois J. Math.*, 21:429–490, 1977.
2. K. Appel, W. Haken, and J. Koch. Every planar map is four colorable. part II. reducibility. *Illinois J. Math.*, 21:491–567, 1977.
3. C. Berge. *Graphes et hypergraphes*. Dunod, Paris, second edition, 1973.

4. G. Chartrand and H.V. Kronk. The point arboricity of planar graphs. *J. Lond. Math. Soc.*, 44:612–616, 1969.
5. R. Cori. *Un code pour les graphes planaires et ses applications*, volume 27. Société mathématique de france, Paris, 1975.
6. H. de Fraysseix and P. Ossona de Mendez. Intersection graphs of Jordan arcs. In *Stirin 1997 Proc.* DIMATIA-DIMACS. (accepted).
7. H. de Fraysseix and P. Ossona de Mendez. Stretchability of Jordan arc contact systems. Technical Report 98-387, KAM-DIMATIA Series, 1998.
8. H. de Fraysseix, P. Ossona de Mendez, and J. Pach. Representation of planar graphs by segments. *Intuitive Geometry*, 63:109–117, 1991.
9. H. de Fraysseix, P. Ossona de Mendez, and P. Rosenstiehl. On triangle contact graphs. *Combinatorics, Probability and Computing*, 3:233–246, 1994.
10. B. Descartes. Solution to advanced problem no. 4256. *Am. Math. Mon.*, 61:352, 1954.
11. P. Erdös, A.L. Rubin, and H. Taylor. Choosability in graphs. In *Proc. West-Coast Conference on Combinatorics, Graph Theory and Computing*, volume XXVI, pages 125–157, Arcata, California, 1979.
12. I. Fáry. On straight line representation of planar graphs. *Acta Scientiarum Mathematicarum (Szeged)*, 11:229–233, 1948.
13. H. Grötzsch. Ein Dreifarbensatz für dreikreisfrei Netze auf der Kugel. *Wissenchaftliche Zeitschrift der Martin-Luther-Universität Halle-Wittenberg. Mathematisch-Naturwissenschaftliche Reihe*, 8:109–120, 1958/1959.
14. P.J. Heawood. Map color theorem. *Q.J. Pure Appl. Math.*, 24:332–338, 1890.
15. H. Heesch. Untersuchungen zum Vierfarbenproblem. Hochschulskriptum 8 10/a/b, Bibliographisches Institut, Mannheim, 1969.
16. D.S. Johnson and H.O. Pollak. Hypergraph planarity and the complexity of drawing Venn diagrams. *Journal of Graph Theory*, 11(3):309–325, 1987.
17. R.P. Jones. *Colourings of Hypergraphs*. PhD thesis, Royal Holloway College, Egham, 1976.
18. J.B. Kelly and L.M. Kelly. Paths and circuits in critical graphs. *Am. J. Math.*, 76:786–792, 1954.
19. L. Lovasz. On chromatic numbers of finite set systems. *Acta Math. Acad. Sci. Hung.*, 19:59–67, 1968.
20. B. Mohar and P. Rosenstiehl. A flow approach to upward drawings of toroidal maps. In *proc. of Graph Drawing '94*, pages 33–39, 1995.
21. J. Mycielski. Sur le coloriage des graphes. *Colloq. Math.*, 3:161–162, 1955.
22. G. Ringel and J.W.T. Youngs. Solution of the Heawood map coloring problem. volume 60 of *Proc. Nat. Acad. Sci.*, pages 438–445, 1968.
23. N. Robertson, D.P. Sanders, P.D. Seymour, and R. Thomas. A new proof of the four color theorem. *Electron. Res. Announc. Amer. Math. Soc.*, 2:17–25, 1996. (electronic).
24. N. Robertson, D.P. Sanders, P.D. Seymour, and R. Thomas. The four color theorem. *J. Comb Theory*, B(70):2–44, 1997.
25. P. Rosenstiehl and R.E. Tarjan. Rectilinear planar layout and bipolar orientation of planar graphs. *Discrete and Computational Geometry*, 1:343–353, 1986.
26. S.K. Stein. Convex maps. In *Proc. Amer. Math. Soc.*, volume 2, pages 464–466, 1951.
27. E. Steinitz and H. Rademacher. *Vorlesung über die Theorie der Polyeder*. Springer, Berlin, 1934.
28. G. Szekeres and H.S. Wilf. An inequality for the chromatic number of a graph. *J. Comb. Theory*, 4:1–3, 1968.

29. C. Thomassen. Every planar graph is 5-choosable. *Journal of Combinatorial Theory*, B(62):180–181, 1994.
30. V.G. Vizing. Coloring the vertices of a graph in prescribed colors. *Mtody Diskret. Anal. v Teorii Kodov i Schem*, 29:3–10, 1976. (In russian).
31. M. Voigt. A not 3-choosable planar graph without 3-cycles. *Discrete Math.*, 146:325–328, 1995.
32. M. Voigt and B. Wirth. On 3-colorable non 4-choosable planar graphs. *Journal of Graph Theory*, 24:233–235, 1997.
33. K. Wagner. Bemerkungen zum Vierfarbenproblem. *Jber. Deutsch. Math. Verein*, 46:26–32, 1936.
34. T.R.S. Walsh. Hypermaps versus bipartite maps. *J. Combinatorial Theory*, 18(B):155–163, 1975.
35. A.T. White. *Graphs, Groups and Surfaces*, volume 8 of *Mathematics Studies*, chapter 13, pages 205–210. North-Holland, Amsterdam, revised edition, 1984.
36. A.A. Zykov. Hypergraphs. *Uspeki Mat. Nauk*, 6:89–154, 1974.

Classifying Discrete Temporal Properties

Thomas Wilke*

Institut für Informatik und Praktische Mathematik
Christian-Albrechts-Universität zu Kiel, D-24098 Kiel, Germany
tw@informatik.uni-kiel.de, http://www.informatik.uni-kiel.de/~tw/

Abstract. This paper surveys recent results on the classification of discrete temporal properties, gives an introduction to the methods that have been developed to obtain them, and explains the connections to the theory of finite automata, the theory of finite semigroups, and to first-order logic.

The salient features of temporal logic[1] are its modalities, which allow it to express temporal relationships. So it is only natural to investigate how and how much each individual modality contributes to the expressive power of temporal logic. One would like to be able to answer questions like: Can a given property be expressed without using the modality "next"? What properties can be expressed using formulas where the nesting depth in the modality "until" is at most 2?

This survey reports on recent progress on answering such questions, presenting results from the papers [3], [2], [14], [11], and [16] and the thesis [17].

The results fall into three categories: (A) characterizations of fragments of future temporal logic, where a fragment is determined by which future modalities (modalities referring to the future and present only) are allowed in building formulas; (B) characterizations of symmetric fragments, where with each modality its symmetric past/future counterpart is allowed; (C) characterization of the levels of the until hierarchy, where the nesting depth in the "until" modality required to express a property in future temporal logic determines its level.

An almost complete account of the results from category (A) will be given in Sections 2 through 4, including full proofs. These results can be obtained with a reasonable effort in an automata-theoretic framework and the methods used to obtain them are fundamental to the whole subject, in particular, to the results from categories (B) and (C). The results from these two categories are presented in Sections 5 and 6 without going into details of the proofs, which would require a thorough background in finite semigroup theory.

In computer science applications, temporal formulas are interpreted in (finite or infinite) sequences (colored discrete linear orderings), which are nothing else than words (strings or ω-words). Therefore, the set of models of a temporal formula—the property defined by it—can be viewed as a formal language, in

---

* Part of the research reported here was conducted while the author was postdoc at DIMACS as part of the Special Year on Logic and Algorithms.

[1] I use "temporal logic" as a synonym for "propositional linear-time temporal logic."

fact, a regular language. In other words, characterizing a fragment of temporal logic amounts to characterizing a certain class of regular languages.

There is a long tradition of classifying regular languages, going back to as early as 1965, when Schützenberger in the seminal paper of the field, [12], characterized the star-free languages as being exactly the ones whose minimal DFA's are counter-free. Given that temporal logic and star-free expressions have the same expressive power (which was only realized much later [5,4,10]), Schützenberger's result also marked the first step in classifying discrete temporal properties: it gave an effective characterization of the class of all regular languages expressible in temporal logic. After an introductory section with terminology and notation, this survey starts off in Section 2 with a new, brief proof that every language recognized by a counter-free DFA is expressible in future temporal logic.

This paper only deals with strings, but most of the results have been extended to ω-words. The reader is referred to the respective original papers.

1 Basic Terminology and Notation

We interpret temporal formulas in strings and use standard notation with regard to strings. The positions of a string of length n are indexed by $0, \ldots, n-1$. When u is a string of length n and $0 \leq i \leq j \leq n$, then $u(i, j)$ denotes the string $u(i)u(i+1)\ldots u(j-1)$. Further, $u(i, *)$ denotes the suffix $u(i, n)$.

A temporal formula over some alphabet Σ is built from the logical constants $\top$ (true) and $\bot$ (false) and the elements of Σ using the boolean connectives $\neg$ (negation), $\wedge$ (conjunction), and $\vee$ (disjunction) and the *temporal modalities* X (next), F (eventually), and U (until). All connectives and modalities are unary except for $\wedge$, $\vee$, and U, which are binary and written in infix notation. The set of all temporal formulas is denoted by TL.

A *fragment* of temporal logic is a subset of TL obtained by allowing only the use of certain temporal modalities in the construction of formulas. When l is a list of temporal modalities, then TL$[l]$ denotes the respective fragment. For instance, TL[F] stands for the class of all temporal formulas which can be built from alphabet symbols and the logical constants using boolean connectives and F as the only temporal modality.

Given a temporal formula φ and a string u, one defines what it means for φ to hold in u, denoted $u \models \varphi$. This definition is inductive, where, in particular,

— for every symbol a, $u \models a$ if $u(0) = a$,

— $u \models X\varphi$ if $|u| > 1$ and $u(1, *) \models \varphi$,

— $u \models F\varphi$ if there exists i with $0 < i < |u|$ such that $u(i, *) \models \varphi$, and

— $u \models \varphi U \psi$ if there exists i with $0 < i < |u|$ such that $u(j, *) \models \varphi$ for every $j \in \{1, \ldots, i-1\}$ and $u(i, *) \models \psi$.

Note that $\bot U \varphi$ has the same meaning as $X\varphi$ for any temporal formula φ, and $\top U \varphi$ has the same meaning as $F\varphi$, which means X and F can be derived from U. Sometimes, we will also use the temporal modality G (always), which is another derived modality: it stands for $\neg F \neg$.

The two modalities F and U have so-called *stutter-invariant counterparts* (for an explanation of the terminology, see Section 4), denoted F_{sf} and U_{sf}, respectively. Their meaning is defined just as above except that i is allowed to be 0 and 0 must also be considered for j. In this regard, the modalities X, F, and U will be referred to as *strict modalities*.

Given a temporal formula φ over some alphabet Σ and an alphabet Γ, we write $\mathcal{L}_\Gamma(\varphi)$ for the set $\{u \in \Gamma^+ \mid u \models \varphi\}$ and say $\mathcal{L}_\Gamma(\varphi)$ is the language over Γ *defined* by φ. (Observe that if Γ is an arbitrary alphabet, φ an arbitrary formula, and ψ the formula obtained from φ by replacing every alphabet symbol not from Γ by $\bot$, then $\mathcal{L}_\Gamma(\varphi) = \mathcal{L}_\Gamma(\psi)$. This means one can always assume that a defining formula only uses symbols from the alphabet of the language in question.) A language is said to be *expressible* in temporal logic (or TL-expressible) if there is a temporal formula that defines it. Similarly, when $\boldsymbol{F}$ is a fragment of temporal logic, a language is expressible in $\boldsymbol{F}$ if there exists a formula in $\boldsymbol{F}$ that defines it.

A *deterministic finite automaton (DFA)* is a tuple $\boldsymbol{A} = (\Sigma, Q, q_I, \delta, F)$ where Σ is a finite alphabet, Q a finite set of *states*, $q_I \in Q$ the *initial state*, $\delta \colon Q \times A \to Q$ the *transition function*, and $F \subseteq Q$ the set of *final states*. The *extended transition function* of $\boldsymbol{A}$, denoted δ^*, is defined by $\delta^*(q, \epsilon) = q$ for $q \in Q$ and $\delta^*(q, ua) = \delta(\delta^*(q, u), a)$ for $q \in Q$, $u \in \Sigma^*$, and $a \in \Sigma$. The language *recognized* by $\boldsymbol{A}$, denoted $\mathcal{L}(\boldsymbol{A})$, is defined by $\mathcal{L}(\boldsymbol{A}) = \{u \in \Sigma^+ \mid \delta^*(q_I, u) \in F\}$. Given a regular language L, the minimal DFA for L is denoted by $\boldsymbol{A}_L$.

When u denotes a string, then u^ρ denotes the *reverse* of u, i.e., if u is of length n, then $u^\rho = u(n-1)u(n-2)\ldots u(0)$. Accordingly, when L denotes a language, then L^ρ denotes the reverse of L, i.e., the language $\{u^\rho \mid u \in L\}$.

2   Full Temporal Logic

It is easy to see that every language expressible in temporal logic is a regular language, i.e., recognizable by a DFA. This raises the question what regular languages are exactly the ones that are expressible in temporal logic. Recall that the minimal DFA recognizing a given regular language is a canonical object to consider when one is interested in classifying a regular language. So more concretely, one can ask for a structural property of DFA's that is enjoyed by the minimal DFA of a given regular language if and only if the language is expressible in temporal logic.

The adequate property is known as counter-freeness. Given a DFA $\boldsymbol{A}$, a sequence $q_0, \ldots, q_{m-1}$ of distinct states is a *counter* for a string u if $m > 1$ and $\delta^*(q_i, u) = q_{i+1}$ for $i < m$ where, by convention, $q_m = q_0$. A DFA is *counter-free* if it does not have a counter.

Theorem 1. [10,4] *A regular language L is expressible in TL if and only if $\boldsymbol{A}_L$ is counter-free.*

This theorem is a simple consequence of two fundamental results: in 1971, McNaughton and Papert [10] proved that counter-free DFA's recognize exactly the languages that are expressible in first-order logic; in 1980, Gabbay, Pnueli, Shelah, and Stavi [4] showed that temporal logic is as expressive as first-order logic.[2] The latter result is an improvement of a result of Kamp [5] from 1968 that says that temporal logic with future as well as past operators is as expressive as first-order logic in Dedekind-complete orderings.

The difficult implication in Theorem 1 is the one that asserts that a regular language L is expressible in temporal logic if $\mathbf{A}_L$ is counter-free. For this part of the theorem only a few direct proofs have been presented thus far. There is a journal paper by Cohen, Perrin, and Pin [1], Maler's thesis [8], and an accompanying conference paper by Maler and Pnueli [9]. Cohen et al. as well as Maler and Pnueli use some kind of decomposition theory (for finite semigroups or for finite automata); the proof presented below, from [17], avoids such theories.

We need more terminology and notation. A *pre-automaton* is a triple (Σ, Q, δ) where Σ is a finite alphabet, Q a finite set of states, and $\delta \colon Q \times \Sigma \to Q$ a transition function. In other words, a pre-automaton is a DFA without initial and final states. The terminology and notation we have introduced for DFA's transfers to pre-automata in a straightforward way (if applicable). For instance, the extended transition function of a pre-automaton and the property of being counter-free are defined in exactly the same way as for DFA's.

Given a set Q, we view the set Q^Q of all functions on Q as a finite semigroup with composition as product operation. Given $\alpha, \beta \colon Q \to Q$, we write $\alpha\beta$ for the composition of α and β, i.e., for the function given by $q \mapsto \beta(\alpha(q))$. For $\alpha \colon Q \to Q$ and $Q' \subseteq Q$, we write $\alpha[Q']$ for the image of Q' under α, i.e., for $\{\alpha(q) \mid q \in Q'\}$.

Let $\mathbf{A} = (\Sigma, Q, \delta)$ be a pre-automaton. For every string $u \in \Sigma^*$ we define its *transformation*, denoted $u^{\mathbf{A}}$, as follows. For every $q \in Q$ we set $u^{\mathbf{A}}(q) = \delta^*(q, u)$, and we let $S_A = \{u^{\mathbf{A}} \mid u \in \Sigma^+\}$. Clearly, this set is closed under functional composition, that is, it is a subsemigroup of Q^Q. It is called the *transformation semigroup* of $\mathbf{A}$. For every $\alpha \colon Q \to Q$, we set $L_\alpha^A = \{u \in \Sigma^+ \mid u^{\mathbf{A}} = \alpha\}$. Further, $\tilde{L}_\alpha^A$ denotes $L_\alpha^A \cup \{\epsilon\}$ if $\alpha = id_Q$ and else L_α^A.— Observe that if a pre-automaton as above is counter-free and u is a string such that $u^{\mathbf{A}}[Q] = Q$, then $u^{\mathbf{A}} = id_Q$.

Proof of Theorem 1, from a counter-free DFA to a temporal formula, [17]. We prove that for every pre-automaton $\mathbf{A} = (\Sigma, Q, \delta)$ and every $\alpha \in S_A$ the language L_α^A is expressible in temporal logic, which is obviously enough. The proof goes by induction on $|Q|$ in the first place and then on $|\Sigma|$: in the induction step, we will consider pre-automata with the same state space but over a smaller alphabet as well as pre-automata with a smaller state space but over a much larger alphabet.

We distinguish two cases. First, assume there is no symbol $a \in \Sigma$ such that $a^{\mathbf{A}}[Q] \subsetneq Q$. Then $a^{\mathbf{A}} = id_Q$ for every $a \in \Sigma$, which means $S_A = \{id_Q\}$. This implies $L_\alpha^A = \Sigma^+$ for every $\alpha \in S_A$, and Σ^+ is obviously expressible in temporal logic.

---

[2] In [4], the authors interpreted temporal logic and first-order logic in ω-words. It is, however, obvious that their result is also valid for strings.

Second, assume $b \in \Sigma$ is such that $b^A[Q] \subsetneq Q$. Let $Q' = b^A[Q]$, $\Gamma = \Sigma \setminus \{b\}$, and let B be the pre-automaton which results from A by restricting it to the symbols from Γ. Further, let $U_0 = \Gamma^* b$, $\Delta = \{u^A \mid u \in U_0\}$, and set $C = (\Delta, Q', \delta')$ where $\delta'(q, \alpha) = \alpha(q)$ for every $q \in Q'$ and $\alpha \in \Delta$. Finally, let $h: U_0^+ \to \Delta^+$ be the function defined by $h(u_0 \ldots u_{n-1}) = u_0^A \ldots u_{n-1}^A$ for $u_0, \ldots, u_{n-1} \in U_0$.

Let $\alpha \in S_A$. We want to show that L_α^A is TL-expressible. To this end, we first partition L_α^A according to how many b's occur in a string; we set

$$L_0 = L_\alpha^A \cap \Gamma^+ , \qquad L_1 = L_\alpha^A \cap \Gamma^* b \Gamma^* , \qquad L_2 = L_\alpha^A \cap \Gamma^* b \Sigma^* b \Gamma^* .$$

Then $L_\alpha^A = L_0 \cup L_1 \cup L_2$. Next, we observe that

$$L_0 = L_\alpha^B , \qquad L_1 = \bigcup_{\alpha = \beta b^A \beta'} \overbrace{\tilde{L}_\beta^B b \tilde{L}_{\beta'}^B}^{L_{\beta,\beta'}} , \qquad L_2 = \bigcup_{\alpha = \beta b^A \gamma \beta'} \overbrace{\tilde{L}_\beta^B b h^{-1}(L_\gamma^C) \tilde{L}_{\beta'}^B}^{L_{\beta,\gamma,\beta'}} ,$$

where $\beta, \beta' \in S_B \cup \{id_Q\}$, and $\gamma \in S_C$. Further, we see that

$$L_{\beta,\beta'} = \tilde{L}_\beta^B b \Sigma^* \cap \Gamma^* b \tilde{L}_{\beta'}^B, \quad L_{\beta,\gamma,\beta'} = \tilde{L}_\beta^B b \Sigma^* \cap \Gamma^* b h^{-1}(L_\gamma^C) \Gamma^* \cap \Sigma^* b \tilde{L}_{\beta'}^B, \quad (1)$$

for $\beta, \beta' \in S_B \cup \{id_Q\}$, and $\gamma \in S_C$.

By induction hypothesis, we know that all L_β^B with $\beta \in S_B$ and all L_γ^C with $\gamma \in S_C$ are TL-expressible. It is now a manageable "programming task" to show that under these assumptions all the sets that are intersected on the right-hand sides of the equations in (1) are TL-expressible, which means L_α^A is TL-expressible, as temporal logic is closed under disjunction (union) and conjunction (union). Lemmas 1 and 2 below provide the details. □

Lemma 1. *Let Σ be an alphabet, $b \in \Sigma$, and $\Gamma = \Sigma \setminus \{b\}$. Assume $L \subseteq \Sigma^+$ and $L' \subseteq \Gamma^+$ are TL-expressible. Then so are $\Gamma^* b L$, $\Gamma^* b(L + \epsilon)$, $\Sigma^* b L'$, $\Sigma^* b(L' + \epsilon)$, $L' b \Sigma^*$, and $(L' + \epsilon)b\Sigma^*$.*

Proof. First, let φ and ψ be formulas over Σ and Γ, respectively, such that $\mathcal{L}_\Sigma(\varphi) = L$ and $\mathcal{L}_\Gamma(\psi) = L'$. Then

$$\Gamma^* b L = \mathcal{L}_\Sigma(\neg b \, \mathsf{U}_{\mathsf{sf}} \, (b \wedge \mathsf{X}\varphi)) , \qquad \Sigma^* b L' = \mathcal{L}_\Sigma(\mathsf{F}_{\mathsf{sf}}(b \wedge \mathsf{G}\neg b \wedge \mathsf{X}\psi)) .$$

The defining formulas for $\Gamma^* b(L + \epsilon)$ and $\Sigma^* b(L' + \epsilon)$ can be obtained in a similar fashion.

Second, we show by induction that for every temporal formula φ over Γ there exists a temporal formula φ^+ such that $\mathcal{L}_\Sigma(\varphi^+) = \mathcal{L}_\Gamma(\varphi)b\Sigma^*$. We can simply set

$$a^+ = a \wedge \mathsf{F}b , \qquad\qquad (\neg\varphi)^+ = \neg\varphi^+ \wedge \neg b \wedge \mathsf{F}b ,$$
$$(\varphi \wedge \psi)^+ = \varphi^+ \wedge \psi^+ , \qquad\qquad (\varphi \, \mathsf{U} \, \psi)^+ = (\varphi^+ \wedge \neg b) \, \mathsf{U} \, (\psi^+ \wedge \neg b) ,$$

where a stands for an arbitrary element of Γ.

Clearly, $\mathcal{L}_\Sigma(\varphi^+ \vee b) = (\mathcal{L}_\Gamma(\varphi) + \epsilon)b\Sigma^*$. □

Lemma 2. *Let Σ, Δ be alphabets, $b \in \Sigma$, $\Gamma = \Sigma \setminus \{b\}$, and $U_0 = \Gamma^* b$. Further, let $h_0 \colon U_0 \to \Delta$ be an arbitrary function and $h \colon U_0^+ \to \Delta^+$ be defined by $h(u_0 \ldots u_{n-1}) = h_0(u_0) \ldots h_0(u_{n-1})$ for $u_0, \ldots, u_{n-1} \in U_0$. For every $d \in \Delta$, let $L_d = \{u \in \Gamma^+ \mid h_0(ub) = d\}$. Assume $L \subseteq \Delta^+$ is expressible in temporal logic and also L_d for every $d \in \Delta$. Then $h^{-1}(L)\Gamma^*$ is expressible in temporal logic.*

Proof. We show by induction that for every temporal formula φ over Δ there exists a temporal formula $\varphi^\#$ over Σ such that $h^{-1}(\mathcal{L}_\Delta(\varphi))\Gamma^* = \mathcal{L}_\Sigma(\varphi^\#)$. For $d \in \Delta$, we either have $h^{-1}(\mathcal{L}_\Delta(d))\Gamma^* = L_d b \Sigma^*$ or $h^{-1}(\mathcal{L}_\Delta(d))\Gamma^* = (L_d + \epsilon)b \Sigma^*$. Thus, the induction basis follows from the previous lemma and the assumption that the languages L_d are TL-expressible. For the induction step, we can set

$$(\neg\varphi)^\# = \neg\varphi^\# \wedge \mathsf{F_{sf}}b \,, \qquad (\varphi \wedge \psi)^\# = \varphi^\# \wedge \psi^\# \,,$$
$$(\varphi \mathbin{\mathsf{U}} \psi)^\# = \psi^\# \vee (\varphi^\# \wedge (b \to \mathsf{X}\varphi^\#) \mathbin{\mathsf{U}} (b \wedge \mathsf{X}\psi^\#)) \,. \qquad \square$$

The above proofs are constructive, i.e., following these proofs one can actually construct a temporal formula defining the language recognized by a given counter-free automaton. A closer analysis of the constructions sketched in the proofs yields the following quantitative statement. (Recall that for every preautomaton with n states, the cardinality of its transformation semigroup is $2^{\mathcal{O}(n \log n)}$.)

Corollary 1. *For every counter-free DFA with at most n states and at most m symbols in the alphabet, there exists a temporal formula of size $m \, 2^{2^{\mathcal{O}(n \log n)}}$ which defines the language recognized by the DFA.*

3 Strict Fragments

The three basic temporal modalities are X, F, and U. So if we determine fragments of TL by disallowing the use of some of these modalities we obtain eight different fragments. Obviously, some of these have the same expressive power. For instance, the modality X as well as the modality F can be expressed using U only. Thus, all fragments that allow U have the expressive power of full temporal logic:

$$\mathrm{TL}[\mathsf{U}] = \mathrm{TL}[\mathsf{X}, \mathsf{U}] = \mathrm{TL}[\mathsf{F}, \mathsf{U}] = \mathrm{TL}[\mathsf{X}, \mathsf{F}, \mathsf{U}] = \mathrm{TL} \,. \tag{2}$$

By abuse of notation we use an expression like $\mathrm{TL}[\mathsf{X}, \mathsf{U}]$ to refer to the specific fragment of TL as well as to the class of languages expressible in this fragment.

The identities in (2) are the only ones that hold: $\mathrm{TL}[\mathsf{X}]$ and $\mathrm{TL}[\mathsf{F}]$ are incomparable in terms of expressive power and both are stronger than $\mathrm{TL}[\,]$ and weaker than $\mathrm{TL}[\mathsf{X}, \mathsf{F}]$, which in turn is weaker than full temporal logic.

The aim of this section is to provide structural properties that exactly characterize each of these fragments, just as counter-freeness characterizes expressibility in full temporal logic.

3.1   Forbidden Patterns

We need a convenient way to describe structural properties of DFA's and therefore borrow the notion of "forbidden pattern" from Cohen, Perrin, and Pin [1].[3]

For brevity in notation, given a transition function $\delta\colon Q \times \Sigma \to Q$, we define a product $Q \times \Sigma^* \to Q$ by setting $q\,u = \delta^*(q, u)$ for $q \in Q$ and $u \in \Sigma^*$. Given a set N, an N-*labeled digraph* is a tuple (V, E) where V is an arbitrary set and E a subset of $V \times N \times V$. The *transition graph* of a DFA $A = (\Sigma, Q, q_I, \delta, F)$ is the Σ^+-labeled digraph (Q, E) where $E = \{(q, u, q\,u) \mid q \in Q \text{ and } u \in \Sigma^+\}$. So the transition graph of any DFA is an infinite graph. (It has infinitely many edges, but only finitely many vertices.)

A *pattern* is a labeled digraph whose vertices are *state variables*, usually denoted p, q, ... , and whose edges are labeled with *variables for labels* of two different types: variables for nonempty strings, usually denoted u, v, ... , and variables for symbols, usually denoted a, b, ... In addition, a pattern comes with *side conditions* stating which state variables are to be interpreted by distinct states. We will draw patterns just as we draw graphs. Consider, for instance, Figure 1. In this figure, as well as in all subsequent figures depicting patterns, we adopt the convention that all states drawn solid must be distinct.

We say a Σ^+-labeled digraph *matches a pattern* if there is an assignment to the variables obeying the type constraints and the side conditions so that the digraph obtained by replacing each variable by the value assigned to it is a subgraph of the given digraph.

3.2   Classification Theorem

Using the notion of a forbidden pattern, we can now characterize all fragments:

Theorem 2. [10,4,1,3,11] *Let L be a regular language and F one of the fragments* TL[ ], TL[X], TL[F], TL[X, F], *or* TL. *Then L is expressible in F if and only if the transition graph of A_{L^ρ} does not match the pattern(s) for F depicted in Figures 1–6.*

Observe that in Figures 1 and 6 the connected graphs are viewed as different patterns (any of which must not occur), whereas Figure 2 shows only one pattern, which happens to be not connected.

The characterizations given in Theorem 2 for TL[ ] and TL[X] are easy to obtain; the characterization for TL is correct because of Theorem 1. The characterization for TL[X, F] was first obtained by Cohen et al. [1]. An alternative proof and a characterization for TL[F] were given in [3], using the same technique for both fragments. In the following two subsections, this technique is demonstrated.

---

[3] To be precise, what is called a "forbidden pattern" here is referred to as a "forbidden configuration" by Cohen et al.

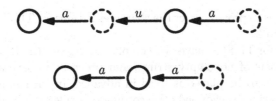

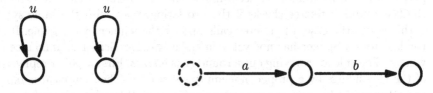

Fig. 1. Patterns forbidden for TL[]

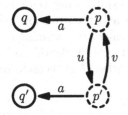

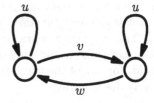

Fig. 2. Pattern forbidden for TL[X]

Fig. 3. Pattern forbidden for stutter invariance

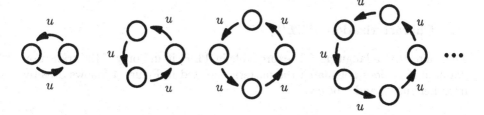

Fig. 4. Pattern forbidden for TL[F]

Fig. 5. Pattern forbidden for TL[X, F]

Fig. 6. Patterns forbidden for TL

3.3   Ehrenfeucht-Fraïssé Games

Ehrenfeucht-Fraïssé (EF) games are a standard tool in mathematical logic to tackle questions about the expressive power of a logic. They allow one to reduce such questions to questions about the existence of strategies in specific two-player games, abstract away syntactical peculiarities, and thus represent the combinatorial core of the problems. In our situation, we will use specifically tailored EF games to prove correct the characterizations for TL[F] (and TL[X, F]) given in Theorem 2.

An EF game for TL[F] is played by two players, *Spoiler* (male) and *Duplicator* (female), on a pair of nonempty strings and proceeds in several rounds. The number of rounds to be played is fixed in advance. In each round, a prefix of each of the two strings is chopped off according to a rule explained below so that the outcome of a round is a new pair of strings or an early win for one of the players if the other cannot act according to the rule. Before each round and after the last round, a referee checks if the two strings start with the same symbol. If this is not the case, the referee calls Spoiler the winner of the game. If after the last round Spoiler has not yet won the game, Duplicator is announced the winner. The rule for carrying out a round is as follows. First, Spoiler replaces one of the two strings by a proper, nonempty suffix of it. Then Duplicator replaces the other string by a proper, nonempty suffix of it. If Spoiler cannot follow this rule because both strings have no proper, nonempty suffix (i. e., if both strings are of length 1), he looses, and if Duplicator cannot reply according to the rules because the other string is of length 1, then Spoiler wins.

The idea behind the game is that Spoiler tries to exhibit a difference between the two strings the game starts with whereas Duplicator tries to show they are similar. This can also be phrased in a formal way: Spoiler has a winning strategy in a k-round game if and only if there is a formula φ of "F depth" at most k that holds for one of the two strings but not for the other. The theorem that we will use is the following.

Theorem 3. [3] *Let L be a language. Then L is expressible in TL[F] if and only if there exists a number k such that for every pair (u, v) with $u \in L$ and $v \notin L$, Spoiler has a winning strategy in the k-round game on (u, v).*

3.4   Characterization of TL[F]

The claim that a language L is expressible in TL[F] if and only if the transition graph of $\mathbf{A}_{L^\rho}$ does not match the pattern depicted in Figure 4 follows directly from Lemmas 3 and 4 below.

Lemma 3. *Let L be a regular language such that the transition graph of $\mathbf{A}_{L^\rho}$ matches the pattern depicted in Figure 4. Then L is not expressible in TL[F].*

Proof. Let $\mathbf{A}_{L^\rho} = (\Sigma, Q, q_I, \delta, F)$ and assume a, u, and v are chosen so that the pattern in Figure 4 is matched. By minimality of $\mathbf{A}_{L^\rho}$, there exist $x, y \in \Sigma^*$ such that $x(uv)^l uay \in L^\rho$ iff $x(uv)^l ay \notin L^\rho$, for every $l \geq 0$. We show that for $l \geq k \geq 0$ and any choice of strings $x, y \in \Sigma^*$, $u, v \in \Sigma^+$, Duplicator wins the k-round game on $(x(uv)^l uay)^\rho$ and $(x(uv)^l ay)^\rho$. Thus, by Theorem 3, L cannot be expressible in TL[F].

First of all, observe that playing on the first $|ay|$ positions of the two strings does not help Spoiler to win the game: Duplicator will simply copy Spoiler's moves. It is therefore sufficient to show that Duplicator wins the k-round game on $(x(uv)^{l+1}u')^\rho$ and $(x(uv)^l u')^\rho$ for $l \geq k \geq 0$ and any choice of strings $x \in \Sigma^*$, $u, u', v \in \Sigma^+$ where u' is a prefix of u.

The proof of this claim is by induction on k. The induction base, $k = 0$, is trivial. For the inductive step, assume $k > 0$. Write s and t for $(x(uv)^{l+1}u')^\rho$ and $(x(uv)^l u')^\rho$. First, suppose Spoiler removes a prefix of length i from t. Then Duplicator replies by removing a prefix of length $i + |uv|$ from s, and the remaining strings will be identical. Second, assume Spoiler removes a prefix from s, say of length i. If $i > |uv|$, then Duplicator removes the prefix of length $i - |uv|$ from t, and the remaining strings will be identical. If $i \leq |uv|$, then Duplicator removes the prefix of length i from t, and the induction hypothesis applies for the following reason. The remaining strings are $(x(uv)^{l+1}u'')^\rho$ and $(x(uv)^l u'')^\rho$ with $u'' \in \Sigma^+$ a prefix of u, or $(xu(vu)^l v')^\rho$ and $(xu(vu)^{l-1}v')^\rho$ with v' a prefix of v, or $(x(uv)^l u'')^\rho$ and $(x(uv)^{l-1}u'')^\rho$ with $u'' \in \Sigma^+$ a prefix of u.   □

For the other direction we need some more notation and terminology. First, we write $\mathrm{SCC}(q)$ for the *strongly connected component* of a node q in a given digraph. Second, given a DFA $A = (\Sigma, Q, q_I, \delta, F)$ and a string $u \in \Sigma^*$, the *rank* of u (with respect to A), denoted $\mathrm{rk}(u)$, is the cardinality of the set $\{\mathrm{SCC}(q_I\,u(0,0)), \ldots, \mathrm{SCC}(q_I\,u(0, |u| - 2))\}$.

Lemma 4. *Let A be a DFA over some alphabet Σ whose transition graph does not match the pattern depicted in Figure 4. Then $\mathcal{L}(A)^\rho$ is expressible in* TL[F].

Proof. We prove that if u and v are nonempty strings over Σ such that $q_I\,u \neq q_I\,v$, then Spoiler wins the $(\mathrm{rk}(u) + \mathrm{rk}(v))$-round game on u^ρ and v^ρ, by induction on $\mathrm{rk}(u) + \mathrm{rk}(v)$.

Write $u = u'a$ and $v = v'b$ for appropriate $a, b \in \Sigma$. If $a \neq b$, then Spoiler wins immediately. So in the rest, assume $a = b$. Write p and q for $q_I\,u'$ and $q_I\,v'$. Clearly, $\mathrm{SCC}(p) \neq \mathrm{SCC}(q)$ in the transition graph of A, because otherwise it would match the pattern depicted in Figure 4. There are three situations that we distinguish.

1. Neither $\mathrm{SCC}(p)$ is reachable from $\mathrm{SCC}(q)$ nor vice versa.
2. $\mathrm{SCC}(p)$ is reachable from $\mathrm{SCC}(q)$, but $\mathrm{SCC}(q)$ is not reachable from $\mathrm{SCC}(p)$.

3. The same as 2. with the roles of p and q exchanged.

First, assume we are in situation 1. Then it is not possible that q_I belongs to both $\mathrm{SCC}(p)$ and $\mathrm{SCC}(q)$, say it does not belong to $\mathrm{SCC}(p)$. Let i be minimal such that $q_I\,u(0,i) \in \mathrm{SCC}(p)$ and set $p' = q_I\,u(0,i)$. Spoiler replaces u^ρ by $u(0,i)^\rho$. Duplicator either looses immediately (because v is of length 1) or she replies by removing a prefix of v^ρ, say she replaces v^ρ by $v(0,j)^\rho$. Set $q' = q_I\,v(0,j)$. If we had $p' = q'$, then $\mathrm{SCC}(q)$ would be reachable from $\mathrm{SCC}(p)$ — a contradiction. Hence, $p' \neq q'$. By the minimality of i, we also have $\mathrm{SCC}(q_I\,u(0,i-1)) \neq \mathrm{SCC}(p)$, which means $\mathrm{rk}(u(0,i)) < \mathrm{rk}(u)$ and, in particular, $\mathrm{rk}(u(0,i)) + \mathrm{rk}(v(0,j)) < \mathrm{rk}(u) + \mathrm{rk}(v)$, so that the induction hypothesis applies. Spoiler wins the remaining game with one round less.

Second, assume we are in situation 2. Choose i as above. Spoiler does the same as before. Duplicator either looses immediately or she removes a prefix from v^ρ, say she replaces v^ρ by $v(0,j)^\rho$. If we had $q_I\,u(0,i) = q_I\,v(0,j)$, then $\mathrm{SCC}(q)$ would be reachable from $\mathrm{SCC}(p)$ — a contradiction. Just as before, we can apply the induction hypothesis. Situation 3 is symmetric to situation 2. □

Exactly the same technique works for proving the correctness of the characterization of $\mathrm{TL}[\mathsf{X},\mathsf{F}]$. In EF games for this fragment, the additional temporal modality is accounted for by an additional type of round, so-called X rounds. In such a round, Spoiler first chops off the first symbol of one the two strings and Duplicator then chops off the first symbol of the other string. For details, see [3].

4 Stutter-Invariant Fragments

In Section 1 we have defined the so-called stutter-invariant counterparts of F and U, namely F_{sf} and U_{sf}. In this section, we will obtain effective characterizations for the *stutter-invariant fragments*, $\mathrm{TL}[\mathsf{F}_{\mathsf{sf}}]$ and $\mathrm{TL}[\mathsf{U}_{\mathsf{sf}}]$. (Observe that $\mathrm{TL}[\mathsf{U}_{\mathsf{sf}}] = \mathrm{TL}[\mathsf{F}_{\mathsf{sf}}, \mathsf{U}_{\mathsf{sf}}]$ and $\mathrm{TL}[\mathsf{X}, \mathsf{F}_{\mathsf{sf}}] = \mathrm{TL}[\mathsf{X}, \mathsf{F}]$.)

Strings u and v are *stutter-equivalent* if they both belong to a language of the form $a_0^+ a_1^+ \ldots a_k^+$ for some k and appropriate symbols a_i. We use $\equiv_{st}$ to denote stutter equivalence, and it is easy to see that $\equiv_{st}$ is in fact an equivalence relation. A language is *stutter-invariant* if whenever u and v are stutter-equivalent strings, then either u and v belong to this language or u and v do not belong to it, i.e., if this language is a union of stutter equivalence classes.

Lamport [7] observed that every language expressible in $\mathrm{TL}[\mathsf{F}_{\mathsf{sf}}, \mathsf{U}_{\mathsf{sf}}]$ is stutter-invariant. This explains why F_{sf} and U_{sf} are called stutter-invariant. Below, we prove that the converse of Lamport's observation holds true as well, in the following sense.

Theorem 4. [3,17] *Let F be one of the stutter-invariant fragments $\mathrm{TL}[\mathsf{F}_{\mathsf{sf}}]$ and $\mathrm{TL}[\mathsf{U}_{\mathsf{sf}}]$ and let F' be its strict counterpart, $\mathrm{TL}[\mathsf{F}]$ respectively $\mathrm{TL}[\mathsf{U}]$. Assume L is an arbitrary language. Then L is expressible in F if and only if L is expressible in F' and stutter-invariant.*

Observe that a regular language L is stutter-invariant if and only if the transition graph of $\mathbf{A}_{L^\rho}$ (or, equivalently, of $\mathbf{A}_L$) does not match the pattern depicted in Figure 3. Thus, the above theorem (together with the classification theorem from the previous section) immediately leads to characterizations of $\mathrm{TL}[\mathsf{F}_{\mathsf{sf}}]$ and $\mathrm{TL}[\mathsf{U}_{\mathsf{sf}}]$ in terms of forbidden configurations.

Using the characterization results we have obtained so far, one can prove:

Corollary 2. *For every fragment (strict or stutter-invariant) F of temporal logic, the following problem is PSPACE-complete. Given a temporal formula φ, decide whether φ is equivalent to a formula in F?*

The upper bound follows from the fact that in polynomial time one can check whether or not the transition graph of a DFA matches a fixed pattern. The lower bound is obtained by a reduction to TL satisfiability.

The proof of Theorem 4 makes use of the notion of a stutter-free string, which is defined as follows. A string u is *stutter-free* if $u(i) \neq u(i+1)$ for all $i < |u| - 1$. Clearly, every equivalence class of $\equiv_{st}$ contains exactly one stutter-free string. As a consequence of Lamport's observation, we note:

Lemma 5. *Let L be a stutter-invariant language over some alphabet Σ and $\varphi \in \mathrm{TL}[\mathsf{F_{sf}}, \mathsf{U_{sf}}]$ a formula over Σ such that $u \models \varphi$ iff $u \in L$, for $u \in \Sigma^+$ stutter-free. Then φ defines L.* $\square$

So Theorem 4 will follow once we have established the following lemma.

Lemma 6. *Let $\boldsymbol{F}$ and $\boldsymbol{F}'$ be as in Theorem 4, and assume $\varphi \in \boldsymbol{F}'$. Then there exists $\varphi' \in \boldsymbol{F}$ such that $u \models \varphi$ iff $u \models \varphi'$, for $u \in \Sigma^+$ stutter-free.*

Proof. The proof is an inductive definition of φ', which works in both situations. The base case is trivial. In the induction step, negation and disjunction can be dealt with easily. What remains are formulas whose outermost connective is F or U. We set

$$
\varphi' = \begin{cases} \displaystyle\bigvee_{a,b\in\Sigma:\, a\neq b} \left(a \wedge \mathsf{F_{sf}}(b \wedge \mathsf{F_{sf}}\psi')\right) , & \text{for } \varphi = \mathsf{F}\psi, \\[2ex] \displaystyle\bigvee_{a,b\in\Sigma:\, a\neq b} \left(a \wedge \left(a \, \mathsf{U_{sf}} \, (b \wedge (\psi' \, \mathsf{U_{sf}} \, \chi'))\right)\right) , & \text{for } \varphi = \psi \, \mathsf{U} \, \chi. \end{cases}
$$

We prove only that the second choice is correct; the proof that the first choice is correct is even simpler. First, assume $u \models \varphi$. Then there exists $i > 0$ such that $u(i, *) \models \chi$ and $u(j, *) \models \psi$ for $j \in \{1, \ldots, i - 1\}$. By induction hypothesis, this means $u(i, *) \models \chi'$ and $u(j, *) \models \psi'$ for $j \in \{1, \ldots, i - 1\}$. Clearly, we have $u \models u(0) \wedge u(0) \, \mathsf{U_{sf}} \, (u(1) \wedge \psi' \, \mathsf{U_{sf}} \, \chi')$, which is a disjunct of φ'.

Second, assume $u \models \varphi'$ and let a and b be symbols for which the corresponding disjunct holds. If $u \models a \wedge a \, \mathsf{U_{sf}} \, (b \wedge \psi' \, \mathsf{U_{sf}} \, \chi')$, then $u(0) = a$ and $u(1) = b$, since u is assumed to be stutter-free. But then $u(1, *) \models \psi' \, \mathsf{U_{sf}} \, \chi'$, which implies, by induction hypothesis, $u(1, *) \models \psi \, \mathsf{U_{sf}} \, \chi$, which, in turn, implies $u \models \psi \, \mathsf{U} \, \chi$. $\square$

This completes the first part of this survey. We have seen how every fragment (determined by which modalities are allowed in forming formulas) of future temporal logic can be characterized in an effective, concise way by describing structural properties of DFA's.

5   Past Modalities and Symmetric Fragments

Thus far, we have only dealt with temporal modalities that refer to the future (and possibly the present) only. But, of course, each of the modalities considered has a symmetric past counterpart: S (since) goes with U, P (eventually in the past) goes with F, Y (previously) goes with X.

Adding past modalities does not increase the expressive power of temporal logic, i. e., $\mathrm{TL} = \mathrm{TL}[\mathsf{U}, \mathsf{S}]$. This is easy to see because for every temporal formula (with future and past modalities) one can still find a counter-free DFA recognizing the language defined by the formula. Similarly, $TL[U_{sf}] = TL[U_{sf}, S_{sf}]$, because even with past stutter-invariant modalities one can only express stutter-invariant languages. Clearly, $TL[X] = TL[X, Y]$. But the expressive power of any other fragment is increased by adding the corresponding past modalities. Nevertheless, we have:

Theorem 5 (Decidability of Symmetric Fragments [16]). *For each of the fragments* $TL[F_{sf}, P_{sf}]$, $TL[F, P]$, *and* $TL[X, Y, F, P]$ *it is decidable whether or not a given temporal property can be expressed in it.*

This theorem is based on similar structural characterizations as the ones given in Theorem 2 for the future fragments of temporal logic. There is, however, a fundamental difference. Instead of looking at the minimal DFA for a given language, one considers its syntactic semigroup, which, by definition, is symmetric in the sense that the syntactic semigroup of the reverse of a language is the reverse of the syntactic semigroup of the language, and thus better suited for investigating symmetric fragments. The proofs get more involved and require non-trivial finite semigroup theory. On the other hand, they also reveal interesting connections to first-order logic.

Remember that Kamp's theorem says that temporal logic (with future modalities only or with both) is as expressive as first-order logic. In this statement, a string $u \in \Sigma^+$ of length n is viewed as a structure in the signature with a binary predicate $<$, for the order relation on the positions, and unary predicates P_a, for each alphabet symbol a a separate predicate.

A simple induction shows that every temporal formula is equivalent to a first-order formula, even to a first-order formula that uses at most three variables. In view of Kamp's theorem, this means that temporal logic and first-order logic with three variables have the same expressive power. Reducing the number of variables to two leads to $TL[F, P]$ and $TL[X, Y, F, P]$, respectively:

Theorem 6 (Kamp's Theorem for Smaller Fragments [16]).
1. A language is expressible in $TL[F, P]$ *if and only if it is expressible in first-order logic with two variables.*
2. A language is expressible in $TL[X, Y, F, P]$ *if and only if it is expressible in first-order logic with two variables when the signature is extended by the built-in predicate* suc *for successor.*

There are more connections to first-order logic and to formal language theory. First, the languages expressible in $TL[F, P]$ are exactly the unambiguous languages in the sense of Schützenberger [13]. Second, the languages expressible in $TL[F, P]$ and $TL[X, Y, F, P]$ are exactly the languages expressible by a Σ_2 and, at same time, a Π_2 formula (over the respective signature). For details, see [16].

6   Until Hierarchy

Which temporal modalities are needed to express a given temporal property is the first question to ask when one is interested in studying the expressive power of the temporal modalities themselves, but there are other, equally important ones, and some prominent ones are concerned with the "until hierarchy" of future temporal logic. The "until" modality is special in several respects. First, it is complete in the sense that no other modality is needed to express all temporal properties. Second, it is the only binary modality. The last fact is crucial; it makes formulas hard to read, especially, when nesting occurs. So the question is whether or not nesting of "until" is necessary, even when the other modalities can be used for free.[4] Using an appropriate Ehrenfeucht-Fraïssé game with additional types of rounds corresponding to X and U, one can actually show that the more nesting is allowed, the more one can express:

Theorem 7 (Strictness of Until Hierarchy [3]). *Let $\Sigma = \{a, b, c\}$ be a three-element alphabet and define φ_n, $n \geq 0$, by $\varphi_0 = a$ and $\varphi_{n+1} = a \wedge X(b \, U \, \varphi_n)$. Then $F\varphi_n$ is of until nesting depth n, but $\mathcal{L}_\Sigma(\varphi_n)$ is not definable by a formula of until nesting depth $< n$.*

We even have:

Theorem 8 (Computability of Until Depth [14]). *Given a temporal formula φ, one can compute the minimal until nesting depth required to express the language defined by φ.*

The proof of this theorem, just as the proof of Theorem 5, makes heavy use of finite semigroup theory. A key ingredient of the proof is the so-called semidirect product/substitution principle, which, in rough outline, states that if two fragments of temporal logic, say F and G, are characterized by classes V and W of finite semigroups, then the fragment which is obtained by substituting formulas of G into formulas of F is characterized by the semidirect product of V and W. Applied to the until hierarchy, this principle says that the k-th level of the hierarchy is characterized by a k-th power of the class of semigroups that characterizes level 1. (Observe that a formula of until depth at most k can be written as a k-fold substitution of formulas of depth at most 1, and vice versa.) For details, see [14] or [15].

7   Conclusion

The results presented in this survey show that there are intimate connections between temporal logic, the theory of finite automata, the theory of finite semigroups, and first-order logic. The classification of discrete temporal properties has been accomplished to a great extent. A problem that is still open is the decidability of the combined until/since hierarchy, where a property is classified according to the nesting depth in U and S required to express it using future as well as past modalities. Note that this hierarchy is known to be strict, see [3].

---

[4] In the literature, other binary modalities (such as "at next" [6] or "as long as" [7]) have been occasionally used, and these operators are as powerful as "until." In fact, nesting depth with regard to any of these two operators is exactly the same as nesting depth with respect to "until."

References

1. Joëlle Cohen, Dominique Perrin, and Jean-Eric Pin. On the expressive power of temporal logic. *J. Comput. System Sci.*, 46(3):271–294, 1993.
2. Kousha Etessami, Moshe Y. Vardi, and Thomas Wilke. First-order logic with two variables and unary temporal logic. In *Proceedings 12th Annual IEEE Symposium on Logic in Computer Science*, pages 228–235, Warsaw, Poland, 1997.
3. Kousha Etessami and Thomas Wilke. An until hierarchy for temporal logic. In *Proceedings 11th Annual IEEE Symposium on Logic in Computer Science*, pages 108–117, New Brunswick, N. J., 1996.
4. Dov M. Gabbay, Amir Pnueli, Saharon Shelah, and Jonathan Stavi. On the temporal analysis of fairness. In *Conference Record of the 12th ACM Symposium on Principles of Programming Languages*, pages 163–173, Las Vegas, Nev., 1980.
5. Johan Anthony Willem Kamp. *Tense Logic and the Theory of Linear Order*. PhD thesis, University of California, Los Angeles, Calif., 1968.
6. Fred Kröger. *Temporal Logic of Programs*. Springer, Berlin, 1987.
7. Leslie Lamport. Specifying concurrent program modules. *ACM Trans. Programming Lang. Sys.*, 5(2):190–222, 1983.
8. Oded Maler. *Finite Automata: Infinite Behavior, Learnability and Decomposition*. PhD thesis, The Weizmann Institute of Science, Rehovot, Israel, 1990.
9. Oded Maler and Amir Pnueli. Tight bounds on the complexity of cascaded decomposition of automata. In *Proceedings of the 31st Annual Symposium on Foundations of Computer Science*, vol. II, pages 672–682, St. Louis, Miss., 1990.
10. Robert McNaughton and Seymour Papert. *Counter-Free Automata*. MIT Press, Cambridge, Mass., 1971.
11. Doron Peled and Thomas Wilke. Stutter-invariant temporal properties are expressible without the next-time operator. *Inform. Process. Lett.*, 63(5):243–246, 1997.
12. Marcel P. Schützenberger. On finite monoids having only trivial subgroups. *Inform. and Computation*, 8:190–194, 1965.
13. Marcel P. Schützenberger. Sur le produit de concatenation non ambigu. *Semigroup Forum*, 13:47–75, 1976.
14. Denis Thérien and Thomas Wilke. Temporal logic and semidirect products: An effective characterization of the until hierarchy. In *Proceedings of the 37th Annual Symposium on Foundations of Computer Science*, pages 256–263, Burlington, Vermont, 1996.
15. Denis Thérien and Thomas Wilke. Temporal logic and semidirect products: An effective characterization of the until hierarchy. Technical report 96-28, DIMACS, Piscataway, N. J., 1996.
16. Denis Thérien and Thomas Wilke. Over words, two variables are as powerful as one quantifier alternation: $FO^2 = \Sigma_2 \cap \Pi_2$. In *Proceedings of the Thirtieth Annual ACM Symposium on Theory of Computing*, pages 41–47, Dallas, Texas, 1998.
17. Thomas Wilke. Classifying discrete temporal properties. Habilitationsschrift (postdoctoral thesis), Kiel, Germany, 1998.

Circuit Complexity of Testing Square-Free Numbers

Anna Bernasconi[1] and Igor Shparlinski[2]

[1] Institut für Informatik, Technische Universität München
D-80290 München, Germany
bernasco@informatik.tu-muenchen.de
[2] School of MPCE, Macquarie University
Sydney, NSW 2109, Australia
igor@mpce.mq.edu.au

Abstract. In this paper we extend the area of applications of the Abstract Harmonic Analysis to the field of Boolean function complexity. In particular, we extend the class of functions to which a spectral technique developed in a series of works of the first author can be applied. This extension allows us to prove that testing square-free numbers by unbounded fan-in circuits of bounded depth requires a superpolynomial size. This implies the same estimate for the integer factorization problem.

1 Introduction

In recent years spectral techniques based on the Abstract Harmonic Analysis on the hypercube have been shown to represent a very useful tool for obtaining lower complexity bounds. Various links between Fourier coefficients of Boolean functions and their complexity characteristics have been studied in a number of works, see [2,3,4,5,8,14,19,20,23,24]. In particular, these spectral techniques have been successfully applied to the parity function and to threshold functions.

However, a limitation of such approach to the study of Boolean function complexity is that, besides the results for parity and threshold functions, spectral methods have provided lower bounds for specially constructed Boolean functions, which are not related to any particular number theoretic or combinatorial problem. In fact, there are very few known examples of functions coming from natural number theoretic or combinatorial problems for which the spectral techniques have produced non-trivial results. The only examples we are aware of are the lower bounds on integer multiplication [13] and on the complexity of computing the discrete logarithm [10,25]. There are also some very interesting results about determinants [15,16].

In this paper we pursue two purposes:

o extend the area of applications of the spectral techniques to the study of Boolean function complexity;
o obtain the first non-trivial lower bound on the circuit complexity of testing square-free numbers.

C. Meinel and S. Tison (Eds.): STACS'99, LNCS 1563, pp. 47–56, 1999.
© Springer-Verlag Berlin Heidelberg 1999

To this aim, we first provide a generalization of the spectral technique developed in [2,3] for getting lower bounds on the size complexity of Boolean functions computed by constant-depth circuits.

We then apply the generalized technique to evaluate the complexity of the Boolean function which decides whether a given $(n+1)$-bit odd integer is square-free, that is the function for which

$$f(x_1,\dots,x_n) = \begin{cases} 1, & \text{if } 2x+1 \text{ is square-free,} \\ 0, & \text{if } 2x+1 \text{ is square-full,} \end{cases} \tag{1}$$

where $2x+1 = x_1 \dots x_n 1$ is the bit representation of $2x+1$, $0 \leq x \leq 2^n - 1$ (if necessary we add several leading zeros).

More precisely, we provide an estimate for the Fourier coefficients of (1) and derive a complexity lower bound showing that this function does not belong to the complexity class $\mathbf{AC}^o$.

In [10,25], some lower bounds are obtained for the function deciding if a given integer x is a quadratic residues modulo p. Here we show that some of the techniques used in [10,25] can be applied to the function (1). This approach is based on the uniformity of distribution of square-free numbers with some fixed binary digits. For the quadratic residuacity a similar property has been established by using the very powerful Weil estimate. Here we use a sieve method.

Notice that our estimate compliments the results of [26] on polynomial representations of the Boolean function deciding whether a given integer x is square-free. Moreover, it provides the first non-trivial lower bound on the circuit complexity of a number theoretic problem which is closely related to the integer factorization problem.

Some results of this paper have recently been generalized in [6]. Several more relevant results can also be found in [1,7].

2   Basic Definitions

Let $\mathfrak{B}_n = \{0,1\}^n$ denote the n dimensional Boolean cube.

We will use the notation $|f|$ to denote the number of strings accepted by the function f, that is $|f| = |\{w \in \mathfrak{B}_n \mid f(w) = 1\}|$. Moreover, p_f denotes the probability that the function f takes the value 1 (over the uniform distribution), that is $p_f = |f|/2^n$.

Given a binary string $w \in \mathfrak{B}_n$, we denote with $|w|$ the **Hamming weight** of w, which is the number of ones in w.

An **unbounded fan-in Boolean circuit** $\mathcal{C}$ with input variables $x_1,\dots,x_n$, consists of several levels of AND, OR and NOT gates. The gates at the bottom level accept values from the input variables $x_1,\dots,x_n$. Each of the other gates may accept output values from any number of gates of the previous levels. The only top level gate contains the output $\mathcal{C}(x_1,\dots,x_n)$. For a more detailed description, see [9,20,24].

The number of levels is called the **depth** of the circuit, the number of gates is called the **size**.

The class of $\mathbf{AC^o}$ **circuits** consists of circuits whose size is bounded by a polynomial in n, and whose depth is bounded by a constant.

A **restriction** ρ is a mapping of the set of the subscripts of input variables $x_1, \ldots, x_n$ to the set $\{0, 1, \star\}$, where

o $\rho(i) = 0$ means that we substitute the value 0 for x_i;
o $\rho(i) = 1$ means that we substitute the value 1 for x_i;
o $\rho(i) = \star$ means that x_i remains a variable.

Given a function f depending on n binary variables, we will denote by f_ρ the function obtained from f by applying the restriction ρ; f_ρ will be a function of the variables x_i for which $\rho(x_i) = \star$, $1 \leq i \leq n$.

The subscripts i and the corresponding variables x_i are called **fixed** if $\rho(i) = 0, 1$, and **free** if $\rho(i) = \star$.

We recall that an integer x is called **square-free** if there is no prime p such that $p^2 | x$. Otherwise x is called **square-full**.

Throughout the paper we denote by $\log x$ the binary logarithm of x.

3 Abstract Harmonic Analysis and Circuits

We give some background on abstract harmonic analysis on the hypercube. We refer to [20,24] for a more detailed exposition.

We consider the space $\mathcal{F}$ of all the two-valued functions on $\mathfrak{B}_n$. The domain of $\mathcal{F}$ is a locally compact Abelian group and the elements of its range, that is 0 and 1, can be added and multiplied as complex numbers. The above properties allow one to analyze $\mathcal{F}$ by using tools from harmonic analysis. This means that it is possible to construct an orthogonal basis set of Fourier transform kernel functions for $\mathcal{F}$. The kernel functions of the Fourier transform are defined in terms of a group homomorphism from $\mathfrak{B}_n$ to the direct product of n copies of the multiplicative subgroup $\{\pm 1\}$ on the unit circle of the complex plane. The functions $Q_w(x) = (-1)^{w_1 x_1} (-1)^{w_2 x_2} \ldots (-1)^{w_n x_n} = (-1)^{w^T x}$ are known as **Fourier transform kernel functions**, and the set $\{Q_w \mid w \in \mathfrak{B}_n\}$ is an orthogonal basis for $\mathcal{F}$.

We can now define the **Abstract Fourier Transform** of a Boolean function f as the rational valued function f^* which defines the coordinates of f with respect to the basis $\{Q_w(x) \mid w \in \mathfrak{B}_n\}$, that is

$$f^*(w) = 2^{-n} \sum_{x \in \mathfrak{B}_n} Q_w(x) f(x) = 2^{-n} \sum_{x \in \mathfrak{B}_n} (-1)^{w^T x} f(x).$$

Then

$$f(x) = \sum_{w \in \mathfrak{B}_n} Q_w(x) f^*(w) = \sum_{w \in \mathfrak{B}_n} (-1)^{w^T x} f^*(w)$$

is the **Fourier expansion** of f.

It is interesting to note that the zero-order Fourier coefficient, that is the coefficient related to the all zeros string, is equal to the probability that the function takes the value 1, while the other Fourier coefficients measure the correlation between the function and the parity of subsets of its input bits (see [19] for more details).

As a consequence of the orthogonality of the functions Q_w, it is also possible to derive a very useful identity, the **Parseval identity:**

$$\sum_{v\in\mathfrak{B}_n} (f^*(v))^2 = 2^{-n} \sum_{v\in\mathfrak{B}_n} f(v) = f_0^* , \qquad (2)$$

where f_0^* denotes the zero-order Fourier coefficient.

We finally present an interesting application of harmonic analysis to circuit complexity which is due to [19].

Lemma 1. *Let f be a Boolean function on n variables computable by a Boolean circuit of depth d and size M, and let ϑ be any integer. Then*

$$\sum_{|w|>\vartheta} (f^*(w))^2 \le \frac{1}{2} M \, 2^{-0.05\vartheta^{1/d}} ,$$

where the sum is taken over all strings $w \in \mathfrak{B}_n$ of cardinality $|w| > \vartheta$.   □

4   A Technique to Prove Lower Bounds on the Size/Depth of Circuits

In [2] and [3] a new technique has been developed with the aim of proving lower bounds on the size-complexity of Boolean functions presenting a rather strong combinatorial structure. This technique is based both on the abstract harmonic analysis on the hypercube, and on the spectral characterization of the size-depth trade-off of Boolean circuits which has been given in Lemma 1.

Let $f : \mathfrak{B}_n \to \{0,1\}$ be a Boolean function depending on n variables. Now, let k, $1 \le k \le n$, be the smallest integer such that f has the following property: for any subfunction f_ρ depending on k variables, $p_{f_\rho} = p_f$, where $p_{f_\rho} = |f_\rho|/2^k$. In this case, we say that the function f is **of level** k (see [3] for more details).

Then, if f is computable by a circuit of constant depth d and size M, it is possible to derive a lower bound on the size M of such a circuit, which depends both on the probability p_f and on the level k:

$$M \ge (p_f - p_f^2) \, 2^{0.05(n-k)^{1/d}+1} .$$

Notice that this result can be viewed as a generalization of the exponential lower bound for the size of constant depth circuits computing the parity function [9,14]. Indeed, parity and its complement are the only two non-constant Boolean functions of level 1, see [2].

The above lower bound can be proved by combining Lemma 1 with some results of [3].

The paper [3] also gives a complete characterization of functions of level k. A Boolean function $f : \mathfrak{B}_n \rightarrow \{0,1\}$ is of level k if and only if $f_0^* = p_f$ and $f^*(w) = 0$ for any string w such that $0 < |w| \leq n - k$.

We now show how the above technique can be generalized in order to be applied also to functions which present such combinatorial structure only in an "approximate sense".

A Boolean function $f : \{0.1\}^n \rightarrow \{0,1\}$ is called δ-**approximately of level** k if

$$|p_{f_\rho} - p_f| \leq \delta$$

for any subfunction f_ρ depending on at least k variables.

In the following theorem we derive a spectral characterization of functions δ-approximately of level k.

Theorem 1. *Let $f : \mathfrak{B}_n \rightarrow \{0,1\}$ be δ-approximately of level k. Then,*

$$|f^*(w)| \leq \delta$$

for any string w such that $0 < |w| \leq n - k$.

Proof. Let $\mu = (\mu_1, \mu_2, \ldots, \mu_n)$ be a Boolean string such that $0 < |\mu| = n - \ell \leq n - k$. Moreover, let $\mathcal{I} = \{i \mid \mu_i = 1\}$.

For any string $u \in \{0,1\}^{n-\ell}$, let $f_{\rho_{\mu,u}}$ denote the subfunction defined by the restriction $\rho_{\mu,u}$ that assigns to the variables x_i such that $i \in \mathcal{I}$, the $(n - \ell)$ values taken from the string u, and leaves free the other ℓ variables. Then, we have

$$f^*(\mu) = \frac{1}{2^n} \sum_{w \in \mathfrak{B}_n} (-1)^{\mu^T w} f(w) = \frac{1}{2^n} \sum_{w \in \mathfrak{B}_n} (-1)^{\sum_{i \in \mathcal{I}} w_i} f(w)$$

$$= \frac{1}{2^n} \sum_{u \in \mathfrak{B}_{n-\ell}} (-1)^{|u|} \sum_{v \in \mathfrak{B}_\ell} f_{\rho_{\mu,u}}(v) = \frac{1}{2^n} \sum_{u \in \mathfrak{B}_{n-\ell}} (-1)^{|u|} |f_{\rho_{\mu,u}}|.$$

For any $u \in \mathfrak{B}_{n-\ell}$, the subfunction $f_{\rho_{\mu,u}}$ depends on $\ell \geq k$ variables and, since f is δ-approximately of level k, we have

$$\left| |f_{\rho_{\mu,u}}| - 2^\ell p_f \right| \leq 2^\ell \delta.$$

Thus, we get

$$|f^*(\mu)| = \frac{1}{2^n} \left| 2^\ell p_f \sum_{u \in \mathfrak{B}_{n-\ell}} (-1)^{|u|} + \sum_{u \in \mathfrak{B}_{n-\ell}} (-1)^{|u|} \left(|f_{\rho_{\mu,u}}| - 2^\ell p_f \right) \right|$$

$$= \frac{1}{2^{n-\ell}} \left| \sum_{u \in \mathfrak{B}_{n-\ell}} (-1)^{|u|} \left(p_{f_{\rho_{\mu,u}}} - p_f \right) \right| \leq \frac{1}{2^{n-\ell}} \sum_{u \in \mathfrak{B}_{n-\ell}} \delta,$$

and the result immediately follows. $\qquad\square$

We are now able to state and prove a theorem which provides a lower bound on the size required by a depth d circuit to compute functions which are δ-approximately of level k.

Theorem 2. *Let $f : \mathfrak{B}_n \to \{0,1\}$ be a function δ-approximately of level k. If f is computable by a circuit of constant depth d and size M, then*

$$M \geq 2^{0.05(n-k)^{1/d}+1} \left(p_f - p_f^2 - \delta^2 2^{(n-k)\log n} \right).$$

Proof. An application of Lemma 1 yields the following inequality:

$$M \geq 2^{0.05\vartheta^{1/d}+1} \sum_{|w|>\vartheta} (f^*(w))^2 .$$

Let us choose $\vartheta = n - k$. Then, by using the Parseval identity (2) we obtain

$$\sum_{|w|>n-k} (f^*(w))^2 = \sum_{w\in\mathfrak{B}_n} (f^*(w))^2 - (f_0^*)^2 - \sum_{1\leq|w|\leq n-k} (f^*(w))^2$$

$$= p_f - p_f^2 - \sum_{1\leq|w|\leq n-k} (f^*(w))^2 ,$$

where, as before, f_0^* denotes the zero-order Fourier coefficient.

We are now left with the evaluation of the sum of the squares of the Fourier coefficients of order less or equal to our threshold $n - k$. From Theorem 1 it follows that

$$\sum_{1\leq|w|\leq n-k}^{n-k} (f^*(w))^2 \leq \delta^2 \sum_{j=1}^{n-k} \binom{n}{j} \leq \delta^2\, 2^{(n-k)\log n} ,$$

where we have applied the inequality

$$\sum_{j=1}^{\ell} \binom{n}{j} \leq n^\ell .$$

Therefore, we obtain

$$\sum_{|w|>n-k} (f^*(w))^2 \geq p_f - p_f^2 - \delta^2\, 2^{(n-k)\log n}$$

and the result follows. □

Note that such a lower bound turns out to be meaningful provided that

$$\delta^2\, 2^{(n-k)\log n} = o(p_f) .$$

5   Circuit Complexity of Testing Square-Free Numbers

First of all we need a result about the uniformity of distribution of odd square-free numbers with some fixed binary digits.

Let ρ be a restriction on the set $\{1, \ldots, n\}$. We denote by $\mathcal{N}_\rho(n)$ the set of integers x, $0 \le x \le 2^n - 1$ such that $x_i = \rho(i)$ for all fixed subscripts $i \in \{1, \ldots, n\}$, where $x_1 \ldots x_n 1$ is the bit representation of $2x + 1$. We also denote by $S_\rho(n)$ the number of $x \in \mathcal{N}_\rho(n)$ for which $2x + 1$ is square-free.

Lemma 2. *For any restriction* ρ *with* $r \le n^{1/2}/3 - 1$ *fixed positions,*

$$S_\rho(n) = \frac{8}{\pi^2} 2^{n-r} + O(2^{n-r-n/3(r+1)}).$$

Proof. Let $T_\rho(n, m)$ be the number of $x \in \mathcal{N}_\rho(n)$ with $m^2 | 2x + 1$. From the inclusion-exclusion principle it follows that

$$S_\rho(n) = \sum_{\substack{1 \le m \le 2^{(n+1)/2} \\ m \equiv 1 \pmod 2}} \mu(m) T_\rho(n, m),$$

where $\mu(m)$ is the Möbius function. We recall that $\mu(1) = 1$, $\mu(m) = 0$ if m is square-full and $\mu(m) = (-1)^{\nu(m)}$ otherwise, where $\nu(m)$ is the number of prime divisors of $m \ge 2$.

Let t be the length of the largest substring of free positions. It is obvious that the elements of $\mathcal{N}_\rho(n)$ can be separated into 2^{n-r-t} groups such that in each group the numbers are of the form $2^s z + a$, $0 \le z \le 2^t - 1$, for some integers s and a.

For an odd integer $m \ge 1$, each such group contains $2^t/m^2 + O(1)$ numbers which are congruent to zero modulo m^2. Taking into account that $t \ge n/(r+1)$, we then obtain

$$T_\rho(n, m) = 2^{n-r}/m^2 + O(2^{n-r-n/(r+1)}).$$

Put $K = 2^{2n/3(r+1)}$. Applying the above asymptotic formula for $m \le K$ and the trivial bound $T_\rho(n, m) \le 2^n/m^2 + 1$ for $m > K$, we obtain

$$S_\rho(n) = \sum_{\substack{1 \le m \le K \\ m \equiv 1 \pmod 2}} \mu(m) \left(\frac{2^{n-r}}{m^2} + O(2^{n-r-n/(r+1)}) \right) + O\left(\sum_{\substack{K < m \le 2^{(n+1)/2} \\ m \equiv 1 \pmod 2}} \frac{2^n}{m^2} \right).$$

From Theorem 287 of [17] we derive

$$\sum_{\substack{1 \le m \le K \\ m \equiv 1 \pmod 2}} \frac{\mu(m)}{m^2} = \sum_{m \equiv 1 \pmod 2} \frac{\mu(m)}{m^2} + O(K^{-1})$$

$$= \sum_{m=1}^{\infty} \frac{\mu(m)}{m^2} - \sum_{m \equiv 0 \pmod 2} \frac{\mu(m)}{m^2} + O(K^{-1})$$

$$= \frac{3}{4} \sum_{m=1}^{\infty} \frac{\mu(m)}{m^2} + O(K^{-1}) = \frac{3}{4} \zeta(2)^{-1} + O(K^{-1}) = \frac{8}{\pi^2} + O(K^{-1}).$$

Therefore, $S_\rho(n) = 8\pi^{-2} 2^{n-r} + O\left(2^{n-r-n/(r+1)}K + 2^n K^{-1}\right)$. Finally, since for $r \leq n^{1/2}/3 - 1$ the first term in the 'O'-symbol dominates, the result follows. $\square$

At this point we are able to derive our main result, namely a lower bound on the Boolean circuit complexity of testing square-free numbers.

Theorem 3. *Assume that the Boolean function f given by (1) is computed by an unbounded fan-in Boolean circuit C of depth d and of size M. Then, for sufficiently large n,*

$$d \log \log M \geq 0.5 \log n + O(\log \log n).$$

Proof. Put $k = n - \lfloor n^{1/2} \log^{-2} n \rfloor - 1$. It follows from Lemma 2 that, for sufficiently large n, f is δ-approximately of level k with $p_f = 8/\pi^2$ and $\delta = \exp(-Cn^{1/2} \log^2 n)$, where $C > 0$ is some absolute constant. Applying Theorem 2 we derive the desired statement. $\square$

In particular, if the depth d is a constant, then the size turns out to be super-polynomial $M \geq \exp(cn^\gamma)$, for some constants $c > 0$ and $\gamma > 0$. In particular, this means that testing square-free numbers, and thus integer factorization, cannot be done by a circuit of the class $\mathbf{AC}^o$. This result has recently been improved in [1], where it is shown that for any prime p, testing square-free numbers as well as primality testing and testing co-primality of two given integers cannot be computed by $\mathbf{AC}^o[p]$ circuits, that is, $\mathbf{AC}^o$ circuits enhanced by MOD_p gates.

Apparently the result of Lemma 2 can be improved by means of some more sophisticated sieve methods (see for instance [18]). This may possibly improve the constant 0.5 in Theorem 3.

6   Concluding Remarks

It would be very interesting to obtain analogous results for other Boolean functions related to number theoretic problems, for example for Boolean functions deciding primality or the parity of the number of prime divisors of the input x. Unfortunately, sieve techniques even more advanced than those used in Lemma 2 are still not powerful enough to produce such results, even under the assumption of the Extended Riemann Hypothesis.

We also remark that that some elementary number theoretic considerations have been used in [26] to obtain a very tight lower bound on the sensitivity of the function which decides whether its input x is a square-free integer.

Recall that the **sensitivity**, $\sigma(f)$, of a Boolean function $f : \mathfrak{B}_n \to \{0,1\}$ (which is also known as the *critical complexity*) is defined as the largest integer $s \leq n$ such that there is a binary vector $x \in \mathfrak{B}_n$ for which $f(x) \neq f(x^{(i)})$ for s values of i, $1 \leq i \leq n$, where $x^{(i)}$ is the vector obtained from x by flipping its ith coordinate. The **average sensitivity** is defined in a similar way as the average values of s taken over all $x \in \mathfrak{B}_n$. Thus

$$\sigma(f) = \max_{x \in \mathfrak{B}_n} \sum_{i=1}^{n} \left| f(x) - f(x^{(i)}) \right| \quad \text{and} \quad \sigma_{av}(f) = \frac{1}{2^n} \sum_{x \in \mathfrak{B}_n} \sum_{i=1}^{n} \left| f(x) - f(x^{(i)}) \right|.$$

In [26] it has been shown that for the function $g(x)$, deciding if an n-bit integer x is square-free, the bound $\sigma(g) \geq \lfloor n/60 \rfloor$ holds.

This sensitivity is of interest because it can be used to obtain lower bounds for the CREW PRAM complexity of a Boolean function f (see [11,12,22,27]), that is the complexity on a *parallel random access machine* with an unlimited number of all-powerful processors, such that simultaneous reads of a single memory cell by several processors are permitted, but simultaneous writes are not. In particular, from the above bound on $\sigma(g)$ one immediately concludes that the CREW PRAM complexity of g is at least $0.5 \log n + O(1)$, see [22].

It is also known that the average sensitivity can be expressed via the Fourier coefficients of f and related to the formula size complexity of f and to the degree of the polynomial approximation of f over the reals, see [2,5,8,21]. Applying our results, one can derive the estimate $\sigma_{av}(f) \geq c\, n^{1/2} \log^{-2} n$ for the function f given by (1), where $c > 0$ is an absolute constant. However, using a more direct approach, it is shown in [7] that in fact

$$\sigma_{av}(f) \geq \frac{4}{9\pi^2} n + o(n).$$

This bound implies several other results about various complexity characteristics of f, such as the formula size, the average decision tree depth and the degrees of exact and approximative polynomial representations of this function over the reals. It should be noted that for some of the above applications it is very essential to get better that the $n^{1/2}$-lower bound for $\sigma_{av}(f)$.

References

1. E. Allender, M. Saks and I. E. Shparlinski, 'A lower bound for primality', *Preprint*, 1998, 1–11.
2. A. Bernasconi, 'On the complexity of balanced Boolean functions', *Lect. Notes in Comp. Sci.*, Springer-Verlag, Berlin, **1203** (1997), 253–263.
3. A. Bernasconi, 'Combinatorial properties of classes of functions hard to compute in constant depth', *Lect. Notes in Comp. Sci.*, Springer-Verlag, Berlin, **1449** (1998), 339–348.
4. A. Bernasconi and B. Codenotti, 'Measures of Boolean function complexity based on harmonic analysis', *Lect. Notes in Comp. Sci.*, Springer-Verlag, Berlin, **778** (1994), 63–72.
5. A. Bernasconi, B. Codenotti and J. Simon, 'On the Fourier analysis of Boolean functions', *Preprint* (1996), 1–24.
6. A. Bernasconi, C. Damm and I. E. Shparlinski, 'Circuit and decision tree complexity of some number theoretic problems', *Tech. Report 98-21*, Dept. of Math. and Comp. Sci., Univ. of Trier, 1998, 1–17.
7. A. Bernasconi, C. Damm and I. E. Shparlinski, 'On the average sensitivity of testing square-free numbers', *Preprint*, 1998, 1–11.
8. R. B. Boppana, 'The average sensitivity of bounded-depth circuits', *Inform. Proc. Letters*, **63** (1997), 257–261.
9. R. B. Boppana and M. Sipser, 'The complexity of finite functions', *Handbook of Theoretical Comp. Sci.*, Vol. A, Elsevier, Amsterdam (1990), 757–804.

10. D. Coppersmith and I. E. Shparlinski, 'On polynomial approximation of the discrete logarithm and the Diffie–Hellman mapping', *J. Cryptology* (to appear).
11. M. Dietzfelbinger, M. Kutyłowski and R. Reischuk, 'Feasible time-optimal algorithms for Boolean functions on exclusive-write parallel random access machine', *SIAM J. Comp.*, **25** (1996), 1196–1230.
12. F. E. Fich, 'The complexity of computation on the parallel random access machine', *Handbook of Theoretical Comp. Sci., Vol. A*, Elsevier, Amsterdam (1990), 757–804.
13. M. Furst, J. Saxe and M. Sipser, 'Parity, circuits, and the polynomial time hierarchy', *Math. Syst. Theory*, **17** (1984), 13–27.
14. M. Goldmann, 'Communication complexity and lower bounds for simulating threshold circuits', *Theoretical Advances in Neural Computing and Learning*, Kluwer Acad. Publ., Dordrecht (1994), 85–125.
15. D. Grigoriev and M. Karpinski, 'An exponential lower bound for depths 3 arithmetic circuits' *Proc. 30 ACM Symp. on Theory of Comp.* (1998), 577–582.
16. D. Grigoriev and A. Razborov, 'Exponential lower bounds for depths 3 arithmetic circuits in algebras of functions over finite fields' *Proc. 39 IEEE Symp. on Found. of Comp. Sci.*, 1998 (to appear).
17. G. H. Hardy and E. M. Wright, *An introduction to the number theory*, Oxford Univ. Press, Oxford (1965).
18. D. R. Heath-Brown, 'The least square-free number in an arithmetic progression', *J. Reine Angew. Math.*, **332** (1982), 204–220.
19. N. Linial, Y. Mansour and N. Nisan, 'Constant depth circuits, Fourier transform, and learnability', *Journal of the ACM*, **40** (1993), 607-620.
20. Y. Mansour, 'Learning Boolean functions via the Fourier transform', *Theoretical Advances in Neural Computing and Learning*, Kluwer Acad. Publ., Dordrecht (1994), 391–424.
21. N. Nisan and M. Szegedy, 'On the degree of Boolean functions as real polynomials', *Comp. Compl.*, **4** (1994), 301–313.
22. I. Parberry and P. Yuan Yan, 'Improved upper and lower time bounds for parallel random access machines without simultaneous writes', *SIAM J. Comp.*, **20** (1991), 88–99.
23. R. Raz, 'Fourier analysis for probabilistic communication complexity', *Comp. Compl.*, **5** (1995), 205–221.
24. V. Roychowdhry, K.-Y. Siu and A. Orlitsky, 'Neural models and spectral methods', *Theoretical Advances in Neural Computing and Learning*, Kluwer Acad. Publ., Dordrecht (1994), 3–36.
25. I. E. Shparlinski, *Number theoretic methods in cryptography: Complexity lower bounds*, Birkhäuser, to appear.
26. I. E. Shparlinski, 'On polynomial representations of Boolean functions related to some number theoretic problems', *Electronic Colloq. on Comp. Compl.*, http://www.eccc.uni-trier.de/eccc/, TR98-054, 1998, 1–13.
27. I. Wegener, *The complexity of Boolean functions*, Wiley-Teubner Series in Comp. Sci., Stuttgart (1987).

Relating Branching Program Size and Formula Size over the Full Binary Basis*

Martin Sauerhoff[1], Ingo Wegener[1], and Ralph Werchner[2]

[1] FB Informatik, LS 2,
Univ. Dortmund, 44221 Dortmund, Germany
sauerhof/wegener@ls2.cs.uni-dortmund.de
[2] Frankfurt, Germany
Ralph.Werchner@frankfurt.netsurf.de

Abstract. Circuit size, branching program size, and formula size of Boolean functions, denoted by $C(f)$, $BP(f)$, and $L(f)$, are the most important complexity measures for Boolean functions. Often also the formula size $L^*(f)$ over the restricted basis $\{\vee, \wedge, \neg\}$ is considered. It is well-known that $C(f) \leq 3\,BP(f)$, $BP(f) \leq L^*(f)$, $L^*(f) \leq L(f)^2$, and $C(f) \leq L(f) - 1$. These estimates are optimal. But the inequality $BP(f) \leq L(f)^2$ can be improved to $BP(f) \leq 1.360\,L(f)^\beta$, where $\beta = \log_4(3 + \sqrt{5}) < 1.195$.

1 Introduction

Circuits, branching programs and formulas are the most important and well-studied computation models for Boolean functions $f \in B_n$, i.e., $f\colon \{0,1\}^n \to \{0,1\}$. For circuits and formulas it is most natural to use the full binary basis B_2, but for formulas also the restricted basis $\{\vee, \wedge, \neg\}$ is of interest. For the sake of convenience we define these computation models and the corresponding complexity measures.

Definition 1.
(1) A circuit over $X_n = \{x_1, \ldots, x_n\}$ is defined as a sequence $G_1, \ldots, G_c$ of gates. A gate G_j is a triple (op_j, I_j^1, I_j^2) where $op_j \in B_2$ and $I_j^1, I_j^2 \in \{0, 1, x_1, \ldots, x_n, G_1, \ldots, G_{j-1}\}$. The inputs are also considered as functions. At gate G_j the operation op_j is applied to the functions represented at the inputs I_j^1 and I_j^2. The circuit complexity $C(f)$ of $f \in B_n$ is the minimal number of gates to compute f.

(2) A formula over X_n is a circuit where each gate may be used at most once as input of another gate (i.e., the underlying graph is a tree if the inputs may be duplicated). The formula size $L(f)$ of $f \in B_n$ is the minimal number of inputs (or leaves) of a formula representing f. By $L^*(f)$ we denote the formula size for formulas where only $\vee$, $\wedge$ and $\neg$ are allowed and negations are given for free.

---

* The first and second author have been supported by DFG grant We 1066/8-1.

C. Meinel and S. Tison (Eds.): STACS'99, LNCS 1563, pp. 57–67, 1999.
© Springer-Verlag Berlin Heidelberg 1999

(3) A branching program (BP) over X_n consists of a directed acyclic graph $G = (V, E)$ and a labelling of the nodes and edges of G. The graph has two sinks, labelled by 0 and 1, and the inner nodes (non-sink nodes) have outdegree 2 and get labels from X_n. For each inner node one of the outgoing edges gets the label 0 and the other one the label 1. Each node v represents a Boolean function $f_v \in B_n$. For the of computation of $f_v(a)$, $a \in \{0, 1\}^n$, we follow a path in G starting at v and leading to one of the sinks. At nodes labelled by x_i the outgoing edge labelled by a_i is chosen. Then $f_v(a)$ is equal to the label of the reached sink. The branching program complexity $BP(f)$ of $f \in B_n$ is the minimal number of inner nodes to compute f.

The consideration of these computation models does not need a further motivation, since they are well-established. The following relations between the complexity measures are well-known (see, e. g., [8]).

Theorem 1. *For all $f \in B_n$,*
(1) $C(f) \leq 3\,BP(f)$,
(2) $BP(f) \leq L^*(f)$,
(3) $L^*(f) \leq L(f)^2$, and
(4) $C(f) \leq L(f) - 1$.

These estimates are optimal. Krapchenko [6] has proved that the parity of n variables has an L^*-complexity of n^2. From Theorem 1 we obtain the estimate $BP(f) \leq L(f)^2$. The relationship between branching programs and formulas has been studied in another context. The famous result of Barrington [2] states that formulas of depth $O(\log n)$, i. e., NC^1-functions, can be represented by polynomial-size branching programs of width 5. Cleve [4] and Cai and Lipton [3] have improved this simulation with respect to the size. They also have considered more general types of branching programs. Their simulations focussed on formulas with a given depth. With respect to formula size, these results have implications only if the formulas are well-balanced. The main result of this paper is the new estimate

$$BP(f) \leq 1.360\,L(f)^\beta, \quad \text{where } \beta = \log_4(3 + \sqrt{5}) < 1.195. \tag{1}$$

In Section 2, we present our simulation and the analysis is performed in Section 3. In Section 4, we prove that the analysis of our simulation is optimal and we discuss whether our simulation itself is optimal.

2   Simulating Formulas by Branching Programs

The task is to construct a branching program (of small size) for a given formula F. Let $F = F_l \otimes F_r$, where $\otimes$ is the operation at the root of the underlying tree of the formula and F_l, F_r are the two sub-formulas of F. For the ease of notation, we use the same name for a formula and its represented function. It is obvious that $BP(F) = BP(\overline{F})$, since it is sufficient to negate the sinks. Therefore, it is sufficient to consider the two cases $\otimes = \wedge$ and $\otimes = \oplus$.

Algorithm 1.
We describe a recursive procedure for the construction of a branching program $G(F)$ for the function represented by a formua F over $X_n = \{x_1, \ldots, x_n\}$. By size$(F)$ we denote the number of inner nodes of the constructed BP. The recursion stops for subformulas with one leaf which are replaced by BPs with one inner node, i. e., size$(x_i) = 1$. Our construction for the two cases $F = F_l \wedge F_r$ and $F = F_l \oplus F_r$ is shown in Fig. 1.

Case 1: $F = F_l \wedge F_r$.
The BP combines the BP $G(F_l)$ for F_l and the BP $G(F_r)$ for F_r. The 1-sink of $G(F_l)$ is replaced by the source of $G(F_r)$. Obviously,

$$\text{size}(F) = \text{size}(F_l) + \text{size}(F_r). \tag{2}$$

Case 2: $F = F_l \oplus F_r$.
Here we combine $G(F_l)$ and a BP $G(F_r, \overline{F_r})$ with two sources representing F_r and $\overline{F_r}$. The 0-sink of $G(F_l)$ is replaced by the F_r-source of $G(F_r, \overline{F_r})$ and the 1-sink of $G(F_l)$ is replaced by the $\overline{F_r}$-source of $G(F_r, \overline{F_r})$. Using the evaluation rule of BPs it follows that the 1-sink is reached iff $F_l(a) = 0$ and $F_r(a) = 1$ or $F_l(a) = 1$ and $\overline{F_r}(a) = 1$ which means $F_r(a) = 0$. Hence, we realize F. We also may interchange the roles of F_l and F_r. Of course, we choose the better alternative. Hence,

$$\text{size}(F) = \min\{\text{size}(F_l) + \text{size}(F_r, \overline{F_r}), \text{size}(F_r) + \text{size}(F_l, \overline{F_l})\}. \tag{3}$$

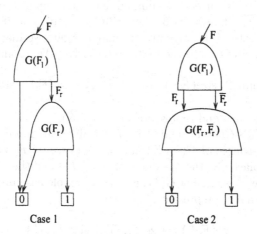

Fig. 1. BPs for F where $F = F_l \wedge F_r$ and $F = F_l \oplus F_r$.

Now we are faced with a new problem. How do we obtain a BP for the pair $(F_r, \overline{F_r})$ and in general for $(F, \overline{F})$? Again we consider the two cases $F = F_l \wedge F_r$ and $F = F_l \oplus F_r$ which are illustrated in Fig. 2. As terminal case we get size$(x_i, \overline{x_i}) = 2$.

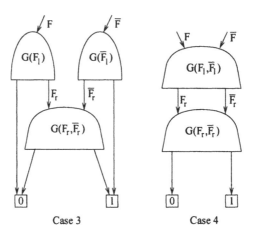

Fig. 2. BPs for $(F, \overline{F})$ where $F = F_l \wedge F_r$ and $F = F_l \oplus F_r$.

Case 3: $F = F_l \wedge F_r$ and representation of $(F, \overline{F})$.
We use the BPs $G(F_l)$, $G(\overline{F_l})$, and $G(F_r, \overline{F_r})$. The BP $G(\overline{F_l})$ is obtained by copying $G(F_l)$ and negating the sinks. As F-source we choose the source of $G(F_l)$ and as $\overline{F}$-source the source of $G(\overline{F_l})$. The 1-sink of $G(F_l)$ is replaced by the F_r-source of $G(F_r, \overline{F_r})$ and the 0-sink of $G(\overline{F_l})$ is replaced by the $\overline{F_r}$-source of $G(F_r, \overline{F_r})$. It is easy to verify that we represent $(F, \overline{F})$. For the computation of $F = F_l \wedge F_r$, we first evaluate F_l. If $F_l(a) = 0$, then $F(a) = 0$. If $F_l(a) = 1$, then $F(a) = F_r(a)$ and F is correctly evaluated. For the computation of $\overline{F} = \overline{F_l} \vee \overline{F_r}$, we first evaluate $\overline{F_l}$. If $\overline{F_l}(a) = 0$, then $\overline{F}(a) = \overline{F_r}(a)$, and if $\overline{F_l}(a) = 1$, then $\overline{F}(a) = 1$. Hence, also $\overline{F}$ is correctly evaluated. Again we may interchange the roles of F_l and F_r and choose the better alternative. We obtain

$$\text{size}(F, \overline{F}) = \min\{2 \cdot \text{size}(F_l) + \text{size}(F_r, \overline{F_r}), 2 \cdot \text{size}(F_r) + \text{size}(F_l, \overline{F_l})\}. \quad (4)$$

Case 4: $F = F_l \oplus F_r$ and representation of $(F, \overline{F})$.
We combine the BPs $G(F_l, \overline{F_l})$ and $G(F_r, \overline{F_r})$ as illustrated in Fig. 2. As F-source we choose the F_l-source of $G(F_l, \overline{F_l})$ and as $\overline{F}$-source its $\overline{F_l}$-source. The 0-sink of $G(F_l, \overline{F_l})$ is replaced by the F_r-source of $G(F_r, \overline{F_r})$ and the 1-sink of $G(F_l, \overline{F_l})$ is replaced by the $\overline{F_r}$-source of $G(F_r, \overline{F_r})$. By a simple case inspection as above it can be shown that we represent $(F, \overline{F})$ and

$$\text{size}(F, \overline{F}) = \text{size}(F_l, \overline{F_l}) + \text{size}(F_r, \overline{F_r}). \quad (5)$$

We have not obtained any new case and our construction is complete.   □

Our main interest in this paper is to prove that the construction in Algorithm 1 leads to a branching program which is small with respect to the size of the given formula. Nevertheless, it is worth mentioning that the algorithm is efficient. The following result follows directly from the description of the algorithm.

Proposition 1. *The construction of the branching program described in Algorithm 1 can be performed in linear time with respect to the size of the resulting branching program. If we are only interested in the size of the resulting branching program, this can be computed in linear time with respect to the size of the given formula.*

3 Estimating the Size of the Constructed Branching Program

In this section we estimate size(F), the size of the branching program constructed by Algorithm 1, with respect to the size of the given formula F. By the description of the algorithm it follows that we have to consider size(F) and size$(F, \overline{F})$ simultaneously. We do not know of a standard technique for the analysis.

It has turned out to be reasonable to bound $\varphi(\text{size}(F), \text{size}(F, \overline{F}))$ for some suitable function $\varphi \colon \mathbb{R}_+^2 \to \mathbb{R}_+$. We have chosen

$$\varphi(s, t) := \max(a_1 s + a_2 t, a_1' s + a_2' t, a_1'' s + a_2'' t), \tag{6}$$

Then we can prove the bound size$(F) \leq \alpha l^\beta$, where l is the number of leaves of F, $\alpha := \varphi(1, 2)/\varphi(1, 1)$ and $\beta := \log_4(3 + \sqrt{5}) = \log_2(\sqrt{3 + \sqrt{5}})$. Having this result (which is proven later) in mind we see that we do not have to care about constant factors of φ. Moreover, the parameters in the definition of φ are chosen in order to minimize the upper bound. First let

$$p_1 := \frac{1}{2}\left(3 + \sqrt{5}\right)^{1/2} = 2^{\beta-1}, \quad p_2 := \left(2 + \sqrt{5}\right)\left(3 + \sqrt{5}\right)^{-1/2} = \left(2 + \sqrt{5}\right)2^{-\beta},$$

$$q_1 := \frac{1}{4}\left(3 + \sqrt{5}\right) = 4^{\beta-1}, \quad q_2 := \frac{1}{2}\left(1 + \sqrt{5}\right).$$

Then we define $a_1, a_2, a_1', a_2', a_1'', a_2''$ as the unique solution of the the following system of linear equations.

$$a_1 p_1 + a_2 p_2 = 1, \qquad a_1' p_1 + a_2' p_2 = 1,$$
$$a_1' q_1 + a_2' q_2 = 1, \qquad a_1'' q_1 + a_2'' q_2 = 1,$$
$$a_2'' = 0, \qquad a_1 + 2a_2 = 2^{-\beta}(3a_1' + 4a_2').$$

One can verify that $a_1, a_2, a_1', a_2', a_1'', a_2''$ are nonnegative.

Lemma 1. *Let $s, t, \tilde{s}, \tilde{t}$ be nonnegative real numbers with*

$$s \leq t \leq 2s, \quad \tilde{s} \leq \tilde{t} \leq 2\tilde{s}, \quad \varphi(s, t) \leq m^\beta, \quad \text{and} \quad \varphi(\tilde{s}, \tilde{t}) \leq \tilde{m}^\beta.$$

Then we have

$$\varphi(s + \tilde{s}, 2s + \tilde{t}) \leq (m + \tilde{m})^\beta \quad \text{if } s < \tilde{s} \text{ or } s = \tilde{s} \wedge \tilde{t} \leq t,$$

and

$$\varphi(s + \tilde{t}, t + \tilde{t}) \leq (m + \tilde{m})^\beta \quad \text{if } \tilde{t} < t \text{ or } \tilde{t} = t \wedge s \leq \tilde{s}.$$

Proof. We only consider φ on the region of all (s,t) with $s \le t \le 2s$. This region is shown in Fig. 3. One can verify that the set of all (s,t) with $\varphi(s,t) = 1$ consists of three segments which meet in $p = (p_1, p_2)$ and $q = (q_1, q_2)$, as shown in Fig. 3. Therefore, we divide the considered region into three sectors by the lines $t = (p_2/p_1)s$ and $t = (q_2/q_1)s$. We number the sectors from bottom to top by I, II, and III. Note that φ is linear within each sector.

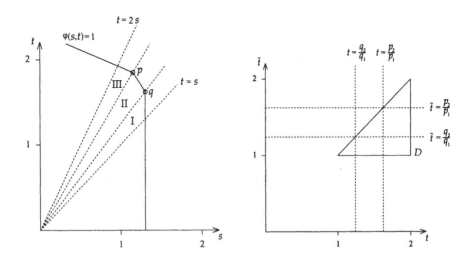

Fig. 3. Function φ.   **Fig. 4.** Triangle D.

We start with the proof of the first inequality. From now on assume that $s, \tilde{s}, t, \tilde{t}$ fulfill the condition for the first claim of the lemma. It is sufficient to show that $\mu(s, \tilde{s}, t, \tilde{t})$ is nonnegative, where $\mu \colon \mathbb{R}_+^4 \to \mathbb{R}$ is defined by

$$\mu(s, \tilde{s}, t, \tilde{t}) := \left(\varphi(s,t)^{1/\beta} + \varphi(\tilde{s}, \tilde{t})^{1/\beta} \right)^{\beta} - \varphi(s + \tilde{s}, 2s + \tilde{t}). \tag{7}$$

Denote by $\varphi_1(s,t)$ the partial derivative of $\varphi(s,t)$ with respect to s and, accordingly, denote by $\varphi_2(s,t)$ the partial derivative with respect to t. μ is monotonically increasing in $\tilde{s}$ since it is continuous and, on every line in $\tilde{s}$-direction, $\partial\mu/\partial\tilde{s}$ is defined on all but at most four points with

$$\frac{\partial\mu}{\partial\tilde{s}}(s, \tilde{s}, t, \tilde{t}) = \left(1 + \left(\frac{\varphi(s,t)}{\varphi(\tilde{s}, \tilde{t})} \right)^{1/\beta} \right)^{\beta-1} \cdot \varphi_1(\tilde{s}, \tilde{t}) - \varphi_1(s + \tilde{s}, 2s + \tilde{t})$$

$$\ge \varphi_1(\tilde{s}, \tilde{t}) - \varphi_1(s + \tilde{s}, 2s + \tilde{t}) \ge 0. \tag{8}$$

The last inequality follows from the fact that, in the positive quadrant, moving in the direction of the vector $(1,2)$ never increases φ_1. The function φ_1 is constant within each of the Sectors I, II, and III, with the smallest value in I and the largest value in III. Since the lines bounding the sectors have a slope of at most 2, the last inequality is correct. Thus, we can further restrict ourselves to the case of $s = \tilde{s}$, $\tilde{t} \leq t$. Since μ is multiplicative, we can even assume $s = \tilde{s} = 1$. Now define $\hat{\mu}$ on the triangle $D \subseteq \mathbb{R}_+^2$ with the corners $(1,1), (2,1), (2,2)$ by

$$\hat{\mu}(t, \tilde{t}) := \mu(1, 1, t, \tilde{t}) = \left(\varphi(1, t)^{1/\beta} + \varphi(1, \tilde{t})^{1/\beta} \right)^{\beta} - \varphi(2, 2 + \tilde{t}). \tag{9}$$

It remains to show that $\hat{\mu}$ is nonnegative on D. As shown in Fig. 4, the four lines $t = q_2/q_1$, $t = p_2/p_1$, $\tilde{t} = q_2/q_1$, and $\tilde{t} = p_2/p_1$ partition D into six regions, which are triangles and rectangles. Restricted to each of these regions, $\hat{\mu}$ is a "linear transformation" of the function $\psi \colon \mathbb{R}_+^2 \to \mathbb{R}_+$ defined by $\psi(z, \tilde{z}) := \left(z^{1/\beta} + \tilde{z}^{1/\beta} \right)^{\beta}$. This means, there are linear functions l_1, l_2, l_3 such that

$$\hat{\mu}(t, \tilde{t}) = l_1(t, \tilde{t}) + \psi(l_2(t), l_3(\tilde{t})). \tag{10}$$

Note that also $\varphi(2, 2 + \tilde{t})$ is linear on the intervals $[1, q_2/q_1]$, $[q_2/q_1, p_2/p_1]$ and $[p_2/p_1, 2]$, since $2 + q_2/q_1 = 2 p_2/p_1$. Since ψ is concave on $\mathbb{R}^2$ (the matrix of second order partial derivatives is negative semi-definite), $\hat{\mu}$ is concave on each of the six regions of D. Thus the minimum value of $\hat{\mu}$ on D appears on a corner of one of the six regions.

It is $\hat{\mu}(q_2/q_1, q_2/q_1) = 0$. One can easily verify that $\hat{\mu}$ is positive on the other nine corners of the regions. Thus, μ is nonnegative on D and we have proved the first inequality of the lemma.

The second claim of the lemma is proved analogously. Assume that $s, \tilde{s}, t, \tilde{t}$ fulfill the condition of the second claim. Setting

$$\nu(s, \tilde{s}, t, \tilde{t}) := \left(\varphi(s, t)^{1/\beta} + \varphi(\tilde{s}, \tilde{t})^{1/\beta} \right)^{\beta} - \varphi(s + \tilde{t}, t + \tilde{t}) \tag{11}$$

we need to show that $\nu(s, \tilde{s}, t, \tilde{t}) \geq 0$. Since

$$\frac{\partial \nu}{\partial t}(s, \tilde{s}, t, \tilde{t}) = \left(1 + \left(\frac{\varphi(\tilde{s}, \tilde{t})}{\varphi(s, t)} \right)^{1/\beta} \right)^{\beta - 1} \cdot \varphi_2(s, t) - \varphi_2(s + \tilde{t}, t + \tilde{t})$$

$$\geq \varphi_2(s, t) - \varphi_2(s + \tilde{t}, t + \tilde{t}) \geq 0, \tag{12}$$

ν is monotonically increasing in t. Again, the last inequality follows from the fact that moving in direction of $(1,1)$ never increases φ_2. Thus we can restrict ourselves to $t = \tilde{t} = 1$ and $s \leq \tilde{s}$. Define $\hat{\nu}$ on the triangle $D' \subseteq \mathbb{R}_+^2$ with the corners $(0.5, 0.5), (0.5, 1), (1, 1)$ by

$$\hat{\nu}(s, \tilde{s}) := \nu(s, \tilde{s}, 1, 1) = \left(\varphi(s, 1)^{1/\beta} + \varphi(\tilde{s}, 1)^{1/\beta} \right)^{\beta} - \varphi(s + 1, 2). \tag{13}$$

The four lines $s = p_1/p_2$, $s = q_1/q_2$, $\tilde{s} = p_1/p_2$, and $\tilde{s} = q_1/q_2$ divide D' into six regions. On each of those regions $\hat{\nu}$ is a linear transformation of ψ, and thus concave (note that even $\varphi(s + 1, 2)$ is linear on every region since $p_1/p_2 + 1 = 2 q_1/q_2$). Evaluating $\hat{\nu}$ on the ten corners of the regions shows that $\hat{\nu}(p_1/p_2, p_1/p_2) = 0$, $\hat{\nu}(0.5, 0.5) = 0$, and $\hat{\nu} > 0$ on the other corners. This completes the proof of the second claim. $\qquad \square$

Lemma 2. *Let F be a formula of size l. Then we have*

$$\varphi(\text{size}(F), \text{size}(F, \overline{F})) \leq \varphi(1,2)\, l^\beta,$$

where $\beta := \log_4(3 + \sqrt{5}) < 1.1943$.

Proof. We prove the lemma by induction on l. For $l = 1$ the claim holds since $\text{size}(F) \leq 1$ and $\text{size}(F, \overline{F}) \leq 2$ in this case.

Let $l > 1$ and let $F = F_l \otimes F_r$, where F_l and F_r are sub-formulas of size k and m, respectively. Set $\overline{\alpha} := \varphi(1,2)$. By the induction hypothesis we have

$$\varphi(\text{size}(F_l), \text{size}(F_l, \overline{F_l})) \leq (\overline{\alpha}^{1/\beta}\, k)^\beta, \tag{14}$$

$$\varphi(\text{size}(F_r), \text{size}(F_r, \overline{F_r})) \leq (\overline{\alpha}^{1/\beta}\, m)^\beta. \tag{15}$$

We distinguish two cases according to $\otimes$.

Case 1: $\otimes = \wedge$.

W. l. o. g., assume that $\text{size}(F_l) < \text{size}(F_r)$ or $\text{size}(F_l) = \text{size}(F_r) \wedge \text{size}(F_r, \overline{F_r}) \leq \text{size}(F_l, \overline{F_l})$. Then the Cases 1 and 3 in Section 2 together with the first inequality of Lemma 1 yield

$$\varphi(\text{size}(F), \text{size}(F, \overline{F})) \leq \varphi(\text{size}(F_l) + \text{size}(F_r), 2\, \text{size}(F_l) + \text{size}(F_r, \overline{F_r}))$$
$$\leq (\overline{\alpha}^{1/\beta}k + \overline{\alpha}^{1/\beta}m)^\beta = \overline{\alpha}\, l^\beta. \tag{16}$$

Case 2: $\otimes = \oplus$.

W. l. o. g., assume that $\text{size}(F_r, \overline{F_r}) < \text{size}(F_l, \overline{F_l})$ or $\text{size}(F_r, \overline{F_r}) = \text{size}(F_l, \overline{F_l}) \wedge \text{size}(F_l) \leq \text{size}(F_r)$. Then the Cases 2 and 4 in Section 2 together with the second part of Lemma 1 yield

$$\varphi(\text{size}(F), \text{size}(F, \overline{F})) \leq \varphi(\text{size}(F_l) + \text{size}(F_r, \overline{F_r}), \text{size}(F_l, \overline{F_l}) + \text{size}(F_r, \overline{F_r}))$$
$$\leq (\overline{\alpha}^{1/\beta}k + \overline{\alpha}^{1/\beta}m)^\beta = \overline{\alpha}\, l^\beta. \tag{17}$$

$\square$

Theorem 2. *Let $\alpha = \varphi(1,2)/\varphi(1,1)$ and $\beta = \log_4(3 + \sqrt{5})$. Then $BP(f) \leq \alpha L(f)^\beta$, $\alpha < 1.3592$, and $\beta < 1.195$.*

Proof. Let F be an optimal formula for f. By the preceding lemma it holds that $\varphi(\text{size}(F), \text{size}(F, \overline{F})) \leq \varphi(1,2)\, L(f)^\beta$. Using the fact that $\varphi(1,y)$ is monotonically increasing in y and that, for $s, t, c \in \mathbb{R}_+$, $\varphi(cs, ct) = c\varphi(s,t)$, we have

$$\text{size}(F) = \varphi(\text{size}(F), \text{size}(F, \overline{F}))\, \varphi(1, \text{size}(F, \overline{F})/\text{size}(F))^{-1}$$
$$\leq \varphi(1,2)\, L(f)^\beta\, \varphi(1,1)^{-1} = \alpha\, L(f)^\beta. \tag{18}$$

The estimates for α and β follow by numerical calculations. $\square$

4   On the Optimality of the Simulation

The first question is whether the analysis of our simulation is optimal. This is indeed the case, as the following example shows.

Definition 2. *The function Alternating Tree,* AT_k*, is defined on* $n = 2^k$ *variables, for an odd number* k*. Let* $\text{AT}_1(x_1, x_2) := x_1 \oplus x_2$ *and*

$$\text{AT}_k(u, v, x, y) := (\text{AT}_{k-2}(u) \wedge \text{AT}_{k-2}(v)) \oplus (\text{AT}_{k-2}(x) \wedge \text{AT}_{k-2}(y))$$

for disjoint variable vectors u*,* v*,* x*, and* y *of length* 2^{k-2}*.*

Theorem 3. *The formula size of* AT_k *equals* $n = 2^k$*. The size of the branching program constructed by Algorithm 1 for the optimal formula obtained from the definition of* AT_k *equals*

$$c_1 \, r^{(k-1)/2} + c_2 \, s^{(k-1)/2}, \quad \text{where}$$

$c_1 = \frac{1}{10} \left(15 + 7\sqrt{5} \right)$*,* $c_2 = \frac{1}{10} \left(15 - 7\sqrt{5} \right)$*,* $r = 3 + \sqrt{5}$*, and* $s = 3 - \sqrt{5}$*.*

This bound is at least $1.340 \, n^\beta - o(1)$*.*

Proof. The result on the formula size is obvious. Because of the symmetry in the definition of AT_k we are able to analyse the size of the resulting BP. Let S_k be the BP size for AT_k and T_k be the size for $(\text{AT}_k, \overline{\text{AT}_k})$. It is easy to check by case inspection that $S_1 = 3$ and $T_1 = 4$. The definition of AT_k can be abbreviated by $F = (F_1 \wedge F_2) \oplus (F_3 \wedge F_4)$, where F_1, F_2, F_3, and F_4 are the same formulas on different variable sets. Therefore, $F_l = F_r$ in the Cases 2 and 3 in Algorithm 1 and the two terms from which we have to take the minimal one are equal. Hence, by the case inspection in Algorithm 1,

$$\text{size}(F) = \text{size}(F_1 \wedge F_2) + \text{size} \left((F_3 \wedge F_4), (\overline{F_3 \wedge F_4}) \right)$$
$$= \text{size}(F_1) + \text{size}(F_2) + 2 \cdot \text{size}(F_3) + \text{size}(F_4, \overline{F_4}), \quad \text{and}$$
$$\text{size}(F, \overline{F}) = \text{size} \left((F_1 \wedge F_2), (\overline{F_1 \wedge F_2}) \right) + \text{size} \left((F_3 \wedge F_4), (\overline{F_3 \wedge F_4}) \right)$$
$$= 2 \cdot \text{size}(F_1) + \text{size}(F_2, \overline{F_2}) + 2 \cdot \text{size}(F_3) + \text{size}(F_4, \overline{F_4}). \quad (19)$$

Since $F_1 = F_2 = F_3 = F_4 = \text{AT}_{k-2}$ for $F = \text{AT}_k$, this leads to

$$S_k = 4S_{k-2} + T_{k-2}, \quad \text{and} \quad T_k = 4S_{k-2} + 2T_{k-2}. \quad (20)$$

The exact solution of these linear difference equations follows by standard techniques (see [5]). $\qquad \square$

This result proves that our analysis is optimal, but it says nothing about the optimality of the simulation. For this problem it would be interesting to know the branching program complexity of the alternating tree function. In this situation we have to complain about the lack of powerful lower bound techniques for the branching program complexity of explicitly defined Boolean functions. The most powerful technique due to Nečiporuk [7] always gives even larger bounds on the formula size and is useless for our purposes. Otherwise there are only bounds of quasilinear size like the bounds of Babai, Pudlák, Rödl, and Szemerédi [1]. They consider symmetric Boolean functions and for these functions the best known upper bounds on the formula size are much larger than their lower bounds on the branching program size. We finish the paper with the observation that the largest trade-off between branching program and formula size can already be proved if we restrict ourselves to read-once formulas.

Definition 3. *A read-once formula is a formula whose leaves are labelled by different variables.*

Theorem 4. *Let $L(l)$ be the class of Boolean functions whose formula size equals l and let $BP(l)$ be the class of all functions with the maximal branching program complexity of all functions in $L(l)$. Then $BP(l)$ also contains a function $f \in L(l)$ representable by a read-once formula of size l.*

Proof. Let $f \in BP(l)$. Then we define f^* by replacing the leaves of an optimal formula F for f by $x_1, \ldots, x_l$. Obviously, $f^* \in L(l)$. It is sufficient to prove $BP(f^*) \geq BP(f)$. This is also easy. Let G be an optimal branching program for f^*. Then we may reverse the replacement of the leaves of F by $x_1, \ldots, x_l$ and obtain a branching program for f of size $BP(f^*)$.   □

Hence, we have reduced our problem to the following one.

Problem 1. Determine the maximal branching program complexity of functions representable by read-once formulas of size l.

Because of Theorem 3, also the following problem is of interest.

Problem 2. Determine the branching program complexity of the alternating tree function.

References

1. L. Babai, P. Pudlák, V. Rödl, and E. Szemerédi. Lower bounds to the complexity of symmetric Boolean functions. *Theoretical Computer Science*, 74:313–323, 1990.
2. D. A. Barrington. Bounded-width polynomial-size branching programs recognize exactly those languages in NC^1. *Journal of Computer and System Sciences*, 38:150–164, 1989.
3. J. Cai and R. J. Lipton. Subquadratic simulations of circuits by branching programs. In *Proc. of the 30th IEEE Symp. on Foundations of Computer Science (FOCS)*, 568–573, 1989.
4. R. Cleve. Towards optimal simulations of formulas by bounded-width programs. *Computational Complexity*, 1:91–105, 1991.
5. R. L. Graham, D. E. Knuth, and O. Patashnik. *Concrete Mathematics*. Addison-Wesley, 1994.

6. V. M. Krapchenko. Complexity of the realization of a linear function in the class of π-circuits. *Math. Notes Acad. Sci. USSR*, 10:21–23, 1971.
7. È. I. Nečiporuk. A Boolean function. *Soviet Mathematics Doklady*, 7(4):999–1000, 1966.
8. I. Wegener. *The Complexity of Boolean Functions*. Wiley-Teubner, 1987.

Memory Organization Schemes for Large Shared Data: A Randomized Solution for Distributed Memory Machines

(Extended Abstract)

Alexander E. Andreev[1], Andrea E. F. Clementi[2], Paolo Penna[2], and
José D. P. Rolim[3]

[1] LSI Logic, (CA) USA
andreev@lsil.com
[2] Dipartimento di Matematica, "Tor Vergata" University of Rome
{lastname}@mat.uniroma2.it
[3] Centre Universitaire d'Informatique, University of Geneva, CH,
rolim@cui.unige.ch

Abstract. We address the problem of organizing a set T of shared data
into the memory modules of a *Distributed Memory Machine* (DMM) in
order to minimize memory access conflicts during read operations.

In this paper we present a new randomized scheme that, with high probability, performs any set of r unrelated read operations on the shared
data set T in $O(\log r + \log \log |T|)$ parallel time *with no memory con-
flicts* and using $O(r)$ processors. The set T is distributed into m DMM
memory modules where m is polynomial in r and *logarithmic* in T, and
the overall size of the shared memory used by our scheme is not larger
than $(1 + 1/\log|T|)|T|$ (this means that there is "almost" no data repli-
cation). The memory organization scheme and most part of all the computations of our method *do not depend* on the read requests, so they can
be performed once and for all during an off-line phase. This is a relevant
improvement over the previous deterministic method recently given in [1]
when "real-time" applications are considered.

1 Introduction

Consider a shared-memory synchronous parallel machine in which a set of p processors can access to a set of b *memory modules* (also called *banks*) in parallel,
provided that a memory module is not accessed by more than one processor
simultaneously. The processors are connected to the memory modules through
a switching network. This parallel model, commonly referred to as *Distributed
Memory Machine* (DMM) or *Module Parallel Computer*, is considered more realistic than the PRAM model and it has been the subject of several studies in
the literature [8,9,14,15,20]. In an EREW PRAM, each of the p processors can in
fact access any of the N memory words, provided that a word is not accessed by
more than one processor simultaneously. To ensure such connectivity, the total

C. Meinel and S. Tison (Eds.): STACS'99, LNCS 1563, pp. 68–77, 1999.
© Springer-Verlag Berlin Heidelberg 1999

number of the switching elements must be $\Theta(pN)$. For large shared memory, constructing a switching network of this complexity is very expensive or even impossible in practice.

One standard way to reduce the complexity of the switching network is to organize the shared memory in *modules* [11], each of them containing several words. A processor switches between modules and not between individual words. So, if the total number of modules is $b << N$ we then need only $\Theta(pb)$ switching elements to realize the interconnection network.

There are two fundamental problems that always arise when the DMM model is adopted. The first one concerns the construction of feasible switching networks and related routing algorithms that must guarantee a full connectivity between processors and memory modules with the best achievable delay. Several randomized and deterministic solutions of this problem have been derived over the last years (see [12] for a good survey).

Once the routing problem is efficiently solved, the shared data have to be distributed (and, if necessary, replicated) among the set of memory modules so that processors can access them avoiding simultaneous reading accesses on the same memory module. This second problem is sometimes referred to as the *granularity problem*. The importance of this problem lies in the fact that reading conflicts on the same shared-memory module (and, in general, any operation that generates conflicts in the use of shared external devices) can only be solved at the cost of a significant time delay. So, the *memory contention*, i.e. the maximum number of shared accesses simultaneously mapped into the same module, is one of the most important factors of the overall time complexity of a DMM algorithm.

In this paper, we address the granularity problem by assuming that we have at hand a sufficiently good solution for the routing problem and thus processors and memory modules can be thought as being ideally connected by a complete bipartite graph. We also assume that our DMM model is provided by *memory interleaving* technology [11] that allows any processor to send access requests to more than one memory module simultaneously, provided that each of these modules is not used by other processors.

Based on the above assumptions, several works have been devoted to the design of efficient solutions for the granularity problem. In particular, randomized solutions have been presented in [8,14,20], while less efficient but deterministic solutions have been introduced in [15,21]. Most of these works concern the problem of simulating a PRAM algorithm on a DMM model. Further relevant applications of the granularity problem concern the design of parallel systems for Private Information Retrieval (PIR) on public-accessible databases [3,4,6], parallel routing-table computations for IP lookup [16,17].

Let T be the table of binary data to be shared and r be the number of parallel read accesses to be satisfied. The performance analysis of any solution of the granularity problem on the DMM model should mainly address the following aspects:

1) The total size of the shared memory required to implement the original table T; 2) The *memory contention*, i.e., the maximum number of simultaneous

access requests that a single memory module must satisfy; 3) The parallel time complexity. This time complexity depends on both the *local computations* performed by each processor and on the memory contention. As discussed before, it is reasonable to assume that the latter represents a dominant factor; 4) The number of processors.

Clearly, the role of the first aspect becomes crucial when large data tables have to be shared (this is the case, for instance, of public databases accessible, say, on WWW servers). On the other hand, in the case of PRAM simulations, it is reasonable to assume that the number of shared data is relatively small.

To our knowledge, the best randomized solution for the granularity problem in the case of PRAM simulation on the DMM model has been introduced by Karp *et al* in [8]. They indeed derived a randomized method that simulates a PRAM with $p \log \log p \log^* p$ processors by using a DMM with p processors with optimal expected delay time $O(\log \log p \log^* p)$ per step of simulation. The memory contention is $O(\log \log p)$. Each of the shared data is replicated in $O(\log \log p)$ copies and mapped to p memory modules by means of $O(\log \log p)$ *hash* functions (see also [9]). The randomized simulation of the PRAM algorithm consists of a sequence of consecutive phases to be executed on the DMM model. Each phase simulates $O(p^{1/10})$ steps of the PRAM algorithm and all the variables (observe that this number is at most $O(p^{11/10})$) used during these steps are distributed into the memory modules by using new randomly selected hash functions. This random distribution is called the *cleaning up* task. Observe also that this solution requires the *a priori* knowledge of the data used during the generic phase and which must be assigned to the memory modules.

The above solution can be considered efficient for the problem of simulating PRAM algorithms where, as already remarked, the number of sharing data is relatively small and it is possible to define the set of data actually used by the program or by a part of it [15]. On the other hand, this randomized solution turns out to be less efficient when: *i*) the number of shared data is significantly larger than the number of available processors, *ii*) it is not possible to determine which is the set of shared data requests that will be performed in the next phase.

These are the typical situations that arise in the case of concurrent accesses to large data structures such as public database available on WWW servers and IP routers databases where "on-line" read requests have to be satisfied in "real time". We emphasize that an application of Karp *et al*'s method for this version of the granularity problem would imply a new assignment of the sharing data to the memory modules (during the cleaning up procedure) for each new set of read requests.

One solution for the problem of performing read accesses on a large database using the DMM model has been recently given by Andreev *et al* [1]. The algorithm performs r arbitrary read requests on a database T of size $N = 2^n$ within the following performances[1]. 1) The total size of the shared memory required to implement the original table T is $2^n(1 + \frac{1}{n})$ to represent the input function and $O(r^3 \log(r^3 n))$ to perform extra algebraic computations; 2) There is no memory

---

[1] We here use a definition of *processor* which is different from that used in [1].

contention, i.e., each memory module receives at most one read request during the algorithm; 3) The worst-case parallel time is logarithmic in r and n; 4) The number of processors is $O(r^5 n)$.

Andreev et al's method have interesting applications for the *direct-sum problem* in circuit complexity [1]. On the other hand, one negative aspect lies in a rather expensive *setup* procedure INIT whose goal is to select the correct addresses of the memory modules that will be considered during the algorithm[2]. This procedure is a sequence of non trivial search and comparing operations inside a matrix M of $O(r^3)$ elements from the finite field $GF(q)$ (where $q = O(r^3 n)$) that allows to select one row of M that satisfies a certain algebraic property. More importantly, INIT depends on the particular values of the input requests, so it must be run for every new set of r requests. This setup procedure performed at "run time" yields an overhead cost that makes the overall algorithm not useful for real time applications.

1.1   Our Result

We provide a randomized version of Andreev et al's algorithm that solves the above problem and requires neither the execution of INIT nor to implement the relative matrix M. The processor programs are thus simpler and more suitable for real time applications.

Given any error probability $0 < \delta < 1/2$, our new randomized version performs with probability at least $(1 - \delta)$ any set of r read requests on T of $N = 2^n$ bits within the following complexity. 1) $2^n(1 + \frac{1}{n})$ memory size (no additional shared memory for extra-computation); 2) there is no memory contention; 3) $O(\log r + \log n + \log(1/\delta))$ parallel time; 4) $O(r((1/\delta)r^3 n))$ processors. (As we will see later, the $(r^3 n)$ factor refers to the amount of the simple *xor* gates of fan-in 2 that are required to parallelize the task of one read request); 5) $O(\log r + \log \delta)$ random bits.

Observe that in case of error, the algorithm does not fail to compute the function but it just might have memory contention greater than 0.

The advantage of our randomized solution is not only in the above performances. In fact, the distribution of T into the memory modules *does not depend on the set of the r read requests*, so it can be done off-line in a pre-processing phase. The use of randomness in our algorithm is required only to select a set of $O(r^2)$ elements from a finite field with uniform distribution. Furthermore, the computation of the memory module addresses which have to be used to satisfy the r read requests is simple and involves only elementary linear algebra on finite fields (more precisely, it is required to compute the set of points that belong to a line specified by one of its points and the parameter of its direction). Finally the number of memory modules in which the table is organized is polynomial in r and *logarithmic* in the size of T. As discussed before, the fact that both the number of processors and the number of memory modules are logarithmic in the

---

[2] A detailed description of the main ideas of this algorithm is given in Section 2.

number of sharing data makes our solution more efficient in the case of large database applications than those proposed for PRAM simulations. A relevant example of such applications is the parallel implementation of *Private Information Retrieval* (PIR) systems [3,4,6] on the DMM model. Due to the lack of space, this application will be described in the full version of this paper.

2   Description of the Algorithm

Let T be the binary database to be shared. Let $|T| = 2^n$ for some integer $n > 0$, we then consider T as the output table of a finite Boolean function $f : \{0,1\}^n \to \{0,1\}$. According to this terminology, our problem is that of computing the function f on a set of r unrelated inputs. As stated in the Introduction, we will adopt the DMM parallel model. The time required by any processor to perform a query to a shared memory module is denoted as extime. In our case, each shared memory module contains one Boolean subfunction (which is stored by means of its *output table*): processors can specify the input of one of these Boolean functions and get one output bit.

Finally, in order to run randomized algorithms, we assume that a public *pseudo-random generator* of bits is available to all processors.

Let $X_1, \ldots, X_r$ be a set of inputs for function f. Our first step consists of splitting the input space $\{0,1\}^n$ in the direct product of two subspaces:

$$\{0,1\}^n = \{0,1\}^{4k} \times \{0,1\}^{n-4k}$$

(the correct choice of k will be given later). The first subspace is here considered as the finite field $GF(q)^4$ where $q = 2^k$. It follows that f can be written as $f(A, B)$, where $A \in GF(q)^4$, $B \in \{0,1\}^{n-4k}$, and our problem is now to compute the set of values $f(A_1, B_1), f(A_2, B_2), \ldots, f(A_r, B_r)$ for arbitrary pairs (A_i, B_i), $i = 1, \ldots, r$.

We need here some algebraic definitions. Consider the set $GF(q)^4$ as a 4-dimensional linear space. Let $l(A, u)$ be the line passing through A and parallel to the vector $h(u) = (1, u, u^2, u^3)$. Notice that the parameter u determines the direction of the line. Let U be any subset of $GF(q)$ (the correct choice of this subset is crucial for our randomized algorithm, and it will be given in Section 4); the term $SL_4(U)$ denotes the set of all lines $l(A, u)$ such that $A \in GF(q)^4$ and $u \in U$. We also define the set of points $l^{\#}(A, u) = l(A, u) \setminus \{A\}$. For any $A \in GF(q)^4$, consider the function $f_A : \{0,1\}^{n-4k} \to \{0,1\}$ defined as $f_A(B) = f(A, B)$. Furthermore, for each line $l \in SL_4(U)$, define the function $g_l : \{0,1\}^{n-4k} \to \{0,1\}$ as

$$g_l = \bigoplus_{A \in l} f_A$$

These functions give our representation of f in the shared memory, i.e., each of them is stored in one single memory module of the DMM. Notice that the construction (more precisely, the size and the structure of the function tables) is independent from f and from the sequence (A_i, B_i), $i = 1, \ldots, r$. So, it can be performed in a preliminary phase once and for all.

A processor can call one of such functions by paying a special time cost denoted as extime. In what follows, we define a system of pairwise independent "computations" of f.

For any $A \in l$, it is easy to prove that

$$f_A(B) = g_l(B) \bigoplus \left(\bigoplus_{A^* \in l \setminus \{A\}} f_{A^*}(B) \right) .$$

Informally speaking, the idea of our solution is that we can compute f on a given input (A_i, B_i) without using the memory module that contains f_{A_i}.

If we consider a set $\{u_1, \ldots, u_r\}$ of elements from $GF(q)$ then we can compute f on (A_i, B_i), $i = 1, \ldots, r$, by applying the following simple procedure:

- **Procedure** ALG_1.
- **input:** f (stored in the shared memory modules by means of functions g_l and f_A);
  (A_i, B_i), $i = 1, \ldots, r$;
  $\{u_1, \ldots, u_r\}$ $(u_i \in GF(q))$ $i = 1, \ldots, r$;
- **begin**
- **for any** $i = 1, \ldots, r$ **do**
-   • **begin**
    • read $g_{l(A_i, u_i)}(B_i)$;
    • **for any** $A^* \in l^\#(A_i, u_i)$ **read** $f_{A^*}(B_i)$;
    • **end**
- **for any** $i = 1, \ldots, r$ **compute**

$$f(A_i, B_i) = g_{l(A_i, u_i)}(B_i) \bigoplus \left(\bigoplus_{A^* \in l^\#(A_i, u_i)} f_{A^*}(B_i) \right) ; \qquad (1)$$

- **end.**

The system in Eq. 1 suggests us a way to avoid memory contention in ALG_1: it suffices to find a set of elements $\{u_1, \ldots, u_r\}$ such that any function of type f_A or g_l (and so any memory module) can participate only in one equation of the system.

In the deterministic algorithm presented by Andreev et al in [1], this task is solved by means of a rather expensive deterministic procedure $INIT$ that considers a suitable matrix $M(i, j)$ of $r^2 \times r$ distinct elements from $GF(q)$ and then computes the first row $\{u_1, \ldots, u_r\}$ of M for which the following property holds

$$\text{for any} \quad j_1 \neq j_2 \rightarrow l^\#(A_{j_1}, u_{j_1}) \bigcap l^\#(A_{j_2}, u_{j_2}) = \emptyset . \qquad (2)$$

This task (in particular, that of checking the above property) is expensive in terms of number of processors, parallel time, and requires non trivial algebraic operations and comparison in $GF(q)$. Furthermore, we emphasize that the output of the procedure INIT in fact depends on A_i $(i = 1, \ldots, r)$ hence it must be performed for every possible values of such prefixes.

In the next section, we will give an algebraic lemma that allows us to avoid the procedure $INIT$ by using a suitable random choice of the sequence $\{u_1, \ldots, u_r\}$.

3   Avoiding Memory Contention via Randomness

The randomized algorithm that, on any input sequence $A_1, \ldots, A_r$, returns an output sequence $u_1, \ldots, u_r$ satisfying Property 2 enjoys of the following result.

Lemma 1. *Let $c \geq 1$ and let U be any subset of $GF(q)$ such that $|U| \geq cr^3$. Let*

$$M = \{u_{i,j}, \ i = 1, \ldots, cr^2; \ j = 1, \ldots, r\}$$

be a matrix of pairwise distinct elements from U. The probability that a randomly chosen index i_0 satisfies the property

$$\text{for any } j_1 \neq j_2, \ l^{\#}(A_{j_1}, u_{i_0,j_1}) \bigcap l^{\#}(A_{j_2}, u_{i_0,j_2}) \ = \ \emptyset$$

is at least $1 - \frac{1}{2c}$.

Proof. Let us define the subset

$$BAD = \{i \in \{1, \ldots, cr^2\} \mid \exists j_1(i) \neq j_2(i), l^{\#}(A_{j_1}, u_{i,j_1(i)}) \cap l^{\#}(A_{j_2}, u_{i,j_2(i)}) \neq \emptyset\}.$$

Assume that for some $U \subseteq GF(q)$ with $|U| \geq cr^3$ and for some matrix M of cr^3 distinct elements of U we have that

$$|BAD| \geq \frac{cr^2}{2c} = \frac{r^2}{2}. \tag{3}$$

For any $i \in BAD$, consider two indexes $j_1(i) \neq j_2(i)$ for which

$$l^{\#}(A_{j_1(i)}, u_{i,j_1(i)}) \cap l^{\#}(A_{j_2(i)}, u_{i,j_2(i)}) \neq \emptyset \tag{4}$$

(observe that, from the definition of BAD, these two indexes must exist).

From the condition of the lemma $u_{i,j_1(i)} \neq u_{i,j_2(i)}$ and Eq. 4, we easily have that $A_{j_1(i)} \neq A_{j_2(i)}$. Since the number of possible pairs $(A_{j_1(i)}, A_{j_2(i)})$ with $(A_{j_1(i)} \neq A_{j_2(i)})$ is

$$\frac{1}{2}r(r-1) < \frac{1}{2}r^2 \leq |BAD|$$

then at least two different i_1 and i_2 exist for which $A_{j_1(i_1)} = A_{j_1(i_2)}$ and $A_{j_2(i_1)} = A_{j_2(i_2)}$. Let $A_1 = A_{j_1(i_1)} = A_{j_1(i_2)}$, $A_2 = A_{j_2(i_1)} = A_{j_2(i_2)}$, and also define

$$C_1 = l(A_1, u_{i_1,j_1(i_1)}) \bigcap l(A_2, u_{i_1,j_2(i_1)}), \ C_2 = l(A_1, u_{i_2,j_1(i_2)}) \bigcap l(A_2, u_{i_2,j_2(i_2)}).$$

Consider now the four vectors

$$V_1 = C_1 - A_1 , \qquad V_2 = A_2 - C_1 , \qquad V_3 = C_2 - A_2 , \qquad V_4 = A_1 - C_2 .$$

It is easy to verify that they are linearly dependent, i.e. $V_1 + V_2 + V_3 + V_4 = 0$. Furthermore, we have that

$$V_1 \parallel \boldsymbol{h}(u_1), \qquad V_2 \parallel \boldsymbol{h}(u_2), \qquad V_3 \parallel \boldsymbol{h}(u_3), \qquad V_4 \parallel \boldsymbol{h}(u_4), \qquad \text{where}$$
$$u_1 = u_{i_1,j_1(i_1)}, u_2 = u_{i_1,j_2(i_1)}, u_3 = u_{i_2,j_2(i_2)}, u_4 = u_{i_2,j_1(i_2)} .$$

At least two of the above vectors V_1, V_2, V_3, V_4 are not zero. It follows that vectors $h(u_i), i = 1, \ldots, 4$ should be linearly dependent. But this is not true: these vectors constitute the well known *Wandermonde* determinant which is always positive for pairwise distinct values of u_i, $i = 1, \ldots, 4$. This implies that $|BAD| < \frac{cr^2}{2c}$ and the lemma is proved.

$\square$

Informally speaking, this lemma states that if we randomly choose a row of M then, with high probability, the r elements contained in this row can be used to compute the system in Eq. 1 avoiding reading conflicts on memory modules.

4 The Global DMM Algorithm and Its Performance Analysis

In what follows, we give an overall description of all the tasks performed by the global algorithm denoted as ALG. ALG receives as input an integer parameter $c \geq 1$, two positive integers n and r ($1 \leq r \leq 2^n$), the output table T of a Boolean function $f : \{0,1\}^n \rightarrow \{0,1\}$, and a set of r inputs $\{X_i = (A_i, B_i), i = 1, \ldots, r\}$.

1. **The Pre-Processing Task: The Shared Memory Partition.** Let

$$k = \lceil \log c + 3 \log r + \log n \rceil , \quad \text{and } q = 2^k$$

Consider the field $GF(q)$ using its standard binary representation. Define U as the first cr^3 elements in $GF(q)$ (any ordering of the field works well). Then, we store the subtables f_A (for any $A \in GF(q)^4$) and g_l (for any $l \in SL_4(U)$) in the shared memory modules. Note that this memory structure depends only on n and r, so if some values in the Table T of f will change (and/or some input X), we just need to update some of the subtables but we do not need to update the memory organization scheme.

2. **The Randomized Procedure RAND.** Consider the $(cr^2 \times r)$-matrix M where

$$M(i,j) \text{ is the } ((i-1)cr^2 + j)\text{-th element of } U$$

(note that we don't need to store M in the shared memory).
Choose uniformly at random an index i_R from the set $1, \ldots, cr^2$ and

$$\text{for any } j = 1, \ldots, r, \text{ return } u_j = M(i_R, j) .$$

3 **The procedure ALG_1.** Apply procedure ALG_1 using $(u_1, \ldots, u_r)$.

We now analyse the costs of ALG by assuming that the pre-processing task has been already done (as already observed, this task depends only on r and n). However, we remark that even this task can be efficiently parallelized since the number of subtables is $|GF(q)^4| + |SL_4(U)|$ which is polynomial in n and r.

Since we have defined $k = \lceil \log c + 3 \log r + \log n \rceil$ and $q = 2^k$, it follows that

$$|SL_4(U)| = \frac{|GF(q)^4||U|}{|GF(q)|} = 2^{4k} \frac{|U|}{2^k} \leq 2^{4k} \frac{1}{n} .$$

The total size of the shared memory used to implement the f is the thus the following

$$\mathsf{mem}(ALG) = |GF(q)^4|2^{n-4k} + |SL_4(U)|2^{n-4k} \leq \left(1 + \frac{1}{n}\right) 2^n .$$

Assume that we have r processors $\{p_1, \ldots, p_r\}$.

- In the procedure RAND we select an element $i_R \in \{1, \ldots, cr^2\}$ by making $\lceil \log r + \log c \rceil$ calls of the public pseudo-random generator. Then p_j $(j = 1, \ldots, r)$ returns the element u_j by computing the $((i-1)r+j)$-th element of U as specified by the procedure RAND. This computation in $GF(q)$ requires $O(k)$ time using $O(k^2)$ processors.

- The third phase requires the computation of procedure ALG_1. For any $i = 1, \ldots, r$, p_i computes the function in Eq. 1.

Assume now that the sequence $u_1, \ldots, u_r$ verifies the property in Eq. 2 (from Lemma 1, this happens with probability greater than $(1 - 1/(2c))$). In this case, each shared memory module receives at most one query, so the memory contention is 0. It follows that the task of each processor p_j can be performed in $O(\log r + \log c + \log n) + \mathsf{ex}$ time parallel time using a number of Boolean gates of fan-in two that satisfies the bound $O(|l^\#(A, u)|) = O(2^k) = O(cr^3 n)$

Finally, we have proved the following theorem.

Theorem 2. *Given any $c \geq 1$, the algorithm ALG computes with probability at least $(1 - 1/(2c))$ any n-input Boolean function f on a set of r inputs, within the following complexity*

- *$(1 + \frac{1}{n})2^n$ memory size (no additional shared memory for extra-computation);*
- *$O(\log r + \log n + \log c) + \mathsf{ex}$ time parallel time (with no memory contention);*
- *r processors each of them having $O(cr^3 n)$ Boolean gates of fan-in 2;*
- *$O(\log r + \log c)$ random bits.*

References

1. A.E. Andreev, A.E.F. Clementi, J. D.P. Rolim (1996), On the parallel computation of Boolean functions on unrelated inputs. Proc. of the *IV IEEE Israel Symposium on Theory of Computing and Systems (ISTCS'96)*, IEEE, pp. 155–161.

2. Chin F. (1986), Security Problems on Inference Control for SUM, MAX and MIN Queries. *J. of ACM*, 33(3), pp. 451–464.
3. Chor B., and Gilboa N. (1997), Computationally Private Information Retrieval. *Proc. of ACM STOC*, p. 304–313.
4. Chor B., Goldreich O., Kushilevitz E., and Sudan M. (1995), Private Information Retrieval. *Proc of IEEE FOCS*, pp. 41-50.
5. Dobkin D., Jones A. K., Lipton R.J. (1979), Secure Databases: Protection Against User Influence, *ACM Trans. on Database Systems*, 4(1), pp. 97–106.
6. Gertner Y., Goldwasser S., and Malkin T. (1998), A Random Server Model for Private Information Retrieval. *Technical Report* MIT-LCS-TR-715. To appear on *Proc. RANDOM '98*
7. Gertner Y., Ishai Y., Kushilevitz E., and Malkin T. (1998), Protecting Data Privacy in Private Information Retrieval Schemes. *Proc. of ACM STOC*.
8. Karp R. M., Luby M., and Meyer auf der Heide F. (1996), Efficient PRAM Simulation on a Distributed Memory Machine. *Algoritmica*, 16, pp. 517-542 (Extended Abstract in *ACM STOC 1992*).
9. Karlin A. and Upfal E. (1986), Parallel hashing - an efficient implementation of shared memory. *Proc. of ACM STOC*, 160-168.
10. Kruskal C.P., Rudolph L., and Snir M. (1990), A Complexity Theory of Efficient Parallel Algorithms. *Theoret. Comput. Sci*, 71, p. 95–132.
11. Kumar V., Grama A., Gupta A., and Karypis G. (1995), *Introduction to Parallel Computing*. Benjamin/Cummings Publ. Company.
12. T. Leighton (1992), *Introduction to parallel algorithms and architectures: arrays, trees, hypercubes*. Morgan Kaufmann Publishers, san Mateo, CA.
13. Liu Z., Li X., and You J. (1992), On storage schemes for parallel array access. *Proc. ACM ICS*, pp. 282–291.
14. Mehlhorn K. and Vishkin U. (1984), Randomized and Deterministic Simulation of PRAM by Parallel Machines with Restricted Granularity of Parallel Memories. *ACTA Informatica*, 21, pp. 339–374.
15. Pietracaprina A., and F. P. Preparata (1993), A Practical Constructive Scheme for Deterministic Shared-Memory Access. *Proc of ACM SPAA*, p. 100–109.
16. Pluris Inc. (1998), Pluris Massively Parallel Routing. *Technical Report* available at *www.pluris.com/wp/index.html*.
17. Pluris Inc. (1998), Parallel Routing, *Technical report* available at *www.pluris.com*.
18. Tannenbaum A.(1994), *Computer Networks*. Prenctice Hall, III Edition.
19. Ullman J. D. (1982) *Principles of Database Systems*. II edition.
20. Upfal E. (1984), Efficient Schemes for Parallel Communication. *J. of the ACM*, 31 (3), pp. 507–517.
21. Upfal E. and Wigderson A. (1987), How to share memory in a distributed system, *J. of the ACM*, 34, pp. 116–127.

The Average Time Complexity to Compute Prefix Functions in Processor Networks*

Andreas Jakoby

Medizinische Universität zu Lübeck
Institut für Theoretische Informatik, Wallstraße 40, 23560 Lübeck, Germany
jakoby@informatik.mu-luebeck.de

Abstract. We analyze the average time complexity of evaluating all prefixes of an input vector over a given algebraic structure $\langle \Sigma, \otimes \rangle$. As a computational model networks of finite controls are used and a complexity measure for the average delay of such networks is introduced. Based on this notion, we then define the average case complexity of a computational problem for arbitrary strictly positive input distributions. We give a complete characterization of the average complexity of prefix functions with respect to the underlying algebraic structure $\langle \Sigma, \otimes \rangle$ resp. the corresponding Moore-machine M. By considering a related reachability problem for finite automata it is shown that the complexity only depends on two properties of M, called *confluence* and *diffluence*. We prove optimal lower bounds for the average case complexity. Furthermore, a network design is presented that achieves the optimal delay for all prefix functions and all inputs of a given length while keeping the network size linear. It differs substantially from the known constructions for the worst case.

1 Introduction

The parallel prefix problem is a fundamental task with a lot of applications such as addition of binary numbers or solving linear recurrences. To each such problem one associates a specific semigroup that describes the algebraic structure of the problem. The complexity of the parallel prefix problem has extensively been investigated in the circuit model. The analysis of the expected time of computing the parallel prefix function goes back to the analysis of Burks, Goldstine, and von Neumann 1946. In [5] it is pointed out that the speed of the arithmetic-logic unit of a processor mainly depends on the speed of the adder used – which can significantly be accelerated in the average. The authors have explored the expected length of a carry chain and shown a logarithmic upper and lower bound for these chains. The results of Burks et. al. have been improved by different authors (see for example [13,4,10,1]). Based on these results several models for computing the sum of two n-bit binary numbers efficiently in the average case have been analyzed in the literature. Expected case upper bounds of order $O(\log \log n)$ for

---

* supported by DFG Research Grant Re 672-2

C. Meinel and S. Tison (Eds.): STACS'99, LNCS 1563, pp. 78–89, 1999.

the addition and other prefix problems have also been obtained in [12] and [10]. Reif has observed that circuit depth $O(\log\log n)$ is sufficient if one allows a small portion of input vectors for which a wrong result may be obtained. He has also introduced a circuit that supervises these errors and corrects them - but this requires one gate of unbounded fanin (see also [6,7]). In contrast to Reif's circuit in [10] we have presented a standard circuit that correctly computes the addition within an asymptotic minimum time for all inputs.

Every prefix function for which the underlying algebraic structure is a semi-group can obviously be computed in logarithmic depth. Ladner and Fischer have shown how this can be done in parallel using only linear circuit size [11]. Further, they have shown that each function that can be expressed by a Moore-machine can be transformed into a semigroup $\langle \Sigma, \otimes \rangle$ where $|\Sigma|$ is exponential in the size of the Moore-machine. Snir has obtained exact bounds for the tradeoff between the depth and the size of prefix circuits [14]. Bilardi and Preparata have studied time-size tradeoffs for this problem. They have given a complete characterization of the semigroups with respect to this question [2] by providing tight lower and upper bounds. The complexity depends on algebraic properties of the semigroup. It is shown that only two cases are possible: within a wide range either the product of time and size grows only linearly or as $n \log n$. In [3] the same authors have studied constant depth unbounded-fanin circuits and classify semigroups according to the property of having linear size prefix circuits.

In [9] we have defined an average measure for the delay of Boolean circuits called *time*. The idea is to take advantage of favorable cases in which the value of a prefix can be computed faster than within the trivial lower bound of log-arithmic depth. We have shown that for several semigroups like the Boolean semigroup with the OR or the AND-operator or for the semigroup corresponding to the addition of binary numbers one can compute all prefixes with an average delay of order $\log\log n$ [9,10]. Furthermore, the circuit size can be kept linear. That means there is an exponential speedup from the worst case to the average case. On the other hand, for functions like PARITY or MAJORITY no speedup is possible. In [10] we have also shown that there is a polynomial time decidable algebraic property, called *confluence*, that decides whether an average delay of order $\log\log n$ can be achieved or not.

To generalize these results to arbitrary parallel machines with restricted de-gree like EREW-PRAMs, CREW-PRAMs, or networks of RAMs we will extend the circuit model to processor networks where each computational unit consists of a finite control. In contrast to the circuit model the nodes of this networks can compute interim results which simplify the analysis of the lower bounds and extensions to more general models. On the other hand the results can be applied to the circuit model by using a statistical estimation (see [8]).

Based on [10] the relation between the average case complexity of the com-putation of parallel prefix problems in a processor network and the structure of the underlying groupoid will be studied in more detail. Using the notion of **con-fluence** and **diffluence** a complete characterization will be given saying that only three substantially different cases are possible. Either the average delay is constant, or is of order $\log \log n$, or it is of logarithmic order, that means equal to the worst case. To obtain these results we translate the prefix functions into a reachability problem. It will be shown that the reachable sets depend on algebraic properties of the underlying Moore-machine. For further investigations see [8].

2   Definitions

2.1   The Prefix Functions

Define $w := w[1]w[2]\ldots \in \Sigma^*$ and for $i \leq j \in \mathbb{N}$ $w[i;j] := w[i]\ldots w[j]$. Further let λ be the empty word. A prefix function can be defined by the transition function of a Moore-machine. Thus, we define:

Definition 1 *A **Moore-machine** $M = (Q, \Sigma_{\mathrm{I}}, \Sigma_{\mathrm{O}}, q_0, \delta, \gamma)$ is a 6-tuple where Q denotes a set of states, Σ_{I} the input alphabet, Σ_{O} the output alphabet, $q_0 \in Q$ the starting state, $\delta : Q \times \Sigma_{\mathrm{I}} \to Q$ the transition function and $\gamma : Q \to \Sigma_{\mathrm{O}}$ the output function of M. To extend the transition function to strings $w \in \Sigma_{\mathrm{I}}^*$ we define for $q \in Q$ and $w \in \Sigma_{\mathrm{I}}^*$: $\delta(q, w) = q$ if $w = \lambda$ and $\delta(\delta(q, w[1]), w[2; |w|])$ elsewhere. Further, define the transformation function*

$$T_{M,q} := \gamma(q)\gamma(\delta(q, w[1]))\gamma(\delta(q, w[1;2]))\ldots\gamma(\delta(q, w)) \ .$$

For easier notion we assume that $\gamma(q_0) = \lambda$. The transition function of M is given by $T_M(w) := T_{M,q_0}(w)$. Furthermore define $\delta(Q', w) := \{\delta(q, w) | q \in Q'\}$, $\delta(q, W) := \{\delta(q, w) | w \in W\}$, and $\delta(Q', W) := \bigcup_{w \in W} \delta(Q', w)$ for $Q' \subseteq Q$ and $W \subseteq \Sigma_{\mathrm{I}}^*$. For a Moore-machine $M = (Q, \Sigma_{\mathrm{I}}, \Sigma_{\mathrm{O}}, q_0, \delta, \gamma)$ and a state $q \in Q$ define $M_q := (Q, \Sigma_{\mathrm{I}}, \Sigma_{\mathrm{O}}, q, \delta, \gamma)$.

*A function $f : \Sigma_{\mathrm{I}}^* \to \Sigma_{\mathrm{O}}^*$ is called a **prefix function** if there exists Moore-machine M with $T_M \equiv f$. The set of all prefix function is denoted by $\mathcal{F}_{\mathbf{prefix}}$.*

Analogously, $\mathcal{F}_{\mathrm{prefix}}$ can be defined using Mealy-machines or deterministic general sequential machine, respectively. On the other hand we can characterize prefix functions by an algebraic structure $\langle \Sigma, \otimes \rangle$, called groupoid, and an homomorphism h. Note that in contrast to a semigroup or a group the binary operator $\otimes$ of a groupoid is not necessarily associative.

Definition 2 *Let $\Sigma, \Sigma_{\mathrm{I}}, \Sigma_{\mathrm{O}}$ be finite alphabets with $\Sigma_{\mathrm{I}} \subseteq \Sigma$. For a **groupoid** $\langle \Sigma, \otimes \rangle$ (i.e. an algebraic structure with a binary operation $\otimes : \Sigma \times \Sigma \to \Sigma$) and an input $w \in \Sigma_{\mathrm{I}}^*$ let $\otimes(w) := w[1] \otimes \ldots \otimes w[|w|]$. Then define for $i \in \mathbb{N}$ and $w \in \Sigma_{\mathrm{I}}^*$ $\mathrm{PP}_{\langle \Sigma, \otimes \rangle, \Sigma_{\mathrm{I}}}^{[i]} := \otimes(w[1;i])$, if $|w| \geq i$, and $\mathrm{PP}_{\langle \Sigma, \otimes \rangle, \Sigma_{\mathrm{I}}}^{[i]} := \lambda$, elsewhere. Further, let $\mathrm{PP}_{\langle \Sigma, \otimes \rangle, \Sigma_{\mathrm{I}}}(w) := \otimes(w[1]) \otimes (w[1;2]) \otimes (w[1;3])\ldots$. For easier notion define $\mathrm{PP}_{\langle \Sigma, \otimes \rangle} := \mathrm{PP}_{\langle \Sigma, \otimes \rangle, \Sigma}$.*

A straightforward analysis leads to the following relation:

Proposition 1 *A function $f : \Sigma_{\mathrm{I}}^* \to \Sigma_{\mathrm{O}}^*$ is a prefix function iff there exists a groupoid $\langle \Sigma, \otimes \rangle$ and an homomorphism $h : \Sigma^* \to \Sigma_{\mathrm{O}}^*$ with $f \equiv h \circ \mathrm{PP}_{\langle \Sigma, \otimes \rangle, \Sigma_{\mathrm{I}}}$.*

In order to index the corresponding Moore-machine of a groupoid define:

Definition 3 *For an algebra $\langle \Sigma, \otimes \rangle$, an input alphabet $\Sigma_I \subseteq \Sigma$, and a homomorphism $h : \Sigma^* \to \Sigma_O^*$ define the Moore-machine $M_{\langle \Sigma, \otimes \rangle, \Sigma_I, h} := (\Sigma \cup \{q_0\}, \Sigma_I, \Sigma_O, q_0, \delta, h)$ with $h(q_0) := \lambda$. Further, define for all $q \in \Sigma \cup \{q_0\}$ and $x \in \Sigma_I$: $\delta(q, x) := q \otimes x$, if $q \neq q_0$, and $\delta(q_0, x) := x$, elsewhere.*

A Moore-machine M will be called **minimal** for the prefix function $f \equiv T_M$ iff for any pair of states $q_1, q_2 \in Q$ holds $T_{M, q_1} \not\equiv T_{M, q_2}$. Analogously, a groupoid $\langle \Sigma, \otimes \rangle$ with a homomorphism h is called **minimal** for $f = h \circ \mathrm{PP}_{\langle \Sigma, \otimes \rangle, \Sigma_I}$ if for any pair $x, y \in \Sigma$ there exists a string $w \in \Sigma_I^*$ such that $h(\otimes(xw)) \neq h(\otimes(yw))$. Note that the corresponding Moore-machine $M_{\langle \Sigma, \otimes \rangle, \Sigma_I, h}$ of a minimal algebra is also minimal. In the following only minimal algebras and Moore-machines will be investigated. To mark the corresponding minimal algebra and homomorphism resp. Moore-machine of a prefix function f we write $\langle \Sigma_f, \otimes_f \rangle$, h_f, and $M_f = (Q_f, \Sigma_I, \Sigma_O, q_{0,f}, \delta_f, \gamma_f) := M_{\langle \Sigma, \otimes \rangle, \Sigma_I, h}$.

2.2   Confluence and Diffluence

Using the corresponding Moore-machine M_f of a prefix function f we will now define some basic features, which allow us to analyze the average time of a network. Define the **reachability set** $R_{M,q}(t) := \delta(q, \Sigma^t)$ as the set of states that can be reached by M in exactly $t \in \mathbb{N}$ steps starting at state $q \in Q$. Further let $R_M := R_{M, q_0}$. It is easy to see that these reachability sets share the following property for all $t_1, t_2 \in \mathbb{N}$ and $q \in Q$: $R_{M,q}(t_1) = R_{M,q}(t_2)$ iff for all $i \in \mathbb{N}$ holds $R_{M,q}(t_1 + i) = R_{M,q}(t_2 + i)$. Since there are at most $2^{|Q|}$ different possibilities for $R_{M,q}(t)$. This property implies the existence of numbers $\tau, \pi \in \{1, 2, \ldots, 2^{|Q|}\}$ with $R_M(\tau) = R_M(\tau + i \cdot \pi)$ for all $i \in \mathbb{N}$. Let us call the smallest τ the **start of periodicity** $\tau(M)$, and the smallest π the **period** $\pi(M)$ of M. For a prefix function $f \in \mathcal{F}_{\mathrm{prefix}}$ define $R_f := R_{M_f}$, $\tau(f) := \tau(M_f)$, and $\pi(f) := \pi(M_f)$.

To construct a parallel machine that computes a prefix function f efficiently in the average, we will use a search strategy for local substrings, that determine the output of $f^{[i]}$, that means the ith output position of f, independently from the concrete prefix of the input. So we define:

Definition 4 *Let q_1, q_2 be two states of a Moore-machine M. A string $w \in \Sigma_I^*$ with $\delta(q_1, w) = \delta(q_2, w)$ is called a **confluence** of these states. Further, w is a confluence for a subset $Q' \subseteq Q$ if $|\delta(Q', w)| = 1$. M has a t-**confluence** if $R_M(t)$ is confluent. A $\tau(M)$-confluence is called a **canonical** confluence of M. A function $f \in \mathcal{F}_{\mathrm{prefix}}$ is called t-confluent if M_f is t-confluent. A Moore-machine M (resp. a function $f \in \mathcal{F}_{\mathrm{prefix}}$) is called **strictly (t, ℓ)-confluent** if each string $w \in \Sigma_I^{\geq \ell}$ is a confluence of $R_M(t)$ (resp. of $M_f(t)$). Define $\mathcal{F}_{\mathrm{conf}}$ resp. $\mathcal{F}_{\mathrm{str\text{-}conf}}$ as the set of all canonical confluent or strictly $(\tau(f), \ell)$-confluent functions $f \in \mathcal{F}_{\mathrm{prefix}}$ for some $\ell \in \mathbb{N}^+$.*

The parameters of a strictly confluent prefix function can be bounded as follows:

Lemma 1 *Let $f \in \mathcal{F}_{\text{str-conf}}$ be strictly (t, ℓ)-confluent for some $t \geq \tau(f)$ and $\ell \in \mathbb{N}$, then f is also strictly $(\tau(f), |\Sigma|^2/2)$-confluent.*

For easier notion we will call the functions $f \in \mathcal{F}_{\text{str-conf}}$ **strictly confluent**. To analyze the lower bounds a property complementary to the confluence is needed, which is called diffluence.

Definition 5 *Let q_1, q_2 be two states of a Moore-machine M. A string $w \in \Sigma_{\text{I}}^+$ is called a **diffluence** of these states if $\{q_1, q_2\} = \delta(\{q_1, q_2\}, w)$. $w \in \Sigma_{\text{I}}^+$ is called a diffluence of a subset $Q' \subseteq Q$ if w is a diffluence for a pair $q_1, q_2 \in Q'$. Q' is called **strictly diffluent** iff each string $w \in \Sigma_{\text{I}}^*$ can be completed to an diffluence for Q', i.e. $\forall w \in \Sigma_{\text{I}}^* \; \exists u \in \Sigma_{\text{I}}^* : \{q_1, q_2\} = \delta(\{q_1, q_2\}, wu)$. M is (strictly) t-**diffluent** if $R_M(t)$ is (strictly) diffluent. M is called (strictly) **canonical diffluent** if M is (strictly) $\tau(M)$-diffluent. $f \in \mathcal{F}_{\text{prefix}}$ is called (strictly) t-diffluent, if M_f is (strictly) t-diffluent. Define $\mathcal{F}_{\text{diff}}$ resp. $\mathcal{F}_{\text{str-diff}}$ as the set of all (strictly) canonical diffluent functions $f \in \mathcal{F}_{\text{prefix}}$.*

Figure 1 gives four examples of prefix functions. The first graph shows the transition graph of the carry propagation f_{carry}, that is the basic groupoid of the addition. Figure 1.b to 1.d are the transition graphs of the parity function f_{parity}, the negation f_{not}, and the prefix function $w[1]^{w[i]}$. The corresponding groupoid $\langle \Sigma, \otimes_{\text{carry}} \rangle$ of the carry propagation is given by $\Sigma_{\text{I}} := \Sigma := \{\text{pro}, \text{gen}, \text{del}\}$ and $x \otimes_{\text{carry}} y := x$, if $y = \text{pro}$, and $x \otimes_{\text{carry}} y := y$, elsewhere. The corresponding groupoids for f_{parity}, f_{not}, and $w[1]^{w[i]}$ can be defined analogously. Investigating these transition graphs it can be seen that $f_{\text{carry}} \in \mathcal{F}_{\text{conf-diff}} := \mathcal{F}_{\text{conf}} \cap \mathcal{F}_{\text{diff}}$, $f_{\text{parity}}, w[1]^{w[i]} \in \mathcal{F}_{\text{str-diff}}$, and $f_{\text{not}} \in \mathcal{F}_{\text{str-conf}}$. A general relationship between $\mathcal{F}_{\text{str-conf}}$, $\mathcal{F}_{\text{conf-diff}}$, and $\mathcal{F}_{\text{str-diff}}$ is given in the following theorem:

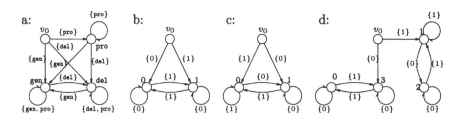

Fig. 1. The transition graphs of the Moore-machines of four prefix functions.

Theorem 1 *$\mathcal{F}_{\text{prefix}}$ is a disjunct union of $\mathcal{F}_{\text{str-conf}}$, $\mathcal{F}_{\text{conf-diff}}$, and $\mathcal{F}_{\text{str-diff}}$.*

For a function $f \in \mathcal{F}_{\text{conf}}$ let $\alpha(f)$ be the minimum multiple of $\pi(f)$ such that there exists a canonical confluence of M_f of length $\alpha(f)$. For $f \in \mathcal{F}_{\text{diff}}$ let $\beta(f)$ be the minimal length of a canonical diffluence of M_f such that $\beta(f) \bmod \pi(f) = 0$. Further, we assume that $\beta(f) \geq \alpha(f)$ if $f \in \mathcal{F}_{\text{conf-diff}}$.

For associative operators $\otimes$ – like $\otimes_{\text{carry}}$ for the addition – we get $\tau(\text{PP}_{\langle \Sigma, \otimes \rangle}) \leq |\Sigma|$ and $\pi(\text{PP}_{\langle \Sigma, \otimes \rangle}) = 1$. Furthermore, it holds that $\alpha(\text{PP}_{\langle \Sigma, \otimes \rangle}) = 1$ if $\text{PP}_{\langle \Sigma, \otimes \rangle} \in \mathcal{F}_{\text{conf}}$ and $\beta(\text{PP}_{\langle \Sigma, \otimes \rangle}) = 1$ if $\text{PP}_{\langle \Sigma, \otimes \rangle} \in \mathcal{F}_{\text{diff}}$.

2.3 Probability Distributions

To investigate the expected time of a machine we consider **families of probability distributions** $\mu_{\Sigma_{\text{I}}} := \mu_{\Sigma_{\text{I}},1}, \mu_{\Sigma_{\text{I}},2}, \ldots$ with $\mu_{\Sigma_{\text{I}},n} : \Sigma_{\text{I}}^n \rightarrow \mathbb{R}$ where $\mu_{\Sigma_{\text{I}},n}(w) \neq 0$ iff $|w| = n$. Furthermore, we bound the computational complexity of these probability distributions by restricting ourself to those which can be approximated by binomial distributions. It is shown in [10,8] that this restriction covers the *strictly positive* distributions that are described by distribution generating circuits [9] or finite irreducible Markov chains.

Definition 6 *Let $p \in (0;1)$. Define $\mathcal{B}_{\Sigma_{\text{in}},n}[p]$ as the set of all families of distributions $\mu_{\Sigma_{\text{in}},p}^{\text{bin}} := \mu_{\Sigma_{\text{in}},1,p}^{\text{bin}}, \mu_{\Sigma_{\text{in}},1,p}^{\text{bin}}, \ldots$ with $\mu_{\Sigma_{\text{in}},n,p}^{\text{bin}} \Sigma^* \rightarrow [0;1]$, such that for all $w \in \Sigma_{\text{I}}^n$ $\mu_{\Sigma_{\text{in}},n,p}^{\text{bin}}(w) \geq p^n$ and 0 elsewhere. Furthermore it holds for any $\mu_{\Sigma_{\text{I}},n,p}^{\text{bin}}$-distributed random variable W: $\forall x \in \Sigma_{\text{I}} \; \forall uv \in \Sigma_{\text{I}}^{n-1} : \; \Pr[Y = x \mid W = uYv] \geq p$.*

To measure the probability of a canonical confluence of a $\mu_{\Sigma_{\text{I}},n,p}^{\text{bin}}$-distributed random variable W we define the **rank of confluence** of $f \in \mathcal{F}_{\text{prefix}}$ and W as $p_f := \mathcal{S}_f \cdot p^{\alpha(f)}$ – where $\mathcal{S}_f$ is the set of all canonical confluences of length $\alpha(f)$ of f. Further define the **rank of diffluence** of f, W and a pair of diffluent states $q_1, q_2 \in R_f(\tau(f))$ as $\widehat{p}_f(q_1, q_2) := \mathcal{O}_f(q_1, q_2) \cdot p^{\beta(f)}$ and $\widehat{p}_f := \max_{q_1, q_2 \in R(\tau(f))} \widehat{p}_f(q_1, q_2)$ where $\mathcal{O}_f(q_1, q_2)$ denotes the set of all diffluences of length $\beta(f)$ of q_1 and q_2.

2.4 Networks of Finite Controls

As the parallel model we will consider networks of finite controls, which are closely related to the model of cellular automata: A **network of finite control (NFC)** $N_{\text{NET}} := (K, K_{\text{in}}, K_{\text{out}}, E)$ of **degree** d is a DAG $G_{N_{\text{NET}}} = (K, E)$ of degree d where the nodes are weighted by synchronised finite controls $K^{[i]} := (Q, \Sigma, \Sigma, \Delta, \Gamma, q_{0,i})$. The **input nodes** of N_{NET} are given by $K_{\text{in}} \subseteq K$ and the **output nodes** of N_{NET} by $K_{\text{out}} \subseteq K$. The input of an NFC is given by the starting states of the input nodes and the output by the final states of the output nodes. The transition function $\Delta : Q \times \Sigma^d \rightarrow \Sigma$ of node $K^{[i]}$ is a function over the state of $K^{[i]}$ and the "outputs" of the direct predecessors $K^{[i_1]}, \ldots, K^{[i_{d_i}]}$ of $K^{[i]}$ where $i_j < i_{j+1}$. Let $q_{0,i}, q_{1,i}, \ldots$ be the sequence of states of $K^{[i]}$. Then it holds that for each $t \geq 0$: $q_{t+1,i} = \Delta(q_{t,i}, \Gamma(q_{t,i_1}), \ldots, \Gamma(q_{t,i_{d_i}}))$.

Define the computation time **time**$_{N_{\text{NET}}}(w)$ of an NFC N_{NET} on input $w \in \Sigma^*$ as the minimum t such that either for all output nodes $K^{[i]}$ holds $q_{t-1,i} = q_{t,i}$. The output $f_{N_{\text{NET}}}(w)$ of N_{NET} on input w is given by output of the output nodes at step $\text{time}_{N_{\text{NET}}}(w)$. For a function $f \in \mathcal{F}_{\text{prefix}}$ let $\mathcal{M}_{\text{NET}}^{f,n}$ denote the set of all NFCs which compute f on inputs of length n.

Like for the circuit model, the domain of a single NFC is bounded to inputs of the same length. So we are mostly interested in **families** of NFCs, where different NFCs have a different number of input nodes. A family of NFCs of a function f is an infinite sequence $N_{\text{NET}}^1, N_{\text{NET}}^2, \ldots$ of NFCs with $N_{\text{NET}}^n \in \mathcal{M}_{\text{NET}}^{f,n}$. We call a family of NFCs **constructible** if there exists a polynomial time bounded DTM that *outputs* N_{NET}^n on input 1^n for all $n \in \mathbb{N}$. For a prefix function f let $\mathcal{M}_{\text{NET}}^f$ denote the set of all constructible families of NFCs for f.

Definition 7 *For an NFC N_{NET}^n with input alphabet Σ_{I} and a distributions $\mu_{\Sigma_{\text{I}},n}$ define the* **expected time** $\text{etime}(N_{\text{NET}}^n, \mu_{\Sigma_{\text{I}},n}) := \sum_{w \in \Sigma_{\text{I}}^n} \mu_{\Sigma_{\text{I}},n}(w) \cdot \text{time}_{N_{\text{NET}}^n}(w)$. *For a family of NFCs* $M := N_{\text{NET}}^1, N_{\text{NET}}^2, \ldots$ *and a family of distributions* $\mu_{\Sigma_{\text{I}}} := \mu_{\Sigma_{\text{I}},1}, \mu_{\Sigma_{\text{I}},2}, \ldots$ *we write* $\text{etime}(M, \mu_{\Sigma_{\text{I}}}) = f$ *iff for all* $n \in \mathbb{N}$ *holds* $\text{etime}(N_{\text{NET}}^n, \mu_{\Sigma_{\text{I}},n}) = f(n)$.

Further, we extend this definition of the expected time to sets of families of NFCs $\mathcal{M}$ and sets of families of distributions $\mathcal{D}$ in a natural way. For example:
$$\text{etime}(\mathcal{M}, \mathcal{D}_{\Sigma_{\text{I}}}) \leq f \quad :\Longleftrightarrow \quad \exists M \in \mathcal{M} \; \forall \mu_{\Sigma_{\text{I}}} \in \mathcal{D} \; : \quad \text{etime}(M, \mu_{\Sigma_{\text{I}}}) \leq f$$
$$\text{etime}(\mathcal{M}, \mathcal{D}_{\Sigma_{\text{I}}}) \geq f \quad :\Longleftrightarrow \quad \forall M \in \mathcal{M} \; \exists \mu_{\Sigma_{\text{I}}} \in \mathcal{D} \; : \quad \text{etime}(M, \mu_{\Sigma_{\text{I}}}) \geq f \; .$$
In the following we will show sharp bounds for the expected time behavior of NFCs computing prefix functions. The lower bounds follow from some arguments based only on restrictions of the data flow. Hence, we can translate these bounds also to other parallel models like EREW-PRAMS or PRAM-networks of bounded degree.

3   Preliminaries

Before we analyze how the existence of a canonical confluence or a canonical diffluence influence the expected delay of a NFC in detail, we will show upper and lower bounds for the probability that the maximal length of a non-confluent substring of a $\mathcal{B}_{\Sigma_{\text{In}},n}[p]$-distributed random variable W passes a given bound. More precisely, we will analyze the expected length of $\text{pro}_f(w, i) := \max_{j \leq i}\{i - j \mid w[j; i]$ is not a confluence for $R_f(j)\}$ and $\text{pro}_f(w) := \max_{i \in [1;|w|]} \text{pro}_f(w, i)$.

For easier notion let $\text{llog} := \log_2 \log_2$. By partitioning the input string into blocks of length $\alpha(f)$ we can show:

Lemma 2 *Let W be a $\mu_{\Sigma_{\text{I}},n,p}^{\text{bin}}$-distrib. rand. variable with $\mu_{\Sigma_{\text{I}},n,p}^{\text{bin}} \in \mathcal{B}_{\Sigma_{\text{I}},n}[p]$, then it holds for any function $f \in \mathcal{F}_{\text{conf}}$ and for any i, ℓ with $\tau(f) + \ell \cdot \alpha(f) \leq i$: $\Pr[\text{pro}_f(W, i) > (\ell + 1) \cdot \alpha(f)] \leq (1 - p_f)^\ell$. If further $\tau(f) + \ell \cdot \alpha(f) \leq n$: $\Pr[\text{pro}_f(W) \geq (2\ell - 1) \cdot \alpha(f)] \leq \lceil (n - \tau(f))/(\alpha(f) \cdot \ell) \rceil \cdot \exp(-\ell \cdot p_f)$.*

If we partition the input string into blocks of length $\beta(f)$ we can analyze the maximum length of a diffluent substring:

Lemma 3 *Let $f \in \mathcal{F}_{\text{diff}}$ and W be a $\mu_{\Sigma_{\text{I}},n,p}^{\text{bin}}$-distributed random variable. Then it holds for $\varepsilon := (\text{llog}_2 n + \log_2 \beta(f) - \text{llog}_2(\widehat{p}_f^{-1}))/\log_2 n$:*
$$\Pr[\text{pro}_f(W) \geq \left\lfloor (1 - \varepsilon) \cdot \log_2 n \; / \; \log_2(\widehat{p}_f^{-1}) \right\rfloor \cdot \beta(f)] \geq p^{|\Sigma_f|}/2 \; .$$

4 Upper and Lower Bounds

To prove the upper bound of the expected delay of a NFC computing a prefix function we will use a network design based on the circuit presented in [10]. The construction of a single finite controls is based on a technique of Ladner and Fischer. In [11] they have shown that for any prefix function f resp. for each Moore-machine M_f there exists a groupoid $\langle \overline{\Sigma}_f, \overline{\otimes}_f \rangle$ with binary associative operator $\overline{\otimes}_f$ and two homomorphisms g_1, g_2 such that $f \equiv g_2 \circ \mathrm{PP}_{\langle \overline{\Sigma}_f, \overline{\otimes}_f \rangle, g_1(\Sigma_\mathrm{I})} \circ g_1$. Using this construction the size of $\overline{\Sigma}_f$ is exponential in $|Q_f|$ and therefore at least exponential in $|\Sigma_\mathrm{I}|$. Since both homomorphisms g_1 and g_2 can be computed by an NFC within constant depth we will focus our analysis on a network computing $\mathrm{PP}_{\langle \overline{\Sigma}_M, \overline{\otimes}_M \rangle, g_1(\Sigma_\mathrm{I})}$.

To compute a given prefix function f on inputs of length $n = 2^k - 2$ we will use a NFC N_{NET}^k as illustrated in figure 2. The NFC N_{NET}^k consists of $n = 2^k - 2$ input nodes $\{x_0, \dots, x_{n-1}\}$, n output nodes $\{y_0, \dots, y_{n-1}\}$, and $2^{k+2} - 2^k - 4 \cdot k - 10$ internal nodes. The internal nodes are partitioned into three types of gates: The *upper part*, denoted by A_k, consists of nodes u_i^j with $0 \le j \le k$ and $0 \le i \le 2^{k-j+2} - 3$ where every node u_i^j is the root of a complete binary tree of height j with leaves $x_{i \cdot 2^j}, \dots, x_{(i+1) \cdot 2^j - 1}$. Further, it holds that $u_i^0 := x_i$ and $u_i^j := u_{2i}^{j-1} \overline{\otimes}_f u_{2i+1}^{j-1}$ for all i. As a whole the upper part of N_{NET}^k is a forest of complete binary trees, in which each height up to k appears exactly twice and the heights decrease when going from left to right. The *lower part* B_k consists of nodes v_i^j, w_i^j with $1 \le j \le k+1$ and $0 \le i \le 2^{k-j+2} - 2$ where

$$v_0^j = u_0^{j-1} \qquad\qquad v_{2i+1}^j = v_i^{j+1} \overline{\otimes}_f u_{4i+2}^{j-1} \qquad\qquad y_{2i+1} = w_i^1$$
$$w_i^j = v_i^j \overline{\otimes}_f u_{2i+1}^{j-1} \qquad\qquad v_{2i}^j = w_{i-1}^{j+1} \overline{\otimes}_f u_{4i}^{j-1} \qquad\qquad y_{2i} = v_i^1 .$$

If each pair (v_i^j, w_i^j) is collapsed to a single node the resulting topology of the lower part is a collection of complete binary trees, in which each height up to k appears exactly once and the heights increase when going from left to right.

Lemma 4 *Let $f \in \mathcal{F}_{\mathrm{conf}}$, then it holds for any $n = 2^k - 2$ with $k \in \mathbb{N}$ and for any input $w \in \Sigma_\mathrm{I}^n$ that* $\mathrm{time}_{N_{\mathrm{NET}}^k}(w) \le 3 \cdot \log_2(\mathrm{pro}_f(w)) + 6$.

From Lemmata 2 and 4 we can conclude:

Theorem 2 *Let $f \in \mathcal{F}_{\mathrm{conf}}$ and N_{NET}^k a NFCs that computes f on all inputs of length $n \in [2^{k-1} - 1; 2^k - 2]$, then it holds* $\mathrm{etime}(N_{\mathrm{NET}}^k, \mathcal{B}_{\Sigma_\mathrm{I}, n}[p]) \le \mathbf{3 \cdot llog\ } n + 3 \cdot \log_2 p_f^{-1} + c_f$ *where c_f is a constant depending on f. If further $f \in \mathcal{F}_{\mathrm{str\text{-}conf}}$, then* $\mathrm{time}_{N_{\mathrm{NET}}^k}(w) \le 6 \cdot \log_2(|\Sigma|) + 3$ *for all $w \in \Sigma_\mathrm{I}^n$.*

In the following we investigate the expected delay of NFCs of degree 2 (the general case of degree d follows similarly). Thus, a lower bound for a strictly diffluent prefix function $f \in \mathcal{F}_{\mathrm{str\text{-}diff}}$ follows directly from the bounded fanout:

Theorem 3 *Let $f \in \mathcal{F}_{\mathrm{str\text{-}diff}}$, then it holds for all n with $n > \tau(f) + |\Sigma_f|$ an for a constant c_f depending on f* $\mathrm{etime}(\mathcal{M}_{\mathrm{NET}}^{f,n}, \mathcal{B}_{\Sigma_\mathrm{I}, n}[p]) \ge \frac{1}{2} \cdot \mathbf{log_2\ } n - \frac{\log |\Sigma_f|}{2} \cdot \log_2 p^{-1} - c_f$.

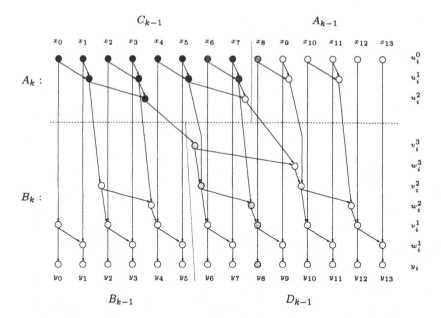

Fig. 2. An average case optimal NFCs N_{NET}^k with $k = 4$ for input vectors of length 14.

Similarly to Theorem 3 it follows from Lemma 3:

Theorem 4 *Let $f \in \mathcal{F}_{\text{diff}}$, then* $\text{etime}(\mathcal{M}_{\text{NET}}^{f,n}, \mathcal{B}_{\Sigma_{\text{I}},n}[p]) \geq \frac{p^{|Q_f|}}{4} \cdot (\text{llog } n - c_f)$ *where* $c_f := |Q_f| \cdot \log_2 p^{-1} + \log_2 \log_2 \hat{p}_f^{-1} + \text{O}(\log_2 \beta(f))$.

Summarizing we can conclude for constant probabilities p – like the uniform distribution with $p = \frac{1}{2}$ – that only three different cases can arise: either a prefix function can be computed with a constant average delay if $f \in \mathcal{F}_{\text{str-conf}}$, or $\text{etime}(\mathcal{M}_{\text{NET}}^{f,n}, \mathcal{B}_{\Sigma_{\text{I}},n}[p]) \in \Theta(\text{llog } n)$ if $f \in \mathcal{F}_{\text{conf-diff}}$, or $\text{etime}(\mathcal{M}_{\text{NET}}^{f,n}, \mathcal{B}_{\Sigma_{\text{I}},n}[p]) \in \Theta(\log n)$ if $f \in \mathcal{F}_{\text{str-diff}}$. Thus, it follows for the examples illustrated in figure 1 that the negation f_{not} can be computed within a constant delay, the carry propagation f_{carry} with in an expected delay of order $\text{llog } n$, whereas the parity function f_{parity} as well as $w[1]^{w[i]}$ requires networks with an expected delay of $\Theta(\log n)$.

5   The Unbounded Fanout Case

Investigating the example function $w[1]^{w[i]}$ from subsection 2.2, it is easy to see that it can be computed by an NFC with unbounded fanout for all inputs in constant time. To find a characterization of the prefix function according to the average time complexity in processor networks with unbounded out-degree we have to examine the structure of a given Moore-machine more carefully.

Definition 8 *A subset $Q' \subseteq Q$ is called a* **component** *of M if Q' is a maximal strongly connected component of the transition graph G_M. For $q \in Q$ let $\boldsymbol{Q}[\boldsymbol{q}]$ denote the component that contains q. A component Q' is called* **closed** *if any state that is reachable from a state of Q' belongs to Q'. Let $\boldsymbol{Q}_M^*$ be the set of all states of Q belonging to a closed component of M.*

Applying the confluence and diffluence properties to the closed components of the transition graph of an Moore-machine we define:

Definition 9 *A Moore-machine M is* **(strictly) suffix inherent confluent** *if for all $q \in R_M(\tau(M)) \cap \boldsymbol{Q}_M^*$ M_q is (strictly) canonical confluent. $f \in \mathcal{F}_{\text{prefix}}$ is (strictly) suffix inherent confluent if M_f is (strictly) suffix inherent confluent. Define $\mathcal{F}_{\text{si-conf}}$ resp. $\mathcal{F}_{\text{si-str-conf}}$ as the set of all functions $f \in \mathcal{F}_{\text{prefix}}$ such that M_f is (strictly) suffix inherent confluent. M is* **(strictly) suffix inherent diffluent** *if there exists at least one state $q \in R_M(\tau(M)) \cap \boldsymbol{Q}_M^*$ such that M_q is (strictly) canonical diffluent. $f \in \mathcal{F}_{\text{prefix}}$ is called (strictly) suffix inherent diffluent if M_f is (strictly) suffix inherent diffluent. Define $\mathcal{F}_{\text{si-diff}}$ resp. $\mathcal{F}_{\text{si-str-diff}}$ as the set of all (strictly) suffix inherent diffluent functions $f \in \mathcal{F}_{\text{prefix}}$. Further define $\mathcal{F}_{\text{si-conf-diff}} := \mathcal{F}_{\text{si-conf}} \cap \mathcal{F}_{\text{si-diff}}$.*

For the four examples of subsection 2.2 it holds $f_{\text{carry}} \in \mathcal{F}_{\text{si-conf-diff}}$, $f_{\text{parity}} \in \mathcal{F}_{\text{si-str-diff}}$, and $f_{\text{not}}, w[1]^{w[i]} \in \mathcal{F}_{\text{si-str-conf}}$. According to Theorem 1 we can show:

Theorem 5 *$\mathcal{F}_{\text{prefix}}$ is a disj. union of $\mathcal{F}_{\text{si-str-conf}}$, $\mathcal{F}_{\text{si-conf-diff}}$, and $\mathcal{F}_{\text{si-str-diff}}$.*

The lower bounds for fanout unbounded NFCs follow analogously to the lower bounds of fanout bounded Networks. Additional to the fanout bounded case we have to consider the probability that a prefix of an input maps the starting state of the corresponding Moore-machine into a state of a closed component. This yields an additional factor of $p^{|\Sigma_f|}$.

To compute a prefix function efficiently in the average, we will use a network which method of working can be subdivided into two steps. In the first step we will determine the values $\boldsymbol{\text{run}_f(w)} := \min\{2^k - 1 \mid \delta(q_0, w[1; 2^k - 1]) \in \mathcal{Q}_{M_f}^*\}$ and $q_{CC} := \delta(q_0, w[1; \text{run}_f(w)])$. This can be done within an expected delay of $O(\tilde{p}(f))$ with $\tilde{p}(\boldsymbol{f}) := p^{|Q_f|} \cdot \exp(p^{|Q_f|})/(\exp(p^{|Q_f|}) - 1) + p^{-1/2 \cdot |Q_f|}$ using a network design as illustrated in figure 3. In the second step we will a network as presented in the previous section.

Theorem 6 *Let $f \in \mathcal{F}_{\text{prefix}}$ and $N_{\text{NET}} \in \mathcal{M}_{\text{NET}}^{f,n}$ an NFC of indegree d then it holds*

$$\text{etime}(N_{\text{NET}}, \mathcal{B}_{\Sigma_I, n}[p]) \geq \Omega\left(\frac{p^{2 \cdot |\Sigma_f|}}{\log_2 d}\right) \cdot \begin{cases} \textbf{llog}\, n & \text{if } f \in \mathcal{F}_{\text{si-conf-diff}} \\ \log_2 n & \text{if } f \in \mathcal{F}_{\text{si-str-diff}} \end{cases}.$$

Furthermore it holds

$$\text{etime}(\mathcal{M}_{\text{NET}}^{f,n}, \mathcal{B}_{\Sigma_I, n}[p]) \leq \begin{cases} 3 \cdot \textbf{llog}\, n + O(\tilde{p}(f)) & \text{if } f \in \mathcal{F}_{\text{si-conf-diff}} \\ 6 \cdot \log_2 |Q_f| + O(\tilde{p}(f)) & \text{if } f \in \mathcal{F}_{\text{si-str-conf}} \end{cases}.$$

Hence, the classification of the average time complexity to compute prefix functions in processor networks is similar in both cases. The unbounded fanout of a

input

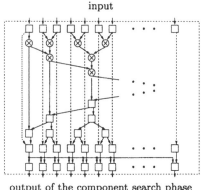

output of the component search phase

Fig. 3. Unbounded outdegree NFC computing $\text{run}_f(w)$.

network enables the NFC to distribute some local knowledge to the whole network within one step. So the existence of a strict diffluence that is only caused by a prefix of constant length does not determine the computation time of the network completely.

6   Conclusions

Our results characterize the average case complexity of prefix functions. They allow a significant speedup in the expected delay for two types of networks: networks with bounded and with unbounded fanout. Using the presented algebraic properties and network designs we can generate a family of NFCs that achieves the shown bounds for all prefix functions. Since the proofs of the lower bounds do not make use of the restricted power of finite state machines the results can simply be translated to other more powerful parallel models like EREW-PRAMs, CREW-PRAMs, and networks of arbitrary powerful control units. On the other hand the lower bounds can not be applied to CRCW-PRAMs or other models with unbounded fanin like ub-circuits. For example, it is known that the addition of two n-bit numbers can be computed be a ub-circuit of polynomial size and constant depth.

Acknowledgment: I thank Gerhard Buntrock, Karin Genther, Kathrin Hauke, Maciej Liśkiewicz, Rüdiger Reischuk, Christian Schindelhauer, Ingo Wegener, and Stephan Weis for fruitful discussions and helpful hints to the literature.

References

1. R. Beigel, B. Gasarch, M. Li, L. Zhang, *Addition in* $\log_2 n + O(1)$ *Steps on Average: A Simple Analysis*, ECCC, TR96-051, 1996.

2. G. Bilardi, F. Preparata, *Size-Time Complexity of Boolean Networks for Prefix Computations*, J. of the ACM 36, 1989, 362-382.
3. G. Bilardi, F. Preparata, *Characterization of Associative Operations with Prefix Circuits of Constant Depth and Linear Size*, SIAM J. Comput 19, 1990, 246-255.
4. B. E. Briley, *Some New Results on Average Worst Case Carry*, IEEE Trans. Computers, C-22:5, 1973.
5. A. W. Burks, H. H. Goldstine, J. von Neumann, *Preliminary Discussion of the Logical Design of an Electronic Computing Instrument*, John von Neumann *Collected Works*, Band 5, Editor A. H. Taub, 1961.
6. B. Fagin, *Fast Addition of Large Integers*, IEEE Tr. Comp. 41, 1992, 1069-1077.
7. P. Gemmell, M. Horchol, *Tight Bounds on Expected Time to Add Correctly and Add Mostly Correctly*, Information Processing Letters, 1994, 77-83.
8. A. Jakoby, *Die Komplexität von Präfixfunktionen bezüglich ihres mittleren Zeitverhaltens*, PhD Thesis, Med. Universität zu Lübeck, 1998. Shaker Verlag, Aachen 1998, Germany.
9. A. Jakoby, R. Reischuk, C. Schindelhauer, *Circuit Complexity: from the Worst Case to the Average Case*, 26. STOC'94, 1994, 58-67.
10. A. Jakoby, R. Reischuk, C. Schindelhauer, S. Weis, *The Average Case Complexity of the Parallel Prefix Problem*, 21. ICALP'94, 1994, 593-604.
11. R. Ladner, M. Fischer, *Parallel Prefix Computation*, J. of the ACM 27, 1980, 831-838.
12. J. Reif, *Probabilistic Parallel Prefix Computation*, Comp. Math. Applic. 26, 1993, 101-110.
13. G. Reitwiesner, *The Determination of Carry Propagation Length for Binary Addition*, IRE Trans. on Electronic Comput., Vol EC-9, 1960, 35-38.
14. M. Snir, *Depth-Size Trade-offs for Parallel Prefix Computation*, J. Alg. 7, 1986, 185-201.

On the Hardness of Permanent

Jin-Yi Cai [*1], A. Pavan[1], and D. Sivakumar [**2]

[1] Department of Computer Science and Engineering, State University of New York
at Buffalo, Buffalo, NY 14260. {cai, aduri}@cse.buffalo.edu
[2] Department of Computer Science, University of Houston, Houston, TX 77204.
siva@cs.uh.edu

Abstract. We prove that if there is a polynomial time algorithm which
computes the permanent of a matrix of order n for any inverse polynomial fraction of all inputs, then there is a BPP algorithm computing
the permanent for *every* matrix. It follows that this hypothesis implies
$P^{\#P} = BPP$. Our algorithm works over any sufficiently large finite field
(polynomially larger than the inverse of the assumed success ratio), or
any interval of integers of similar range. The assumed algorithm can
also be a probabilistic polynomial time algorithm. Our result is essentially the best possible based on any black box assumption of permanent
solvers, and is a simultaneous improvement of the results of Gemmell
and Sudan [GS92], Feige and Lund [FL92] as well as Cai and Hemachandra [CH91], and Toda (see [ABG90]).

1 Introduction

The *permanent* of an $n \times n$ matrix A is defined as

$$\text{per}(A) = \sum_{\sigma \in S_n} \prod_{i=1}^{n} A_{i,\sigma(i)} \ ,$$

where S_n is the symmetric group on n letters, i.e., the set of all permutations of
$\{1, \ldots, n\}$.

The permanent function has a rich history in combinatorics and in computational complexity theory. All known general algorithms computing the permanent function over the integers take exponential time. In fact, this exponential
time complexity remains true for any general algorithm over any field of characteristic other than two. The best known general computational procedure for
the permanent is a formula due to Ryser, which runs in time $O(n^2 2^n)$ [Rys63].

In terms of computational complexity theory, in 1979, Valiant [Val79] proved
the seminal result that computing the permanent of integer matrices is complete for the counting complexity class #P, and is therefore NP-hard. A decade
later, Toda [Tod89] demonstrated the surprising power of #P; Toda's theorem,

* Supported in part by NSF grant CCR-9634665, and by a Guggenheim Fellowship.
** Supported in part by an NSF CAREER award CCR-9734164.

C. Meinel and S. Tison (Eds.): STACS'99, LNCS 1563, pp. 90–99, 1999.

together with Valiant's result, implies that the permanent is hard for the entire polynomial-time hierarchy.

The permanent has two other fascinating properties that endow it with rich computational structure. Lipton [Lip91] observed that the permanent of a matrix with entries that are low degree polynomials in an indeterminate x is itself a low-degree polynomial. In particular, taking linear polynomials in x as entries, the permanent of an $n \times n$ matrix is a polynomial of degree at most n. It follows that the computation of the permanent of a matrix can be reduced, via polynomial interpolation, to computing the permanent of uniformly distributed random matrices. This property of *random self-reducibility* has the important consequence that the permanent is computationally hard in the worst case if and only if it is hard on the average. The permanent also has the *downward self-reducibility* property, which means that computing the permanent of an $n \times n$ matrix can be reduced (in polynomial time) to the computation of the permanent of $(n-1) \times (n-1)$ matrices, via Laplacian expansion. Lund *et al.* [LFKN90] applied this property together with the random self-reducibility, to obtain the breakthrough result that #P has interactive proof protocols. Very recently, Impagliazzo and Wigderson [IW98] have used these two properties to obtain exciting new connections between computational hardness vs. randomness.

The line of research initiated by Lipton [Lip91] connecting the worst-case and the average-case complexities of the permanent was subsequently pursued by Gemmell et al. [GLRSW91], Gemmell and Sudan [GS92], and by Feige and Lund [FL92]. It is shown in [GS92] that if there is a polynomial time algorithm that can compute the permanent on at least a $(1/2) + (1/\text{poly})$ fraction of all $n \times n$ matrices over the finite field $\mathbf{Z}_p$, then there is a probabilistic polynomial time algorithm that can compute the permanent of every $n \times n$ matrix over $\mathbf{Z}_p$, provided p is sufficiently large with respect to n. Feige and Lund [FL92] improved this and showed a similar result under a weaker hypothesis that assumes an algorithm that can compute the permanent on at least a $(1/2) - 1/n$ fraction of all $n \times n$ matrices over the finite field $\mathbf{Z}_p$. A polynomial reconstruction algorithm (aka. Reed-Solomon decoder) of Berlekamp and Welch [BW] is the crucial technical procedure in both of these papers. The interactive protocol of [LFKN90] is another crucial ingredient in [FL92].

Another result regarding the computational complexity of the permanent is due to Cai and Hemachandra [CH91] and independently due to Toda (see [ABG90]). According to this result, if there is a polynomial time algorithm such that for every matrix of order n, it can enumerate a list of polynomially many values, one of which is the correct value of the permanent, then one can compute the permanent of any matrix in polynomial time.

In this note, we show that if there is a polynomial time algorithm that can compute the permanent on at least an inverse polynomial fraction $(1/n^c)$ of all $n \times n$ matrices over the finite field $\mathbf{Z}_p$, then there is a probabilistic polynomial time algorithm that can compute the permanent of every $n \times n$ matrix over $\mathbf{Z}_p$, provided p is somewhat larger than the inverse of the success ratio of the assumed algorithm. The proof can be extended to the case where the assumed algorithm is also a probabilistic polynomial time algorithm. We also prove the same result for matrices over any finite field, (again assuming that the field size is sufficiently large compared to the inverse of the success ratio), and over integers (with an appropriate restriction on the length of the entry size and a corresponding definition of uniform distribution over such an interval of integers.) These results are a simultaneous improvement of the results in [GS92, FL92] as well as in [CH91] (except replacing probabilistic for deterministic polynomial time algorithm in the latter).

We achieve this improvement by applying an improved Reed-Solomon decoder due to Madhu Sudan [Sud96], building on earlier work by Ar et al. [ALRS92]. Our proof uses a family of intermediate matrices whose entries are univariate polynomials in an indeterminate x; this definition unifies two ideas used in previous papers, and makes the proof simple and completely self-contained.

2   Hardness over $\mathbf{Z}_p$

We begin with a theorem concerning the complexity of permanent over the finite field $\mathbf{Z}_p$, where p is somewhat larger than the inverse of the success ratio of the assumed algorithm.

Theorem 1. *For any constant $k > 0$, if there exists a deterministic polynomial time algorithm $\mathcal{A}$ such that for all n and $p \geq 9n^{2k+2}$, $\mathcal{A}$ computes the permanent of order n over the field $\mathbf{Z}_p$ correctly on greater than $1/n^k$ fraction of inputs, then $\mathrm{P}^{\#\mathrm{P}} = \mathrm{BPP}$.*

Proof. Assume that the success probability of $\mathcal{A}$ is q. Then $q \geq 1/n^k$. Without loss of generality we assume $k \geq 1$. Let M be an $n \times n$ matrix over $\mathbf{Z}_p$ whose permanent we wish to compute. Let $M_{11}, M_{21}, \ldots, M_{n1}$ be the n minors of M. Consider the following matrix polynomial, defined by choosing the matrices B and C (of dimension $n - 1$) uniformly and independently at random,

$$D(x) = \sum_{i=1}^{n} \delta_i(x) M_{i1} + \alpha(x)(B + xC),$$

where each $\delta_i(x)$ is a polynomial of degree $n - 1$ such that

$$\delta_i(x) = \begin{cases} 1 \text{ if } x = i \\ 0 \text{ if } x \neq i \text{ and } 1 \leq x \leq n, \end{cases}$$

and $\alpha(x)$ is a degree-n polynomial that vanishes for $x = 1, 2, \ldots, n$, given by $\alpha(x) = (x - 1)(x - 2) \cdots (x - n)$. Note that $D(i) = M_{i1}$, for $1 \leq i \leq n$.

Let $D = \{D(x) \mid x = n + 1, n + 2, \ldots, p\}$. The permanent of $D(x)$ is a polynomial of degree less than n^2. If we know the polynomial $\mathrm{per}(D(x))$ then we can compute the permanent of M, by Laplacian expansion. If the matrices B and C are chosen uniformly at random, then all the matrices in D are uniformly distributed. Moreover, if the matrices B and C are chosen uniformly and independently, then it can be shown easily that the matrices in D are pairwise independent. So the algorithm $\mathcal{A}$ has success probability close to q on D also. More precisely, the following lemma can be proved by the Chebyshev inequality.

Lemma 1. *The probability (over random choices of B and C) that the algorithm $\mathcal{A}$ correctly computes the permanent on more than $q/2$ fraction of inputs from D is at least $1 - \frac{1}{(p-n)q^2}$.*

Proof. For $i = n+1, n+2, \ldots, p$, define Z_i to be 1 if $\mathcal{A}$ succeeds on $D(i)$, else 0. We know that the expectation $E(Z_i) = q$ for $i = n+1, n+2, \ldots, p$. Define the random variable $Z = \frac{\sum_{i=n+1}^{p} Z_i}{p-n}$ as the average of the Z_i's. Expectation of Z is also q.

$$\begin{aligned} \Pr[Z \leq (q/2)] &= \Pr[Z - E(Z) \leq (-q/2)] \\ &\leq \Pr[|Z - E(Z)| \geq (q/2)] \\ &\leq \frac{\mathbf{Var}(Z)}{(q/2)^2} \leq \frac{1}{(p-n)q^2}. \end{aligned}$$

The last inequality is obtained using the facts that Z_i's are pairwise independent and the variance of a 0-1 random variable is at most $1/4$. So with high probability $\mathcal{A}$ works correctly on more than $q/2$ fraction of inputs from D. ∎

We call a set D (which is defined by choosing B and C) *good* if, $\mathcal{A}$ works correctly on more than $q/2$ fraction of inputs from D. In the following discussion assume that D is good.

For a polynomial f, the *Graph(f)* is the set $\{(i, f(i)) \mid i = 1, 2, \ldots, p\}$. Define a set S as follows, $S = \{(i, \mathcal{A}(D(i))) \mid i = n+1, \ldots, p\}$. Consider the set of all polynomials f of degree less than n^2 such that S intersects the set *Graph(f)* with at least $(p-n)q/2$ points. The polynomial per$(D(x))$ is one among them, since $\mathcal{A}$ computes the permanent correctly on $(p-n)q/2$ matrices from D. We can prove that there exist at most polynomially (in n) many such polynomials f.

Lemma 2. *If $p \geq 9n^{2k+2}$, then there are at most $3/q$ many polynomials f of degree less than n^2 that satisfy $|Graph(f) \cap S| \geq (p-n)q/2$.*

Proof. If there exist more than $3/q$ such polynomials, consider a set $\mathcal{F}$ of $N = \lceil 3/q \rceil$ polynomials among them. For a polynomial f define the set, $S_f = \{i \mid (i, f(i)) \in Graph(f) \cap S\}$. By the inclusion-exclusion principle

$$\begin{aligned} p - n \geq \left| \bigcup_{f \in \mathcal{F}} S_f \right| &\geq \sum_{f \in \mathcal{F}} |S_f| - \sum_{f, f' \in \mathcal{F}, f \neq f'} |S_f \cap S_{f'}| \\ &\geq N \frac{(p-n)q}{2} - \frac{N(N-1)}{2}(n^2 - 1) \end{aligned}$$

The last inequality uses the fact that any two distinct polynomials of degree less than n^2 can agree on no more than $n^2 - 1$ points. If $N = \lceil 3/q \rceil$ then $N - 1 < 3/q$

and $N \geq 3/q$. Since $p \geq 9n^{2k+2}$ and $q \geq 1/n^k$, we have $p - n > 9(n^2 - 1)/q^2$. Thus we obtain the following contradiction

$$
\begin{aligned}
p - n &\geq \frac{N}{2}\left((p-n)q - (N-1)(n^2-1)\right) \\
&> \frac{N}{2}\left((p-n)q - \frac{3(n^2-1)}{q}\right) \\
&\geq \frac{3}{2q}\left((p-n)q - \frac{3(n^2-1)}{q}\right) \\
&= p - n + \frac{1}{2}\left(p - n - \frac{9(n^2-1)}{q^2}\right) \ > \ p - n.
\end{aligned}
$$

∎

Next we show how all these polynomials can be obtained explicitly by a randomized procedure with high probability. To handle a minor technical complication, we will divide this procedure into two cases. We remind the reader that we are working under the assumption that the set D is good, that is, the fraction of inputs from D for which $\mathcal{A}$ works correctly is more than $q/2 = 1/(2n^k)$.

(Case $9n^{2k+2} \leq p < 161n^{3k+2}$):

In this case, we apply the algorithm $\mathcal{A}$ to the matrix $D(x)$ for every $x = n+1, \ldots, p$. Clearly, this can be accomplished in polynomial time. Since D is assumed to be *good*, $\mathcal{A}$ computes the permanent correctly for at least $(p - n)/(2n^k)$ of these $p - n$ matrices. Letting $L = p - n$ and $d = n^2$, we have a list of L pairs $\{(x_j, y_j) \mid j = 1, \ldots, L\}$, such that the x_j's are all distinct, and moreover, there is a polynomial f of degree at most d (namely $\mathrm{per}(D(x))$) whose graph intersects the list on at least $(p - n)/(2n^k)$ places. The condition $p \geq 9n^{2k+2}$ implies that $(p-n)/(2n^k) > \sqrt{2(p-n)n^2} = \sqrt{2Ld}$, for all n.

(Case $p \geq 161n^{3k+2}$):

Pick $L = 40n^{2k+2}$ many values x uniformly and independently from the set $\{n+1, \ldots, p\}$. For the set of (distinct) x's produced by this process, produce the matrix $D(x)$ and apply the algorithm $\mathcal{A}$ to it. Our goal is to argue that with overwhelming probability, for at least $9n^{k+2}$ *distinct* x's (out of the L choices made), $\mathcal{A}$ will give us the correct value of the permanent. (The reason for the choice of 161 and 9 with respect to 40 will be clear shortly.)

Call an x *lucky* if $\mathcal{A}$ computes $\mathrm{per}(D(x))$ correctly. We define the events E_j, $j = 1, \ldots, L$, as follows: the event E_j occurs if the j-th random choice of x is lucky, *and* the j-th random choice of x is distinct from all the previously chosen lucky x's. The worst case for E_j to occur is when $j = L$ and all the previously chosen x's were distinct and lucky, thus making it least likely that the jth choice from $\{n+1, \ldots, p\}$ is a lucky x distinct from all previous points. Even in this case, there are more than $((p - n)/(2n^k)) - L$ distinct lucky x's that have not been picked, and which, if picked next, would cause E_j to occur. Therefore, for every j, and under any condition on the events $E_1, E_2, \ldots, E_{j-1}$, the probability that E_j occurs is at least

$$\frac{\frac{p-n}{2n^k} - L}{p-n} = \frac{1}{2n^k} - \frac{40n^{2k+2}}{p-n} \geq \frac{1}{2n^k} - \frac{1}{4n^k} = \frac{1}{4n^k},$$

since $p \geq 161n^{3k+2}$. With this estimate we will prove that with high probability at least $9n^{k+2}$ out of the L events occur.

The event "at least $9n^{k+2}$ out of L E_j's occur" is stochastically dominated by the event that we obtain at least $9n^{k+2}$ successes in L Bernoulli trials, each having an independent success probability of $1/(4n^k)$. This can be seen by performing the L Bernoulli trials in the following fashion: first, perform the experiment which defines E_j, where an occurrence of E_j is counted as a success in the jth Bernoulli trial; second, for each j, if E_j did not occur, let the jth Bernoulli trial be a success with probability $1/(4n^k) - e_j$, where e_j is the conditional probability of E_j occurring given the previous occurrences (and non-occurrences) of $E_1, E_2, \ldots, E_{j-1}$.

Now the probability of the event that "at least $9n^{k+2}$ successes in L Bernoulli trials with success probability of $1/(4n^k)$" can be shown to be very close to one, using Chernoff bounds. Therefore, with very high probability we have $9n^{k+2}$ distinct x's for which the value of the polynomial $\text{per}(D(x))$ of degree less than n^2 is available. Of course, these x's and the values of $\text{per}(D(x))$ are part of a list of at most L pairs (x_j, y_j). Once again, recalling that $L = 40n^{2k+2}$ and letting $d = n^2$, we notice that $9n^{k+2} > \sqrt{2Ld} = \sqrt{80n^{2k+4}}$.

To summarize, in either case, with high probability, we have a list $\{(x_j, y_j)\}$ of at most L pairs (for some L which is polynomially bounded in n) such that the graph of the polynomial $\text{per}(D(x))$ of degree at most $d = n^2$, intersects the list on more than $\sqrt{2Ld}$ places. Given such a list, the following lemma of Sudan [Sud96], building on earlier work by [ALRS92], shows how one can construct all the polynomials f whose graphs intersect the list on more than $\sqrt{2Ld}$ places. We note that Sudan's procedure is based on bivariate polynomial factoring. Therefore, it can be implemented in randomized polynomial time in L and d and $\log p$, or in deterministic time polynomial in L, d, and p (see [Kal92]).

Lemma 3 ([Sud96]). *Given a sequence of L distinct pairs $\{(x_i, y_i) \mid 1 \leq i \leq L\}$, where x_i and y_i are elements of a field $\mathbf{F}$. Let t and d be integers such that $t > \sqrt{2Ld}$. Then there is a probabilistic polynomial time algorithm that finds all polynomials f of degree at most d such that the number of i's such that $y_i = f(x_i)$ is at least t.*

Sudan's polynomial reconstruction algorithm is very elegant and simple. For the sake of completeness, we will sketch his algorithm here.

Consider all bivariate polynomials of the form $F(x, y) = \sum_{i,j} a_{i,j} x^i y^j$, where $i + jd \leq \sqrt{2Ld}$. Thus there are exactly

$$\sum_{j=0}^{\lfloor\sqrt{2L/d}\rfloor} \left(\lfloor\sqrt{2Ld}\rfloor - jd + 1\right) = \left(\lfloor\sqrt{2L/d}\rfloor + 1\right)\left(\lfloor\sqrt{2Ld}\rfloor + 1\right) - d\binom{\lfloor\sqrt{\frac{2L}{d}}\rfloor + 1}{2},$$

many coefficients $a_{i,j}$. This quantity can be shown to be strictly greater than L as follows:

Let $I = \lfloor \sqrt{2L/d} \rfloor$, then $\sqrt{2L/d} = I + x$, for some $0 \leq x < 1$. Since $\lfloor \sqrt{2Ld} \rfloor + 1 > \sqrt{2Ld}$, the total sum is strictly greater than

$$\left(\sqrt{2L/d} + 1 - x \right) \sqrt{2Ld} - \frac{d}{2} \left(\sqrt{2L/d} + 1 - x \right) \left(\sqrt{2L/d} - x \right),$$

which can be simplified to $L + \sqrt{\frac{Ld}{2}} + \frac{x(1-x)d}{2} > L$.

Now if we set $F(x_i, y_i) = 0$ for all $1 \leq i \leq L$, we have a homogeneous linear equation system in the coefficients $a_{s,t}$, with more unknowns than L, the number of equations, and hence we can find at least one non-trivial bivariate polynomial F. If we substitute $y = f(x)$ where f is a polynomial of degree at most d and passes through at least t points (x_i, y_i), then we have a univariate polynomial $F(x, f(x))$ of degree at most $\sqrt{2Ld}$, but vanishes on $t > \sqrt{2Ld}$ many points. Hence $F(x, f(x))$ is identically 0 in $\mathbf{F}[x]$. Thus as polynomials in $\mathbf{F}(x)[y]$, $y - f(x) \mid F(x, y)$. But since $y - f(x)$ is monic and in fact belongs to $\mathbf{F}[x][y]$, $y - f(x) \mid F(x, y)$ also in $\mathbf{F}[x, y]$.

Clearly $y - f(x)$ is irreducible in $\mathbf{F}[x, y]$, which is a UFD. In probabilistic polynomial time one can factor a polynomial in $\mathbf{F}[x, y]$ [Kal92], then $y - f(x)$ must be one of the the the irreducible factors.

We return to the proof of Theorem 1. Thus we have a randomized procedure that, with high probability, computes a list of at most $3/q = O(n^k)$ polynomials, such that one of them is the permanent of the $(n-1) \times (n-1)$ matrix $D(x)$. The remaining task is to identify the correct polynomial from this list. Two ideas come into play here: First, we can find in deterministic polynomial time a point $v \in \mathbf{Z}_p$, such that all $3/q$ polynomials disagree on v. This is because each pair of polynomials can agree on at most n^2 points, and there are strictly less than $9n^{2k}$ such pairs, and $p \geq 9n^{2k+2}$. Secondly, if we can *somehow* obtain the *correct* value of $\mathrm{per}(D(v))$, then we can eliminate all but at most one polynomial on our list, by cross-checking the values of each polynomial in our list at v against the correct value of $\mathrm{per}(D(v))$. Assuming D is *good*, then with very high probability, the correct polynomial $\mathrm{per}(D(x))$ is on the list of $O(n^k)$ polynomials, thus exactly one polynomial must remain, and the remaining polynomial must be the correct $\mathrm{per}(D(x))$. Furthermore, D is *good* with high probability. Once we have the correct polynomial, by evaluating the correct polynomial $\mathrm{per}(D(x))$ at $x = 1, \dots, n$, we may compute the permanents of the n minors of the matrix M that we started with, and thus also compute $\mathrm{per}(M)$.

What we have achieved is a reduction, using $\mathcal{A}$ as an auxiliary procedure, of the computation of the permanent of $n \times n$ matrices to the computation of the permanent of $(n-1) \times (n-1)$ matrices. The reduction is probabilistic, and has a high probability of success. In fact, the error probability is bounded by the probability that D is not *good*, which is at most $1/((p-n)q^2) = O(1/n^2)$, plus the probability that given a good D still we did not get a sufficient number of distinct points on which $\mathcal{A}$ evaluates correctly, which is exponentially small, plus the failure probability of the factoring algorithm, which also can be made exponentially small. Thus the overall error probability is $O(1/n^2)$, say c/n^2 for some constant $c > 0$. Therefore, if we carry this process through $n, n-1, \dots, K$

for some large constant K, the total error probability is bounded by $\sum_{j=K}^{n}(c/j^2)$, which can be made arbitrarily small, say $\epsilon < 1/2$, by choosing a large enough K. This implies that we may use the *same* procedure recursively to compute the correct value of $\text{per}(D(v))$. The recursion terminates when the order of the matrix becomes less than K, and we can compute the permanent directly.

Finally, to show that $P^{\#P} = BPP$, we need a probabilistic polynomial time algorithm for the permanent with negligible (less than any inverse polynomial) error probability. We can achieve that by repeating the above algorithm for the permanent a sufficiently large polynomial number of times, and taking the majority vote. By Chernoff bound, this will succeed with exponentially small error probability. ∎

3   Some Extensions

The above proof can be seen to work over any finite field as long as the cardinality of the field is at least as large as $9n^{2k+2}$, and also there is a polynomial length representation of field elements.

Theorem 2. *Fix any constant $k > 0$, if there exists a deterministic polynomial time algorithm $\mathcal{A}$ such that for all n, $\mathcal{A}$ computes the permanent of order n over the finite field $\mathbf{F}$ of characteristic other than two correctly on greater than $1/n^k$ fraction of inputs, where $|\mathbf{F}| > 9n^{2k+2}$, and each field element has a representation of bit length at most $n^{O(1)}$, then $P^{\#P} = BPP$.*

Similarly we can extend the proof to the case of integers. Admittedly this is the most interesting case classically. But our proof will reduce this case to the case with a finite field $\mathbf{Z}_p$, for an appropriate prime p.

We must first define properly what is meant to be the uniform distribution of integer $n \times n$ matrices. For our purposes, we will define simply as follows: Consider any bit length ℓ between $\Omega(\log n)$ and $n^{O(1)}$. Then we consider uniform distribution of all integer $n \times n$ matrices where each entry of the matrix is an integer with absolute value bounded by 2^ℓ.

Theorem 3. *For any constant $k > 0$, if there exists a deterministic polynomial time algorithm $\mathcal{A}$ such that for all n, $\mathcal{A}$ computes the permanent of order n over the integers in the interval $[-2^\ell, 2^\ell]$ on greater than $1/n^k$ fraction of inputs, where $(2k+3)\log_2 n \leq \ell \leq n^{O(1)}$, then $P^{\#P} = BPP$.*

We prove this by choosing a prime close to 2^ℓ, and reason in the finite field $\mathbf{Z}_p$. We omit the details.

The proof of our theorems can also be carried out assuming only the existence of a probabilistic polynomial time algorithm $\mathcal{B}$ with the expected success ratio at least inverse polynomial on an inverse polynomial fraction of the inputs. We omit the proof here.

Theorem 4. *The same result holds in the above theorems if we assumed the existence of a probabilistic polynomial time algorithm only.*

Our theorems are an improvement of the results in [GS92, FL92]. They also simultaneously generalize the results in [CH91]. In this latter result, it is assumed that there exists a polynomial time algorithm such that for every matrix of order n, it can enumerate a list of polynomially many values, one of which is the correct value of the permanent. By taking one entry from such a list at random, we obtain a probabilistic polynomial time algorithm with an inverse polynomial success ratio.

In a related development (after this paper was submitted to STACS), Goldreich, Ron, and Sudan published a technical report in ECCC [GRS98] showing that if there is a polynomial time algorithm $\mathcal{B}$ that is able to guess the permanent of a random $n \times n$ matrix on $2n$-bit integers modulo a random n-bit prime with inverse polynomial success probability, then $P^{\#P} = BPP$. To prove this result, they develop algorithmic tools to decode an error correcting code based on the Chinese Remainder Theorem. While their decoding algorithms may be of independent interest, the above result on permanent, which is the main motivation for their algorithm, may be proved directly from our results in this paper.

Specifically, given such an algorithm $\mathcal{B}$, we can show how to produce an algorithm $\mathcal{A}$ that meets the hypothesis of Theorem 3. The idea is to randomly choose $p(n)$ many n-bit primes for a large polynomial $p(n)$, (by choosing polynomially many integers of this size and applying a probabilistic primality test). With high probability, we will be able to find sufficiently many primes p such that the algorithm $\mathcal{B}$ succeeds with a fixed inverse polynomial probability in computing the permanent modulo p on random $n \times n$ matrices. Here "sufficiently many" means that the number of such primes, which we will call *good* primes, is sufficient, via Chinese remaindering, to describe any integer that might be the value of the permanent of $n \times n$ matrices of $2n$-bit integers. Furthermore, by the distribution of primes according to the Prime Number Theorem, with high probability, we will have a sufficiently many good primes in hand, among all the primes generated, which are large enough in value to apply the procedures in the proof of Theorem 1. Of course, we will also have many *bad* primes where the algorithm $\mathcal{B}$ fails to achieve such success rate. The main idea is that, through the use of the random self-reducibility and downward self-reducibility of the permanent (as in the proof of Theorem 1), for any prime, with high probability we will know whether the procedures in the proof of Theorem 1 had succeeded or not. The essential idea is contained in the LFKN protocol. We can amplify the correctness probability of this identification of good primes exponentially close to one, namely $> 1 - e^{-n^c}$ for any constant c. Once we have sufficiently many good primes identified, we may apply the proof ideas of Theorem 1 and for a given matrix M, compute its permanent modulo the good primes, and finally apply an error-free Chinese remaindering to compute the permanent of the matrix M.

The permanent function has played a pivotal role in complexity theory, e.g., see [Lip91, LFKN90, IW98]. Frequently, it is a standard technique to use Yao's XOR Lemma to amplify the unpredictability of some hard function. The theorems presented here has the interesting feature that, for the permanent function, no such amplification is needed if one requires only less than inverse polynomial unpredictability. Perhaps these theorems can be used as an alternative to the standard amplification technique. The advantage is that we do not need any replication of the inputs. This has been an important concern in some recent results due to Impagliazzo and Wigderson [IW98].

References

[ABG90]   A. Amir, R. Beigel, and W. Gasarch. Some connections between query classes and non-uniform complexity. In *Proceedings of the 5th Structure in Complexity Theory*, pages 232–243. IEEE Computer Society, 1990.

[ALRS92]   S. Ar, R. Lipton, R. Rubinfeld, and M. Sudan. Reconstructing algebraic functions from mixed data. In *Proc. 33rd FOCS*, pages 503–512, 1992.

[BW]   E. Berlekamp and L. Welch. Error correction of algebraic codes. US Patent Number 4,633,470.

[CH91]   J. Cai and L. Hemachandra. A note on enumerative counting. *Information Processing Letters*, 38(4):215–219, 1991.

[FL92]   U. Feige and C. Lund. On the hardness of computing permanent of random matrices. In *Proceedings of 24th STOC*, pages 643–654, 1992.

[GLRSW91]   P. Gemmell, R. Lipton, R. Rubinfeld, M. Sudan, and A. Wigderson. Self-testing/correcting for polynomials and for approximate functions. In *Proceedings of 23rd STOC*, pages 32–42, 1991.

[GS92]   P. Gemmell and M. Sudan. Highly resilient correctors for polynomials. *Information Processing Letters*, 43:169–174, 1992.

[GRS98]   O. Goldreich and D. Ron and M. Sudan. Chinese remaindering with errors. ECCC Technical Report TR 98-062, October 29, 1998. Available at www.eccc.uni-trier.de.

[IW98]   R. Impagliazzo and A. Wigderson. Randomness vs Time, Derandomization under a uniform assumption. Manuscript, 1998. *To appear in FOCS '98*.

[Kal92]   E. Kaltofen. Polynomial factorization 1987–1991. *LATIN '92*, I. Simon (Ed.), LNCS, vol.583, pp294–313, Springer, 1992.

[LFKN90]   C. Lund, L. Fortnow, H. Karloff, and N. Nisan. Algebraic methods for interactive proof systems. In *Proceedings of 31st FOCS*, pages 2–10, 1990.

[Lip91]   R. Lipton. *New directions in testing*, In Distributed Computing and Cryptography, volume 2 of DIMACS Series in Discrete Mathematics and Theoretical Computer Science, pages 191–202. AMS, 1991

[MR95]   R. Motwani and P. Raghavan. *Randomized Algorithms*. Cambridge University Press, 1995.

[Rys63]   H. J. Ryser. *Combinatorial Mathematics*. Carus Mathematical Monograph No 14, Math. Assoc. of America, 1963.

[Sud96]   M. Sudan. Maximum likelihood decoding of Reed-Solomon codes. In *Proceedings of the 37th FOCS*, pages 164–172, 1996.

[Tod89]   S. Toda. On the computational power of PP and $\oplus$P. In *Proceedings of the 30th FOCS*, pages 514–519, 1989.

[Val79]   L. Valiant. The complexity of computing the permanent. *Theoretical Computer Science*, 47(1):85–93, 1979.

One-Sided Versus Two-Sided Error in Probabilistic Computation

Harry Buhrman[1]* and Lance Fortnow[2]**

[1] CWI, PO Box 94079, 1090 GB Amsterdam, The Netherlands,
buhrman@cwi.nl,
http://www.cwi.nl/~buhrman
[2] University of Chicago, Department of Computer Science,
1100 E. 58th St., Chicago, IL 60637
fortnow@cs.uchicago.edu,
http://www.cs.uchicago.edu/~fortnow

Abstract. We demonstrate how to use Lautemann's proof that **BPP** is in Σ_2^p to exhibit that **BPP** is in $\mathbf{RP}^{\mathbf{PromiseRP}}$. Immediate consequences show that if **PromiseRP** is easy or if there exist quick hitting set generators then $\mathbf{P} = \mathbf{BPP}$. Our proof vastly simplifies the proofs of the later result due to Andreev, Clementi and Rolim and Andreev, Clementi, Rolim and Trevisan.

Clementi, Rolim and Trevisan question whether the promise is necessary for the above results, i.e., whether $\mathbf{BPP} \subseteq \mathbf{RP}^{\mathbf{RP}}$ for instance. We give a relativized world where $\mathbf{P} = \mathbf{RP} \neq \mathbf{BPP}$ and thus the promise is indeed needed.

1 Introduction

Andreev, Clementi and Rolim [ACR98] show how given access to a quick hitting set generator, one can approximate the size of easily describable sets. As an immediate consequence one gets that if quick hitting set generators exist then $\mathbf{P} = \mathbf{BPP}$. Andreev, Clementi, Rolim and Trevisan [ACRT97] simplify the proof and apply the result to simulating **BPP** with weak random sources.

Much earlier, Lautemann [Lau83] gave a proof that $\mathbf{BPP} \subseteq \Sigma_2^p = \mathbf{NP}^{\mathbf{NP}}$, simplifying work of Gács and Sipser [Sip83]. Lautemann's proof uses two simple applications of the probabilistic method to get the existence results needed. As often with the case of the probabilistic method, the proof actually shows that the overwhelming number of possibilities fulfill the needed requirements. With this observation, we show that Lautemann's proof puts **BPP** in the class $\mathbf{RP}^{\mathbf{PromiseRP}[1]}$. Since quick hitting set generators derandomize **PromiseRP** problems, we get the existence of quick hitting set generators implies $\mathbf{P} = \mathbf{BPP}$. This greatly simplifies the proofs of Andreev, Clementi and Rolim [ACR98] and Andreev, Clementi, Rolim and Trevisan [ACRT97].

---

* Partially supported by the European Union through NeuroCOLT ESPRIT Working Group Nr. 8556, and HC&M grant nr. ERB4050PL93-0516.
** Supported in part by NSF grant CCR 92-53582.

The difference between **RP** and **PromiseRP** is subtle but important. In the class **RP** we require the probabilistic Turing machine to either reject always or accept with probability at least one-half for all inputs. In **PromiseRP** we only need to solve instances where the machine rejects always or accepts with probability at least one-half.

A survey paper by Clementi, Rolim and Trevisan [CRT98] asks whether we can remove the promise in our result, i.e., whether $\mathbf{BPP} \subseteq \mathbf{RP}^{\mathbf{RP}}$. We give a relativized counterexample to this conjecture by exhibiting an oracle A such that $\mathbf{P}^A = \mathbf{RP}^A$ but $\mathbf{P}^A \neq \mathbf{BPP}^A$. Since virtually all the techniques used in derandomization relativize, this means that new techniques will be required to collapse **BPP** in this way.

2   Definitions

We assume the reader familiar with the standard notions of Turing machines, and deterministic, nondeterministic and probabilistic polynomial-time computation. We let Σ represent the binary alphabet $\{0,1\}$.

A quick hitting set generator finds strings in large easily describable sets.

Definition 1. *A quick δ-hitting set generator is a polynomial-time computable function h mapping 1^n to a set of strings of length n such that for all n if $f : \Sigma^n \to \{0,1\}$ is a function computed by circuits of at most n gates and $\Pr_{x \in \Sigma^n}(f(x) = 1) \geq \delta$ then $f(x) = 1$ for some x in $h(1^n)$.*

Andreev, Clementi and Rolim [ACR98] show that for any $\delta, \delta' > 0$, if quick δ-hitting set generators exist than so do δ'-hitting set generators. We will drop δ in this case.

We have many variations of probabilistic complexity classes. In this paper, we will concern ourselves with **RP**, **BPP**, **PromiseRP** and **PromiseBPP**.

Definition 2. *A language L is in the class **RP** if there exists a probabilistic polynomial-time Turing machine such that for all $x \in \Sigma^*$,*

- *If x is in L then $\Pr(M \text{ accepts } x) \geq 1/2$, and*
- *If x is not in L then $\Pr(M \text{ accepts } x) = 0$.*

Sometimes the class **RP** is denoted simply by **R**.

Definition 3. *A language L is in the class **BPP** if there exists a probabilistic polynomial-time Turing machine such that for all $x \in \Sigma^*$,*

- *If x is in L then $\Pr(M \text{ accepts } x) \geq 2/3$, and*
- *If x is not in L then $\Pr(M \text{ accepts } x) \leq 1/3$.*

Languages in **RP** require machines M that fulfill the requirements of Definition 2 for all inputs. Sometimes we would like to consider probabilistic machines restricted to inputs where the desired requirements hold. We use **PromiseRP** to describe these problems. This does not form a class per se, but we can formally define the notions of **PromiseRP** being easy and oracle access to **PromiseRP**.

Definition 4. *We say that a language A is* **RP**-*consistent with a probabilistic polynomial-time Turing machine M if for all $x \in \Sigma^*$,*

- *x is in A if $\Pr(M \text{ accepts } x) \geq 1/2$, and*
- *x is not in A if $\Pr(M \text{ accepts } x) = 0$.*

Note that A may be arbitrary for x such that $0 < \Pr(M \text{ accepts } x) < 1/2$.

Definition 5. *We say* **PromiseRP** *is easy if for every probabilistic polynomial-time Turing machine M there is a set A in* **P** *that is* **RP**-*consistent with M.*

Using repetition we can reduce the error in Definitions 2-5 to $2^{-q(|x|)}$ for any polynomial q.

Contrast Definition 5 to Definition 2. In particular we have **PromiseRP** is easy implies **P** = **RP**. The converse is not so simply provable, relativized counterexamples easily follow from known results on generic oracles [IN88]. The oracle we develop in Section 4 also gives a relativizable counterexample.

Definition 6. *For any relativizable complexity class C, L is in $C^{\text{PromiseRP}}$ if there is a probabilistic polynomial-time Turing machine M such that L is in C^A for all A* **RP**-*consistent with M.*

We can also define $C^{\text{PromiseRP}[k]}$ if we allow only k queries to A in Definition 6. We can use the notation **PromiseBPP** in a similar manner.

One might want to require in Definition 6 that L be in C^A via a fixed machine depending only on M. Grollmann and Selman [GS88] show that this restriction does not affect Definition 6. For completeness we give a proof of the equivalence of the two definitions in Section 5.

It is not hard to see that there is an easy connection between hitting set generators and **PromiseRP**.

Fact 1 *If there are quick hitting set generator then* **PromiseRP** *is easy.*

3   One-Sided Promise Gives BPP

Theorem 1.
$$\mathbf{BPP} \subseteq \mathbf{RP}^{\text{PromiseRP}[1]}$$

Proof: We basically use the proof of Lautemann [Lau83] that **BPP** is in Σ_2^p to prove Theorem 1.

Let L be a language in **BPP** and M a probabilistic polynomial-time Turing machine accepting L with an error of 2^{-n} on inputs of length n. Let $q(n)$ be the maximum number of coin tosses on any computation path of M on any input of length n. Note $q(n)$ is bounded by a polynomial in n.

Let A be the set of pairs $\langle x, r \rangle$ such that $|r| = q(|x|)$ and $M(x)$ using r as its random coins will accept. Note that A is computable in deterministic polynomial time. We now define the set B as:

$$B = \{\langle x, z_1, \ldots, z_{q(|x|)} \rangle \mid |z_1| = \cdots = |z_{q(|x|)}| = q(|x|) \text{ implies there is some}$$
$$w \in \Sigma^{q(|x|)} \text{ such that } \langle x, w \oplus z_1 \rangle \notin A \wedge \cdots \wedge \langle x, w \oplus z_{q(|x|)} \rangle \notin A\}.$$

Here $u \oplus v$ for $|u| = |v|$ is the bitwise parity of u and v.

Note we have $B \in \mathbf{NP}$. First we will show that L is in $\mathbf{RP}^{B[1]}$. Our $\mathbf{RP}^B$ algorithm on input x with $n = |x|$ simply chooses $z_1, \ldots, z_{q(n)}$ independently at random from $\Sigma^{q(n)}$ and then accepts if $\langle x, z_1, \ldots, z_{q(n)} \rangle$ is not in B.

If x is in L then consider a fixed w and i, $1 \leq i \leq q(n)$. The probability that $\langle x, w \oplus z_i \rangle$ is not in A is at most 2^{-n}. Since the z_i's are chosen independently, the chance that $\langle x, w \oplus z_i \rangle$ is not in A for every z_i, $1 \leq i \leq q(n)$ is at most $2^{-nq(n)}$. Since there are $2^{q(n)}$ possible w's we have

$$\Pr(\langle x, z_1, \ldots, z_{q(n)} \rangle \in B) \leq 2^{-n}.$$

Now suppose that x is not in L. Fix $z_1, \ldots, z_{q(n)}$ and i, $1 \leq i \leq q(n)$. If we choose w at random, the probability that $w \oplus z_i$ is in A is at most 2^{-n}. The probability that $w \oplus z_i$ is in A for some i is at most $q(n)2^{-n}$ which for sufficiently large n is much smaller than $1/2$. Thus for every $z_1, \ldots, z_{q(n)}$ of strings of length $q(n)$, $\langle x, z_1, \ldots, z_{q(|x|)} \rangle$ is in B.

Now we wish to show that L is in $\mathbf{RP}^{\mathbf{PromiseRP}[1]}$. Let C be any set such that C and B agree on tuples where the w is chosen at random and the acceptance probability is either zero or greater than one-half.

More specifically $\langle x, z_1, \ldots, z_{q(|x|)} \rangle$ is in C if

1. $|z_i| = q(|x|)$ for each i, $1 \leq i \leq q(|x|)$, and
2. the number of w of length $q(|x|)$ such that

$$\langle x, w \oplus z_1 \rangle \notin A \wedge \cdots \wedge \langle x, w \oplus z_{q(|x|)} \rangle \notin A$$

is greater than $2^{q(|x|)-1}$.

The tuple $\langle x, z_1, \ldots, z_{q(|x|)} \rangle$ is not in C if

1. $|z_i| = q(|x|)$ for each i, $1 \leq i \leq q(|x|)$, and
2. there are no w of length $q(|x|)$ such that

$$\langle x, w \oplus z_1 \rangle \notin A \wedge \cdots \wedge \langle x, w \oplus z_{q(|x|)} \rangle \notin A.$$

The set C can be arbitrary for all other inputs.

The proof above that L is in $\mathbf{RP}^{B[1]}$ also shows that L is in $\mathbf{RP}^{C[1]}$. $\square$

In the proof of Theorem 1, if x is in L and the z_i are badly chosen then the number of w such that

$$\langle x, w \oplus z_1 \rangle \notin A \wedge \cdots \wedge \langle x, w \oplus z_{q(|x|)} \rangle \notin A$$

might be nonzero yet small. This is why we need **PromiseRP** instead of just **RP** for this proof. Theorem 3 shows that any relativizable proof would need to use **PromiseRP**.

From Theorem 1 and its proof we get the following two corollaries.

Corollary 1. *If* **PromiseRP** *is easy then* **P** = **BPP** *and* **PromiseBPP** *is easy.*

Corollary 2 (Andreev-Clementi-Rolim). *If quick hitting set generators exist then* **P** = **BPP**.

The proof of Theorem 1 only uses the set A restricted to the inputs of the form $\langle x, r \rangle$. Thus we can use **PromiseBPP** is easy instead of just **P** = **BPP** in Theorem 1 and Corollaries 1 and 2.

Andreev, Clementi and Rolim [ACR98] prove the following stronger result to get Corollary 2.

Theorem 2 (Andreev-Clementi-Rolim). *For any $\epsilon > 0$, there is a polynomial-time algorithm that, given access to a quick hitting set generator, and given as input a circuit C returns a value D such that*

$$\left| \Pr_{x \in \Sigma^n} (C(x) = 1) - D \right| \leq \epsilon.$$

We should note that Theorem 2 also follows from Theorem 1. One just need notice that distinguishing the possibilities that $\Pr_{x \in \Sigma^n}(C(x) = 1) \geq D + \epsilon$ and $\Pr_{x \in \Sigma^n}(C(x) = 1) \leq D - \epsilon$ is a **PromiseBPP** question.

4   RP Can Be Easy without BPP Being Easy

In this section we show that Theorem 1 cannot be improved to show that **P** = **R** implies **P** = **BPP** using relativizing techniques.

Theorem 3. *There exists a relativized world where* **P** = **RP** $\neq$ **BPP**.

Define the following function $tower(0) = 2$, $tower(n + 1) = 2^{tower(n)}$, i.e. $tower(n)$ is an exponential tower of $n + 1$ 2's. We will use a special type of generic (see [FFKL93] for an overview) to prove the theorem.

Definition 7. *A **BPP**-generic oracle G is a type of generic oracle that is only defined at length n such that $n = tower(m)$ for some m. Moreover at these lengths it will always be the case that at most $1/3$ or more than $2/3$ of the strings of length n are in G. We will call oracles that satisfy these requirements oracles that are **BPP**-promise.*

The oracle that fulfills the conditions of Theorem 3 will be **QBF** $\oplus$ G for G a **BPP**-generic. Here **QBF** is the **PSPACE**-complete set of true quantified boolean formulae. The following lemma shows that the second part of Theorem 3 is fulfilled.

Lemma 1. *Let G be a **BPP**-generic. $\mathbf{P^{QBF \oplus G}} \neq \mathbf{BPP^{QBF \oplus G}}$.*

Proof: This follows because G is generic and the condition that **P** $\neq$ **BPP** can be met under the **BPP** promise of G. $\square$

The more difficult part is to show that $\mathbf{P^{QBF \oplus G}} = \mathbf{R^{QBF \oplus G}}$. We will need the following notion of categoricity.

Definition 8. *A polynomial time nondeterministic machine M is categorically* **R** *if for all* **BPP**-*promise oracles B it is the case that for all x* $M^{\mathbf{QBF} \oplus B}(x)$ *has either more than 1/2 of its paths accepting or none. We will also call these machines categorical.*

The idea is to show that if M is categorical then there is a polynomial time (relative to **QBF**) algorithm that computes for all x whether $M(x)$ accepts or rejects. The core of this proof will be an argument from Nisan [Nis91].

The proof of Theorem 3 follows from Lemmas 2 and 3. Lemma 2 says that if we have a machine $M(x)$ that is categorically **R** and we only consider oracles A such that at most 1/6 or at least 5/6 of the strings of length n are in A then $M^{\mathbf{QBF} \oplus A}(x)$ can be decided in polynomial time relative to $\mathbf{QBF} \oplus A$.

Lemma 2. *Fix an input x and let* $n = |x|$. *Let* $M(x)$ *be a categorical machine. For any set A that only contains strings of length n with the promise that either at most 1/6 or at least 5/6 of the strings of length n are in A, there exists a deterministic strategy that determines* $M^{\mathbf{QBF} \oplus A}(x)$, *querying only a fixed polynomial number of strings in A. Moreover this strategy can be computed in a fixed polynomial time relative to* $\mathbf{QBF} \oplus A$.

Proof We follow the lines of the proof of Nisan [Nis91]. Suppose M runs in time $p(n)$. Call any B that fulfills the 1/6, 5/6 promise $\mathbf{BPP}_2$-promise. Fix A to be any $\mathbf{BPP}_2$-promise oracle.

The deterministic strategy to determine $M^{\mathbf{QBF} \oplus A}(x)$ works as follows. Let S_1 contain all the oracles B such that $M^{\mathbf{QBF} \oplus B}(x)$ accepts:

$$S_1 = \{B \mid \Pr(M^{\mathbf{QBF} \oplus B}(x) \text{ accepts}) > 0\}$$

Let S_0 contain all the $\mathbf{BPP}_2$-promise oracles such that $M(x)$ rejects:

$$S_0 = \{C \mid C \text{ is } \mathbf{BPP}_2 \text{ -promise and } \Pr(M^{\mathbf{QBF} \oplus C}(x) \text{ accepts}) = 0\}$$

Let B_1 a set in S_1. Fix any accepting path π of $M^{\mathbf{QBF} \oplus B_1}(x)$ with queries $q_1, \ldots, q_p(n)$ on it and let $b_1, \ldots b_{p(n)}$ be such that $B_1(q_i) = b_i$. Next query $q_1, \ldots, q_{p(n)}$ to A and let $a_1, \ldots a_{p(n)}$ be the answers (i.e. $a_i = A(q_i)$). If for all i it holds that $a_i = b_i$ we know that $M^{\mathbf{QBF} \oplus A}(x)$ accepts and we are done. So assume that this is not the case.

At this point we have the following claim:

Claim. For all $C \in S_0$ at least half of the computation paths of $M^{\mathbf{QBF} \oplus C}(x)$ query a string in $Q = q_1, \ldots, q_{p(n)}$.

Proof Suppose this is not true and that there is a $C \in S_0$ such that less than half of the computation paths of $M^{\mathbf{QBF} \oplus C}(x)$ query a string in Q. Consider the oracle C' which is defined as follows. For all $x \notin Q$, $C'(x) = C(x)$ and for $q_i \in Q$, $C'(q_i) = b_i$. (i.e. C' equals C except for the queries in Q where it equals B_1). Since C was $\mathbf{BPP}_2$-promise it follows that C' is $\mathbf{BPP}$-promise. Since $M^{\mathbf{QBF} \oplus C'}(x)$ has at least one accepting path π and it is categorical it follows that at least 1/2 of its paths are accepting. On the other hand since

$M^{\mathbf{QBF} \oplus C}(x)$ has no accepting paths and more than half of the computation paths do not query anything in Q it follows that less than $1/2$ of the paths changed and hence that $M^{\mathbf{QBF} \oplus C'}(x)$ still rejects. A contradiction. $\square$

Next adjust S_0 and S_1 such that they only contain oracles that agree with $A(q_1), \ldots, A(q_{p(n)})$ and repeat the above construction. It follows that in each round we learn the answer to a new query that is queried on at least half of the computation paths. Suppose after $2p(n)$ rounds we have not yet encountered a proof that $M^{\mathbf{QBF} \oplus A}(x)$ accepts. Either all the queries on all the paths of $M^{\mathbf{QBF} \oplus A}(x)$ have been queried or the current S_0 is empty. Let E be the set of queries made to A in all the rounds. We will have that $M^{\mathbf{QBF} \oplus A}(x)$ accepts if and only if $M^{\mathbf{QBF} \oplus (A \cap E)}(x)$ has an accepting path.

To choose the set B_1 in each round we need remember the oracle queries previously made to A. It is not hard to see then that this construction can be carried out in **PSPACE** and reducible to **QBF**. $\square$

Let D be the deterministic strategy that comes out of Lemma 2. The next lemma shows that this strategy also works for **BPP**-promise oracles.

Lemma 3. *For any* **BPP**-*promise oracle A. Let D be the strategy as described in Lemma 2. D will compute correctly $M^{\mathbf{QBF} \oplus A}(x)$.*

Proof Suppose that D does not compute $M^{\mathbf{QBF} \oplus A}(x)$ for some **BPP**-promise A. Suppose that A contains at most $1/3$ of the strings of length n. The case where A contains more than $2/3$ of the strings of length n can be handled similarly.

Suppose D accepted but did not find an accepting path of $M^{\mathbf{QBF} \oplus A}(x)$. This could only have happened if the final S_0 was empty. Let E be a minimal subset of A consistent with D's queries to A such that $M^{\mathbf{QBF} \oplus E}(x)$ rejects. Since S_0 is empty, E must contain at least $\frac{2^n}{6}$ strings. Removing any string y from E not queried by D will cause $M^{\mathbf{QBF} \oplus (E - \{y\})}(x)$ to accept with probability at least one-half. Thus every string in E not queried by D must occur on at least half of the computation paths of $M^{\mathbf{QBF} \oplus E}(x)$ which cannot happen by a simple counting argument.

Thus the only way the strategy can make an error is when D rejects whereas $M^{\mathbf{QBF} \oplus A}(x)$ accepts. Let $Q = q_1, \ldots, q_{2p(n)^2}$ be the queries made by D. and let $R = r_1, \ldots, r_{p(n)}$ be the queries on some accepting path of $M^{\mathbf{QBF} \oplus A}(x)$. Consider the following set A'. For all $q \in Q$ set $A'(q) = A(q)$, and for $r \in R$ set $A'(r) = A(r)$. For all the other strings x set $A'(x) = 0$. It now follows that A' contains at most a polynomial number of strings of length n and is **BPP**$_2$-promise. Moreover since $M^{\mathbf{QBF} \oplus A'}(x)$ has an accepting path it follows that $M^{\mathbf{QBF} \oplus A'}(x)$ accepts. But since all the queries made by D will be the same for A and A' it follows that D still rejects contradicting Lemma 2. $\square$

Proof (of Theorem 3) By Lemma 1 it follows that $\mathbf{P}^{\mathbf{QBF} \oplus G} \neq \mathbf{BPP}^{\mathbf{QBF} \oplus G}$. Let M be any categoric machine that runs in time $p(n)$. let x be any string of length l and let m be the biggest m such that $tower(m) \leq p(n)$. Set $n = tower(m)$. Query all the relevant strings in G of length strictly less than n. Since G is only defined at lengths that are a tower of 2's it follows that the previous relevant length is so small that one can query all those strings in polynomial time. Next apply Lemma 3 and use **QBF** to compute $M^{\mathbf{QBF} \oplus A}(x)$. The last possibility is that $M^{\mathbf{QBF} \oplus G}$ happens to be an **R** machine but it is not categoric. This however can not happen since the genericity of G will diagonalize against such non-categoric machines. (See [BI87]) □

Theorem 3 in combination with Theorem 1 gives a relativized world where **PromiseRP** is not easy but **P** = **RP**. This corollary also follows from work of Impagliazzo and Naor [IN88].

Heller [Hel86] exhibits a relativized world where **BPP** = **NEXP**. One might suspect that the techniques of Heller and those used in the proof of Theorem 3 may lead to an oracle A where $\mathbf{P}^A = \mathbf{RP}^A$ and $\mathbf{BPP}^A = \mathbf{NEXP}^A$. We show this cannot happen.

Theorem 4. *In all relativized worlds, if* **P** = **RP** *and* **NP** ⊆ **BPP** *then* **P** = **BPP**.

Proof Zachos [Zac88] shows that if **NP** ⊆ **BPP** then **NP** = **RP**. We then have $\mathbf{P} = \mathbf{NP} = \Sigma_2^p$ and thus **P** = **BPP**. These arguments all relativize. □

5   Relativizing to PromiseRP

Definition 6 may allow the machine that exhibits L in $\mathcal{C}^{\mathbf{PromiseRP}}$ to depend on A instead of just the underlying probabilistic machine. Grollmann and Selman [GS88] give a general result that implies that disallowing this dependence does not change the class $\mathcal{C}^{\mathbf{PromiseRP}}$. For completeness we give a proof of this result.

For simplicity we will show the equivalence for the class $\mathbf{P}^{\mathbf{PromiseRP}}$. The proof works similarly for many other natural classes such as $\mathbf{RP}^{\mathbf{PromiseRP}}$, $\mathbf{NP}^{\mathbf{PromiseRP}}$, $\mathbf{RP}^{\mathbf{PromiseRP}[1]}$, $\mathbf{P}^{\mathbf{PromiseBPP}}$, etc.

Theorem 5 (Grollmann-Selman). *For every language L and the following are equivalent:*

1. *L is in $\mathbf{P}^{\mathbf{PromiseRP}}$, i.e., there exists a probabilistic polynomial-time Turing machine M such that for all A **RP**-consistent with M, there is a polynomial-time oracle Turing machine N such that $L = L(N^A)$.*
2. *There exist a probabilistic polynomial-time Turing machine M and a polynomial-time oracle Turing machine N such that for all A **RP**-consistent with M, $L = L(N^A)$.*

Proof: (2) is more restrictive than (1). We have to show that (1) implies (2). Fix L in $\mathbf{P}^{\mathbf{PromiseRP}}$ and a M that witnesses this.

Let D be the set of x such that $M(x)$ accepts with probability zero or probability at least one-half. Let E be the set of x such that $M(x)$ accepts with probability at least one-half. We have that A is **RP**-consistent with M if and only if $A \cap D = E$.

Let us assume that (2) fails for M, i.e., for every polynomial-time Turing machine N there is an A such that $A \cap D = E$ and $L \neq L(N^A)$. We will create a set B with $B \cap D = E$ such that for all polynomial-time Turing machines N, $L \neq L(N^B)$. This contradicts that fact that M witnesses L in $\mathbf{P}^{\overline{\mathbf{PromiseRP}}}$.

Let $N_1, N_2, \ldots$ be an enumeration of the polynomial-time oracle Turing machines.

We create B in stages, in each stage we give a partial setting of whether some strings are or are not in B. Let B_0 be the oracle where all strings in E are put in B_0 and all strings in $D - E$ are put out of B_0. Let $m_0 = 0$.

Our goal at stage i will be to guarantee that for any oracle A extending B_i, $L \neq L(N_i^A)$. At the end of stage i we will have all strings of length less than m_i defined in B_i and only the strings in D of length greater than i will be defined.

Stage $i + 1$:

Claim. There exists an **RP**-consistent A extending B_i such that $L \neq L(N_i^A)$.

Proof: Suppose not. Create machine N^C that simulates N_i^C except that on oracle queries of length less than m_i, N will answer them according to B_i. Let C be any **RP**-consistent language. Then N^C will simulate N_i^F where

$$F = (B_i \cap \Sigma^{<m_i}) \cup (C \cap \Sigma^{\geq m_i}).$$

Since $C \cap D = E$ we have that F extends B_i. By the assumption that the claim fails we have $L(N^C) = L(N_i^F) = L$. We now have that $L(N^C) = L$ for all **RP**-consistent C contradicting the assumption that (2) fails. $\square$

Fix an **RP**-consistent A and an x such that $x \in L \Leftrightarrow x \notin L(N_i^A)$. Let m_{i+1} be one more than length the longest oracle query made by $N_i^A(x)$ and let B_{i+1} be the extension of B_i where all strings of length less than m_{i+1} are set according to A. $\square$

Acknowledgments

We thank Noam Nisan for bringing the "**PromiseRP** is easy implies $\mathbf{P} = \mathbf{BPP}$" problem to our attention. We thank Luca Trevisan for suggesting the alternate definition for relativized **PromiseRP** (Definition 6) and the possibility that the two definitions are equivalent.

We thank Luca Trevisan, Noam Nisan, Stuart Kurtz, John Rogers and Steve Fenner for many helpful discussions. Dieter van Melkebeek provided many helpful suggestions on earlier drafts. We also thank Ronald de Wolf for the use of his notes.

References

[ACR98]   A. Andreev, A. Clement, and J. Rolim. A new derandomization method. *Journal of the ACM*, 45(1):179–213, Januari 1998.

[ACRT97] A. Andreev, A. Clement, J. Rolim, and L. Trevisan. Weak random sources, hittings sets, and BPP simulations. In *Proceedings of the 38th IEEE Symposium on Foundations of Computer Science*, pages 264–272. IEEE, New York, 1997.

[BI87]   M. Blum and R. Impagliazzo. Generic oracles and oracle classes. In *Proceedings of the 28th IEEE Symposium on Foundations of Computer Science*, pages 118–126. IEEE, New York, 1987.

[CRT98]   A. Clementi, J. Rolim, and L. Trevisan. Recent advances towards proving BPP = P. *Bulletin of the European Association for Theoretical Computer Science*, 64:96–103, February 1998.

[FFKL93]   S. Fenner, L. Fortnow, S. Kurtz, and L. Li. An oracle builder's toolkit. In *Proceedings of the 8th IEEE Structure in Complexity Theory Conference*, pages 120–131. IEEE, New York, 1993.

[GS88]   J. Grollmann and A Selman. Complexity measures for public-key cryptosystems. *SIAM Journal on Computing*, 17:309–355, 1988.

[Hel86]   H. Heller. On relativized exponential and probabilistic complexity classes. *Information and Computation*, 71:231–243, 1986.

[IN88]   R. Impagliazzo and M. Naor. Decision trees and downward closures. In *Proceedings of the 3rd IEEE Structure in Complexity Theory Conference*, pages 29–38. IEEE, New York, 1988.

[Lau83]   C. Lautemann. BPP and the polynomial hierarchy. *Information Processing Letters*, 17(4):215–217, 1983.

[Nis91]   N. Nisan. CREW PRAMSs and decision trees. *SIAM Journal on Computing*, 20(6):999–1007, December 1991.

[Sip83]   M. Sipser. A complexity theoretic approach to randomness. In *Proceedings of the 15th ACM Symposium on the Theory of Computing*, pages 330–335. ACM, New York, 1983.

[Zac88]   S. Zachos. Probabilistic quantifiers and games. *Journal of Computer and System Sciences*, 36:433–451, 1988.

An Optimal Competitive Strategy for Walking in Streets

Christian Icking, Rolf Klein, and Elmar Langetepe

FernUniversität Hagen, Praktische Informatik VI, D-58084 Hagen.*

Abstract. We present an optimal strategy for searching for a goal in a street which achieves the competitive factor of $\sqrt{2}$, thus matching the best lower bound known before. This finally settles an interesting open problem in the area of competitive path planning many authors have been working on.

Key words: Computational geometry, autonomous robot, competitive strategy, LR-visibility, on-line navigation, path planning, polygon, street.

1 Introduction

In the last decade, the path planning problem of autonomous mobile systems has received a lot of attention in the communities of robotics, computational geometry, and on-line algorithms; see e. g. Rao et al. [14], Blum et al. [4], and the upcoming surveys by Mitchell [13] and Berman [3].

Among the basic problems is searching for a goal in an unknown environment. One is interested in strategies that are correct, in that the goal will always be reached whenever this is possible, and in performance guarantees that allow us to relate the length of the robot's path to the length of the shortest path from start to goal, or to other measures of the complexity of the scene.

It is well known that there are some differences between the outdoor setting, where the robot has to circumnavigate a set of compact obstacles in order to get to the target, and the indoor setting where the obstacles are situated in a—not necessarily rectangular—room whose walls may further impede the robot; see e. g. Angluin et al. [1]. Therefore, it is reasonable to study the indoor problem in its most simple form, that is, where the walls of the room are the only obstacles the robot has to cope with.

Suppose a point-shaped mobile robot equipped with a 360° vision system is placed inside a room whose walls are modeled by a simple polygon. Neither the floorplan nor the position of the target point are known to the robot. As the robot moves around it can build a partial map of those parts that have so far been visible. Also, it will recognize the target point on sight.

It is quite easy to see that in arbitrary simple polygons no strategy can guarantee a search path at most a constant times as long as the shortest path from start to goal. The question arose if there are subclasses of polygons for which a

---

* This work was supported by the Deutsche Forschungsgemeinschaft, grant Kl 655/8-3.

C. Meinel and S. Tison (Eds.): STACS'99, LNCS 1563, pp. 110–120, 1999.

constant performance ratio can be achieved. To this end Klein [9] introduced the concept of *streets*. A polygon P with two distinguished vertices s and t is called a street if the two boundary chains leading from s to t are mutually weakly visible, i. e. if each point on one of the chains can see at least one point of the other. Equivalently, from each s-to-t path inside P each point of the polygon is at least once visible. Klein provided a competitive strategy for searching for the target point, t, of a street, starting from s. He proved an upper bound of 5.72 for the ratio of the length of the robot's path over the length of the shortest path from s to t in P. Also, it was shown that no strategy can achieve a competitive ratio of less than $\sqrt{2} \approx 1.41$. This lower bound applies to randomized strategies, too.

Since then, the street problem has attracted considerable attention. Some research was devoted to structural properties. Tseng et al. [17] have shown how to report all pairs of vertices (s, t) of a given polygon for which it is a street; for star-shaped polygons many of such vertex pairs exist. Das et al. [5] have improved on this result by giving an optimal linear time algorithm. Ghosh and Saluja [6] have described how to walk an unknown street incurring a minimum number of turns.

Other research addressed the gap between the $\sqrt{2}$ lower bound and the first upper bound of 5.72 known for the class of street polygons. The upper bound was lowered to 4.44 in Icking [7], then to 2.61 in Kleinberg [10], to 2.05 in López-Ortiz and Schuierer [11], to 1.73 in López-Ortiz and Schuierer [12], to 1.57 in Semrau [16], and to 1.51 in Icking et al. [8].

But it has remained open, until now, if $\sqrt{2}$ is really the largest lower bound, and how to design an optimal strategy for searching the target in a street; compare the open problems mentioned in Mitchell [13].

In this paper both questions are finally answered. We introduce a new strategy and prove that the search path it generates, in any particular street, is at most $\sqrt{2}$ times as long as the shortest path from s to t. This result makes the street problem one of the few problems in on-line navigation whose competitive complexity is precisely known (the only other example we are aware of is the result by Baeza-Yates et al. [2] on multiway search).

One might wonder if this paper is but another small step in a chain of technical improvements. We do not think so, for the following reason. Unlike many approaches discussed in previous work, the optimal strategy we are presenting here is not an artifact. Rather, its definition is well motivated by backward reasoning.

The crucial subproblem can be parametrized by a single angle, ϕ. For each possible value of ϕ a lower bound can be established, see Sect. 3.1. For the maximum value $\phi = \pi$ the existence of a strategy matching this bound is obvious. We state a requirement in Sect. 3.2 that would allow us to extend an optimal strategy from a given value of ϕ to smaller values. From this requirement we can infer how the strategy should proceed; see Sect. 3.3.

After this work was finished, and made publicly available via the Internet, we learned that Schuierer and Semrau [15] have simultaneaously and independently studied the same strategy. However, their analytic approach is quite different from our proof.

2   Definitions and Known Properties

We briefly repeat necessary definitions and known facts, mostly from [9].

A simple polygon P is considered as a room, the edges are opaque walls. Two points are mutually *visible*, i. e. see each other, if the connecting line segment is contained within P. As usual, two sets of points are said to be *mutually weakly visible* if each point of one set can see at least one point of the other set.

Definition 1. A simple polygon P in the plane with two distinguished vertices s and t is called a *street* if the two boundary chains from s to t are weakly mutually visible; see Fig. 1. Streets are sometimes also denoted as *LR-visible polygons* [5, 17], where L denotes the left and R the right boundary chain from s to t.

A strategy for searching a goal in an unknown street is an on-line algorithm for a mobile system (robot), modeled by a point, that starts at vertex s, moves around inside the polygon and eventually arrives at the goal t. The robot is equipped with a vision system that provides the visibility polygon, $vis(x)$, for the actual position, x, at each time, and everything which has been visible is memorized. When the goal becomes visible the robot goes there and its task is accomplished.

Compared to the shortest path, SP, from s to t inside P, it seems clear that most of the time a detour is unavoidable. Our aim is to bound that detour.

Definition 2. *A strategy for searching a goal in a street is* competitive with factor c *(or c-competitive, for short) if its path is never longer than c times the length of the shortest path from s to t.*

The shortest path from the startpoint s to the goal t inside a simple polygon P, which only turns at reflex[1] vertices of P, is a useful guide for any strategy. At each time, either the next vertex on the shortest path to t is known and there is no question where to go. Or there is some uncertainty, but we will see that only two candidates remain for the next vertex on the shortest path to t. Each part of the polygon which has never been visible is called a *cave*, and each cave is hidden behind a reflex vertex. Such a reflex vertex v that causes a cave is called *left* reflex vertex if its adjacent segments on P lie to the left of the ray from the actual position of the robot through v, and analogously for *right* vertices.

First, we consider the situation at the beginning. From the startpoint s we order clockwise around s the set of the left and right reflex vertices, obviously they appear in the same clockwise order on the boundary of P. As seen from s, let v_l be the clockwise most advanced left reflex vertex and v_r the counterclockwise most advanced right reflex vertex, see Fig. 1.

---

[1] A *reflex vertex* is one whose internal angle exceeds $180°$.

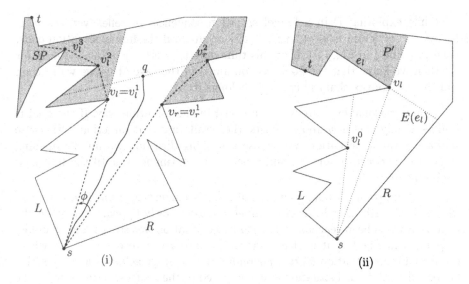

Fig. 1. Typical situations in streets.

If they exist, vertex v_l belongs to the left chain L, and v_r belongs to the right chain R. Only the following situations occur. If both, v_l and v_r, exist, see Fig. 1 (i), then the goal has to be in one of the caves behind v_r and v_l, thus SP passes over either v_r or v_l. If there is no vertex v_r, see Fig. 1 (ii), then the goal has to be inside the cave behind v_l, and the robot moves straight towards v_l along SP. We proceed correspondingly if only v_r exists.

To prove these properties assume that w.l.o.g. the goal is inside the cave behind a left reflex vertex v_l^0, see Fig. 1 (ii), and v_l^0 appears before v_l clockwise from s. Then the boundary of P inside the cave P' behind vertex v_l would belong to the right chain R. The extension, $E(e_l)$, of the invisible clockwise adjacent edge, e_l, of v_l cannot hit the left chain L. Therefore no point inside the cave P' can see a part of L, a contradiction to the street property.

By the same arguments we can prove that the counterclockwise angle $\phi \geq 0$ between sv_r and sv_l is always smaller than π, see Fig. 1 (i). Therefore in the vicinity of s the robot should always walk into the triangle $v_l s v_r$ to avoid unnecessary detours to v_r and v_l.

Now we look at the general situation. We assume that a strategy has led the robot to an actual position somewhere in the polygon. We will see that the properties discussed for the start essentially remain valid. Vertices v_l and v_r are defined as before, i.e. $v_l \in L$ is the clockwise most advanced left reflex vertex and $v_r \in R$ the counterclockwise most advanced right reflex vertex.

There is no reason for a strategy to loose the current v_l or v_r out of sight, so we assume that v_l and v_r are always visible, as long as they exist. As already discussed above, the only non-trivial case is if both, v_l and v_r, actually exist. We call this a *funnel situation*. The angle, ϕ, between the directions from the actual position to v_l and to v_r is called the *opening angle*, it is always smaller than π.

While exploring P in a funnel situation sequences of reflex vertices $v_l \in \{v_l^1, v_l^2, \ldots, v_l^m\}$ and $v_r \in \{v_r^1, v_r^2, \ldots, v_r^n\}$ occur until the funnel situation ends, see e.g. point q in Fig. 1 (i). If at this time only $v_l = v_l^m$ exists (analogously for v_r) then we know that the goal t is contained in the cave of v_l, we walk to v_l, and the left convex chain $v_l^1 v_l^2 \ldots v_l^m$ belongs to SP.

So any reasonable strategy will proceed in the following way. If the goal is visible or only one of v_l and v_r exists, then walk into that direction. Otherwise we have a funnel situation, we choose a walking direction within the opening angle, i.e. between v_l and v_r, and repeat this continuously until the first case applies again.

It is important to note that at the robot's current position is a vertex of the shortest path SP whenever a funnel situation newly appears *and* when the next vertex has been reached after the funnel situation was solved. For example, at point q in Fig. 1 (i) it is clear that we have to go to vertex $v_l^3 \in SP$ where the next funnel situation starts. Therefore, if a strategy achieves a competitive factor c in each funnel situation (i.e. compared to the shortest path between the two visited vertices of SP) then it achieves the same factor in arbitrary streets.

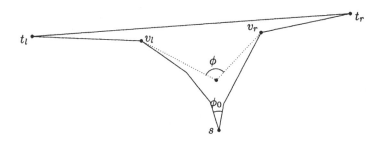

Fig. 2. A funnel.

As a consequence, we can restrict our attention to very special polygons, the so-called *funnels*. A funnel consists of two chains of reflex vertices with a common start point s, see Fig. 2 for an example. The two reflex chains end in vertices t_l and t_r, respectively, and the line segment $t_l t_r$ closes the polygon. A funnel polygon represents a funnel situation in which the goal t lies arbitrarily close behind either t_l or t_r, and the strategy will know which case applies only when the line segment $t_l t_r$ is reached. For analyzing a strategy, both cases have to be considered and the worse of them determines the competitive factor. Other funnel situations which end with a smaller opening angle or where the goal is further away from t_l or t_r will produce a smaller detour.

Since the walking direction is always within the opening angle, ϕ is *strictly increasing*. It starts at the angle, ϕ_0, between the two edges adjacent to s, and reaches, but never exceeds, $180°$ when finally the goal becomes visible. By this property, it is natural to take the opening angle ϕ for *parameterizing a strategy*.

We can further restrict ourselves to consider only funnels with initial opening angle $\phi_0 \geq 90°$. As was shown in [8, 16], any strategy which achieves a factor $\geq \sqrt{2}$ for all funnels with $\phi_0 \geq 90°$ can be adapted to the general case without changing its factor in the following way. First, we walk along the angular bisector of the current pair v_l and v_r until an opening angle of $\pi/2$ is reached. Then we proceed with the given strategy.

3   A Strategy Which Always Takes the Worst Case into Account

3.1   A Generalized Lower Bound

We start with a generalized lower bound for initial opening angles $\geq 90°$. For an arbitrary angle ϕ, let $K_\phi := \sqrt{1 + \sin \phi}$.

Lemma 1. *Assume an initial opening angle $\phi_0 \geq \frac{\pi}{2}$. Then no strategy can guarantee a smaller competitive factor than K_{ϕ_0}.*

Proof. We take an isosceles triangle with angle ϕ_0 at vertex s and other vertices t_l and t_r. The goal becomes visible only when the line segment $t_l t_r$ is reached. If this happens to the left of its midpoint m then the goal may be to the right, and vice versa. In any case the path length is at least the distance from s to m plus the distance from m to t_l. For the ratio, c, of the path length to the shortest path we obtain by simple trigonometry $c \geq \cos \frac{\phi_0}{2} + \sin \frac{\phi_0}{2} = \sqrt{1 + \sin \phi_0} = K_{\phi_0}$. □

For $\phi_0 = \frac{\pi}{2}$, we have the well-known lower bound of $\sqrt{2}$ stemming from a rectangular isoceles triangle [9]. Remark also that the bound K_{ϕ_0} also applies for any non-symmetric situation, since at the start the funnel is unknown except for the two edges adjacent to s and it may turn into a nearly symmetric case immediately after the start. This means in other words that for an initial opening angle ϕ_0 a competitive factor of K_{ϕ_0} is always the best we can hope for.

In the following we will develop a strategy which achieves exactly this factor.

3.2   Sufficient Requirements for an Optimal Strategy

In a funnel with opening angle π the goal is visible and there is a trivial strategy that achieves the optimal competitive factor $K_\pi = 1$. So we look backwards to decreasing angles.

Let us assume for the moment that the funnel is a triangle, and that we have a strategy with a competitive factor of K_{ϕ_2} for all triangular funnels with initial opening angle ϕ_2. How can we extend this to initial opening angles ϕ_1 with $\pi \geq \phi_2 > \phi_1 \geq \frac{\pi}{2}$?

Starting with an angle ϕ_1 at point p_1 we walk a certain path of length w until we reach an angle of ϕ_2 at point p_2 from where we can continue with the known strategy; see Fig. 3. The left and right reflex vertices, v_l and v_r as defined in Sect. 2, do not change.

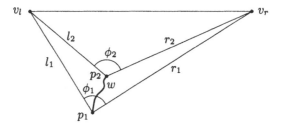

Fig. 3. Getting from angle ϕ_1 to ϕ_2.

Let l_1 and l_2 denote the distances from p_1 resp. p_2 to v_l at the left side and r_1 and r_2 the corresponding distances at the right. If $t = v_l$ then the path length from p_1 to t is not greater than $w + K_{\phi_2}l_2$. If now $K_{\phi_1}l_1 \geq w + K_{\phi_2}l_2$ holds and the analogous inequality $K_{\phi_1}r_1 \geq w + K_{\phi_2}r_2$ for the right side, which can also be expressed as

$$w \leq \min(K_{\phi_1}l_1 - K_{\phi_2}l_2, K_{\phi_1}r_1 - K_{\phi_2}r_2), \tag{1}$$

then we have a competitive factor not bigger than K_{ϕ_1} for triangles with initial opening angle ϕ_1.

Note that condition (1) is additive in the following sense. If it holds for a path w_{12} from ϕ_1 to ϕ_2 and for a continuing path w_{23} from ϕ_2 to ϕ_3 then it is also true for the combined path $w_{12} + w_{23}$ from ϕ_1 to ϕ_3. This will turn out to be very useful: if (1) holds for arbitrarily small, successive steps w then it is also true for all bigger ones.

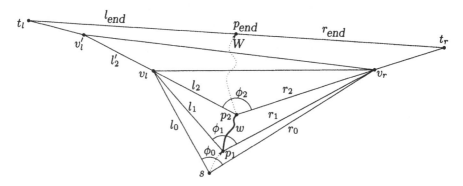

Fig. 4. When reaching p_2, the most advanced visible point to the left jumps from v_l to v'_l.

Now let us go further backwards and observe what happens if the current v_l or v_r change. We assume that condition (1) holds for path w from p_1 to p_2 and that v_l changes at p_2; see Fig. 4. The visible left chain is extended by l'_2. Nothing changes on the right side of the funnel, and for the left side of the funnel we have

$$w \leq K_{\phi_1} l_1 - K_{\phi_2} l_2 \leq K_{\phi_1}(l_1 + l'_2) - K_{\phi_2}(l_2 + l'_2). \tag{2}$$

The last inequality holds because $K_\phi = \sqrt{1 + \sin\phi}$ is decreasing with increasing ϕ. Here, $l_1 + l'_2$ and $l_2 + l'_2$ are the lengths of the shortest paths from p_1 and p_2 to v'_l, respectively. But (2) in fact means that (1) remains valid even if changes of v_l or v_r occur.

Under the assumption that (1) holds for all small steps where v_l and v_r do not change we can make use of the additivity of (1) and obtain the following for the path length, W, from an initial opening angle ϕ_0 to the point p_{end} where the line segment $t_l t_r$ is reached; see Fig. 4.

$$W \leq \min \left(K_{\phi_0}(\text{length of left chain}) - K_\pi l_{end}, \right.$$
$$\left. K_{\phi_0}(\text{length of right chain}) - K_\pi r_{end} \right)$$

But, since $K_\pi = 1$, this inequality exactly means that we have a competitive factor not bigger than K_{ϕ_0}. It only remains to find a curve that fulfills (1) for small steps.

3.3   Developing the Curve and Checking the Requirements

One could try to fulfill condition (1) by analyzing, for fixed p_1, ϕ_1, and ϕ_2, which points p_2 meet that requirement. To avoid this tedious task, we argue as follows. For fixed ϕ_2, the point p_2 lies on a circular arc through v_l and v_r. While p_2 moves along this arc, the length l_2 is strictly increasing while r_2 is strictly decreasing. Therefore, we maximize our chances to fulfill (1) if we require

$$K_{\phi_2} l_2 - K_{\phi_1} l_1 = K_{\phi_2} r_2 - K_{\phi_1} r_1 \quad \text{or} \quad K_{\phi_2}(l_2 - r_2) = K_{\phi_1}(l_1 - r_1). \tag{3}$$

In other words: if we start with initial values ϕ_0, l_0, r_0, we have a fixed constant $A := K_{\phi_0}(l_0 - r_0)$ and for any $\phi_0 \leq \phi \leq \pi$ with corresponding lengths l_ϕ and r_ϕ we want that

$$K_\phi(l_\phi - r_\phi) = A. \tag{4}$$

In the symmetric case $l_0 = r_0$ this condition means that we walk along the bisector of v_l and v_r. Otherwise condition (4) defines a nice curve which can be determined in the following way. We choose a coordinate system with horizontal axis $v_l v_r$, the midpoint being the origin. We scale such that the distance from v_l to v_r equals 1.

W.l.o.g. let $l_0 > r_0$. For any $\phi_0 \leq \phi < \pi$ the corresponding point of the curve is the intersection of the hyperbola and the circle given by

$$\frac{X^2}{\left(\frac{A}{2K_\phi}\right)^2} - \frac{Y^2}{\left(\frac{1}{2}\right)^2 - \left(\frac{A}{2K_\phi}\right)^2} = 1 \quad \text{and} \quad X^2 + \left(Y + \frac{\cot\phi}{2}\right)^2 = \frac{1}{4\sin^2\phi}.$$

Solving the equations gives, after some transformations, the following solutions.

$$X(\phi) = \frac{A}{2} \cdot \frac{\cot \frac{\phi}{2}}{1 + \sin \phi} \sqrt{\left(1 + \tan \frac{\phi}{2}\right)^2 - A^2} \tag{5}$$

$$Y(\phi) = \frac{1}{2} \cot \frac{\phi}{2} \left(\frac{A^2}{1 + \sin \phi} - 1\right) \tag{6}$$

Since $A < \sqrt{1 + \sin \phi}$ holds, the functions $X(\phi)$ and $Y(\phi)$ are well defined and continuous and the curve is contained in the triangle defined by ϕ_0, l_0, r_0.

Fig. 5 shows how these curves look like for all possible values of ϕ and A and also for $l_0 \leq r_0$. All points with an initial opening angle of $\frac{\pi}{2}$ lie on the lower half circle. Two cases can be distinguished. For $A \leq 1$ the curves can be

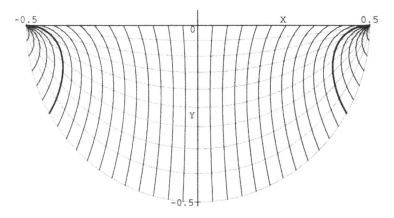

Fig. 5. The curves fulfilling condition (4) for all values of ϕ and A.

continuously completed to an endpoint on the line $v_l \, v_r$ with $X(\pi) = \pm\frac{A}{2}$ and $Y(\pi) = 0$ where also (4) is fulfilled. For $A > 1$ the curves end up in v_l and v_r, resp., with parameter $\phi = \arcsin \sqrt{A^2 - 1} < \pi$. The curves for the limiting case $A = 1$ are emphasized with a thick line in Fig. 5.

For lack of space we give only a sketch of the proof that the given curve fulfills condition (1). Because of the additive property of (1) it is sufficient to verify this for very small intervals. The arc length of the curve from angle ϕ to $\phi + \epsilon$ has to be compared to the right side of (1). Because of (3) the min can be dropped, and (1) transforms to

$$\int_{\phi}^{\phi + \epsilon} \sqrt{X'(\phi)^2 + Y'(\phi)^2} \, d\phi \leq K_\phi l_\phi - K_{\phi + \epsilon} l_{\phi + \epsilon} \qquad \text{for all } \epsilon > 0.$$

This can now be verified by inserting (5) and (6) and by making use of well-known facts from planar geometry and analysis.

To summarize, our strategy for searching a goal in an unknown street works as follows.

Strategy *WCA* (worst case aware):
If the initial opening angle is less than 90° walk along the angular bisector of v_l and v_r until a right opening angle is reached; see the end of Sect. 2. Depending on the actual parameters ϕ_0, l_0, and r_0, walk along the corresponding curve given by (5) and (6) until one of v_l and v_r changes. Switch over to the curve corresponding to the new parameters ϕ_1, l_1, and r_1. Continue until the line $t_l\,t_r$ is reached.

Theorem 1. *By using strategy WCA we can search a goal in an unknown street with a competitive factor of at most $\sqrt{2}$. This is optimal.*

4   Conclusions

We have developped a competitive strategy for walking in streets which guarantees an optimal factor of at most $\sqrt{2}$ in the worst case, thereby settling an old open problem.

Furthermore, the strategy behaves even better for an initial opening angle $\phi_0 > 90°$ in which case an optimal factor $K_{\phi_0} = \sqrt{1 + \sin\phi_0}$ between 1 and $\sqrt{2}$ is achieved.

The idea for this strategy comes from the generalized lower bound in Lemma 1 and from the two conditions (1) and (3), which are not strictly necessary for the optimal competitive factor but turn out to be very useful. Therefore, we do not claim that this is the only optimal strategy. It would be interesting if there are substantially different but also optimal strategies.

References

[1] D. Angluin, J. Westbrook, and W. Zhu. Robot navigation with range queries. In *Proc. 28th Annu. ACM Sympos. Theory Comput.*, pages 469–478, May 1996.

[2] R. Baeza-Yates, J. Culberson, and G. Rawlins. Searching in the plane. *Inform. Comput.*, 106:234–252, 1993.

[3] P. Berman. On-line searching and navigation. In A. Fiat and G. Woeginger, editors, *Competitive Analysis of Algorithms*. Springer-Verlag, 1998.

[4] A. Blum, P. Raghavan, and B. Schieber. Navigating in unfamiliar geometric terrain. *SIAM J. Comput.*, 26(1):110–137, Feb. 1997.

[5] G. Das, P. Heffernan, and G. Narasimhan. LR-visibility in polygons. *Comput. Geom. Theory Appl.*, 7:37–57, 1997.

[6] S. K. Ghosh and S. Saluja. Optimal on-line algorithms for walking with minimum number of turns in unkn. streets. *Comput. Geom. Theory Appl.*, 8:241–266, 1997.

[7] C. Icking. *Motion and Visibility in Simple Polygons*. PhD thesis, FernUniversität Hagen, 1994.

[8] C. Icking, A. López-Ortiz, S. Schuierer, and I. Semrau. Going home through an un-
     known street. Technical Report 228, Dep. of Comp. Science, FernUniversität Ha-
     gen, 1998. http://wwwpi6.fernuni-hagen.de/Publikationen/tr228.pdf, submitted.
[9] R. Klein. Walking an unknown street with bounded detour. *Comput. Geom.
     Theory Appl.*, 1:325–351, 1992.
[10] J. M. Kleinberg. On-line search in a simple polygon. In *Proc. 5th ACM-SIAM
     Sympos. Discrete Algorithms*, pages 8–15, 1994.
[11] A. López-Ortiz and S. Schuierer. Going home through an unknown street. In *Proc.
     4th Workshop Algorithms Data Struct.*, volume 955 of *Lecture Notes Comput. Sci.*,
     pages 135–146. Springer-Verlag, 1995.
[12] A. López-Ortiz and S. Schuierer. Walking streets faster. In *Proc. 5th Scand.
     Workshop Algorithm Theory*, volume 1097 of *Lecture Notes Comput. Sci.*, pages
     345–356. Springer-Verlag, 1996.
[13] J. S. B. Mitchell. Geometric shortest paths and network optimization. In J.-
     R. Sack and J. Urrutia, editors, *Handbook of Computational Geometry*. Elsevier
     Science Publishers B.V. North-Holland, Amsterdam, 1998.
[14] N. S. V. Rao, S. Kareti, W. Shi, and S. S. Iyengar. Robot navigation in un-
     known terrains: introductory survey of non-heuristic algorithms. Technical Report
     ORNL/TM-12410, Oak Ridge National Laboratory, 1993.
[15] S. Schuierer and I. Semrau. An optimal strategy for searching in unknown streets.
     Institut für Informatik, Universität Freiburg, 1998.
[16] I. Semrau. Analyse und experimentelle Untersuchung von Strategien zum Finden
     eines Ziels in Straßenpolygonen. Diploma thesis, FernUniversität Hagen, 1996.
[17] L. H. Tseng, P. Heffernan, and D. T. Lee. Two-guard walkability of simple poly-
     gons. *Internat. J. Comput. Geom. Appl.*, 8(1):85–116, 1998.

An Optimal Strategy for Searching in Unknown Streets*

Sven Schuierer and Ines Semrau

Institut für Informatik, Universität Freiburg
Am Flughafen 17, D-79110 Freiburg, Germany
{schuiere,semrau}@informatik.uni-freiburg.de

Abstract. We consider the problem of a robot searching for an unknown, yet visually recognizable target in a *street*. A *street* is a simple
polygon with start and target on the boundary so that the two boundary
chains between them are weakly mutually visible. We are interested in
the ratio of the search path length to the shortest path length which
is called the *competitive ratio* of the strategy. We present an optimal
strategy whose competitive ratio matches the known lower bound of $\sqrt{2}$,
thereby closing the gap between the lower bound and the best known
upper bound.

1 Introduction

A fundamental problem in robot motion planning is to search for a target in
an unknown environment. We consider a robot with an on-board vision system
that can identify the target on seeing it. The robot's information consists of the
local visibility maps it has obtained so far. Thus the search strategy performs
on-line, and the method of competitive analysis as introduced by Sleator and
Tarjan [19] can be applied to measure its quality. A strategy is *c-competitive* or
has a *competitive ratio* (or *factor*) of c if its cost does not exceed the cost of
an optimal solution times c. In our context, the distance traveled by the robot
must not exceed c times the shortest path length. Competitive on-line searching
has been investigated in various settings such as searching in special classes of
simple polygons [2,3,5,15,16,17] and among convex obstacles [8].

   We restrict ourselves to searching in so-called *streets* which were introduced
by Klein as the first environment that allows searching with a *constant* competitive ratio [9]. Klein presents the strategy *lad* which is based on the idea
of minimizing the *local absolute detour*. He gives an upper bound on its competitive ratio of ≈ 5.71. The upper bound on the competitive factor was later
improved by Icking to ≈ 4.44 [4].

   A number of other strategies have been presented since by Kleinberg [10],
López-Ortiz and Schuierer [12,13,14] and Semrau [18]. The currently best known
competitive ratio is ≈ 1.514 [7]. In this paper, we present an optimal strategy
with competitive ratio $\sqrt{2} \approx 1.41$. The same strategy was independently discovered and analysed by Icking et al. [6]. In the next section, we give an outline of

* This research is supported by the DFG-Project "Diskrete Probleme", No. Ot 64/8-2.

how to search in streets and point out the subproblem that must be solved. In section 3, an optimal strategy for this problem is presented.

2 Searching in a Street

The robot, modeled as a point, is located at the start position s in a simple polygon P. It has to find the target t. Both points, s and t, are vertices of P. Together with the polygon, they form a street subject to the following definition.

Definition 1. *A simple polygon P in the plane with two distinguished vertices s and t on its boundary is called a* street *if the two boundary chains from s to t are weakly mutually visible.*

An example for a street is depicted in Figure 1(a). We briefly summarize some facts about searching in streets (see [9,12,18]). Due to the simple lower bound example shown in Figure 1(b), there is no strategy with a competitive ratio less than $\sqrt{2}$ [9]. If a strategy moves to the left or right before seeing t, then t can be placed on the opposite side, thus forcing the robot to travel more than $\sqrt{2}$ times the diagonal.

Crucial for search strategies is how they behave in *funnels*, shown as shaded areas in Figure 1(a). A *funnel* consists of two reflex chains induced by vertices of the street polygon. A reflex chain is a polygonal chain all of whose vertices have an interior angle larger than π. Klein shows that if a strategy achieves a competitive factor c in funnels, then it can be embedded in a so-called *high level strategy* to provide a c-competitive strategy for searching in streets [9]. Outside of funnels, the high level strategy takes the shortest path as depicted in Figure 1(a).

To examine the funnel situation, we introduce some notations. The two reflex chains start in a common point p_1 and end in the vertices v_n and w_n, cf. Figure 2(a). A strategy to search in a funnel will know whether to go to v_n or

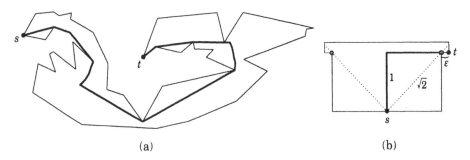

(a)   (b)

Fig. 1. (a) An optimal search path in a street (b) A lower bound for searching in rectilinear streets

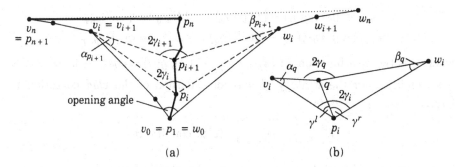

Fig. 2. A funnel and its notations

w_n before reaching the line segment $v_n w_n$ [9], where we assume that no three vertices of the street are collinear.

For a path point p_i in a funnel, we denote the most advanced visible point on the left chain by v_i and the most advanced visible point on the right chain by w_i. The last path point in the funnel is denoted by p_{n+1} which equals v_n or w_n. The path point before going straight to p_{n+1} is denoted by p_n.

We define the visibility angle of a point p_i to be the angle between the line segments from p_i to v_i and to w_i having a value of $2\gamma_i$ which is always $< \pi$. The visibility angle of p_1 is called the *opening angle* of the funnel.

For a point q in the triangle $v_i p_i w_i$, we denote the angle between $v_i p_i$ and $v_i q$ by α_q and the angle between $w_i p_i$ and $w_i q$ by β_q; cf. Figure 2(b). Note that the value of the visibility angle of q is $2\gamma_q = 2\gamma_i + \alpha_q + \beta_q$. Furthermore, we define $\delta_i := \gamma_i - \pi/4$, $\gamma_{n+1} := \gamma_n$, and $v_0 := w_0 := p_1$. The length of a line segment between two points a and b is denoted by ab and its length by $|ab|$.

As proved by López-Ortiz and Schuierer, a c-competitive strategy for funnels with opening angle $\geq \pi/2$ can be extended by their strategy *clad* (continuous lad) to a c-competitive strategy for arbitrary funnels [14]. Thus, we only consider funnels with opening angle $\geq \pi/2$.

3   The Strategy *glad*

We shall define a strategy to traverse funnels by specifying a condition which must be fulfilled by two path points p_i and p_{i+1}. In the following, these two points are always chosen such that on the path between them (without end-points) no new vertex becomes visible.

3.1   Outline and Correctness of the Strategy

The strategy, *glad* (generalized clad), can be regarded as a generalization of the strategy *clad* proposed by López-Ortiz and Schuierer [14] where every two path points p_i and p_{i+1}, $i < n$, fulfill the condition $|p_i v_i| - |p_{i+1} v_i| = |p_i w_i| - |p_{i+1} w_i|$. For a path point p_i and a point q with visibility angle $2\gamma_q$ in the triangle $v_i p_i w_i$ we define

$$l_{p_i,q} := |p_iv_i|g(\gamma_i) - |qv_i|g(\gamma_q) \qquad D^l_{p_i,q} := |p_iq|/l_{p_i,q}$$
$$r_{p_i,q} := |p_iw_i|g(\gamma_i) - |qw_i|g(\gamma_q) \qquad D^r_{p_i,q} := |p_iq|/r_{p_i,q}$$

where g is the *glad* function $g : [\pi/4, \pi/2] \to [0,1]$, $g(\gamma) = \cos(\gamma - \pi/4)$.

The condition for the strategy *glad* is an extension of the *clad* condition to $l_{p_i,p_{i+1}} = r_{p_i,p_{i+1}}$, $i < n$, or

$$D^l_{p_i,p_{i+1}} = D^r_{p_i,p_{i+1}} \quad (\textit{glad} \text{ condition}).$$

Two segments of length $l_{p_i,p_{i+1}} = r_{p_i,p_{i+1}}$ are depicted in Figure 3.

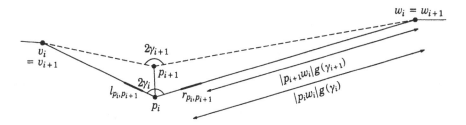

Fig. 3. The points p_i and p_{i+1} fulfill the *glad* condition

It is possible to obtain an outline of the *glad* path if only a few point pairs have to fulfill the *glad* condition; cf. Figure 4. Every point pair with no new visible vertex on the connecting path fulfills the *glad* condition, e.g. point pairs (p_1, p_2) and (p_1, p_3). Note that if (p_i, p_{i+1}) and (p_{i+1}, p_{i+2}) fulfill the *glad* condition, then (p_i, p_{i+2}) also does.

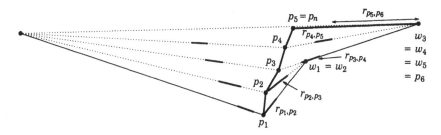

Fig. 4. A polygonal path passing through a finite number of path points

Before stating Lemma 1, we give the domains (if not restricted further in the text) of some frequently used variables:

$$\tfrac{\pi}{4} \le \gamma_i < \gamma_{i+1} < \tfrac{\pi}{2} \qquad 0 \le \delta_i = \gamma_i - \tfrac{\pi}{4} < \tfrac{\pi}{4} \qquad \alpha_q + \beta_q \le \pi - 2\gamma_i.$$

Lemma 1. *Let q be a point in the triangle $v_i p_i w_i$. The segment $p_i q$ divides the visibility angle of p_i into γ^l and γ^r as in Figure 2(b). Then*

$$D^l_{p_i,q} = \frac{\sin \alpha_q}{\sin(\gamma^l + \alpha_q)\cos\delta_i - \sin\gamma^l \cos\left(\delta_i + \frac{\alpha_q + \beta_q}{2}\right)}$$

$$D^r_{p_i,q} = \frac{\sin \beta_q}{\sin(\gamma^r + \beta_q)\cos\delta_i - \sin\gamma^r \cos\left(\delta_i + \frac{\alpha_q + \beta_q}{2}\right)}.$$

Proof. Let the value of the visibility angle of q be $2\gamma_q$. Dividing numerator and denominator of $D^l_{p_i,q}$ by $|p_i q|$ and applying the law of sines in the triangle $v_i p_i w_i$ yields

$$D^l_{p_i,q} = \frac{|p_i q|}{|p_i v_i| g(\gamma_i) - |q v_i| g(\gamma_q)} = \frac{\sin \alpha_q}{\sin(\pi - \gamma^l - \alpha_q) g(\gamma_i) - \sin\gamma^l g\left(\gamma_i + \frac{\alpha_q + \beta_q}{2}\right)}$$

$$= \frac{\sin \alpha_q}{\sin(\gamma^l + \alpha_q)\cos\delta_i - \sin\gamma^l \cos\left(\delta_i + \frac{\alpha_q + \beta_q}{2}\right)}.$$

The proof for $D^r_{p_i,q}$ is analogous.   □

Before analyzing the strategy, we show that the *glad* path approaches a point on $v_n w_n$. Therefore, we examine the path in one triangle $v_i p_i w_i$. *In the remaining paper we assume that the points p_i, v_i, and w_i always satisfy the condition $|p_i v_i| \le |p_i w_i|$.* Of course, the analysis is analogous if $|p_i v_i| > |p_i w_i|$.

Lemma 2. *The* glad *path approaches $v_i w_i$ in a triangle $v_i p_i w_i$ as long as no new vertex becomes visible. Every path point p_k lies in the interior of the triangle $v_i p_i w_i$ to the left of or on the bisector of p_i's visibility angle (that is, $\gamma^l \le \gamma^r$), and $l_{p_i,p_k} = r_{p_i,p_k} > 0$.*

The proof is omitted due to space limitations.

Lemma 2 implies that the *glad* path approaches a point on $v_n w_n$ since if a new vertex becomes visible on the path in the triangle $v_i p_i w_i$, then the continuing path approaches $v_{i+1} w_{i+1}$ of the next triangle. Two paths are exemplarily shown in Figure 5.

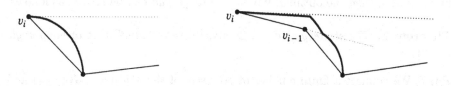

Fig. 5. Two *glad* paths

3.2   Analysis of the Strategy

We analyze the competitive ratio of the strategy *glad* by approximating the path with polygonal paths. This idea (Lemma 3 and Theorem 1) is taken from López-Ortiz and Schuierer [14]; Lemma 3 is only extended by the *glad* function. With this lemma, the ratio of the length of a polygonal path (connecting a finite number of points p_i) to the shortest path length can be estimated by regarding every segment $p_i p_{i+1}$ and a corresponding portion of the shortest path. An example is depicted in Figure 4.

We denote the shortest path in the funnel from a to b by $S(a,b)$, the *glad* path by $G(a,b)$, and the length of a path X by $|X|$.

Lemma 3. *Let p_i, $1 \leq i \leq n$, be path points in a funnel with opening angle $\geq \pi/2$ and let g be the glad function. If $p_{n+1} = v_n$ and $D^l_{p_i,p_{i+1}} \leq c$ or $p_{n+1} = w_n$ and $D^r_{p_i,p_{i+1}} \leq c$, for all $1 \leq i \leq n$, then the length of the polygonal path through the points p_i, $1 \leq i \leq n+1$, is at most $c\,|S(p_1, p_{n+1})|$.*

Proof. Let Π denote the polygonal path. We have to estimate $|\Pi|/|S(p_1,p_{n+1})|$ and assume $p_{n+1} = w_n$. The case for $p_{n+1} = v_n$ is symmetric. If no new vertex is visible in p_{i+1}, then $|w_i w_{i+1}| = 0$, else $|w_i w_{i+1}| > 0$. In either case, $|w_i w_{i+1}| = |p_{i+1} w_{i+1}| - |p_{i+1} w_i|$, $0 \leq i < n$; cf. Figure 2(a). This equation together with $|p_1 w_0| = 0$ and $|p_{n+1} w_n| = 0$ yields

$$\sum_{i=1}^{n} r_{p_i,p_{i+1}}$$

$$= \sum_{i=1}^{n} \left[|p_i w_i| g(\gamma_i) - |p_{i+1} w_i| g(\gamma_{i+1}) \right] - |p_1 w_0| g(\gamma_1) + |p_{n+1} w_n| g(\gamma_{n+1})$$

$$= \sum_{i=0}^{n-1} \left(|p_{i+1} w_{i+1}| - |p_{i+1} w_i| \right) g(\gamma_{i+1}) \leq \sum_{i=0}^{n-1} |w_i w_{i+1}| = |S(p_1, p_{n+1})|.$$

By the inequality $(a_1 + a_2)/(b_1 + b_2) \leq \max\{a_1/b_1, a_2/b_2\}$, which holds for b_1, $b_2 > 0$, and the above lower bound on $|S(p_1, p_{n+1})|$ it follows that

$$\frac{|\Pi|}{|S(p_1,p_{n+1})|} \leq \frac{\sum_{i=1}^{n} |p_i p_{i+1}|}{\sum_{i=1}^{n} r_{p_i,p_{i+1}}} \leq \max_{1 \leq i \leq n} \frac{|p_i p_{i+1}|}{r_{p_i,p_{i+1}}} = \max_{1 \leq i \leq n} D^r_{p_i,p_{i+1}} = c;$$

here, $r_{p_i,p_{i+1}} > 0$ is required in the second inequality, for $1 \leq i \leq n$, which is ensured by Lemma 2.   □

If we approximate the *glad* path by polygonal paths and estimate their lengths by Lemma 3, then the optimality of the strategy *glad* can be proved as follows.

Theorem 1. *The strategy* glad *is $\sqrt{2}$-competitive in funnels with opening angle $\geq \pi/2$.*

Proof. We connect a finite number of points p_i of the *glad* path $G(p_1, p_n)$ and the point p_{n+1} to a polygonal path. The length of the *glad* path $G(p_1, p_{n+1})$ is the supremum of the lengths of all such polygonal paths. Let M_α be the set of all polygonal paths Π for which α, $\alpha \in (0, \pi/2)$, is an upper bound on $\alpha_{p_{i+1}}$.

For the supremum, it suffices to consider the polygonal paths from a single set M_α since every polygonal path outside this set can be extended to one contained in it by inserting additional points.

To estimate the length of a polygonal path $\Pi \in M_\alpha$ by Lemma 3, we first show the upper bound $c := \sqrt{2}/(1 - \sin(\alpha))$ for all values $D^l_{p_i,p_{i+1}}$ and $D^r_{p_i,p_{i+1}}$, $1 \leq i \leq n$. For $i < n$, the points p_i and p_{i+1} satisfy the *glad* condition. We prove in Lemma 5 that for two such points the values $D^l_{p_i,p_{i+1}}$ and $D^r_{p_i,p_{i+1}}$ are at most $\sqrt{2}/(1 - \sin \alpha_{p_{i+1}})$ implying an upper bound of c on them as $\alpha_{p_{i+1}} \leq \alpha < \pi/2$. For $i = n$, we obtain (cf. Figure 4)

$$D^r_{p_n,p_{n+1}} = \frac{|p_n p_{n+1}|}{|p_n w_n|g(\gamma_n) - 0} = \frac{|p_n w_n|}{|p_n w_n|g(\gamma_n)} < \frac{1}{g(\pi)} = \sqrt{2} < c.$$

Analogously, $D^l_{p_n,p_{n+1}} < c$. By Lemma 3, every polygonal path from the set M_α has a length of at most $c|S(p_1, p_{n+1})|$. Since every set M_α, $\alpha \in (0, \pi/2)$, yields an upper bound on the length of the *glad* path, we obtain

$$|G(p_1, p_{n+1})| \leq \inf_{\alpha \in (0, \pi/2)} \frac{\sqrt{2}}{1 - \sin(\alpha)} \cdot |S(p_1, p_{n+1})| = \sqrt{2}\,|S(p_1, p_{n+1})|.$$

$\square$

It remains to prove the bound $\sqrt{2}/(1 - \sin \alpha_{p_{i+1}})$ for $D^l_{p_i,p_{i+1}}$ $(= D^r_{p_i,p_{i+1}})$ if p_i and p_{i+1} satisfy the *glad* condition. From now on, we only deal with the path point p_{i+1} and write $\alpha_{p_{i+1}}$ and $\beta_{p_{i+1}}$ without indices. Since p_{i+1} lies in the interior of the triangle $v_i p_i w_i$ by Lemma 2, i.e. α and β are positive, we define $z := \beta/\alpha$, and $z > 0$ holds. We need the following function definitions and the properties thereafter which hold for the path point p_{i+1}.

Definition 2. Let $\delta_i := \gamma_i - \pi/4$, $x \in (0, \pi/2)$, $\gamma_i \in [\pi/4, \pi/2)$, $z\alpha > 0$. The functions T, A, A^*, B, B^*, and D^*, whose (additional) arguments z, γ_i, and (if depending on) α are omitted for the sake of readability, are defined as follows.

$$T(x) := \frac{\cos x - \cos \frac{\alpha + z\alpha}{2} + \tan \delta_i \sin \frac{\alpha + z\alpha}{2}}{\sin x}$$

$$A := 1 - \frac{T(z\alpha)}{\tan \gamma_i} \quad \text{and} \quad A^* := 1 - \frac{\tan \delta_i}{\tan \gamma_i}$$

$$B := \frac{T(z\alpha) - T(\alpha)}{\sin(2\gamma_i)} \quad \text{and} \quad B^* := 2 \frac{1 - z}{1 + z} \frac{\tan \delta_i}{\sin(2\gamma_i)}$$

$$D^* := \frac{\sqrt{1 + \tan^2 \gamma^l}}{\cos \delta_i + \frac{1 \pm z}{2} \tan \gamma^l \sin \delta_i}.$$

The functions A^* and B^* are closely related to their corresponding limit functions $A_0 := \lim_{\alpha \to 0} A$ and $B_0 := \lim_{\alpha \to 0} B$. In particular,

$$A^* = A_0 + (1 - z)/(2z) \cdot \tan \delta_i / \tan \gamma_i \quad \text{and} \quad B^* = B_0 \cdot 4z/(1 + z)^2. \tag{1}$$

The function D^* results from transforming $D^l_{p_i,p_{i+1}}$ by

$$\sin(y+x)\cos\delta_i - \sin y \cos\left(\delta_i + \tfrac{\alpha+z\alpha}{2}\right) = \sin x \cos\delta_i \left[\cos y + T(x)\sin y\right] \quad (2)$$

to

$$D^l_{p_i,p_{i+1}} = \frac{1}{\cos\delta_i\left(\cos\gamma^l + T(\alpha)\sin\gamma^l\right)} = \frac{\sqrt{1+\tan^2\gamma^l}}{\cos\delta_i\left(1+T(\alpha)\tan\gamma^l\right)}$$

and replacing $T(\alpha)$ by its limit function $\tfrac{1+z}{2}\tan\delta_i$ for $\alpha \to 0$.

Lemma 4. *Let p_i and p_{i+1} be two path points with $\gamma_{i+1} > \gamma_i$, and let $\delta_i := \gamma_i - \pi/4$. The segment $p_i p_{i+1}$ divides the visibility angle of p_i into γ^l and γ^r. Then, with the functions from Definition 2,*

(a) $z \in (0,1]$

(b) $\tan\gamma^l = A\tan\gamma_i/(A+B)$

(c) $A + B > 0$ *and* $A > 0$

(d) $A \le A^*$ *and* $B \ge B^*$

(e) $D^l_{p_i,p_{i+1}} \le D^*/(1-\sin\alpha)$.

Proof. (a): We have already stated $z > 0$. The proof for the upper bound, $z \le 1$, is by contradiction. Assume $z > 1$. We show that this implies $D^r_{p_i,p_{i+1}} \ne D^l_{p_i,p_{i+1}}$, hence z must be ≤ 1. The assumption $z > 1$ is equivalent to $\beta > \alpha$ and yields for the expressions from Lemma 1

$$1/D^l_{p_i,p_{i+1}} > \cos(\gamma^l - \delta_i) \quad \text{and} \quad 1/D^r_{p_i,p_{i+1}} < \cos(\gamma^r - \delta_i).$$

By Lemma 2 we have $0 < \gamma^l \le \gamma_i$ and $\gamma_i \le \gamma^r < 2\gamma_i$, for which the inequality $\cos(\gamma^l - \delta_i) \ge \cos(\gamma^r - \delta_i)$ easily can be shown. Hence, $1/D^l_{p_i,p_{i+1}} > 1/D^r_{p_i,p_{i+1}}$.

(b): Starting with $D^l_{p_i,p_{i+1}} = D^r_{p_i,p_{i+1}}$ and solving for $\tan\gamma^l$ yields the claim by using $\gamma^r = 2\gamma_i - \gamma^l$ and (2).

The proofs of (c)–(e) are based on $T(\alpha) \le \lim_{\alpha \to 0} T(\alpha) = \tfrac{1+z}{2}\tan\delta_i$ and $T(z\alpha) \ge \lim_{\alpha \to 0} T(z\alpha) = \tfrac{1+z}{2z}\tan\delta_i$ which can easily be shown. From this follows that $A + B \ge 1 - T(z\alpha) + T(z\alpha) - T(\alpha) > 0$, $A \le A_0$, and $B \ge B_0 \ge 0$, where A_0 and B_0 again denote the limit functions of A and B for $\alpha \to 0$. $A_0 \le A^*$ and $B_0 \ge B^*$ follows from (1) and $B_0 \ge 0$ and proves (d). From $A + B > 0$, $\tan\gamma^l > 0$, and Lemma 4(b), we obtain $A > 0$. The proof of (e) is omitted due to space limitations. $\qquad\square$

Now we prove an upper bound for $D^l_{p_i,p_{i+1}}$ $(= D^r_{p_i,p_{i+1}})$, p_i and p_{i+1} being two points satisfying the *glad* condition. Replacing the occurrences of γ^l in $D^l_{p_i,p_{i+1}}$ via Lemma 4(b) yields $D^l_{p_i,p_{i+1}}$ as a function of α, z and γ_i. A numerical analysis of this function (with respect to the conditions $A > 0$ and $z \in (0,1]$ from Lemma 4) suggests that the function achieves its maximal value $\sqrt{2}$ for $z = 1$. In the following analysis of the function, we show the upper bound $\sqrt{2}/(1-\sin\alpha)$ which is sufficient for our purpose.

Lemma 5. *For two points p_i and p_{i+1} satisfying the* glad *condition, $D^l_{p_i,p_{i+1}} = D^r_{p_i,p_{i+1}} \leq \sqrt{2}/(1 - \sin\alpha)$.*

Proof. Because of Lemma 4(e), it suffices to show that $D^* \leq \sqrt{2}$. Let f_1 and f_2 denote the numerator and denominator of D^*, respectively. We use $A + B > 0$ from Lemma 4(c) and prove $(A + B)^2 \left(2f_2^2 - f_1^2\right) \geq 0$ which implies $2 \geq f_1^2/f_2^2$, hence $\sqrt{2} \geq f_1/f_2 = D^*$. In addition to the functions f_1 and f_2, we define the two functions

$$Q(\gamma_i, z) := (1 - z)(1 - \tan\gamma_i)$$
$$R(\gamma_i, z) := 2\sin(2\gamma_i) - (1 + z)\cos(2\gamma_i)\tan\gamma_i$$

Since $-\cos(2\gamma_i)/\sin(2\gamma_i) = (\tan^2\gamma_i - 1)/(2\tan\gamma_i)$ and $z \leq 1$,

$$B^* \cdot R(\gamma_i, z) \geq (1 - z)\tan\delta_i(1 + \tan^2\gamma_i). \tag{3}$$

Furthermore, since $\tan\delta_i = (\tan\gamma_i - 1)/(1 + \tan\gamma_i)$,

$$A^* \tan\gamma_i(1 - \tan\gamma_i) = \tan\delta_i(-\tan^2\gamma_i - 1). \tag{4}$$

Now, using $\cos\delta_i = (\cos\gamma_i + \sin\gamma_i)/\sqrt{2}$, $\sin\delta_i = (\sin\gamma_i - \cos\gamma_i)/\sqrt{2}$ and replacing $\tan\gamma^l$ via Lemma 4(b), we obtain that $(A + B)^2\left(2f_2^2 - f_1^2\right)$ equals

$$(A + B)^2(\cos\gamma_i + \sin\gamma_i)^2 + 2(A + B)(\sin^2\gamma_i - \cos^2\gamma_i)(1 + z)/2 \cdot A\tan\gamma_i +$$
$$(1 + z)^2/4 \cdot A^2\tan^2\gamma_i(\sin\gamma_i - \cos\gamma_i)^2 - (A + B)^2 - A^2\tan^2\gamma_i.$$

If we subtract $2B^2 \sin\gamma_i \cos\gamma_i$, use $(1 + z)^2/4 \geq z$, and simplify the resulting expression, we obtain

$$(A + B)^2\left(2f_2^2 - f_1^2\right) \geq A^2 \cdot \tan\gamma_i \, Q(\gamma_i, z) + AB \cdot R(\gamma_i, z);$$

since $Q(\gamma_i, z) \leq 0$ and $R(\gamma_i, z) \geq 0$ because of $z \leq 1$, $\tan\gamma_i \geq 1$ and $\cos(2\gamma_i) \leq 0$, we obtain by Lemma 4(c) and (d)

$$\geq A\big(A^* \tan\gamma_i \cdot Q(\gamma_i, z) + B^* \cdot R(\gamma_i, z)\big)$$

and use (3) and (4) to finally get

$$\geq A(1 - z)\big(A^* \tan\gamma_i(1 - \tan\gamma_i) + \tan\delta_i(1 + \tan^2\gamma_i)\big) = 0.$$

$\square$

Theorem 1 completes the proof of Lemma 5, the optimality of the strategy *glad*. Together with the lower bound $\sqrt{2}$, we have the following result.

Theorem 2. *For robots with vision system, the strategy* glad, *combined with the strategy* clad *and embedded into the high level strategy, is a $\sqrt{2}$-competitive strategy for target searching in streets, and this is optimal.*

4 Conclusions

We have solved the target searching problem for streets by presenting an optimal strategy for funnels. The strategy consists of two parts and changes from *clad* to *glad* upon reaching the visibility angle $\pi/2$. Regarding the strategy and the path, two questions arise. Is it possible to traverse a funnel optimally without changing the strategy? And is it possible to traverse a funnel with a path consisting only of line segments? Concerning the second question, the strategy *clad* can be replaced without losing optimality by following the bisector of the first visibility angle until a new vertex becomes visible and repeating this until the visibility angle $\pi/2$ is reached [11,18]. For the remaining part of the funnel, López-Ortiz already shows that an *optimal* path of line segments cannot be obtained if the robot changes its direction only on seeing a new vertex [11]. It is questionable if optimality can be achieved here with a finite number of additional direction changes.

References

1. A. Blum, P. Raghavan, and B. Schieber. Navigating in unfamiliar geometric terrain. In *Proc. 23rd ACM Symp. on Theory of Computing*, pages 494–503, 1991.
2. A. Datta, Ch. Hipke, and S. Schuierer. Competitive searching in polygons—beyond generalized streets. In *Proc. Sixth International Symposium on Algorithms and Computation*, pages 32–41. LNCS 1004, 1995.
3. A. Datta and Ch. Icking. Competitive searching in a generalized street. In *Proc. 10th ACM Symp. on Computational Geometry*, pages 175–182, 1994.
4. Ch. Icking. *Motion and Visibility in Simple Polygons*. PhD thesis, FernUniversität Hagen, 1994.
5. Ch. Icking and R. Klein. Searching for the kernel of a polygon: A competitive strategy. In *Proc. 11th ACM Symp. on Computational Geometry*, pages 258–266, 1995.
6. Ch. Icking, R. Klein and E. Langetepe. An Optimal Competitive Strategy for Walking in Streets. *This Proceedings.*
7. Ch. Icking, A. López-Ortiz, S. Schuierer, and I. Semrau. Going home through an unknown street. Manuscript, 1998. Submitted to *Comptutational Geometry: Theory and Applications.*
8. B. Kalyanasundaram and K. Pruhs. A competitive analysis of algorithms for searching unknown scenes. *Computational Geometry: Theory and Applications*, 3:139–155, 1993.
9. R. Klein. Walking an unknown street with bounded detour. *Computationl Geometry: Theory and Applications*, 1:325–351, 1992.
10. J. M. Kleinberg. On-line search in a simple polygon. In *Proc. 5th ACM-SIAM Symp. on Discrete Algorithms*, pages 8–15, 1994.
11. A. López-Ortiz. *On-line Target Searching in Bounded and Unbounded Domains*. PhD thesis, Department of Computer Science, University of Waterloo, 1996.
12. A. López-Ortiz and S. Schuierer. Going home through an unknown street. S. G. Akl, F. Dehne, and J.-R. Sack, editors, In *Proc. of 4th Workshop on Data Structures and Algorithms*, LNCS 955, pages 135–146, 1995.

13. A. López-Ortiz and S. Schuierer. Simple, efficient and robust strategies to traverse streets. In Ch. Gold and J.-M. Robert, editors, *Proc. Seventh Canadian Conference on Computational Geometry*, pages 217–222. Université Laval, 1995.
14. A. López-Ortiz and S. Schuierer. Walking streets faster. In *Proc. 5th Scandinavian Workshop on Algorithm Theory*, pages 345–356. LNCS 1097, 1996
15. A. López-Ortiz and S. Schuierer. Generalized streets revisited. In M. Serna J. Diaz, editor, *Proc. 4th European Symposium on Algorithms*, pages 546–558. LNCS 1136, 1996.
16. A. López-Ortiz and S. Schuierer. Position-independent near optimal searching and on-line recognition in star polygons. In *Proc. 5th Workshop on Algorithms and Data Structures*, pages 284–296. LNCS 1272, 1997.
17. A. López-Ortiz and S. Schuierer. The Exact Cost of Exploring Streets with a CAB. In *Proc. 10th Canadian Conf. on Computational Geometry*, 1998.
18. I. Semrau. Analyse und experimentelle Untersuchung von Strategien zum Finden eines Ziels in Straßenpolygonen. Masters thesis (Diplomarbeit), FernUniversität Hagen, 1996.
19. D. D. Sleator and R. E. Tarjan. Amortized efficiency of list update and paging rules. *Communications of the ACM*, 28:202–208, 1985.

Parallel Searching on m Rays[*]

Mikael Hammar[1], Bengt J. Nilsson[1], and Sven Schuierer[2]

[1] Department of Computer Science, Lund University, Box 118, S-221 00 Lund,
Sweden,
{Bengt.Nilsson, Mikael.Hammar}@cs.lth.se
[2] Institut für Informatik, Am Flughafen 17, Geb. 051, D-79110 Freiburg, Germany,
schuiere@informatik.uni-freiburg.de

Abstract. We investigate parallel searching on m concurrent rays. We
assume that a target t is located somewhere on one of the rays; we are
given a group of m point robots each of which has to reach t. Furthermore, we assume that the robots have no way of communicating over
distance. Given a strategy S we are interested in the competitive ratio
defined as the ratio of the time needed by the robots to reach t using S
and the time needed to reach t if the location of t is known in advance.
If a lower bound on the distance to the target is known, then there is a
simple strategy which achieves a competitive ratio of 9 — independent
of m. We show that 9 is a lower bound on the competitive ratio for two
large classes of strategies if $m \geq 2$.
If the minimum distance to the target is not known in advance, we show
a lower bound on the competitive ratio of $1 + 2(k + 1)^{k+1}/k^k$ where
$k = \lceil \log m \rceil$. We also give a strategy that obtains this ratio.

1 Introduction

Searching for a target is an important and well studied problem in robotics. In
many realistic situations the robot does not possess complete knowledge about
its environment, for instance, the robot may not have a map of its surroundings,
or the location of the target may be unknown [3, 4, 5, 6, 9, 10, 11, 13, 15, 16].

The search of the robot can be viewed as an *on-line* problem since the robot's
decisions about the search are based only on the part of its environment that
it has seen so far. We use the framework of *competitive analysis* to measure the
performance of an on-line search strategy S [19]. The *competitive ratio* of S is
defined as the maximum of the ratio of the distance traveled by a robot using S
to the optimal distance from its starting point to the target, over all possible
locations in the environment of the target.

A problem with paradigmatic status in this framework is searching on m
concurrent rays. Here, a point robot or—as in our case—a group of point robots
is imagined to stand at the origin of m concurrent rays. One of the rays contains
the target t whose distance to the origin is unknown. A robot can detect t only
if it stands on top of it. It can be shown that an optimal strategy for one robot is

---

[*] This research is supported by the DFG-Project "Diskrete Probleme", No. Ot 64/8-1.

C. Meinel and S. Tison (Eds.): STACS'99, LNCS 1563, pp. 132–142, 1999.
© Springer-Verlag Berlin Heidelberg 1999

to visit the rays in cyclic order, increasing the step length each time by a factor of $m/(m-1)$ if it starts with a step length of 1 [1, 7]. The competitive ratio C_m achieved by this strategy is given by

$$C_m = 1 + 2\frac{m^m}{(m-1)^{m-1}}.$$

The lower bound for searching in m rays has proven to be a very useful tool for proving lower bounds for searching in a number of classes of simple polygons, such as star-shaped polygons [12], generalized streets [6, 14], HV-streets [5], θ-streets [5, 8], and k-spiral polygons [18].

In this paper we are interested in obtaining upper and lower bounds for the competitive ratio of *parallel searching* on m concurrent rays.

Assume that a group of m point robots searches for the target. Neither the ray containing the target nor the distance to the target are known. Now all the robots have to reach the target and the only way two robots can communicate is if they meet, that is, they have no communication device. Baeza-Yates and Schott investigate searching on the real line, that is, the case $m = 2$ [2]. They present two strategies both of which achieve a competitive ratio of 9. They also consider searching for a target line in the plane with multiple robots and present symmetric and asymmetric strategies. However, the question of optimality, that is, corresponding lower bounds, is not considered.

In this paper we investigate search strategies for parallel searching on m concurrent rays. If a lower bound on the distance to the target is known, then there is a simple strategy that achieves a competitive ratio of 9—independent of m. We show that even in the case $m = 2$ there is a matching lower bound of 9 on the competitive ratio of two large classes of strategies. Moreover, we show that, for all strategies, a lower bound of 9 for $m = 2$ implies a lower bound of 9 for $m > 2$—as is to be expected.

If the minimum distance to the target is not known in advance, then we show a lower bound on the competitive ratio of $1 + 2(k+1)^{k+1}/k^k$ where $k = \lceil \log m \rceil$. We also present a strategy that achieves this competitive ratio.

The paper is organized as follows. In the next section we present some definitions and preliminary results. In particular, we present three strategies to search on the line ($m = 2$), each with a competitive ratio of 9 and show that one of them can also be used to search on m rays with the same competitive ratio. In Section 3 we show a matching lower bound of 9 for strategies that are monotone or symmetric. Finally, in Section 4 we present an optimal algorithm to search on m rays if no lower bound on the distance to the target is known in advance.

2 Preliminaries

In the following we consider the problem of a group of m robots searching for a target of unkown location on m rays in parallel. The robots have the same maximal speed which we assume w.l.o.g. to be 1 distance unit per time unit. If the robots have unbounded speed, then the time to find the target (both off-line and on-line) can be made arbitrarily small. The speed of a robot may be positive (if it moves away from the origin) or negative (if it moves towards the origin).

Let S be a strategy for parallel searching on m rays and $T_S(D)$ the maximum time the group of robots needs to find and reach a target placed at a distance of D if it uses strategy S. Since the maximum speed of a robot is one, the time needed to reach the target if the position of the target is known is D time units. The competitive ratio is now defined as the maximum of $T_S(D)/D$, over all $D \geq 0$. In some applications a lower bound D_{min} on the distance to the target may be known. If such a lower bound exists, then we assume without loss of generality that $D_{min} = 1$. It will turn out that the existence of D_{min} leads to a drastically lower competitive ratio if $m > 2$.

We define different classes of possible strategies to search on m rays in parallel. We say a strategy is *monotone* if, at all times, all the robots (that do not know the location of the target) have non-negative speed. We say a strategy is *full speed* if all the robots travel at a speed of 1 or -1 at all times. We say a strategy is *symmetric* if, at all times, all the robots (that do not know the location of the target) have the same speed.

We illustrate the different types of strategies for $m = 2$. The optimal monotone strategy is for each robot to travel at a speed of $1/3$ on each ray. After one robot has found the target, it runs back to fetch the other. This leads to a competitive ratio of 9. This strategy is described in [2]. In the next section we show a lower bound of 9 on the competitive ratio of monotone strategies. The optimal (full-speed) symmetric strategy is for each robot to double the distance that has been explored before and then to return to the origin. This strategy can only be applied if a lower bound on the distance to the target is known. It achieves a competitive ratio of 9. Again this strategy is described in [2] and we show a lower bound of 9 on the competitive ratio of symmetric strategies in the next section. Finally, an asymmetric strategy is for both robots to walk together and to use the optimal strategy for one robot to search on two rays. This again yields a competitive ratio of 9.

3   With Lower Bound on the Minimum Distance

In this section we assume that a lower bound on the distance from the origin to the target of $D_{min} = 1$ is known.

Initially we study the parallel search problem on two rays and prove a lower bound on the competitive ratio for monotone strategies. We can view the two rays as being the positive and negative parts of the real line with the two robots initially placed at the origin. As time passes the robots move continuously and monotonically with some speed along the line until one of them finds the target. This robot now travels at full speed to the other robot and communicates to it the location of the target and they both return to this target point.

Let $v_1(T)$ and $v_2(T)$ be the *average* speeds of the two robots at time T, i.e., the distance of the robot to the origin at this time divided by the time. It is clear that a search strategy is completely specified by the two average speed functions. We can prove the following lower bound.

Lemma 1. *There is no monotone strategy that achieves a better competitive ratio than 9 to search on two rays in parallel.*

Next we look at symmetric strategies. It is easy to show that only symmetric full speed strategies need to be considered. In the following we show a lower bound for symmetric strategies. We start with the simpler case when $m = 2$ and consider the general version later.

Lemma 2. *Let $\mathcal{Z}$ be the set of infinite positive sequences $Z = (z_0, z_1, z_2, \ldots)$ with $z_{2k} > z_{2k-2}$ and $z_{2k+1} > z_{2k-1}$, for all $k \geq 1$. If S is a symmetric strategy, then the competitive ratio C_S of S is at least*

$$\inf_{Z \in \mathcal{Z}} \sup_{k \geq 1} 1 + 2F_k(Z)$$

where

$$F_k(Z) = \max \left\{ \frac{z_{2k} + z_{2k+1}}{z_{2k-1}}, \frac{z_{2k+2} + z_{2k+3}}{z_{2k+2} - z_{2k}} \right\}.$$

Proof. Since the strategy is symmetric, the two robots will use the same local strategy to search its own ray. Furthermore, as we mentioned above we can assume that it is a full speed strategy. We can model a full speed strategy for a robot by saying that it first moves a distance x_0 forward along the ray at full speed, then it moves a distance y_0 backwards at full speed, then a distance x_1 forward, a distance y_1 backward, and so on. When one of the robots finds the target, it runs back at full speed until it meets the other robot, and they both run to the target at full speed.

The proof uses an adversary to place the target point in order to maximize the competitive ratio.

We say that a robot is in *step* k when it moves forward and backward the $k + 1^{\text{st}}$ time. Let L_k denote the distance to the origin of the turning point where the robot begins step k and let U_k denote the distance to the origin of the turning point where the robot starts to move backwards during step k. We have that

$$L_k = L_{k-1} + x_{k-1} - y_{k-1} = \sum_{i=0}^{k-1} x_i - y_i,$$

$$U_k = L_k + x_k = x_k + \sum_{i=0}^{k-1} x_i - y_i = y_k + \sum_{i=0}^{k} x_i - y_i.$$

The total time that the robot has travelled when it completes step k is

$$T_k = \sum_{i=0}^{k} x_i + y_i.$$

First of all, we can assume that $y_k \leq U_k$ for all $k \geq 0$, that is, a robot always stays on the same ray, since if this does not hold, we can exchange the strategy for an equivalent one where, if the two robots meet at the origin, they exchange places and continue on their own ray instead of the other robot's ray.

Secondly, we can assume that $U_{k-1} < U_k$, since otherwise, the strategy will not explore any new part of the ray during step k, and we can exchange the strategy for another equivalent one, where the assumption holds. In particular, we can assume that $y_k < x_{k+1}$, for all $k \geq 0$.

Assume that the target is placed at distance D with $1 = D_{min} \leq D$ and $U_{k-1} < D \leq U_k$, for some $k \geq 0$. This means that one of the robots will find the target during step k. If T_R again denotes the time for the robot that found the target to reach the second robot at full speed, then the competitive ratio for this placement is given by

$$C_D = \frac{T_{k-1} + (D - L_k) + 2T_R}{D} = 1 + 2\frac{\sum_{i=0}^{k-1} y_i + T_R}{D}.$$

We will only consider two possible placements for the target and let the adversary choose the one that maximizes the competitive ratio. For the first placement of the target, we assume that the other robot is reached while it is still in step k. (This is the best case for the strategy.) Since we place the target in the interval $]U_{k-1}, U_k]$, the ratio C_k^1 of the time needed by the strategy to the optimal time is given by

$$C_k^1 = \sup_{D \in]U_{k-1}, U_k]} \left\{ 1 + 2\frac{\sum_{i=0}^{k-1} y_i + T_R}{D} \right\} = 1 + 2\frac{\sum_{i=0}^{k-1} y_i + U_k}{U_{k-1}}.$$

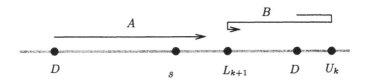

Fig. 1. Robot A misses robot B if $D + L_{k+1} > U_k - D + U_k - L_{k+1}$.

In the second possible placement the robot that finds the target just fails to reach the other robot during step k. Since both robots travel at full speed, the earliest they can meet is during the step $k + 1$. In order for robot A to miss robot B the distance $D + L_{k+1}$ from the point where A finds the target to the forward turning point of B at the beginning of step $k + 1$ has to be larger than the distance from the point D on B's ray to B's backward turning point at U_k plus the distance from U_k to L_{k+1} (see Figure 1), that is,

$$L_{k+1} + D > U_k - D + U_k - L_{k+1} = U_k - D + y_k,$$

i.e., when $D > y_k$. The placement of the target is therefore restricted to the interval $]y_k, U_k]$, which implies that the adversary places the target right after y_k. The ratio C_k^2 of the time needed by the strategy to the optima is given by

$$C_k^2 = \sup_{D \in]y_k, U_k]} \left\{ 1 + 2 \frac{\sum_{i=0}^{k-1} y_i + T_R}{D} \right\} = 1 + 2 \frac{\sum_{i=0}^{k} y_i + U_{k+1}}{y_k}.$$

It may happen that independently of where the target is placed in the interval $]U_{k-1}, U_k]$ the robot that finds the target will never miss the other robot during step k. This means that $y_k \geq U_k$ because placing the target on the point U_k and requiring that the other robot is met during step k implies that the two robots meet at the origin. On the other hand, we know from before that $y_k \leq U_k$, and hence, that $y_k = U_k$.

Now, we consider the situation if the adversary places the target right after the point U_k. We obtain a competitive ratio of

$$1 + 2 \frac{\sum_{i=0}^{k} y_i + T_R}{U_k} = 1 + 2 \frac{\sum_{i=0}^{k} y_i + U_{k+1}}{y_k} = C_k^2.$$

Hence, C_k^2 is a lower bound for the competitive ratio independent of the fact whether the robot that finds the target can miss or cannot miss the other robot in step k. Hence, the best competitive ratio for any symmetric strategy is bounded below by

$$C = \sup_{k \geq 0} \max\{C_k^1, C_k^2\}.$$

If we make the variable substitutions $z_{2k} = \sum_{i=0}^{k} y_i$ and $z_{2k+1} = U_{k+1}$, for all $k \geq 0$, and let $Z = (z_0, z_1, \ldots)$, we can express $\max\{C_k^1, C_k^2\}$, $k \geq 2$ as the functional

$$1 + 2F_{k-1}(Z) = 1 + 2 \max \left\{ \frac{z_{2k-2} + z_{2k-1}}{z_{2k-3}}, \frac{z_{2k} + z_{2k+1}}{z_{2k} - z_{2k-2}} \right\}$$

where $z_{2i} > z_{2i-2}$ and $z_{2i+1} > z_{2i-1}$, for all $i \geq 1$. This proves the claim. □

Now we show the main result on symmetric strategies.

Theorem 1. *There is no symmetric strategy that achieves a better competitive ratio than 9 to search on two rays in parallel.*

Proof. **(Sketch)** By Lemma 2 the competitive ratio C_S of a symmetric search strategy S satisfies $C_S \geq \inf_{Z \in \mathcal{Z}} \sup_{k \geq 1} 1 + 2F_k(Z)$. Let $a = \varlimsup_{n \to \infty} (z_n)^{1/n}$. Using methods by Schuierer [17] it can be shown that there exist two positive numbers γ_0 and γ_1 such that

$$\sup_{0 \leq k < \infty} F_k(Z) \geq \sup_{0 \leq k < \infty} F_k(\gamma_0, \gamma_1 a, \gamma_0 a^2, \ldots)$$

where $\gamma_0, \gamma_1 > 0$. Let $\gamma = \gamma_1 / \gamma_0$; then,

$$F_k(\gamma_0, \gamma_1 a, \gamma_0 a^2, \dots) = \max\left\{\frac{a}{\gamma} + a^2, \frac{a^2 + \gamma a^3}{a^2 - 1}\right\}$$

which is minimized for $\gamma = a - 1/a$ and $a = \sqrt{2}$ yielding a value of 4, that is, $\sup_{0 \le k < \infty} 1 + 2F_k(Z) \ge 9$, for all $Z \in \mathcal{Z}$ as claimed. $\qquad\square$

Theorem 2. *There is no monotone or symmetric strategy that achieves a better competitive ratio than 9 to search on m rays in parallel.*

Proof. We consider monotone strategies first. It is obvious that if the robots have further information about the location of the target, then the competitive ratio for a strategy that exploits this information does not increase. Assume that the robots know that the target is on one of the two first rays. They can all explore these rays monotonically in common. Consider now the strategy we get by taking the furthest robot from the origin on each of the two rays. This strategy is a monotone strategy for two robots on two rays and by Lemma 1 no such strategy can do better than a competitive ratio of 9.

For symmetric strategies we can argue in a similar manner. Once a robot has found the target, the competitive ratio is bounded below by the time it takes to fetch all the other robots and to go back to the target. This ratio is bounded below by the time it takes for the robot that found the target to fetch *one* other robot and for it to go to the target. By Theorem 1 this competitive ratio is bounded below by 9. $\qquad\square$

Now, there is a strategy that achieves a competitive ratio of 9 to search on m rays in parallel. The strategy is known as the *doubling strategy* and goes as follows. Each robot starts by going one unit at full speed on its ray and then goes back to the origin. Then they each go two units, four units, and so on, on their corresponding ray, always doubling the distance travelled and repeatedly going back to the origin. Once a robot finds the target, it goes back at full speed to the origin and waits there until the other robots reach it. It then communicates the location of the target to the other robots and they all move at full speed to the location. The competitive strategy of the doubling strategy, if the target is at distance D from the origin is

$$C \le \sup_{D \in]2^{k-1}, 2^k]}\left\{\frac{2\sum_{i=0}^{k} 2^i + D}{D}\right\} \le 1 + \sup_{D \in]2^{k-1}, 2^k]}\left\{\frac{2\sum_{i=0}^{k} 2^i}{D}\right\}$$

$$\le 1 + \frac{2\sum_{i=0}^{k} 2^i}{2^{k-1}} = 1 + 2\frac{2^{k+1} - 1}{2^{k-1}} = 9 - \frac{1}{2^{k-2}} \le 9.$$

We have proved the following theorem.

Theorem 3. *The doubling strategy achieves a competitive ratio of 9 to search on m rays in parallel given a lower bound on the distance to the target.*

4   Without Lower Bound on the Minimum Distance

In this section we consider the problem of a group of m robots searching for a target of unkown location on m rays in parallel and no lower bound on the distance from the origin to the target is known.

We begin by presenting a strategy that achieves a competitive ratio of

$$1 + 2\frac{(k+1)^{k+1}}{k^k},$$

where $k = \lceil \log m \rceil$. We then show that, in fact, no strategy can do better.

4.1   The Strategy

The optimal strategy is a monotone strategy where all the robots move, one on each ray, with a constant speed v. When one robot finds the target it searches for a robot at full speed to tell it where the target is located. Then they both go at full speed to search for two more robots and tell them the location of the target, and so on. After each step the number of robots that know the location of the target is doubled. Once all robots know the location, they all move to the target. Suppose the target is on some ray and at distance D from the origin. The strategy consists of *steps*. Step i starts when 2^i robots know the location to the target and ends when 2^{i+1} robots know the location to the target; that is, in step i the 2^i robots that currently know the position of the target chase 2^i of those robots that do not. Let T_i denote the time it takes to complete step i. It takes any of the robots D/v time to find the target, and when all robots know the location of the target, it takes them time T_F to go to the target. Hence, the competitive ratio of the strategy is

$$C \;=\; \frac{D/v + \sum_{i=0}^{k-1} T_i + T_F}{D},$$

where $k = \lceil \log m \rceil$.

We can show that $\frac{D/v + \sum_{i=0}^{k-1} T_i + T_F}{D} = 1 + 2\frac{(k+1)^{k+1}}{k^k}$ if the speed is set to $v = \frac{1}{2k+1}$. With this given we have proved the following result.

Theorem 4. *There is a monotone strategy that achieves a competitive ratio of*

$$1 + 2\frac{(k+1)^{k+1}}{k^k},$$

where $k = \lceil \log m \rceil$, to search on m rays in parallel, if no lower bound on the distance to the target is known.

4.2   Lower Bound

We consider first the parallel search problem for monotone strategies and prove a lower bound on the competitive ratio for these strategies. As time passes the robots move continuously and monotonically with some speed along the rays until one of them has found the target. This robot now travels at full speed to one of the other robots and communicates to him the location of the target, they both travel to other robots to communicate the location of the target, and so on. When all robots know where the target is they all go to this target point.

Let $v_1(T), v_2(T), \ldots, v_m(T)$ be the average speeds of the m robots at time T, i.e., the coordinate position of the robot on its ray at this time divided by the time. It is clear that a search strategy is completely specified by the m average speed functions. We have the following lower bound.

Lemma 3. *There is no monotone strategy that achieves a better competitive ratio than*

$$1 + 2\frac{(k+1)^{k+1}}{k^k},$$

where $k = \lceil \log m \rceil$, to search on m rays in parallel.

We are now in a position to prove the following theorem.

Theorem 5. *There is no strategy whatsoever that achieves a better competitive ratio than*

$$1 + 2\frac{(k+1)^{k+1}}{k^k},$$

where $k = \lceil \log m \rceil$, to search on m rays in parallel.

Proof. Any strategy can be given by the average speed functions $v_1(T), \ldots, v_m(T)$ of the m robots. (In conjunction with information about whether a robot switches ray, at some point in time.) Given these average speed functions, an adversary can extract information about the time length that a robot moves monotonically along a ray. (This includes also the time that a robot stands still at the origin.) Let T_i, for $1 \leq i \leq m$ denote the time that robot i moves monotonically along a ray. Each T_i is greater than 0 since either the robot stands still or moves along some ray. If $T_\epsilon = \min_{1 \leq i \leq m}\{T_i\}$, then we let the adversary place the target on some ray, say ray 1, at distance $D = v_1(T_D)T_D$, such that $T_D < \frac{T_\epsilon}{1+2\frac{(k+1)^{k+1}}{k^k}}$, for $k = \lceil \log m \rceil$. If the strategy uses more than T_ϵ time, then the competitive ratio is trivially bounded from below by

$$1 + 2\frac{(k+1)^{k+1}}{k^k},$$

for $k = \lceil \log m \rceil$. If, on the other hand, the strategy uses less than T_ϵ time, then the strategy is monotone in the interesting time interval and we can apply Lemma 3, proving our claim.   □

5 Conclusions

We considered search strategies for parallel search on m concurrent rays. We show that a straight forward generalization of the so called doubling strategy, from searching on the line to searching on m concurrent rays, yields a competitive ratio of 9 if a minimum distance from the origin to the target is known in advance. Furthermore, we prove that 9 is a lower bound on the competitive ratio for both monotone and symmetric strategies in this case.

We also prove a lower bound of

$$1 + 2\frac{(k+1)^{k+1}}{k^k}$$

on the competitive ratio, if a minimum distance from the origin to the target is not known in advance. Finally, we give a search strategy that achieves this ratio regardless of whether such a minimum distance is known or not, giving us an optimal search strategy in the latter case.

References

[1] R. Baeza-Yates, J. Culberson, and G. Rawlins. Searching in the plane. *Information and Computation*, 106:234–252, 1993.

[2] R. Baeza-Yates and R. Schott. Parallel searching in the plane. *Comput. Geom. Theory Appl.*, 5:143–154, 1995.

[3] Margrit Betke, Ronald L. Rivest, and Mona Singh. Piecemeal learning of an unknown environment. In *Sixth ACM Conference on Computational Learning Theory (COLT 93)*, pages 277–286, July 1993.

[4] K-F. Chan and T. W. Lam. An on-line algorithm for navigating in an unknown environment. *International Journal of Computational Geometry & Applications*, 3:227–244, 1993.

[5] A. Datta, Ch. Hipke, and S. Schuierer. Competitive searching in polygons—beyond generalized streets. In *Proc. Sixth Annual International Symposium on Algorithms and Computation*, pages 32–41. LNCS 1004, 1995.

[6] A. Datta and Ch. Icking. Competitive searching in a generalized street. In *Proc. 10th Annu. ACM Sympos. Comput. Geom.*, pages 175–182, 1994.

[7] S. Gal. *Search Games*. Academic Press, 1980.

[8] Ch. Hipke. Online-Algorithmen zur kompetitiven Suche in einfachen Polygonen. Master's thesis, Universität Freiburg, 1994.

[9] Christian Icking and Rolf Klein. Searching for the kernel of a polygon: A competitive strategy. In *Proc. 11th Annu. ACM Sympos. Comput. Geom.*, pages 258–266, 1995.

[10] R. Klein. Walking an unknown street with bounded detour. *Comput. Geom. Theory Appl.*, 1:325–351, 1992.

[11] J. M. Kleinberg. On-line search in a simple polygon. In *Proc. of 5th ACM-SIAM Symp. on Discrete Algorithms*, pages 8–15, 1994.

[12] A. López-Ortiz. *On-line Searching on Bounded and Unbounded Domains*. PhD thesis, Department of Computer Science, University of Waterloo, 1996.

[13] A. López-Ortiz and S. Schuierer. Going home through an unknown street. In S. G. Akl, F. Dehne, and J.-R. Sack, editors, *Proc. 4th Workshop on Algorithms and Data Structures*, pages 135–146. LNCS 955, 1995.

[14] A. López-Ortiz and S. Schuierer. Generalized streets revisited. In M. Serna J. Diaz, editor, *Proc. 4th European Symposium on Algorithms*, pages 546–558. LNCS 1136, 1996.

[15] A. Mei and Y. Igarashi. Efficient strategies for robot navigation in unknown environment. In *Proc. of 21st Intl. Colloquium on Automata, Languages and Programming*, pages 630–641, 1994.

[16] C. H. Papadimitriou and M. Yannakakis. Shortest paths without a map. In *Proc. 16th Internat. Colloq. Automata Lang. Program.*, volume 372 of *Lecture Notes in Computer Science*, pages 610–620. Springer-Verlag, 1989.

[17] S. Schuierer. Lower bounds in on-line geometric searching. In *Proc. of the 11th Intern. Symp. on Fundamentals of Computation Theory*, pages 429–440. LNCS 1279, 1997.

[18] S. Schuierer. On-line searching in geometric trees. In *Proc. 9th Canadian Conf. on Computational Geometry*, pages 135–140, 1997.

[19] D. D. Sleator and R. E. Tarjan. Amortized efficiency of list update and paging rules. *Communications of the ACM*, 28:202–208, 1985.

A Logical Characterisation of Linear Time on Nondeterministic Turing Machines[*]

Clemens Lautemann, Nicole Schweikardt, and Thomas Schwentick

Institut für Informatik / FB 17
Johannes Gutenberg-Universität D–55099 Mainz
{cl|nisch|tick}@informatik.uni-mainz.de

Abstract. The paper gives a logical characterisation of the class NTIME(n) of problems that can be solved on a nondeterministic Turing machine in linear time. It is shown that a set L of strings is in this class if and only if there is a formula of the form $\exists f_1 \cdots \exists f_k \exists R_1 \cdots \exists R_m \forall x \varphi$ that is true exactly for all strings in L. In this formula the f_i are unary function symbols, the R_i are unary relation symbols and φ is a quantifier-free formula. Furthermore, the quantification of functions is restricted to *non-crossing, decreasing* functions and in φ no equations in which different functions occur are allowed. There are a number of variations of this statement, e.g., it holds also for $k = 3$. From these results we derive an Ehrenfeucht game characterisation of NTIME(n).

1 Introduction

Since Fagin's seminal result that NP is the class of problems that can be described by an existential second-order (ESO) formula [6] there have been several characterisations of subclasses of NP by sublogics of ESO [7, 14, 9]. Lynch showed that all problems in NTIME(n), i.e., all problems that can be solved in linear time on nondeterministic Turing machines, can also be expressed by a monadic ESO formula in the presence of a built-in addition relation. Although NTIME(n) is a relatively robust class, in order to capture the notion of linear time as used in algorithm design, Turing machines seem far too restrictive: apparently, simple operations, such as traversing a tree, cannot be done. Therefore, a number of alternative models have been proposed to capture this notion. Notably, in a series of papers ([9, 8, 10]), Grandjean introduced and investigated linear–time classes DLIN and NLIN, based on determistic and nondeterministic random access machines. NLIN contains NTIME(n) as a subclass but it is not known whether this inclusion is strict. Grandjean proved that Lynch's logic even captures (at least) all the languages in NLIN, hence indicating that this logic probably does not exactly characterize NTIME(n). He also showed that a set L is in NLIN if and only if it is the set of models of formulas of the form $\exists f_1 \cdots \exists f_k \forall x \, \varphi$, where φ is quantifier–free and the second–order quantifiers $\exists f_i$ range over unary functions.

---

[*] full paper:
ftp://ftp.informatik.uni-mainz.de/pub/publications/misc/Schwentick/stacs99_NTimeN_full.ps

C. Meinel and S. Tison (Eds.): STACS'99, LNCS 1563, pp. 143–152, 1999.
© Springer-Verlag Berlin Heidelberg 1999

In the present paper, we give an *exact* characterisation of NTIME(n). The motivation behind exact logical characterisations of (presumably) weaker and weaker complexity classes is the hope that they might enable lower bound proofs by methods of Finite Model Theory like Ehrenfeucht games (cf. [5]). Such games have been successfully used in non-expressibility results for similar logics (for a survey see, e.g. [17]).

Our characterisation is obtained by restrictiing Grandjean's logic[1] for NLIN. The main difference concerns the function quantifiers. In our logic, quantification of functions is restricted to *non–crossing* functions, i.e., functions on $\{1,..,n\}$ whose graph, when drawn in the upper half plane with vertices on the line, has no crossing arcs. Such functions have, in different guises, been used before to describe computations, e.g., they play an important part in the lower bound proof of [15], and in the separation of DTIME(n) from NTIME(n), proved in [16]. In the form of *matchings*, they were used in a logical characterisation of context–free languages in [13]. In fact, we make use of the close connection between context–free languages and NTIME(n), which is expressed in the theorem of [3], stating that a set is in NTIME(n) iff it is the projective image of the intersection of three context–free languages. To be more precise, we show that a set of strings can be recognised by a nondeterministic Turing machine in linear time iff it is the set of models of a formula $\exists f_1 \cdots \exists f_k \exists R_1 \cdots \exists R_m \forall x \varphi$, where φ is a quantifier–free formula (with certain syntactical restrictions), the second-order quantifiers $\exists R_i$ range over unary relations, and the function quantifiers $\exists f_i$ range over decreasing non–crossing functions only. By restricting the number, k, of function variables, we obtain a strict hierarchy of classes from $k = 0$ to $k = 3$: $k = 0$ characterises the regular languages [4], $k = 1$ the context–free languages, and for $k \geqslant 3$, we obtain NTIME(n). Using the lower bound from [15], it can be seen that the class of languages defined by formulas with $k = 2$ function quantifiers lies strictly between context–free languages and NTIME(n).

Our logic is fairly robust, allowing a number of variations. If we want to show that some set is contained in NTIME(n), we can do this by using a rather liberal syntax. On the other hand, in order to show that some set is *not* contained in NTIME(n), the more constrained our formula class, the better. We present an Ehrenfeucht game for NTIME(n), based on our most restrictive characterisation, in which the players play only three rounds of rather restricted moves.

2   Preliminaries

2.1   Strings and Structures

Let Σ be an alphabet (i.e. a finite nonempty set). By ϵ we denote the *empty* string, Σ^* is the class of all finite strings over Σ, and $\Sigma^+ := \Sigma^* \setminus \{\epsilon\}$. By $|w|$ we denote the *length* of a string $w \in \Sigma^*$. The signature τ_Σ associated to an alphabet Σ consists of two constant symbols *min* and *max*, unary function symbols s and

---

[1] Note however, that Grandjean's encoding of strings as structures is different from ours, which is the straightforward one.

p, and a unary relation symbol W_σ for each letter $\sigma \in \Sigma$. With each string $w = w_1 \cdots w_n \in \Sigma^+$ we associate the τ_Σ-structure $\underline{w} := \langle [n], 1, n, s, p, (W_\sigma)_{\sigma \in \Sigma} \rangle$, where $[n]$ is an abbreviation for $\{1, .., n\}$, $s(i) := i + 1$ for $i < n$, $s(n) := n$, $p(i) := i - 1$ for $i > 1$, $p(1) := 1$, and, for every $\sigma \in \Sigma$, $W_\sigma := \{i \in [n] \, / \, w_i = \sigma\}$.

2.2 Formulas

We consider structures with unary functions, unary relations and constants. Consequently, terms are built from variables, constants, and function symbols, and an atomic formula is either a term equality or a unary relation symbol (also called a *predicate*) applied to a term. We will consistently use g, f, f_i for function symbols, R, Q, R_i, Q_i, W_σ for unary relation symbols. Our logics will be fragments of existential second-order logic, obtained by both syntactic and semantic restrictions. If Φ is a class of formulas, we write $\forall x\, \Phi$ for the class of all those formulas of the form $\forall x\, \varphi$, where $\varphi \in \Phi$. Analogously we use notations like $\exists f_1 \cdots f_k\, \Phi$, $\exists \overline{f}\, \Phi$, $\exists R_1 \cdots R_m\, \Phi$. Thus, as an example, $\psi \in \exists f_1 f_2 f_3 \exists \overline{R}\, FO$ means that ψ is of the form $\exists f_1 \exists f_2 \exists f_3 \exists R_1 \cdots \exists R_m \forall x\, \varphi$, for some $m \geqslant 0$, where φ is a first-order formula. Our semantic restrictions concern the scope of the function quantifiers $\exists f_i$. Let, for every n, F_n be a class of functions on $[n]$, $F := \bigcup_{n \geqslant 1} F_n$. For a formula φ and a string w we write $\underline{w} \models^F \exists f_1 \cdots \exists f_k\, \varphi$ iff there are functions $f_1^w, .., f_k^w \in F_{|w|}$ such that $\langle \underline{w}, f_1^w, .., f_k^w \rangle \models \varphi$. Accordingly, if $\psi = \exists f_1 \cdots \exists f_k\, \varphi$ we write $Mod^F(\psi)$ for the F-model set of ψ, i.e., the set $\{w \, / \, \underline{w} \models^F \psi\}$, and for a class Ψ of formulas, $MOD^F(\Psi) := \{Mod^F(\psi) \, / \, \psi \in \Psi\}$.

2.3 Non-crossing Functions

Let $f : [n] \rightarrow [n]$ be a *non-increasing* function, i.e., $f(j) \leqslant j$, for all $j \in [n]$. We call f *non-crossing*, iff for all $j, j' \in [n]$ such that $f(j) < j' \leqslant j$, it holds that $f(j') \geqslant f(j)$. Let NNC denote the class of all non-increasing, non-crossing functions, and DNC the class of all *decreasing* non-crossing functions (where additionally to $f \in NNC$ we require that $f(j) < j$, for all $j > 1$). Instead of $f \in NNC$ ($f \in DNC$) we shall also say f *is nnc* (f *is dnc*). With regard to expressive power, in our context, the difference between NNC and DNC is immaterial, as the following lemma shows.

Lemma 1. *Let $\Psi_k := \exists f_1 \cdots f_k \exists \overline{R}\, \forall x\, FO$. Then $MOD^{NNC}(\Psi_k) = MOD^{DNC}(\Psi_k)$.*

For the inclusion "$\subseteq$" we represent a function $f_i \in NNC$ by a function $f_i' \in DNC$ together with a set I_i as follows:

$$I_i(j) \leftrightarrow f_i(j) = j, \quad \text{and} \quad f_i'(j) = \begin{cases} f_i(j) & \text{if } f_i(j) \neq j \\ p(j) & \text{if } f_i(j) = j. \end{cases}$$

Lemma 1 allows us to use either class, depending on which suits our purposes best. The more restrictive class DNC has some particularly useful properties.

Lemma 2. *A function $f : [n] \rightarrow [n]$ is dnc iff $f(1) = 1$, and for every $j > 1$ there is a number $q \geqslant 0$ such that $f(j) = f^q(j - 1)$. Figure 1 (a) gives an illustration.*

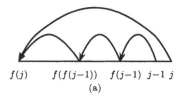

$$f(j) \qquad f(f(j-1)) \quad f(j-1) \;\; j-1 \;\; j$$
(a)

Fig. 1. An illustration of Lemma 2.

3   The Main Result

Let Φ_k be the class of quantifier–free formulas over the signature τ_Σ which have as free variables at most one FO variable x, at most k unary function symbols $f_1, .., f_k$, and an arbitrary number of unary relation symbols, and in which no term equation contains occurrences of more than one of the symbols $f_1, .., f_k$. Let $\Phi_* = \bigcup_{k \geqslant 0} \Phi_k$.

Theorem 1. $\mathrm{NTIME}(n) = MOD^{DNC}(\exists \overline{f} \, \exists \overline{R} \, \forall x \, \Phi_*)$.

The proof is given in the next subsections. We first show in Subsection 3.1 that every context–free language is the DNC–model set of a formula in $\exists f \, \exists \overline{R} \, \forall x \, \Phi_1$. Together with the fact that every language in $\mathrm{NTIME}(n)$ is the image under a projection of the intersection of three context–free languages (c.f., [3]) this proves the inclusion from left to right (even with three function variables), see Proposition 2. For the other direction, in Subsection 3.2 (Proposition 3), we first transform every formula in $\exists \overline{f} \, \exists \overline{R} \, \forall x \, \Phi_*$ into one whose atoms are all of a very restricted form, without changing its DNC–model set. Finally, in Subsection 3.3 (Proposition 4), we construct, for every such formula, a nondeterministic, linear time Turing machine which evaluates the formula on its input string.

3.1   Expressing Derivations

Concerning context–free grammars, we use the standard notation of [11]. Let us briefly recall the notion of a *derivation tree* for a context–free grammar $G = (V, \Sigma, P, S)$. Every $\sigma \in \Sigma$ is a derivation tree for $\sigma \Rightarrow_G^* \sigma$. If $\alpha_i \in \Sigma \cup V$, $A \to \alpha_1 \cdots \alpha_r$ is a production in P, and T_i is a derivation tree for $\alpha_i \Rightarrow_G^* u_i \in \Sigma^+$, then $A(T_1, .., T_r)$ is a derivation tree for $A \Rightarrow_G^* u_1 \cdots u_r$, where by $A(T_1, .., T_r)$ we denote the ordered tree whose root is labelled with A and has r subtrees $T_1, .., T_r$.

Proposition 1. *If a language $L \subseteq \Sigma^+$ is context–free, then $L = Mod^{DNC}(\psi)$ for a τ_Σ–formula $\psi \in \exists f \, \exists \overline{R} \, \forall x \, \Phi_1$.*

Proof. As the constructions of the proof of Lemma 1 do not essentially change the structure of the FO formulas, it suffices to find a formula ψ such that $L = Mod^{NNC}(\psi)$. Let L be generated by some grammar $G = (V, \Sigma, P, S)$ in *quadratic double Greibach normal form*, i.e., by a grammar whose productions are of the form $A \to \alpha$ where $\alpha \in \Sigma \cup \Sigma\Sigma \cup \Sigma V \Sigma \cup \Sigma VV \Sigma$ ([2], Theorem 3.3). We have to find a formula ψ such that for all $w \in \Sigma^+$ we have $\underline{w} \models^{NNC} \psi$ iff $S \Rightarrow_G^* w$.

Let $w \in L$ be of length n, and let T be a derivation tree for $S \Rightarrow_G^* w$. As G is in quadratic double Greibach normal form, the leftmost and the rightmost child of each internal node of T are leaves, labelled with terminal symbols; and every position in w corresponds to a leftmost or to a rightmost child of an internal node of T. We can thus represent T by an nnc-function f^T and sets $Q_{A \to \alpha}^T$ for each production $A \to \alpha$ in P as follows: For all $j \in [n]$ we define

$$f^T(j) \quad := \quad \begin{cases} i \text{ if } i \text{ corresponds to the leftmost, } j \text{ to the right-} \\ \quad \text{most child of the same internal node of } T, \\ \\ j \text{ if no such } i \text{ exists.} \end{cases}$$

$Q_{A \to \alpha}^T(j) \iff j$ corresponds to the rightmost child of an internal node of T which is associated to the production $A \to \alpha$, i.e., which is labelled with A, has exactly $|\alpha|$ children, the ith of which is labelled with the ith letter in α (for all $1 \leqslant i \leqslant |\alpha|$).

As one can easily see, $\langle \underline{w}, f^T, (Q_{A \to \alpha}^T)_{(A \to \alpha) \in P} \rangle$ satisfies the formula

$$\varphi_{tree} := \forall x \; \varphi_{disjoint} \wedge \varphi_{start} \wedge \bigwedge_{(A \to \alpha) \in P} (Q_{A \to \alpha}(x) \to \varphi_{A \to \alpha}), \text{ where}$$

$$\varphi_{disjoint} := \bigwedge_{q' \neq q \in P} \neg (Q_q(x) \wedge Q_{q'}(x))$$

$$\varphi_{start} := x{=}max \; \to \; (fx{=}min \wedge \bigvee_{(S \to \alpha) \in P} Q_{S \to \alpha}(x))$$

$$\varphi_{A \to \sigma} := W_\sigma(x) \wedge fx{=}x$$

$$\varphi_{A \to \sigma\tau} := W_\sigma(fx) \wedge W_\tau(x) \wedge fx{=}px{\neq}x$$

$$\varphi_{A \to \sigma B\tau} := W_\sigma(fx) \wedge W_\tau(x) \wedge fx{=}pfpx{\neq}fpx \wedge \bigvee_{(B \to \beta) \in P} Q_{(B \to \beta)}(px)$$

$$\varphi_{A \to \sigma CB\tau} := W_\sigma(fx) \wedge W_\tau(x) \wedge fx{=}pfpfpx{\neq}fpfpx \wedge$$
$$(\bigvee_{(B \to \beta) \in P} Q_{B \to \beta}(px)) \wedge (\bigvee_{(C \to \gamma) \in P} Q_{C \to \gamma}(pfpx)).$$

We thus obtain $\underline{w} \models^{NNC} \psi$, where $\psi := \exists f (\exists Q_q)_{q \in P} \; \varphi_{tree}$. For the opposite direction let w be a string of length n, let f be an nnc-function on $[n]$, and let $(Q_{A \to \alpha})_{(A \to \alpha) \in P}$ be subsets of $[n]$ such that φ_{tree} is satisfied by $\langle \underline{w}, f, (Q_{A \to \alpha})_{(A \to \alpha) \in P} \rangle$. We have to show that $S \Rightarrow_G^* w$, i.e., that $w \in L$. For all $j \in [n]$ such that $Q_{A \to \alpha}(j)$ for some $(A \to \alpha) \in P$ we define a tree $T(j)$ as follows:

$$\begin{aligned} &\text{If} \quad Q_{A \to \sigma}(j) \quad \text{then } T(j) := A(\sigma), \\ &\text{if} \quad Q_{A \to \sigma\tau}(j) \quad \text{then } T(j) := A(\sigma, \tau), \\ &\text{if} \quad Q_{A \to \sigma B\tau}(j) \quad \text{then } T(j) := A(\sigma, T(p(j)), \tau), \\ &\text{if} \quad Q_{A \to \sigma CB\tau}(j) \quad \text{then } T(j) := A(\sigma, T(pfp(j)), T(p(j)), \tau). \end{aligned}$$

By a straightforward induction on the depth of $T(j)$ one can easily show that if $Q_{A \to \alpha}(j)$, then $T(j)$ is a derivation tree for $A \Rightarrow_G^* w_{f(j)} \cdots w_j$. As φ_{start}

guarantees that $f(n) = 1$ and that $Q_{S \to \alpha}(n)$ for some production $S \to \alpha$ in P, we conclude that $T(n)$ is a derivation tree for $S \Rightarrow_G^* w$, and hence $w \in L$.   □

Let Σ and $\tilde{\Sigma}$ be alphabets. A mapping $h : \tilde{\Sigma} \to \Sigma$ is called a *projection*. We extend h to map strings in the canonical way: $h(w_1 \cdots w_n) := h(w_1) \cdots h(w_n)$. If $L \subseteq \tilde{\Sigma}^*$ then the set $h(L) := \{h(w) \, / \, w \in L\}$ is called the *projection* of L under h.[2]

Theorem 2 ([3], Theorem 4.1). *A language L is in* NTIME(n) *if and only if L is a projection of the intersection of three context-free languages.*

Hence, from Proposition 1 we obtain the following:

Proposition 2. *If a language $L \subseteq \Sigma^+$ is in* NTIME(n)*, then $L = \mathrm{Mod}^{DNC}(\psi)$ for a τ_Σ-formula $\psi \in \exists f_1 f_2 f_3 \, \exists \overline{R} \, \forall x \, \Phi_3$.*

Proof. From Theorem 2 we obtain an alphabet $\tilde{\Sigma}$, context–free languages $L_1, L_2, L_3 \subseteq \tilde{\Sigma}^+$, and a projection $h : \tilde{\Sigma} \to \Sigma$ such that $L = h(L_1 \cap L_2 \cap L_3)$. W.l.o.g., $\Sigma \cap \tilde{\Sigma} = \emptyset$. Proposition 1 provides formulas $\psi_i \in \exists f \, \exists \overline{R} \, \forall x \, \Phi_1$ over the signature $\tau_{\tilde{\Sigma}}$ such that $L_i = \mathrm{Mod}^{DNC}(\psi_i)$ for $i \in \{1, 2, 3\}$. We define a formula

$$\psi := (\exists W_{\tilde{\sigma}})_{\tilde{\sigma} \in \tilde{\Sigma}} \left(\psi_1 \wedge \psi_2 \wedge \psi_3 \right) \wedge \left(\forall x \bigvee_{\tilde{\sigma} \in \tilde{\Sigma}} \left(W_{\tilde{\sigma}}(x) \wedge W_{h(\tilde{\sigma})}(x) \wedge \bigwedge_{\tilde{\sigma}' \neq \tilde{\sigma}} \neg W_{\tilde{\sigma}'}(x) \right) \right),$$

which holds for a string $w \in \Sigma^+$ iff there exists a string $\tilde{w} \in \tilde{\Sigma}^+$ of the same length, such that $\tilde{w}$ satisfies ψ_1, ψ_2, and ψ_3, and $h(\tilde{w}) = w$. Hence we obtain $\mathrm{Mod}^{DNC}(\psi) = h(L_1 \cap L_2 \cap L_3) = L$. Furthermore, ψ can easily be transformed into a formula in $\exists f_1 f_2 f_3 \, \exists \overline{R} \, \forall x \, \Phi_3$.   □

3.2   Simplifying Formulas

In this subsection we will prove the following proposition.

Proposition 3. *For every formula $\psi \in \exists f_1 \cdots f_k \, \exists \overline{R} \, \forall x \, \Phi_k$ there is a formula $\psi' \in \exists f_1 \cdots f_k \, \exists \overline{R} \, \forall x \, \Phi_k$ such that $\mathrm{Mod}^{DNC}(\psi') = \mathrm{Mod}^{DNC}(\psi)$, and the atoms of ψ' are of the following forms:*

- *$x{=}min$, $x{=}max$, $f_i x{=}f_i^q px$ (where $i \in [k]$ and $q \geqslant 0$),*
- *$Q(x)$, $Q(sx)$, $Q(px)$, $Q(f_i x)$ (where $i \in [k]$ and Q is a unary relation symbol).*

Proof. (sketch) The proof proceeds in several steps:

1. We replace every equational atom $tx{=}t'x$ by a new relational atom $Q_{\{t,t'\}}(x)$ (with the intended meaning that $Q_{\{t,t'\}}(j) \leftrightarrow t(j) = t'(j)$), and we replace $t\mu = y$ (for $\mu \in \{min, max\}$ and an arbitrary term y) by the new relational atom $Q_{t\mu}(y)$ (with the intended meaning that $Q_{t\mu}(j) \leftrightarrow t(\mu) = j$).

---

[2] In the literature, projections are sometimes called *length-preserving homomorphisms*.

2. We eliminate all predicates $Q_{\{t,t'\}}$, where $\{t,t'\} \neq \{f_i, f_i^q p\}$ by replacing them with adequate formulas. (In this step we make essential use of the special properties of dnc-functions).

3. We take the conjunction of the resulting formula with formulas that define the predicates $Q_{\{f_i, f_i^q p\}}$ and $Q_{t\mu}$. After this step, all equations of the resulting formula are of the desired form.

4. We replace every relational atom $Q(tx)$ by a new relational atom $Q_t(x)$ (with the intended meaning that $Q_t(j) \leftrightarrow Q(t(j))$) and take the conjunction of the resulting formula with formulas that define the predicates Q_t.

5. A similar construction is performed to replace atoms of the form $Q(tmin)$ and $Q(tmax)$.

More details can be found in the full version of the paper.   □

3.3   Evaluating Formulas

We now conclude the proof of Theorem 1 by showing how a nondeterministic Turing machine can evaluate formulas of the form given in Proposition 3.

Proposition 4. *Let ψ be of the form $\exists f_1 \cdots \exists f_k \exists R_1 \cdots \exists R_m \forall x\, \varphi$, where φ is a quantifier–free formula in which all atoms are of one of the following forms:*

- *$x{=}min$, $x{=}max$, $f_i x{=}f_i^q px$ (where $i \in [k]$ and $q \geqslant 0$),*
- *$Q(x)$, $Q(sx)$, $Q(px)$, $Q(f_i x)$ (where $i \in [k]$ and Q is a unary relation symbol).*

Then there is a nondeterministic, linear time Turing machine which accepts precisely those strings w for which $\underline{w} \models^{DNC} \psi$.

Proof. The machine, M, will scan its input from left to right, guessing the values of all the relations, and evaluating φ accordingly. Since writing down (and reading) the values $f_i(j)$ would take too long ($\Omega(\lg n)$ steps per input position) M represents these values indirectly, by the movements of pushdown heads. To this end, we equip M with k pushdown tapes, one for each of the function variables f_i. When scanning the input at position j, pushdown tape i will hold information about $f_i(j), f_i^2(j), .., 1$, in this order. More precisely, on input w, M proceeds in $n = |w|$ "metasteps", where in the jth metastep it looks at the jth input symbol w_j. It maintains three variables, P^-, P, P^+, where $P^- = (\langle r_1^-, .., r_m^- \rangle, \sigma^-)$ contains the information about the previous position, $j{-}1$. r_l^- is intended to be the truth value of R_l, σ^- the input symbol, w_{j-1}, at that position. P and P^+ contain the same information about the current and the next position, respectively. The details are given in the full paper.   □

Remark 1. It should be noted that in the proof of Proposition 4 M scans its input only once, from left to right, and uses as many pushdown tapes as ψ has function variables. In particular, for $k = 1$, the formula can be evaluated by a pushdown automaton, i.e, the language is context–free.

4   Discussion and Further Results

A logical characterisation of a complexity class can expose what is typical for that class. Our characterisation shows that, in some sense, non–crossing functions capture the essence of linear time on nondeterministic Turing machines. NTIME(n) allows some variation in the precise definition of Turing machines. This is reflected in the logic: we can vary both our syntactical and semantical restrictions quite considerably, without changing the class of model sets. Some of these variations lead to interesting insights: we obtain a strict hierarchy of classes within NTIME(n) and a characterisation of the class by an Ehrenfeucht game, which looks considerably easier to play for the duplicator than the game one obtains directly from Theorem 1.

4.1   Variations

The following proposition subsumes several possible variations of Theorem 1, where by Φ'_k we denote the class of all Boolean combinations of atoms of the forms $x=min$, $x=max$, $f_i x=x$, $f_i x=f_i^q px$, $Q(x)$, $Q(gx)$, for $g \in \{s, p, f_1, .., f_k\}$ and unary relation symbols Q.

Proposition 5. *For $F \in \{DNC, NNC\}$ we have*

$$\text{NTIME}(n) = MOD^F(\exists \overline{f} \, \forall x \, \Phi_*) = MOD^F(\exists \overline{f} \, \exists \overline{R} \, \forall x \, \Phi_*) = MOD^F(\exists f_1 f_2 f_3 \, \exists \overline{R} \, \forall x \, \Phi'_3).$$

From the results of [13] one obtains a different logic for NTIME(n). There, over the relational signature $\{<, W_\sigma \mid \sigma \in \Sigma\}$, it was shown that[3] CFL $=$ $MOD^{MATCH}(\exists M \, FO)$, where $MATCH$ is essentially the class of all injective, partial dnc-functions. Together with the theorem of Book and Greibach (Theorem 2), it follows that a language L is in NTIME(n) iff $L = Mod^{MATCH}(\psi)$ for a formula $\psi := \exists M_1 \exists M_2 \exists M_3 \exists R_1 \cdots \exists R_m(\varphi_1 \wedge \varphi_2 \wedge \varphi_3)$, where $<$ and M_i are the only binary relation symbols occurring in the FO–formula φ_i (for $i \in \{1, 2, 3\}$).

4.2   Separations

As stated in Remark 1, the number of function symbols corresponds to the number of pushdown tapes needed for evaluating a formula (with respect to the function class $F \in \{DNC, NNC\}$). As a consequence, we obtain a strict hierarchy of classes from $k=0$ to $k=3$ by restricting the number k of function variables, as illustrated in the following picture. The class $MOD^F(\exists f_1 f_2 \, \exists \overline{R} \, \forall x \, \Phi_2)$ is separated from CFL by the language $\{ww \mid w \in \{0,1\}^+\}$, and NTIME($n$) is separated from $MOD^F(\exists f_1 f_2 \, \exists \overline{R} \, \forall x \, \Phi_2)$ by the result of [15], which says that the language SMT (*sparse matrix transposition*) cannot be accepted with two pushdown tapes in time $o(n \log n)$, but with three pushdown tapes it can (even deterministically) be accepted in linear time.

---

[3] By CFL we denote the class of all context–free languages.

$$MOD^F(\exists \overline{f}\,\exists \overline{R}\,\forall x\,\Phi_*) = MOD^F(\exists f_1 f_2 f_3\,\exists \overline{R}\,\forall x\,\Phi_3) = NTIME(n)$$
$$|$$
$$MOD^F(\exists f_1 f_2\,\exists \overline{R}\,\forall x\,\Phi_2)$$
$$|$$
$$MOD^F(\exists f\,\exists \overline{R}\,\forall x\,\Phi_1) = \qquad CFL$$
$$|$$
$$MOD^F(\exists \overline{R}\,\forall x\,\Phi_0) = \qquad REG$$

4.3 Games

Ehrenfeucht games have been proved a useful tool for showing inexpressibility results in Finite Model Theory. For a general introduction to Ehrenfeucht games we refer to the textbooks of Immerman [12] and of Ebbinghaus and Flum [5]. Our game for NTIME(n) makes use of the idea of Ajtai and Fagin [1] that in Ehrenfeucht games for existential second-order logic the duplicator can choose the second structure after the spoiler has selected relations (or, in our case, functions) for the first structure. From each of the possible logical characterisations of NTIME(n) one can derive a corresponding Ehrenfeucht game. We are going to describe here one variant that looks particularly easy to play for the duplicator. In particular, after choosing functions and colourings, both players have to select only one position in each string. A closer inspection of the proof of Proposition 1 shows that in the characterisation of NTIME(n) we can restrict ourselves to *special* nnc–functions f which, apart from being nnc, have the following properties:

- $f(f(j)) = f(j)$ for all $j \in [n]$,
- for every j there is at most one $j' \neq j$ with $f(j') = j$, i.e., f is one–one, if we ignore loops $f(j) = j$, and
- the *width* of each arc $(f(j), j)$ is at most 2, where by *width* we denote the number of arcs lying on the surface beneath that arc. To be precise, $(f(j), j)$ has width 0 if $f(j) = j$ or $j-1$, width 1 if $f(j) = f(j-1)-1$, and width 2 if $f(j) = f(f(j-1)-1)-1$.

The game for a set L of strings consists of the following three rounds:

1. The spoiler chooses a number $m \geqslant 0$. Afterwards, the duplicator chooses a string $w \in L$. In the following, let n denote the length of w.
2. The spoiler chooses *special* nnc–functions f_1, f_2, f_3 on $[n]$, and colours w with m colours. Afterwards, the duplicator chooses a string $w' \notin L$, *special* nnc–functions f_1', f_2', f_3' on $[n']$, and m colours on w'. (Here, n' denotes the length of w'.)
3. The spoiler chooses $j' \in [n']$. Afterwards, the duplicator chooses $j \in [n]$.

The duplicator wins the game iff the following conditions are satisfied:

- $j = 1$ iff $j' = 1$, and $j = n$ iff $j' = n'$,
- the colour of position j in w is the same as the colour of position j' in w'. The corresponding statements hold for the positions $j-1$ and $j'-1$, and for $f_i(j)$ and $f_i'(j')$ (for $i \in \{1, 2, 3\}$).

- The arc $(f_i(j), j)$ has the same width as the arc $(f_i'(j'), j')$, and it is a loop iff $(f_i'(j'), j')$ is a loop (for $i \in \{1, 2, 3\}$).

Proposition 6. *A set L of strings is in* NTIME(n) *iff the spoiler has a winning strategy in the game for L.*

Acknowledgements We would like to thank Malika More and Arnaud Durand for stimulating discussions.

References

[1] M. Ajtai and R. Fagin. Reachability is harder for directed than for undirected finite graphs. *Journal of Symbolic Logic*, 55(1):113–150, 1990.

[2] J. Autebert, J. Berstel, and L. Boasson. Context-free languages and pushdown automata. *Handbook of Formal Languages*, 2:111–174, 1997.

[3] R. Book and S. Greibach. Quasi-realtime languages. *Math. Syst. Theory*, 4:97–111, 1970.

[4] J.R. Büchi. Weak second-order arithmetic and finite automata. *Zeitschrift für Mathematische Logik und Grundlagen der Mathematik*, 6:66–92, 1960.

[5] H.D. Ebbinghaus and J. Flum. *Finite Model Theory*. Springer-Verlag, New York, 1995.

[6] R. Fagin. Generalized first-order spectra and polynomial-time recognizable sets. In R.M. Karp, editor, *Complexity of Computation*, volume 7 of *SIAM-Proceedings*, pages 43–73. AMS, 1974.

[7] E. Grädel. Capturing complexity classes by fragments of second order logic. *Theoretical Computer Science*, 101:35–57, 1992.

[8] E. Grandjean. Invariance properties of RAMs and linear time. *Computational Complexity*, 4:62–106, 1994.

[9] E. Grandjean. Linear time algorithms and NP-complete problems. *SIAM Journal of Computing*, 23:573–597, 1994.

[10] E. Grandjean. Sorting, linear time and the satisfiability problem. *Annals of Mathematics and Artificial Intelligence*, 16:183–236, 1996.

[11] J. Hopcroft and J. Ullman. *Introduction to Automata Theory, Languages, and Computation*. Addison-Wesley Publishing Company, 1979.

[12] N. Immerman. *Descriptive and Computational Complexity*. Springer-Verlag, New York, 1998.

[13] C. Lautemann, T. Schwentick, and D. Thérien. Logics for context-free languages. In *Proceedings of the Annual Conference of the EACSL*, volume 933 of *Lecture Notes in Computer Science*, pages 205–216, 1994.

[14] J. F. Lynch. The quantifier structure of sentences that characterize nondeterministic time complexity. *Computational Complexity*, 2:40–66, 1992.

[15] W. Maass, G. Schnitger, E. Szemerédi, and G. Turán. Two tapes versus one for off-line turing machines. *Computational Complexity*, 3:392–401, 1993.

[16] W. Paul, N. Pippenger, E. Szemerédi, and W. Trotter. On determinism versus nondeterminism and related problems. In *Proc. 24th Ann. Symp. Found. Comput. Sci.*, pages 429–438, 1983.

[17] T. Schwentick. Padding and the expressive power of existential second-order logics. In *11th Annual Conference of the EACSL, CSL '97*, pages 461–477, 1997.

Descriptive Complexity of Computable Sequences

Bruno Durand[1], Alexander Shen[2], and Nikolai Vereshagin[3*]

[1] LIP, Ecole Normale Supérieure de Lyon, 46 Allée d'Italie, 69364 Lyon Cedex 07, France. `Bruno.Durand@ens-lyon.fr`
[2] Institute of Problems of Information Transmission, Moscow, Russia. `shen@mccme.ru`
[3] Dept. of Mathematical Logic and Theory of Algorithms, Moscow State University, Vorobjevy Gory, Moscow 119899, Russia. `ver@mech.math.msu.su`

Abstract. Our goal is to study the complexity of infinite binary recursive sequences. We introduce several measures of the quantity of information they contain. Some measures are based on size of programs that generate the sequence, the others are based on the Kolmogorov complexity of its finite prefixes. The relations between these complexity measures are established. The most surprising among them are obtained using a specific two-players game.

1 Introduction

The notion of Kolmogorov entropy (=complexity) for finite binary strings was introduced in the 60ies independently by Solomonoff, Kolmogorov and Chaitin [7,4,1]. There are different versions (plain Kolmogorov entropy, prefix entropy, etc. see [8] for the details) that differ from each other not more than by an additive term logarithmic in the length of the argument. In the sequel we are using plain Kolmogorov entropy $K(x|y)$ as defined in [4], but similar results can be obtained for prefix complexity.

When an infinite 0-1-sequence is given, we may study the entropy (=complexity) of its finite prefixes. If prefixes have high complexity, the sequence is random (see [5] for details and references); if prefixes have low complexity, the sequence is computable. In the sequel, we study the latter type.

Let $K(x)$, $K(x|y)$ denote the plain Kolmogorov entropy (complexity) of a binary string x and the conditional Kolmogorov entropy (complexity) of x when y (some other binary string) is known. That is, $K(x)$ is the length of the shortest program p that prints x; $K(x|y)$ is the length of the shortest program that prints x given y as input. (For details see [5] or [9].)

Let $\omega_{1:n}$ denote first n bits (= n-prefix) of the sequence ω.

Let us recall the following criteria of computability of ω in terms of entropy of its finite prefixes.

---

* The work was done while visiting LIP, Ecole Normale Supérieure of Lyon.

C. Meinel and S. Tison (Eds.): STACS'99, LNCS 1563, pp. 153–162, 1999.

(a) ω is computable if and only if $K(\omega_{1:n}|n) = \mathcal{O}(1)$. This result is attributed in [6] to A.R. Meyer (see also [9,5]).
(b) ω is computable if and only if $K(\omega_{1:n}) < K(n) + \mathcal{O}(1)$ [2].
(c) ω is computable if and only if $K(\omega_{1:n}) < \log_2 n + \mathcal{O}(1)$ [2].

These results provide criteria of the computability of infinite sequences. For example, (a) can be reformulated as follows: sequence ω is computable if and only if $M(\omega)$ is finite, where

$$M(\omega) = \max_n K(\omega_{1:n}|n) = \max_n \min_p \{l(p) \mid p(n) = \omega_{1:n}\}.$$

($l(p)$ stands for the length of program p; $p(n)$ denotes its output on n).

Therefore, $M(\omega)$ can be considered as a complexity measure for ω: $M(\omega)$ is finite iff ω is computable.

Another straightforward approach is to define entropy (complexity) of a sequence ω as the length of the shortest program computing ω:

$$K(\omega) = \min\{l(p) \mid \forall n \ p(n) = \omega_{1:n}\},$$

(and by definition $K(\omega) = \infty$ if ω is not computable.)

The difference between $K(\omega)$ and $M(\omega)$ can be explained as follows: $M(\omega) \leq m$ means that for every n there is a program p_n of size at most m that computes $\omega_{1:n}$ given n; this program may depend on n. On the other hand, $K(\omega) \leq m$ means that there is a one such program that works for all n. Thus, $M(\omega) \leq K(\omega)$ for all ω, and one can expect that $M(\omega)$ may be significantly less than $K(\omega)$. (Note that the known proofs of (a) give no bounds of $K(\omega)$ in terms of $M(\omega)$.)

Indeed, theorem 3 shows that there is no computable bound for $K(\omega)$ in terms of $M(\omega)$: for any computable function $\alpha(m)$ there exist computable infinite sequences $\omega^0, \omega^1, \omega^2 \ldots$ such that $M(\omega^m) \leq m + \mathcal{O}(1)$ and $K(\omega^m) \geq \alpha(m) - \mathcal{O}(1)$.

The situation changes surprisingly when we compare "almost all" versions of $K(\omega)$ and $M(\omega)$ defined in the following way:

$$K_\infty(\omega) = \min\{l(p) \mid \forall^\infty n \ p(n) = \omega_{1:n}\}$$
$$M_\infty(\omega) = \limsup_n K(\omega_{1:n}|n) = \min\{m \mid \forall^\infty n \exists p \ (l(p) \leq m \text{ and } p(n) = \omega_{1:n})\},$$

($\forall^\infty n$ stands for "for all but finitely many n"). It is easy to see that $M_\infty(\omega)$ is finite only for computable sequences. Indeed, if $M_\infty(\omega)$ is finite, then $M(\omega)$ is also finite, and the computability of ω is implied by Meyer's theorem.

Surprisingly, it turns out that $K_\infty(\omega) \leq 2M_\infty(\omega) + \mathcal{O}(1)$ (theorem 5) so the difference between K_∞ and M_∞ is not so large as between K and M. We stress that this result is rather strange because a multiplicative constant 2 appears, and has no intuitive meaning taking into account that all the six complexity measures ("entropies") mentioned above are "well calibrated" in the following sense: there are $\Theta(2^m)$ sequences whose entropy does not exceed m. –In the general theory of Kolmogorov complexity, additive constants often appear, but not multiplicative ones. As theorem 6 shows, *this bound is tight*.

It is interesting also to compare K_∞ and M_∞ with K and M, as well as with relativized versions of K. For any oracle A one may consider a relativized Kolmogorov complexity K^A allowing programs to access the oracle. Then $K^A(\omega)$ is defined in a natural way. By $K'(\omega)$ [or $K''(\omega)$] we mean $K^A(\omega)$ where $A = \mathbf{0}'$ [or $\mathbf{0}''$]. The results of this comparison are shown by a diagram (Fig. 1).

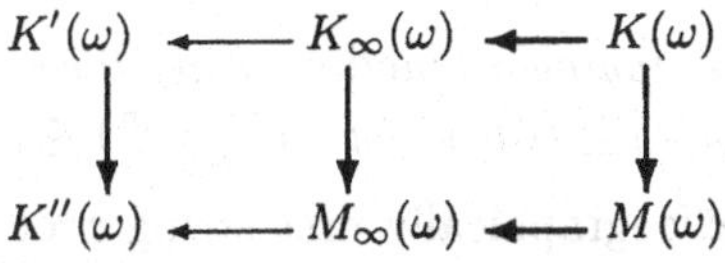

$$
\begin{array}{ccccc}
K'(\omega) & \longleftarrow K_\infty(\omega) & \longleftarrow K(\omega) \\
\downarrow & \downarrow & \downarrow \\
K''(\omega) & \longleftarrow M_\infty(\omega) & \longleftarrow M(\omega)
\end{array}
$$

Fig. 1. Relations between different complexity measures for infinite sequences

Arrows go from the bigger quantity to the smaller one (up to $\mathcal{O}(1)$-term, as usual). Bold arrows indicate inequalities that are immediate consequences of the definitions. Other arrows are provided by Theorem 1 ($K'(\omega) \le K_\infty(\omega) + \mathcal{O}(1)$) and Theorem 4 ($K''(\omega) \le M_\infty(\omega) + \mathcal{O}(1)$).

As we have said, $K_\infty(\omega) \le 2M_\infty(\omega) + \mathcal{O}(1)$, so K_∞ and M_∞ differ only by a bounded factor. If we ignore such a difference, we get a simplified diagram

$$K''(\omega) \longleftarrow K'(\omega) \longleftarrow K_\infty(\omega), M_\infty(\omega) \longleftarrow M(\omega) \longleftarrow K(\omega)$$

where $X \leftarrow Y$ means that $X = \mathcal{O}(Y)$.

On the last diagram no arrow could be inverted. Indeed, $K''(\omega)$ is finite while $K'(\omega)$ is infinite for a sequence ω that is $\mathbf{0}''$-computable but not $\mathbf{0}'$-computable. Therefore the first arrow cannot be inverted. The second one cannot be inverted for similar reasons: $K'(\omega)$ is finite while $K_\infty(\omega)$ and $M_\infty(\omega)$ are infinite for a sequence that is $\mathbf{0}'$-computable but not computable. Theorem 2 shows that $K_\infty(\omega)$ and $M_\infty(\omega)$ could be small while $M(\omega)$ is large. Finally, Theorem 3 shows that $M(\omega)$ could be small while $K(\omega)$ is large.

These diagrams and the statements we made about them do not tell us whether the inequalities $K_\infty(\omega) \le M(\omega) + \mathcal{O}(1)$ and $K'(\omega) \le M_\infty(\omega) + \mathcal{O}(1)$ are true. The first one is not true, as Theorem 6 implies. We don't know whether the second one is true.

Other open questions: (1) is it possible to reverse the second arrow ($K'(\omega) \leftarrow K_\infty(f), M_\infty(f)$) for *computable* sequences? (2) what can be said about similar notions for finite strings? in particular, is $\limsup_n K(x|n)$ equal to $K'(x) + \mathcal{O}(1)$ or not?[1]

2   Theorems and Proofs

Theorem 1. $K'(\omega) < K_\infty(\omega) + \mathcal{O}(1)$.

Proof. Let $p(n) = \omega_{1:n}$ for almost all n. The following program q (with access to $\mathbf{0}'$) computes $\omega_{1:n}$ given n: For $k = n, n+1, \ldots$ find out (using $\mathbf{0}'$) whether

---

[1] It was shown recently by the third author that $\limsup_n K(x|n) = K'(x) + \mathcal{O}(1)$.

(a) $p(k)$ is defined and is a binary string of length k; (b) $p(m)$ is consistent with $p(k)$ for all $m > k$; consistency means that either $[p(m)$ has length m and has prefix $p(k)]$ or $[p(m)$ is undefined]. As soon as k satisfying both (a) and (b) is found, print the first n bits of $p(k)$.

Obviously, $q(n) = \omega_{1:n}$ for all n and the bit length of q is $\mathcal{O}(1)$ longer than that of p. $\square$

Theorem 2. *For any computable function $\alpha(m)$ there exist infinite sequences $\omega^0, \omega^1, \ldots$ such that $M(\omega^m) \geq \alpha(m)$ while $K_\infty(\omega^m) \leq m + \mathcal{O}(1)$.*

Proof. Let x_m be the lexicographically first string x of length $\alpha(m)$ such that $K(x|\alpha(m)) \geq \alpha(m)$. (Such a string exists since the number of programs of length less than k is less than 2^k.)

Now let $\omega^m = x_m 0000 \ldots$. By definition, $M(\omega^m) \geq K(x_m|\alpha(m)) \geq \alpha(m)$. On the other hand, $K_\infty(\omega^m) \leq m + \mathcal{O}(1)$. Indeed, the set $\{x \mid K(x|l(x)) < l(x)\}$ is enumerable. Consider the program p_m that having input n performs n steps of enumeration of this set. Then the program p_m finds the first string x_m^n of length m that was not encountered, and outputs first n bits of the sequence $x_m^n 0000 \ldots$. If n is large enough then $x_m^n = x_m$ and p outputs $\omega_{1:n}^m$. It remains to note that the length of p_m is $\log m + \mathcal{O}(1)$. $\square$

Theorem 3. *For any computable function $\alpha(m)$ there exist infinite sequences $\omega^0, \omega^1, \ldots$ such that $K(\omega^m) \geq \alpha(m)$ while $M(\omega^m) \leq m + \mathcal{O}(1)$.*

Proof. Let c be a constant (to be specified later). The set $E = \{\langle x, k \rangle \mid K(x) < \alpha(k) + c\}$ is enumerable. Consider the process of its enumeration. Let $s(m)$ be the time (step number) when all pairs of type $\langle x, m \rangle$ with a given m have been appeared in E. Now let $\omega^m = 0^{s(m)} 1111 \ldots$.

Let us prove that $K(\omega^m) > \alpha(m) - \mathcal{O}(1)$. Assume that $p(n) = \omega_{1:n}^m$ for all n. Given p we can find the first 1 in ω^m and hence $s(m)$. Thus $K(s(m)) \leq K(\omega^m) + \mathcal{O}(1)$. On the other hand, given $s(m)$ we can find the (lexicographically) first string x_m of entropy $\alpha(m)$ or more, therefore, $\alpha(m) \leq K(x_m) \leq K(s(m)) + \mathcal{O}(1)$. Hence $\alpha(m) \leq K(\omega^m) + \mathcal{O}(1)$.

Let us prove now that $M(\omega^m) \leq m + \mathcal{O}(1)$. Let the program q on input n output n zeros. Then $q(n) = \omega_{1:n}^m$ for all $n \leq s(m)$.

Consider the program p_m that on input n does n steps of enumeration of the set E, finds the number $s(m,n)$ of the last step among them when a new pair of type $\langle x, m \rangle$ with a given m has been appeared, and then outputs the first n bits of the sequence $0^{s(m,n)} 1111111 \ldots$. If $n \geq s(m)$, then p_m outputs the correct prefix of ω^m.

Thus, for any n, either p_m or q (given n) outputs $\omega_{1:n}^m$. It remains to note that the length of p_m is $\log m + \mathcal{O}(1)$. $\square$

These theorems 2 and 3 can be reinforced using a technique presented in [3]: they are true for any computable infinite family of distinct sequences $\omega^0, \omega^1, \ldots$ (the family itself should be computable). Anyways these *pathological* cases are rare: the difference between $K(x)$ and $K''(x)$ can be huge but this concerns only an exponentially small portion of strings x of a given size.

Theorem 4.
$$K''(\omega) < M_\infty(\omega) + \mathcal{O}(1).$$

Proof. Let $m = M_\infty(\omega) + 1$. Consider the set $T = \{x \mid K(x|l(x)) < m\}$. By definition, all sufficiently long prefixes of ω belong to T. The set T is enumerable. For each n there are at most 2^m strings of length n in T. A string $x \in T$ is called "good" if there is a sequence ξ such that x is a prefix of ξ and all prefixes of ξ longer than x belong to T (in other words, if x lies on the infinite path in T). It is easy to see that König's lemma allows to express the statement "x is good" as $\forall\exists$-statement. Therefore, the set $\overline{T}$ of all good strings is $\mathbf{0}''$-decidable.

This set can be represented as an union of non-overlapping infinite paths: consider all the strings in order of increasing length; if a string in $\overline{T}$ is found that is not already included in one of the paths, take a path that starts with it (if there are many of them, choose the lexicographically first, i.e., turn to the left when possible). The number of different paths does not exceed 2^m. This decomposition process is $\mathbf{0}''$-effective, i.e., there is an $\mathbf{0}''$-algorithm that gives k-bit prefix of path number i for given k and i. Appending i (considered as m-bit string) to that algorithm, we get a $\mathbf{0}''$-program that gives k-bit prefixes of i-th path for all k (this program needs also m to construct T and $\overline{T}$, but m is given implicitly as the length of i). Since one of the paths goes along ω, we conlude that $K''(f) \leq m + \mathcal{O}(1) = M_\infty(\omega) + \mathcal{O}(1)$. $\square$

The next two theorems provide the connection between K_∞ and M_∞.

Theorem 5. $K_\infty(\omega) < 2M_\infty(\omega) + \mathcal{O}(1)$.

Theorem 6. *There is a sequence ω^m of infinite strings such that $M(\omega^m) \leq m + \mathcal{O}(1)$ and $K_\infty(\omega^m) \geq 2m$ (hence $M_\infty(\omega^m), M(\omega^m) = m + \mathcal{O}(1), K_\infty(\omega^m) = 2m + \mathcal{O}(1)$).*

Proof. (The original proof of theorem 5 was simplified significantly by An. A. Muchnik.) First, let us define a game that is relevant to both theorems 5 and 6 and may be interesting in its own right.

Let k, l be integer parameters. The (k,l)-*game* is played by two players called the Man (M) and the Nature (N). On its moves, N builds a binary rooted tree. More specifically, during its move N adds a binary string to a finite set T (initally empty). On his moves, M may color certain binary strings using colors from the set $\{1, 2, \ldots, l\}$ (several colors may be attached to the same string; attached colors cannot be removed later).

The game stops after a finite number of moves if

(1) T is not a tree (that is, there are $x \in T$ and $y \notin T$ such that y is a prefix of x); in this case M wins, or
(2) for some n the number of strings of length n in T (the number of nodes having depth n) exceeds k; in this case M also wins, or
(3) there are two different strings of the same length colored by the same color; in this case N wins.

Otherwise the game lasts indefinitely long, and the winner is determined as follows. Let T be the ultimate tree (formed by all strings included in T at all steps). An infinite 0-1-sequence is called an *infinite branch* of T if $\omega_{1:n} \in T$ for all n.

M wins if for any infinite branch β there exists a color c such that all but finitely many nodes of β are colored by c (and, may be, by other colors). Otherwise N wins.

(One may give the following interpretation to this game. The tree built by Nature is the tree of all breeds of animals, and nodes at height n are breeds existing at time n. The coloring is giving names to breeds. Thus Man is required to give stable names to all eternal breeds.)

We will use also a modified version of this game where the rule (1) is omitted and the definition of an infinite branch is changed as follows: sequence ω is an infinite branch if all but finitely many prefixes of ω are in T. (Obviously, the modified game is more difficult for M than the original one.)

The following two lemmas play a key role in the proof of theorems 5 and 6.

Lemma 1. *For any k, there is a computable winning strategy for M in the modified (k, k^2)-game (the winning algorithm has k as an input).*

Lemma 2. *N has a computable winning strategy in the (k, l)-game if $l < k^2/4$.*

Before proving these lemmas, let us finish the proof of theorems 5 and 6 using them.

Theorem 5 requires us to prove that $K_\infty(\omega) < 2M_\infty(\omega) + \mathcal{O}(1)$.

Fix ω. Let $T = \{x \mid \mathrm{K}(x|l(x)) \le M_\infty(\omega)\}$. Then for any n the set T has no more than $k = 2^{M_\infty(\omega)+1}$ strings of length n. According to our assumption, $\omega_{1:n} \in T$ for all but finitely many n. Thus ω is an infinite branch in T. Consider now the following strategy for N in modified (k, k^2)-game: N just enumerates T (ignoring M's replies). M can defeat this strategy using his computable strategy that exists according to lemma 1.

Since both M and N are using computable strategies, the set $C = \{\langle x, p \rangle \mid$ node x gets color p at some stage$\}$ is enumerable. As M wins, there is a color p that is attached to $\omega_{1:n}$ for all sufficiently large n. Each color can be considered as binary string of length $2(M_\infty(\omega) + 1)$, since there are at most k^2 colors.

The following algorithm computes $\omega_{1:n}$ given n and p. First find the value $k = 2^{M_\infty(\omega)+1} = 2^{l(p)/2}$. Second, enumerate C until a pair $\langle x, p \rangle$ appears with $l(x) = n$, i.e., until some node x having depth n gets color p. Then return x. For all sufficiently large n this algorithm will return $\omega_{1:n}$ (since the infinite branch ω has color p assigned).

The program q to compute $\omega_{1:n}$ given n for almost all n consists of the above algorithm with the string p appended. Thus, the length of q is $2M(\omega) + \mathcal{O}(1)$, and the theorem 5 (modulo lemma 1) is proved.

Now let us derive theorem 6 from lemma 2. We need to prove that there exist infinite sequences $\omega^0, \omega^1, \dots$ such that $M(\omega^m) \le m + \mathcal{O}(1)$ and $K_\infty(\omega^m) \ge 2m$.

For any fixed m consider the following strategy for M. He enumerates all triples $\langle p, n, x \rangle$ such that $p(n) = x$; if it turns out that $l(x) = n$ and $l(p) < 2m$, he assigns color p to string x. This strategy may be performed by an algorithm having m as an input.

Let $k = 2^{m+1}$, $l = 2^{2m} - 1$. Since $l < k^2/4$, the lemma 2 guarantees that N could defeat this strategy using its own computable strategy. Therefore, there exists an algorithm A that given m generates a tree T^m which has an infinite branch ω that is not properly colored, i.e., there is no p of length less than $2m$ such that $p(n) = \omega_{1:n}$ for almost all n. In other words, $K_\infty(\omega) \geq 2m$.

On the other hand, $M(\omega) \leq m + \mathcal{O}(1)$. Indeed, let n be a natural number. Let us describe a program of size $m + \mathcal{O}(1)$ that computes $\omega_{1:n}$. Consider an algorithm B that for a given string q of length $m + 1$ and for any n uses A to generate T^m and waits until q nodes (here q is identified with its ordinal number among all strings of length $m + 1$) at height n appear. Then B outputs the node that appeared last. Since $\omega_{1:n} \in T^m$, for some q the output will be equal to $\omega_{1:n}$. The string q appended to B constitutes a program to compute $\omega_{1:n}$ given n. This program has size $m + \mathcal{O}(1)$.

Theorem 6 is proved (modulo lemma 2)

Now we have to prove lemmas 1 and 2.

Recall that lemma 1 says that for any k, there is a computable winning strategy for M in the modified (k, k^2)-game (the winning algorithm has k as an input).

Proof. (Using An. Muchnik's argument.) Let M use k^2 colors indexed by pairs (a, b), where a and b are natural numbers in range $1..k$. Let us explain how the color (a, b) is assigned. (Different colors are assigned independently.) Observing the growing set T, M looks for all pairs of strings u and v such that:

(a) u has number a if we count all the (already appeared) strings in T in the lexicographic order;
(b) v has number b if we count all the (already appeared) strings in T in the reverse lexicographic order;
(c) u is a prefix of v.

After such a pair of strings is found, any prefix of u gets color (a, b) unless some other string of the same length already has this color (and M is prohibited to use (a, b) again on that level). Then M looks for another pair of strings u and v with the same properties, etc.

We need to prove that this strategy guarantees that any infinite branch will be colored uniformly starting at some point. Let T be the set of all strings that N gives (at all steps). Let ω be an infinite branch, so $\omega_{1..n} \in T$ for all sufficiently large n. For these n let a_n denote the lexicographic number of $\omega_{1..n}$ in the set T_n of all strings of length n that are in T, and let b_n denote the inverse lexicographic number of $\omega_{1..n}$ in T_n. Let $a = \limsup a_n$ and $b = \limsup b_n$. We claim that for sufficiently large n the string $\omega_{1..n}$ will have color (a, b).

Indeed, consider a pair (u, v) that satisfies the conditions listed above. Let us prove first that for sufficiently long sequences only prefixes of ω have chance to get colored with color (a, b). Indeed, for large enough n we have $a_n \leq a$, so sufficiently long strings u are "on the right of ω" or are prefixes of ω. ("On the right of ω means that u follows the prefix of ω having the same length, in the lexicographic order.) For the same reasons all sufficiently long strings v are on the left of ω or are prefixes of ω. Therefore, the only chance for u to be a prefix of v (if both are long enough) is when both u and v are prefixes of ω. Therefore, no other long strings (except prefixes of ω) could get color (a, b).

According to the definition of a and b there are infinitely many n such that $a_n = a$ and infinitely many m such that $b_m = b$. Choose a pair of such n and m; assume that $n \leq m$. The strings $u = \omega_{1..n}$ and $v = \omega_{1..m}$ will be discovered after all strings of length n and m appear in the enumeration of T since they will have correct ordinal numbers. And all prefixes of u will get color (a, b) unless some other vertex of the same length already has this color. (And this is possibly only for short strings, as we have seen). Since u may be arbitrarily long, all sufficiently long prefixes of ω will get color (a, b). Lemma 1 is proved.

Lemma 2 says that N has a computable winning strategy in (k, l)-game of $l < k^2/4$.

Proof. Let $m = k/2$. First we introduce some terminology. We consider finite trees T with m distinguished leaves at the height equal to height of the tree. Those distinguished leaves are called *tops* of the tree. The m paths from the root to m tops are called *trunks* of the tree. All the nodes that belong to the trunks are called *trunk nodes*; other are called *side nodes*.

We call a tree T' an *extension of a tree* T if (a) $T \subset T'$; (b) T' does not contain new vertices on the levels that exist in T (i.e., any string is $T' - T$ is longer than any string in T); (c) all trunks of T' continue those of T (that is, jth trunk of T' continues jth trunk of T for all $j \leq m$).

First N builds any tree T_0 of width m that has m trunks. Then N continues all the m trunks of T_0 (for example, by adding, for any top v, nodes $v0$, $v00$, and so on) and waits until M starts to color nodes on the trunks (otherwise he looses). More specifically, N waits until there exists h_1 such that the nodes at height h_1 on all m trunks are colored. We call those nodes *special* ones. The colors of special nodes are be pairwise different, as the special nodes are at the same height (otherwise M looses). Let h_2 be the height of trunks when M colors the last special node ($h_2 \geq h_1$).

N has just forced M to use m different colors and has constructed a finite tree of width m. However, we wish (for the next iteration) that the nodes colored in m different colors do not belong to trunks at the expense of increasing the width of the tree by 1. This is done as follows. Once N has forced M to color m special nodes at the same height h_1, it chooses one the trunks and cuts it (this means that N will not continue that trunk). Then N takes the father of the special node on that trunk and starts from the father another trunk instead of the cut trunk. The nodes lying on the cut trunk from the height h_1 to h_2 become side nodes. Thus at least one side node is colored. Call that node a *distinguished* node. After that N still grows m trunks in parallel (continuing $m - 1$ non-cut trunks and the trunk having a branch with the distinguished node) until M colors m nodes on m trunks at a new height $h_3 > h_2$.

Call those nodes the *new special nodes*. Now N chooses a trunk whose new special node is colored in a color different from the color of the distinguished node, cuts it and starts a new trunk from its node at height $h_3 - 1$. We thus obtain the second side node colored in a color different from the color of the distiguished node. Call that side node also a distiguished node. Thus we have two distinduished side nodes having different colors.

This process is repeated m times. Each time N cuts a trunk whose special node is colored in a color different from the colors of the existing distiguished nodes (such a special node exists while the number of distinguished nodes is less than m). After m repetitions we have a tree of width $m + 1$ that has m distinguished side nodes colored in m different colors.

The described strategy will be denoted by S_1. Its starting point may be any tree T with m trunks. It either terminates and constructs an extension T' of T such that $T' - T$ is colored in m different colors, or wins. The set $T' - T$ has width $m + 1$.

Now let us describe the induction step. Assume X is a subset of a tree T. Let colors(X) [sidecolors(X)] denote the set of colors of all nodes [all side nodes] in X.

Assume we have a strategy S_i $(i < m)$ for N with the following properties. Starting from any tree T with m trunks it constructs a finite extension T' of T such that the difference $T' - T$ has width $m + i$ and $|\,\text{sidecolors}(T' - T)| \geq im$.

Our goal is to define a strategy S_{i+1} satisfying the same conditions (for increased value of i). We define first an auxilliary strategy $\tilde{S}_{i+1}$ that, starting from any tree T with m trunks, constructs a finite extension T' of T such that the difference $T' - T$ has width $m + i$, $|\,\text{colors}(T' - T)| \geq (i+1)m$, and $|\,\text{sidecolors}(T' - T)| \geq im$ (or $\tilde{S}_{i+1}$ wins).

The strategy $\tilde{S}_{i+1}$ given a tree T works as follows. Apply S_i starting from T. Wait until S_i terminates. Let T_1 be the continuation of T constructed by S_i. Then $|\,\text{sidecolors}(T_1 - T)| \geq im$. Apply S_i starting from T_1. Wait until S_i constructs a continuation T_2 of T_1 with $|\,\text{sidecolors}(T_2 - T_1)| \geq im$. Applying S_i many times, we get $T_1, T_2, T_3, \ldots$. Wait until there exist j and s such that $j \leq s$ and all the nodes along all the trunks inside $T_j - T_{j-1}$ at step s are colored and each trunk has its own color (if no such j and s exist, the startegy $\tilde{S}_{i+1}$ never terminates and wins). Let $T' = T_s$. The tree T_s has im different colors on side nodes in $T_j - T_{j-1}$ and m new colors on nodes on m trunks.

Now we are able to define the strategy S_{i+1}. Starting from a tree T it works as follows. Apply $\tilde{S}_{i+1}$ starting from T. Wait until it terminates. Let T_1 denote the resulting tree. The set colors($T_1 - T$) has at least $(i + 1)m$ colors. The problem, however, is that some of them may be used for trunk nodes only. In this case choose a trunk of T_1 that has a node colored in a color $c \in \text{colors}(T_1 - T) - \text{sidecolors}(T_1 - T)$. Let j be the number of that trunk. We add to T_1 a new branch starting from the jth top of T and declare this branch a new trunk of T_1; the old jth trunk is not a trunk anymore. This operation increases the width of

$T_1 - T$ to $m+i+1$. The gain is that the set $\mathsf{sidecolors}(T_1 - T)$ has got a new color c. So $|\mathsf{sidecolors}(T_1 - T)| \geq im+1$ now. If it happens that the set $\mathsf{sidecolors}(T_1 - T)$ already has at least $(i+1)m$ colors, we stop. Otherwise, we apply once more the strategy $\tilde{S}_{i+1}$ starting from T_1. We get T_2 such that $|\mathsf{colors}(T_2 - T_1)| \geq (i+1)m$. As $|\mathsf{sidecolors}(T_1 - T)| < (i+1)m$, the set $\mathsf{colors}(T_2 - T_1)$ has at least one color that does not belong to $\mathsf{sidecolors}(T_1 - T)$. We choose again a color c from $\mathsf{colors}(T_2 - T_1) - \mathsf{sidecolors}(T_1 - T)$, choose a trunk node in $T_2 - T_1$ colored by c, make a new trunk from the top of T_1 lying on that trunk and thus get $\mathsf{sidecolors}(T_2 - T) \geq \mathsf{sidecolors}(T_1 - T) + 1 \geq im+2$. Repeating this trick at most m times, we obtain an extension T' such that $\mathsf{sidecolors}(T' - T) \geq (i+1)m$ and the width of $T' - T$ is at most $m + i$.

The induction step is described. Note that the strategy $\tilde{S}_m$ wins in the $2m, (m^2 - 1)$-game. $\square$

References

1. G.J. Chaitin. "On the length of programs for computing finite binary sequences: statistical considerations," *J. of ACM*, 16:145–159, 1969.
2. G.J. Chaitin. "Information-theoretic characterizations of recursive infinite strings," *Theor. Comp. Sci.*, 2:45–48, 1976.
3. B. Durand, S. Porrot. Comparison between the complexity of a function and the complexity of its graph. In MFCS'98. *Lecture Notes in Computer Science*, 1998.
4. A.N. Kolmogorov. "Three approaches to the quantitative definition of information." *Problems of Information Transmission*, 1(1):1–7, 1965.
5. M. Li, P. Vitányi. An Introduction to Kolmogorov Complexity and its Applications. Second edition. Springer Verlag, 1997.
6. D.W. Loveland. "A variant of Kolmogorov concept of Complexity", *Information and Control*, 15:510–526, 1969.
7. R.J. Solomonoff. "A formal theory of inductive inference, part 1 and part 2," *Information and Control*, 7:1–22, 224-254, 1964.
8. V.A. Uspensky, A.Kh. Shen'. "Relations between varieties of Kolmogorov complexities," *Math. Systems Theory*, 29:271–292, 1996.
9. A.K. Zvonkin, L.A. Levin. "The complexity of finite objects and the development of the concepts of information and randomness by means of theory of algorithms." *Russian Math. Surveys*, 25(6):83–124, 1970.

Complexity of Some Problems
in Universal Algebra
Extended Abstract*

Clifford Bergman[1] and Giora Slutzki[2]

[1] Department of Mathematics, Iowa State University, Ames, Iowa 50011
cbergman@iastate.edu
[2] Department of Computer Science, Iowa State University, Ames, Iowa 50011
slutzki@cs.iastate.edu

Abstract. In this paper we consider the complexity of several problems involving finite algebraic structures. Given finite universal algebras **A** and **B**, these problems ask: (1) Do **A** and **B** satisfy precisely the same identities? (2) Do they satisfy the same quasi-identities? and (3) Do **A** and **B** have the same set of term operations?

In addition to the general case in which we allow arbitrary (finite) algebras, we consider each of these problems under the restrictions that all operations are unary, and that **A** and **B** have cardinality two. We briefly discuss the relationship of these problems to algebraic specification theory.

There are several relationships between mathematical structures that might be considered "fundamental". First and foremost is certainly the isomorphism relation. Questions about isomorphic structures occur throughout mathematics and apply to universal algebras, topological spaces, graphs, partially ordered sets, etc. Many other relationships are more specialized. For example, given two graphs **G** and **H**, one may wish to know whether **H** is a subgraph of **G**, or perhaps a minor of **G**.

Properly formulated, questions about these relationships give rise to complexity questions. Generally speaking, we must impose some sort of finiteness assumption on the structures in question so that notions of computational complexity make sense. The complexity of various isomorphism problems have received a great deal of attention. The graph isomorphism problem has been intensively studied, partly because its exact relationship to the classes **P** and **NP** is still unknown, and partly because it provides a paradigm for other problems of unknown complexity status. In this case, both graphs are assumed to have finitely many vertices and finitely many edges. With a similar formulation, the isomorphism problem for algebras has the same complexity as does graph isomorphism. More generally, Kozen [10] showed that the isomorphism problem

---

* The complete vesion of this paper is available from
http://www.math.iastate.edu/cbergman/papers.html

C. Meinel and S. Tison (Eds.): STACS'99, LNCS 1563, pp. 163–172, 1999.
© Springer-Verlag Berlin Heidelberg 1999

for finitely presented algebras has this same complexity. See [2, 12] for further discussion and references on the isomorphism problem.

In this paper we consider the complexity of three relationships that arise from considerations in universal algebra. Any algebraic structure satisfies certain identities and fails to satisfy others. Roughly speaking, an identity is an equality between two expressions built from the operations of the algebra. Examples of identities are the associative law (which involves one binary operation) and DeMorgan's law (two binary and one unary operation). Identities are one of the primary organizing tools in algebra.

Given two algebras **A** and **B**, we may ask whether they satisfy precisely the same set of identities. Notice that this is a far weaker notion than isomorphism. For example, any algebra satisfies the same identities as each of its direct powers. Nevertheless, if **A** and **B** satisfy the same identities, then they will be constrained to behave in a similar way. One of our problems, called VAR-EQUIV, is this: Given two finite algebras of the same finite similarity type, determine whether they satisfy the same identities.

This problem has implications for several areas of computer science. *Formal algebraic specifications* are expressions in a mathematical language which describe the properties and/or input-output behavior that a software system must exhibit, without putting any restrictions on the way in which these properties are implemented. This *abstraction* makes formal specifications extremely useful in the process of developing software systems where it serves as a reference point for users, implementers, testers and writers of instruction manuals. Formal specifications have been applied successfully in deployment of sophisticated software systems, see [16], especially the references there.

Mathematically, formal algebraic specifications are firmly grounded on algebraic concepts, especially ideas, notions and methods from universal algebra [4]. The relationship between implementation and equational specification corresponds, in algebraic terms, to the relationship between an algebra and a set of identities satisfied by the algebra. Thus, two algebras that satisfy the same identities correspond to a pair of implementations with precisely the same specification. The computational complexity of these problems, in the universal algebraic framework, is thus quite relevant to the body of research in formal specification theory, and to the construction of supporting tools such as theorem provers and model checkers.

Generalizing the notion of identity, we arrive at a quasi-identity. We shall leave a precise definition for Sect. 1, but crudely speaking, a quasi-identity involves a conjunction of identities and an implication. An example is the left-cancellation law (for, say, a semigroup). In direct analogy with the previous problem we can ask for the complexity of the following. Given two finite algebras of the same finite similarity type, determine whether they satisfy exactly the same quasi-identities. This notion too extends to algebraic specification theory, since "conditional specifications" take the form of quasi-identities.

Our third problem involves the term operations of an algebra. Although an algebra may be endowed with only finitely many basic operations, we can construct many more by composing the basic ones in various combinations. These are called the term operations of the algebra. Two algebras (presumably of different similarity types) are called term-equivalent if they have the same universe and exactly the same set of term operations. In universal algebra, term-equivalent algebras are considered the same, "for all practical purposes". The problem we call TERM-EQUIV is that of determining whether two finite algebras are term-equivalent. Returning once again to the realm of specification theory, in this problem we are asking whether a pair of implementations for two entirely different specifications have the property that they exhibit the same input-output behavior.

Each of these three problems makes sense for arbitrary finite algebras with an arbitrary (but finite) set of basic operations. In addition to this most general formulation we consider, for each of the three problems, two more restricted settings that, experience tells us, may result in different complexities. The first is to require that all basic operations on our algebras be unary. In the second, we only consider algebras of cardinality two.

We would like to thank Joel Berman, Gary Leavens and Ross Willard for many helpful discussions on this and related topics.

1 Preliminaries

We shall assume that the reader is familiar with the fundamental definitions and concepts of universal algebra. Our primary reference for this material is [13]. Other good references are [3], especially for the material on quasivarieties, and [7]. Although a bit dated, Taylor's survey in [15] is particularly readable.

We use the notation $\mathbf{V}(\mathbf{A})$ (respectively $\mathbf{Q}(\mathbf{A})$) for the variety (quasivariety) generated by an algebra $\mathbf{A}$, and $\mathrm{Clo}(\mathbf{A})$ for the clone of term operations on $\mathbf{A}$. We write $\mathbf{A} \sim \mathbf{B}$ to indicate that the algebras $\mathbf{A}$ and $\mathbf{B}$ have the same similarity type. Throughout the paper, all algebras are assumed to have finite similarity type.

We assume that the reader is familiar with the most common notions of complexity theory. Our notation and definitions for complexity classes comes from [8]. In particular, we use the names $\mathbf{L}$ and $\mathbf{NL}$ for the classes of languages decidable in logarithmic and nondeterministic logarithmic space, respectively. Most of our problems require a pair of algebras as input. Let us be more specific as to the form we assume the input will take. The underlying set of an algebra can be assumed to be $\{0, 1, \ldots, n - 1\}$ for some positive integer n. In fact, this set can be represented in the input by its cardinality, which requires $\log n$ bits of storage. (All logarithms will be to the base 2.) A k-ary operation on this set is represented as a table of values, in other words, a k-dimensional array with both the indices and entries coming from $\{0, 1, \ldots, n - 1\}$. Notice that this can be represented in the input stream using $n^k \cdot \log n$ bits.

2   Discussion of the Problems

In this paper we shall consider three equivalence relations on algebraic structures. First, given two algebras $\mathbf{A}$ and $\mathbf{B}$ of the same similarity type, is $\mathbf{V(A)} = \mathbf{V(B)}$? This is equivalent to asking whether $\mathbf{A}$ and $\mathbf{B}$ satisfy exactly the same identities. Note that this only makes sense if the two algebras have the same similarity type. It was shown in [9] that this problem is decidable. We shall denote this problem VAR-EQUIV. Thus

$$\text{VAR-EQUIV} = \{\,(\mathbf{A}, \mathbf{B}) : \mathbf{A} \sim \mathbf{B} \,\&\, \mathbf{V(A)} = \mathbf{V(B)}\,\}\ .$$

We have an analogous problem for quasivarieties:

$$\text{QVAR-EQUIV} = \{\,(\mathbf{A}, \mathbf{B}) : \mathbf{A} \sim \mathbf{B} \,\&\, \mathbf{Q(A)} = \mathbf{Q(B)}\,\}\ .$$

The assertion $(\mathbf{A}, \mathbf{B}) \in$ QVAR-EQUIV is equivalent to $\mathbf{A}$ and $\mathbf{B}$ satisfying exactly the same quasi-identities. Surprisingly, even though the logical form of a quasi-identity is much more complicated than that of an identity, QVAR-EQUIV has a relatively low computational complexity compared to VAR-EQUIV. Note that QVAR-EQUIV $\subseteq$ VAR-EQUIV as sets.

The third problem we shall consider is term-equivalence. Two algebras $\mathbf{A}$ and $\mathbf{B}$ are *term-equivalent* if and only if they have the same underlying set and $\mathrm{Clo}(\mathbf{A}) = \mathrm{Clo}(\mathbf{B})$. For this problem, we do not require that $\mathbf{A}$ and $\mathbf{B}$ have the same similarity type, but we do require that they have the same universe:

$$\text{TERM-EQUIV} = \{\,(\mathbf{A}, \mathbf{B}) : A = B \,\&\, \mathrm{Clo}(\mathbf{A}) = \mathrm{Clo}(\mathbf{B})\,\}\ .$$

It was shown in [1] that TERM-EQUIV is complete for **EXPTIME**.

There are several restrictions of these problems which are of interest and which turn out to have a lower complexity. In particular, we can bound either the cardinality of the underlying sets or the ranks of the operations of the algebras. For example, it was shown in [11] that TERM-EQUIV is complete for **PSPACE** when restricted to *unary algebras*, that is, algebras in which every operation has rank 1. For each of our three problems, we shall consider, in addition to the general case, the subcases obtained by considering only unary algebras and only 2-element algebras. We shall denote the subcase by appending a superscript '1' or subscript '2' to the problem. To be precise, let us define

$$U = \{\,\mathbf{A} : \mathbf{A} \text{ is a unary algebra}\,\}$$
$$T = \{\,\mathbf{A} : A = \{0,1\}\,\}$$

then $\mathrm{X}^1 = \mathrm{X} \cap (U \times U)$ and $\mathrm{X}_2 = \mathrm{X} \cap (T \times T)$ for X any of TERM-EQUIV, VAR-EQUIV, or QVAR-EQUIV.

Our results for each of these nine problems can be summarized in Table 1. In this table, the first row concerns the subcase consisting of 2-element algebras, the second of unary algebras and the third, the general case. Each of the nine entries gives the smallest complexity class known to contain the problem, and a superscript '*' indicates that the result is sharp, i.e., the problem is complete for the given complexity class.

Table 1. Summary of results

	QVAR-EQUIV	TERM-EQUIV	VAR-EQUIV
card2	**L**	**NL**	**L**
unary	**NP**	**PSPACE***	**PSPACE**
general	**NP***	**EXPTIME***	**2-EXPTIME**

3   The Quasivariety Problems

We begin with the problems that ask whether two algebras generate the same quasivariety. It is sometimes convenient to work with an asymmetric variant of this. We write

$$\text{QVAR-MEM} = \{ (\mathbf{A}, \mathbf{B}) : \mathbf{A} \sim \mathbf{B} \ \& \ \mathbf{B} \in \mathbf{Q}(\mathbf{A}) \} \ .$$

Since the '$\mathbf{Q}$' operator has the usual properties of closure, we obviously have

$$(\mathbf{A}, \mathbf{B}) \in \text{QVAR-EQUIV} \iff (\mathbf{A}, \mathbf{B}), (\mathbf{B}, \mathbf{A}) \in \text{QVAR-MEM} \ . \tag{1}$$

It follows that for an instance of size s, membership in QVAR-EQUIV can be tested with two calls to an algorithm for QVAR-MEM, both using inputs of size s. In a natural way, we also have the restricted problems QVAR-MEM^1 and QVAR-MEM$_2$ consisting of pairs of unary and two-element algebras, respectively.

Theorem 1. QVAR-MEM $\in$ **NP**.

Proof. Let $\mathbf{A}$ and $\mathbf{B}$ be a pair of similar, finite algebras. We wish to determine whether $\mathbf{B} \in \mathbf{Q}(\mathbf{A})$. Here is a nondeterministic algorithm. For each unordered pair $\{a, b\}$ of distinct elements of B, guess a function $\psi_{\{a,b\}} \colon B \to A$ such that $\psi_{\{a,b\}}(a) \neq \psi_{\{a,b\}}(b)$. Test whether $\psi_{\{a,b\}}$ is a homomorphism. If it is not, then reject. But if every $\psi_{\{a,b\}}$ passes the homomorphism test, then accept.

Since $\mathbf{A}$ is finite, $\mathbf{Q}(\mathbf{A}) = \mathbf{SP}(\mathbf{A})$. It is a simple matter to verify that our algorithm accepts the pair $(\mathbf{A}, \mathbf{B})$ if and only if $\mathbf{B} \in \mathbf{Q}(\mathbf{A})$. To bound the running time, Let s denote the size of the input. A function ψ from B to A can be guessed in time on the order of $|B| \cdot |A|$, which is at most s^2. The verification that ψ is a homomorphism also takes time in $O(s^2)$. The total number of functions we need to construct is $\binom{|B|}{2} \leq s^2$. Thus the total running time lies in $O(s^4)$. $\square$

Corollary 1. *The following problems lie in* **NP**: QVAR-EQUIV, QVAR-MEM^1, QVAR-MEM$_2$, QVAR-EQUIV^1 *and* QVAR-EQUIV$_2$.

Theorem 2. QVAR-MEM^1, QVAR-MEM, SUBALG^1, SUBALG *and* QVAR-EQUIV *are all complete for* **NP**.

Let us point out that Theorem 2 does not include QVAR-EQUIV[1]. The exact complexity of QVAR-EQUIV[1] is open.

Now we turn to the problem QVAR-EQUIV$_2$. Suppose that $\mathbf{A}$ and $\mathbf{B}$ are two-element algebras. We claim that $\mathbf{B} \in \mathbf{Q}(\mathbf{A})$ if and only if $\mathbf{B} \cong \mathbf{A}$. To see this, note that every two-element algebra is simple. Since $|B| = |A|$, we obtain

$$\mathbf{B} \in \mathbf{Q}(\mathbf{A}) = \mathbf{SP}(\mathbf{A}) \implies \mathbf{B} \in \mathbf{S}(\mathbf{A}) \implies \mathbf{B} \cong \mathbf{A} \ .$$

The converse, that $\mathbf{B} \cong \mathbf{A} \implies \mathbf{B} \in \mathbf{Q}(\mathbf{A})$ is trivial.

Theorem 3. QVAR-EQUIV$_2$, QVAR-MEM$_2$ $\in \mathbf{L}$.

Proof. As we argued in the previous paragraph, $(\mathbf{B}, \mathbf{A}) \in$ QVAR-MEM$_2$ if and only if $\mathbf{B} \cong \mathbf{A}$. There are only two bijections from B to A, and each of these can be tested to see if it is a homomorphism. The testing requires just a couple of counters, which has a space bound that is logarithmic in the size of the input. Thus QVAR-MEM$_2$ $\in \mathbf{L}$. Now apply assertion (1) to deduce that QVAR-EQUIV$_2$ $\in \mathbf{L}$. $\qquad\square$

4   The Variety Problems

The problem VAR-EQUIV asks: if $\mathbf{A}$ and $\mathbf{B}$ are two algebras of the same similarity type, is $\mathbf{V}(\mathbf{A}) = \mathbf{V}(\mathbf{B})$? As with quasivarieties, it is convenient to introduce an auxiliary problem, VAR-MEM.

$$\text{VAR-MEM} = \{\, (\mathbf{A}, \mathbf{B}) : \mathbf{A} \sim \mathbf{B} \ \& \ \mathbf{B} \in \mathbf{V}(\mathbf{A}) \,\} \ .$$

Unlike the situation for quasivarieties, the relationship between VAR-EQUIV and VAR-MEM is clear-cut. We have

$$\begin{aligned}
(\mathbf{A}, \mathbf{B}) \in \text{VAR-EQUIV} &\iff (\mathbf{A}, \mathbf{B}), (\mathbf{B}, \mathbf{A}) \in \text{VAR-MEM}, \\
(\mathbf{A}, \mathbf{B}) \in \text{VAR-MEM} &\iff (\mathbf{A}, \mathbf{A} \times \mathbf{B}) \in \text{VAR-EQUIV} \ .
\end{aligned} \tag{2}$$

The second equivalence follows from the fact that both $\mathbf{A}$ and $\mathbf{B}$ are homomorphic images of $\mathbf{A} \times \mathbf{B}$.

We begin with the two-element problem. The crucial point is the following theorem.

Theorem 4. *Let $\mathbf{A}$ and $\mathbf{B}$ be two-element algebras of the same similarity type. Then $\mathbf{V}(\mathbf{A}) = \mathbf{V}(\mathbf{B})$ if and only if $\mathbf{A} \cong \mathbf{B}$.*

Theorem 5. VAR-EQUIV$_2$ $\in \mathbf{L}$.

Proof. From Theorem 4, testing whether $(\mathbf{A}, \mathbf{B}) \in$ VAR-EQUIV is equivalent to testing $\mathbf{A} \cong \mathbf{B}$. Arguing as we did at the end of Sect. 3, there are only two possible isomorphisms to test. This can be done deterministically in logarithmic space. $\qquad\square$

In order to proceed to the remaining two problems, we need some more detailed information on the relationship between clones, terms and varieties. Let $\mathbf{A} = \langle A, F \rangle$ be an algebra of cardinality n and let m be a positive integer. It is always the case that the set $\mathrm{Clo}_m(\mathbf{A})$ forms a subalgebra of $\mathbf{A}^{(A^m)}$, that we denote $\mathbf{Clo}_m(\mathbf{A})$. Notice that we follow the usual typographic convention and print 'Clo' in boldface when it is to be used as an algebra. Since varieties are closed under the formation of both powers and subalgebras, it follows that both $\mathbf{A}^{(A^m)}$ and $\mathbf{Clo}_m(\mathbf{A})$ lie in $\mathbf{V}(\mathbf{A})$.

Every term on $\mathbf{A}$ can be viewed as a tree in a natural way. Let us write $\mathrm{ht}(t)$ for the height of the tree corresponding to the term t. If one thinks about the natural way to construct the set $\mathrm{Clo}_m(\mathbf{A})$, we see that for every m-ary term t, there is a term t' such that $t^{\mathbf{A}} = (t')^{\mathbf{A}}$ and $\mathrm{ht}(t') < n^{(n^m)}$. In the special case that $\mathbf{A}$ is a unary algebra, we can do better. Since every m-ary term operation (for any m) is essentially unary, the bound on the height of t' can be reduced to n^n. Using these remarks, we have the following Theorem.

Theorem 6. *Let $\mathbf{A}$ and $\mathbf{B}$ be finite algebras of the same similarity type. Assume that the cardinalities of $\mathbf{A}$ and $\mathbf{B}$ are n and m respectively. Then the following are equivalent.*

(i) $\mathbf{B} \in \mathbf{V}(\mathbf{A})$.
(ii) For every pair of terms s and t, each of height at most $n^{(n^m)}$, if $\mathbf{A}$ satisfies the identity $s \approx t$ then so does $\mathbf{B}$.
(iii) $\mathbf{B}$ is a homomorphic image of the algebra $\mathbf{Clo}_m(\mathbf{A})$.

If $\mathbf{A}$ and $\mathbf{B}$ are unary algebras, then the bound $n^{(n^m)}$ in (ii) can be reduced to n^n.

Theorem 6 suggests an approach that can be used to test the condition $\mathbf{B} \notin \mathbf{V}(\mathbf{A})$: simply *guess* an identity ϵ an check to see whether $\mathbf{A}$ satisfies ϵ while $\mathbf{B}$ fails to satisfy ϵ. This approach seems to be quite effective—at least for unary algebras. For in this case, we have the improved bound n^n in part (ii) of the theorem.

Let us fix a set $F = \{f_1, \ldots, f_k\}$ of operation symbols, each of rank 1. Also, let us add an additional unary operation symbol f_0 which will always be interpreted as the identity operation. This has no effect on the algebras, but will save us a subscript in our analysis. A typical term over F is of the form $f_{i_\ell} f_{i_{\ell-1}} \cdots f_{i_2} f_{i_1}(x)$ where $i_1, i_2, \ldots, i_\ell \in \{0, 1, \ldots, k\}$. The height of this term is ℓ. Since each term involves only one variable, every identity is of one of two possible forms:

$$s(x) \approx t(x) \quad \text{or} \quad s(x) \approx t(y) .$$

Notice that the second of these is quite degenerate since it requires that the term operations corresponding to s and t both be constant, and in fact, the same constant. Nevertheless, it must be considered in the analysis.

Now suppose that $\mathbf{A}$ and $\mathbf{B}$ are algebras of type F and of cardinalities n and m respectively. Algorithm 1 is a nondeterministic algorithm that accepts the pair $(\mathbf{A}, \mathbf{B})$ if and only if $\mathbf{B} \notin \mathbf{V}(\mathbf{A})$.

1. $s^A \leftarrow t^A \leftarrow f_0^{\mathbf{A}}; \quad s^B \leftarrow t^B \leftarrow f_0^{\mathbf{B}}.$
2. for $i = 1$ to n^n do
3.   guess $j, \ell \in \{0, 1, \ldots, k\}$
4. $s^A \leftarrow f_j^{\mathbf{A}} \circ s^A; \quad s^B \leftarrow f_j^{\mathbf{B}} \circ s^B; \quad t^A \leftarrow f_\ell^{\mathbf{A}} \circ t^A; \quad t^B \leftarrow f_\ell^{\mathbf{B}} \circ t^B$
5.   if $\big((\forall x, y \in A)\,(s^A(x) = t^A(y))$ and $(\exists x, y \in B)\,(s^B(x) \neq t^B(y))\big)$ or
       $\big((\forall x \in A)\,(s^A(x) = t^A(x))$ and $(\exists x \in B)\,(s^B(x) \neq t^B(x))\big)$
       then **accept**.

Algorithm 1. Testing $(\mathbf{A}, \mathbf{B}) \notin \text{VAR-MEM}^1$

How much space is used by this algorithm? Let $p = \max(n, m)$. Each of the four unary operations can be represented as a vector of length p. Each such vector requires $p \log(p) \leq p^2$ bits. We also need space for the counter i, which ranges from 0 to n^n. Since $\log(n^n) = n \log(n) \leq p^2$, i requires another p^2 bits. It follows that the total amount of space required is on the order of p^2 bits.

What is the size of the input? The algebra $\mathbf{A}$ requires $\log(n) + kn \log(n) > n$ bits. Similarly $\mathbf{B}$ requires at least m bits. The total input size is at least $n + m > p$ bits. It follows that our algorithm's space requirements are bounded above by the square of the size of the input.

Theorem 7. VAR-MEM^1 *and* VAR-EQUIV^1 *lie in* **PSPACE**.

Proof. The above algorithm can be used to test whether $(\mathbf{A}, \mathbf{B})$ lies in the *complement* of VAR-MEM^1. Since the algorithm is nondeterministic, we get $\text{VAR-MEM}^1 \in$ **co-NPSPACE**. But from Savitch's Theorem [14] **NPSPACE** = **PSPACE** and since every deterministic class is closed under complements, **co-PSPACE** = **PSPACE**. Thus $\text{VAR-MEM}^1 \in$ **PSPACE**. Now it follows from (2) that $\text{VAR-EQUIV}^1 \in$ **PSPACE** as well. $\qquad\square$

It is not clear whether this is the best possible bound for these two problems. We leave it as an open question.

Problem 1. Are either VAR-MEM^1 or VAR-EQUIV^1 **PSPACE**-complete?

One might hope to apply the same techniques used above to the unrestricted problem, VAR-EQUIV. Unfortunately, the resources needed to evaluate an arbitrary term in a given algebra jump dramatically as soon as we allow a binary operation. Our approach instead is to try to construct the homomorphism guaranteed by Theorem 6(*iii*). The best we seem to be able to do is the following hyperexponential bound.

Theorem 8. $\text{VAR-EQUIV}, \text{VAR-MEM} \in$ **2-EXPTIME**.

5   Term-Equivalence

Recall that the algebras **A** and **B** are term-equivalent if $A = B$ and $\mathrm{Clo}(\mathbf{A}) = \mathrm{Clo}(\mathbf{B})$. From a universal algebraic standpoint, term-equivalent algebras are generally interchangeable. When considering the complexity of the problem TERM-EQUIV, it is convenient, once again, to consider a slightly different problem. Thus we define the problem CLO-MEM to consist of all pairs (F, g) in which $F \cup \{g\}$ is a set of operations on some finite set A and $g \in \mathrm{Clo}^A(F)$. The problem CLO-MEM^1 is similar, but we require all of the operations in $F \cup \{g\}$ to be unary. For the problem CLO-MEM$_2$, the set A has cardinality 2.

Historically, CLO-MEM was the first problem, of those discussed in this paper, to be considered in the literature. Kozen proved in 1977 [11], that CLO-MEM^1 is complete for **PSPACE**. In 1982, Friedman proved that CLO-MEM is complete for **EXPTIME**, [6]. However, that manuscript was never published. A proof of this result appears in [1].

Let $\mathbf{A} = \langle A, F \rangle$ and $\mathbf{B} = \langle B, G \rangle$. Then

$$(\mathbf{A}, \mathbf{B}) \in \text{TERM-EQUIV}$$

$$\Updownarrow \tag{3}$$

$$A = B \ \& \ (\forall g \in G)(\forall f \in F)\left((F, g), (G, f) \in \text{CLO-MEM} \right)$$

Conversely, if $F \cup \{g\}$ is a set of operations on A, then

$$(F, g) \in \text{CLO-MEM} \iff (\langle A, F \rangle, \langle A, F \cup \{g\} \rangle) \in \text{TERM-EQUIV} \ . \tag{4}$$

Of course, similar relationships hold for the unary and two-element variants of these problems.

Now if follows easily from (4) that CLO-MEM is log-space reducible to TERM-EQUIV. This is true for the general, unary and two-element variants of the problems. However, a reduction in the other direction is a bit problematic. Let us first consider the general case. Given a pair $(\mathbf{A}, \mathbf{B})$ of size S, (3) tells us that we can test $(\mathbf{A}, \mathbf{B}) \in$ TERM-EQUIV by making several calls to an algorithm for CLO-MEM. The input to each such call will certainly have size at most S, hence will run in time at most $2^{p(S)}$ for some polynomial p. Furthermore, there will clearly be at most S such calls. Hence a bound on the running time for TERM-EQUIV will be $S \cdot 2^{p(S)}$ which is still exponential in S. Combining our observations we conclude that the (general) problem TERM-EQUIV is complete for **EXPTIME**.

For CLO-MEM^1 and CLO-MEM$_2$, we need to argue a bit differently, since we will be interested in a space-bound. Let C denote one of these two problems, and let T denote the corresponding term-equivalence problem. Suppose we have an algorithm for C that runs in space $f(x)$ on an input of size x. As is commonplace, we assume that f is a monotonically increasing function. In applying (3) to test $(\mathbf{A}, \mathbf{B}) \in$ T, the first call to C will require space bounded above by $O(f(S))$. But subsequent calls to C can reuse the same space. Hence, the total space requirement for T is on the order of $\log S + f(S)$. (The $\log S$ term accounts for some counters.)

For the specific case $C = \text{CLO-MEM}^1$, the function f is a polynomial, so we conclude that TERM-EQUIV^1 is complete for **PSPACE**. When $C = \text{CLO-MEM}_2$, it turns out that $f(S)$ is on the order of $\log S$. From this we get our final Theorem.

Theorem 9. CLO-MEM_2 *and* TERM-EQUIV_2 *lie in* **NL**.

References

[1] C. Bergman, D. Juedes, and G. Slutzki, *Computational complexity of term-equivalence*, Inter. Jour. Algebra and Computation (1997), to appear.

[2] S. Burris, *Computers and universal algebra: some directions*, Algebra Universalis **34** (1995), 61–71.

[3] S. Burris and H. P. Sankappanavar, *A course in universal algebra*, Springer-Verlag, New York, 1981.

[4] H. Ehrig and B. Mahr, *Fundamentals of algebraic specification 1*, Vol. 6 of EATCS Monographs on Theoretical Computer Science [5], 1985, Equations and initial semantics.

[5] ———, *Fundamentals of algebraic specification 2*, EATCS Monographs on Theoretical Computer Science, vol. 21, Springer-Verlag, Berlin, 1990, Module specifications and constraints.

[6] H. Friedman, *Function composition and intractability I*, manuscript, March 1982.

[7] G. Grätzer, *Universal algebra*, second ed., Springer-Verlag, New York, 1979.

[8] D. Johnson, *A catalog of complexity classes*, Handbook of Theoretical Computer Science (J. van Leeuwen, ed.), vol. A, Elsevier, Amsterdam, 1990, pp. 69–161.

[9] J. Kalicki, *On comparison of finite algebras*, Proc. Amer. Math. Soc. **3** (1952), 36–40.

[10] D. Kozen, *Complexity of finitely presented algebras*, Proceedings of the Ninth Annual Symposium on the Theory of Computing, ACM, 1977, pp. 164–177.

[11] ———, *Lower bounds for natural proof systems*, 18th Annual Symposium on Foundations of Computer Science (Providence, R.I., 1977), IEEE Comput. Soc., Long Beach, CA, 1977, pp. 254–266.

[12] L. Kučera and V. Trnková, *The computational complexity of some problems in universal algebra*, Universal Algebra and its Links with Logic, Algebra, Combinatorics and Computer Science (P. Burmeister et. al., ed.), Heldermann Verlag, 1984, pp. 261–289.

[13] R. McKenzie, G. McNulty, and W. Taylor, *Algebras, lattices, varieties*, vol. I, Wadsworth & Brooks/Cole, Belmont, CA, 1987.

[14] W. J. Savitch, *Relationships between nondeterministic and deterministic tape complexities*, J. Computer and Systems Sciences **4** (1970), no. 2, 177–192.

[15] W. Taylor, *Equational logic*, Houston Jour. Math. (1979), iii+83, survey.

[16] J. Wing, *A specifier's introduction to formal methods*, Computer (1990), 8–24.

New Branchwidth Territories

Ton Kloks[1]*, Jan Kratochvíl**[2], and Haiko Müller[3]

[1] Department of Mathematics and Computer Science
Vrije Universiteit
1081 HV Amsterdam, The Netherlands
kloks@cs.vu.nl
[2] Department of Applied Mathematics and DIMATIA
Charles University, Malostranské nám. 25
118 00 Praha 1, Czech Republic
honza@kam.ms.mff.cuni.cz
[3] Friedrich-Schiller-Universität Jena
Fakultät für Mathematik und Informatik
07740 Jena, Germany
hm@minet.uni-jena.de

Abstract. We give an algorithm computing the branchwidth of interval graphs in time $O(n^3 \log n)$. This method generalizes to permutation graphs and, more generaly, to trapezoid graphs. In contrast, we show that computing branchwidth is NP-complete for splitgraphs and bipartite graphs.

1 Introduction

The research into the branchwidth of fundamental graph classes commenced in [6]. The authors of this paper showed that the branchwidth problem can be solved in polynomial time for planar graphs. In this paper we present some further developments.

The major reason for studying graph parameters like branchwidth are the fast algorithms one can obtain for problems when restricted to graphs for which this parameter is not too large. Treewidth and branchwidth are closely related connectivity measures introduced by Robertson and Seymour. The treewidth parameter has drawn most of the attention until now. The two parameters differ by at most a small constant factor (more precisely, branchwidth$(G) \leq$ treewidth$(G) + 1 \leq \frac{3}{2}$branchwidth(G)).

Our motivation for studying branchwidth is twofold. The 'fast' algorithms one obtains for various NP-complete problems are of mere theoretical interest in many cases, because of the huge constants involved. These constants appear in two stages of the algorithms. In the first stage one needs to construct a

* The first author acknowledges support of DIMATIA Charles University, where he held a visiting position in 1997/8.
** The second author acknowledges partial support of Czech Research grants GAČR 201/1996/0194 and GAUK 1996/194.

branch- or tree-decomposition of the graph with a small width. In this paper we show that the complexity of obtaining an optimal branch- or tree-decomposition can differ enormously. In the second stage one solves the seemingly difficult (i.e., NP-complete) problem at hand using the decomposition. The complexity of the second stage usually depends heavily on the width of the decomposition, and although the width of the tree- or branch-decomposition differs only by a constant factor, this can make or break the application. Therefore, obtaining efficient algorithms for the branchwidth of a graph is of independent interest.

The complexity of the treewidth problem restricted to special graph classes was considered in various papers (see, e.g., [4]). In this paper we investigate the computational complexity of the branchwidth problem for some of the most fundamental classes of graphs (split graphs, bipartite graphs and interval graphs.) It should be noted that though the existence of an efficient algorithm for interval graphs is not unexpected, it is somewhat surprising that this algorithm is by no means straightforward and its correctness requires a nontrivial proof. We show that our efficient algorithm for interval graphs can be generalized to a parameterized class of graphs called d-trapezoid graphs. As far as we know, besides [6] and [1], these are the first results dealing with the computational complexity of the branchwidth problem.

2   Preliminaries

The notion of branchwidth was introduced in [5]. We think it is more convenient to work with a more relaxed (yet equivalent) version of the branch-decomposition tree.

Definition 1. *A pair (T, τ) is a relaxed branch-decomposition if T is a tree with vertices of degree at most three and τ is a surjective mapping which maps every leaf of T to an edge of E. (Hence every edge of E is represented by at least one leaf). The* order *of an edge e in T is the number of vertices x of G such that there are leaves t_1 and t_2 in different components of $T - e$ with x incident both with $\tau(t_1)$ and with $\tau(t_2)$. The* width *of (T, τ) is the maximum order of an edge of T. The* branchwidth *of G, $\beta(G)$, is the minimum width over all relaxed branch-decompositions of G.*

Consider a relaxed branch-decomposition (T, τ). For a vertex v we denote by T_v the smallest subtree T_v of T that contains all leaves λ such that $\tau(\lambda)$ is incident with v.

3   Splitgraphs and Bipartite Graphs

A graph $G = (V, E)$ is called a *splitgraph* if V can be split into an independent set I of G and a clique C of G. Such a graph is also denoted as $G = (I, C, E)$.

Since C is a clique in G, $\beta(G) \geq \lceil \frac{2}{3}|C| \rceil$. It is easy to see that $\beta(G) = 2k$ if $|C| = 3k$ and there is a partition of C into three sets C_1, C_2 and C_3, each of cardinality k, such that for every vertex $i \in I$ there is an index $j \in \{1, 2, 3\}$ with $N(i) \cap C_j = \emptyset$.

We will show that it is NP-complete to decide for a given hypergraph (X, S) whether there is an *admissible coloring*, i.e., a 3-coloring of X such that the color classes are of equal size and every subset $s \in S$ contains at most two colors. The NP-completeness of the branchwidth problem restricted to split graphs follows by setting $C = X$ and $I = S$ with $N(s) = s$ for every $s \in S$.

The reduction is from GRAPH 3-COLORABILITY, see problem [GT4] in [3]. Let $G = (V, E)$ be an instance of GRAPH 3-COLORABILITY with $|V| = n > 0$ and $|E| = m$. We choose a vertex $v_0 \in V$ and a set W such that $|W| = 2n$ and $V \cap W = \{v_0\}$. The hypergraph $H = (X, S)$ is defined by $X = (V \cup W) \times \{1, 2, 3\}$ and

$$S = \{\{(v, i), (v', j), (v', k)\} : \{v, v'\} \in E \text{ and } \{i, j, k\} = \{1, 2, 3\}\} \cup$$
$$\{\{(w, i) : w \in W \text{ and } i \neq j\} : j \in \{1, 2, 3\}\}$$

Lemma 1. *If G is 3-colorable then H has an admissible coloring.*

Proof. Let $f : V \to \{1, 2, 3\}$ be a proper coloring of G. We define an admissible coloring $c : X \to \{1, 2, 3\}$ by

$$c(v, i) = j \text{ if and only if } v \in V \text{ and } f(v) + i \equiv j \pmod{3} \text{ and}$$
$$c(w, i) = c(v_0, i) \text{ for all } w \in W.$$

$\square$

Lemma 2. *For every admissible coloring c of H we have $c(w_1, i) = c(w_2, i)$ for all $w_1, w_2 \in W$ and $i \in \{1, 2, 3\}$.*

Proof. First we observe that $|X| = 9n - 3$ for $|V| = n$. We assume that there exist two elements $w_1, w_2 \in W$ such that $c(w_1, i) \neq c(w_2, i)$ for some $i \in \{1, 2, 3\}$. Then every vertex (w, j), $w \in W$ and $j \in \{1, 2, 3\}$ is colored either with color $c(w_1, i)$ or with color $c(w_2, i)$, since $\{(w, 1), (w, 2)\} : w \in W\}$, $\{(w, 2), (w, 3)\} : w \in W\}$, and $\{(w, 3), (w, 1)\} : w \in W\}$ are hyperedges of H. This implies that the union of these two color classes contains at least $6n$ vertices of H contradicting the fact that the three color classes have equal cardinality. $\square$

Lemma 3. *If c is an admissible coloring of H and $\{u, v\}$ is an edge of G such that all three colors appear on $(v, 1)$, $(v, 2)$, and $(v, 3)$, then all three colors appear on $(u, 1)$, $(u, 2)$, and $(u, 3)$.*

Proof. For simplicity we assume $c(v, i) = i$ for $i = 1, 2, 3$ and, on the contrary, $c(u, 1) = c(u, 2)$. If $c(u, 1) \neq 3$ then on the hyperedge $\{(u, c(u, 1)), (v, 3 - c(u, 1)), (v, 3)\}$ appear all three colors. Otherwise, if $c(u, 1) = 3$ then on the hyperedge $\{(u, c(u, 3)), (v, 3 - c(u, 3)), (u, 3)\}$ appear all three colors. $\square$

Lemma 4. *If G is connected and H has an admissible coloring then G has a proper 3-coloring.*

Proof. It follows from Lemma 2 that all three colors appear at $(v_0, 1)$, $(v_0, 2)$, and $c(v_0, 2)$. Now an inductive argument using Lemma 3 shows that all three colors appear at $(v, 1)$, $(v, 2)$, and $(v, 3)$ for every vertex $v \in V$. We define a 3-coloring $f : V \rightarrow \{1, 2, 3\}$ of G by $f(v) = c(v, 3)$ for all $v \in V$. For every egde $\{u, v\} \in E$ we have $f(u) \neq f(v)$ since $c(u, 3) \in \{c(v, 1), c(v, 2)\}$ by Lemma 3.   □

Theorem 1. *The branchwidth problem is NP-complete when restricted to split graphs.*

Proof. The proof is a direct consequence of Lemmas 1 and 4.   □

It is easy to see that if H is a proper subdivision of an arbitrary graph G then $\beta(H) = \max(1, \beta(G))$. It follows that the branchwidth problem is also NP-complete when restricted to bipartite graphs.

4   Interval Graphs

An ordering $\mathcal{X} = (X_1, \dots, X_\ell)$ of the maximal cliques of G is called a *consecutive clique arrangement* (cca for short) if for every vertex of G, the maximal cliques containing this vertex occur consecutively in the ordering. Equivalently, $X_{i-1} \cap X_{i+1} \subseteq X_i$ for all $i = 2, \dots, \ell - 1$. It is well known that G is an interval graph if and only if G has a cca. There is a linear time algorithm that either constructs a cca or detects that the input is not an interval graph [2].

By $[a, b]$ we denote the interval $\{a, a + 1, \dots, b\}$. Let $X_{a,b}$ be shorthand for $\bigcup_{i=a}^{b} X_i$. A *fragmentation* of $\mathcal{X}$ is a partition $\mathcal{F}$ of $[1, \ell]$ into intervals $[a_1, b_1]$, $[a_2, b_2], \dots, [a_t, b_t]$.

Definition 2. *Let $\mathcal{X} = (X_1, \dots, X_\ell)$ be a cca of G. We say that $X_{a,b}$ is a k-fragment if and only if*

(i) $|X_{a,b}| \leq \frac{3}{2}k$,
(ii) $|X_{a-1} \cap X_{a,b}| \leq k$ and $|X_{b+1} \cap X_{a,b}| \leq k$
(iii) $|X_{a,b} \cap (X_{a-1} \cup X_{b+1})| \leq k$ or $|X_{a,b}| + |X_{a-1} \cap X_{b+1}| \leq 2k$.

A k-fragmentation of $\mathcal{X}$ is a partition $\mathcal{F}$ of $[1, \ell]$ into intervals such that for every $[a, b] \in \mathcal{F}$ the set $X_{a,b}$ is a k-fragment.

For an interval graph G given with a cca we show in this section that $\beta(G) \leq k$ if and only if the cca has a k-fragmentation. For the proof of the "only if" part of this claim we need some lemmas.

Suppose G has a branch-decomposition (T, τ) of width at most k. Let α be a node of T. We call the connected components of $T - \alpha$ the *branches* of T with respect to α (shortly the α-branches). For every vertex $x \in V(G)$, we denote by T_x the subtree of T with leaves $\alpha \in V(T)$ such that $x \in \tau(\alpha)$. For every clique $X \subseteq V(G)$, let $T(X)$ be the subtree of T induced by $\bigcap_{x \in X} V(T_x)$. It follows from the Helly property that $T(X) \neq \emptyset$ for every clique X of G.

Definition 3. *A graph G is* saturated, *if $|V(T(X))| = 1$ for every maximal clique X of G and every relaxed branch-decomposition (T, τ) of width $\beta(G)$.*

If G is a saturated interval graph with cca $(X_1, \ldots, X_\ell)$ and branch-decomposition (T, τ), we denote by $T(i)$ the unique vertex of $T(X_i)$.

Lemma 5. *Let $(X_1, \ldots, X_\ell)$ be a cca of an interval graph G. There exists a saturated interval graph G' with cca $(X_1', \ldots, X_\ell')$ such that $\beta(G') = \beta(G)$ and $X_i' \cap V(G) = X_i$ for $i = 1, 2, \ldots, \ell$.*

Proof. Let $k = \beta(G)$. We choose an interval graph G' with $\beta(G') = k$ obtained from G by adding a maximal number of new vertices which appear in exactly one maximal clique of G' each. We will show that G' is saturated. Let $(X_1', \ldots, X_\ell')$ be the cca of G' such that $X_i' \cap V(G) = X_i$ for $i = 1, 2, \ldots, \ell$. Since $|X_i'| \leq \frac{3}{2}k$ for every i, such a graph G' exists.

We consider a relaxed branch-decomposition (T', τ') of G' of width k. Suppose for the contrary that $T'(X_i')$ has more than one vertex for some particular i. Then all the trees $T_x, x \in X_i'$, contain a common edge of T', say e (and hence $|X_i'| \leq k$). Let G'' be the interval graph obtained from G' by adding a new vertex x to X_i'.

We define a branch-decomposition (T'', τ'') as follows: Take a copy of $K_{1,3}$ with central vertex α and leaves α_1, α_2, and α_3, disjoint with T'. Subdivide the edge e of T' by a new extra node and identify this node with α_3. Let $n_1 = \lfloor \frac{1}{2}|X_i'| \rfloor$ and $n_2 = \lceil \frac{1}{2}|X_i'| \rceil$. Subdivide the edge $\{\alpha, \alpha_1\}$ by $n_1 - 1$ new vertices and pend leaves $\alpha_{1,j}, j = 1, 2, \ldots, n_1 - 1$, on them (one on each). Similarly, create $n_2 - 1$ leaves $\alpha_{2,j}$ along the edge $\{\alpha, \alpha_2\}$. The resulting tree is T''. Partition X_i' into two almost equal parts Y_1 and Y_2 of sizes $|Y_1| = n_1$ and $|Y_2| = n_2$. Define τ'' so that its restriction to $\{\alpha_1\} \cup \{\alpha_{1j} : j = 1, 2, \ldots, n_1 - 1\}$ is a bijection onto the set of edges $\{(x, y) : y \in Y_1\}$, and similarly the restriction to $\{\alpha_2\} \cup \{\alpha_{2j} : j = 1, 2, \ldots, n_2 - 1\}$ is a bijection onto the set of edges $\{(x, y) : y \in Y_2\}$. For other leaves u of T'', $\tau''(u) = \tau'(u)$.

Clearly (T'', τ'') is a branch-decomposition of G'' of width k contradicting the choice of G'. $\qquad\square$

From now on, we will assume that G itself is saturated. (If we show that G' allows a k fragmentation, the inherited fragmentation of G is a k-fragmentation as well.) Note that assuming $\beta(G) \geq 2$, $T(i)$ is never a leaf of the decomposition.

Lemma 6. *If G has a relaxed branch-decomposition (T, τ) of width $k = \beta(G)$ such that $T(i) = T(h) \neq T(j)$ for some $i < j < h$, then G has a decomposition $(\bar{T}, \bar{\tau})$ of the same width such that $\bar{T}(i) \neq \bar{T}(h)$ and such that no two clique-representatives are unified in $(\bar{T}, \bar{\tau})$ unless they were unified in (T, τ).*

Proof. Let α and δ be the two neighbors of $T(i)$ that are not in the $T(i)$-branch containing $T(j)$. Let T^* be the subtree rooted at $T(i)$ containing α and δ.

Take two copies T_1^* and T_2^* of T^*. We will trim the leaves of T_1^* and T_2^*. In T_1^* we leave only those leaves that are mapped by τ onto edges $\{x, y\}$ where both x and y belong to $X_{1,j}$, and in T_2^* we leave the leaves mapped onto edges $\{x', y'\}$ where both x' and y' belong to $X_{j,\ell}$. Remove from T the subtrees at α and δ which do not contain $T(i)$ and replace them by T_1^* and T_2^*, respectively. Denote this decomposition $(\bar{T}, \bar{\tau})$.

Note first that $(\bar{T}, \bar{\tau})$ is again a relaxed branch-decomposition for G. Indeed, if an edge $\{x, y\} \in E(G)$ has both endpoints $x, y \in X_{1,j}$ and there was a leaf $v \in T^*$ such that $\tau(v) = (x, y)$, then there remains a copy $\bar{v}$ of v in T_1^* such that $\bar{\tau}(\bar{v}) = (x, y)$. If such a v lies in the $T(i)$-branch that contains $T(j)$ then $\bar{v} = v$. Similarly for edges $\{x, y\}$ with $x, y \in X_{j,\ell}$.

Next we show that the width of $(\bar{T}, \bar{\tau})$ does not exceed k. The orders of edges in the $T - T^*$ subtree of $\bar{T}$ did not change. Let u be the neighbor of $T(i)$ on the path to $T(j)$. If $x \in V(G)$ is such that $\bar{T}_x$ contains the edge $\{\alpha, T(i)\}$, then $x \in \bar{\tau}(v)$ for some leaf $v \in T_1^*$ and $x \in \bar{\tau}(v')$ for another leaf $v' \in \bar{T} - T_1^*$. If $v \in T - T^*$ then T_x contains the edge $\{T(i), u\}$. If $v' \in T_2^*$ then necessarily $x \in X_j$ and again T_x contains the edge $\{T(i), u\}$, since T_x contains a leaf in T^* and $T(j) \in T_x$. In any case, the order of the edge $\{\alpha, T(i)\}$ in $\bar{T}$ does not exceed the order of the edge $\{T(i), u\}$ in T. Similarly for the edge $\{\delta, T(i)\}$. An analogous argument shows that the orders of edges within T_1^* (T_2^*) did not increase.

Denote by u_1^* (u_2^*) the copy of vertex $u \in T^*$ in T_1^* (in T_2^*, respectively). (In particular, $T(i)_1^* = \alpha$ and $T(i)_2^* = \delta$.) Observe that

$$\bar{T}(m) = \begin{cases} T(m) & \text{if } T(m) \notin T^* \\ T(m)_1^* & \text{if } T(m) \in T^* \text{ and } m \le j, \text{ and} \\ T(m)_2^* & \text{if } T(m) \in T^* \text{ and } m \ge j. \end{cases}$$

This is because $\bar{T}(m)$ (defined as above) belongs to $\bar{T}_x$ for every $x \in X_m$, and $|\bar{T}(X_m)| = 1$. Therefore $\bar{T}(i) = \alpha \ne \bar{T}(h) = \delta$ and $\bar{T}(m) = \bar{T}(m')$ only if $T(m) = T(m')$. □

Corollary 1. *For the graph G with a cca $\mathcal{X}$, there exists a fragmentation $\mathcal{I}$ and a relaxed branch-decomposition (T, τ) of width $k = \beta(G)$ such that $T(i) = T(j)$ if and only if i, j belong to the same interval of $\mathcal{I}$.*

Proof. For example, every decomposition with the maximum number of distinct $T(i)$'s defines such a fragmentation. □

Assume now that a fragmentation $\mathcal{I}$ as in Corollary 1 is fixed. We will further consider the interval supergraph $G^{\mathcal{I}}$ of G such that $\{X_{a,b} : [a, b] \in \mathcal{I}\}$ is the set of maximal cliques of $G^{\mathcal{I}}$. This graph has again branchwidth $k = \beta(G)$ (in a branch-decomposition satisfying Corollary 1, for every $x \in X_i$, $y \in X_j$ s.t. i, j belong to the same interval of the fragmentation - i.e., $T(i) = T(j)$ - and $xy \notin E(G)$, pend a leaf v_{xy} to a new vertex subdividing an edge incident with $T(i)$ which is contained in $T_x \cap T_y$, and set $\tau(v_{xy}) = (x, y)$). Again, in every relaxed branch-decomposition (T, τ) of width k, the cliques $X_i^{\mathcal{I}}$ have unique representatives $T(i)$, $i = 1, 2, \ldots, t$. For the sake of brevity we will now assume that $G = H^I$ from some graph H. In other words, we assume that G is saturated and G has a relaxed decomposition of width k in which the representatives of the maximal cliques are all distinct.

Definition 4. *Let X_i, X_j, and X_h be maximal cliques of a saturated graph G with relaxed branch-decomposition (T, τ). We say that the representatives of these cliques are in* claw-position *if there exists a node α of T such that every branch of T with respect to α contains one of the vertices $T(i)$, $T(j)$ and $T(h)$. We say that they are in (i, j, h)-path-position, if $T(i)$ and $T(h)$ lie in different branches of T with respect to $T(j)$.*

Note that for pairwise distinct $T(i)$, $T(j)$ and $T(h)$, exactly one of the following statements is true:

- $T(i)$, $T(j)$, and $T(h)$ are in claw-position,
- $T(i)$, $T(j)$, and $T(h)$ are in (i, j, h)-path-position,
- $T(i)$, $T(j)$, and $T(h)$ are in (j, i, h)-path-position,
- $T(i)$, $T(j)$, and $T(h)$ are in (i, h, j)-path-position.

For $i < j < h$, cliques with representatives in the last two mentioned positions are said to be in *wrong-order positions*.

Lemma 7. *If G has a relaxed branch-decomposition (T, τ) of width $k = \beta(G)$ such that clique representatives are distinct and $T(i-1), T(i)$ and $T(i+1)$ are in $(i-1, i+1, i)$-path-position for some i, then G has a decomposition $(\bar{T}, \bar{\tau})$ of the same width such that $\bar{T}(i-1), \bar{T}(i), \bar{T}(i+1)$ are in claw-position and such that no two clique representatives coincide and no other triple $(j-1, j, j+1)$ in wrong-order position is created.*

Proof. We rebuild the decomposition (T, τ) similarly as in the proof of Lemma 6. This time T^* is the tree rooted in $T(i+1)$ that does not contain $T(i)$ and $\bar{T}$ is constructed by deleting T^* from T, adding two new vertices α, δ adjacent to $T(i+1)$ and adding T_1^* rooted in α and T_2^* rooted in δ (again, only the leaves mapped by τ onto edges with both endpoints in $X_{1,i}$ are left in T_1^*, and T_2^* contains only copies of those leaves of T^* which are mapped by τ onto edges with both endpoints in $X_{i,\ell}$). Clearly, $(\bar{T}, \bar{\tau})$ is a relaxed branch-decomposition of G of width k, all clique representatives are still distinct and $\bar{T}(i-1) = T(i-1)_1^* \in T_1^*$, $\bar{T}(i) = T(i)$ and $\bar{T}(i+1) = \delta$ are in claw-position.

It remains to show that no other triple $(j-1, j, j+1)$ with clique-representatives in wrong-order position was created. For the contrary, suppose that $\bar{T}(j-1), \bar{T}(j)$ and $\bar{T}(j+1)$ are in wrong-order position but $T(j-1), T(j), T(j+1)$ were not. It follows that two of $\bar{T}(j-1), \bar{T}(j), \bar{T}(j+1)$ are in T_1^* and the last one is in T_2^*, or vice versa (two of $\bar{T}(j-1), \bar{T}(j), \bar{T}(j+1)$ are in T_2^* and the third one is in T_1^*). But $\bar{T}(h)$ is in T_1^* (T_2^*) only if $h < i$ ($h > i$, respectively), and thus the indices of the newly created triple with clique representatives in wrong-order position cannot be consecutive. $\square$

Corollary 2. *Every cca of a saturated interval graph G has a fragmentation $\mathcal{I}$ and a relaxed branch-decomposition (T, τ) of $G^{\mathcal{I}}$ of width $\beta(G)$ such that all representatives $T(i), i = 1, 2, \ldots, t$ are distinct and for every $i = 2, 3, \ldots, t - 1$, the representatives $T(i - 1), T(i)$ and $T(i + 1)$ are either in claw-position or in $(i - 1, i, i + 1)$-path-position.*

Proof. Obviously, Lemma 7 has a symmetric variant that kills a triple $(T(i - 1), T(i), T(i+1))$ in $(i, i-1, i+1)$-path-position. Therefore any relaxed branch-decomposition with minimum number of triples $(T(i-1), T(i), T(i+1))$ in wrong-order position has actually no triples $(T(i - 1), T(i), T(i + 1))$ in wrong-order position. □

The "only if" part of Theorem 2 below now follows from the following two lemmas.

Lemma 8. *There exists a relaxed branch-decomposition of width k for $G[X_p \cup X_q \cup X_r]$ in which the representatives of cliques X_p, X_q, X_r, $p < q < r$ are in claw-position if and only if $|X_i| \leq \frac{3}{2}k$ (for $i = p, q, r$) and $|X_q \cap (X_p \cup X_r)| \leq k$.*

Proof. Suppose $G[X_p \cup X_q \cup X_r]$ has a decomposition (T, τ). The inequalities $|X_i| \leq \frac{3}{2}k$ are obvious. Let u be the vertex of T such that $T(p), T(q)$ and $T(r)$ lie in different u-branches, and let e_p, e_q, e_r be the edges incident with u on the paths towards $T(p), T(q), T(r)$, respectively. Set $V_{ij} = \{x \in X_p \cup X_q \cup X_r : e_i, e_j \in T_x\}$ for $i, j = p, q, r$. Then $X_p \cap X_q \subseteq V_{pq}$ (since for $x \in X_p \cap X_q$, $T(p), T(q) \in T_x$ and e_p, e_q lie on the unique path connecting $T(p)$ and $T(q)$ in T) and similarly, $X_r \cap X_q \subseteq V_{rq}$. Therefore the order of e_q is at least $|V_{pq} \cup V_{rq}|$ and $|X_q \cap (X_p \cup X_r)| \leq k$ follows.

On the other hand, suppose X_p, X_q, X_r satisfy the conditions. Take a vertex u adjacent to x_p, x_q and x_r and add leaves mapping onto the edges of the cliques near the vertices x_p, x_q, x_r so that $T(i) = x_i$ for $i = p, q, r$. Since $|X_i| \leq \frac{3}{2}k$ (for $i = p, q, r$), this can be done so that the orders of the edges incident with the clique representatives are $\leq k$. It only remains to show that the orders of the edges incident with u are small as well. For every $i \neq j \in \{p, q, r\}$ and $x \in X_i \cap X_j$, the tree T_x contains the path x_i, u, x_j, and so the order of the path $x_i - u$ is $|X_i \cap (X_j \cup X_h)|$ (where j, h are such that $\{i, j, h\} = \{p, q, r\}$). For $i = q$, $|X_q \cap (X_p \cup X_r)| \leq k$ by the assumption. For $i = p$, $X_p \cap (X_q \cup X_r) \subseteq X_p \cap X_q$ (since $X_p \cap X_r \subseteq X_q$ as $p < q < r$) and hence $|X_q \cap (X_p \cup X_r)| \leq |X_p \cap X_q| \leq |X_q \cap (X_p \cup X_r)| \leq k$. Similarly for $i = r$. □

Lemma 9. *There exists a relaxed branch-decomposition of width k for $G[X_p \cup X_q \cup X_r]$ in which the representatives of cliques X_p, X_q, X_r, $p < q < r$ are in (p, q, r)-path-position if and only if $|X_i| \leq \frac{3}{2}k$ (for $i = p, q, r$), $|X_i \cap X_j| \leq k$ (for $i \neq j$, $i, j = p, q, r$) and $|X_q| + |(X_p \cap X_r)| \leq 2k$.*

Proof. Suppose a decomposition (T, τ) exists. Call e_1, e_2 and e_3 the outgoing edges of $T(q)$ such that e_1 is on the path towards $T(p)$ and e_3 is on the path towards $T(r)$.

Let α be the number of vertices x of X_q such that $e_1, e_3 \in T_x$, β the number of vertices such that $e_1, e_2 \in T_x$ and γ the number of vertices such that $e_2, e_3 \in T_x$. Then we can find a suitable assignment exactly when we can find numbers α, β and γ satisfying:

1. $\alpha \geq |X_p \cap X_r|$, $\alpha + \beta \geq |X_p \cap X_q|$, $\alpha + \gamma \geq |X_q \cap X_r|$
2. $\alpha + \beta + \gamma = |X_q|$
3. $\alpha + \beta \leq k$, $\alpha + \gamma \leq k$ and $\beta + \gamma \leq k$

It is easy to see that this system of inequalities has a solution exactly when the given restrictions are satisfied. $\qquad\square$

Theorem 2. *Let $\mathcal{X} = (X_1, \ldots, X_\ell)$ be an arbitrary cca of an interval graph G. For $k \geq 2$, $\beta(G) \leq k$ if and only if a k-fragmentation of $\mathcal{X}$ exists.*

Proof. The "only if" part follows from Lemma 8 and Lemma 9. We proceed with the converse. Let $\mathcal{F} = \{[a_i, b_i] : i \in [1, t]\}$ be a k-fragmentation of $\mathcal{X}$ such that $a_i - 1 = b_{i-1}$ for every $i = 2, 3, \ldots, t$. We will show that this implies $\beta(G) \leq k$. For simplicity let $X_0 = X_{\ell+1} = \emptyset$.

Following the constructions in Lemmas 8 and 9, we choose branch-decompositions (T_i, τ_i) of $G[X_{a_i, b_i}]$ such that T_i contains edges e_i and f_i with

$$e_i \in \bigcap_{x \in X_{a_i-1} \cap X_{a_i}} E(T_{ix}) \quad \text{and} \quad f_i \in \bigcap_{x \in X_{b_i+1} \cap X_{b_i}} E(T_{ix}).$$

If the *claw condition* $|X_{a,b} \cap (X_{a-1} \cup X_{b+1})| \leq k$ is fulfilled then $e_i = f_i$, and if the *path condition* $|X_{a,b}| + |X_{a-1} \cap X_{b+1}| \leq 2k$ is fulfilled then e_i and f_i are adjacent. If for X_{a_i, b_i} the path condition holds, then we subdivide e_i and f_i by additional vertices c_i and d_i, respectively. If for X_{a_i, b_i} the claw condition holds, then we subdivide $e_i = f_i$ by an additional vertex, which is adjacent to an additional leaf. In this case $c_i = d_i$ is this leaf. Now we obtain T by adding edges $\{d_{i-1}, c_i\}$ for $i \in [2, t]$, and we define $\tau(\lambda) = \tau_i(\lambda)$ if the leaf λ of T is also a leaf of T_i.

Obviously (T, τ) is a relaxed branch decomposition of G the width of (T, τ) is at most k. $\qquad\square$

Below we present a procedure to check whether the branchwidth of G is at most k, i.e., we check whether $[1, \ell]$ has a k-fragmentation. We use a boolean array $A[a, b]$ to indicate whether $[a, b]$ has a k-fragmentation. We show that this procedure can be implemented such that it runs in $O(n^3)$ time. We use a boolean vertex versus clique matrix with the consecutive ones property for rows. For each vertex x let $f(x)$ be the index of the first clique containing x and $l(x)$ be the index of the last clique containing x. Clearly the functions $f(x)$ and $l(x)$ can be computed in $O(n^2)$ time. Notice that now $X_{a,b}$ can be computed in $O(n)$ time for every pair a, b, since $x \notin X_{a,b}$ iff $l(x) < a$ or $f(x) > b$. Since there are $O(n^2)$ of these unions, all these can be computed in $O(n^3)$ time. We obtain the following theorem.

Theorem 3. *There exists an $O(n^3 \log n)$ algorithm to compute the branchwidth of an interval graph.*

The result can be extended to d-trapezoid graphs. We restrain from giving the details.

Input: A graph G with a cca $(X_1, \ldots, X_\ell)$ and an integer k.
Output: A statement whether the branchwidth of G is at most k.

```
begin
    X₀ ← ∅;   Xₗ₊₁ ← ∅;
    for d ← 0 to ℓ do
        for b ← d + 1 to ℓ do
            begin
                a ← b − d;
                if |Xₐ₋₁ ∩ Xₐ| > k or |Xᵦ ∩ Xᵦ₊₁| > k
                    then A[a, b] ← FALSE
                else
                    begin
                        if |Xₐ,ᵦ| ≤ ³⁄₂k and
                           (|Xₐ,ᵦ ∩ (Xₐ₋₁ ∪ Xᵦ₊₁)| ≤ k or
                            |Xₐ,ᵦ| + |Xₐ₋₁ ∩ Xᵦ₊₁| ≤ 2k)
                            then A[a, b] ← TRUE
                        else
                            begin
                                A[a, b] ← FALSE;
                                for c ← a to b − 1 do
                                    A[a, b] ← A[a, b] or (A[a, c] and A[c + 1, b])
                            end
                    end
            end
    end
    if A[1, ℓ] then output "β(G) ≤ k" else output "β(G) > k"
end
```

5   Concluding Remarks

One of the basic questions we could not settle thus far is the complexity of branchwidth for cocomparability graphs. Also for the subclass of cobipartite graphs, the complexity is an open problem.

References

1. Bodlaender, H. and D. M. Thilikos, Constructive linear time algorithms for branchwidth, *Proceedings ICALP'97*, Springer–Verlag, LNCS 1256, (1997), pp. 627–637.
2. Booth, K. and G. Lueker, Testing for the consecutive ones property, interval graphs, and graph planarity testing using PQ-tree algorithms, *J. of Computer and System Sciences* **13**, (1976), pp. 335–379.

3. Garey, M. R. and D. S. Johnson, *Computers and intractability, a guide to the theory of NP-completeness*, Freeman, San Francisco 1979.
4. Kloks, T., *Treewidth–Computations and Approximations*, Springer–Verlag, LNCS 842, 1994.
5. Robertson, N. and P. D. Seymour, Graph minors X: Obstructions to tree-decomposition, *Journal of Combinatorial Theory, Series B* **52**, (1991), pp. 153–190.
6. Seymour, P. D. and R. Thomas, Call routing and the ratcatcher, *Combinatorica* **14**, (1994), pp. 217–241.

Balanced Randomized Tree Splitting with Applications to Evolutionary Tree Constructions

Ming-Yang Kao[1], Andrzej Lingas[2], and Anna Östlin[2]

[1] Department of Computer Science, Yale University, New Haven, CT 06520, USA,
Kao-Ming-Yang@cs.yale.edu
[2] Department of Computer Science, Lund University, Box 118, S-221 00 Lund,
Sweden
{Andrzej.Lingas,Anna.Ostlin}@dna.lth.se

Abstract. We present a new technique called *balanced randomized tree splitting*. It is useful in constructing unknown trees recursively. By applying it we obtain two new results on efficient construction of evolutionary trees: a new upper time-bound on the problem of constructing an evolutionary tree from experiments, and a relatively fast approximation algorithm for the maximum agreement subtree problem for binary trees for which the maximum number of leaves in an optimal solution is large. We also present new lower bounds for the problem of constructing an evolutionary tree from experiments and for the problem of constructing a tree from an ultrametric distance matrix.

1 Introduction

Several of the known efficient algorithms for trees rely on their excellent separator properties. It is well known that each tree contains a vertex whose removal splits it into components of balanced size. Unfortunately, finding such a vertex usually requires the knowledge of the tree. In this paper, we consider a more general situation when the tree is unknown and we can obtain some partial information on its topology at some cost. More precisely, the partial information is in the form of the topological subtree induced by a subset of the leaves and the cost corresponds to the time taken by the construction of the subtree in a given model. We introduce an efficient randomized technique of balanced splitting of an unknown tree termed *balanced randomized tree splitting*. It can be used to construct an unknown tree recursively. Our technique seems to be especially useful in the efficient construction of evolutionary trees.

The problem of constructing evolutionary trees is central in computational biology and has been studied in several papers [1, 2, 3, 4, 5, 6, 7, 8, 9, 10, 11, 12, 13], An *evolutionary tree* is a tree where the leaves represent species and internal nodes represent common ancestors, see Fig. 1 and 2 for two examples. There are many different approaches to the problem of constructing an evolutionary tree depending, among other things, on what kind of data that is available.

A well known variant of the problem of constructing an evolutionary tree is based on experiments. An *experiment* determines how three species are related

C. Meinel and S. Tison (Eds.): STACS'99, LNCS 1563, pp. 184–196, 1999.
© Springer-Verlag Berlin Heidelberg 1999

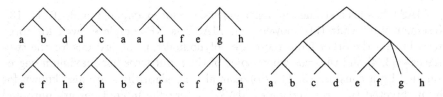

Fig. 1. An example of experiment results and the corresponding evolutionary tree.

in the evolutionary tree, i.e., returns the topological subtree (without weights on the edges) for the three species. Fig. 1 shows an example of a tree and the outcome of experiments. Note that there are two different types of trees one can get from an experiment.

The problem of constructing evolutionary trees using experiments have been studied by Kannan, Lawler, and Warnow [8]. In [8], they present algorithms for binary evolutionary trees. The fastest of them runs in time $O(n \log n)$ and performs at most $4n \log_2 n$ experiments. Experiments are expensive and hence it is important to minimize their number. The algorithm that performs the smallest number of experiments presented in [8] runs in time $O(n^2)$ and performs at most $n \log_2 n$ experiments. For trees not restricted to be binary Kannan *et al.* present an $O(n^2)$-time algorithm, which performs $O(dn \log n)$ experiments, where d is the maximum degree in the tree, and for the unrestricted degree case they present an $O(n^2)$-time algorithm, that performs $O(n^2)$ experiments [8].

By using our new technique of balanced randomized tree splitting, we show that an evolutionary tree for n species can be determined, using experiments, in expected time $O(nd \log n \log \log n)$, where d is the maximum degree in the tree. This is a dramatic improvement over the quadratic time-bound of Kannan *et al.* causing only a slight, practically negliable, increase in the number of experiments.

In [8] it is also shown that the lower bound on the number of experiments required to construct the tree in worst case is $\Omega(n^2)$. In this paper we show an $\frac{1}{3} \max\{(\deg(u))^2 + \sum_{v \in IV(T) - \{u\}} (\deg(v) - 1)^2, \Omega(n \log n)\}$ lower bound on the number of experiments, where u is an internal vertex of maximum degree and $IV(T)$ is the set of internal vertices in the tree. In particular, if the tree contains $\Omega(n/d)$ vertices of degree $\Omega(d)$, $\Omega(n(\log n + d))$ experiments are required. We derive analogous lower bounds for the number of entries in the so called ultrametric distance matrix (see Sect. 3.2) that every algorithm constructing the tree from it has to access.

Another popular variant of the problem of constructing an evolutionary tree is called the *maximum agreement subtree problem* (MAST for short), or maximum homeomorphic subtree problem (MHT for short). Given k rooted leaf labeled trees $T_1, ..., T_k$, each with n leaves uniquely labeled by elements from an n-element set A, MAST asks for a rooted tree T with the maximum number of leaves uniquely labeled by some elements from A, such that for $i = 1, ..., k$, T is homeomorphic to the subtree of T_i induced by the leaves of T. See Fig. 2 for an example.

MAST has been studied in many papers, see for example [3, 5, 6, 10, 11, 13]. Keselman and Amir [11] showed that MAST is NP-complete, even for three trees, but on the other hand there are polynomial time algorithms for the two-tree case [3, 4] and the case where some of the input trees has maximum degree bounded by a constant [3, 11]. Polynomial time algorithms are also known for the undirected two-tree version of MAST, where the input trees are unrooted [4, 10, 13]. The best known algorithm, due to Farach, Przytycka, and Thorup, for MAST for k n-leaf trees, some of which has maximum degree bounded by d, runs in time $O(kn^3 + n^d)$ [3].

In practice it is often the case that the input trees are both binary and very similar, and only a small fraction of the leaves has to be removed to find an agreement subtree. For this reason, we have considered the problem of an efficient approximation of MAST in the case when the number of leaves in the MAST tree is a large fraction of n. For large n, such an approximation can be useful as test of whether or not the optimal solution is sufficiently large, and so enough interesting, to run the more costly exact algorithm to produce it. In this paper, we present an algorithm which for k input n-leaf binary trees, admitting an agreement subtree on at least βn leaves, where $\beta \in (0.8, 1]$, constructs an agreement subtree whose expected number of leaves is at least $0.4\beta n$ in time $O(kn^{9/5} \log^{6/5} n)$. This approximation algorithm is our second example of successful application of the technique of balanced randomized tree splitting.

Sect. 2 presents some general results on balanced randomized tree splitting. In Sect. 3 the upper bound on constructing a tree from experiments is shown and the lower bounds are presented. Sect. 4 presents the approximation algorithm for MAST.

2   Balanced Randomized Splitting of an Unknown Tree

To find a balanced splitting of the unknown tree T, we randomly pick a sample of its leaves and build the topological subtree T' induced by the sample. The removal of the vertices of T' from T splits T into components of balanced number of leaves with high probability.

Given a tree T and a subset S of leaves of T, the *topological subtree T' of T induced by S*, denoted by $T \parallel S$, is the minimum size tree that has S as the set

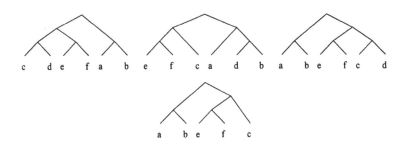

Fig. 2. Three binary input trees with six species yield a maximum agreement subtree with five species.

of its leaves and is homeomorphic to the subtree of T composed of paths in T joining each pair of leaves in S.

For an edge (u, w) of T, let $T_{u,w}$ denote the subtree of T rooted at u and composed of all nodes reachable from w via u.

For a pair of vertices w_1, w_2 of T, let $T(w_1, w_2)$ denote the forest composed of all distinct subtrees $T_{u,w}$ where w is an internal vertex on the path joining w_1 with w_2 in T and u is a vertex of T outside this path.

A *component* of $T - T'$ is either (1) a subtree $T_{u,w}$ where $w \in V(T')$ and u does not belong to any path in T joining w with an adjacent vertex in T' or (2) a forest $T(w_1, w_2)$ where (w_1, w_2) is an edge of T'.

Theorem 1. *Let T be a tree on n leaves. Let $2 \leq n_0 < n$ and $k \geq 2$. For any sample S of at least $2k \frac{n}{n_0} \log n$ leaves chosen uniformly at random among the n leaves of T, and the topological subtree T' of T induced by S, each of the components of $T - T'$ contains less than n_0 leaves with probability at least $1 - n^{-k}$.*

Proof. Note that T has at least two edges. For a pair of vertices w_1, w_2 of T, suppose that $T(w_1, w_2)$ has at least n_0 leaves of T. The probability that none of the leaves is chosen in S is not greater than $(1 - \frac{n_0}{n})^{n \frac{2k \log n}{n_0}}$, i.e., it is not greater than $\frac{1}{n^{2k}}$. Since T has at most n vertices, there are less than n^2 forests $T(w_1, w_2)$. Consequently, the probability that there is such a forest $T(w_1, w_2)$ with at least n_0 leaves of T none of which is chosen to S is less than $\frac{1}{n^{2k-2}}$ which is less than $\frac{1}{n^k}$ since $k \geq 2$. Each of the components of $T - T'$ is included in a forest of the form $T(w_1, w_2)$, therefore we conclude that each of them contains less than n_0 leaves with probability larger than $1 - n^{-k}$. $\square$

Note that Theorem 1 implies the existence of a single vertex in T' whose removal from T partitions the leaves of T into balanced size components with high probability.

Corollary 1. *Let T be a tree on n leaves. Let $2 \leq n_0 \leq n/2$ and $k \geq 2$. For any sample S of at least $2k \frac{n}{n_0} \log n$ leaves chosen uniformly at random among the n leaves of T, the topological subtree T' of T induced by S contains a vertex of T whose removal disconnects T into subtrees none of which contains more than $n/2 + n_0$ leaves with probability at least $1 - n^{-k}$. Given T' and the size of the components of $T - T'$, such a vertex can be found in time $O((n/n_0) \log n)$.*

Proof. Transform T' to an auxiliary tree T^*, by breaking each edge $e = (w_1, w_2)$ into two edges (w_1, w_e), (w_e, w_2), where w_e is a new vertex uniquely associated with e, and adding for each component C of $T - T'$ a unique leaf l_C. If C is of the form $T_{u,w}$ then l_C is adjacent to the vertex w in T'. Otherwise, if C is of the form $T(w_1, w_2)$ then l_C is adjacent to w_e where $e = (w_1, w_2)$. For each leaf l_C of T^*, set its weight to the number of leaves of T in C. For each of the remaining vertices w of T^*, set the weight w to zero.

To prove the thesis, it is sufficient to find a vertex of T' whose removal disconnects T^* into subtrees none of which contains vertices of total weight greater than $n/2 + n_0$.

By Theorem 1, each vertex of T^* has weight not exceeding n_0, which does not exceed $n/2$ by our assumptions, with probability at least $1 - n^{-k}$.

Root T^* at a vertex of T'. If for each child of the root the total weight of descendant vertices is not greater than $n/2$ then we are done. Otherwise, walk down the tree always in direction of the child for which the total weight of descendant vertices is larger than $n/2$. If there is no such child, stop. Note that since each leaf of T^* has weight not larger than $n/2$, the above procedure stops at an internal vertex v of T^*. Also, by the definition of v, its removal disconnects T^* into subtrees each containing vertices of total weight not greater than $n/2$. If v is a vertex of T', we are done. Otherwise, we choose the parent w of v in the rooted T^*. Since v in this case has only two children, one of which is a leaf of weight less than n_0, it follows that w satisfies the thesis in this case.

Given T' and the size of the components of $T - T'$, the above procedure can be easily implemented in time linear in the size of T'.   □

Corollary 1 is mostly interesting in the situation when the number of leaves in the components of T induced by the removal of the vertices in T' can be determined faster than it takes to build the whole T.

3   Improved Bounds on Building Trees from Experiments

3.1   An Upper Bound

We later use the result from Sect. 2 to derive a new upper time-bound on the construction of an evolutionary tree from experiments, depending on the maximum degree of the constructed tree. The following procedure will yield our upper bound.

Algorithm BuildTree(L)
Input: A set L of species for which experiments can be made.
Output: An evolutionary tree for L.
1. Pick a random sample S of species from L of size $8\lceil \log |L| \rceil$;
2. Build an evolutionary tree T' for S by using the quadratic-time algorithm from [8];
3. Determine the components $C_1, ..., C_j$ of $T - T'$ where T is the evolutionary tree to construct;
4. Augment each of the components of $T - T'$ of the form $T(x, y)$ by a single species labeling a leaf of T' lying below the lowest common ancestor of x and y (keeping unchanged the remaining components);
5. **For** each augmented component C_i **do**
   if C_i contains at most $\log n$ species then construct the evolutionary tree for C_i by using the quadratic-time algorithm
   else BuildTree(C_i)
6. Hang the evolutionary trees recursively computed for the augmented components C_i at T' as follows:
   if C_i is of the form $T_{x,y}$ **then** link the root of the evolutionary tree for C_i to y;

if C_i has been originally of the form $T(x, y)$ **then** identify the root of the evolutionary tree for the augmented C_i with the lowest common ancestor of x, y, and identify its leaf labeled by the additional species with the other vertex in $\{x,\ y\}$.

The next fact is useful for analyzing the time complexity of BuildTree(L).

Fact 1 (see [8]) *Let v be an internal node of the evolutionary tree for a set U of species. Let $W \subseteq U$, and let $W_1, ..., W_q$ be the splitting of W into non-empty components induced by v. For any $u \in U - W$, we can determine whether u belongs to any of the components of U induced by v that is a superset of one of the components $W_1, ..., W_q$, and if so, also the index of the component, in time $O(q)$ by performing $\lceil \frac{q}{2} \rceil$ experiments.*

Theorem 2. *An evolutionary tree for n species can be determined in expected $O(nd \log n \log \log n)$ time, where d is the maximum degree in the tree.*

Proof. Let L be the set of n species. We use BuildTree(L) to prove the theorem. The correctness of this procedure follows from the correctness of combining the evolutionary trees recursively computed for the augmented components in step 6.

As for the expected running time of BuildTree(L), it is easily seen to be dominated by steps 2, 3 and the "if" part of step 5.

To estimate the total work (including recursive calls) required in step 2 and the "if" part of step 5 note that each internal node of T can appear in at most one of the evolutionary trees T' built for the samples. Let $n_1, ..., n_q$ be the sizes of the samples drawn by the algorithm. We have $\sum_{i=1}^{q} n_i \leq n$, and $n_i \leq 8\lceil \log n \rceil$ for $i = 1, ..., q$. Hence, the total work for constructing the evolutionary trees T' for the samples (step 2) does not exceed

$$O\left(\max_{\sum_i n_i \leq n, n_i \leq 8\lceil \log n \rceil} \left(\sum n_i^2\right)\right) = O\left(\frac{n}{\log n} \log^2 n\right) = O(n \log n)$$

Analogously, the total time taken by the construction of the evolutionary trees for the components of size not greater than $\log n$ (the "if" part of step 5) is $O((n/\log n) \log^2 n) = O(n \log n)$.

In order to derive our upper bound on the expected time required by step 3, we consider the following method of determining the components of $T - T'$ proceed as follows.

First, we find a (separator) vertex v of T' whose removal disconnects T' into subtrees T_i', $i = 1, ..., q$, none of which has more than $\frac{2}{3}$ of the vertices of T'. Now, we can determine the leaf sets of the subtrees of the form $T_{u,v}$, where (u, v) is an edge of T, by performing at most $\lceil d/2 \rceil n$ experiments in time $O(dn)$ by Fact 1. In this way, in particular we can determine the components of $T - T'$ of the form $T_{u,v}$. Let v_1 through v_l be the neighbors of v in T', and L_1 through L_l be the leaf sets assigned to the branches (v, u_1) through (v, u_l) by the aforementioned experiments. Note that for $i = 1, ..., l$, L_i is just the leaf set of a subtree $T_{u,v}$ where u is the second vertex on the path from v to u_i in T. Let T'_i be the subtree of T' corresponding to the branch (v, u_i), i.e., the maximal subtree of T' including u_i and excluding v. For $i = 1, ..., l$, we analogously find a vertex separator v_i of T'_i and perform at most $\lceil d/2 \rceil |L_i|$ experiments in order to split L_i into the subsets of leaf sets of the subtrees of the form T_{u,v_i}. Importantly, note that if $u = v$ then the subset of L_i assigned to the branch (v_i, v) yields the component $T(v, v_i)$. By proceeding in this way recursively, we can determine all the components of $T - T'$.

A recursive separator partition of T' in the form of a tree of vertex separators of T' can be found in time linear in the size of T', i.e., in time $O(\log n)$. Hence, the total time taken by finding such separator partitions including the recursive calls of the BuildTree procedure is easily seen to be $O(n)$.

In order to estimate the expected total work taken by determining the components of the tree T restricted to the current component with respect to the evolutionary tree T' for the sample for the current component let us make the following observation.

Since the recursive separator partition of T' has depth logarithmic in the size of T', each species labeling a leaf of T takes part in $O(\log \log n)$ leaf sets that are partitioned by the experiments in order to determine the components of $T - T'$.

For a species s, let $n_1, ..., n_h$ be the sizes of the components that s belongs to during the performance of the algorithm, where $n_1 = n$, $n_i > \log n$ for $i = 1, ..., h-1$, and $n_h \leq \log n$. Let $h(n) = \log n - \log \log n$.

By Theorem 1 it follows that the probability that for $i = 1, ..., \min(h(n), h) - 1$, $n_{i+1} \leq n_i/2$ is at least $\prod_{i=1}^{\min(h(n),h)-1}(1 - \frac{1}{n_i}) \geq \prod_{i=1}^{h(n)-1}(1 - \frac{1}{\log n}) \geq e^{-1}$. Consequently, $\text{Prob}(h \leq h(n)) \geq e^{-1}$ holds. Hence, by Markov inequality the expected value of h is at most $eh(n)$.

Thus, the expected number of leaf sets s belongs to during our algorithm is $O(\log n \log \log n)$. For each leaf set species s belongs to, s has to participate in $O(nd)$ experiments (determining the components of the tree T restricted to the current component with respect to the splitting trees). The expected total work and the expected total number of experiments are therefore $O(nd \log n \log \log n)$.

□

3.2   Lower Bounds

We shall show a lower bound on the number of experiments required by any algorithm constructing an evolutionary tree from them. We do this by firstly showing a lower bound on the number of entries in the matrix that has to be accessed by any algorithm constructing a tree from the so called ultrametric distance matrix.

A *distance matrix* D for n species is a $n \times n$ matrix where all elements D_{ij} are non-negative real numbers. The value D_{ij} represents the distance between the species i and j. For example D_{ij} could represent the distance between DNA sequences for species i and j. If it is possible to construct a tree, with edge weights, such that the distance between every pair of leaves equals the corresponding value in the matrix, then the matrix is said to be *additive*. If the tree can be rooted in such a way that the distance from the root to all leaves are equal, then the matrix is called *ultrametric*.

Culberson and Rudnicki gave an $O(n^2)$-time algorithm for the problem of constructing a tree from an additive distance matrix [1], and showed it to be optimal. When the degree in the tree is bounded by d, their algorithm runs in time $O(dn\log_d n)$.

Our lower bound on constructing a tree from an ultrametric distance matrix is based on the following simple lemma.

Lemma 2. *Let T be a tree realizing an ultrametric distance matrix M. Let v be an internal vertex of T, and let $C_1, ... C_k$ be the connected components of T resulting from removing v. For $i, j \in \{1, ..., k\}$, $i \neq j$, any tree-realizability algorithm that constructs T on the basis of M without any a priori knowledge on M has to access at least one entry of M corresponding to a leaf in C_i and a leaf in C_j.*

Proof. Suppose otherwise, i.e., that a tree-realizability algorithm doesn't access any entry of M corresponding to a leaf in C_i and a leaf in C_j. Let m be the minimum distance between a leaf in C_i and a leaf in C_j. Transform M to another tree-realizable matrix M' by decreasing the distance between each leaf in C_i and each leaf in C_j by $m - \epsilon$ so in the tree T' realizing M' the components C_i and C_j are forced to form the same component with respect to the vertex corresponding to v. The algorithm would output the tree T instead of the correct one, i.e., T'.   □

Theorem 3. *Let T be a tree realizing an ultrametric distance $n \times n$ matrix M, and let u be an internal vertex of T of maximum degree. Any tree-realizability algorithm that constructs T on the basis of M without any a priori knowledge on M has to access $\max\{(\deg(u))^2 + \sum_{v \in IV(T) - \{u\}} (\deg(v) - 1)^2, \Omega(n \log n)\}$ entries in M.*

Proof. The $\Omega(n \log n)$ bound directly follows from the corresponding information-theoretic lower bound on the number of experiments determining the topology of evolutionary subtrees on three leaves necessary to find out the topology of the whole evolutionary tree due to Kannan, Lawler and Warnow [8]. The lower bound of Kannan *et al.* is valid even in case of binary evolutionary trees. Since the topology of a tree on three leaves can be determined by accessing three entries of M respectively indexed by the leaves, we conclude that $\Omega(n \log n)$ accesses to the entries of M are required in order to determine the topology of T.

To prove the first lower bound, root T at u. For each non-leaf vertex v of the rooted T, at least $(\deg(v) - 1)^2$ entries of M corresponding to pairs of leaves a, b where a and b are in different subtrees rooted at the children of v have to be accessed by Lemma 2 (for $v = u$, the corresponding number is at least

$(\deg(u))^2)$. Now, it is sufficient to note that such a pair a, b can only be counted at the lowest common ancestor of a and b at the rooted T.   □

Corollary 2. *For a tree T realizing an ultrametric distance matrix M and having its leaves pending from $n/(d-1)$ internal vertices of degree d each, any tree-realizability algorithm that constructs T on the basis of M without any a priori knowledge on M has to access $\Omega(n(\log n + d))$ entries of M.*

Since ultrametric distance matrices are a special case of distance matrices the above lower bounds hold also for distance matrices in general. These lower bounds can be also easily translated into corresponding ones for algorithms constructing evolutionary trees from experiments on three species.

As pointed out in [8], an experiment on three species could be implemented using an ultrametric distance matrix, by looking at its three entries. Hence, if we could construct the evolutionary tree by performing less than, say, $g(n)$ experiments, then we could construct it by looking at less than $3 \cdot g(n)$ entries in the matrix. Combining this Kannan's *et al.* observation with our lower bounds for ultrametric distance matrices, we obtain the following theorem.

Theorem 4. *Let T be the evolutionary tree for a set of n species, and let u be an internal vertex of T of maximum degree. Any algorithm that constructs T on the basis of experiments without any a priori knowledge on T has to perform at least $\frac{1}{3} \max\{(\deg(u))^2 + \sum_{v \in IV(T) - \{u\}}(\deg(v) - 1)^2, \Omega(n \log n)\}$ experiments. In particular, if T contains $\Omega(n/d)$ vertices of degree $\Omega(d)$, $\Omega(n(\log n + d))$ experiments are required.*

4   An Approximation Algorithm for MAST

In practical applications of the maximum agreement subtree problem, one has typically a large number of binary (or, at least bounded degree) input trees and looks for a large agreement subtree covering the majority of the input species. Although there are known polynomial-time algorithms for MAST when some of the input trees have maximum degree bounded by a constant [3, 11], they are not practical as they require $\Omega(kn^3)$ time for k input trees, each on n leaves. For this reason, it seems worthy to develop faster approximation algorithms for the variants of MAST occuring in practise. They could be used to test whether the input instance has large enough agreement subtree worthy to run a costly exact algorithm on it or not. In this section, we present an efficient method for verifying whether or not the binary input trees have a large maximum agreement subtree based on our technique of balanced randomized tree splitting. Because of space considerations and for the sake of presentation clarity, we omit an extension of our method to include input trees of constantly bounded degree. Our method for binary trees is specified as follows.

   Algorithm MAST$(T_1, ..., T_k)$
   Input: binary trees $T_1, ..., T_k$, each on n leaves uniquely labeled with 1 through n
   Output: an agreement subtree U of $T_1, ..., T_k$

1. Randomly pick a subset L of $\lceil n^{3/5} \log^{2/5} n \rceil$ leaf labels (species) in $\{1, 2, ..., n\}$.

2. For $i = 1, ..., k$, set H_i to $T_i \parallel L$.

3. Find the maximum agreement subtree J for $H_1, ..., H_k$ by using the algorithm from [3].

4. For each edge (x, y) of J and $i = 1, ..., k$, determine the set $L_i[x, y]$ of leaf labels in the forest $T_i(x, y)$. Set $L[x, y]$ to $\bigcap_{i=1}^{k} L_i[x, y]$.

5. For each edge (x, y) of J, if $L[x, y] \neq \emptyset$ and y is a descendant of x in J then construct the maximum agreement subtree $J[x, y]$ for the trees of $T_i^y \parallel (L[x, y] \cup \{a_1, ..., a_{|L[x,y]|+1}\})$, $i = 1, ..., k$, where T_i^y is the tree obtained from T_i by deleting all descendants of y and instead hanging at y a binary tree (the same for $i = 1, ..., k$) with $|L[x, y]| + 1$ leaves uniquely labeled with the elements a_1 through $a_{|L[x,y]|+1}$ different from the leaf labels of T.

6. Expand J to the tree U as follows:

   for each edge (x, y) of J, where y is a descendent of x, if $L[x, y] \neq \emptyset$ then identify x with the root of $J[x, y]$ and y with an arbitrary leaf of $J[x, y]$ with a label of the form a_l.

7. Set U to the subtree of U induced by the leaves labeled with leaf labels of T and output it.

Note that the maximum agreement subtree $J[x, y]$ always contains a leaf with a label of the form a_l since the binary tree with $|L[x, y]| + 1$ leaves uniquely labeled with the elements a_1 through $a_{|L[x,y]|+1}$ is an agreement subtree for the same trees including more than half of the possible leaf labels. Hence, the following lemma easily follows from the construction of U.

Lemma 3. *The tree U output by $MAST(T_1, ..., T_k)$ is an agreement subtree for the input trees $T_1, ..., T_k$.*

The analysis of time complexity of $MAST(T_1, ..., T_k)$ is a bit more complicated.

Lemma 4. *For the binary input trees $T_1, ..., T_k$, each on n leaves, $MAST(T_1, ..., T_k)$ runs in $O(kn^{9/5} \log^{6/5} n)$ time with probability at least $1 - n^{-1}$.*

Proof. Steps 1, 2 are easily seen to take $O(n^{3/5} \log^{2/5} n)$-time or $O(kn)$-time, respectively. By using the algorithm for MAST from [3], Step 3 can be implemented in time $O(k(n^{3/5} \log^{2/5} n)^3)$. In Step 4, the sets $L_i[x, y]$ can be determined by standard tree searches in time $O(kn)$. Note that by Theorem 1 each of them is of size $O(n^{2/5} \log^{3/5} n)$ with probability $1 - n^{-1}$. Consequently, each of the sets $L[x, y]$ is of the size $O(n^{2/5} \log^{3/5} n)$ with probability $1 - n^{-1}$. The sets $L[x, y]$ can be computed in Step 4 by sorting lexicographically the sets $L_i[x, y]$ in total time $O(kn)$. By [3], the highly probable $O(n^{2/5} \log^{3/5} n)$ bound on the size of $L[x, y]$, and the n upper bound on the sum of the sizes of $L[x, y]$, the agreement subtrees $J[x, y]$ can be computed in time $O((n^{3/5}/\log^{3/5} n)(k(n^{2/5} \log^{3/5} n)^3)$ with probability at least $1 - n^{-1}$ in Step 5. The final expansion and pruning steps take $O(n)$ time. $\qquad\square$

The analysis of approximation properties of $\text{MAST}(T_1, ..., T_k)$ is more involved. To begin with, we need the following probabilistic lemma.

Lemma 5. *Let T be a tree on n leaves and let $\beta \in (0.8, 1]$. Next, let T' be the topological subtree of T induced by a sample of at least $(2\beta - 1)\lceil n^{3/5} \log^{2/5} n \rceil$ leaves of T chosen from $\beta \lceil n^{3/5} \log^{2/5} n \rceil$ leaves which are chosen uniformly at random from the n leaves of T. The probability that there is a set of at most $(1 - \beta)\lceil n^{3/5} \log^{2/5} n \rceil$ edges (x, y) such that the forests $T(x, y)$ totally contain at least $(1 - 0.4\beta)n$ leaves of T is $2^{-\Omega(n^{3/5} \log^{2/5} n)}$.*

Proof. Each of the edges (x, y) of T' is determined by at most three leaves of the sample, two to determine the lower endpoint (x, y) as their lowest common ancestor, and one additional to determine the higher endpoint. On the other hand, in the best case, $(1 - \beta)\lceil n^{3/5} \log^{2/5} n \rceil + 1$ leaves in the sample can already determine all the $(1 - \beta)\lceil n^{3/5} \log^{2/5} n \rceil$ edges (x, y). For this reason, we can bound the aforementioned probability from above, for $y = n^{3/5} \log^{2/5} n$, by

$$\sum\nolimits_{(1-\beta)y < x \leq 3(1-\beta)y} \binom{n}{\lceil x + (1-\beta)y \rceil} \binom{\lceil n - (1-0.4\beta)n - x - (1-\beta)y \rceil}{\lceil \beta y - x - (1-\beta)y \rceil} \bigg/ \binom{n}{\lfloor \beta y \rfloor}.$$

By straightforward calculations using the Stirling approximation formula, we obtain the lemma thesis.   □

Theorem 5. *let $\beta \in (0.8, 1]$. If the number of leaves in a maximum agreement subtree of the trees $T_1, ..., T_k$ is at least βn then the expected number of leaves in the agreement subtree U produced by $MAST(T_1, ..., T_k)$ is at least $0.4\beta n$.*

Proof. Let S be a maximum agreement subtree of $T_1, ..., T_k$. Suppose first that at least $\beta \lceil n^{3/5} \log^{2/5} n \rceil$ elements of L belong to S. It follows that there is an agreement subtree for H_i's of size not less than $\beta \lceil n^{3/5} \log^{2/5} n \rceil$ induced by $L \cap S$. Hence, a maximum agreement subtree for H_i's is of size at least $\beta \lceil n^{3/5} \log^{2/5} n \rceil$. It might contain at most $\lceil n^{3/5} \log^{2/5} n \rceil - \beta \lceil n^{3/5} \log^{2/5} n \rceil$ leaf labels outside S. Hence, it contains at least $(2\beta - 1)\lceil n^{3/5} \log^{2/5} n \rceil$ leaf labels in $S \cap L$.

Let S' be the agreement subtree for H_i's induced by the restriction of the leaf labels in the maximum agreement subtree J for H_i's to $S \cap L$. Clearly, S' is also an agreement subtree for T_i's. It can also be obtained from S by the restriction of the leaf labels in S to $S \cap L$. It follows that S' is a topological subtree of both J and S.

Each maximal subtree of J with leaf labels in $L - S$ can have its root on at most one path in J that one-to-one corresponds to a single edge of S'. It follows that the number of edges in S' that do not appear in J is bounded from above by the number of leaf labels of J outside S, i.e., it is at most $(1 - \beta)\lceil n^{3/5} \log^{2/5} n \rceil$

If an edge (x, y) of S' is also an edge of J then $\text{MAST}(T_1, ..., T_k)$ in Step 5 will find a super-tree of a maximum agreement subtree for the leaf labels of $S'(x, y)$. We conclude by Lemma 5 that under the preliminary assumption that at least $\beta \lceil n^{3/5} \log^{2/5} n \rceil$ elements of L belong to S, the agreement subtree U produced by $\text{MAST}(T_1, ..., T_k)$ contains at least $n - (1 - 0.4\beta)n$ leaf labels with very high probability. Now it is sufficient to note that the preliminary supposition holds with the probability at least $\frac{1}{2}$ in order to get the lemma thesis.   □

5 Final Remarks

The technique of balanced randomized tree splitting presented in this paper should be useful in finding efficient algorithms for several other problems involving construction of unknown trees (both within computational biology as well as outside it).

Our approximation algorithm for MAST presumably yields much better approximation than that stated in Theorem 5. We conjecture that the expected number of leaves in the agreement subtree produced by it is at least (the minimum number of edges in S' that appear in J times the expected size of a forest $S'(x, y)$, which is) $\frac{7}{8}\beta n$.

References

[1] J. C. Culbertson and P. Rudnicki. A fast algorithm for constructing trees from distance matrices. *Information Processing Letters*, 30(4):215–220, 1989.

[2] M. Farach, S. Kannan, and T. J. Warnow. A robust model for finding optimal evolutionary trees. *Algorithmica*, 13(1/2):155–179, 1995.

[3] M. Farach, T. M. Przytycka, and M. Thorup. Computing the agreement of trees with bounded degrees. In P. Spirakis, editor, *Lecture Notes in Computer Science 979: Proceedings of the Third Annual European Symposium on Algorithms*, pages 381–393. Springer-Verlag, New York, NY, 1995.

[4] M. Farach and M. Thorup. Fast comparison of evolutionary trees (extended abstract). In *Proceedings of the 5th Annual ACM-SIAM Symposium on Discrete Algorithms*, pages 481–488, 1994.

[5] C. R. Finden and A. D. Gordon. Obtaining common pruned trees. *Journal of Classification*, 2:255–276, 1985.

[6] L. Gąsieniec, J. Jansson, A. Lingas, and A. Östlin. On the complexity of computing evolutionary trees. In *Lecture Notes in Computer Science 1276: Proceedings of the 3rd Annual International Computing and Combinatorics Conference*, pages 134–145, 1997.

[7] M. R. Henzinger, V. King, and T. J. Warnow. Constructing a tree from homeomorphic subtrees, with applications to computational biology. In *Proceedings of the 7th Annual ACM-SIAM Symposium on Discrete Algorithms*, pages 333–340, 1996.

[8] S. K. Kannan, E. L. Lawler, and T. J. Warnow. Determining the evolutionary tree using experiments. *Journal of Algorithms*, 21:26–50, 1996.

[9] M. Y. Kao. Tree contractions and evolutionary trees. *SIAM Journal on Computing*, 1998. To appear.

[10] M. Y. Kao, T. W. Lam, T. M. Przytycka, W. K. Sung, and H. F. Ting. General techniques for comparing unrooted evolutionary trees. In *Proceedings of the 29th Annual ACM Symposium on Theory of Computing*, pages 54–65, 1997.

[11] D. Keselman and A. Amir. Maximum agreement subtree in a set of evolutionary trees – metrics and efficient algorithms. In *Proceedings of the 35th Annual IEEE Symposium on the Foundations of Computer Science*, pages 758–769, 1994. To appear in SIAM Journal on Computing.

[12] C. Phillips and T. J. Warnow. The asymmetric median tree - a new model for building consensus trees. In *Proceedings of the 7th Annual Symposium on Combinatorial Pattern Matching*, pages 234–252, 1996.

[13] M. Steel and T. J. Warnow. Kaikoura tree theorems: Computing the maximum agreement subtree. *Information Processing Letters*, 48:77–82, 1993.

Treewidth and Minimum Fill-In of Weakly Triangulated Graphs

Vincent Bouchitté and Ioan Todinca

LIP-École Normale Supérieure de Lyon
46 Allée d'Italie, 69364 Lyon Cedex 07, France
{Vincent.Bouchitte, Ioan.Todinca}@ens-lyon.fr

Abstract. We use the notion of potential maximal clique to characterize the maximal cliques appearing in minimal triangulations of a graph. We show that if these objects can be listed in polynomial time for a class of graphs, the treewidth and the minimum fill-in are polynomially tractable for these graphs. Finally we show how to compute in polynomial time the potential maximal cliques of weakly triangulated graphs.

1 Introduction

The notion of *treewidth* was introduced by Robertson and Seymour in [17]. It plays a major role in graph algorithm design. Indeed, it has been shown that many classical NP-hard problems become polynomial and even linear when restricted to graphs with small treewidth. These algorithms use a tree decomposition or a *triangulation* of the input graph, which is a chordal supergraph, i.e. all the cycles with at least four vertices of the supergraph have a chord. Computing the treewidth consists in finding a triangulation of minimum cliquesize. A related probem is the *minimum fill-in* problem, which consists in finding a triangulation of a graph such that the number of added edges is minimum. This parameter is used in sparse matrix factorization.

When computing the treewidth or the minimum fill-in, we are looking for triangulations of a graph. In both cases we can restrict to triangulations minimal by inclusion, that we call *minimal triangulations*. Also both problems are NP-complete. Nevertheless, these parameters can be computed in polynomial time for several classes of graphs such as chordal bipartite graphs [12, 3], circle and circular-arc graphs [8, 19, 15], AT-free graphs with polynomial number of separators [14] and HHD-free graphs [2]. Most of these algorithms use the fact that these classes of graphs have a polynomial number of *minimal separators*. It was conjectured in [10, 11] that the treewidth and the minimum fill-in should be tractable in polynomial time for all the graphs having a polynomial number of minimal separators. The conjecture is still open.

A potential maximal clique of a graph is a vertex set which induces a maximal clique in some minimal triangulation of the graph. We show here that if one can list in polynomial time all the potential maximal cliques of some class of graphs, then the treewidth and the minimum fill-in of those graphs can be computed in

C. Meinel and S. Tison (Eds.): STACS'99, LNCS 1563, pp. 197–206, 1999.
© Springer-Verlag Berlin Heidelberg 1999

polynomial time. This notion is related to the work of [1], from which we can easily deduce that the potential maximal cliques of the previously cited classes of graphs can be listed in polynomial time.

The class of *weakly triangulated graphs*, introduced in [6], is a class of graphs with polynomial number of separators, probably the only one for which the treewith and minimum fill-in problems were still open. We give an algorithm computing the potential maximal cliques of these graphs. Consequently, the treewidth and the minimum fill-in of weakly triangulated graphs are computable in polynomial time.

2   Chordal Graphs and Minimal Separators

Throughout this paper we consider connected, simple, finite, undirected graphs.

A graph H is *chordal* (or *triangulated*) if every cycle of length at least four has a chord. A *triangulation* of a graph $G = (V, E)$ is a chordal graph $H = (V, E')$ such that $E \subseteq E'$. H is a *minimal triangulation* if for any intermediate set E'' with $E \subseteq E'' \subset E'$, the graph (V, E'') is not triangulated.

A subset $S \subseteq V$ is an *a, b-separator* for two nonadjacent vertices $a, b \in V$ if the removal of S from the graph separates a and b in different connected components. S is a *minimal a, b-separator* if no proper subset of S separates a and b. We say that S is a *minimal separator* of G if there are two vertices a and b such that S is a minimal a, b separator. Notice that a minimal separator can be strictly included in another. We denote by Δ_G the set of all minimal separators of G.

Let G be a graph and S a minimal separator of G. We note $\mathcal{C}_G(S)$ the set of connected components of $G - S$. A component $C \in \mathcal{C}_G(S)$ is *full* if every vertex of S is adjacent to some vertex of C. We denote by $\mathcal{C}_G^*(S)$ the set of all full components of $G - S$. For the following lemma, we refer to [5].

Lemma 1. *A set S of vertices of G is a minimal a, b-separator if and only if a and b are in different full components of S.*

If $C \in \mathcal{C}(S)$, we say that $(S, C) = S \cup C$ is a *block* of S. A block (S, C) is called *full* if C is a full component of S.

Definition 1. *Two separators S and T cross, denoted by $S \sharp T$, if there are some distinct components C and D of $G - T$ such that S intersects both of them. If S and T do not cross, they are called parallel, denoted by $S \| T$.*

It is easy to prove that these relations are symmetric. Remark that for any couple of parallel separators S and T, T is contained in some block (S, C) of S.

A *clique* of G is a complete subgraph of G. In [4] Dirac showed that all the minimal separators of a chordal graph are cliques. Using the fact that a separator cannot separate two adjacent vertices we deduce the following lemma.

Lemma 2. *Let G be a graph, S a minimal separator and Ω a clique of G. Then Ω is included in some block of S. In particular, the minimal separators of a chordal graph are pairwise parallel.*

Definition 2. *Let G be a graph. The* treewidth *of G, denoted by $tw(G)$, is the minimum, over all triangulations H of G, of $\omega(H) - 1$, where $\omega(H)$ is the the maximum cliquesize of H.*

Definition 3. *The* minimum fill-in *of a graph G, denoted by $mfi(G)$, is the smallest value of $|E(H) - E(G)|$, where the minimum is taken over all triangulations H of G.*

In other words, computing the treewidth of G means finding a triangulation with smallest cliquesize, while computing the minimum fill-in consists in finding a triangulation with smallest number of edges. In both cases we can restrict our work to minimal triangulations.

Let $S \in \Delta_G$ be a minimal separator. We denote by G_S the graph obtained from G by *completing* S, i.e. by adding an edge between every pair of non-adjacent vertices of S. If $\Gamma \subseteq \Delta_G$ is a set of separators of G, G_Γ is the graph obtained by completing all the separators of Γ. The results of [13], concluded in [16], establish a strong relation between the minimal triangulations of a graph and its minimal separators.

Theorem 1. *Let $\Gamma \in \Delta_G$ be a maximal set of pairwise parallel separators of G. Then $H = G_\Gamma$ is a minimal triangulation of G and $\Delta_H = \Gamma$.*

Let H be a minimal triangulation of a graph G. Then Δ_H is a maximal set of pairwise parallel separators of G and $H = G_{\Delta_H}$.

In other terms, every minimal triangulation of a graph G is obtained by considering a maximal set Γ of pairwise parallel separators of G and completing the separators of Γ. The minimal separators of the triangulation are exactly the elements of Γ.

It is important to know that the elements of Γ, who become the separators of H, have strictly the same behavior in H as in G. Indeed, the connected components of $H - S$ are exactly the same in $G - S$, for every $S \in \Gamma$. Moreover, the full components are the same in the two graphs, that is $C_H^*(S) = C_G^*(S)$.

3   Potential Maximal Cliques and Maximal Sets of Neighbor Separators

The previous theorem gives a characterization of the minimal triangulations of a graph by means of minimal separators, but it needs a global look over the set of minimal separators. Therefore, it gives no algorithmic information about how we should construct a minimal triangulation in order to optimize its cliquesize or the fill-in. In [1] we introduced the notion of "maximal sets of neighbor separators", and we showed how these sets are related to the maximal cliques of any triangulation of a graph that are called here "potential maximal cliques". We will give a new tool to recognize the potential maximal cliques of a graph, which will also help to compute all the potential maximal cliques of weakly triangulated graphs.

Definition 4. *A vertex set Ω of a graph G is called a* potential maximal clique *if there is a minimal triangulation H of G such that Ω is a maximal clique of H.*

If K is a vertex set of G, we denote by $\Delta(K)$ the minimal separators of G included in K.

Definition 5. *A family $\mathcal{S}$ of minimal separators of a graph G is called* maximal set of neighbor separators *if there is a potential maximal clique Ω of G such that $\mathcal{S} = \Delta(\Omega)$. We also say that $\mathcal{S}$ borders Ω in G.*

Definition 6. *Let G be a graph and $\mathcal{S} \subseteq \Delta_G$ a set of pairwise parallel separators such that for any $S \in \mathcal{S}$, there is a block $(S, C(S))$ containing all the separators of $\mathcal{S}$. Suppose that $\mathcal{S}$, ordered by inclusion, has no greatest element. We define the* piece between *the elements of $\mathcal{S}$ by*

$$P(\mathcal{S}) = \bigcap_{S \in \mathcal{S}} (S, C(S))$$

Notice that for any $S \in \mathcal{S}$ the block of S containing all the separators of $\mathcal{S}$ is unique : if $T \in \mathcal{S}$ is not included in S, there is a unique connected component of $G - S$ containing $T - S$.

The two following theorems, proved in [1], allow us to recognize a potential maximal clique of a graph :

Theorem 2. *Let Ω be a vertex set of G and suppose that $\Delta(\Omega)$ has a maximum element S, i.e. every T in $\Delta(\Omega)$ is included in S. Then Ω is a potential maximal clique if and only if Ω is some block (S, C) and $G_S[\Omega]$ is a clique.*

Theorem 3. *Let Ω be a vertex set of G and suppose that $\Delta(\Omega)$ ordered by inclusion has no greatest element. Then Ω is a potential maximal clique if and only if $\Omega = P(\Delta(\Omega))$ and $G_{\Delta(\Omega)}[\Omega]$ is a clique.*

Notice that if we consider a minimal triangulation H of G such that Ω is a maximal clique of H, then the separators of G bordering Ω are also minimal separators of H, as shown in the following lemma :

Lemma 3. *Let H be a minimal triangulation of a graph G and T be a minimal separator of G such that $H[T]$ is a clique. Then T is also a minimal separator of H.*

We are going to give a strong characterization of potential maximal cliques, which does not use minimal separators. We need some easy observations on theorems 2 and 3.

Proposition 1. *Let Ω be a potential maximal clique of G and let $S \in \Delta(\Omega)$. Then S is strictly contained in Ω and $\Omega - S$ is in a full component of S.*

Proposition 2. *Let Ω be a potential maximal clique of a graph G and let a be any vertex of $V - \Omega$. There is a minimal separator $S \subset \Omega$ that separates a and $\Omega - S$.*

Let now K be a set of vertices of a graph G. We denote by $C_1(K), \ldots, C_p(K)$ the connected components of $G - K$. We denote by $S_i(K)$ the vertices of K adjacent to at least one vertex of $C_i(K)$. When no confusion is possible we will simply speak of C_i and S_i. If $S_i(K) = K$ we say that $C_i(K)$ is a *full component* of K.

Lemma 4. *Let Ω be a potential maximal clique of a graph G and let $\Delta(\Omega)$ be the maximal set of neighbor separators bordering Ω. Then the elements of $\Delta(\Omega)$ are exactly the sets $S_i(\Omega)$.*

Proof. We prove that for any i, $1 \leq i \leq p$, S_i is a minimal a, b-separator for some $a \in C_i$ and $b \in \Omega - S_i$. Proposition 2 tells us that there is some minimal separator $S \in \mathcal{S}$ that separates a from $\Omega - S$; recall that $\Omega - S$ is not empty. Since every vertex in S_i has a neighbor in C_i, if S does not contain a vertex $x \in S_i$, S cannot separate $x \in \Omega - S$ from a so we get $S_i \subseteq S$. By proposition 1, $\Omega - S$ is in a full component of S, and therefore of S_i. Let b be a vertex of $\Omega - S$. Then a and b are in different full components of S_i, so S_i is a minimal a, b-separator by lemma 1. We have to prove now that for any minimal separator $S \subseteq \Omega$, there is some $i, 1 \leq i \leq p$ such that $S = S_i$. We have that $\Omega - S \neq \emptyset$ and $\Omega - S$ is in some full component of S. Let C be another full component of S. Then C is a connected component of $G - \Omega$, let us say C_i. It follows that $S \subseteq S_i$. Let $x \in S_i - S$, since C_i is a full component of S_i, we must have $x \in C$ contradicting $C = C_i$. So we get $S = S_i$. We conclude that the separators of G included in Ω are exactly the sets S_i. □

We also give a "sufficient condition" to characterize the potential maximal cliques, which is somehow the dual of lemma 4.

Theorem 4. *Let $K \subseteq V$ be a set of vertices. We denote by $\mathcal{S}$ the set of all $S_i(K)$. K is a potential maximal clique if and only if :*

1. *K has no full components.*
2. *$G_{\mathcal{S}}[K]$ is a clique.*

Moreover, $\mathcal{S}$ is the maximal set of neighbor separators bordering K.

Proof. We prove the "only if" part. Suppose that K is a potential maximal clique of G. By lemma 4, the maximal set of neighbor separators bordering K is $\mathcal{S}$. By theorems 2 and 3, K is a clique in the graph $G_{\mathcal{S}}$. It remains to show that K has no full components. Let C_i be any connected component of $G - K$. Then S_i are the neighbors of C_i in K. Since K is a potential maximal clique and S_i is a separator contained in K, we have that S_i is strictly contained in K, by proposition 1. Therefore, C_i is not a full component of K.

We prove now the "if" part. Let us show at first that $S_i, 1 \leq i \leq p$, is a minimal separator. S_i is clearly a separator and C_i is a full component of S_i. Let

x be a vertex of $K - S_i$. We show that x belongs to a full component of S_i different from C_i. We denote by C_x the connected component of $G - S_i$ containing x. For any $y \in S_i$, y must have a neighbor in C_x. This is true if x and y are adjacent in G. If x and y are not adjacent, by the second condition of the theorem, x and y belong to a same S_j. C_j being a full component of S_j, there is a path in G connecting x to y entirely contained in C_j except from x and y, we deduce that $C_j \subseteq C_x$. It follows that y has a neighbor in C_x since it has a neighbor in C_j. S_i is a minimal separator of G according to lemma 1.

Now, given two distinct separators S_i and S_j, we have to show that they are parallel. We prove that $K - S_i$ is in a connected component of $G - S_i$. Let $x, y \in K - S_i$. If x and y are adjacent they are clearly in the same component of $G - S_i$. Otherwise, since $G_S[K]$ is a clique, they are in a same S_k, so they are connected via C_k. So S_j intersects only the component of $G - S_i$ containing $K - S_i$ and then $S_i \| S_j$. Therefore S consists in a set of pairwise parallel separators.

We have to show that any separator of G included in K is an element of S. Consider a minimal triangulation H of G such that all the elements of S are separators of H. We know that K is a clique in H. Now let $U \subseteq K$ be any minimal separator of G. Notice that U must be strictly included in K, otherwise K would have two full components in G contradicting our choice of K. Clearly U is a clique in H, so by lemma 3, it is a minimal separator of H. Since K is a clique in H, it must be included in some full block of U. Let (U, C) be another full block of U in H, and consequently in G. We have that C is a connected component of $G - U$ and U separates C and $K - U$. We deduce that C is also a connected component of $G - K$, let us say C_i. By definition of S_i, we have $U \subseteq S_i$. Suppose there exists a vertex $x \in S_i - U$, since x has a neighbor in C, the connected component C of $G - U$ would contain x contradicting $C = C_i$. So we have $U = S_i$ and $U \in S$.

We want to prove that S satisfies the conditions of theorems 2 or 3. Remark that for any $y \in V - K$, y is in some connected component C_i of $G - K$ and the separator $S_i \in S$ separates y from $K - S_i$. Suppose now that S has an element S, maximum by inclusion. Let (S, C) be the block of S containing K. By the previous remark, for any $y \in V - K$, S separates y and $K - S$, so $y \notin (S, C)$. It follows that $(S, C) = K$, so S satisfies all the conditions of theorem 2. Now if S does not have an element maximum by inclusion, clearly K is contained in the piece between the separators of S in G. By the previous remark, $P_G(S)$ does not contain any $y \in V - K$, so $K = P_G(S)$ and therefore we are under the conditions of theorem 3. It follows that S forms a maximal set of neighbor separators of G, bordering the set K.   □

4   Triangulating Blocks

In this section we prove that the potential maximal cliques of a graph are sufficient to compute its treewidth and its minimum fill-in.

Let $B = (S, C)$ be a block of the graph G. The graph $R(S, C) = G_S[S \cup C]$ is called the *realization* of the block B. The following proposition, proved in [14], gives a relation between the treewidth and the minimum fill-in of a graph and some minimal triangulations of its realizations.

Proposition 3. *Let G be a non-complete graph. Then*

$$tw(G) = \min_{S \in \Delta_G} \max_{C \in \mathcal{C}(S)} tw(R(S,C))$$

$$mfi(G) = \min_{S \in \Delta_G} (fill(S) + \sum_{C \in \mathcal{C}(S)} mfi(R(S,C)))$$

where $fill(S)$ is the number of non-edges of S.

We want to give a characterization of the minimal triangulations of a realization $R(S,C)$ using the potential maximal cliques Ω with $S \subset \Omega$ and $\Omega \subseteq (S,C)$ and the minimal triangulations of the realizations of some blocks (S_i, C_i), strictly included in (S,C). We will compute the treewidth and the minimum fill-in by dynamic programming on blocks.

   The minimal triangulations of the realizations of non-full blocks are easy reducible to the case of full blocks. For the following proposition, see [14] :

Proposition 4. *Let (S,C) be a non-full block of G and let S^* be the vertices of S adjacent in G to at least one vertex of C. Then*

$$tw(R(S,C)) = \max(|S| - 1, tw(R(S^*,C)))$$

$$mfi(R(S,C)) = fill(S) + mfi(R(S^*,C))$$

   It remains to express the treewidth and the minimum fill-in of realizations of full blocks from realizations of smaller blocks. For this let us give first a characterization of minimal triangulations of a graph using a potential maximal clique and the realizations of some blocks. The proof has been omitted due to space restriction.

Theorem 5. *Let H be a minimal triangulation of G and let Ω be a maximal clique of H. For each connected component $C_i, 1 \leq i \leq p$ of $G - \Omega$, let S_i be the vertices of Ω having a neighbor in C_i. Then $H_i = H[S_i \cup C_i]$ are minimal triangulations of the realizations $R(S_i, C_i)$.*

   Conversely, let Ω be a potential maximal clique of G. For each connected component $C_i, 1 \leq i \leq p$ of $G - \Omega$, let H_i be a minimal triangulation of $R(S_i, C_i)$. Then $H = (V, E(H))$ with $E(H) = \bigcup_{i=1}^{p} E(H_i) \cup \{\{x, y\} | x, y \in \Omega\}$ is a minimal triangulation of G.

   One can also prove that for any minimal triangulation $H(S,C)$ of the realization of any full block $R(S,C)$, there is a maximal clique Ω of $H(S,C)$ such that $S \subset \Omega \subseteq (S,C)$ and Ω is a potential maximal clique of G. We deduce :

Proposition 5. *Let (S,C) be a full block of G. Then*

$$tw(R(S,C)) = \min_{S \subset \Omega \subseteq (S,C)} \max(|\Omega| - 1, tw(R(S_i, C_i)))$$

$$mfi(R(S,C)) = \min_{S \subset \Omega \subseteq (S,C)} fill(S) + \sum mfi(R(S_i, C_i))$$

Propositions 3, 4 and 5 give us a dynamic programming algorithm which, using the list of all potential maximal cliques of a graph G, computes the treewidth and the minimum fill-in of G. The algorithm is clearly polynomial in the number of vertices and the number of potential maximal cliques of G.

5   Weakly Triangulated Graphs

We consider now two non-adjacent vertices x, y of G. Let G' be the graph obtained from G by adding the edge $\{x, y\}$. We will show in this section that the potential maximal cliques of G can be computed from the minimal separators of G and the potential maximal cliques of G'. We will use this technique to compute all the potential maximal cliques of any weakly triangulated graph.

Let once again Ω be a potential maximal clique of G. Let $C_1, \ldots, C_p$ be the connected components of $G - \Omega$ and let S_i be the set of vertices of Ω having at least a neighbor in C_i. We want to describe the behaviour of Ω and S in the graph G'. We denote by $C'_1, \ldots, C'_q$ the connected components of $G' - \Omega$ and by $S'_1, \ldots, S'_q$ the neighborhoods of C'_i in G'. From theorem 4, it follows that Ω is a clique in $G'_{\{S'_1, \ldots, S'_q\}}$. If Ω has no full component in G', then Ω is a potential maximal clique of G'. We deduce :

Theorem 6. *Let Ω be a potential maximal clique of G. Let x, y be two non-adjacent vertices of G and let $G' = G \cup \{x, y\}$. Two cases are possible :*

1. *Ω can be written as $S_1 \cup \{x\}$, $S_1 \cup \{y\}$ or $S_1 \cup S_2$, where S_1, S_2 are minimal x, y-separators of G.*
2. *Ω is a potential maximal clique of G'.*

The weakly triangulated graphs were introduced in [6]. A graph G is called *weakly triangulated* if neither G nor its complement $\overline{G}$ have an induced cycle with strictly more than four vertices. This class contains the chordal graphs, the chordal bipartite graphs and the distance hereditary graphs.

We denote by $N(x)$ the neighbors of the vertex x. We say that two vertices x, y of a graph G form a *two-pair* if their common neighbors $N(x) \cap N(y)$ form an x, y-separator. It was proved in [7] that every weakly triangulated graph that is not a clique has a two-pair. Spinrad and Sritharan gave in [18] an algorithm recognizing the weakly triangulated graphs, based on the following theorem:

Theorem 7. *Let $G = (V, E)$ be a graph and let $\{x, y\}$ be a two-pair of G. Let $G' = (V, E')$ be the graph obtained from G by adding the edge $\{x, y\}$. Then G is weakly triangulated if and only if G' is weakly triangulated.*

Notice that a clique is a weakly triangulated graph. The recognition algorithm considers an input graph G and, while G has a two-pair $\{x, y\}$, it adds the edge between x and y to G. At the end of the loop, either G became a clique, in which case the initial graph was weakly triangulated by theorem 7, or G is not a clique and it has no two-pair, in which case the input graph could not be weakly triangulated.

We denote by $\bar{e}$ the number of edges of $\overline{G}$. Let now $G = (V, E)$ be a weakly triangulated graph and let $f_1 = \{x_1, y_1\}, \ldots, f_{\bar{e}} = \{x_{\bar{e}}, y_{\bar{e}}\}$ be the edges added to G by the recognition algorithm in this order. We denote by G_i the graph $(V, E \cup \{f_1, f_2, \ldots, f_i\})$, with $0 \leq i \leq \bar{e}$ (so $G_0 = G$ and $G_{\bar{e}}$ is a clique). We will describe the minimal separators, respectively the potential maximal cliques of G_i using the minimal separators, respectively the potential maximal cliques of G_{i+1}, for any $i < \bar{e}$.

It is known that a weakly triangulated graph has at most $\bar{e}$ minimal separators (Kloks, [9]). Indeed, it is easy to check that if x, y is a two-pair of G and $G' = (V, E \cup \{xy\})$, then any minimal separator S of G, different from $S_{xy} = N_G(x) \cap N_G(y)$, is also a minimal separator of G'. So for any $i < p$, G_i has at most one more minimal separator than G_{i+1}. We deduce:

Proposition 6. *A weakly triangulated graph G has at most $\bar{e}$ minimal separators, where $\bar{e}$ is the number of edges of $\overline{G}$.*

Notice that if x and y form a two-pair of G, then S_{xy} is the unique x, y-minimal separator of G. In particular, if Ω is a potential maximal clique of G, we can not have two x, y-minimal separators S_1 and S_2 with $\Omega = S_1 \cup S_2$. So we can refine the results of theorem 6 :

Proposition 7. *Let Ω be a potential maximal clique of G. Let x, y be a two-pair of G and let $G' = G \cup \{xy\}$. Let $S_{xy} = N_G(x) \cap N_G(y)$. Two cases are possible :*

1. *Ω can be written as $S_{xy} \cup \{x\}$ or $S_{xy} \cup \{y\}$.*
2. *Ω is a potential maximal clique of G'.*

Proposition 8. *A weakly triangulated graph G has at most $2\bar{e} + 1$ potential maximal cliques.*

Proof. We consider the sequence of graphs $G_0 = G, G_1, \ldots, G_{\bar{e}}$ previously defined. Since G_{i+1} is obtained from G_i by adding an edge between a two-pair, by proposition 7 G_i has at most two more potential maximal cliques than G_{i+1}. The graph $G_{\bar{e}}$, which is a clique, has a unique potential maximal clique. $\square$

Theorem 8. *The treewidth and the minimum fill-in of weakly triangulated graphs can be computed in polynomial time.*

Clearly, all potential maximal cliques of a weakly triangulated graph can be listed in polynomial time, and therefore the treewidth and the minimum fill-in are computable in polynomial time.

6 Conclusions

We still do not know if the treewidth and the minimum fill-in are polynomially tractable for all graphs with a polynomial number of minimal separators. A way to prove this conjecture would be to answer the following question: does there exists a polynomial P such that for any graph G, one can compute a sequence of graphs $G_0 = G, G_1, \ldots, G_p$ where G_p is a clique, G_{i+1} is obtained from G_i by adding an edge and, for all $i, 1 \leq i \leq p$, $|\Delta_{G_i}| \leq P(|\Delta_G|, |V(G)|)$?

References

[1] V. Bouchitté and I. Todinca. Minimal triangulations for graphs with "few" minimal separators. In *Proceedings 6th Annual European Symposium on Algorithms (ESA'98)*, volume 1461 of *Lecture Notes in Computer Science*, pages 344–355. Springer-Verlag, 1998.

[2] H.J. Broersma, E. Dahlhaus, and T. Kloks. Algorithms for the treewidth and minimum fill-in of HHD-free graphs. In *Workshop on Graphs (WG'97)*, volume 1335 of *Lecture Notes in Computer Science*, pages 109–117. Springer-Verlag, 1997.

[3] M. S. Chang. Algorithms for maximum matching and minimum fill-in on chordal bipartite graphs. In *ISAAC'96*, volume 1178 of *Lecture Notes in Computer Science*, pages 146–155. Springer-Verlag, 1996.

[4] G.A. Dirac. On rigid circuit graphs. *Abh. Math. Sem. Univ. Hamburg*, 21:71–76, 1961.

[5] M. C. Golumbic. *Algorithmic Graph Theory and Perfect Graphs*. Academic Press, New York, 1980.

[6] R. Hayward. Weakly triangulated graphs. *J. Combin. Theory ser. B*, 39:200–208, 1985.

[7] R. Hayward, C. Hoàng, and F. Maffray. Optimizing weakly triangulated graphs. *Graphs Combin.*, 5:339–349, 1989.

[8] T. Kloks. Treewidth of circle graphs. In *Proceedings 4th Annual International Symposium on Algorithms and Computation (ISAAC'93)*, volume 762 of *Lecture Notes in Computer Science*, pages 108–117. Springer-Verlag, 1993.

[9] T. Kloks, 1998. Private communication.

[10] T. Kloks, H.L. Bodlaender, H. Müller, and D. Kratsch. Computing treewidth and minimum fill-in: All you need are the minimal separators. In *Proceedings First Annual European Symposium on Algorithms (ESA'93)*, volume 726 of *Lecture Notes in Computer Science*, pages 260–271. Springer-Verlag, 1993.

[11] T. Kloks, H.L. Bodlaender, H. Müller, and D. Kratsch. Erratum to the ESA'93 proceedings. In *Proceedings Second Annual European Symposium on Algorithms (ESA'94)*, volume 855 of *Lecture Notes in Computer Science*, page 508. Springer-Verlag, 1994.

[12] T. Kloks and D. Kratsch. Treewidth of chordal bipartite graphs. *J. Algorithms*, 19(2):266–281, 1995.

[13] T. Kloks, D. Kratsch, and H. Müller. Approximating the bandwidth for asteroidal triple-free graphs. In *Proceedings Third Annual European Symposium on Algorithms (ESA'95)*, volume 979 of *Lecture Notes in Computer Science*, pages 434–447. Springer-Verlag, 1995.

[14] T. Kloks, D. Kratsch, and J. Spinrad. On treewidth and minimum fill-in of asteroidal triple-free graphs. *Theoretical Computer Science*, 175:309–335, 1997.

[15] T. Kloks, D. Kratsch, and C.K. Wong. Minimum fill-in of circle and circular-arc graphs. *J. Algorithms*, 28(2):272–289, 1998.

[16] A. Parra and P. Scheffler. Characterizations and algorithmic applications of chordal graph embeddings. *Discrete Appl. Math.*, 79(1-3):171–188, 1997.

[17] N. Robertson and P. Seymour. Graphs minors. II. Algorithmic aspects of treewidth. *J. of Algorithms*, 7:309–322, 1986.

[18] J. Spinrad and R. Sritharan. Algorithms for weakly triangulated graphs. *Discrete Applied Mathematics*, 59:181–191, 1995.

[19] R. Sundaram, K. Sher Singh, and C. Pandu Rangan. Treewidth of circular-arc graphs. *SIAM J. Discrete Math.*, 7:647–655, 1994.

Decidability and Undecidability of Marked PCP

Vesa Halava[1], Mika Hirvensalo[2,1*], and Ronald de Wolf[3,4]

[1] Turku Centre for Computer Science, Lemminkäisenkatu 14 A, 4th floor, FIN-20520, Turku, Finland, vehalava@cs.utu.fi
[2] Department of Mathematics, University of Turku, FIN-20014, Turku, Finland. mikhirve@utu.fi
[3] CWI, P.O. Box 94079, Amsterdam, The Netherlands, rdewolf@cwi.nl
[4] University of Amsterdam

Abstract. We show that the *marked* version of the Post Correspondence Problem, where the words on a list are required to differ in the first letter, is decidable. On the other hand, PCP remains undecidable if we only require the words to differ in the first *two* letters. Thus we locate the decidability/undecidability-boundary between marked and 2-marked PCP.

1 Introduction: PCP and Marked PCP

The Post Correspondence Problem (PCP) [6] is one of the most useful undecidable problems, because it can be simply described and many other problems can easily be reduced to it, particularly problems in formal language theory. The general form of the problem is as follows. An instance of PCP is a four-tuple $I = (\Sigma, \Delta, g, h)$, consisting of a finite *source alphabet* $\Sigma = \{a_1, \ldots, a_n\}$, a finite *target alphabet* Δ and two homomorphisms $g, h : \Sigma^* \to \Delta^*$ ($g(ab) = g(a)g(b)$ and $h(ab) = h(a)h(b)$ whenever $a, b \in \Sigma^*$). It is enough to define $g, h : \Sigma \to \Delta^*$, the extension is just concatenation. PCP is the following decision problem:

Given $I = (\Sigma, \Delta, g, h)$, is there an $x \in \Sigma^+$ such that $g(x) = h(x)$?

In other words, we have two lists of words $g(a_1), \ldots, g(a_n)$ and $h(a_1), \ldots, h(a_n)$ and we want to decide if there is a correspondence between them: are there $a_{i_1}, \ldots, a_{i_k} \in \Sigma$ such that $g(a_{i_1}) \ldots g(a_{i_k}) = h(a_{i_1}) \ldots h(a_{i_k})$?

The general form of this problem is undecidable [6], the reason being that the two morphisms together can simulate the computation of a Turing machine on a specific input. Examining restricted versions of PCP allows one to determine the exact boundary between decidability and undecidability. For instance, the problem becomes trivially decidable (but NP-complete) if we ask for the existence of a solution x of length at most some fixed k [2, p. 228]. If we restrict to g, h which have to be *injective* (g is injective if $x \neq y \Rightarrow g(x) \neq g(y)$), the problem remains undecidable [4]. Also PCP(7), where we restrict to $n = 7$, is

---

* Supported by the Academy of Finland under grant 14047.

C. Meinel and S. Tison (Eds.): STACS'99, LNCS 1563, pp. 207–216, 1999.
© Springer-Verlag Berlin Heidelberg 1999

still undecidable [5], but PCP(2) is decidable [1]. As far as we know, decidability or undecidability is still open for $2 < n < 7$.

A further restriction which we will examine in this paper is to have g and h *marked*, which we formally define as follows. If z is a string, we use $Pref_k(z)$ to denote the prefix of length k of z ($Pref_k(z) = z$ if $|z| \leq k$). A homomorphism g is k-*marked* if $g(a)$ and $g(b)$ are nonempty and have $Pref_k(g(a)) \neq Pref_k(g(b))$ whenever $a \neq b \in \Sigma$. An instance $I = (\Sigma, \Delta, g, h)$ of PCP is k-*marked* if both g and h are k-marked, and k-marked PCP is the PCP decision problem restricted to k-marked instances. We will abbreviate 1-marked to *marked*. If I is marked then $g(a)$ and $g(b)$ start with a different letter whenever $a \neq b \in \Sigma$, which implies that $|\Sigma| \leq |\Delta|$. Without loss of generality we may assume $\Sigma \subseteq \Delta$. Markedness clearly implies injectivity: suppose g is marked and $x \neq y \in \Sigma^+$, let $x = zax'$ and $y = zby'$, a and b being the first letter where x and y differ. Because of markedness we have $g(a) \neq g(b)$, hence $g(x) = g(z)g(a)g(x') \neq g(z)g(b)g(y') = g(y)$, so g is injective. The converse does not hold. Consider for instance $\Sigma = \Delta = \{1, 2\}$, $g(1) = 11$, $g(2) = 12$, then g is injective but not marked.

The proof of decidability of PCP(2) in [1] is based on a reduction from arbitrary instances of PCP(2) to marked instances of *generalized* PCP(2). [1] then prove by means of extensive case analysis that marked generalized PCP(2) is decidable. In particular marked PCP(2) is decidable. Here we prove that marked PCP is decidable for *any* alphabet size. We will in fact show that marked PCP is in **EXPTIME** (the class of languages that can be recognized in time upper bounded by $2^{p(N)}$ for some polynomial p of the input size N).

As stated above, PCP can be used for establishing the boundaries between decidability and undecidability. The main result of this paper is decidability of marked PCP. How much can we weaken the markedness condition before we lose decidability? We will show in Section 3 that 2-marked PCP is undecidable, thus locating the decidability/undecidability-boundary between 1-markedness and 2-markedness.

In another direction, we can weaken the markedness condition by only requiring g and h to be *prefix* morphisms (g is prefix if no $g(a_i)$ is a prefix or another $g(a_j)$) or even *biprefix* (g is biprefix if no $g(a_i)$ is a prefix or suffix of another $g(a_j)$). It turns out that biprefix PCP is undecidable [8].[1]

2   Marked PCP Is Decidable

2.1   A Simpler Decision Problem

We would like to give a decision method for marked PCP. First we give an algorithm for the following simpler problem, which also occurs in [1, Section 6]:

> Given marked $I = (\Sigma, \Delta, g, h)$ and $a \in \Delta$, are there $x, y \in \Sigma^+$ such that $g(x) = h(y)$ and $g(x)$ starts with a?

---

[1] Clearly, a marked morphism is prefix. Both marked and biprefix PCP are special cases of injective PCP, but 2-marked PCP is not. See also at the end of Section 3.

We do not look for $g(x) = h(x)$ here but only for $g(x) = h(y)$, and we additionally require that $g(x)$ starts with some specific $a \in \Delta$. For example, if I has

$$
\begin{array}{llll}
g(a_1) = a_1 & g(a_2) = a_2 & g(a_3) = a_3a_4 & g(a_4) = a_4 \\
h(a_1) = a_1a_3 & h(a_2) = a_4a_2 & h(a_3) = a_3a_3 & h(a_4) = a_2a_2
\end{array}
$$

then for $a = a_1$, a solution would be $x = a_1a_3a_2$ and $y = a_1a_2$.

The next algorithm decides the problem.

1. Set $G = H = \emptyset$, $i = j = 1$.
2. If there are $x_1, y_1 \in \Sigma$ such that $g(x_1)$ and $h(y_1)$ start with a, then set
   $x = x_1$, $y = y_1$
   else goto 4.
3. (a) If $g(x) = h(y)$, then print "solution $x = x_1 \ldots x_i$ and $y = y_1 \ldots y_j$" and
       terminate.
   (b) If $g(x)$ is not a prefix of $h(y)$ nor vice versa, then goto 4.
   (c) If $g(x)s = h(y)$, then do the following.
       If $s \in G$ then goto 4; else set $i = i + 1$ and $G = G \cup \{s\}$.
       If there is an x_i such that $g(x_i)$ and s start with the same letter, then
       set $x = xx_i$ and goto 3; else goto 4.
   (d) If $g(x) = h(y)s$, then analogous to previous step.
4. Print "no solution" and terminate.

Informally, we are building $x = x_1 \ldots x_i$ and $y = y_1 \ldots y_j$, trying to achieve $g(x) = h(y)$. We add on a new x_{i+1} as long as $g(x)$ is a proper prefix of $h(y)$ (i.e., $g(x)s = h(y)$ for some suffix s), and add on a new y_{j+1} if $h(y)$ is a proper prefix of $g(x)$. Note that at each point such x_{i+1} or y_{j+1} are unique (if they exist) because of markedness; if they do not exist we know there is no solution. We keep track of the suffixes we have seen so far in the sets G and H. Because the number of possible suffixes is finite, either the process terminates with a solution, or at some point a suffix is encountered for the second time, in which case we know the process will cycle forever and there is no solution.

The solutions produced by this algorithm are of minimal length. Note carefully that the whole procedure is deterministic, because g and h are marked. Furthermore, if N is the length of the instance I given as input (i.e., the number of bits needed to describe the instance), then this procedure runs in time polynomial in N. Namely, each $g(a_i)$ and $h(a_i)$ can have length at most N, and hence can have at most $N - 1$ proper suffixes. Since there are only $2n = O(N)$ different $g(a_i)$ and $h(a_i)$, there are only $O(N^2)$ different suffixes, hence the loop of the algorithm can be repeated at most $O(N^2)$ times. This loop itself takes $O(N)$ steps, because (1) to check if $g(x) = h(y)$ or $g(x)s = h(y)$ or $g(x) = h(y)s$, we only need to check the way $g(x)$ and $h(y)$ have been changed by the addition of the previous x_i or y_j, and (2) searching for a new x_i (in step c) or y_j (in step d) can be done in $O(n) = O(N)$ steps. Therefore the whole procedure runs in $O(N^3)$ steps.

2.2   Reducing to Simpler Instances

Consider an instance $I = (\Sigma, \Delta, g, h)$ of marked PCP: we have two marked homomorphisms $g, h : \Sigma^+ \to \Delta^+$, where $\Sigma = \{a_1, \ldots, a_n\} \subseteq \Delta$, and we want to decide if there is an $x \in \Sigma^+$ such that $g(x) = h(x)$. Below we describe an approach to decide I by reducing it to an equivalent but simpler instance I' of marked PCP ("equivalent" meaning that I has a solution iff I' has one).

Suppose $\Delta = \{a_1, \ldots, a_l\}$, $l \geq n$. We can run the procedure of the previous section for every $a_i \in \Delta$, yielding pairs of (minimal-length) solutions $(u_1, v_1), \ldots, (u_l, v_l)$ where $u_i, v_i \in \Sigma^+$ and $g(u_i) = h(v_i)$ starts with a_i, or non-existence of solutions for certain i. At most n of the a_i can have a solution. Without loss of generality assume $1, \ldots, m \leq n$ are the i that have a solution. We can turn this into a new instance $I' = (\Sigma', \Delta, g', h')$ of PCP, where $\Sigma' = \{a_1, \ldots, a_m\}$, $g'(a_i) = u_i$ and $h'(a_i) = v_i$. Note that g' and h' are marked, so I' is an instance of marked PCP. Also, since the procedure of the previous section runs in $O(N^3)$ steps and has to be run n times here, I' can be built from I in $O(N^4)$ steps. The reduction from I to I' preserves equivalence:

Lemma 1. *If I and I' are as above, then I and I' are equivalent.*

Proof. Note that every solution x to I must be built up from u_i and v_i: there must be $i_1, \ldots, i_k$ such that $x = u_{i_1} \ldots u_{i_k} = v_{i_1} \ldots v_{i_k}$. This is easy to see from the example in Figure 1. Here $u_1 = a_5 a_3 a_1$ and $v_1 = a_5 a_3$ is a solution to the simpler problem for a_1, similarly $(a_2 a_4, a_1 a_2)$ is a solution for a_6 and $(a_6 a_3, a_4 a_6 a_3)$ is a solution for a_2. Here $x = a_5 a_3 a_1 a_2 a_4 a_6 a_3$ is a solution to I, $x' = a_1 a_6 a_2$ is a solution to I', related by $x = g'(x')$.

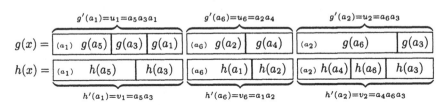

Fig. 1. How a solution to I translates to I' and vice versa

In general, by construction, if x' is a solution to I' then $x = g'(x') = h'(x')$ is a solution to I. And conversely, for every solution x to I there is a solution x' to I' such that $x = g'(x') = h'(x')$. Thus I and I' are equivalent.   □

If we could prove that I' is somehow simpler than I, then we could repeat the procedure, reduce to simpler and simpler equivalent instances I'', I''', ..., and eventually decide I. There are at least two ways in which I' can be simpler than I: $|\Sigma'| < |\Sigma|$ ($m < n$) or $\sigma(I') < \sigma(I)$, where σ measures the "suffix complexity" of an instance $I = (\Sigma, \Delta, g, h)$ [1]:

$$\sigma(I) = |\cup_{a \in \Sigma} \{x \mid x \text{ is a proper suffix of } g(a)\}|$$
$$+ |\cup_{a \in \Sigma} \{x \mid x \text{ is a proper suffix of } h(a)\}|$$

If $n = m$, we would like I' to be simpler than I in the sense that $\sigma(I') < \sigma(I)$. The following lemma shows that I' at least cannot be *more* complex than I:

Lemma 2. *If I and I' are as above, then $\sigma(I') \leq \sigma(I)$.*

Proof. Define the following four sets:

$$G = \cup_{a \in \Sigma}\{x \mid x \text{ is a proper suffix of } g(a)\}$$
$$G' = \cup_{a \in \Sigma'}\{x \mid x \text{ is a proper suffix of } g'(a)\}$$
$$H = \cup_{a \in \Sigma}\{x \mid x \text{ is a proper suffix of } h(a)\}$$
$$H' = \cup_{a \in \Sigma'}\{x \mid x \text{ is a proper suffix of } h'(a)\}$$

We will define an injective function $p : G' \to H$. Let $u \in G'$, so u is a proper suffix of some specific $g'(a_i) = u_i = x_1 \ldots x_c$ generated by the procedure of the previous section. Let x_r be the first letter of u, and s be the shortest suffix of some $h(y_t)$ due to which x_r was added to u_i in the procedure of the previous section, so s is a prefix of $g(x_r)$ (see Figure 2) or vice versa. Define p as $p(u) = s$.

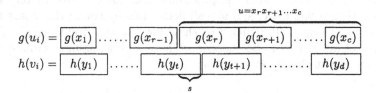

Fig. 2. The suffix s corresponding to u

We will show p is injective. If $u, u' \in G'$ and $p(u) = p(u')$, then u and u' are associated with the same suffix $s = p(u)$, hence u and u' must start with the same x_r and (by determinism of the procedure of the previous section) continue in the same way, giving $u = u'$. Thus p is injective, which implies $|G'| \leq |H|$.

Similarly we can define an injective function from H' to G, which proves $|H'| \leq |G|$. It now follows that $\sigma(I') = |G'| + |H'| \leq |G| + |H| = \sigma(I)$. □

2.3   The Algorithm

We will here give a method to decide if a given instance $I = (\Sigma, \Delta, g, h)$ of marked PCP has a solution. The idea is to make a sequence of equivalence-preserving reductions $I_0 = I, I_1, I_2, \ldots$, such that once in a while a reduction from I_i to I_{i+1} simplifies the instance (makes the source alphabet or the suffix complexity smaller). We will show that either this sequence of reductions reaches an I_j which has source alphabet of size 1 or σ equal to 0 (so I_j is decidable), or the sequence will repeat itself after a while and start cycling. Such cycles are detectable, and we will show that every I leading to such a cycle is easily decidable.

So suppose the sequence of reductions does not reach an I_j with alphabet of size 1 or $\sigma(I_j) = 0$. Then it must get "stuck" at a certain source alphabet size and σ. That is, there exist a k, m and z such that all I_i in the infinite sequence $I_k, I_{k+1}, I_{k+2}, \ldots$ have source alphabet of size m and have $\sigma(I_i) = z$. Now this sequence must repeat itself after a while, for otherwise there would be infinitely many distinct instances with the same alphabet and σ-value, contradicting the next lemma.

Lemma 3. *Let $\Sigma = \{a_1, \ldots, a_m\} \subseteq \Delta$ be finite sets and z be a positive natural number. There exist only finitely many distinct instances $I = (\Sigma, \Delta, g, h)$ of PCP that satisfy $\sigma(I) \leq z$.*

Proof. An instance $I = (\Sigma, \Delta, g, h)$ is completely specified by giving the $2m$ words $g(a_1), \ldots, g(a_m), h(a_1), \ldots, h(a_m) \in \Delta^+$. Note that if one of those words has length $> z+1$, then this word has more than z proper suffixes and $\sigma(I) > z$. Accordingly, each of the $2m$ words can have length at most $z + 1$. There are $\sum_{i=1}^{z+1} |\Delta|^i \leq |\Delta|^{z+2}$ such words. Thus there are at most $|\Delta|^{(z+2)2m}$ choices for $2m$ such words, and hence finitely many different I that satisfy $\sigma(I) \leq z$. $\square$

This lemma shows that if the procedure does not converge to very simple instances then it will cycle, and we can detect this by noting that some I_k and I_r $(k < r)$ are equal. It remains to show how we can decide such "cycling" instances of marked PCP. So suppose we have a cycle, assume without loss of generality that it already starts at I_0:

$$I_0 \to I_1 \to \cdots \to I_{r-1} \to I_r = I_0,$$

where $I_i = (\Sigma, \Delta, g_i, h_i)$. By the proof of Lemma 1, for every solution x_i to some I_i, there is a solution x_{i+1} to I_{i+1} such that $x_i = g_{i+1}(x_{i+1}) = h_{i+1}(x_{i+1})$. Suppose x_0 is a solution to I_0 of minimal length. There must exist some solution x_r to I_r such that

$$x_0 = g_1 g_2 \ldots g_r(x_r)$$
$$x_0 = h_1 h_2 \ldots h_r(x_r)$$

Since the g_i and h_i cannot be length-decreasing, we have $|x_0| \geq |x_r|$. But x_0 was chosen to be a minimal-length solution to I_0 and x_r is also a solution to $I_r = I_0$, hence $|x_0| = |x_r|$. This implies that $g_0(= g_r)$ and $h_0(= h_r)$ map the letters occurring in x_r to letters. But then the first letter of x_r is already a solution, hence $|x_0| = |x_r| = 1$. Thus I_0 has a solution iff I_0 has a 1-letter solution (i.e., there is an $a \in \Sigma_0$ such that $g_0(a) = h_0(a)$), and this is trivially decidable.

Below we summarize this analysis in an algorithm and a theorem:

Decision procedure for marked PCP

1. Set $\mathcal{I} = \emptyset$, $i = 0$, $I_0 = I$.
2. Set $i = i + 1$.
3. Reduce I_{i-1} to I_i in the way stated above.
4. If I_i has source alphabet of size 1 or $\sigma = 0$, then decide I_i, print the outcome and terminate.

5. If I_i is simpler than I_{i-1} (smaller source alphabet or σ) then set $\mathcal{I} = \emptyset$ and goto 2.
6. If $I_i \in \mathcal{I}$ then there is a cycle and we can decide I_i by checking if it has a 1-letter solution, print the outcome and terminate;
else set $\mathcal{I} = \mathcal{I} \cup \{I_i\}$ and goto 2.

Theorem 1. *Marked PCP is decidable.*

2.4 Complexity Analysis

Let us analyze the complexity of this algorithm. Let N be the length of the input instance I. Each reduction from I_i to I_{i+1} can be done in $O(N^4)$ steps. How many different reductions do we need to make? For a fixed alphabet size $|\Sigma| \leq |\Delta| = m$ and suffix complexity z, we can make at most $m^{(z+2)2m}$ reductions before detecting a cycle (proof of Lemma 3). Since $m = O(N)$ and $z = O(N^2)$, this gives an upper bound of $2^{O(\log N \cdot N^3)}$ on the number of reductions for fixed alphabet size and suffix complexity. Alphabet size and suffix complexity cannot increase during the process. There are at most $n = O(N)$ different alphabet sizes and at most $\sigma(I) = O(N^2)$ different suffix complexities possible, so we have to make no more than $O(N^3) \cdot 2^{O(\log N \cdot N^3)}$ reductions. Since the set $\mathcal{I}$ can contain at most $2^{O(\log N \cdot N^3)}$ instances, the test $I_i \in \mathcal{I}$ in step 6 can be performed in $2^{O(\log N \cdot N^3)}$ steps. Thus the whole algorithm works in $2^{O(\log N \cdot N^3)}$ steps, which means that marked PCP is in **EXPTIME**.

3 2-Marked PCP Is Undecidable

Here we will show that if we weaken the condition of markedness, by only requiring the morphisms to be 2-marked, then PCP becomes undecidable again.

Consider the following semi-group S_7 with set of 5 generators $\Gamma = \{a, b, c, d, e\}$ and 7 relations:

$$S_7 = \langle a, b, c, d, e \mid R \rangle$$
$$R = \{ac = ca, ad = da, bc = cb, bd = db, eca = ce, edb = de, cca = ccae\}$$

Tzeitin [10] (see also [7, p. 445]) proved that the following problem for this semi-group is undecidable:

Given $u, v \in \Gamma^+$, is $u = v \in S_7$?

Note that the set of 7 left-hand-sides of R is 2-marked, and similarly for the set of 7 right-hand-sides of R. We will reduce this problem to 2-marked PCP. We use a slight modification of the standard reduction from word problems to PCP, involving an alphabet with some underlined letters in order to ensure 2-markedness.

Define the source alphabet as

$$\Sigma = \Gamma \cup \underline{\Gamma} \cup \{B, E, \#, \underline{\#}, r_1, r_2, \ldots, r_7, \underline{r_1}, \underline{r_2}, \ldots, \underline{r_7}\},$$

where $\underline{\Gamma} = \{\underline{a}, \underline{b}, \underline{c}, \underline{d}, \underline{e}\}$, and $r_1, \ldots, r_7$ are the 7 relations in R and $\underline{r_1}, \ldots, \underline{r_7}$ are their underlined versions (considered as single letters), so $r_1 = [ac = ca]$, $\underline{r_1} = [\underline{ac} = \underline{ca}]$ etc. Define the target alphabet as

$$\Delta = \Gamma \cup \underline{\Gamma} \cup \{B, E, \#, \underline{\#}\}.$$

B and E will mark the beginning and end of expressions, respectively. Given $u, v \in \Gamma^+$, g and h are defined by Table 1:

	B	E	$\#$	$\underline{\#}$	a	$\ldots$	e	$\underline{a}$	$\ldots$	$\underline{e}$	$[s = t]$	$[\underline{s} = \underline{t}]$
g	$Bu\#$	E	$\#$	$\underline{\#}$	$\underline{a}$	$\ldots$	$\underline{e}$	a	$\ldots$	e	$\underline{t}$	s
h	B	$\underline{\#}vE$	$\#$	$\underline{\#}$	a	$\ldots$	e	$\underline{a}$	$\ldots$	$\underline{e}$	s	$\underline{t}$

Table 1. Definition of g and h

Note that the constructed instance $I = (\Sigma, \Delta, g, h)$ is an instance of 2-marked PCP. The following lemma shows that the reduction preserves equivalence with Tzeitin's problem:

Lemma 4. *Let u, v, I be as above. Then $u = v \in S_7$ iff I has a solution.*

Proof.

$\Longrightarrow$: Suppose $u = v \in S_7$. Then there is a sequence $u = u_1 \rightarrow u_2 \rightarrow \cdots \rightarrow u_k = v$, where $u_i = u'su''$ and $u_{i+1} = u'tu''$, and $s = t \in R$ or $t = s \in R$. We construct a solution to I by induction on k.

If $k = 1$, then $u = v \in \Gamma^+$. Now $x = Bu\#\underline{u}E$ is a solution to I.

Now let $I' = (\Sigma, \Delta, g', h')$ be the instance of 2-marked PCP corresponding to $u = u_{k-1} \in S_7$. By the induction hypothesis we can assume that I' has a minimal-length solution x'. It is easy to see that every solution must begin with B and end with E, so $x' = ByE$, and therefore $g'(By) = w\#\underline{u_{k-1}}$ and $h'(By) = w$ for some w. Note that since I and I' only differ in the assignment $h(E)$ and $h'(E)$, and E cannot occur in y (because x' is minimal), we also have $g(By) = w\#\underline{u_{k-1}}$ and $h(By) = w$. We distinguish two cases. Firstly, $u_{k-1} = u'su''$ and $v = u_k = u'tu''$, where $r = [s = t]$ is one of the 7 relations. Then it is easily verified that $x = By\underline{\#}\underline{u'ru''}\#\underline{u'tu''}E$ is a solution to I. Secondly, if $u_{k-1} = u'tu''$ and $v = u_k = u'su''$, then $x = By\underline{\#}\underline{u'tu''}\#\underline{u'ru''}E$ is a solution. This completes the induction step.

$\Longleftarrow$: Suppose I has a solution x. We can assume x is of minimal length. This x must be of the form $Bx_1x_2\ldots x_mE$, where $x_i \in \Sigma$, so $g(Bx_1\ldots x_mE) = Bu\#g(x_1\ldots x_m)E = h(Bx_1\ldots x_mE) = Bh(x_1\ldots x_m)\#vE$. Ignoring the underlining, $g(x) = h(x)$ must be of the form $Bu_1\#u_2\#\ldots\#u_{k-1}\#u_kE$, where $u_i \in \Gamma^*$, $u_1 = u$ and $u_k = v$. We will show that $u_i = u_{i+1} \in S_7$ for every $1 \leq i \leq k-1$, from which $u = v \in S_7$ follows.

Because $\#$ occurs in $h(x_1\ldots x_m)$, there must be some least i such that $x_i = \#$, and hence $u = h(x_1\ldots x_{i-1})$. Since there is no underlining in u, it follows that

$x_1, \ldots, x_{i-1}$ must have been chosen from $a, \ldots, e, r_1, \ldots, r_7$. Let $x_1 \ldots x_{i-1} = w_1 r_{i_1} w_2 r_{i_2} \ldots w_l$, with $w_i \in \Gamma^*$ and $r_i = [s_i = t_i] \in \{r_1, \ldots, r_7\}$. Then $u = h(w_1 r_{i_1} w_2 r_{i_2} \ldots w_l) = w_1 s_{i_1} w_2 s_{i_2} \ldots w_l$. See Figure 3 for illustration.

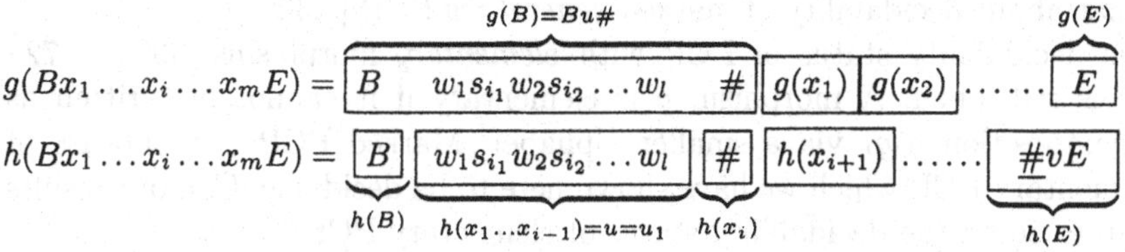

$$g(Bx_1 \ldots x_i \ldots x_m E) = \boxed{B \quad \overbrace{w_1 s_{i_1} w_2 s_{i_2} \ldots w_l}^{g(B)=Bu\#} \quad \#} \; \boxed{g(x_1)} \; \boxed{g(x_2)} \ldots \ldots \boxed{E}^{g(E)}$$

$$h(Bx_1 \ldots x_i \ldots x_m E) = \underbrace{\boxed{B}}_{h(B)} \; \underbrace{\boxed{w_1 s_{i_1} w_2 s_{i_2} \ldots w_l}}_{h(x_1 \ldots x_{i-1})=u=u_1} \; \underbrace{\boxed{\#}}_{h(x_i)} \; \boxed{h(x_{i+1})} \ldots \ldots \underbrace{\boxed{\#vE}}_{h(E)}$$

Fig. 3. Picture leading to $u = v$

Note that $g(x_1 \ldots x_{i-1}) = g(w_1 r_{i_1} w_2 r_{i_2} \ldots w_l) = w_1 t_{i_1} w_2 t_{i_2} \ldots w_l$. But now, since we must have $g(x_1 \ldots x_m E) = h(x_{i+1} \ldots x_m E)$, there must be a least $j > i$ such that $x_j \in \{\#, \underline{\#}\}$ and $h(x_{i+1} \ldots x_{j-1}) = g(x_1 \ldots x_{i-1}) = w_1 t_{i_1} w_2 t_{i_2} \ldots w_l$. The latter string (without underlining) is u_2. Note that $u_1 = u_2 \in S_7$, because $u_1 (= u)$ and u_2 only differ by u_2 having t_i where u_1 has s_i.

Continuing this reasoning, we can show that for every two words $u_i, u_{i+1} \in \Gamma^*$ occurring in $g(x) = h(x)$ separated by $\#$, ignoring underlining, we must have $u_i = u_{i+1} \in S_7$ (some of the words u_i and u_{i+1} may actually already be equal in Σ^+). Hence $u = v \in S_7$, since $g(x)$ starts with $u_1 = u$ and ends with $u_k = v$. □

Together with Tzeitin's result, the above lemma implies:

Theorem 2. 2-*Marked PCP is undecidable.*

To end this section, we note that 2-marked PCP is not a special case of injective PCP. For example, the morphism defined by $g(1) = 23$, $g(2) = 2$, $g(3) = 3$ is 2-marked but not injective. We can combine k-markedness and injectivity by calling a morphism g *strongly* k-*marked* if g is both k-marked and prefix (i.e., no $g(a_i)$ is a prefix of another $g(a_j)$). This clearly implies injectivity. It follows from a construction of Ruohonen [8] that strongly 5-marked PCP is undecidable: the biprefix instances of PCP constructed there to show undecidability of biprefix PCP are also 5-marked. Decidability of strongly k-marked PCP for $1 < k < 5$ is still open.

4 Conclusion and Future Work

We can investigate the boundary between decidability and undecidability by examining which restrictions on the Post Correspondence Problem render the problem decidable. We have shown here that restricting PCP to *marked* morphisms gives us decidability. On the other hand, 2-marked PCP is still undecidable.

The following questions are left open by this research:

- Is exponential time the best we can do when deciding marked PCP, or is there a polynomial-time algorithm for the problem?
- What about decidability of strongly k-marked PCP for $1 < k < 5$?
- What about decidability of marked *generalized* PCP [1,3]?
- The decidability status of PCP with *elementary* morphisms [9, pp. 72–77] is still open. A morphism g is elementary if it cannot be written as a composition $g_2 g_1$ via a smaller alphabet. Marked PCP is a subcase of elementary PCP which we have shown here to be decidable. Can our results help to settle the decidability status of elementary PCP?

Acknowledgment

We would like to thank Tero Harju, Juhani Karhumäki, and John Tromp for reading and commenting this paper, and Harry Buhrman for some discussions. The second author would like to thank the CWI for its hospitality during the summer of 1998, when part of this work was done.

References

1. A. Ehrenfeucht, J. Karhumäki, and G. Rozenberg. The (generalized) Post correspondence problem with lists consisting of two words is decidable. *Theoretical Computer Science*, 21(2):119–144, 1982.
2. M. Garey and D. Johnson. *Computers and Intractability: A Guide to the Theory of NP-Completeness*. W. H. Freeman, 1979.
3. T. Harju, J. Karhumäki, and D. Krob. Remarks on generalized Post correspondence problem. In *Proceedings of 13th STACS*, volume 1046 of *Lecture Notes in Computer Science*, pages 39–48. Springer-Verlag, 1996.
4. Y. Lecerf. Récursive insolubilité de l'équation générale de diagonalisation de deux monomorphismes de monoïdes libres $\phi x = \Psi x$. *Comptes Rendus Acad. Sci. Paris*, 257:2940–2943, 1963.
5. Y. Matiyasevich and G. Sénizergues. Decision problems for semi-Thue systems with a few rules. In *Proceedings of the 11th IEEE Symposium on Logic in Computer Science*, pages 523–531, 1996.
6. E. L. Post. A variant of a recursively unsolvable problem. *Bulletin of the American Mathematical Society*, 52:264–268, 1946.
7. G. Rozenberg and A. Salomaa, editors. *Handbook of Formal Languages*, volume 1. Springer-Verlag, Berlin, 1997.
8. K. Ruohonen. Reversible machines and Post's correspondence problem for biprefix morphisms. *Journal of Information Processing and Cybernetics (EIK)*, 21(12):579–595, 1985.
9. A. Salomaa. *Jewels of Formal Language Theory*. Pitman, 1981.
10. G. C. Tzeitin. Associative calculus with an unsolvable equivalence problem. *Tr. Mat. Inst. Akad. Nauk*, 52:172–189, 1958. In Russian.

On Quadratic Word Equations

John Michael Robson[1] and Volker Diekert[2] *

[1] LaBRI, Université Bordeaux I
351, cours de la Libération, F-33405 Talence Cedex, France
[2] Institut für Informatik, Universität Stuttgart
Breitwiesenstr. 20-22, D-70565 Stuttgart, Germany

Abstract. We investigate the satisfiability problem of word equations where each variable occurs at most twice (quadratic systems). We obtain various new results: The satisfiability problem is NP-hard (even for a single equation). The main result says that once we have fixed the lengths of a possible solution, then we can decide in linear time whether there is a corresponding solution. If the lengths of a minimal solution were at most exponential, then the satisfiability problem of quadratic systems would be NP-complete. (The inclusion in NP follows also from [21])
In the second part we address the problem with regular constraints: The uniform version is PSPACE-complete. Fixing the lengths of a possible solution doesn't make the problem much easier. The non-uniform version remains NP-hard (in contrast to the linear time result above). The uniform version remains PSPACE-complete.

1 Introduction

A major result in combinatorics on words states that the existential theory of equations over free monoids is decidable. This result was obtained by Makanin [14], who showed that the satisfiability of word equations with constants is decidable. For the background we refer to [19], to the corresponding chapter in the Handbook of Formal Languages, [4], or to the forthcoming [5]. There are also two volumes in the Springer lecture notes series dedicated to word equations and related topics: [23] and [2]. Makanin's Algorithm is the construction of a finite search graph. It's finiteness proof is probably among the most complex proofs in theoretical computer science. The algorithm was implemented in 1987 at Rouen by Abdulrab, see [1].
In 1990 Schulz showed an important generalization: Makanin's result remains true when adding regular constraints, [22]. Thus, we may specify for each word variable x a regular language L_x and we are only looking for solutions where the value of each variable x is in L_x. Having this form it was also possible to extend Makanin's result to free partially commutative monoids (known also as trace monoids), see [6, 17]

---

* This work was partially supported by the French German project PROCOPE, the authors acknowledge this support and the valuable comments of the anonymous referees

The inherent complexity of the satisfiability problem of word equations is however not well-understood. The known lower bound follows already from the unary case. A system of word equations over a unary alphabet of constants is equivalent (under logspace reductions) to an instance of linear integer programming – and vice versa. It is a well-known classical fact that linear integer programming is NP-complete, see e.g. [8]. Thus, the satisfiability problem for a single equation over a binary alphabet of constants is NP-hard. For the upper bound, a first analysis in the works of Jaffar and Schulz showed a 4-NEXPTIME result [5, 9, 24]. By Kościelski and Pacholski [10, Cor. 4.6] this went down to 3-NEXPTIME. The present state of the art is due to [7]. Gutiérrez showed that the problem is in EXPSPACE, in particular, it is in 2-DEXPTIME. Another recent and very interesting result is due to Rytter and Plandowski [21]. It shows that the minimal solution of a word equation is highly compressible in terms of Lempel-Ziv encodings. It is conjectured that the length of a minimal solution is at most exponential in the denotational length of the equation. If this were true, then the Lempel-Ziv encoding has polynomial length and, following [21], the satisfiability problem for word equation with constants would become NP-complete. At the moment we believe we are far away from proving this audacious conjecture.

The objects of interest here are quadratic systems, i.e., systems of word equations where each variable occurs at most twice. In combinatorial group theory these systems have been introduced by [12], see also [13]. They play an important rôle in the classification of closed surfaces and basic ideas of how to handle quadratic equations go back to [18]. The explicit statement of an algorithm for the solution of quadratic systems of word equations appears in [16].

We obtain various new results concerning quadratic systems. We show that the satisfiability problem is NP-hard (even for a single equation). The main result of the paper states that once we have fixed the lengths of a possible solution, then we can decide in linear time whether there is a corresponding solution. As a corollary we can say that if the lengths of a minimal solution of solvable quadratic systems were at most exponential, then the satisfiability problem would be NP-complete. The conclusion of containment in NP follows also from [21], but the method here is more direct and yields a much simpler approach to the special situation of quadratic systems.

In the second part we address the problem with regular constraints. The uniform version is PSPACE-complete. We also show that fixing the lengths of a possible solution doesn't make the problem much easier. The non-uniform version remains NP-hard (in contrast to the linear time result above). The uniform version remains PSPACE-complete.

Due to lack of space this extended abstract does not contain all proofs. They will appear elsewhere.

2   Notations and Preliminaries

Let A be an alphabet of constants and let Ω be a set of variables. As usual, $(A \cup \Omega)^*$ means the free monoid over the set $A \cup \Omega$; the empty word is denoted by ε. A *word equation* $L = R$ is a pair $(L, R) \in (A \cup \Omega)^* \times (A \cup \Omega)^*$, and a *system* of word equations is a set of equations $\{L_1 = R_1, \ldots, L_k = R_k\}$. A *solution* is a homomorphism $\sigma: (A \cup \Omega)^* \to A^*$ leaving the letters of A invariant such that $\sigma(L_i) = \sigma(R_i)$ for all $1 \leq i \leq k$. A solution $\sigma: \Omega \to A^*$ is called *minimal*, if the sum $\sum_{x \in \Omega} |\sigma(x)|$ is minimal.

A system of word equations is called *quadratic*, if each variable occurs at most twice. In the present paper we consider only quadratic systems.

3   Quadratic Equations

Quadratic systems are, in principle, easy to solve by using Nielsen transformations. The standard algorithm is from [16]; it uses non-deterministic linear space and works as follows: The first step is to guess which variables can be replaced by the empty word. Then we may assume that the first equation is of the form

$$x \cdots = y \cdots$$

where $x \neq y$ and y is a variable. Moreover, we may also assume that x is a prefix of y. Then we replace all occurrences of y (at most two) by xy, and we cancel x on the left of the first equation. Having done this, we guess whether y can be replaced by the empty word. Then we repeat the process. The size of the quadratic system never increases, but the length of a minimal solution decreases in each round. Hence, the non-deterministic algorithm will find a solution, if there is any. A non-redundant execution of the algorithm will go through at most exponentially many different systems. Thus, there is an doubly exponential upper bound on the the length of the minimal solution. This seems to be quite an overestimation. We have the following conjecture.

Conjecture   The length of a minimal solution of a solvable quadratic system of word equations is at most polynomial in the input size.

The value of Theorem 2 would already increase, if only the following much weaker conjecture were true.

Conjecture (weak form)   The length of a minimal solution of a solvable quadratic system of word equations is at most exponential in the input size.

The first result of the present paper shows that the satisfiability problem of word equations remains NP-hard, even in the restricted case of quadratic systems. In fact, based on the conjectures above, we strongly believe that it is NP-complete in this case, see Corollary 1.

Theorem 1. *Let* $|A| \geq 2$. *The following problem is NP-hard.*
INSTANCE: A quadratic word equation.
QUESTION: Is there a solution $\sigma: \Omega \longrightarrow A^*$?

Proof. We give a reduction from 3-SAT. Let $F = C_0 \wedge \cdots \wedge C_{m-1}$ be a propositional formula in 3-CNF over a set of variables Ξ. Each clause has the form

$$C_i = (\tilde{X}_{3i} \vee \tilde{X}_{3i+1} \vee \tilde{X}_{3i+2})$$

where the $\tilde{X}_j$ are literals. We can assume that every variable has both positive and negative occurrences.

First we construct a quadratic system of words equations using word variables

$$c_i, d_i, \ 0 \leq i \leq m - 1,$$
$$x_j, \ 0 \leq j \leq 3m - 1,$$
$$y_X, z_X, \text{ for each } X \in \Xi.$$

We use the constants $a, b, a \neq b$. For each clause C_i we have two equations:

$$c_i x_{3i} x_{3i+1} x_{3i+2} = a^{3m} \quad \text{and} \quad c_i d_i = a^{3m-1}.$$

Now let $X \in \Xi$. Consider the set of positions $\{i_1, \ldots, i_k\}$ where $X = \tilde{X}_{i_1} = \cdots = \tilde{X}_{i_k}$ and the set of positions $\{j_1, \ldots, j_n\}$ where $\bar{X} = \tilde{X}_{j_1} = \cdots = \tilde{X}_{j_n}$. We deal with the case $k \leq n$; the case $n \leq k$ is symmetric. With each X we define two more equations:

$$y_X z_X = b \text{ and } x_{i_1} \cdots x_{i_k} y_X a^n b x_{j_1} \cdots x_{j_n} z_X = a^n b a^n b.$$

The formula is satisfiable if and only if the quadratic system has a solution. Next, a system of k word equations $L_1 = R_1, \ldots, L_k = R_k$, $k \geq 1$ with $R_1 \cdots R_k \in \{a, b\}^*$ is equivalent to a single equation temporarily using a third constant c: $L_1 c \cdots L_{k-1} c L_k = R_1 c \cdots R_{k-1} c R_k$. Finally, we can eliminate the use of the third letter c without increasing the number of occurrences of any variable by the well-known technique of coding the three letters as aba, $abba$ and $abbba$ and replacing each occurrence of a variable x by axa, for a classical reference see [15].

The following theorem is the main result of the paper. In a slightly different form it appeared first in an unpublished manuscript of the first author [20].

Theorem 2. *There is a linear time algorithm to solve the following problem (on a unit cost RAM).*
INSTANCE: A quadratic system of word equations with a list of natural numbers $b_x \in \mathbb{N}$, $x \in \Omega$, written in binary.
QUESTION: Is there a solution $\sigma: \Omega \longrightarrow A^*$ such that $|\sigma(x)| = b_x$ for all $x \in \Omega$?

Proof. In a linear time preprocessing we can split the system into equations each containing a maximum of three variable occurrences: to see this let $x_1 \cdots x_g = x_{g+1} \cdots x_d$ be a word equation of the system with $1 \leq g < d$, $x_i \in A \cup \Omega$ for $1 \leq i \leq d$. Then the equation is equivalent to:

$$\begin{array}{ll}
x_1 = y_1, & x_{g+1} = y_{g+1}, \\
y_1 x_2 = y_2, & y_{g+1} x_{g+2} = y_{g+2}, \\
\quad \vdots & \quad \vdots \\
y_{g-1} x_g = y_g, & y_{d-1} x_d = y_d, \\
\end{array}$$
$$y_g = y_d.$$

Here $y_1, \ldots, y_d$ denote new variables, each of them occurring exactly twice. After the obvious simplification of equations with only one variable or constant on each side, we obtain a system where each equation has the form $z = xy$, $x, y, z \in A \cup \Omega$. In fact, using (for the first time) that the lengths b_x are given, we may assume that $b_x \neq 0$ for all variables $x \in \Omega$ and that each equation has the form $z = xy$, where z is a variable. If m denotes the number of equations, then we can define the input size of a problem instance E as:

$$d(E) = m + \sum_{x \in \Omega} \log_2(b_x).$$

For $x \in A \cup \Omega$ let $|x| = b_x$ if $x \in \Omega$, and $|x| = 1$ if $x \in A$. We are looking for a solution σ such that $|\sigma(x)| = |x|$ for all $x \in A \cup \Omega$. A variable $z \in \Omega$ is called *doubly defined*, if E contains two equations $z = xy$ and $z = uv$. Let $dd(E)$ be the number of doubly defined variables. Define $c = 0.55$ and $k = 3/\ln(1/c)$ (≈ 5.01). Finally, we define the weight of the instance E as follows

$$W(E) = |\Omega| + dd(E) + k \sum_{x \in \Omega} \ln |x|.$$

We start the algorithm with the assumption that $|z| = |x| + |y|$ for all equations $z = xy$ and $|x| \neq 0$ for all variables x. Since $W(E) \in O(d(E))$, it is enough to show how to reduce the weight by at least 1 in constant time. If $dd(E) = 0$, then the system is solvable and we are done. Hence let $dd(E) > 0$. Consider a doubly defined variable z and its two equations:

$$z = xy,$$
$$z = uv.$$

If $|x| = |u|$, then we either have an immediate contradiction or we can eliminate at least one variable, namely z. Therefore, without restriction, $0 < |u| < |x|$. Let w be a new variable with $|w| = |x| - |u|$. We replace the equations $z = xy$ and $z = uv$ by:

$$x = uw,$$
$$v = wy.$$

If $|w| \leq c|z|$ we have reduced $\sum_{x \in \Omega} \ln |x|$ by at least $\ln(1/c)$ while increasing $dd(E)$ by at most 1. So the weight has been reduced by at least 2 and we are done with this step.

Hence in what follows we assume $|w| > c|z|$.

If $x = v$, then u and y are conjugates. Hence for some $\alpha \geq 0$ and new variables r, s we can write:

$$u = rs, w = (rs)^\alpha r, \text{ and } y = sr$$

Note that the values of $\alpha, |r|$, and $|s|$ can be calculated in constant time from $|x|$ and $|u|$. Since the system is quadratic, there are no other occurrences of x.

Hence we can replace the equations $x = uw$ and $v = wy$ by:

$$u = rs,$$
$$y = sr.$$

The overall effect is that z and x have been replaced by r and s. The number of doubly defined variables may be greater by at most 1, but we have $|r| \leq |u| < |z| - |w| \leq (1 - c)|z|$ and $|s| < |x|$, so that $\sum_{x \in \Omega} \ln |x|$ has been reduced by at least $\ln(1/(1 - c))$. Hence again we have reduced $W(E)$ by more than 1. Let us make a comment here: In a minimal solution we must have $\alpha = 0$. But this contradicts the assumption $|w| > c|z| > (1 - c)|z| > |u|$. Hence, the case $x = v$ is not possible for a minimal solution at this stage of the algorithm.

We are in the case $x \neq v$ and $|w| > c|z|$. If neither x nor v has a second definition, then $\sum_{x \in \Omega} \ln |x|$ has not increased but the number of doubly defined variables has decreased by 1 thus decreasing $W(E)$ by 1. Hence we may assume that there is an equation

$$x = pq.$$

If p is long, which means here $|p| \geq c|z|$, then we return to the original situation:

$$z = xy,$$
$$x = pq,$$
$$z = uv.$$

We introduce a new variable r with $|r| = |z| - |p|$ and we replace the first two equations $z = xy$ and $x = pq$ by:

$$z = pr,$$
$$r = qy.$$

Since $|r| \leq (1 - c)|z|$ and $|x| \geq c|z|$ we have reduced $\sum_{x \in \Omega} \ln |x|$ by at least $\ln(c/(1 - c))$ while leaving $dd(E)$ unchanged (since r is not doubly defined). So $W(E)$ has decreased by more than $k \ln(c/(1-c)) > 1$ and we are done. Therefore, the situation is as follows:

$$x = uw,$$
$$x = pq,$$
$$v = wy.$$

We have $|u| < (1-c)|z|, |p| < c|z|$, and $|w| \geq c|z|$. If $|u| = |p|$, then again, either there is an immediate contradiction or we can eliminate at least one variable. Assume first $|u| < |p|$. Then we introduce a new variable r with $|r| = |p| - |u|$ and we replace the equations $x = uw$ and $x = pq$ by:

$$p = ur,$$
$$w = rq.$$

The other (and final) case is in fact symmetric. If $|p| < |u|$, then we introduce r with $|r| = |u| - |p|$ and we replace the equations $x = uw$ and $x = pq$ by:

$$u = pr,$$
$$q = rw.$$

Overall, the number of doubly defined variables may have increased by at most 2. The variables z and x are replaced by r and w. But, in each case, we have $|r| < c|z|$ and $|w| < |x|$. Hence $\sum_{x \in \Omega} \ln |x|$ has decreased by at least $\ln(1/c)$ and so the net decrease in $W(E)$ is at least 1.

Hence, in all cases, we have decreased $W(E)$ by at least 1 in $O(1)$ arithmetic operations.

Remark 1. The method above yields a most general solution in the following sense. Let E be an instance to the problem of Theorem 2 and assume that E is solvable. Then we produce in linear time a quadratic system over a set of variables Γ (but without doubly defined variables) such that the set of solutions satisfying the length constraints is in a canonical one-to-one correspondence with the set of mappings $\psi : \Gamma \longrightarrow A^*$ where $|\psi(x)| = |b_x|$ for $x \in \Gamma$.

Corollary 1. *If the conjecture (weak form) above is true, then the satisfiability problem for quadratic systems of word equations is NP-complete.*

Remark 2. Given Theorem 1, Corollary 1 follows also from a recent work of Rytter and Plandowski [21]. They have shown that if the lengths $b_x, x \in \Omega$, are given in binary as part of the input together with a word equation (not necessarily quadratic), then there is a deterministic polynomial time algorithm for the satisfiability problem. Their method is based on Lempel-Ziv encodings and technically involved. Our contribution shows that the situation becomes much simpler for quadratic systems. In particular, we can reduce polynomial time to linear time; and our method is fairly straightforward using variable splitting. In view of the conjectures above it is not clear that the use of Lempel-Ziv encodings can improve the running time for deciding the satisfiability of quadratic systems. The most difficult part is apparently to get an idea of the lengths b_x for $x \in \Omega$. Once these lengths are known (or fixed), the corresponding satisfiability problem for quadratic systems of word equation becomes extremely simple.

4   Regular Constraints

There is an interesting generalization of Makanin's result. The generalization is due to Schulz [22] and says that if a word equation is given with a list of regular languages $L_x \subseteq A^*, x \in \Omega$, then one can decide whether there is a solution $\sigma : \Omega \longrightarrow A^*$ such that $\sigma(x) \in L_x$ for all $x \in \Omega$. In the following we shall assume that regular languages are specified by non-deterministic finite automata (NFA). In the uniform version the NFA are part of the input. In the non-uniform version the NFA are restricted such that each is allowed to have at most k states, where k is a fixed constant being not part of the input. Using a recent result of Gutiérrez [7] one can show that the uniform satisfiability problem of word equations with regular constraints can be solved in EXPSPACE (more precisely in DSPACE $(2^{o(d^3)})$, if d denotes the input size), see [5]. So, from the general case it is not really clear whether adding regular constraints makes the satisfiability problem of word equations harder. We give here however some evidence that, indeed, it does. Restricted to quadratic systems the uniform satisfiability problem with regular constraints becomes PSPACE complete.

The non-uniform version is NP-hard and it remains NP-hard, even if the lengths $b_x, x \in \Omega$, are given in unary as part of the input. This is in sharp contrast to Theorem 2. Having regular constraints it is also easy to find examples where the length of a minimal solution increases exponentially; the next example is of this kind. Note however that this refers to the uniform version of the problem

Example 1. (modified from one by [3]) Let $n > 0$. Consider the following word equation with regular constraints:

$$A = \{a, b\}, \quad \Omega = \{x_i \mid 0 \leq i \leq n\},$$
$$L_{x_i} = a, \text{ for } i \leq 1,$$
$$L_{x_i} = aA^*a \backslash (A^*b^i A^*) \text{ for } i > 1,$$
$$x_n b^n x_n = x_0 b x_0 \ b^2 x_1 b x_1 \ \cdots \ b^n x_{n-1} b^{n-1} x_{n-1}$$

Theorem 3. *The following problem is PSPACE-complete.*
INSTANCE: A quadratic system of word equations with a list of regular constraints $L_x \subseteq A^*$, $x \in \Omega$.
QUESTION: Is there a solution $\sigma \colon \Omega \longrightarrow A^*$ such that $\sigma(x) \in L_x$ for all $x \in \Omega$?
Moreover, the problem remains PSPACE-complete, if the input is given together with a list of numbers b_x, $x \in \Omega$ (a number $b \in \mathbb{N}$ resp.), written in binary, and if we ask for a solution satisfying in addition the requirement $|\sigma(x)| = b_x$ ($|\sigma(x)| = b$ resp.) for all $x \in \Omega$?

Proof. The PSPACE-hardness follows directly from a well-known result on regular sets. Let $L_1, \ldots L_n$ be regular languages specified by NFA. Then the emptiness problem $L_1 \cap \cdots \cap L_n = \emptyset$ is PSPACE-complete, [11]. If the intersection is not empty, then there is a witness of at most exponential length. Let b be this upper bound on the length of a witness. Using a new letter c such that $c \notin A$, we can ask whether the intersection

$$L_1 c^* \cap \cdots \cap L_n c^*$$

contains a word of length b. (Instead of using a new letter we may also use some coding provided $|A| \geq 2$.) The quadratic system is given by n variables $x_1, \ldots, x_n$ and regular constraints $L_{x_i} = L_i c^*$ for $1 \leq i \leq n$. The equations are trivial: $x_1 = x_2$, $x_2 = x_3$, $\cdots$, $x_{n-1} = x_n$.
The PSPACE algorithm for the uniform satisfiability problem is a modification of the proof of Theorem 2.

Theorem 4. *Let $r \geq 4$ be a fixed constant, which is not part of the input. The following problem is NP-complete.*

INSTANCE: A quadratic system of word equations with a list of natural numbers $b_x \in \mathbb{N}$ written in binary, a list of regular constraints $L_x \subseteq A^*$, $x \in \Omega$, such that each language can be specified by some NFA of at most r states, and $|A| \geq 2$.
QUESTION: Is there a solution $\sigma: \Omega \longrightarrow A^*$ such that $|\sigma(x)| = b_x$ and $\sigma(x) \in L_x$ for all $x \in \Omega$?
Moreover, the problem remains NP-hard, if the numbers b_x, $x \in \Omega$, are written in unary, $|A| = 2$, and the system is a single equation.

5   Conclusion

Problems of satisfiability of quadratic word equations, with or without regular constraints, are apparently simple subcases of general problems known to be decidable. However there are a number of interesting questions still open. In the three cases studied (no constraints, uniform constraints and non-uniform constraints), we have only hardness results with no close upper bounds for the general problem where no information is given on the lengths of the solution. It would be very interesting to find a proof (or disproof!) of the conjectures of Section 3 on the minimal solution length.

References

[1] Habib Abdulrab and Jean-Pierre Pécuchet. Solving word equations. *J. Symbolic Computation*, 8(5):499–521, 1990.

[2] Habib Abdulrab and Jean-Pierre Pécuchet, editors. *Proceedings of Word Equations and Related Topics (IWWERT '91)*, volume 677 of *Lect. Notes Comp. Sci.* Springer-Verlag, Berlin, Heidelberg, New York, 1993.

[3] Thierry Arnoux. Untersuchungen zum Makaninschen Algorithmus. Diplomarbeit 1613, Universität Stuttgart, 1998.

[4] Christian Choffrut and Juhani Karhumäki. Combinatorics of words. In G. Rozenberg and A. Salomaa, editors, *Handbook of Formal Languages, Vol. 1*, pages 329–438. Springer-Verlag, Berlin, Heidelberg, New York, 1997.

[5] Volker Diekert. Makanin's Algorithm. In Jean Berstel and Dominique Perrin, editors, *Algebraic Combinatorics on Words*. Cambridge University Press. A preliminary version is on the web:
http://www-igm.univ-mlv.fr/ berstel/Lothaire/index.html.

[6] Volker Diekert, Yuri Matiyasevich, and Anca Muscholl. Solving trace equations using lexicographical normal forms. In P. Degano, R. Gorrieri, and A. Marchetti-Spaccamela, editors, *Proc. of the 24th ICALP, Bologna, 1997*, number 1256 in Lect. Notes Comp. Sci., pages 336–347. Springer-Verlag, Berlin, Heidelberg, New York, 1997.

[7] Claudio Gutiérrez. Satisfiability of word equations with constants is in exponential space. In *Proc. of the 39th Ann. Symp. on Foundations of Computer Science, FOCS 98*, pages 112–119, Los Alamitos, California, 1998. IEEE Computer Society Press.

[8] John E. Hopcroft and Jeffrey D. Ullman. *Introduction to Automata Theory, Languages, and Computation.* Addison-Wesley, Reading, Mass., 1979.

[9] Joxan Jaffar. Minimal and complete word unification. *J. Assoc. Comput. Mach.*, 37(1):47–85, 1990.

[10] Antoni Kościelski and Leszek Pacholski. Complexity of Makanin's algorithm. *J. Assoc. Comput. Mach.*, 43(4):670–684, 1996. Preliminary version in *Proc. of the 31st Annual IEEE Symposium on Foundations of Computer Science*, Los Alamitos, 1990.

[11] Dexter Kozen. Lower bounds for natural proof systems. In *Proc. of the 18th Ann. Symp. on Foundations of Computer Science, FOCS 77*, pages 254–266. IEEE Computer Society Press, 1977.

[12] Roger C. Lyndon. Equations in free groups. *Transactions of the American Mathematical Society*, 96, 1960.

[13] Roger C. Lyndon and Paul E. Schupp. *Combinatorial Group Theory.* Springer-Verlag, Berlin, Heidelberg, New York, 1977.

[14] Gennadiĭ S. Makanin. The problem of solvability of equations in a free semigroup. *Mat. Sb.*, 103(2):147–236, 1977. In Russian; English translation in: *Math. USSR Sbornik, 32*, 129–198, 1977.

[15] Andrei A. Markov. *Theory of Algorithms.* Academy of Sciences of the USSR, 1954. English translation, Add. Wesley 1981.

[16] Yuri Matiyasevich. A connection between systems of word and length equations and Hilbert's Tenth Problem. *Sem. Mat. V. A. Steklov Math. Inst. Leningrad*, 8:132–144, 1968. In Russian; English translation in: *Seminars in Mathematics, V. A. Steklov Mathematical Institute, 8*, 61–67, 1970.

[17] Yuri Matiyasevich. Some decision problems for traces. In Sergej Adian and Anil Nerode, editors, *Proceedings of the 4th International Symposium on Logical Foundations of Computer Science (LFCS'97), Yaroslavl, Russia, July 6–12, 1997*, number 1234 in Lect. Notes Comp. Sci., pages 248–257. Springer-Verlag, Berlin, Heidelberg, New York, 1997.

[18] Jakob Nielsen. Die Isomorphismen der allgemeinen, unendlichen Gruppe mit zwei Erzeugenden. *Mathematische Annalen*, 78, 1918.

[19] Dominique Perrin. Equations in words. In H. Ait-Kaci and M. Nivat, editors, *Resolution of equations in algebraic structures, Vol. 2*, pages 275–298. Academic Press, 1989.

[20] John Michael Robson. Word equations with at most 2 occurrences of each variable. Preprint, LaBRI, Université de Bordeaux I, 1998.

[21] Wojciech Rytter and Wojciech Plandowski. Application of Lempel-Ziv encodings to the solution of words equations. In Kim G. Larsen et al., editors, *Proc. of the 25th ICALP, Aarhus, 1998*, number 1443 in Lect. Notes Comp. Sci., pages 731–742. Springer-Verlag, Berlin, Heidelberg, New York, 1998.

[22] Klaus U. Schulz. Makanin's algorithm for word equations: Two improvements and a generalization. In K.-U. Schulz, editor, *Proceedings of Word Equations and Related Topics, 1st International Workshop, IWWERT'90, Tübingen, Germany*, volume 572 of *Lect. Notes Comp. Sci.*, pages 85–150. Springer-Verlag, Berlin, Heidelberg, New York, 1992.

[23] Klaus U. Schulz, editor. *Proceedings of Word Equations and Related Topics (IWWERT '90)*, volume 572 of *Lect. Notes Comp. Sci.* Springer-Verlag, Berlin, Heidelberg, New York, 1992.

[24] Klaus U. Schulz. Word unification and transformation of generalized equations. *Journal of Automated Reasoning*, 11(2):149–184, 1993.

Some Undecidability Results Related to
the Star Problem in Trace Monoids*
(Extended Abstract)**

Daniel Kirsten

Dresden University of Technology
Department of Computer Science
Institute of Software Engineering I
D-01062 Dresden, Germany
Daniel.Kirsten@inf.tu-dresden.de
http://www.inf.tu-dresden.de/~dk11

Abstract. This paper deals with decision problems related to the star problem in trace monoids, which means to determine whether the iteration of a recognizable trace language is recognizable. Due to a theorem by RICHOMME from 1994 [18], we know that the star problem is decidable in trace monoids which do not contain a C4-submonoid. It is not known whether the star problem is decidable in C4. In this paper, we show undecidability of some related problems: Assume a trace monoid which contains a C4. Then, it is undecidable whether for two given recognizable languages K and L, we have $K \subseteq L^*$, although we can decide $K^* \subseteq L$. Further, we can not decide recognizability of $K \cap L^*$ as well as universality and recognizability of $K \cup L^*$.

1 Introduction

Free partially commutative monoids, also called trace monoids, were introduced by CARTIER and FOATA in 1969 [2]. In 1977, MAZURKIEWICZ proposed these monoids as a potential model for concurrent processes [14], which marks the beginning of a systematic study of trace monoids by mathematicians and theoretical computer scientists, see e.g., [5,6,7].

One main stream in trace theory is the study of recognizable trace languages, which can be considered as an extension of the well studied concept of regular languages in free monoids. A major step in this research is OCHMAŃSKI's PhD thesis from 1984 [17]. Some of the results concerning regular languages in free monoids can be generalized to recognizable languages in trace monoids. However, there is one major difference: The iteration of a recognizable trace language does not necessarily yield a recognizable language. This fact raises the so called star

---

* This work has been supported by the postgraduate program "Specification of discrete processes and systems of processes by operational models and logics" of the German Research Community (Deutsche Forschungsgemeinschaft).

** See http://orchid.inf.tu-dresden.de/gdp/publikation.html for a complete version [12].

C. Meinel and S. Tison (Eds.): STACS'99, LNCS 1563, pp. 227–236, 1999.
© Springer-Verlag Berlin Heidelberg 1999

problem: Given a recognizable language L, is L^* recognizable? In general, it is not known whether the star problem is decidable. The main result after a stream of publications dealing with this problem is a theorem stated by RICHOMME in 1994, saying that the star problem is decidable in trace monoids which do not contain a particular submonoid called C4 [18]. It is not known whether the star problem is decidable in trace monoids with a C4-submonoid. It is even unknown for finite trace languages.

In this paper, we consider some decision problems for recognizable trace languages which are related to the star problem. If we have two recognizable languages K and L in a trace monoid with a C4-submonoid, then it is undecidable whether K is a subset of L^* and whether $K \cup L^*$ yields the complete monoid. Further, recognizability of $K \cup L^*$ and $K \cap L^*$ is undecidable.

The paper is organized as follows. After this introduction, I explain some concepts from algebra, formal language theory, and trace theory. We deal with recognizable sets, rational sets, and relations between them. Then, we discuss some decision problems concerning recognizable and rational trace languages and their solutions as far as known.

In Section 3, we establish a method to define two recognizable trace languages $\mathbb{P}$ and $\mathbb{R}$ from a given instance of POST's Correspondence Problem. We examine properties of $\mathbb{R}$, $\mathbb{P}$, and $\mathbb{P}^*$. In Section 4, we use these properties to develop the main results. In Section 5, we compare the new results to known results.

2   Formal Definitions

2.1   Monoids, Languages, and Traces

I briefly introduce basic notions from algebra and trace theory. By $\mathbb{N}$, we denote the set of natural numbers including zero. Assume two monoids $\mathbb{M}_1$ and $\mathbb{M}_2$. We denote their Cartesian Product by $\mathbb{M}_1 \times \mathbb{M}_2$. We denote its elements by $\binom{p}{q}$. For $p \in \mathbb{M}_1$ and $M \subseteq \mathbb{M}_2$, we denote by $\binom{p}{M}$ the set of all pairs $\binom{p}{q}$ for $q \in M$.

By an *alphabet*, we mean a finite set of symbols called *letters*. Assume an alphabet Σ. We denote the *free monoid over* Σ by Σ^*. We denote the *empty word* by λ. For every word $w \in \Sigma^*$, we call the number of letters of w the *length* of w, and denote it by $|w|$. For every $n \in \mathbb{N}$, we denote by $\Sigma^{\leq n}$ the set of words $w \in \Sigma^*$ with $|w| \leq n$. Accordingly, we use the notions $\Sigma^{<n}$, $\Sigma^{\geq n}$, and $\Sigma^{>n}$.

We call a binary relation I over Σ an *independence relation* iff I is irreflexive and symmetric. For every pair of letters a and b with aIb, we say that a and b are *independent*, otherwise a and b are *dependent*. We call the pair (Σ, I) an *independence alphabet*. We call two words $w_1, w_2 \in \Sigma^*$ equivalent iff we can transform w_1 into w_2 by exchanging independent adjacent letters which we denote by $w_1 \sim_I w_2$. For instance, if a and c are independent letters, *baacbac*, *bacabac* and *bcaabca* are mutually equivalent words.

The relation $\sim_I$ is a congruence. For every $w \in \Sigma^*$, we denote by $[w]_I$ the equivalence class of w. Therefore, we can define a monoid with the sets $[w]_I$ as elements. For $w_1, w_2 \in \Sigma^*$, we define the product of $[w_1]_I$ and $[w_2]_I$ by $[w_1w_2]_I$.

We denote this monoid by $\mathbb{M}(\Sigma, I)$ and call it the *trace monoid* over Σ and I. We call its elements, i.e., the equivalence classes $[w]_I$, *traces* and its subsets *trace languages* or shortly *languages*. The function $[\,]_I$ is a homomorphism from the free monoid Σ^* to $\mathbb{M}(\Sigma, I)$. As long as no confusion arises, we omit the index I at $[\,]_I$.

If I is the empty relation over Σ, then the trace monoid $\mathbb{M}(\Sigma, I)$ is the free monoid Σ^*. If I is the biggest irreflexive relation over Σ, i.e., two letters a and b are independent iff a and b are different, then the trace monoid $\mathbb{M}(\Sigma, I)$ is the free commutative monoid over Σ. Opposed to this very brief introduction, we formally define P3 and C4.

Lemma 1. *Assume two disjoint alphabets Σ_1 and Σ_2, and assume the independence relation $I = \Sigma_1 \times \Sigma_2 \cup \Sigma_2 \times \Sigma_1$. The trace monoid $\mathbb{M}(\Sigma_1 \cup \Sigma_2, I)$ is isomorphic to the monoid $\Sigma_1^* \times \Sigma_2^*$. An isomorphism maps every letter $a \in \Sigma_1$ to $\binom{a}{\lambda}$, and every letter $b \in \Sigma_2$ to $\binom{\lambda}{b}$.* $\square$

This lemma is an application of a method by FLIESS to transform arbitrary trace monoids into (sub)monoids of Cartesian Products of free monoids, see Chapter 1 in [7]. Iff one of the alphabets Σ_1 and Σ_2 is a doubleton, and the other one is a singleton, we denote by P3 both the monoid $\Sigma_1^* \times \Sigma_2^*$ and the independence alphabet $(\Sigma_1 \cup \Sigma_2, I)$ with I from Lemma 1. Iff both alphabets are doubletons, we accordingly use the notion C4. The notions P3 and C4 abbreviate *path of 3 letters* and *cycle of 4 letters*, respectively.

Assume an independence alphabet (Σ, I). A trace $t \in \mathbb{M}(\Sigma, I)$ is called *connected* iff for every non-empty traces t_1 and t_2 with $t = t_1 t_2$, there are a letter a in t_1 and a letter b in t_2 such that a and b are dependent. A trace $\binom{u}{v}$ in P3 or C4 is connected iff $u = \lambda$ or $v = \lambda$. A trace language L is called *connected* iff every trace in L is connected.

2.2   Recognizable Sets

I introduce the concept of recognizability as far as we use it in this paper, for a more general overview I recommend [1,8].

Definition 1. *Assume a monoid $\mathbb{M}$. An $\mathbb{M}$-automaton is a triple $\mathcal{A} = [Q, h, F]$, where Q is a finite monoid, h is a homomorphism $h : \mathbb{M} \to Q$, and F is a subset of Q. The language of an $\mathbb{M}$-automaton $\mathcal{A}$ is defined by $L(\mathcal{A}) = h^{-1}(F)$.* $\square$

If $L(\mathcal{A}) = L$, then we say is an $\mathbb{M}$-automaton for L. We call a set $L \subseteq \mathbb{M}$ a *recognizable* language over $\mathbb{M}$ iff there is an $\mathbb{M}$-automaton for L. We denote the class of all recognizable languages over $\mathbb{M}$ by $\mathrm{REC}(\mathbb{M})$. Calling the triple $[Q, h, F]$ an $\mathbb{M}$-automaton is due to COURCELLE [4]. The next theorem is a classic one, you find the proof in, e.g., [1,8].

Theorem 1. *Assume a monoid $\mathbb{M}$. The class $\mathrm{REC}(\mathbb{M})$ contains the empty set, $\mathbb{M}$ itself, and it is closed under union, intersection, complement and inverse homomorphisms.* $\square$

Theorem 2. *Assume a trace monoid* $\mathrm{IM}(\Sigma, I)$. *The class* $\mathrm{REC\,IM}(\Sigma, I)$ *contains all finite subsets of* $\mathrm{IM}(\Sigma, I)$ *and is closed under monoid product and iteration of connected trace languages.*   □

The proof of the closure under monoid product originates from FLIESS. Closure under iteration of connected trace languages is due to OCHMAŃSKI, CLERBOUT and LATTEUX, and MÉTIVIER [17,3,15]. Chapter 6 of [7] is a recent survey on recognizable trace languages including a convenient proof of Theorem 2.

If a trace monoid $\mathrm{IM}(\Sigma, I)$ is a free monoid, then every trace is connected. Thus, Theorem 2 includes the classic result that recognizable languages in free monoids are closed under iteration.

The following result is widely known as MEZEI's theorem [1,8].

Theorem 3. *Assume two monoids* IM, IM'. *A set L is recognizable in* $\mathrm{IM} \times \mathrm{IM}'$ *iff there are an* $n \in \mathrm{IN}$, *sets* $L_1, \ldots, L_n \in \mathrm{REC}(\mathrm{IM})$, *and sets* $L_1', \ldots, L_n' \in \mathrm{REC}(\mathrm{IM}')$ *such that* $L = (L_1 \times L_1') \cup \ldots \cup (L_n \times L_n')$.   □

Let us shortly mention the notion of rational sets. Assume some monoid IM. The set of rational expressions $\mathrm{REX}(\mathrm{IM})$ is the smallest set which contains the symbol $\emptyset$, the elements in IM and is closed as follows: For the expressions $r, r_1, r_2 \in \mathrm{REX}(\mathrm{IM})$, the expressions r^*, $(r_1 \cup r_2)$, and $(r_1 r_2)$ belong to $\mathrm{REX}(\mathrm{IM})$. Every rational expression r defines a language $L(r)$ as usual.

We have KLEENE's classic result which asserts that in free monoids the recognizable sets and the rational sets coincide [21]. In trace monoids, we have just one direction due to a more general result by MCKNIGHT [1,8]: Every recognizable trace language is rational. Moreover, we can transform every automaton into a rational expression which defines the same language. However, there are rational trace languages which are not recognizable unless the underlying trace monoid is a free monoid. See Chapter 5 in [7] for more information on rational trace languages. We continue with a well-known example.

Example 1. We consider the alphabets $\Sigma = \{a, b\}$, $\Sigma_1 = \{a\}$, $\Sigma_2 = \{b\}$ and the monoids Σ^* and $\Sigma_1^* \times \Sigma_2^*$. We define the language L in Σ^* by the rational expression $(ab)^*$. Hence, L is rational and recognizable. We apply the homomorphism $[]$ on L, we get $[L] = \left\{ \binom{a^n}{b^n} \,\middle|\, n \in \mathrm{IN} \right\}$. We show that $[L]$ is not recognizable by the closure properties of recognizable sets. By applying the inverse homomorphism $[]^{-1}$ on $[L]$, we get $[[L]]^{-1} = \{w \mid |w|_a = |w|_b\}$. This language is not recognizable. If we assume that $[[L]]^{-1}$ is recognizable, its intersection with the recognizable language defined by $a^* b^*$ would also be recognizable. But, this intersection yields $\{a^n b^n \mid n \in \mathrm{IN}\}$, which is not recognizable.

However, $[L]$ is rational, because it is the iteration of a singleton.   □

2.3 Some Decision Problems for Trace Languages

The following decision problems arise:

Recognizability Problem: Can we decide whether the language of a rational expression is a recognizable language?

Star Problem: Can we decide whether the iteration of a recognizable language yields a recognizable language?

In 1987 and 1992, SAKAROVITCH proved the following theorem [19,20].

Theorem 4. *Assume a trace monoid* $\mathbb{M}(\Sigma, I)$. *The following three assertions are equivalent:*

- (Σ, I) *does not contain an P3-subalphabet.*
- *The rational languages of* $\mathbb{M}(\Sigma, I)$ *form an (effective) Boolean algebra.*
- *We can decide whether the language of a rational expression yields a recognizable language.*   □

During the recent 15 years, many papers have dealt with the star problem. Only partial results have been achieved. I give just a brief survey about their history. The star problem in the free monoid is trivial due to KLEENE, and it is decidable in free commutative monoids due to GINSBURG and SPANIER [10,11]. In 1984, OCHMAŃSKI examined recognizable trace languages in his PhD thesis and stated the star problem. During the 80's, OCHMAŃSKI, CLERBOUT and LATTEUX, and MÉTIVIER independently proved that the iteration of a connected recognizable trace language is recognizable [17,3,15]. SAKAROVITCH's solution of the recognizability problem in 1992 (cf. Theorem 4 above) implies the decidability of the star problem in trace monoids which do not contain a P3-submonoid. The attempt to extend SAKAROVITCH's characterization to the star problem failed, just in the same year, GASTIN, OCHMAŃSKI, PETIT and ROZOY showed the decidability of the star problem in P3 [9]. During the subsequent years, MÉTIVIER and RICHOMME showed beside other results the decidability of the star problem for languages containing at most four traces as well as for finite sets containing at most two connected traces [16]. In 1994, RICHOMME proved the following theorem [18].

Theorem 5. *The star problem is decidable in trace monoids which do not contain a C4-submonoid.*   □

The star problem in trace monoids with a C4-submonoid remains open.

3   A Tricky Language

In this section, we show a method to derive two recognizable trace languages from a given instance of POST's Correspondence Problem (PCP). We examine how properties of the iteration of one of the defined languages depend on the existence or non-existence of a solution of the underlying PCP instance.

An instance of the PCP consists of two alphabets Υ and Σ, and two homomorphisms $\alpha, \beta : \Upsilon^* \to \Sigma^*$. We call the letters of Υ indices. It is well known that it is undecidable whether a given instance of a PCP has a solution, i.e., whether there is some word $w \in \Upsilon^+$ with $\alpha(w) = \beta(w)$. We assume that for $i \in \Upsilon$, we have $\alpha(i) \neq \lambda$ and $\beta(i) \neq \lambda$. This restriction of the PCP is also undecidable.

3.1 Definition of $\mathbb{R}$ and $\mathbb{P}$

We define the languages $\mathbb{R}$ and $\mathbb{P}$. We assume an instance of the PCP, consisting of Υ, Σ, α, and β. We call it the *underlying PCP instance*. We denote the number of letters of Υ by k and treat Υ as $\{i_1, \ldots, i_k\}$.

We enrich the alphabet Υ by nine letters, we set $\Gamma = \{i_1, \ldots, i_k, a_1, \ldots, a_9\}$, while we assume $a_1, \ldots, a_9 \notin \Sigma$. For m, n with $1 \le m < n \le 9$, we abbreviate the word $a_n a_{n+1} \ldots a_m$ by $a_{n..m}$, e.g., we write $a_{3..5}$ instead of $a_3 a_4 a_5$.

We need a function $\gamma : \Upsilon^* \to \Gamma^*$ to "code" words in Υ^*. We set $\gamma(\lambda) = a_{1..9}$. For $w \in \Upsilon^*$ and $i \in \Upsilon$, we set $\gamma(wi) = \gamma(w) i a_{1..9}$. For instance, we have $\gamma(i_6 i_2) = a_{1..9} i_6 a_{1..9} i_2 a_{1..9}$. Obviously, γ is not a homomorphism.

Definition 2. *The language* $\mathbb{R} \subseteq (\Gamma^* \times \Sigma^*)$ *is defined by* $\mathbb{R} = \gamma(\Upsilon^+) \times \Sigma^*$. □

We denote the complement of $\mathbb{R}$ by $\overline{\mathbb{R}}$, i.e., $\overline{\mathbb{R}} = (\Gamma^* \times \Sigma^*) \setminus \mathbb{R}$. Consequently, $\overline{\mathbb{R}}$ yields the language $(\Gamma^* \setminus \gamma(\Upsilon^+)) \times \Sigma^*$. I recommend to read the following definition just briefly, now, and to study the details when we apply it.

Definition 3. *The language* $\mathbb{P} \subseteq (\Gamma^* \times \Sigma^*)$ *is defined as the union of the sets:*

$$\mathbb{P}_{1,1} = \bigcup_{i \in \Upsilon} \binom{a_{1..9}\, i\, a_1}{\Sigma < |\alpha(i)|} \qquad \mathbb{P}_{1,2} = \bigcup_{i \in \Upsilon} \binom{a_{2..9}\, i\, a_1}{\Sigma \le |\alpha(i)|} \qquad \mathbb{P}_{1,3} = \left\{ \binom{a_{2..9}}{\lambda} \right\}$$

$$\mathbb{P}_{2,1} = \left\{ \binom{a_{1..2}}{\lambda} \right\} \qquad \mathbb{P}_{2,2} = \bigcup_{i \in \Upsilon} \binom{a_{3..9}\, i\, a_{1..2}}{\Sigma |\alpha(i)|} \qquad \mathbb{P}_{2,5} = \left\{ \binom{a_{4..9}}{\lambda} \right\}$$

$$\mathbb{P}_{2,3} = \bigcup_{i \in \Upsilon} \binom{a_{3..9}\, i\, a_{1..3}}{\Sigma |\alpha(i)| \setminus \{\alpha(i)\}} \qquad \mathbb{P}_{2,4} = \bigcup_{i \in \Upsilon} \binom{a_{4..9}\, i\, a_{1..3}}{\Sigma |\alpha(i)|}$$

$$\mathbb{P}_{3,1} = \bigcup_{i \in \Upsilon} \binom{a_{1..9}\, i\, a_{1..4}}{\Sigma > |\alpha(i)|} \qquad \mathbb{P}_{3,2} = \bigcup_{i \in \Upsilon} \binom{a_{5..9}\, i\, a_{1..4}}{\Sigma |\alpha(i)|} \qquad \mathbb{P}_{3,3} = \left\{ \binom{a_{5..9}}{\lambda} \right\}$$

$$\mathbb{P}_{4,1} = \bigcup_{i \in \Upsilon} \binom{a_{1..9}\, i\, a_{1..5}}{\Sigma < |\beta(i)|} \qquad \mathbb{P}_{4,2} = \bigcup_{i \in \Upsilon} \binom{a_{6..9}\, i\, a_{1..5}}{\Sigma \le |\beta(i)|} \qquad \mathbb{P}_{4,3} = \left\{ \binom{a_{6..9}}{\lambda} \right\}$$

$$\mathbb{P}_{5,1} = \left\{ \binom{a_{1..6}}{\lambda} \right\} \qquad \mathbb{P}_{5,2} = \bigcup_{i \in \Upsilon} \binom{a_{7..9}\, i\, a_{1..6}}{\Sigma |\beta(i)|} \qquad \mathbb{P}_{5,5} = \left\{ \binom{a_{8..9}}{\lambda} \right\}$$

$$\mathbb{P}_{5,3} = \bigcup_{i \in \Upsilon} \binom{a_{7..9}\, i\, a_{1..7}}{\Sigma |\beta(i)| \setminus \{\beta(i)\}} \qquad \mathbb{P}_{5,4} = \bigcup_{i \in \Upsilon} \binom{a_{8..9}\, i\, a_{1..7}}{\Sigma |\beta(i)|}$$

$$\mathbb{P}_{6,1} = \bigcup_{i \in \Upsilon} \binom{a_{1..9}\, i\, a_{1..8}}{\Sigma > |\beta(i)|} \qquad \mathbb{P}_{6,2} = \bigcup_{i \in \Upsilon} \binom{a_9\, i\, a_{1..8}}{\Sigma |\beta(i)|} \qquad \mathbb{P}_{6,3} = \left\{ \binom{a_9}{\lambda} \right\}$$ □

This is a neat little rip. We remark that $\mathbb{P}_{1,1}, \ldots, \mathbb{P}_{6,3}$ are mutually disjoint.

By using the letters $a_1, \ldots, a_9$, we control the concatenation of the traces in $\mathbb{P}$. We can consider two kinds of traces in $\mathbb{P}^*$: "well-formed" traces, i.e., traces in $\mathbb{R}$, and "trash" traces. To examine the intersection $\mathbb{R} \cap \mathbb{P}^*$, we just have to examine the "well-formed" traces in $\mathbb{P}^*$. Thereby, we are able to show connections between the existence of a solution of the underlying PCP instance and properties of $\mathbb{R} \cap \mathbb{P}^*$.

3.2   Properties of $\mathbb{R}$, $\mathbb{P}$, and $\mathbb{P}^*$

The first important property of $\mathbb{R}$, $\overline{\mathbb{R}}$ and $\mathbb{P}$ is recognizability. We can effectively construct $(\Gamma^* \times \Sigma^*)$-automata for $\mathbb{R}$, $\overline{\mathbb{R}}$ and $\mathbb{P}$ from the underlying PCP-instance. See [12] for details. We examine $\mathbb{P}^*$. We are mainly interested in traces in $\mathbb{P}^*$ whose first component is a word from $\gamma(\Upsilon^+)$.

Lemma 2. *For every $w \in \Upsilon^+$, we have assertions (1) and (2). If w is not a solution of the underlying PCP instance, we further have assertion (3).*

$$(1)\ \left(\begin{smallmatrix} \gamma(w) \\ \Sigma^* \setminus \{\alpha(w)\} \end{smallmatrix}\right) \subseteq \mathbb{P}^* \quad (2)\ \left(\begin{smallmatrix} \gamma(w) \\ \Sigma^* \setminus \{\beta(w)\} \end{smallmatrix}\right) \subseteq \mathbb{P}^* \quad (3)\ \left(\begin{smallmatrix} \gamma(w) \\ \Sigma^* \end{smallmatrix}\right) \subseteq \mathbb{P}^* \qquad \square$$

At this point, we somehow firmly feel that something very unpleasant will happen in the case that w is a solution of the underlying PCP instance.

Proof: We show (1). We assume some $w \in \Upsilon^+$ and some $u \neq \alpha(w)$ from Σ^*. We have to show $\left(\begin{smallmatrix} \gamma(w) \\ u \end{smallmatrix}\right) \in \mathbb{P}^*$. We branch into three cases, depending on whether $|u| < |\alpha(w)|$, $|u| = |\alpha(w)|$, or $|u| > |\alpha(w)|$. We treat $w = j_1 \dots j_n$ for some $n \geq 1$ and some $j_1, \dots, j_n \in \Upsilon$. Consequently, $\alpha(w) = \alpha(j_1) \dots \alpha(j_n)$.

- Case 1: $|u| < |\alpha(w)|$
  We factorize u into $u_1, \dots, u_n$ such that $|u_1| < |\alpha(j_1)|$, and for $l \in \{2, \dots, n\}$, we have $|u_l| \leq |\alpha(j_l)|$. At this point, we need the assumption $\alpha(j_1) \neq \lambda$. We define t_1, t_{n+1}, and t_l for $l \in \{2, \dots, n\}$:

$$t_1 = \left(\begin{smallmatrix} a_{1..9}\, j_1\, a_1 \\ u_1 \end{smallmatrix}\right) \in \mathbb{P}_{1,1} \quad t_l = \left(\begin{smallmatrix} a_{2..9}\, j_l\, a_1 \\ u_l \end{smallmatrix}\right) \in \mathbb{P}_{1,2} \quad t_{n+1} = \left(\begin{smallmatrix} a_{2..9} \\ \lambda \end{smallmatrix}\right) \in \mathbb{P}_{1,3}$$

  It is a straightforward verification that $t_1 \dots t_{n+1}$ yields $\left(\begin{smallmatrix} \gamma(w) \\ u \end{smallmatrix}\right)$.

- Case 2: $|u| = |\alpha(w)|$
  We factorize u into $u_1, \dots, u_n$ such that $|u_l| = |\alpha(j_l)|$ for $l \in \{1, \dots, n\}$. Because $u \neq \alpha(w)$, there is some $z \in \{1, \dots, n\}$ with $u_z \neq \alpha(j_z)$. We define the following traces for $l \in \{1, \dots, z-1\}$ and $m \in \{z+1, \dots, n\}$:

$$t_0 = \left(\begin{smallmatrix} a_{1..2} \\ \lambda \end{smallmatrix}\right), t_l = \left(\begin{smallmatrix} a_{3..9}\, j_l\, a_{1..2} \\ u_l \end{smallmatrix}\right), t_z = \left(\begin{smallmatrix} a_{3..9}\, j_z\, a_{1..3} \\ u_z \end{smallmatrix}\right), t_m = \left(\begin{smallmatrix} a_{4..9}\, j_m\, a_{1..3} \\ u_m \end{smallmatrix}\right), t_{n+1} = \left(\begin{smallmatrix} a_{4..9} \\ \lambda \end{smallmatrix}\right)$$

  They belong to $\mathbb{P}_{2,1}, \dots, \mathbb{P}_{2,5}$, respectively. Their concatenation yields $\left(\begin{smallmatrix} \gamma(w) \\ u \end{smallmatrix}\right)$.

We can show the remaining case $|u| > |\alpha(w)|$ as Case 1 using traces in $\mathbb{P}_{3,1}$, $\mathbb{P}_{3,3}$, and $\mathbb{P}_{3,3}$, instead. We can prove assertion (2) in the same way using traces in $\mathbb{P}_{4,1}, \dots, \mathbb{P}_{6,3}$. If w is not a solution of the underlying PCP instance, i.e., if $\alpha(w) \neq \beta(w)$, then (1) and (2) imply (3). See [12] for details. $\square$

Corollary 1. *If the PCP instance has no solution, we have $\mathbb{R} \subseteq \mathbb{P}^*$.* $\square$

This corollary is an obvious conclusion from Lemma 2. We need some kind of opposite to Lemma 2 and Corollary 1. We show that some traces in $\mathbb{R}$ do not belong to $\mathbb{P}^*$ if the underlying PCP has a solution. Together with Corollary 1, we obtain a strong tool, which will allow us to proceed straightforward proofs of the main goals of this paper.

Lemma 3. *Assume some $w \in \Upsilon^+$ with $\alpha(w) = \beta(w)$. We have $\binom{\gamma(w)}{\alpha(w)} \notin \mathbb{P}^*$.* $\square$

Proof: (sketch) We assume some word $w \in \Upsilon^+$ with $\alpha(w) = \beta(w)$ such that $\binom{\gamma(w)}{\alpha(w)} \in \mathbb{P}^*$, and show a contradiction. Thus, we assume an integer $n \geq 1$ and traces $t_1, \ldots, t_n \in \mathbb{P}$ such that $t_1 \ldots t_n = \binom{\gamma(w)}{\alpha(w)}$. The first component of t_1 is either λ, or a word starting with a_1. Consequently, we have $t_1 \in \mathbb{P}_{1,1}, \mathbb{P}_{2,1}, \mathbb{P}_{3,1}, \mathbb{P}_{4,1}, \mathbb{P}_{5,1}$, or $\mathbb{P}_{6,1}$, i.e., we have to branch into six cases.

Assume $t_1 \in \mathbb{P}_{1,1}$. We want to determine traces $t_2, \ldots, t_n \in \mathbb{P}$ such that $t_1 \ldots t_n = \binom{\gamma(w)}{\alpha(w)}$. To obtain the word $\gamma(w)$ as first component, we are forced to select traces $t_2, \ldots, t_{n-1}$ from $\mathbb{P}_{1,2}$ and to finish with $t_n = \binom{a_2 \cdot 9}{\lambda} \in \mathbb{P}_{1,3}$. Thus, we have $t_1 \ldots t_n \in \mathbb{P}_{1,1}\mathbb{P}_{1,2}^*\mathbb{P}_{1,3}$. Then, we cannot achieve that the second component of $t_1 \ldots t_n$ yields $\alpha(w)$. We defined the sets $\mathbb{P}_{1,1}$ and $\mathbb{P}_{1,2}$ in a way that the second component of $t_1 \ldots t_n$ is properly shorter than $\alpha(w)$.

Let us try to build $\binom{\gamma(w)}{\alpha(w)}$ from traces $t_1, \ldots, t_n \in \mathbb{P}$ by starting with a trace $t_1 \in \mathbb{P}_{2,1}$. Then, we can show $t_1 \ldots t_n \in \mathbb{P}_{2,1}\mathbb{P}_{2,2}^*\mathbb{P}_{2,3}\mathbb{P}_{2,4}^*\mathbb{P}_{2,5}$. Then, the length of the second component of $t_1 \ldots t_n$ is exactly the length of $\alpha(w)$. However, there is one trace from $\mathbb{P}_{2,3}$ among $t_1, \ldots, t_n$. This trace causes an error, i.e., because of this trace, the second component of $t_1 \ldots t_n$ is different from $\alpha(w)$.

The attempt to start with a trace $t_1 \in \mathbb{P}_{3,1}, \mathbb{P}_{4,1}, \mathbb{P}_{5,1}$, or $\mathbb{P}_{6,1}$ fails in a similar way. Note that $\alpha(w) = \beta(w)$. See [12] for a presentation of this proof in a more formal way. $\square$

4   Main Results

Now, we are able to prove the following theorem.

Theorem 6. *The following four assertions are equivalent:*

(1) The underlying PCP instance has no solution.
(2) $\mathbb{R} \subseteq \mathbb{P}^*$
(3) $\overline{\mathbb{R}} \cup \mathbb{P}^*$ is recognizable.
(4) $\mathbb{R} \cap \mathbb{P}^*$ is recognizable. $\square$

Proof: We have (1)→(2) by Corollary 1. To show (2)→(1), assume a solution w of the PCP instance. Then, $\binom{\gamma(w)}{\alpha(w)} \notin \mathbb{P}^*$ by Lemma 3, but $\binom{\gamma(w)}{\alpha(w)} \in \mathbb{R}$ by Def. 2.

We have (2)→(4), because $\mathbb{R} \cap \mathbb{P}^*$ yields $\mathbb{R}$ which is recognizable. We have (2)→(3), because $\overline{\mathbb{R}} \cup \mathbb{P}^*$ yields the monoid $\Gamma^* \times \Sigma^*$ which is recognizable.

We show (4)→(1). Assume a $\Gamma^* \times \Sigma^*$-automaton $\mathcal{A} = [Q, h, F]$ for $\mathbb{R} \cap \mathbb{P}^*$. Assume a solution w of the PCP instance. For $n \geq 1$, the words w^n are also solutions. Because Q is finite, there are $m > n \geq 1$ with $h\binom{\gamma(w^n)}{\lambda} = h\binom{\gamma(w^m)}{\lambda}$. Then, we have $h\binom{\gamma(w^n)}{\alpha(w^n)} = h\binom{\gamma(w^n)}{\lambda}h\binom{\lambda}{\alpha(w^n)} = h\binom{\gamma(w^m)}{\lambda}h\binom{\lambda}{\alpha(w^n)} = h\binom{\gamma(w^m)}{\alpha(w^n)}$. Hence, either both or none of the traces $\binom{\gamma(w^n)}{\alpha(w^n)}$ and $\binom{\gamma(w^m)}{\alpha(w^n)}$ belong to $\mathbb{R} \cap \mathbb{P}^*$. But, on one hand, $\binom{\gamma(w^n)}{\alpha(w^n)} \notin \mathbb{P}^*$ by Lemma 3. On the other hand, we have

$\left(\begin{smallmatrix}\gamma(w^m)\\\alpha(w^n)\end{smallmatrix}\right) \in \mathbb{R}$ and $\left(\begin{smallmatrix}\gamma(w^m)\\\alpha(w^n)\end{smallmatrix}\right) \in \mathbb{P}^*$ by Lemma 2(1). We can show (3)$\rightarrow$(1) in the same way [12].   □

We deduce the main result of this paper.

Theorem 7. *Assume an independence alphabet* (Σ, I) *which contains a C4-subalphabet. There is no algorithm, whose input are two* $\mathbb{M}(\Sigma, I)$-*automata for languages* K *and* L *in* $\mathbb{M}(\Sigma, I)$ *which decides one of the following properties:*

(1) $K \subseteq L^*$
(2) $K \cup L^* = \mathbb{M}(\Sigma, I)$
(3) $K \cup L^* \in \mathrm{REC}\,\mathbb{M}(\Sigma, I)$
(4) $K \cap L^* \in \mathrm{REC}\,\mathbb{M}(\Sigma, I)$ □

Proof: (sketch) At first, we note that there is no algorithm whose input are two alphabets Σ_1 and Σ_2, and further, two recognizable languages $K, L \subseteq \Sigma_1^* \times \Sigma_2^*$, which decides (1), (3), or (4). By usual coding techniques, we obtain the same result for C4. Then, the generalization to trace monoids with C4-submonoids is obvious. (2) is an immediate consequence of (1). See [12] for details.   □

5   Conclusions and Future Goals

Let us discuss about Theorem 7. Opposed to the undecidability of property (1), we can decide whether $K^* \subseteq L$ in every trace monoid. Given automata for K and L, we can construct a rational expression k for K, and check $L(k^*) \subseteq L$.

Opposed to the undecidability of property (2), we can trivially decide whether L^* yields the complete trace monoid: we have simply to check whether every letter of Σ occurs as a one letter trace in L. Moreover, for recognizable languages L and M, we can decide whether $L^* = M$ [6].

To decide property (3) in Theorem 7 is a special case of the recognizability problem. The star problem is a special case of (3), namely for $K = \emptyset$.

Due to classic results and Theorem 4 by SAKAROVITCH, properties (1) to (4) are decidable in trace monoids which do not contain a P3-submonoid [12].

Let us consider the decision problems in Theorem 7 in arbitrary trace monoids for finite sets K. Then, (1) and (4) are obviously decidable. RICHOMME remarked that property (3) restricted to finite sets K is decidable iff the star problem is decidable. RICHOMME further showed that (2) is decidable for finite sets K as follows. Choose some $n \in \mathbb{N}$ such that for $t \in K$, we have $|t| < n$. Then, $K \cup L^*$ yields the complete monoid iff every trace t with $|t| \leq 2n$ belongs to $K \cup L^*$.

Recently, MARCINKOWSKI and myself examined improvements of Theorem 7, e.g., restrictions to finite sets L and partial generalizations to P3 [13].

6   Acknowledgments

I acknowledge the discussions with my supervisor HEIKO VOGLER as well as with MANFRED DROSTE and DIETRICH KUSKE from the Institute of Algebra. I thank GWÉNAËL RICHOMME and anonymous referees for reading preliminary versions and making helpful remarks.

References

1. J. Berstel. *Transductions and Context-Free Languages*. B. G. Teubner, Stuttgart, 1979.
2. P. Cartier and D. Foata. *Problèmes combinatoires de commutation et réarrangements*, volume 85 of *Lecture Notes in Computer Science*. Springer-Verlag, Berlin, 1969.
3. M. Clerbout and M. Latteux. Semi-commutations. *Information and Computation*, 73:59–74, 1987.
4. B. Courcelle. Basic notions of universal algebra for language theory and graph grammars. *Theoretical Computer Science*, 163:1–54, 1996.
5. V. Diekert. *Combinatorics on Traces*, volume 454 of *Lecture Notes in Computer Science*. Springer-Verlag, Berlin, 1990.
6. V. Diekert and Y. Métivier. Partial commutation and traces. In G. Rozenberg and A. Salomaa, editors, *Handbook of Formal Languages, Vol. 3, Beyond Words*, pages 457–534. Springer-Verlag, Berlin, 1997.
7. V. Diekert and G. Rozenberg, editors. *The Book of Traces*. World Scientific, Singapore, 1995.
8. S. Eilenberg. *Automata, Languages, and Machines, Volume A*. Academic Press, New York, 1974.
9. P. Gastin, E. Ochmański, A. Petit, and B. Rozoy. Decidability of the star problem in $A^* \times \{b\}^*$. *Information Processing Letters*, 44(2):65–71, 1992.
10. S. Ginsburg and E. Spanier. Bounded regular sets. In *Proceedings of the AMS*, volume 17:5, pages 1043–1049, 1966.
11. S. Ginsburg and E. Spanier. Semigroups, presburger formulas and languages. *Pacific Journal of Mathematics*, 16:285–296, 1966.
12. D. Kirsten. Some undecidability results related to the star problem in trace monoids. Technical Report ISSN 1430-211X, TUD/FI98/07, Dresden University of Technology, Department of Computer Science, May 1998.
13. D. Kirsten and J. Marcinkowski. Some more results related to the star problem in trace monoids. (submitted).
14. A. Mazurkiewicz. Concurrent program schemes and their interpretations. DAIMI Rep. PB 78, Aarhus University, 1977.
15. Y. Métivier. Une condition suffisante de reconnaissabilité dans un monoïde partiellement commutatif. *R.A.I.R.O. - Informatique Théorique et Applications*, 20:121–127, 1986.
16. Y. Métivier and G. Richomme. New results on the star problem in trace monoids. *Information and Computation*, 119(2):240–251, 1995.
17. E. Ochmański. *Regular Trace Languages (in Polish)*. PhD thesis, Warszawa, 1984.
18. G. Richomme. Some trace monoids where both the star problem and the finite power property problem are decidable. In I. Privara et al., editors, *MFCS'94 Proceedings*, volume 841 of *Lecture Notes in Computer Science*, pages 577–586. Springer-Verlag, Berlin, 1994.
19. J. Sakarovitch. On regular trace languages. *Theoretical Computer Science*, 52:59–75, 1987.
20. J. Sakarovitch. The "last" decision problem for rational trace languages. In I. Simon, editor, *LATIN'92 Proceedings*, volume 583 of *Lecture Notes in Computer Science*, pages 460–473. Springer-Verlag, Berlin, 1992.
21. S. Yu. Regular languages. In G. Rozenberg and A. Salomaa, editors, *Handbook of Formal Languages, Vol. 1, Word, Language, Grammar*, pages 41–110. Springer-Verlag, Berlin, 1997.

An Approximation Algorithm for MAX p-SECTION

Gunnar Andersson

Royal Institute of Technology
Department of Numerical Analysis and Computing Science
SE-100 44 Stockholm, SWEDEN
Fax: +46 8 790 09 30
gunnar@nada.kth.se

Abstract. We present an approximation algorithm for the problem of partitioning the vertices of a weighted graph into p blocks of equal size so as to maximize the weight of the edges connecting different blocks. The algorithm is based on semidefinite programming and can in some sense be viewed as a generalization of the approximation algorithm by Frieze and Jerrum for the MAX BISECTION problem. Our algorithm, as opposed to that of Frieze and Jerrum, gives better performance than the naive randomized algorithm also for $p > 2$.

1 Introduction

The MAX CUT problem takes as input an undirected graph $G = (V, E)$ with non-negative edge weights. The objective is to find a cut, i.e., a partition of the vertices into two halves, so that the sum of the weights of the edges between the two halves is maximized. This is one of the most studied **NP** optimization problems, and the corresponding decision problem was shown to be **NP**-complete by Karp [6]. Interest has therefore turned to the design of approximation algorithms, and for a long time the best such was a 0.5-approximation algorithm, i.e., the weight of the cut output by the algorithm is at least 0.5 times the optimum cut. This changed dramatically a few years ago with the seminal paper of Goemans and Williamson [4]. They showed how a semidefinite relaxation of the natural integer programming formulation could be combined with an elegant randomized rounding scheme to yield a 0.87856-approximation algorithm for MAX CUT. Obviously this new technique attracted a lot of attention, and since then semidefinite programming has become a standard tool for constructing approximation algorithms.

Frieze and Jerrum [3] extended the approach of Goemans and Williamson and applied it to two interesting generalizations of MAX CUT: The MAX p-CUT problem, where the vertices are to be partitioned into p parts instead of two, and the MAX BISECTION problem, where the vertices are to be partitioned into two halves of equal size. For the MAX p-CUT problem they obtained a $\left(\frac{p-1}{p} + \Theta(p^{-2}\log p)\right)$-approximation algorithm and for the MAX BISECTION problem a 0.651-approximation algorithm.

C. Meinel and S. Tison (Eds.): STACS'99, LNCS 1563, pp. 237–247, 1999.

In this paper we study the MAX p-SECTION problem, which is a generalization of the MAX BISECTION problem: The vertices are to be partitioned into p parts of equal size so as to maximize the weight of the edges connecting different parts. For this problem, the approach of Frieze and Jerrum does not improve on the obvious randomized algorithm: Select a p-section uniformly at random. It is easy to see that this gives a $\frac{p-1}{p}$-approximation algorithm as the probability of an edge being cut is $\frac{p-1}{p}$. Is there a way to improve on this simple algorithm? For some problems, e.g. the MAX E3-SAT problem, it has been shown that there does not exist any approximation algorithm better than the naive randomized algorithm unless $\mathbf{P} = \mathbf{NP}$ [5]. The main contribution of this paper is a $\left(\frac{p-1}{p} + \Theta(p^{-3})\right)$-approximation algorithm for MAX p-SECTION, thus showing that the naive randomized algorithm is not the best possible for this problem. Our algorithm is based on semidefinite programming. It is easy to formulate but the analysis is non-trivial.

Why is it harder to come up with an approximation algorithm for the MAX p-SECTION problem than for the MAX p-CUT problem? What complicates matters is the constraint that all parts of the partition must have equal size. Traditionally, approximation algorithms based on semidefinite programming have been analyzed by evaluating, analytically or numerically, the performance on local configurations (see e.g. the thorough work of Zwick on constraint satisfaction problems [8]). For the MAX p-SECTION problem, this technique of analysis must be amended with a global analysis to show that an even p-section is produced.

2   Preliminaries

Definition 1. *The* MAX p-SECTION *problem is that of finding a partition of the vertices of a graph $G = (V, E)$ with weights w_{ij} associated with the edges into p subsets of equal size so as to maximize the total weight of the edges cut by the partition.*

The special case $p = 2$ is called MAX BISECTION.

We will denote by n the number of vertices in the graph. For a p-section to exist we must have $p \mid n$, so we assume that this is the case.

Definition 2. *An approximation algorithm A for an* **NP** *maximization problem M has performance guarantee $r < 1$ if $A(I) \geq r \cdot \mathrm{opt}_M(I)$ for all instances I of M.*

When this holds, we will refer to A as an r-approximation algorithm.

3   The Main Algorithm

The foundation of our approximation algorithm is a semidefinite relaxation of the MAX p-SECTION problem. It is based on the relaxation used for the MAX E2-LIN MOD p problem by Andersson, Engebretsen and Håstad [2]. To each vertex x_i corresponds a set of p vectors $\{v_j^i\}_{j=0}^{p-1}$ which are the vertices of a regular $(p-1)$-simplex in $\mathbf{R}^n$ centered at the origin. Following [2], we will refer to this object as a *simplicial porcupine*. To the edge (x_i, x_j) corresponds the term $\frac{p-1}{p} - \frac{p-1}{p^2}\sum_{k=0}^{p-1}\langle v_k^i, v_k^j\rangle$ (omitting the weight w_{ij}) in the objective function. If the simplices corresponding to all the vertices x_i shared vertices in $\mathbf{R}^n$, we could solve the MAX p-SECTION problem to optimality using this approach. Alas, this is not the case, and we therefore add inequalities which are valid for such a configuration and which simplify the analysis. The semidefinite program we construct is

$$\text{maximize } \frac{p-1}{p}\sum_{i,j} w_{ij}\left(1 - \frac{1}{p}\sum_{k=0}^{p-1}\langle v_k^i, v_k^j\rangle\right)$$

$$
\begin{aligned}
\text{subject to } & \langle v_k^i, v_k^i\rangle = 1 \text{ for all } i, k, \\
& \langle v_k^i, v_{k'}^i\rangle = \tfrac{-1}{p-1} \text{ for all } i \text{ and all } k \neq k', \\
& \langle v_k^i, v_{k'}^j\rangle \geq \tfrac{-1}{p-1} \text{ for all } i \neq j \text{ and all } k, k', \\
& \langle v_j^i, v_{j+k}^i\rangle = \langle v_{j'}^{i'}, v_{j'+k}^{i'}\rangle \text{ for all } i, i' \text{ and all } j, j', k, \\
& \textstyle\sum_i v_k^i = 0 \text{ for all } k.
\end{aligned}
\tag{1}
$$

The first two constraints guarantee that $\{v_j^i\}_{j=0}^{p-1}$ is a simplicial porcupine, the third and fourth constraints make the solution more symmetric and therefore easier to analyze, and the last constraint encourages an even partition of the variables after the randomized rounding. For $p = 2$ the relaxation is equivalent to the relaxation used by Frieze and Jerrum [3] for the MAX BISECTION problem.

In the remainder of this paper we will analyze the following randomized algorithm for the MAX p-SECTION problem.

Algorithm 1: Approximation algorithm for MAX p-SECTION.
(1)   Solve the above semidefinite program.
(2)   Generate $r \in \mathbf{R}^n$ by choosing each component as $N(0,1)$ independently. For each vertex x_i and each $j = 0, 1, \ldots, p-1$, let $q_{ij} = \frac{1}{p} + c\langle r, v_j^i\rangle$. Now fix i. If all q_{ij} are in $[0, 2/p]$, set $p_{ij} = q_{ij}$ for all j, otherwise set $p_{ij} = 1/p$ for all j.
(3)   For each i, put x_i in part j of the partition with probability p_{ij}.
(4)   Balance the partition so that each part contains n/p vertices. This is described in detail in Section 5 below.

Note that step (3) is well-defined in the sense that $\sum_j p_{ij} = 1$ for all i; this follows from the property $\sum_j v_j^i = 0$ of simplicial porcupines.

The value of the constant c depends on p; a precise expression for c will be given in Lemma 2 below.

An interesting feature of the algorithm is that it involves three randomized passes: First r is chosen as a random vector in $\mathbf{R}^n$ in step (2), then a preliminary partition is constructed from independent coin tosses in step (3), and finally the balancing scheme in step (4) also makes use of randomness.

The running time of the algorithm is dominated by the time it takes to solve the semidefinite relaxation in step (1).

4   Analyzing Local Configurations

In this section we will analyze the performance of the algorithm up to step (3); the adjustments made in step (4) will be analyzed in Section 5.

The intuition behind the algorithm is as follows: If the porcupines $\{v_k^i\}_{k=0}^{p-1}$ and $\{v_k^j\}_{k=0}^{p-1}$ corresponding to the vertices x_i and x_j are almost perfectly mis-aligned, in the sense that $\langle v_k^i, v_k^j \rangle$ is close to $\frac{-1}{p-1}$, then the random variables p_{ik} and p_{jk} will be negatively correlated and the probability that x_i and x_j will be put in the same part of the partition by step (3) will be less than $1/p$. On the other hand, if the two porcupines are almost perfectly aligned, corresponding to the inner products being close to 1, the probability that the vertices will be put in the same part will be greater than $1/p$.

Consider the edge (x_i, x_j). The contribution to the objective function from this edge is

$$\frac{p-1}{p}\left(1 - \frac{1}{p}\sum_{k=0}^{p-1}\langle v_k^i, v_k^j \rangle\right), \tag{2}$$

where we omit the weight w_{ij} from now on as it does not affect the analysis. If we can bound the ratio between this contribution and the probability that x_i and x_j end up in different parts, we have a bound on the performance guarantee after step (3). We therefore set out to do just that.

Denote with $X_{ij}(r)$ the probability that the edge (x_i, x_j) is cut given the random vector r. Then the expected performance guarantee after step (3), which we will denote G, satisfies

$$G \geq \min \frac{\mathrm{E}[X_{ij}(r)]}{\frac{p-1}{p}(1 - \langle v_0^i, v_0^j \rangle)} \tag{3}$$

where the minimum is taken over all possible configurations of the two porcupines $\{v_k^i\}_{k=0}^{p-1}$ and $\{v_k^j\}_{k=0}^{p-1}$, and the expectation is over the choice of r. This follows from $\langle v_0^i, v_0^j \rangle = \langle v_k^i, v_k^j \rangle$ for all k, which is a consequence of the fourth constraint in the semidefinite program (1).

As the porcupines $\{v_k^i\}_{k=0}^{p-1}$ and $\{v_k^j\}_{k=0}^{p-1}$ together span a space of dimension at most $2(p-1)$, we will from now on assume that $r \in \mathbf{R}^{2(p-1)}$ for the sake of convenience. Let $\Omega_{ij} = \{r : |\langle v_k^i, r \rangle| \leq 1/pc$ and $|\langle v_k^j, r \rangle| \leq 1/pc$ for all $k\}$. Note that Ω_{ij} is symmetric; $r \in \Omega_{ij}$ if and only if $-r \in \Omega_{ij}$. This will simplify the analysis later on.

We can write

$$X_{ij}(r) = \begin{cases} 1 - \sum_{k=0}^{p-1} p_{ik}p_{jk} & \text{if } r \in \Omega_{ij}, \\ \frac{p-1}{p} & \text{otherwise,} \end{cases} \tag{4}$$

and we now turn to bounding $E[X_{ij}(r)]$. As the geometry of Ω_{ij} makes it hard to calculate an exact expression, we will settle for a lower bound. This suffices as we seek to bound the performance guarantee from below.

Let $D_c = \{r : |r| \leq 1/pc\}$. Then $D_c \subseteq \Omega_{ij}$ and we obtain

$$
\begin{aligned}
E[X_{ij}(r)] &= \frac{1}{(2\pi)^{p-1}} \int_{\mathbf{R}^{2(p-1)}} X_{ij}(r) e^{-|r|^2/2} dV \\
&\geq \frac{1}{(2\pi)^{p-1}} \int_{D_c} \left(\frac{p-1}{p} - \frac{c}{p} \sum_{k=0}^{p-1} \langle v_k^i + v_k^j, r \rangle - c^2 \sum_{k=0}^{p-1} \langle v_k^i, r \rangle \langle v_k^j, r \rangle \right) e^{-|r|^2/2} dV \\
&= \frac{1}{(2\pi)^{p-1}} \int_{D_c} \left(\frac{p-1}{p} - c^2 \sum_{k=0}^{p-1} \langle v_k^i, r \rangle \langle v_k^j, r \rangle \right) e^{-|r|^2/2} dV \\
&= \frac{1}{(2\pi)^{p-1}} \int_{D_c} \left(\frac{p-1}{p} - \frac{pc^2}{2(p-1)} |r|^2 \langle v_0^i, v_0^j \rangle \right) e^{-|r|^2/2} dV \\
&= \frac{1}{(2\pi)^{p-1}} \int_{S^{2p-3}} dS \int_0^{1/pc} \left(\frac{p-1}{p} - \frac{pc^2}{2(p-1)} \langle v_0^i, v_0^j \rangle s^2 \right) s^{2p-3} e^{-s^2/2} ds \\
&= \frac{1}{2^{p-1}(p-2)!} \int_0^{1/(pc)^2} \left(\frac{p-1}{p} - \frac{pc^2}{2(p-1)} \langle v_0^i, v_0^j \rangle u \right) u^{p-2} e^{-u/2} du.
\end{aligned}
\tag{5}
$$

using symmetry and the formula for the surface area of the hypersphere,

$$
\int_{S^{2p-3}} dS = \frac{2\pi^{p-1}}{(p-2)!}.
\tag{6}
$$

From elementary calculus we have

$$
\int u^m e^{-u/2} du = -2e^{-u/2} \sum_{k=0}^{m} \frac{2^{m-k} m!}{k!} u^k
\tag{7}
$$

which gives the lower bound

$$
E[X_{ij}(r)] \geq \frac{p-1}{p} \left[e^{-u/2} \sum_{k=0}^{p-2} \frac{u^k}{2^k k!} \right]_{1/(pc)^2}^0 - pc^2 \langle v_0^i, v_0^j \rangle \left[e^{-u/2} \sum_{k=0}^{p-1} \frac{u^k}{2^k k!} \right]_{1/(pc)^2}^0.
\tag{8}
$$

We want to estimate the ratio between this lower bound and the contribution to the objective function; $\frac{p-1}{p}(1 - \langle v_0^i, v_0^j \rangle)$. A nice feature of the lower bound on $E[X_{ij}(r)]$ is that the only parameter describing the geometric relation between the two porcupines $\{v_k^i\}_{k=0}^{p-1}$ and $\{v_k^j\}_{k=0}^{p-1}$ is the scalar product $\langle v_0^i, v_0^j \rangle$ which is also present in the objective function.

Lemma 1.

$$\frac{\dfrac{p-1}{p}\left[e^{-u/2}\displaystyle\sum_{k=0}^{p-2}\dfrac{u^k}{2^k k!}\right]_{1/(pc)^2}^{0} - pc^2\langle v_0^i, v_0^j\rangle\left[e^{-u/2}\displaystyle\sum_{k=0}^{p-1}\dfrac{u^k}{2^k k!}\right]_{1/(pc)^2}^{0}}{\dfrac{p-1}{p}\left(1 - \langle v_0^i, v_0^j\rangle\right)} \tag{9}$$

is an increasing function in $\langle v_0^i, v_0^j\rangle$ for $\langle v_0^i, v_0^j\rangle \in \left[-\frac{1}{p-1}, 1\right]$.

Proof. The numerator is the integral of a non-negative function over a subset of $R^{2(p-1)}$ and hence non-negative for all $\langle v_0^i, v_0^j\rangle$ in the interval and, as a special case, for $\langle v_0^i, v_0^j\rangle = 1$. This means that the fraction above can be written as

$$\frac{a - b\langle v_0^i, v_0^j\rangle}{1 - \langle v_0^i, v_0^j\rangle} \tag{10}$$

with $a \geq b > 0$. This function is increasing for $\langle v_0^i, v_0^j\rangle \in \left[-\frac{1}{p-1}, 1\right]$.

By the lemma, we only have to consider the case $\langle v_0^i, v_0^j\rangle = -\frac{1}{p-1}$ when looking for a lower bound on the performance guarantee. This gives the lower bound

$$G \geq \frac{p-1}{p} + \frac{pc^2}{p-1} - e^{-1/2(pc)^2}\left(\frac{p-1}{p}\sum_{k=0}^{p-2}\frac{1}{(2p^2c^2)^k k!} + \frac{pc^2}{p-1}\sum_{k=0}^{p-1}\frac{1}{(2p^2c^2)^k k!}\right). \tag{11}$$

Out first goal is to prove that $G \geq \frac{p-1}{p} + \Theta(p^{-k})$ for some constant k. To that end we need to choose c as a function on p.

Let $c = \alpha/p^{1.5}$. We make the following estimates:

$$\frac{p-1}{p}\sum_{k=0}^{p-2}\frac{1}{(2p^2c^2)^k k!} + \frac{pc^2}{p-1}\sum_{k=0}^{p-1}\frac{1}{(2p^2c^2)^k k!} \leq \{ \text{ if } 2\alpha^2 \leq 1/2 \}$$

$$\leq \frac{p-1}{p}\frac{2}{(p-2)!}\left(\frac{p}{2\alpha^2}\right)^{p-2} + \frac{\alpha^2}{p^2(p-1)}\frac{2}{(p-1)!}\left(\frac{p}{2\alpha^2}\right)^{p-1} \tag{12}$$

$$\leq \frac{3}{(p-2)!}\left(\frac{p}{2\alpha^2}\right)^{p-2} \leq \frac{3p^p}{(2\alpha^2)^{p-2}p!} \leq \frac{3e^p}{\sqrt{2\pi p}(2\alpha^2)^{p-2}}$$

using Stirling's formula. Thus

$$G \geq \frac{p-1}{p} + \frac{1}{p^3}\left(\alpha^2 - \frac{3}{\sqrt{2\pi}}e^{p-0.5\log p-(p-2)\log(2\alpha^2)-p/2\alpha^2}\right). \tag{13}$$

The following lemma is now easily verified and summarizes this section.

Lemma 2. *For $c = 0.2/p^{1.5}$ the performance guarantee G of Algorithm 1 when the cost of the balancing is neglected satisfies*

$$G \geq \frac{p-1}{p} + \frac{1}{30p^3}. \tag{14}$$

5 Balancing the Partition

The partition produced in step (3) of Algorithm 1 is not necessarily an even p-section. In this section we will estimate the decrease in the objective function due to the cost of balancing and provide a balancing scheme. This concludes the analysis of the performance guarantee of Algorithm 1.

5.1 The Distance to an Even p-Section

There are two types of errors that might make the partition uneven:

1. For a fixed j, $\sum_i p_{ij}$ may differ from n/p.
2. The actual number of vertices placed in part j may differ from $\sum_i p_{ij}$.

We start off by analyzing how balanced the partition can be expected to be. To that measure we let $Z_j = \sum_i p_{ij}$. For i fixed, step (2) of the algorithm sets $p_{ij} = 1/p$ for all j if any $q_{ij} = \frac{1}{p} + c\langle r, v_j^i \rangle$ is outside $[0, 2/p]$. The interval $[0, 2/p]$ is symmetric around $1/p$ and all q_{ij} have symmetric distribution around $1/p$, hence $\mathrm{E}[Z_j] = n/p$. In order to estimate the cost for balancing we need to study the variance of Z_j:

$$\mathrm{Var}[Z_j] = \mathrm{E}\Big[\sum_i \sum_{i'} p_{ij} p_{i'j}\Big] - \mathrm{E}\Big[\sum_i p_{ij}\Big]^2$$

$$= c^2 \sum_i \sum_{i'} \langle v_j^i, v_j^{i'} \rangle - c^2 \sum_i \sum_{i'} \int_{\boldsymbol{R}^{2(p-1)} - \Omega_{ii'}} \langle v_j^i, r \rangle \langle v_j^{i'}, r \rangle \frac{e^{-|r|^2/2}}{(2\pi)^{p-1}} dV \tag{15}$$

Use $\sum_i v_j^i = 0$ for all j and $|\langle v_j^i, r \rangle| \le |r|$ (by the Cauchy-Schwartz inequality):

$$\mathrm{Var}[Z_j] \le c^2 \sum_i \sum_{i'} \int_{\boldsymbol{R}^{2(p-1)} - \Omega_{ii'}} |r|^2 \frac{e^{-|r|^2/2}}{(2\pi)^{p-1}} dV$$

$$\le c^2 n^2 \int_{|r| \ge 1/pc} |r|^2 \frac{e^{-|r|^2/2}}{(2\pi)^{p-1}} dV = c^2 n^2 e^{-1/2(pc)^2} \sum_{k=0}^{p-1} \frac{1}{k!} \Big(\frac{1}{2c^2 p^2}\Big)^k \tag{16}$$

as $r \in \Omega_{ii'}$ when $|r| \le 1/pc$.

Letting $c = \alpha/p^{1.5}$ and again applying the same kind of estimates as in Sec. 4 we obtain

$$\mathrm{Var}[Z_j] \le \frac{1}{2\sqrt{2\pi}} n^2 e^{p-3.5 \log p - p/2\alpha^2 - (p-2) \log 2\alpha^2}. \tag{17}$$

It is easily verified that choosing $\alpha = 0.2$ results in $\mathrm{Var}[Z_j] \le p^{-25}$ for all $p \ge 2$.

Chebyshev's inequality can now be applied:

$$\Pr\big[|Z_j - \mathrm{E}[Z_j]| \ge an\big] \le \frac{\mathrm{Var}[Z_j]}{(an)^2} \tag{18}$$

The unbalancedness due to the second kind of error is easily estimated. Let W_{ij} be the indicator variable for the event "x_i is put in part j by step (3) of the algorithm". Applying standard Chernoff bounds (see e.g. theorems A.4 and A.13 of [1]) gives

$$\Pr\left[\left|\sum_i W_{ij} - \mathrm{E}[Z_j]\right| > n^{3/4}\right] < e^{-2n^{1/2}} + e^{-pn^{1/2}}. \tag{19}$$

The decrease in the objective function due to correcting this small distortion turns out to be negligible.

5.2   A Balancing Scheme and Its Performance

Suppose that the p parts of the partition have sizes $s_0, s_1, \ldots, s_{p-1}$ after step (3) of Algorithm 1. Consider the following simple balancing scheme:

(1)	$S \leftarrow \emptyset$
(2)	For each i such that $s_i > n/p$:
(3)	Choose T of size $s_i - n/p$ randomly and uniformly from part i.
(4)	$S \leftarrow S \cup T$
(5)	Remove T from part i.
(6)	For each i such that $s_i < n/p$:
(7)	Choose T of size $n/p - s_i$ randomly and uniformly from S.
(8)	$S \leftarrow S - T$
(9)	Add T to part i.

How much will the objective value decrease due to balancing? Clearly the worst case is when none of the vertices being moved in step (9) of the algorithm are endpoints of any cut edges. Next we bound the cost of the balancing.

Lemma 3. *The expected decrease in the expected performance guarantee for part i in the partition is at most* $\max \frac{s_i - n/p}{s_i} \cdot \frac{p}{p-1}$.

Proof. Let ζ_i be the number of cut edges with one endpoint in part i of the partition. Consider the number of such edges after the set S has been formed by repeated applications of step (3). The expected decrease compared to the number prior to the balancing algorithm being run is

$$\zeta_i \frac{\max\{0, s_i - n/p\}}{s_i}. \tag{20}$$

The lemma follows from this and the simple observation that at least a fraction $\frac{p-1}{p}$ of the edges are cut in the optimal p-section.

The analysis from the previous section gives us the tools we need to bound the expected decrease (over the choice of r) in the expected performance guarantee due to balancing.

Theorem 1. *Algorithm 1 has expected performance guarantee $\frac{p-1}{p} + \Theta(p^{-3})$ for $p \geq 3$ when run with $c = 0.2/p^{1.5}$.*

Proof. If we let $a = p^{-8}$ in (18) and combine this equation with (19) we obtain

$$\Pr\left[\max\{s_i - n/p\} \leq p^{-8}n + n^{3/4}\right] \geq 1 - p \cdot (p^{-9} + e^{-2n^{1/2}} + e^{-pn^{1/2}}). \quad (21)$$

Lemma 3 can be applied when $\max\{s_i - n/p\}$ is small; otherwise the decrease in the expected performance guarantee can be bounded from above by 1. Combining Lemma 2 with this analysis shows that the expected performance guarantee is at least

$$\frac{p-1}{p} + \frac{1}{30p^3} - p \cdot (p^{-9} + e^{-2n^{1/2}} + e^{-pn^{1/2}}) - \frac{np^{-8}}{n/p - np^{-8}} \cdot \frac{p}{p-1}. \quad (22)$$

For $p \geq 3$ and large enough n this clearly is $\frac{p-1}{p} + \Theta(p^{-3})$.

We defer the analysis of the special case $p = 2$ (i.e., MAX BISECTION) to the next section.

This shows that Algorithm 1 beats the trivial randomized algorithm.

Remark 1. This performance guarantee, $\frac{p-1}{p} + \Theta(p^{-3})$, is somewhat weaker than the $\frac{p-1}{p} + \Theta(p^{-2} \log p)$ achieved by Frieze and Jerrum [3] for the MAX p-CUT problem. It may be possible to sharpen the bound for MAX p-SECTION but it seems hard to reach $\frac{p-1}{p} + \Theta(p^{-2})$ using the approach taken in this paper.

6   Modifications for MAX BISECTION

One may wonder if Algorithm 1 improves on the algorithm of Frieze and Jerrum when $p = 2$. It turns out that a modified rounding scheme in step (2) improves the performance of Algorithm 1:

> (2')   Generate $r \in \mathbf{R}^n$ by choosing each component as $N(0, 1)$ independently. For each vertex x_i and each $j = 0, 1, \ldots, p - 1$, let $q_{ij} = \frac{1}{p} + c\langle r, v_j^i \rangle$. Now fix i. If all q_{ij} are in $[0, 1]$, set $p_{ij} = q_{ij}$ for all j, otherwise set $p_{ij} = 0$ if $q_{ij} \leq 0$ and $p_{ij} = 1$ if $q_{ij} \geq 1$.

For $p > 2$ this might lead to $\sum_j p_{ij} \neq 1$ for some i, but for $p = 2$ it is a generalization of Frieze and Jerrum's algorithm:

Theorem 2. *Frieze and Jerrum's algorithm for* MAX BISECTION *has the same worst-case behavior as the modified version of Algorithm 1 with* $c = \infty$.

Proof sketch. The semidefinite program (1) is, when $p = 2$, easily seen to be equivalent to the one used by Frieze and Jerrum. Their rounding scheme corresponds to the limit $c \to \infty$ in the modified algorithm above. The greedy balancing scheme they use is equivalent to our probabilistic scheme in the worst case.

Numerical simulations indicate that $c = \infty$ is the best choice in Algorithm 1 so our approach does not provide a better approximation algorithm for MAX BISECTION.

7   Open Problems

– **Lower bounds**
  There are no strong lower bounds on the approximability of the MAX p-SECTION problem, probably because the most successful technique for proving lower bounds, probabilistically checkable proofs (PCP), focuses on local properties of problems. For the MAX CUT problem, it has been shown that there cannot exist any approximation algorithm with performance guarantee better than $16/17$ unless $\mathbf{P} = \mathbf{NP}$ [5,7]. PCPs have been less useful on problems with global constraints than on pure constraint satisfaction problems.
– **Better algorithms for small** p
  In practice, special cases of MAX p-SECTION where p is small — especially MAX BISECTION— are the most important ones. Is it possible to beat the 0.651-approximation algorithm for MAX BISECTION?

Acknowledgments

I am most grateful to Johan Håstad for inspiration and helpful discussions. Lars Engebretsen and Viggo Kann gave comments on earlier versions of this paper.

References

1. Noga Alon and Joel H. Spencer. *The Probabilistic Method.* Wiley, New York, 1992.
2. Gunnar Andersson, Lars Engebretsen, and Johan Håstad. A new way to use semidefinite programming with applications to linear equations mod p. In *Proc. Tenth Ann. ACM-SIAM Symp. on Discrete Algorithms*, 1999.
3. Alan Frieze and Mark Jerrum. Improved approximation algorithms for MAX k-CUT and MAX BISECTION. In *Proc. 4th Conf. on Integer Prog. and Combinatorial Optimization*, volume 920 of *Lecture Notes in Comput. Sci.*, pages 1–13, Berlin, 1995. Springer-Verlag.
4. Michel X. Goemans and David P. Williamson. .878-approximation algorithms for MAX CUT and MAX 2SAT. In *Proc. Twenty-sixth Ann. ACM Symp. on Theory of Comp.*, pages 422–431. ACM, New York, 1994.

5. Johan Håstad. Some optimal inapproximability results. In *Proc. Twenty-nineth Ann. ACM Symp. on Theory of Comp.*, pages 1–10. ACM, New York, 1997.
6. Richard M. Karp. Reducibility among combinatorial problems. In R. Miller and J. Thatcher, editors, *Complexity of Computer Computations*, pages 85–103. Plenum Press, New York, NY, 1972.
7. Luca Trevisan, Gregory B. Sorkin, Madhu Sudan, and David P. Williamson. Gadgets, approximation, and linear programming. In *Proc. of 37th Ann. IEEE Symp. on Foundations of Comput. Sci.*, pages 617–626. IEEE Computer Society, Los Alamitos, 1996.
8. Uri Zwick. Approximation algorithms for constraint satisfaction problems involving at most three variables per constraint. In *Proc. Nineth Ann. ACM-SIAM Symp. on Discrete Algorithms*, pages 201–210. ACM-SIAM, 1998.

Approximating Bandwidth
by Mixing Layouts of Interval Graphs

Dieter Kratsch[1] and Lorna Stewart[2]

[1] Fakultät für Mathematik und Informatik
Friedrich-Schiller-Universität
07740 Jena
Germany
[2] Department of Computing Science
University of Alberta
Edmonton, Alberta, Canada
T6G 2H1

Abstract. We examine the bandwidth problem in circular-arc graphs, chordal graphs with a bounded number of leaves in the clique tree, and k-polygon graphs (fixed k). We show that all of these graph classes admit efficient approximation algorithms which are based on exact or approximate bandwidth layouts of related interval graphs. Specifically, we obtain a bandwidth approximation algorithm for circular-arc graphs that executes in $O(n \log^2 n)$ time and has performance ratio 2, which is the best possible performance ratio of any polynomial time bandwidth approximation algorithm for circular-arc graphs. For chordal graphs with not more than k leaves in the clique tree, we obtain a performance ratio of 2k in $O(k(n + m))$ time, and our algorithm for k-polygon graphs has performance ratio $2k^2$ and runs in time $O(n^3)$.

1 Introduction

A *layout* of a graph $G = (V, E)$ is an assignment of distinct integers from $\{1, \dots, n\}$ to the elements of V. Equivalently, a layout L may be thought of as an ordering $L(1), L(2), \dots, L(n)$ of V, where $|V| = n$. We shall use $<_L$ to denote the ordering of the elements in a layout L. The *width* of a layout L, $b(G, L)$, is the maximum over all edges $\{u, v\}$ of G of $|L(u) - L(v)|$. That is, it is the length of the longest edge in the layout. The *bandwidth* of G, $bw(G)$, is the minimum width over all layouts. A *bandwidth layout* for graph G is a layout satisfying $b(G, L) = bw(G)$.

The problem of finding the bandwidth of a graph has applications in sparse matrix computations. An overview of the bandwidth problem is given in [5]. The minimum bandwidth decision problem (Given a graph $G = (V, E)$ and integer k, is $bw(G) \le k$?) is known to be NP-complete [21], even for trees having maximum degree 3 [12], caterpillars with hairs of length at most 3 [20] and cobipartite graphs [17]. The problem is polynomially solvable for caterpillars with hairs of length 1 and 2 [2], cographs [15], and interval graphs [16,19,22].

To date there was not much known about the approximation hardness of the bandwidth minimization problem for graphs in general. Recently Feige presented

C. Meinel and S. Tison (Eds.): STACS'99, LNCS 1563, pp. 248–258, 1999.
© Springer-Verlag Berlin Heidelberg 1999

an approximation algorithm with performance ratio $O(\log^{9/2} n)$ [11]. Very recently Unger has shown in [23] that assuming P$\neq$NP, there is no polynomial time approximation algorithm with constant performance ratio for the bandwidth minimization problem for graphs, even when the inputs are restricted to a special class of trees known as caterpillars of maximum degree three.

Since the bandwidth minimization problem remains NP-complete for such simple classes of graphs, and since no polynomial time algorithm for approximating the bandwidth of general graphs, or even trees, to within a constant factor exists unless P=NP, it is worthwhile to investigate approximation algorithms for this problem on restricted classes of graphs. Some results in this direction have been presented in [17].

In this paper, we examine the bandwidth problem in circular-arc graphs, chordal graphs with a bounded number of leaves in the clique tree, and k-polygon graphs (fixed k). All of these graph classes admit efficient approximation algorithms which are based on exact or approximate bandwidth layouts of related interval graphs. Specifically, we obtain a bandwidth approximation algorithm for circular-arc graphs that has performance ratio 2 and executes in $O(n \log^2 n)$ time, or performance ratio 4 while taking $O(n)$ time. For chordal graphs with not more than k leaves in the clique tree, we obtain a performance ratio of 2k in $O(k(n + m))$ time, and our algorithm for k-polygon graphs has performance ratio $2k^2$ and runs in time $O(n^3)$.

Finally it is worth mentioning that our approximation algorithm with performance ratio 2 for circular-arc graphs has *optimal* performance ratio, since there is no polynomial time bandwidth approximation algorithm for (unit) circular-arc graphs with performance ratio $2 - \epsilon$ for any $\epsilon > 0$ unless P=NP [23].

2   Preliminaries

For $G = (V, E)$, we will denote $|V|$ as n and $|E|$ as m. We sometimes refer to the vertex set of G as $V(G)$ and the edge set as $E(G)$. We let $N(v)$ denote the set of vertices adjacent to v. The *degree* of a vertex v, $degree(v)$, is the number of vertices adjacent to v. $\Delta(G)$ denotes the maximum degree of a vertex in graph G. The subgraph of $G = (V, E)$ induced by $V' \subseteq V$ will be referred to as $G[V']$.

The following well-known lower bound on the bandwidth of a graph is given in [7].

Lemma 1. *[The degree bound] [7] For any graph G, $bw(G) \geq \Delta(G)/2$.*

The distance in graph $G = (V, E)$ between two vertices $u, v \in V$, $d_G(u, v)$, is the length of a shortest path between u and v in G. For any graph $G = (V, E)$, the dth power of G, G^d, is the graph with vertex set V and edge set $\{\{u, v\}|d_G(u, v) \leq d\}$.

Lemma 2. *[The distance bound] [17] (also attributed in part to [6] in [5]) Let G and H be graphs with the same vertex set V, such that $E(G) \subseteq E(H) \subseteq E(G^d)$ or $E(H) \subseteq E(G) \subseteq E(H^d)$ for an integer $d \geq 1$, and let L be an optimal layout for*

H, *i.e.,* $b(H, L) = bw(H)$. *Then* L *approximates the bandwidth of* G *by a factor of* d, *i.e.,* $b(G, L) \leq d \cdot bw(G)$.

Many references, including [14], contain comprehensive overviews of the many known structural and algorithmic properties of interval graphs.

Definition 1. *A graph* $G = (V, E)$ *is an interval graph if there is a one-to-one correspondence between* V *and a set of intervals of the real line such that, for all* $u, v \in V$, $\{u, v\} \in E$ *if and only if the intervals corresponding to* u *and* v *have a nonempty intersection.*

A set of intervals whose intersection graph is G is termed an *interval model* for G. Many algorithms exist which, given a graph $G = (V, E)$, determine whether or not G is an interval graph and, if so, construct an interval model for it, in $O(n + m)$ time (see, for example, [4,8]). We assume that an interval model is given by a left endpoint and a right endpoint for each interval, namely, *left*(v) and *right*(v) for all $v \in V$. Furthermore, we assume that we are also given a sorted list of the endpoints, and that the endpoints are distinct. We will sometimes blur the distinction between an interval and its corresponding vertex, when no confusion can arise.

Polynomial time algorithms for computing the exact bandwidth of an interval graph have been given in [16,19,22]. For an interval graph with n vertices, Kleitman and Vohra's algorithm solves the decision problem ($bw(G) \leq k$?) in $O(nk)$ time and can be used to produce a bandwidth layout in $O(n^2 \log n)$ time, and Sprague has shown how to implement Kleitman and Vohra's algorithm to answer the decision problem in $O(n \log n)$ time and thus produce a bandwidth layout in $O(n \log^2 n)$ time.

The following two lemmas demonstrate that, for interval graph G, a layout L with $b(G, L) \leq 2 \cdot bw(G)$ can be obtained in time $O(n)$, assuming the sorted interval endpoints are given.

Lemma 3. *Given an interval graph* G, *the layout* L *consisting of vertices ordered by right endpoints of corresponding intervals has* $b(G, L) \leq 2 \cdot bw(G)$.

Proof. Let L be the layout of vertices ordered by right interval endpoints. We first observe that, for all $u, v \in V$ such that $\{u, v\} \in E$ and $u <_L v$, all vertices between u and v in L are adjacent to v. Now consider a longest edge in L, i.e., an edge $\{u, v\}$ such that $|L(u) - L(v)| = b(G, L)$. Assume, without loss of generality, that $u <_L v$. From the previous observation, it must be that $degree(v) \geq L(v) - L(u) = b(G, L)$. Now the degree bound (Lemma 1) implies $bw(G) \geq b(G, L)/2$.   □

Lemma 4. *Given an interval graph* G, *the layout* L *consisting of vertices ordered by left endpoints of corresponding intervals has* $b(G, L) \leq 2 \cdot bw(G)$.

Proof. Consider a set of intervals representing G, and the layout L, ordered by left endpoints. Now, flipping the intervals of the model horizontally results in another interval representation for G, and the ordering of vertices by right endpoints of these intervals is the reversal of L. Thus, this lemma follows from the previous one.   □

We will use the following lemma in subsequent sections of the paper.

Lemma 5. *Let* I *be a set of intervals on the real line corresponding to interval graph* $G = (V, E)$. *Let* p_1 *be a point on the line such that at least one interval endpoint is to the left of* p_1 *and only left endpoints are to the left of* p_1. *Let* p_2 *be a point on the line such that at least one interval endpoint is to the right of* p_2 *and only right endpoints are to the right of* p_2. *Let* C_1 *be the set of all intervals that contain* p_1, *and* C_2 *be the set of all intervals that contain* p_2. *If* L *is a layout for* G *in which vertices are ordered by increasing left endpoints of corresponding intervals or by increasing right endpoints, or if* L *is a layout produced by Kleitman and Vohra's bandwidth algorithm [16], then*
(i) $\forall v \in C_1 : \{v, L(1)\} \in E$, *and*
(ii) $\forall v \in C_2 : \{v, L(n)\} \in E$.

Proof. Part (i) for the left endpoint ordering follows from the fact that $L(1) \in C_1$ and C_1 is a clique. In the other two layouts, $L(1)$ is the interval with smallest right endpoint. This interval is either in C_1 or is contained in all intervals of C_1. Thus, (i) holds for the three layouts.

Part (ii) follows immediately for the right endpoint layout, since $L(n) \in C_2$. In the left endpoint order, $L(n)$ is either in C_2 or contained in all intervals of C_2, implying (ii).

The proof of Part (ii) for Kleitman-Vohra layouts heavily relies on details of the algorithm in [16] and is omitted here.   □

3   Circular-Arc Graphs

Circular-arc graphs are the intersection graphs of arcs on a circle. Thus, a graph $G = (V, E)$ is a circular-arc graph if and only if it has a (not necessarily unique) circular-arc model or representation, consisting of a set of arcs on a circle, such that, for all $u, v \in V$, $\{u, v\} \in E$ if and only if the arcs corresponding to u and v have a nonempty intersection. In such a model, we assume, without loss of generality, that the arc endpoints are distinct, and we label the endpoints from 1 to $2n$ in clockwise order around the circle, starting at an arbitrary endpoint. Thus, each vertex $v \in V$ corresponds to an arc given by its counterclockwise endpoint, $ccw(v)$, and its clockwise endpoint, $cw(v)$. We refer to any segment of the circle by its two endpoints and the direction of traversal, i.e., $[p_1, p_2]_{cw}$ refers to the closed arc covered by a clockwise traversal beginning at p_1 and ending at p_2. The arc $[p_1, p_2]_{ccw}$ is the set of all points in a counterclockwise traversal from p_1 to p_2, and parentheses will indicate that the arc is open at one or both ends. Note that, for any two points (not necessarily arc endpoints) on the circle, p_1 and p_2, the arcs $[p_1, p_2]_{cw}$ and $[p_1, p_2]_{ccw}$ cover the entire circle, and their intersection is $\{p_1, p_2\}$.

Eschen and Spinrad [10] have given an $O(n^2)$ algorithm which determines whether or not an n-vertex graph is a circular-arc graph. If so, the algorithm produces a circular-arc model for the graph. Our algorithms assume that the input circular-arc graph is given as a set of arcs on a circle.

Henceforth, we will refer to a set of $2n$ *scanpoints* on the circle, none of which is an arc endpoint, such that exactly one of these points is between each consecutive pair of arc endpoints. We shall label these points from 1 to $2n$ in clockwise order, beginning at any one.

Our bandwidth approximation algorithm works as follows, for a circular-arc graph G. Roughly speaking, we cut the circular-arc representation in half, to form two equal-sized interval graphs, compute exact or approximate bandwidth layouts for the two interval graphs, and then mix the two layouts to form an approximate bandwidth layout for G.

Let $G = (V, E)$ be a circular-arc graph with corresponding circular-arc representation. The first step is to find a scanpoint p on the circle such that $|C_1 \cup C_2 \cup A| = |C_1 \cup C_2 \cup B|$ where C_1 is the set of arcs that contain scanpoint 1, C_2 is the set of arcs that contain scanpoint p, A is the set of arcs entirely contained in $(1, p)_{cw}$, and B is the set of arcs entirely contained in $(1, p)_{ccw}$. Note that $C_1 \cup C_2 \cup A \cup B = V$. We will use scanpoints 1 and p to cut the circle and create two equal-sized interval graphs.

Procedure FINDp

Let $C_1 \leftarrow C_2 \leftarrow$ all arcs that contain scanpoint 1; $A \leftarrow \emptyset$; $B \leftarrow V \setminus C_1$
$a \leftarrow |C_1|$; $b \leftarrow n$ $\{ a = |C_1 \cup C_2 \cup A|$; $b = |C_1 \cup C_2 \cup B| \}$
$p \leftarrow 1$
repeat until $a = b$ **or** $p = 2n$
    $\{$ **Invariant:** $a \le b\}$
    $\{$ **Variant:** $2n - p\}$
    $p \leftarrow p + 1$
    if the endpoint between $p - 1$ and p is a ccw endpoint (say of arc i) **then**
        $C_2 \leftarrow C_2 \cup \{i\}$
        if $i \notin C_1$ **then**
            $B \leftarrow B \setminus \{i\}$
            $a \leftarrow a + 1$
    if between $p - 1$ and p is a cw endpoint (of arc i) **then**
        $C_2 \leftarrow C_2 \setminus \{i\}$
        if $i \notin C_1$ **then**
            $A \leftarrow A \cup \{i\}$
            $b \leftarrow b - 1$
$\{$ Now C_2 is the set of arcs that contain point $p\}$
$\{|C_1 \cup C_2 \cup A| = |C_1 \cup C_2 \cup B|\}$

Claim. **Procedure** FINDp will terminate with $a = b$.

Proof. It is a matter of routine to verify the stated invariant and variant. If the loop terminates with $p = 2n$ then all arc endpoints will have been examined. For all arcs except those of C_1, a will have been incremented by 1 and b will have been decremented by 1. Let a_i and a_f be the initial and final values, respectively, of variable a, and b_i and b_f the initial and final values, respectively, of variable b.

Upon termination of the loop with $p = 2n$, $a_f = a_i + n - |C_1| = |C_1| + n - |C_1| = n$ and $b_f = b_i - (n - |C_1|) = n - n + |C_1| = |C_1|$. But then $b_f < a_f$ (assuming $C_1 \neq V$), contradicting our invariant.   □

We may assume that A and B will be nonempty; otherwise G can be partitioned into two cliques, one of which must have size at least $n/2$, implying (by Lemma 1) $bw(G) \geq n/2 - 1$. Thus, any layout in which the first and last vertices are not adjacent is a 2-approximation.

We now describe how to construct two interval subgraphs of G by cutting the circle at scanpoints 1 and p. We wish to cut the circle and the arcs of C_1 and C_2 at scanpoints 1 and p, producing two line segments, each with a set of intervals that correspond to an interval graph. However, if any arc, say v, contains both scanpoints 1 and p then it covers one entire part of the circle (i.e. $[1, p]_{cw}$ or $[1, p]_{ccw}$) and appears as two disconnected pieces in the other part. Thus, this second part of the circle may not correspond to an interval subgraph, as vertex v is represented by two disconnected intervals. We eliminate this problem by shrinking v's arc on the circle so that it no longer contains p and thus v is removed from C_2. The altered set of arcs might not represent all of the edges of G; specifically, some edges between v and elements of A (or B) may be missing. Let E' denote edges of G that are not represented by the changed arcs. Note that the sets $C_1 \cup C_2 \cup A$ and $C_1 \cup C_2 \cup B$ remain unchanged.

Now, we can cut the circle and the arcs of C_1 and C_2 at scanpoints 1 and p, producing two line segments, $[1, p]_{cw}$ and $[1, p]_{ccw}$. The arcs of the circular-arc model become intervals on the two lines. Let I_A (respectively I_B) be the resulting set of intervals on the line segment $[1, p]_{cw}$ (respectively $[1, p]_{ccw}$). We may assume that the intervals of $C_1 \cup C_2$ are altered slightly in I_A and in I_B without changing intersections, so that interval endpoints are distinct.

Let $G_A = (V_A, E_A)$ and $G_B = (V_B, E_B)$ be the intersection graphs of I_A and I_B, respectively. Now, G_A and G_B are both interval graphs and (not necessarily induced) subgraphs of G. Furthermore, $|V_A| = |V_B|$, and $E_A \cup E_B \cup E' = E$. Figure 1 illustrates this process.

Our method for obtaining an approximate bandwidth layout for a circular-arc graph is to first compute exact or approximate bandwidth layouts, L_A and L_B, for G_A and G_B, respectively, and then mix the two layouts.

Different methods of computing L_A and L_B yield different approximation bounds and time complexities for our algorithm.

Regardless of how we obtain L_A and L_B, the mixing is done as follows. Let $k = |C_1 \cup C_2 \cup A| = |C_1 \cup C_2 \cup B|$. Given

$$L_A = L_A(1), L_A(2), \ldots, L_A(k)$$

and

$$L_B = L_B(1), L_B(2), \ldots, L_B(k)$$

we begin by producing

$$L_M = L_A(1), L_B(1), L_A(2), L_B(2), \ldots, L_A(k), L_B(k).$$

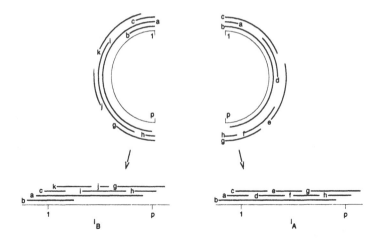

Fig. 1. Cutting the circular-arc model to form two interval graphs

For convenience, we will refer to elements of L_A as having the colour *red* and elements of L_B as having the colour *blue*. Notice that L_M will contain two copies of each vertex of $C_1 \cup C_2$ – one red and one blue. For each $v \in C_1 \cup C_2$, we shall distinguish between the two copies of v in L_M as follows: the red copy will be referred to as v_{red} and the blue as v_{blue}. Each vertex of $A \cup B$ occurs only once in L_M.

From L_M, we produce L by deleting the leftmost copy of each vertex of C_1 and the rightmost copy of each vertex of C_2. Recall that we constructed C_1 and C_2 so that no vertex appears in both. Thus, L is a layout for G.

Lemma 6. *Let $G = (V, E)$ be a circular-arc graph, and let I_A, I_B, G_A, and G_B be constructed as previously described, from a circular-arc model for G. Let L_A and L_B be layouts for G_A and G_B, respectively, satisfying:*
- *$\forall v \in C_1: \{v, L_A(1)\}, \{v, L_B(1)\} \in E$, and*
- *$\forall v \in C_2: \{v, L_A(k)\}, \{v, L_B(k)\} \in E$.*

Let L_M and L be obtained from L_A and L_B as previously described. Then

$$b(G, L) \leq 2 \cdot \max[b(G_A, L_A), b(G_B, L_B)].$$

The proof of the lemma is omitted due to space restrictions.

Theorem 1. *The bandwidth of a circular-arc graph can be approximated to within a factor of four in $O(n)$ time, and to within a factor of two in $O(n \log^2 n)$ time.*

Proof. We have three approximation algorithms for approximating the bandwidth of a circular-arc graph, namely, the algorithm previously described in which:
(i) L_A and L_B are layouts of vertices ordered by left endpoints of intervals,
(ii) L_A and L_B are layouts of vertices ordered by right endpoints of intervals, or
(iii) L_A and L_B are layouts computed by Kleitman and Vohra's algorithm.

Algorithms (i) and (ii) have time complexity $O(n)$, provided the sorted arc endpoints are given, and they output a layout L that satisfies:

$$
\begin{aligned}
b(G, L) &\leq 2 \cdot \max[b(G_A, L_A), b(G_B, L_B)] \\
&\leq 2 \cdot \max[2 \cdot bw(G_A), 2 \cdot bw(G_B)] \\
&= 4 \cdot \max[bw(G_A), bw(G_B)] \\
&\leq 4 \cdot bw(G)
\end{aligned}
$$

Algorithm (iii) requires $O(n \log^2 n)$ time but produces a layout L satisfying:

$$
\begin{aligned}
b(G, L) &\leq 2 \cdot \max[b(G_A, L_A), b(G_B, L_B)] \\
&\leq 2 \cdot \max[bw(G_A), bw(G_B)] \\
&\leq 2 \cdot bw(G)
\end{aligned}
$$

These performance ratios follow from Lemmas 5 and 6, and the fact that any subgraph of graph G has bandwidth not larger than $bw(G)$.   □

4   Chordal Graphs with Clique Trees Having a Bounded Number of Leaves

A graph G is a *chordal* graph if every cycle of length greater than three has a chord. Chordal graphs are exactly the intersection graphs of subtrees in a tree [13]. More precisely, for each chordal graph $G = (V, E)$, there exists a tree T such that the vertices of T correspond to the maximal cliques of G, and the vertices of T corresponding to cliques of G containing any fixed vertex $v \in V$ form a subtree T_v of T. Note the consequence that two vertices of G are adjacent if and only if their corresponding subtrees have nonempty intersection. For a given chordal graph $G = (V, E)$, such a tree, called a *clique tree* for G, will have no more than n nodes and can be constructed in $O(n + m)$ time [3].

We use the idea of mixing layouts of interval graphs, as in the previous section. While a circular-arc graph roughly consists of two interval graphs arranged in a circle, a chordal graph may be thought of as several interval graphs arranged in a tree-like structure. A chordal graph with k leaves in its clique tree may be viewed as a collection of k interval graphs.

Our algorithm is as follows, assuming a clique tree T has been computed for a given chordal graph $G = (V, E)$.

1. Root T at an arbitrary vertex, r.
2. Let k be the number of leaves of T (excluding r). For each root-to-leaf path P_i in T, the collection of subtrees, restricted to P_i, form a set of intervals. Let I_i be this set of intervals in which the left endpoint of each interval is taken to be the one closer to r. Let $G_i = (V_i, E_i)$ be the corresponding interval graph.

3. **for** $i \leftarrow 1$ **to** k **do**

   $L_i \leftarrow$ layout for G_i consisting of V_i ordered by increasing left endpoints of intervals (with ties broken arbitrarily, but the same way in all the L_i's)

4. Mix the L_i's to form L_M, as follows:

   $L_M \leftarrow L_1(1)L_2(1)L_3(1)\ldots L_k(1)L_1(2)L_2(2)\ldots L_k(2)\ldots$

5. For each vertex $v \in V$ that appears in more than one of the G_i's: delete all but the rightmost copy of v from L_M. The result is a layout L for G.

The proof of the following theorem is omitted due to space restrictions.

Theorem 2. *Let $G = (V, E)$ be a chordal graph having a clique tree with at most k leaves. Then a layout L for G satisfying $b(G, L) \leq 2k \cdot bw(G)$ can be computed in $O(k(n + m))$ time by the algorithm described above.*

5   k-Polygon Graphs for Fixed k

We make use of the results of the previous section as follows. We transform any k-polygon graph G into a chordal graph H having a clique tree with at most k leaves, by taking a minimal triangulation of the input graph. We show that there exists such a triangulation which is a subgraph of G^k. Combining these observations with Lemma 2, and the approximation algorithm of the previous section, we obtain an $O(n^3)$ approximation algorithm for the bandwidth of k-polygon graphs which has performance ratio $2k^2$.

A graph $G = (V, E)$ is a k-polygon graph if it is the intersection graph of chords inside a convex k-polygon, where each chord has its endpoints on two different sides of the polygon. A polygon representation, or diagram, for $G = (V, E)$, is a k-sided polygon together with a set of chords such that, for all $u, v \in V$, $\{u, v\} \in E$ if and only if the chords corresponding to u and v cross.

Circle graphs are the intersection graphs of chords inside a circle. Thus, circle graphs are the union of all k-polygon graphs, over all $k \geq 2$. Given a graph $G = (V, E)$, it can be determined in $O(|V|^k)$ time whether or not G is a k-polygon graph and, if so, a polygon representation can be constructed [9]. However, the general problem, given a circle graph, determine the minimum k such that G is a k-polygon graph, remains NP-complete [9].

Our algorithm assumes that a k-polygon representation for the input graph G is provided.

Definition 2. *A triangulation of a graph G is a chordal graph H with the same vertex set as G, such that G is a subgraph of H. A triangulation H of a graph G is called a minimal triangulation of G, if no proper subgraph of H is a triangulation of G.*

Our bandwidth approximation algorithm works as follows. First it computes a minimum triangulation H (i.e., one with the minimum number of edges) for the given graph G using an $O(n^3)$ algorithm for circle graphs presented in [18]. Hence

H is a minimal triangulation of G. Any minimal triangulation of a circle graph, and thus of a k-polygon graph, can be represented by a planar triangulation of a particular convex polygon [18]. Now our algorithm takes the dual graph of the planar triangulation of H (except the exterior face) which is a tree with not more than k leaves and constructs a clique tree of H with not more than k leaves. Finally the algorithm of the previous section is applied to H and its clique tree. For the overall algorithm we obtain the following. (For details we refer to the full version of our paper.)

Theorem 3. *The algorithm described above computes for a k-polygon graph given as a k-polygon representation a layout L satisfying* $b(G, L) \leq 2k^2 \cdot bw(G)$. *It executes in time* $O(n^3)$.

References

1. Ando, K., A. Kaneko, S. Gervacio, The bandwidth of a tree with k leaves is at most $\lceil \frac{k}{2} \rceil$, *Discrete Mathematics* **150** (1996), 403–406.
2. Assmann, S. F., G. W. Peck, M. M. Sysło and J. Zak, The bandwidth of caterpillars with hairs of length 1 and 2, *SIAM J. Algebraic Discrete Methods* **2** (1981), 387–393.
3. Blair, J. R. S., B. Peyton, *An introduction to chordal graphs and clique trees*, in *Graph Theory and Sparse Matrix Computation*, A. George, J. R. Gilbert, J. W. H. Liu (Eds.), The IMA Volumes in Mathematics and its Applications, Volume 56.
4. Booth, K. S. and G. S. Lueker, Testing for the consecutive ones property, interval graphs, and graph planarity using PQ-tree algorithms, *J. Comput. System Sci.* **13** (1976), 335–379.
5. Chinn, P. Z., J. Chvátalová, A. K. Dewdney and N. E. Gibbs, The bandwidth problem for graphs and matrices—a survey, *J. Graph Theory* **6** (1982), 223–254.
6. Chvátalová, J., A. K. Dewdney, N. E. Gibbs and R. R. Korfhage, The bandwidth problem for graphs: a collection of recent results, Research report #24, Department of Computer Science, UWO, London, Ontario (1975).
7. Chvátal, V., A remark on a problem of Harary, *Czech Math. J.* **20** (1970), 109–111.

8. Corneil, D. G., S. Olariu and L. Stewart, The ultimate interval graph algorithm?, *Proceedings of the Ninth Annual ACM-SIAM Symposium on Discrete Algorithms* (San Francisco, 1998), 175–180.
9. Elmallah, E. S. and L. K. Stewart, Polygon graph recognition, *J. Algorithms* **26** (1998), 101–140.
10. Eschen, E. M. and J. P. Spinrad, An $O(n^2)$ algorithm for circular-arc graph recognition, *Proceedings of the Fourth Annual ACM-SIAM Symposium on Discrete Algorithms* (Austin, TX, 1993), 128–137, ACM, New York, 1993.
11. Feige, U., Approximating the bandwidth via volume respecting embeddings, Technical Report CS98-03, Weizmann Insitute of Science, Rehovot, Israel, 1998.
12. Garey, M. R., R. L. Graham, D. S. Johnson and D. E. Knuth, Complexity results for bandwidth minimization, *SIAM J. Appl. Math.* **34** (1978), 477–495.
13. Gavril, F., The intersection graphs of subtrees in a tree are exactly the chordal graphs, *J. Comb. Theory, Ser. B* **16** (1974), 47–56.
14. Golumbic, M. C., Algorithmic Graph Theory and Perfect Graphs, Academic Press, New York, 1980.

15. Jiang, S., The bandwidth problem and bandwidth of cographs, unpublished manuscript, 1992.
16. Kleitman, D. J. and R. V. Vohra, Computing the bandwidth of interval graphs, *SIAM J. Discrete Math.* **3** (1990), pp. 373–375.
17. Kloks, T., D. Kratsch and H. Müller, Approximating the bandwidth for asteroidal triple-free graphs, *Algorithms—ESA '95* (Corfu), 434–447, Lect. Notes Comput. Sci. 979, Springer, Berlin, 1995.
18. Kloks, T., D. Kratsch and C. K. Wong, Minimum fill-in on circle and circular-arc graphs, *J. Algorithms* **28** (1998), 272–289.
19. Mahesh, R., C. Pandu Rangan and A. Srinivasan, On finding the minimum bandwidth of interval graphs, *Inf. Comput.* **95** (1991), 218–224.
20. Monien, B., The bandwidth minimization problem for caterpillars with hair length 3 is NP-complete, *SIAM J. Algebraic Discrete Methods* **7** (1986), 505–512.
21. Papadimitriou, C. H., The NP-completeness of the bandwidth minimization problem, *Computing* **16** (1976), 263–270.
22. Sprague, A. P., An $O(n \log n)$ algorithm for bandwidth of interval graphs, *SIAM J. Discrete Math.* **7** (1994), 213–220.
23. Unger, W., The complexity of the approximation of the bandwidth problem, *Proceedings of the Thirty-ninth Annual IEEE Symposium on Foundations of Computer Science* (Palo Alto, CA, 1998).

Linear Time $\frac{1}{2}$-Approximation Algorithm for Maximum Weighted Matching in General Graphs

Robert Preis *

Universität Paderborn, D-33095 Paderborn, Germany
robsy@uni-paderborn.de

Abstract. A new approximation algorithm for maximum weighted matching in general edge-weighted graphs is presented. It calculates a matching with an edge weight of at least $\frac{1}{2}$ of the edge weight of a maximum weighted matching. Its time complexity is $O(|E|)$, with $|E|$ being the number of edges in the graph. This improves over the previously known $\frac{1}{2}$-approximation algorithms for maximum weighted matching which require $O(|E| \cdot log(|V|))$ steps, where $|V|$ is the number of vertices.

1 Introduction

Graph Matching is a fundamental topic in graph theory. Let $G = (V, E)$ be a graph with vertices V and undirected edges E without multi-edges or self-loops. A matching of G is a subset $M \subset E$, such that no two edges of M are adjacent. A vertex incident to an edge of M is called *matched* and a vertex not incident to an edge of M is called *free*. An enormous amount of work has been done in matching theory in the past. Different types of matchings have been discussed, their existence and properties have been analyzed and efficient algorithms for the calculating of specific matchings have been developed. Many results have been achieved for specific types of graphs like bipartite, planar or other ones.

A central aspect are matchings with high cardinality. A *Maximal Matching* M_{MAX} is a matching which cannot be enlarged by an additional edge without breaking the matching property. A graph may have several different maximal matchings and, especially, maximal matchings of different cardinality. A *Maximum Cardinality Matching* M_{MCM} is a matching of maximum size, i.e. for all matchings $\bar{M}$ of G holds $|M_{\mathsf{MCM}}| \geq |\bar{M}|$. Matchings are also discussed for graphs with edge weights $w : E \to I\!\!R$. For a set $F \subset E$ let $W(F) := \sum_{\{a,b\} \in F} w(\{a,b\})$ be the weight of F. A *Maximum Weighted Matching* M_{MWM} is a matching of highest weight, i.e. for all matchings $\bar{M}$ of G holds $W(M_{\mathsf{MWM}}) \geq W(\bar{M})$.

Many algorithms for the calculating of matchings have been developed in the past. Please consult e.g. [8,9] for the history of matching algorithms. In the following, only the currently fastest algorithms are stated. Simple methods

---

* Supported by DFG/HNI-Graduiertenkolleg "Parallele Rechnernetze in der Produktionstechnik" and DFG Sonderforschungsbereich 376: "Massive Parallelität"

C. Meinel and S. Tison (Eds.): STACS'99, LNCS 1563, pp. 259–269, 1999.

with time complexity $O(|E|)$ can be used to calculate maximal matchings. The fastest algorithm for maximum cardinality matching up to date is by Micali and Vazirani [10] which has a time complexity of $O(|E|\sqrt{|V|})$. In the edge-weighted case, the algorithm by Gabow [3] calculates a maximum weighted matching in time $O(|V| \cdot |E| + |V|^2 log(|V|))$. This time complexity has been improved by Gabow and Tarjan [4] under the assumption of integral costs that are not particularly high: if the weight function $w : E \rightarrow [-N, ..., N]$ assigns only integers between $-N$ and N, their algorithm will run in $O(\sqrt{|V| \cdot \alpha(|E|, |V|)} \cdot log(|V|) \cdot |E| \cdot log(|V| \cdot N))$ time, where α is the inverse of Ackermann's function.

All algorithms discussed so far have super-linear time complexity. Recently, approximation algorithms for matching problems have attracted more and more attention. They have a smaller time complexity than an optimal algorithm and calculate suboptimal solutions. The guaranteed quality is described by an approximation factor which states the worst case loss to an optimal solution, e.g. a factor of $\frac{1}{2}$ guarantees that the solution quality is at least half the value of the optimum solution. It is a simple exercise to prove that any maximal matching M_{MAX} has a cardinality of at least $\frac{1}{2}$ the cardinality of a maximum cardinality matching, i.e. $|M_{MAX}| \geq \frac{1}{2}|M_{MCM}|$. Therefore, any algorithm for maximal matching is an $\frac{1}{2}$-approximation algorithm for maximum cardinality matching.

Augmenting paths are often considered for graph matching, especially for approximating maximum cardinality matching. It is a path of an odd number of edges with alternating edges of M and of $E \backslash M$ and two free vertices as endpoints, i.e. an augmenting path of length l consists of $\frac{l-1}{2}$ edges of M and $\frac{l+1}{2}$ edges of $E \backslash M$. If such a path exists, the cardinality of M can be increased by one by exchanging the matched and unmatched edges of the path. Based on the work by Hopcroft and Karp [6], it can be shown that if the shortest augmenting path with respect to a matching M_l is l, then $|M_l| \geq \frac{l-1}{l+1}|M_{MCM}|$ (see e.g. [7], p.156). Matchings without short augmenting paths can be calculated very fast, e.g. a matching with a shortest path of length $l \geq 5$ can be computed in time $O(|E|)$, resulting in $|M_5| \geq \frac{2}{3}|M_{MCM}|$. If the minimum degree min and maximum degree max of all vertices are considered, it can easily be proven that if the shortest augmenting path has a length $l \geq 5$, then $|M_5| \geq \frac{min}{max+2 \cdot min}|V|$, which implies $|M_5| \geq \frac{1}{3}|V|$ for graphs with a regular degree, i.e. $\frac{2}{3}$ of the vertices are matched.

For the weighted case, the GREEDY-algorithm of Figure 1 calculates a matching M_{GREEDY} with weight $W(M_{GREEDY}) \geq \frac{1}{2}W(M_{MWM})$ which is analyzed by Avis [1]. It requires a time of $O(|E| \cdot log(|V|))$, if the edges are sorted by their weights in a preprocessing step.

1.1   New Result

The new algorithm is similar to GREEDY. It guarantees the same approximation quality, but has only a time complexity of $O(|E|)$. The new algorithm LAM is described in the following sections and we now state the new theorem:

Theorem 1. *Let $G = (V, E)$ be a graph with vertices V and weighted undirected edges E. A matching M_{LAM} of G with an edge weight of at least $\frac{1}{2}$ of the edge weight of a maximum weighted matching can be computed in linear time $O(|E|)$.*

```
GREEDY-Algorithm
    M_GREEDY := ∅;
    WHILE (E ≠ ∅)
        take an edge {a, b} ∈ E with highest weight;
        add {a, b} to M_GREEDY;
        remove all edges incident to a or b from E;
    ENDWHILE
```

Fig. 1. GREEDY:$\frac{1}{2}$-app. alg. for maximum weighted matching in time $O(|E| \cdot log(|V|))$.

Proof. Lemma 4 (Section 3) shows that the algorithm LAM of Figure 4 (Section 2) calculates a matching M_{LAM} with an edge weight of at least $\frac{1}{2}$ the edge weight of a maximum weighted matching and Lemma 5 (Section 4) shows the time complexity of $O(|E|)$. □

According to corollary 2 (Section 2), matching M_{LAM} computed by LAM is also maximal. As stated above, $|M_{MAX}| \geq \frac{1}{2}|M_{MCM}|$, resulting in:

Corollary 1. *The matching M_{LAM} computed by algorithm LAM not only has a weight of at least $\frac{1}{2}$ the weight of a maximum weighted matching, but also has a cardinality of at least $\frac{1}{2}$ the cardinality of a maximum cardinality matching.*

1.2 Our Motivation: Matching for Multilevel Graph Partitioning

Graph Matching has a wide range of application, some of which are discussed by Lovász and Plummer in [9]. Our motivation comes from the use of matchings in a multilevel approach for efficient partitioning of very large graphs. In graph partitioning, the vertices of a graph are partitioned in a fixed number of equally sized parts, such that the number of edges connecting vertices of different parts is minimized. In the weighted case, the sum of weights of crossing edges is minimized. The calculation of a partition with minimum weight of crossing edges is NP-complete, even for the case of partitioning a graph with equal edge weights into two parts [5].

Therefore, efficient heuristics are used to calculate good partitions in a reasonable amount of time and many freely available partitioning libraries like PARTY [11] include several different types of methods, but many heuristics still have a high time complexity when used for very large graphs. The solution is to coarsen the large graph in several levels to smaller and smaller graphs with a similar structure, to partition the smallest graph and to project the here found partition back through the levels to a partition of the original graph.

The single coarsening steps between two levels are usually performed by matchings, i.e. a matching of the graph is calculated and the vertices incident to a matching edge are contracted. It is important to contract those vertices which are connected via an edge of a high weight, because it is very likely that this edge does not cross between parts in a partition with a low weight of crossing edges.

Generally, the use of a maximum weighted matching would benefit the coarsening step the most, but the super-linear time complexity of an optimal algorithm is too high for real examples. Therefore, fast approximation algorithms like the new one in this paper are very useful here.

2   New Linear Time Approximation Algorithm

Figure 2 outlines the new LAM algorithm. It starts with an empty matching M_{LAM} and repeatedly adds an edge $\{a, b\}$ to M_{LAM} and removes all edges incident to a or b, because they cannot be part of the final matching. The key idea is to add *locally heaviest edges* $\{a, b\}$, i.e. an edge with a weight of at least as high as the weight of all adjacent edges remaining in E, i.e. $w(\{a, b\}) \geq w(\{x, y\})$ for any $\{x, y\} \in E$ with $a = x$ or $b = x$. After a locally heaviest edge is removed from E, further edges may become locally heaviest. Note that at least one locally heaviest edge always exist.

Outline of LAM-Algorithm

$M_{LAM} := \emptyset$;
WHILE $(E \neq \emptyset)$
    take a locally heaviest edge $\{a, b\} \in E$;
    add $\{a, b\}$ to M_{LAM};
    remove all edges incident to a or b from E;
ENDWHILE

Fig. 2. Outline of LAM: Linear time $\frac{1}{2}$-app. alg. for maximum weighted matching.

The main problem is to find such an edge. Figure 3 shows the idea. The algorithm starts with an arbitrary edge and checks the remaining adjacent edges. As long as an adjacent edge with higher weight can be found, the algorithm switches to the new edge and repeats the checking procedure until a locally heaviest edge is reached, i.e. the weight increases along the path.

The detailed algorithm LAM is shown in Figure 4. It starts with an empty matching M_{LAM}. The global sets U and R store the unchecked edges ($U = E$ at the start) and the removed edges ($R = \emptyset$ at the start). The main algorithm is a WHILE loop which calls the procedure 'try match' with an arbitrary unchecked edge $\{a, b\}$. This edge is not added to the matching until all adjacent edges to free vertices are checked for higher weight which is managed in the WHILE part of the procedure. Every call of procedure 'try match $(\{a, b\})$' stores its own sets of locally checked edges $C_{\{a,b\}}(a)$ and $C_{\{a,b\}}(b)$, depending on whether the checked edges are incident to a or b. Let C be the union of all locally checked edges, i.e. $C := \bigcup_{\{a,b\} \in E} C_{\{a,b\}}(a) \cup C_{\{a,b\}}(b)$ (we will show later that 'try match' is called not more than once for every edge). As long as a and b are free and at least one of them is incident to an unchecked edge, it is checked and, if it has a higher weight, 'try match' calls itself recursively with the new edge. Recursive calls are repeated

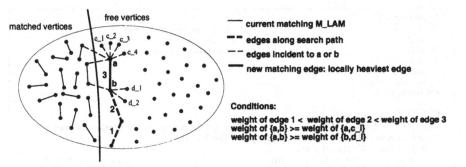

Fig. 3. Starting with an arbitrary edge 1, the path progresses along edges 2 and 3 with higher weight until a locally heaviest edge $\{a, b\}$ is found. (Only edges of the current matching M_{LAM}, edges along the path and edges adjacent to $\{a, b\}$ are shown.)

until a locally heaviest edge is reached. Then, the edge is added to M_{LAM}, the algorithm terminates the current call of 'try match', tracks back the search path by one edge and continues the WHILE loop by checking further adjacent edges.

The WHILE loop terminates if vertices a and/or b were matched in a recursive call or if no further adjacent unchecked edge exists. Then in the IF-ELSE part, $\{a, b\}$ is added to M_{LAM} if a and b are free. In addition, edges incident to matched vertices and checked in the current call are removed, i.e. moved from C to R. If a or b remains free, all edges which were checked in the current call and are incident to two free vertices are unchecked, i.e. moved back from C to U. Lemma 5 will show that only a limited number of times edges are moved back from C to U.

Note that the WHILE loop alternately checks adjacent edges incident to a and incident to b, as long as both types are available. This property will be used in Section 4 to show the linear time requirement. Additionally, removed edges cannot take part in the following matching process, but we need to keep track of how many edges are removed in every part of the algorithm. Unlike in the outline of the algorithm in Figure 2, edges incident to matched vertices are not removed immediately, but removed in those procedure calls of the algorithm where they were checked the last time.

It is fairly obvious that the unchecked edges in U have the property that both incident vertices are free. It holds true for the start. New edges are only moved back to U in the IF-ELSE part, but only if they are incident to two free vertices. Furthermore, an edge $\{a, b\}$ may only be matched at the end of the IF-ELSE part. In this case, a and b have to be free, and the WHILE loop terminated because there was no edge $\{a, c\}$ or $\{b, d\}$ in U, i.e. all edges in U keep the property of being incident to two free vertices.

The central part of the algorithm is the procedure 'try match' and the following lemma shows the number of times it is called:

Lemma 1 ($|E|$ calls). *Procedure 'try match' is called only once for every edge.*

```
LAM-Algorithm
  M_LAM := ∅;                              /* empty matching at the start */
  U := E;                                  /* all edges are unchecked at the start */
  R := ∅;                                  /* no removed edges at the start */
  WHILE (U ≠ ∅)
    take arbitrary edge {a, b} ∈ U;
    try match ({a, b});
  ENDWHILE

PROCEDURE try match ({a, b})
  C_{a,b}(a) := ∅; C_{a,b}(b) := ∅;        /* empty local sets of checked edges at the start */
  WHILE (a is free AND b is free AND (∃{a, c} ∈ U OR ∃{b, d} ∈ U))
    IF (a is free AND ∃{a, c} ∈ U)
      move {a, c} from U to C_{a,b}(a);                    /* move from U to C */
      IF (w({a, c}) > w({a, b}))
        try match ({a, c});                               /* call heavier edge */
      ENDIF
    ENDIF
    IF (b is free AND ∃{b, d} ∈ U)
      move {b, d} from U to C_{a,b}(b);                    /* move from U to C */
      IF (w({b, d}) > w({a, b}))
        try match ({b, d});                               /* call heavier edge */
      ENDIF
    ENDIF
  ENDWHILE
  IF (a is matched AND b is matched)
    move edges C_{a,b}(a) and C_{a,b}(b) to R;             /* move from C to R */
  ELSE IF (a is matched AND b is free)
    move edges C_{a,b}(a) and {{b, d} ∈ C_{a,b}(b)| d is matched} to R;/* move from C to R */
    move edges {{b, d} ∈ C_{a,b}(b)| d is free} back to U;   /* move from C back to U */
  ELSE IF (b is matched AND a is free)
    move edges C_{a,b}(b) and {{a, c} ∈ C_{a,b}(a)| c is matched} to R;/* move from C to R */
    move edges {{a, c} ∈ C_{a,b}(a)| c is free} back to U;   /* move from C back to U */
  ELSE /* a is free AND b is free */
    move edges C_{a,b}(a) and C_{a,b}(b) to R;             /* move from C to R */
    add {a, b} to M_LAM;                        /* new matching edge {a, b} */
  ENDIF
```

Fig. 4. LAM: Linear time $\frac{1}{2}$-app. alg. for maximum weighted matching.

Proof. The procedure 'try match' is only called with an edge $\{a, b\}$ from U, which ensures that both vertices are free. The same edge cannot be the parameter in deeper recursive calls, because the search path only progresses along strictly higher edge weights. Finally, in the IF-ELSE part of 'try match', either a and/or b are already matched or they are matched by including edge $\{a, b\}$ in M_{LAM}. Therefore, after the first call of 'try match($\{a, b\}$)', a and/or b are matched, i.e. $\{a, b\}$ cannot be in U anymore and cannot be called a second time.   □

The following lemma shows the status of the edges:

Lemma 2 (edge status). *At the start all edges are not checked ($U = E$), and there are no checked or removed edges ($R = C = \emptyset$). At any stage, an edge $\{a, b\} \in E$ is either unchecked ($\in U$), checked ($\in C$) or removed ($\in R$), resulting in $E = U \cup C \cup R$. At the end, all edges are removed, i.e. $U = C = \emptyset$ and $R = E$.*

Proof. The status at the start is clear from the algorithm. The edges are only moved between the sets U, C and R, ensuring the stated status throughout the algorithm. The WHILE loop of the main algorithm terminates when U is empty. Besides, all edges $C_{\{a,b\}}(a)$ and $C_{\{a,b\}}(b)$, which have been moved in the WHILE loop of procedure call 'try match ($\{a,b\}$)', are either moved to R or back to U in the IF-ELSE part of the same call. Therefore, C is also empty after termination of all 'try match' calls, resulting, with $E = U \cup C \cup R$, in $E = R$.   □

Lemma 2 ensures that all edges are removed at the end of the algorithm, because at least one of the incident vertices has been matched. Thus:

Corollary 2 (maximal). *The resulting matching M_{LAM} is maximal.*

3 $\frac{1}{2}$ - Approximation Quality

In this section the property of locally heaviest edge for any edge added to the matching and the $\frac{1}{2}$-approximation are shown. A similar proof has been done in [1] to show the $\frac{1}{2}$-approximation for the GREEDY algorithm of Figure 1.

Lemma 3 (locally heaviest edge). *Algorithm LAM starts with an empty matching and an edge $\{a,b\}$ is only added if a and b are free and neither a nor b are adjacent to a free vertex with an edge of higher weight than $\{a,b\}$.*

Proof. An edge $\{a,b\}$ is only added to M_{LAM} in the last ELSE part of 'try match', i.e. a and b are free. Let a be adjacent to a free vertex c (or b adjacent to a free vertex d). According to Lemma 2, edge $\{a,c\}$ ($\{b,d\}$) may be

in R : Impossible, because then either a or c are already matched.
in U : Impossible, because then the WHILE loop would not have terminated.
in C : The weight of $\{a,b\}$ ($\{b,d\}$) is higher than of all other edges in the search path. When edges were checked along the search path, either their weight was not higher than the weight of the corresponding path edge, or the search path progressed along this edge. In the later case, either the recursive call terminated with at least one incident vertex of that edge being matched, or the edge is still in the search path.   □

Lemma 4 ($\frac{1}{2}$-approximation). *Algorithm LAM computes a matching M_{LAM} with at least $\frac{1}{2}$ of the edge weight of a maximum weighted matching M_{MWM}.*

Proof. Compare M_{LAM} to an arbitrary matching M_{MWM}, let V_{LAM} be the matched vertices of the current M_{LAM} and V_{MWM} be the matched vertices of M_{MWM}. Throughout the algorithm we will show that the weight of the current matching M_{LAM} is at least $\frac{1}{2}$ the weight of the edges of M_{MWM} incident to a vertex of V_{LAM}:

$$W(M_{\text{LAM}}) \geq \frac{1}{2}W(\{\{u,v\} \in M_{\text{MWM}} | u \in V_{\text{LAM}} \vee v \in V_{\text{LAM}}\})$$

It holds for the start ($M_{\text{LAM}} := \emptyset$). After adding an edge $\{a,b\}$ to M_{LAM}, $W(M_{\text{LAM}})$ increases by $w(\{a,b\})$, but also the right hand side may increase.

If $\{a,b\} \in M_{\mathsf{MWM}}$, the right hand side only increases by $\frac{1}{2}w(\{a,b\})$. Otherwise, let $\{a,c\},\{b,d\} \in M_{\mathsf{MWM}}$ be the possible edges adjacent to $\{a,b\}$. The choice of matching edge $\{a,b\}$ excluded the possible choice of $\{a,c\}$ and $\{b,d\}$ throughout the rest of the algorithm. These are the only two edges by which the subset of M_{MWM} may increase, i.e. the right hand side may only increase by $\frac{1}{2}(w(\{a,c\}) + w(\{b,d\}))$. If $c \in V_{\mathsf{LAM}}$ ($d \in V_{\mathsf{LAM}}$) before we add edge $\{a,b\}$, then $\{a,c\}$ ($\{b,d\}$) is already in the subset of M_{MWM}. If $c \notin V_{\mathsf{LAM}}$ ($d \notin V_{\mathsf{LAM}}$), i.e. c (d) is free, Lemma 3 insures that $w(\{a,b\}) \geq w(\{a,c\})$ ($w(\{a,b\}) \geq w(\{b,d\})$). Therefore, the value on the right hand side cannot increase by more than $w(\{a,b\})$.

At the end, algorithm LAM terminates with a maximal matching M_{LAM} (Corollary 2), i.e. for all edges $\{u,v\}$ is $u \in V_{\mathsf{LAM}}$ or $v \in V_{\mathsf{LAM}}$. Therefore,

$$W(M_{\mathsf{LAM}}) \geq \frac{1}{2}W(\{\{u,v\} \in M_{\mathsf{MWM}} | u \in V_{\mathsf{LAM}} \lor v \in V_{\mathsf{LAM}}\}) = \frac{1}{2}W(M_{\mathsf{MWM}})\square$$

4   Linear Time Requirement

The time requirement depends on the number of times edges are moved between the sets U of unchecked, C of checked and R of removed edges. There are only three ways in which an edge may be moved: (1) in the WHILE loop of 'try match', previously unchecked edges are checked (moved from U to C), in the IF-ELSE part, the checked edges are either (2) removed (moved from C to R), or (3) unchecked again (moved from C to U). Therefore, once an edge is removed and stored in R, it will not be moved to any other set, ensuring that an edge may be moved from C to R at most $|E|$ times. Furthermore, an edge may be moved several times between U and C, but we will show that every time when edges are moved back from C to U, an almost equal number of edges is moved from C to R, showing that the number of edges moved between U and C is $O(|E|)$.

Lemma 5 (linear time). *Algorithm* LAM *of Figure 4 runs in $O(|E|)$ time.*

Proof. The loop of the main algorithm has at most $|E|$ iterations, because if it makes a call 'try match $(\{a,b\})$' with an edge $\{a,b\} \in U$, at least one of a and b are matched after the completion of the call. As stated in Section 2, an unchecked edge in U is always incident to two free vertices. Therefore, U reduces by at least the edge $\{a,b\}$ in every iteration of the WHILE loop of the main algorithm. Although new edges may be added to U in the IF-ELSE part of 'try match', those edges have previously been moved in the WHILE loop of the same procedure. According to Lemma 1, the 'try match' procedure is not called more than once for every edge $\{a,b\} \in E$. In sum, the time complexity of the algorithm is:

$$O(|E|) + \sum\nolimits_{\{a,b\}\in E} O(\text{try match } (\{a,b\}))$$

The procedure 'try match$(\{a,b\})$' consists of a WHILE loop and an IF-ELSE part. In every WHILE loop, an edge $\{a,c\}$ and/or $\{b,d\}$ is checked and moved from U to the local sets $C_{\{a,b\}}(a)$ and/or $C_{\{a,b\}}(b)$ of checked edges, resulting in a time of $O(|C_{\{a,b\}}(a)| + |C_{\{a,b\}}(b)|)$ for the WHILE loop. The values $|C_{\{a,b\}}(a)|$ and $|C_{\{a,b\}}(b)|$ refer to the maximum sizes of the sets $C_{\{a,b\}}(a)$ and $C_{\{a,b\}}(b)$, which occur after the completion of the WHILE loop. In the IF-ELSE part, the edges of the local sets $C_{\{a,b\}}(a)$ and $C_{\{a,b\}}(b)$ are either removed (moved to R), or they are unchecked again (moved to U), which again leads to a time complexity of $O(|C_{\{a,b\}}(a)| + |C_{\{a,b\}}(b)|)$. Thus, the run time of the algorithm is:

$$O(|E|) + \sum_{\{a,b\}\in E} O(|C_{\{a,b\}}(a)| + |C_{\{a,b\}}(b)|) \tag{1}$$

We will show that the number of check operations is not much larger than the number of remove operations for every call. Let $R_{\{a,b\}}$ be the set of edges removed in the procedure call 'try match($\{a,b\}$)', i.e. $\sum_{\text{try match}}(\{a,b\}) R_{\{a,b\}} = R$. We distinguish between the cases of the IF-ELSE part:

1. **a and b are matched; a and b are free:** In both cases the size of sets $C_{\{a,b\}}(a)$ and $C_{\{a,b\}}(b)$ is equal to the number of removed edges, i.e.

$$|C_{\{a,b\}}(a)| + |C_{\{a,b\}}(b)| = |R_{\{a,b\}}| \tag{2}$$

2. **a is matched, b is free:** In this case, a was matched with an adjacent vertex c in a recursive call of 'try match'. It may be matched in a recursive call just one level deeper as shown in Figure 5(i), but it may also be matched in a deeper level of recursion as result of a loop in the search path as shown in Figure 5(ii). In addition, it may also be matched in a recursive call made from vertex b as shown in Figure 5(iii). We will show, that the number of

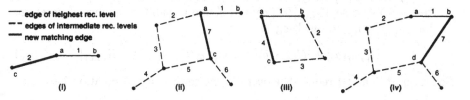

Fig. 5. Vertices a or b are matched in a recursive call. Edges of consecutive recursive calls are shown and numbers indicate the level of recursion, i.e. the edge weight increases with the numbers. (i): a is matched in a recursive call from itself one level deeper. (ii): a is matched in a recursive call from itself several levels deeper, forming a loop. (iii): a is matched in a recursive call from b. (iv): b is matched in a recursive call from a.

uncheck operations is not larger than the number of remove operations. Let us assume, the WHILE loop terminates after i iterations, i.e. a and c were free at the start of every iteration and edge $\{a,c\}$ matched in the i-th iteration. We have to show, that an edge $\{a,c_i\} \in U$ was checked in every iteration i.

According to Lemma 2, $\{a, c\}$ is either in U, C or R at every stage and, especially, at the beginning of every iteration of the WHILE loop of the procedure call 'try match $(\{a, b\})$'. If $\{a, c\}$ is checked, it may either be checked from the procedure 'try match $(\{a, b\})$', or from a procedure call earlier in the search path. When edges were checked along the search path, either their weight was not higher than the weight of the corresponding path edge, or the search path progressed along this edge. In the later case, either the recursive call terminated with at least one incident vertex of that edge being matched, or the edge is still in the search path. Therefore, if $\{a, c\}$ is checked, it cannot have a higher weight than $\{a, b\}$, but this is impossible, because it was matched in iteration i while being the final edge in the search path and $\{a, b\}$ being within the search path. Furthermore, $\{a, c\}$ can also not be removed, because then either a or c is matched. Therefore, $\{a, c\} \in U$ at the start of every iteration, i.e. the condition of the first IF condition in the WHILE loop was true in each iteration and an edge $\{a, c_i\}$ is checked and moved to $C_{\{a,b\}}(a)$, i.e. $|C_{\{a,b\}}(a)| = i$.

In addition, it is clear that $|C_{\{a,b\}}(b)|$ cannot be more than the number of iterations, i.e. $|C_{\{a,b\}}(b)| \leq i$, resulting in

$$|C_{\{a,b\}}(a)| + |C_{\{a,b\}}(b)| \leq 2|C_{\{a,b\}}(a)| \leq 2 \cdot |R_{\{a,b\}}| \tag{3}$$

3. **b is matched, a is free:** This case is similar to the previous one, but not identical, because a is checked first in the WHILE loop. It may happen that b is matched in a recursive call from a as shown in Figure 5(iv). In this case, in the final iteration of the WHILE loop, b was already matched after completion of the recursive calls from a and no further edge was added to $C_{\{a,b\}}(b)$. Consequently, we can only guarantee $i \geq |C_{\{a,b\}}(b)| \geq i - 1$. As in the previous case, $|C_{\{a,b\}}(a)|$ cannot be more than the number of iterations, i.e. $|C_{\{a,b\}}(a)| \leq i$, resulting in

$$|C_{\{a,b\}}(a)| + |C_{\{a,b\}}(b)| \leq 2|C_{\{a,b\}}(b)| + 1 \leq 2 \cdot |R_{\{a,b\}}| + 1 \tag{4}$$

Equations 2, 3 and 4 reduce the overall time complexity of equation 1 to

$$O(|E|) + \sum\nolimits_{\{a,b\} \in E} O(2 \cdot |R_{\{a,b\}}| + 1) = O(|E|),$$

because the total number of removed edges cannot exceed $|E|$.   □

5   Conclusion

The new algorithm LAM has been tested experimentally and compared with other matching heuristics on many large graphs from real applications by Birger Boyens in [2]. It is implemented in the graph partitioning library PARTY [11], where it is used for graph coarsening in the multilevel partitioning approach.

I would like to thank Burkhard Monien and Marco Riedel for many fruitful discussions on graph matching.

References

1. D. Avis. A survey of heuristics for the weighted matching problem. *Networks*, 13:475–493, 1983.
2. B. Boyens. Schrumpfungstechniken zur effizienten Graphpartitionierung. Diplom-Thesis, Universität Paderborn, Germany, June 1998. (in German).
3. H.N. Gabow. Data structures for weighted matching and nearest common ancestors with linking. *ACM-SIAM Symposium on Discrete Algorithms.*, pages 434–443, 1990.
4. H.N. Gabow and R.E. Tarjan. Faster scaling algorithms for general graph-matching problems. *Journal of the ACM*, 38(4):815–853, 1991.
5. M.R. Garey, D.S. Johnson, and L. Stockmeyer. Some simplified NP-complete graph problems. *Theoretical Computer Science*, 1:237–267, 1976.
6. J. Hopcroft and R.M. Karp. An $O(n^{5/2})$ algorithm for maximum matching in bipartite graphs. *SIAM Journal on Computing*, 2:225–231, 1973.
7. M. Karpinski and W. Rytter. *Fast Parallel Algorithms for Graph Matching Problems, Oxford Lecture Series in Math. and its Appl.*. Oxford University Press, 1998.
8. E.L. Lawler. *Combinatorial Optimization: Networks and Matroids*. Holt, Rinehart and Winston, New York, 1976.
9. L. Lovász and M.D. Plummer. *Matching Theory*, volume 29 of *Annals of Discrete Mathematics*. North-Holland Mathematics Studies, 1986.
10. S. Micali and V.V. Vazirani. An $O(\sqrt{|V|} \cdot |E|)$ algorithm for finding maximum matching in general graphs. In *IEEE Annual Symposium on Foundations of Computer Science*, pages 17–27, 1980.
11. R. Preis and R. Diekmann. The PARTY partitioning-library, user guide, version 1.1. Technical Report TR-RSFB-96-024, Universität Paderborn, Sep 1996.

Extending Downward Collapse from 1-versus-2 Queries to j-versus-$j + 1$ Queries

Edith Hemaspaandra[1*], Lane A. Hemaspaandra[2**], and Harald Hempel[3***]

[1] Dept. of Comp. Sci., Rochester Institute of Technology, Rochester, NY 14623, USA.
eh@cs.rit.edu
[2] Dept. of Computer Science, University of Rochester, Rochester, NY 14627, USA.
lane@cs.rochester.edu
[3] Inst. für Informatik, Friedrich-Schiller-Universität Jena, 07743 Jena, Germany.
hempel@informatik.uni-jena.de

Abstract. The above figure shows some classes from the boolean and (truth-table) bounded-query hierarchies. It is well-known that if either collapses at a given level, then all higher levels collapse to that same level. This is a standard "upward translation of equality" that has been known for over a decade. The issue of whether these hierarchies can translate equality *downwards* has proven vastly more challenging. In particular, with regard to the figure above, consider the following claim:

$$P^{\Sigma_k^p}_{m\text{-tt}} = P^{\Sigma_k^p}_{m+1\text{-tt}} \Rightarrow \text{DIFF}_m(\Sigma_k^p) = \text{coDIFF}_m(\Sigma_k^p) = \text{BH}(\Sigma_k^p). \quad (**)$$

This claim, if true, says that equality translates downwards between levels of the bounded-query hierarchy and the boolean hierarchy levels that (before the fact) are immediately below them. Until recently, it was not known whether (**) *ever* held, except in the trivial $m = 0$ case. Then Hemaspaandra et al. [15] proved that (**) holds for all m, whenever $k > 2$. For the case $k = 2$, Buhrman and Fortnow [5] then showed that (**) holds when $m = 1$. In this paper, we prove that for the case $k = 2$, (**) holds for *all* values of m. As Buhrman and Fortnow showed that no relativizable technique can prove "for $k = 1$, (**) holds for all m," our achievement of the $k = 2$ case is unlikely to be strengthened to $k = 1$ any time in the foreseeable future. The new downward translation we obtain tightens the collapse in the polynomial hierarchy implied by a collapse in the bounded-query hierarchy of the second level of the polynomial hierarchy.

* Supported in part by grant NSF-INT-9513368/DAAD-315-PRO-fo-ab. Work done in part while visiting Friedrich-Schiller-Universität Jena.
** Supported in part by grants NSF-CCR-9322513 and NSF-INT-9513368/DAAD-315-PRO-fo-ab. Work done in part while visiting Friedrich-Schiller-Universität Jena.
*** Supported in part by grant NSF-INT-9513368/DAAD-315-PRO-fo-ab. Work done in part while visiting Le Moyne College.

C. Meinel and S. Tison (Eds.): STACS'99, LNCS 1563, pp. 270–280, 1999.
© Springer-Verlag Berlin Heidelberg 1999

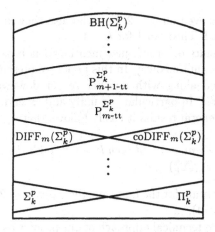

1   Introduction

Does the collapse of low-complexity classes imply the collapse of higher-complexity classes? Does the collapse of high-complexity classes imply the collapse of lower-complexity classes? These questions—known respectively as downward and upward translation of equality—have long been central topics in computational complexity theory. For example, in the seminal paper on the polynomial hierarchy, Meyer and Stockmeyer [20] proved that the polynomial hierarchy displays upward translation of equality (e.g., $P = NP \Rightarrow P = PH$).

The issue of whether the polynomial hierarchy—its levels and/or bounded access to its levels—ever displays *downward* translation of equality has proved more difficult. The first such result regarding bounded access was recently obtained by Hemaspaandra, Hemaspaandra, and Hempel [15], who proved that if for some high level of the polynomial hierarchy one query equals two queries, then the hierarchy collapses down not just to one query to that level, but rather to that level itself. That is, they proved the following result (note: the levels of the polynomial hierarchy [20,26] are denoted in the standard way, namely, $\Sigma_0^p = P$, $\Sigma_1^p = NP$, $\Sigma_k^p = NP^{\Sigma_{k-1}^p}$ for each $k > 1$, and $\Pi_k^p = \{L \mid \overline{L} \in \Sigma_k^p\}$ for each $k \geq 0$).

Theorem 1. *([15])   For each $k > 2$: If $P^{\Sigma_k^p[1]} = P^{\Sigma_k^p[2]}$, then $\Sigma_k^p = \Pi_k^p = PH$.*

This theorem has two clear directions in which one might hope to strengthen it. First, one might ask not just about one-versus-two queries but rather about m-versus-$m + 1$ queries. Second, one might ask if the $k > 2$ can be improved to $k > 1$. Both of these have been achieved. The first strengthening was achieved in a more technical section of the same paper by Hemaspaandra, Hemaspaandra, and Hempel [15]. They showed that Theorem 1 was just the $m = 1$ special case of a more general downward translation result they established, for $k > 2$, between bounded access to Σ_k^p and the boolean hierarchy over Σ_k^p. The second type of strengthening was achieved by Buhrman and Fortnow [5], who showed that Theorem 1 holds even for $k = 2$, but who also showed that no relativizable technique can establish Theorem 1 for $k = 1$.

Neither of the results or proofs just mentioned is broad enough to achieve both strengthenings simultaneously. In this paper we present a new result strong enough to, when used along with another recent downward translation [13], achieve this (and more). In particular, we unify and extend all the above results, and from our more general results it easily follows that:

Corollary 1. *For each $m > 0$ and each $k > 1$ it holds that:* $P^{\Sigma_k^p}_{m\text{-tt}} = P^{\Sigma_k^p}_{m+1\text{-tt}} \Rightarrow$ $\text{DIFF}_m(\Sigma_k^p) = \text{coDIFF}_m(\Sigma_k^p)$.

In particular, we obtain for the first time the cases $(k = 2 \wedge m = 2)$, $(k = 2 \wedge m = 3)$, $(k = 2 \wedge m = 4)$, ...

Informally put, the technical approach of our proof is as follows. In the previous work extending Theorem 1 to the boolean hierarchy (part 1 of Theorem 2), the "coordination" difficulties presented by the fact that boolean hierarchy sets are in effect handled via collections of machines were resolved via using certain lexicographically extreme objects as clear signposts to signal machines with. In the current stronger context that approach fails. Instead, we integrate into the structure of easy-hard-technique proofs (especially those of [15,5]) the so-called "telescoping" normal form possessed by the boolean hierarchy over Σ_k^p (for each k, see [18,6,10,30]), which in concept dates back to Hausdorff's work on algebras of sets, and has often proven useful in "controlling" the complexity-theoretic boolean hierarchies (see, e.g., [18,6,7,3]), and has also been used in the context of the easy-hard technique (see, e.g., [23]). This normal form guarantees that if $L \in \text{DIFF}_m(\Sigma_k^p)$, then there are sets $L_1, L_2, \ldots, L_m \in \Sigma_k^p$ such that $L = L_1 - (L_2 - (L_3 - \cdots (L_{m-1} - L_m) \cdots))$ and $L_1 \supseteq L_2 \supseteq \cdots \supseteq L_{m-1} \supseteq L_m$. (Picture, if you will, an archery target with concentric rings of membership and nonmembership. That is exactly the effect created by this normal form.)

As noted at the end of Section 4, the stronger downward translation we obtain yields a strengthened collapse of the polynomial hierarchy under the assumption of a collapse in the bounded-query hierarchy (by which, throughout this paper, we mean the truth-table bounded-query hierarchy) over NP^{NP}.

We conclude this section with some additional literature pointers. We mention that the proofs of Theorem 1 and all that grew out of it—including this paper—are indebted to, and use (strong) extensions of, the "easy-hard" technique that was invented by Kadin ([17], as further developed by Beigel, Chang, Ogihara, Kadin, and Wagner [27,28,4,8]) to study *upward* translations of equality resulting from the collapse of the boolean hierarchy. We also mention that there is a body of literature showing that equality of exponential-time classes translates downwards in a limited sense: Relationships are obtained with whether *sparse* sets collapse within lower time classes (the classic paper in this area is that of Hartmanis, Immerman, and Sewelson [9], see also [21]; limitations of such results are presented in [1,2,16]). Other than being a restricted type of downward translation of equality, that body of work has no close connection with the present paper due to that body of work's applicability only to sparse sets.

Finally, we mention that downward separation is closely tied to the recently developed theory of "query order" (see the survey [11]).

2   Preliminaries

To explain exactly what we do and how it extends previous results, we now state the previous results in the more general forms in which they were actually established, though in some cases with different notations or statements (see, e.g., the interesting recent paper of Wagner [29] regarding the relationship between "delta notation" and truth-table classes). Before stating the results, we very briefly remind the reader of some definitions/notations, namely the Δ levels of the polynomial hierarchy, truth-table access, symmetric difference classes, and boolean hierarchies. A detailed introduction to the boolean hierarchy, including its motivation and applications, can be found in [6].

Definition 1. *1. As is standard, for each $k \geq 1$, Δ_k^p denotes $\mathrm{P}^{\Sigma_{k-1}^p}$ (see [20]). As is standard, for each $m \geq 0$ and each set A, $\mathrm{P}_{m\text{-tt}}^A$ denotes the class of languages accepted by deterministic polynomial-time machines allowed m truth-table (i.e., non-adaptive) queries to A (see [19]).*

2. For any classes C and D, $C\Delta D = \{L \mid (\exists C \in C)(\exists D \in D)[L = C\Delta D]\}$, where $C\Delta D = (C - D) \cup (D - C)$.

3. ([6,7], see also [10,18])   Let C be any complexity class. We now define the levels of the boolean hierarchy.

(a) $\mathrm{DIFF}_1(C) = C$.

(b) For any $m \geq 1$, $\mathrm{DIFF}_{m+1}(C) = \{L \mid (\exists L_1 \in C)(\exists L_2 \in \mathrm{DIFF}_m(C))[L = L_1 - L_2]\}$.

(c) For any $m \geq 1$, $\mathrm{coDIFF}_m(C) = \{L \mid \overline{L} \in \mathrm{DIFF}_m(C)\}$.

(d) $\mathrm{BH}(C)$, the boolean hierarchy over C, is $\bigcup_{m \geq 1} \mathrm{DIFF}_m$.

The relationship between the levels of the boolean hierarchy over Σ_k^p and bounded access to Σ_k^p is as follows. For each $k \geq 0$ and each $m \geq 0$,

$$\mathrm{P}_{m\text{-tt}}^{\Sigma_k^p} \subseteq \mathrm{DIFF}_{m+1}(\Sigma_k^p) \subseteq \mathrm{P}_{m+1\text{-tt}}^{\Sigma_k^p}$$
$$\subseteq \mathrm{coDIFF}_{m+1}(\Sigma_k^p) \subseteq$$

Now we can state what the earlier papers achieved (and, in doing so, those papers obtained as corollaries the results mentioned above).

Theorem 2. *1. ([15])   Let $m > 0$, $0 \leq i < j < k$, and $i < k - 2$. If $\mathrm{P}^{\Sigma_i^p[1]}\Delta \mathrm{DIFF}_m(\Sigma_k^p) = \mathrm{P}^{\Sigma_j^p[1]}\Delta \mathrm{DIFF}_m(\Sigma_k^p)$, then $\mathrm{DIFF}_m(\Sigma_k^p) = \mathrm{coDIFF}_m(\Sigma_k^p)$.*

2. ([5])   If $\mathrm{P}\Delta\Sigma_2^p = \mathrm{NP}\Delta\Sigma_2^p$, then $\Sigma_2^p = \Pi_2^p = \mathrm{PH}$.

3. ([25,24])   If $\Sigma_i^p\Delta\Sigma_k^p$ is closed under complementation, then the polynomial hierarchy collapses.[1]

---

[1] Selivanov [25,24] establishes only that the hierarchy collapses to a higher level, namely a level that contains Σ_{k+1}^p; thus this result is an upward translation of equality rather than a downward translation of equality.

In this paper, we unify all three of the above results—and achieve the strengthened corollary alluded to above (and stated later as Corollary 2) regarding the relative power of m and $m + 1$ queries to Σ_k^p—by proving the following downward translation of equality result:

Let $m > 0$ and $0 < i < k$. If $\Delta_i^p \Delta \mathrm{DIFF}_m(\Sigma_k^p) = \Sigma_i^p \Delta \mathrm{DIFF}_m(\Sigma_k^p)$, then $\mathrm{DIFF}_m(\Sigma_k^p) = \mathrm{coDIFF}_m(\Sigma_k^p)$.

3   A New Downward Translation of Equality

We first need a definition and a useful lemma.

Definition 2. *For any sets C and D, $C \tilde{\Delta} D = \{\langle x, y \rangle \mid x \in C \Leftrightarrow y \notin D\}$.*

Lemma 1. *C is $\leq_m^p$-complete for $\mathcal{C}$ and D is $\leq_m^p$-complete for $\mathcal{D}$, then $C \tilde{\Delta} D$ is $\leq_m^p$-hard for $\mathcal{C} \Delta \mathcal{D}$.*

Proof: Let $L \in \mathcal{C} \Delta \mathcal{D}$. We need to show that $L \leq_m^p C \tilde{\Delta} D$. Let $\widehat{C} \in \mathcal{C}$ and $\widehat{D} \in \mathcal{D}$ be such that $L = \widehat{C} \Delta \widehat{D}$. Let $\widehat{C} \leq_m^p C$ by f_C, and $\widehat{D} \leq_m^p D$ by f_D. Then $x \in L$ iff $x \in \widehat{C} \Delta \widehat{D}$, $x \in \widehat{C} \Delta \widehat{D}$ iff $(x \in \widehat{C} \Leftrightarrow x \notin \widehat{D})$, $(x \in \widehat{C} \Leftrightarrow x \notin \widehat{D})$ iff $(f_C(x) \in C \Leftrightarrow f_D(x) \notin D)$, and $(f_C(x) \in C \Leftrightarrow f_D(x) \notin D)$ iff $\langle f_C(x), f_D(x) \rangle \in C \tilde{\Delta} D$. ∎

We now state our main result. (Note that as both Δ_i^p and Σ_i^p contain both $\emptyset$ and Σ^*, it is clear that the classes involved in the first equality are at least as large as the classes involved in the second equality.)

Theorem 3. *Let $m > 0$ and $0 < i < k$. If $\Delta_i^p \Delta \mathrm{DIFF}_m(\Sigma_k^p) = \Sigma_i^p \Delta \mathrm{DIFF}_m(\Sigma_k^p)$, then $\mathrm{DIFF}_m(\Sigma_k^p) = \mathrm{coDIFF}_m(\Sigma_k^p)$.*

This result almost follows from Lemma 2—or, to be more accurate, most of its cases are corollaries of Lemma 2. However, the missing cases—which are by far the most challenging ones—need to be established, and Theorem 4 below does exactly that.

Lemma 2. *Let $m > 0$ and $0 < i < k - 1$. If $\Sigma_i^p \Delta \mathrm{DIFF}_m(\Sigma_k^p)$ is closed under complementation, then $\mathrm{DIFF}_m(\Sigma_k^p) = \mathrm{coDIFF}_m(\Sigma_k^p)$.*

We do not prove Lemma 2 here. It was first established in a precursor version of this paper [12], but alternately follows immediately from an even more general version[2] of Lemma 2 [13] that was built over that precursor. Note that Theorem 4 does not rely on Lemma 2.

Theorem 4. *Let $m > 0$ and $k > 1$. If $\Delta_{k-1}^p \Delta \mathrm{DIFF}_m(\Sigma_k^p) = \Sigma_{k-1}^p \Delta \mathrm{DIFF}_m(\Sigma_k^p)$, then $\mathrm{DIFF}_m(\Sigma_k^p) = \mathrm{coDIFF}_m(\Sigma_k^p)$.*

---

[2] Namely: Let $s, m > 0$ and $0 < i < k - 1$. If $\mathrm{DIFF}_s(\Sigma_i^p) \Delta \mathrm{DIFF}_m(\Sigma_k^p)$ is closed under complementation, then $\mathrm{DIFF}_m(\Sigma_k^p) = \mathrm{coDIFF}_m(\Sigma_k^p)$ [13].

Definition 3. *For each $k > 1$, choose any fixed problem that is $\leq_m^p$-complete for Σ_k^p and call it $L'_{\Sigma_k^p}$. Now, having fixed such sets, for each $k > 1$ choose one fixed set $L''_{\Sigma_{k-2}^p}$ that is in Σ_{k-2}^p and one fixed set $L_{\Sigma_{k-1}^p}$ that is $\leq_m^p$-complete for Σ_{k-1}^p and that satisfy[3] $L'_{\Sigma_k^p} = \{x \mid (\exists y \in \Sigma^{|x|})(\forall z \in \Sigma^{|x|})[\langle x, y, z\rangle \in L''_{\Sigma_{k-2}^p}]\}$ and $L_{\Sigma_{k-1}^p} = \{\langle x, y, z\rangle \mid |x| = |y| \wedge (\exists z')[(|x| = |y| = |zz'|) \wedge \langle x, y, zz'\rangle \notin L''_{\Sigma_{k-2}^p}]\}$.*

Proof of Theorem 4 Let $L_{\Sigma_{k-1}^p} \in \Sigma_{k-1}^p$ be as defined in Definition 3, and let $L_{\Delta_{k-1}^p}$ and $L_{\mathrm{DIFF}_m(\Sigma_k^p)}$ be any fixed $\leq_m^p$-complete sets for Δ_{k-1}^p and $\mathrm{DIFF}_m(\Sigma_k^p)$, respectively; such languages exist, e.g., via the standard canonical complete set constructions using enumerations of clocked machines. From Lemma 1 it follows that $L_{\Delta_{k-1}^p} \tilde{\Delta} L_{\mathrm{DIFF}_m(\Sigma_k^p)}$ is $\leq_m^p$-hard for $\Delta_{k-1}^p \Delta \mathrm{DIFF}_m(\Sigma_k^p)$. (Though this is not needed for this proof, we note in passing that it also can be easily seen to be in $\Delta_{k-1}^p \Delta \mathrm{DIFF}_m(\Sigma_k^p)$, and so it is in fact $\leq_m^p$-complete for $\Delta_{k-1}^p \Delta \mathrm{DIFF}_m(\Sigma_k^p)$.) Since $L_{\Sigma_{k-1}^p} \tilde{\Delta} L_{\mathrm{DIFF}_m(\Sigma_k^p)} \in \Sigma_{k-1}^p \Delta \mathrm{DIFF}_m(\Sigma_k^p)$ and by assumption $\Delta_{k-1}^p \Delta \mathrm{DIFF}_m(\Sigma_k^p) = \Sigma_{k-1}^p \Delta \mathrm{DIFF}_m(\Sigma_k^p)$, there exists a polynomial-time many-one reduction h from $L_{\Sigma_{k-1}^p} \tilde{\Delta} L_{\mathrm{DIFF}_m(\Sigma_k^p)}$ to $L_{\Delta_{k-1}^p} \tilde{\Delta} L_{\mathrm{DIFF}_m(\Sigma_k^p)}$ (in light of the latter's $\leq_m^p$-hardness). So, for all $x_1, x_2 \in \Sigma^*$: if $h(\langle x_1, x_2\rangle) = \langle y_1, y_2\rangle$, then $(x_1 \in L_{\Sigma_{k-1}^p} \Leftrightarrow x_2 \notin L_{\mathrm{DIFF}_m(\Sigma_k^p)})$ if and only if $(y_1 \in L_{\Delta_{k-1}^p} \Leftrightarrow y_2 \notin L_{\mathrm{DIFF}_m(\Sigma_k^p)})$. Equivalently, for all $x_1, x_2 \in \Sigma^*$: if $h(\langle x_1, x_2\rangle) = \langle y_1, y_2\rangle$, then

$$(x_1 \in L_{\Sigma_{k-1}^p} \Leftrightarrow x_2 \in L_{\mathrm{DIFF}_m(\Sigma_k^p)}) \text{ iff } (y_1 \in L_{\Delta_{k-1}^p} \Leftrightarrow y_2 \in L_{\mathrm{DIFF}_m(\Sigma_k^p)}). \quad (**)$$

We can use h to recognize some of $\overline{L_{\mathrm{DIFF}_m(\Sigma_k^p)}}$ by a $\mathrm{DIFF}_m(\Sigma_k^p)$ algorithm. In particular, we say that a string x is *easy for length n* if there exists a string x_1 such that $|x_1| \leq n$ and $(x_1 \in L_{\Sigma_{k-1}^p} \Leftrightarrow y_1 \notin L_{\Delta_{k-1}^p})$ where $h(\langle x_1, x\rangle) = \langle y_1, y_2\rangle$.

Let p be a fixed polynomial, which will be exactly specified later in the proof. We have the following algorithm to test whether $x \in \overline{L_{\mathrm{DIFF}_m(\Sigma_k^p)}}$ in the case that (our input) x is an easy string for $p(|x|)$. Guess x_1 with $|x_1| \leq p(|x|)$, let $h(\langle x_1, x\rangle) = \langle y_1, y_2\rangle$, and accept if and only if $(x_1 \in L_{\Sigma_{k-1}^p} \Leftrightarrow y_1 \notin L_{\Delta_{k-1}^p})$ and $y_2 \in L_{\mathrm{DIFF}_m(\Sigma_k^p)}$.[4] This algorithm is not necessarily a $\mathrm{DIFF}_m(\Sigma_k^p)$ algorithm, but it does inspire the following $\mathrm{DIFF}_m(\Sigma_k^p)$ algorithm to test whether $x \in \overline{L_{\mathrm{DIFF}_m(\Sigma_k^p)}}$ in the case that x is an easy string for $p(|x|)$.

Let $L_1, L_2, \ldots, L_m$ be languages in Σ_k^p such that $L_{\mathrm{DIFF}_m(\Sigma_k^p)} = L_1 - (L_2 - (L_3 - \cdots (L_{m-1} - L_m)\cdots))$ and $L_1 \supseteq L_2 \supseteq \cdots \supseteq L_{m-1} \supseteq L_m$ (this can be done, as it is simply the "telescoping" normal form of the levels of the boolean hierarchy over Σ_k^p, see [6,10,30]). For $1 \leq r \leq m$, define L'_r as the language

---

[3] By Stockmeyer's [26] standard quantifier characterization of the polynomial hierarchy's levels, there do exist sets satisfying this definition.

[4] To understand what is going on here, simply note that if $(x_1 \in L_{\Sigma_{k-1}^p} \Leftrightarrow y_1 \notin L_{\Delta_{k-1}^p})$ holds then by equation $(**)$ we have $x \in \overline{L_{\mathrm{DIFF}_m(\Sigma_k^p)}} \Leftrightarrow y_2 \in L_{\mathrm{DIFF}_m(\Sigma_k^p)}$. Note also that both of $x_1 \in L_{\Sigma_{k-1}^p}$ and $y_1 \notin L_{\Delta_{k-1}^p}$ can be very easily tested by a machine that has a Σ_{k-1}^p oracle.

accepted by the following Σ_k^p machine: On input x, guess x_1 with $|x_1| \leq p(|x|)$, let $h(\langle x_1, x \rangle) = \langle y_1, y_2 \rangle$, and accept if and only if $(x_1 \in L_{\Sigma_{k-1}^p} \Leftrightarrow y_1 \notin L_{\Delta_{k-1}^p})$ and $y_2 \in L_r$.

Note that $L_r' \in \Sigma_k^p$ for each r, and that $L_1' \supseteq L_2' \supseteq \cdots \supseteq L_{m-1}' \supseteq L_m'$. We will show that if x is an easy string for length $p(|x|)$, then $x \in \overline{L_{\text{DIFF}_m(\Sigma_k^p)}}$ if and only if $x \in L_1' - (L_2' - (L_3' - \cdots (L_{m-1}' - L_m') \cdots))$.

So suppose that x is an easy string for $p(|x|)$. Define r' to be the unique integer such that (a) $0 \leq r' \leq m$, (b) $x \in L_s'$ for $1 \leq s \leq r'$, and (c) $x \notin L_s'$ for $s > r'$. It is immediate that $x \in L_1' - (L_2' - (L_3' - \cdots (L_{m-1}' - L_m') \cdots))$ if and only if r' is odd.

Let w be some string such that $(\exists x_1 \in (\Sigma^*)^{\leq p(|x|)})(\exists y_1)[h(\langle x_1, x \rangle) = \langle y_1, w \rangle \wedge (x_1 \in L_{\Sigma_{k-1}^p} \Leftrightarrow y_1 \notin L_{\Delta_{k-1}^p})]$, and $w \in L_{r'}$ if $r' > 0$. Note that such a w exists, since x is easy for $p(|x|)$. By the definition of r' (namely, since $x \notin L_s'$ for $s > r'$), $w \notin L_s$ for all $s > r'$. It follows that $w \in L_{\text{DIFF}_m(\Sigma_k^p)}$ if and only if r' is odd.

It is clear, keeping in mind the definition of h, that $x \in \overline{L_{\text{DIFF}_m(\Sigma_k^p)}}$ iff $w \in L_{\text{DIFF}_m(\Sigma_k^p)}$, $w \in L_{\text{DIFF}_m(\Sigma_k^p)}$ iff r' is odd, and r' is odd iff $x \in L_1' - (L_2' - (L_3' - \cdots (L_{m-1}' - L_m') \cdots))$. This completes the case where x is easy, as $L_1' - (L_2' - (L_3' - \cdots (L_{m-1}' - L_m') \cdots))$ in effect specifies a $\text{DIFF}_m(\Sigma_k^p)$ algorithm.

We say that x is *hard for length* n if $|x| \leq n$ and x is not easy for length n, i.e., if $|x| \leq n$ and for all x_1 with $|x_1| \leq n$, $(x_1 \in L_{\Sigma_{k-1}^p} \Leftrightarrow y_1 \in L_{\Delta_{k-1}^p})$, where $h(\langle x_1, x \rangle) = \langle y_1, y_2 \rangle$. Note that if x is hard for $p(|x|)$, then $x \notin L_1'$.

If x is a hard string for length $p(|x|)$, then x induces a many-one reduction from $\left(L_{\Sigma_{k-1}^p} \right)^{\leq p(|x|)}$ to $L_{\Delta_{k-1}^p}$, namely, $f(x_1) = y_1$, where $h(\langle x_1, x \rangle) = \langle y_1, y_2 \rangle$. (Note that f is computable in time polynomial in $\max(|x|, |x_1|)$.) So it is not hard to see that if we choose p appropriately large, then a hard string x for $p(|x|)$ induces Σ_{k-1}^p algorithms for $(L_1)^{=|x|}, (L_2)^{=|x|}, \ldots, (L_m)^{=|x|}$ (essentially since each is in $\Sigma_k^p = \text{NP}^{\Sigma_{k-1}^p}$, $L_{\Sigma_{k-1}^p}$ is $\leq_m^p$-complete for Σ_{k-1}^p, and $\text{NP}^{\Delta_{k-1}^p} = \Sigma_{k-1}^p$), which we can use to obtain a $\text{DIFF}_m(\Sigma_{k-1}^p)$ algorithm for $L_{\text{DIFF}_m(\Sigma_k^p)}$, and thus certainly a $\text{DIFF}_m(\Sigma_k^p)$ algorithm for $\left(\overline{L_{\text{DIFF}_m(\Sigma_k^p)}} \right)^{=|x|}$.

However, there is a problem. The problem is that we cannot combine the $\text{DIFF}_m(\Sigma_k^p)$ algorithms for easy and hard strings into one $\text{DIFF}_m(\Sigma_k^p)$ algorithm for $\overline{L_{\text{DIFF}_m(\Sigma_k^p)}}$ that works all strings. Why? It is too difficult to decide whether a string is easy or hard; to decide this deterministically takes one query to Σ_k^p, and we cannot do that in a $\text{DIFF}_m(\Sigma_k^p)$ algorithm. This is also the reason why the methods from [15] failed to prove that if $\text{P}\Delta\Sigma_2^p = \text{NP}\Delta\Sigma_2^p$, then $\Sigma_2^p = \Pi_2^p$. Recall from the introduction that the latter theorem was proven by Buhrman and Fortnow [5]. We will use their technique at this point. The following lemma, which we will prove after we have finished the proof of this theorem, states a generalized version of the technique from [5]. It has been generalized to deal with arbitrary levels of the polynomial hierarchy and to be useful in settings involving boolean hierarchies.

Lemma 3. *Let $k > 1$. For all $L \in \Sigma_k^p$, there exist a polynomial q and a set $\widehat{L} \in \Pi_{k-1}^p$ such that*

1. *for each natural number n', $q(n') \geq n'$,*
2. *$\widehat{L} \subseteq \overline{L}$, and*
3. *if x is hard for $q(|x|)$, then $x \in \overline{L}$ iff $x \in \widehat{L}$.*

Due to space limitations, we refer the reader to the full version of this paper for the proof of Lemma 3 [12]. From Lemma 3, it follows that there exist sets $\widehat{L_1}, \widehat{L_2}, \ldots, \widehat{L_m} \in \Pi_{k-1}^p$ and polynomials $q_1, q_2, \ldots, q_m$ with the following properties for all $1 \leq r \leq m$:

1. $\widehat{L_r} \subseteq \overline{L_r}$, and
2. if x is hard for $q_r(|x|)$, then $x \in \overline{L_r}$ iff $x \in \widehat{L_r}$.

Take p to be an (easy-to-compute—we may without loss of generality require that there is an ℓ such that it is of the form $n^\ell + \ell$) polynomial such that p is at least as large as all the q_rs, i.e., such that, for each natural number n', we have $p(n') \geq \max\{q_1(n'), \ldots, q_m(n')\}$. By the definition of hardness and condition 1 of Lemma 3, if x is hard for $p(|x|)$ then x is hard for $q_r(|x|)$ for all $1 \leq r \leq m$. As promised earlier, we have now specified p. Define $\widehat{L}_{\text{DIFF}_m(\Sigma_k^p)}$ as follows: On input x, guess r, r even, $0 \leq r \leq m$, and accept if and only if both (a) $x \in L_r$ or $r = 0$, and (b) if $r < m$ then $x \in \widehat{L_{r+1}}$. Clearly, $\widehat{L}_{\text{DIFF}_m(\Sigma_k^p)} \in \Sigma_k^p$. In addition, this set inherits certain properties from the $\widehat{L_r}$s. In particular, in light of the definition of $\widehat{L}_{\text{DIFF}_m(\Sigma_k^p)}$, the definitions of the $\widehat{L_r}$s, and the fact that:

$$x \in \overline{L_{\text{DIFF}_m(\Sigma_k^p)}} \text{ iff for some even } r, \ 0 \leq r \leq m, \text{ we have: } (x \in L_r \text{ or } r = 0) \text{ and } (x \in \overline{L_{r+1}} \text{ or } r = m),$$

we have that the following properties hold: (1) $\widehat{L}_{\text{DIFF}_m(\Sigma_k^p)} \subseteq \overline{L_{\text{DIFF}_m(\Sigma_k^p)}}$, and (2) if x is hard for $p(|x|)$, then $x \in \overline{L_{\text{DIFF}_m(\Sigma_k^p)}}$ iff $x \in \widehat{L}_{\text{DIFF}_m(\Sigma_k^p)}$.

Finally, we are ready to give the algorithm. Recall that $L_1', L_2', \ldots L_m'$ are sets in Σ_k^p such that: (1) $L_1' \supseteq L_2' \supseteq \cdots \supseteq L_{m-1}' \supseteq L_m'$, and (2) if x is easy for $p(|x|)$, then $x \in \overline{L_{\text{DIFF}_m(\Sigma_k^p)}}$ if and only if $x \in L_1' - (L_2' - (L_3' - \cdots (L_{m-1}' - L_m') \cdots))$, and (3) if x is hard for $p(|x|)$, then $x \notin L_1'$. We claim that for all x, $x \in \overline{L_{\text{DIFF}_m(\Sigma_k^p)}}$ iff $x \in (L_1' \cup \widehat{L}_{\text{DIFF}_m(\Sigma_k^p)}) - (L_2' - (L_3' - \cdots (L_{m-1}' - L_m') \cdots))$, which completes the proof of Theorem 4, as Σ_k^p is closed under union.

($\Rightarrow$):   If x is easy for $p(|x|)$, then $x \in L_1' - (L_2' - (L_3' - \cdots (L_{m-1}' - L_m') \cdots))$, and so certainly $x \in (L_1' \cup \widehat{L}_{\text{DIFF}_m(\Sigma_k^p)}) - (L_2' - (L_3' - \cdots (L_{m-1}' - L_m') \cdots))$. If x is hard for $p(|x|)$, then $x \in \widehat{L}_{\text{DIFF}_m(\Sigma_k^p)}$ and $x \notin L_r'$ for all r (since $x \notin L_1'$ and $L_1' \supseteq L_2' \supseteq \cdots$). Thus, $x \in (L_1' \cup \widehat{L}_{\text{DIFF}_m(\Sigma_k^p)}) - (L_2' - (L_3' - \cdots (L_{m-1}' - L_m') \cdots))$.

($\Leftarrow$):   Suppose $x \in (L_1' \cup \widehat{L}_{\text{DIFF}_m(\Sigma_k^p)}) - (L_2' - (L_3' - \cdots (L_{m-1}' - L_m') \cdots))$. If $x \in \widehat{L}_{\text{DIFF}_m(\Sigma_k^p)}$, then $x \in \overline{L_{\text{DIFF}_m(\Sigma_k^p)}}$. If $x \notin \widehat{L}_{\text{DIFF}_m(\Sigma_k^p)}$, then $x \in L_1' - (L_2' - (L_3' - \cdots (L_{m-1}' - L_m') \cdots))$ and so x must be easy for $p(|x|)$ (as $x \in L_1'$, and this is possible only if x is easy for $p(|x|)$). However, this says that $x \in \overline{L_{\text{DIFF}_m(\Sigma_k^p)}}$. ∎

4   Conclusions

We have proven a general downward translation of equality sufficient to yield, as a corollary:

Corollary 2. *For each $m > 0$ and each $k > 1$ it holds that:*

$$P^{\Sigma_k^p}_{m\text{-tt}} = P^{\Sigma_k^p}_{m+1\text{-tt}} \Rightarrow \mathrm{DIFF}_m(\Sigma_k^p) = \mathrm{coDIFF}_m(\Sigma_k^p).$$

Corollary 2 itself has an interesting further consequence. From this corollary, it follows that for a number of previously missing cases (namely, when $m > 1$ and $k = 2$), the hypothesis $P^{\Sigma_k^p}_{m\text{-tt}} = P^{\Sigma_k^p}_{m+1\text{-tt}}$ implies that the polynomial hierarchy collapses to about one level lower in the boolean hierarchy over Σ_{k+1}^p than could be concluded from previous papers. This is because we can, thanks to Corollary 2, when given $P^{\Sigma_k^p}_{m\text{-tt}} = P^{\Sigma_k^p}_{m+1\text{-tt}}$, invoke the powerful collapses of the polynomial hierarchy that are known to follow from $\mathrm{DIFF}_m(\Sigma_k^p) = \mathrm{coDIFF}_m(\Sigma_k^p)$. Regarding what collapses do follow from $\mathrm{DIFF}_m(\Sigma_k^p) = \mathrm{coDIFF}_m(\Sigma_k^p)$, a long line of research started by Kadin and Wagner a decade ago has studied that, and the strongest currently known connection was recently obtained, independently, by Hemaspaandra et al. and by Reith and Wagner [14,22], namely, they proved: For all $m > 0$ and all $k > 0$, if $\mathrm{DIFF}_m(\Sigma_k^p) = \mathrm{coDIFF}_m(\Sigma_k^p)$ then $\mathrm{PH} = \mathrm{DIFF}_m(\Sigma_k^p)\Delta\mathrm{DIFF}_{m-1}(\Sigma_{k+1}^p)$. Putting all the above together, one sees that, for all cases where $m > 1$ and $k > 1$, $P^{\Sigma_k^p}_{m\text{-tt}} = P^{\Sigma_k^p}_{m+1\text{-tt}}$ implies that the polynomial hierarchy collapses to $\mathrm{DIFF}_m(\Sigma_k^p)\Delta\mathrm{DIFF}_{m-1}(\Sigma_{k+1}^p)$; of course, for the case $m = 1$, we already know [15,5] that, for $k > 1$, if $P^{\Sigma_k^p[1]} = P^{\Sigma_k^p[2]}$, then $\Sigma_k^p = \Pi_k^p = \mathrm{PH}$.

Acknowledgments: We thank the referees for helpful comments.

References

[1] E. Allender. Limitations of the upward separation technique. *Mathematical Systems Theory*, 24(1):53–67, 1991.

[2] E. Allender and C. Wilson. Downward translations of equality. *Theoretical Computer Science*, 75(3):335–346, 1990.

[3] R. Beigel. Bounded queries to SAT and the boolean hierarchy. *Theoretical Computer Science*, 84(2):199–223, 1991.

[4] R. Beigel, R. Chang, and M. Ogiwara. A relationship between difference hierarchies and relativized polynomial hierarchies. *Mathematical Systems Theory*, 26(3):293–310, 1993.

[5] H. Buhrman and L. Fortnow. Two queries. In *Proceedings of the 13th Annual IEEE Conference on Computational Complexity*, pages 13–19. IEEE Computer Society Press, June 1998.

[6] J. Cai, T. Gundermann, J. Hartmanis, L. Hemachandra, V. Sewelson, K. Wagner, and G. Wechsung. The boolean hierarchy I: Structural properties. *SIAM Journal on Computing*, 17(6):1232–1252, 1988.

[7] J. Cai, T. Gundermann, J. Hartmanis, L. Hemachandra, V. Sewelson, K. Wagner, and G. Wechsung. The boolean hierarchy II: Applications. *SIAM Journal on Computing*, 18(1):95–111, 1989.

[8] R. Chang and J. Kadin. The boolean hierarchy and the polynomial hierarchy: A closer connection. *SIAM Journal on Computing*, 25(2):340–354, 1996.

[9] J. Hartmanis, N. Immerman, and V. Sewelson. Sparse sets in NP–P: EXPTIME versus NEXPTIME. *Information and Control*, 65(2/3):159–181, 1985.

[10] F. Hausdorff. *Grundzüge der Mengenlehre*. Leipzig, 1914.

[11] E. Hemaspaandra, L. Hemaspaandra, and H. Hempel. An introduction to query order. *Bulletin of the EATCS*, 63:93–107, 1997.

[12] E. Hemaspaandra, L. Hemaspaandra, and H. Hempel. Translating equality downwards. Technical Report TR-657, Department of Computer Science, University of Rochester, Rochester, NY, April 1997.

[13] E. Hemaspaandra, L. Hemaspaandra, and H. Hempel. Downward collapse from a weaker hypothesis. In *Proceedings of the 6th Italian Conference on Theoretical Computer Science*. World Scientific Press, November 1998. To appear.

[14] E. Hemaspaandra, L. Hemaspaandra, and H. Hempel. What's up with downward collapse: Using the easy-hard technique to link boolean and polynomial hierarchy collapses. *SIGACT News*, 29(3):10–22, 1998.

[15] E. Hemaspaandra, L. Hemaspaandra, and H. Hempel. A downward collapse within the polynomial hierarchy. *SIAM Journal on Computing*, 28(2):383–393, 1999.

[16] L. Hemaspaandra and S. Jha. Defying upward and downward separation. *Information and Computation*, 121(1):1–13, 1995.

[17] J. Kadin. The polynomial time hierarchy collapses if the boolean hierarchy collapses. *SIAM Journal on Computing*, 17(6):1263–1282, 1988. Erratum appears in the same journal, 20(2):404.

[18] J. Köbler, U. Schöning, and K. Wagner. The difference and truth-table hierarchies for NP. *RAIRO Theoretical Informatics and Applications*, 21:419–435, 1987.

[19] R. Ladner, N. Lynch, and A. Selman. A comparison of polynomial time reducibilities. *Theoretical Computer Science*, 1(2):103–124, 1975.

[20] A. Meyer and L. Stockmeyer. The equivalence problem for regular expressions with squaring requires exponential space. In *Proceedings of the 13th IEEE Symposium on Switching and Automata Theory*, pages 125–129, 1972.

[21] R. Rao, J. Rothe, and O. Watanabe. Upward separation for FewP and related classes. *Information Processing Letters*, 52(4):175–180, 1994.

[22] S. Reith and K. Wagner. On boolean lowness and boolean highness. In *Proceedings of the 4th Annual International Computing and Combinatorics Conference*, pages 147–156. Springer-Verlag *Lecture Notes in Computer Science #1449*, August 1998.

[23] P. Rohatgi. Saving queries with randomness. *Journal of Computer and System Sciences*, 50(3):476–492, 1995.

[24] V. Selivanov. Two refinements of the polynomial hierarchy. In *Proceedings of the 11th Annual Symposium on Theoretical Aspects of Computer Science*, pages 439–448. Springer-Verlag *Lecture Notes in Computer Science #775*, February 1994.

[25] V. Selivanov. Fine hierarchies and boolean terms. *Journal of Symbolic Logic*, 60(1):289–317, 1995.

[26] L. Stockmeyer. The polynomial-time hierarchy. *Theoretical Computer Science*, 3:1–22, 1977.

[27] K. Wagner. Number-of-query hierarchies. Technical Report 158, Institut für Mathematik, Universität Augsburg, Augsburg, Germany, October 1987.

[28] K. Wagner. Number-of-query hierarchies. Technical Report 4, Institut für Informatik, Universität Würzburg, Würzburg, Germany, February 1989.

[29] K. Wagner. A note on parallel queries and the symmetric-difference hierarchy. *Information Processing Letters*, 66(1):13–20, 1998.

[30] G. Wechsung. On the boolean closure of NP. In *Proceedings of the 5th Conference on Fundamentals of Computation Theory*, pages 485–493. Springer-Verlag *Lecture Notes in Computer Science #199*, 1985. (An unpublished precursor of this paper was coauthored by K. Wagner).

Sparse Sets, Approximable Sets, and Parallel Queries to NP

Vikraman Arvind[1] and Jacobo Torán[2]

[1] Institute of Mathematical Sciences
C. I. T. Campus
Chennai 600 113, India
arvind@imsc.ernet.in

[2] Abteilung Theoretische Informatik,
Universität Ulm,
D-89069 Ulm, Germany
toran@informatik.uni-ulm.de

Abstract. We show that if an NP-complete set or a coNP-complete set is polynomial-time disjunctive truth-table reducible to a sparse set then $FP_{||}^{NP} = FP^{NP}[\log]$. Similarly, we show that if SAT is $O(\log n)$-approximable then $FP_{||}^{NP} = FP^{NP}[\log]$. Since $FP_{||}^{NP} = FP^{NP}[\log]$ implies that SAT is $O(\log n)$-approximable [BFT97], it follows from our result that these two hypotheses are equivalent. We also show that if an NP-complete set or a coNP-complete set is disjunctively reducible to a sparse set of polylogarithmic density then, in fact, P = NP.

1 Introduction

The study of the existence of sparse hard sets for complexity classes has occupied complexity theorists for over two decades. The first results in this area were motivated by the Berman-Hartmanis isomorphism conjecture [BH77] and by the study of connections between uniform and nonuniform complexity classes [KL80]. The focus shifted to proving, for various reducibilities[1] (whose strengths lie between the many-one and the Turing reducibility), that P = NP is equivalent to SAT being reducible to a sparse set via such a reducibility. It is now known (see the recent survey [CO97]) for several reducibilities that P = NP is equivalent to SAT being reducible to a sparse set via such a reducibility: a well-known example here is the result for the case of bounded truth-table reducibility [OW91]. However, it remains a challenging open problem to prove that P = NP if there is a sparse Turing-hard set for NP. Indeed, this question remains open even for reducibilities stronger than the Turing reducibility.

In this paper we consider the question of existence of sparse hard sets for NP w. r. t. *disjunctive truth-table* reductions. We briefly recall some known results: it is shown in [AKM96] that if there is a sparse hard set for NP under disjunctive reductions then PH collapses to Δ_2^p. More recently, it is shown in [CNS96]

---

[1] All reducibilities considered in this paper are polynomial-time computable.

C. Meinel and S. Tison (Eds.): STACS'99, LNCS 1563, pp. 281–290, 1999.
© Springer-Verlag Berlin Heidelberg 1999

that if there are sparse hard sets for NP under the disjunctive reducibility then RP = NP. The proof technique in [CNS96] is based, in turn, on powerful algebraic techniques from [CS95] tailored for application in the area of reductions to sparse sets. With these techniques some long standing conjectures of Hartmanis regarding logspace and NC^1 reductions to sparse sets have recently been settled (the survey [CO97] contains a nice overview of these results).

A Summary of the Results

The main contribution of this paper is to relate the question of the existence of sparse hard sets for NP under disjunctive reductions to other, apparently different, hypotheses in complexity theory considered in the recent work of Buhrman et. al. [BFT97]. Among these are the following.

(1) P = NP.
(2) $FP_{||}^{NP} = FP^{NP}[\log]$.
(3) SAT is $O(\log n)$ approximable.[2]
(4) $(1SAT, SAT)$ has a solution in P.

Clearly, hypothesis (1) implies the others. Furthermore, it is shown that (2) implies (3) in [BFT97]. More recently, Sivakumar [Si98], using algebraic techniques from [ALRS92], has shown that (3) implies (4).

It is known that RP = NP follows from (4) [VV86], and it is an outstanding open problem in structural complexity whether P = NP follows from any of the hypotheses (2), (3) or (4). This is the main motivation for studying them. Cai, Naik and Sivakumar (in the technical report version of [CNS96]) have shown that if SAT is disjunctively reducible to a sparse set then hypothesis (4) holds.

Building on the technique of [CNS96] we prove the following new results:

- If SAT or $\overline{SAT}$ is disjunctively reducible to a sparse set then $FP_{||}^{NP} = FP^{NP}[\log]$.
- For any prime k, if $Mod_k P$ is disjunctively reducible to a sparse set then $(1SAT, SAT)$ has a solution in P.

There are collapse results that follow from hypothesis (2) that are not known to follow from (4). For example, in [JT95] it is shown that if $FP_{||}^{NP} = FP^{NP}[\log]$ then a polylogarithmic amount of nondeterminism can be simulated in polynomial time. Combining this with our result yields as corollary that if SAT or $\overline{SAT}$ is disjunctively reducible to a sparse set of polylogarithmic density then P = NP. With related techniques we observe some consequences of SAT being majority reducible to a sparse set that are discussed in the last section of the paper.

With arguments similar to those we use for proving the above results on sparse sets, but now combined with Sivakumar's technique used in [Si98], we show:

- If SAT is $O(\log n)$ approximable then $FP_{||}^{NP} = FP^{NP}[\log]$.

---

[2] SAT is the canonical NP-complete set, defined as the set of satisfiable Boolean formulas.

This proves that hypotheses (2) and (3) are equivalent, answering an open question in [BFT97]. From these results we conclude that both these hypotheses are at least as weak as SAT being disjunctively reducible to a sparse set.

2 Preliminaries

We fix the alphabet $\Sigma = \{0, 1\}$. The set $\bigcup_{0 \le i \le n} \Sigma^i$ of all strings in Σ^* of length up to n is denoted by $\Sigma^{\le n}$. For any set $A \subseteq \Sigma^*$, $A^{\le n} = A \cap \Sigma^{\le n}$, and $A^{=n} = A \cap \Sigma^n$. χ_A denotes the characteristic function of A. By abuse of notation, let $\chi_A(x_1, x_2, \ldots, x_m)$ denote the function that maps the list of strings $x_1, x_2, \ldots, x_m$ to the m-bit vector whose ith bit is $\chi_A(x_i)$. The length of a string x is denoted by $|x|$, and the cardinality of a set A is denoted by $||A||$. The density function of a set A is defined as $d_A(n) = ||A^{\le n}||$. A set S is *sparse* if its density function is bounded above by a polynomial. A sparse set has *polylog density* if its density function is bounded above by $\log^k n$ for some constant $k > 0$. The complement of a language A is denoted by $\overline{A}$. Let $\langle \cdot, \cdot \rangle$ denote a standard polynomial-time computable, one to one and polynomial-time invertible pairing function which can be extended in a standard fashion to encode arbitrary sequences $(x_1, \ldots, x_k)$ of strings into a string $\langle x_1, \ldots, x_k \rangle$.

All reducibilities in this paper are polynomial-time computable. Apart from the many-one reducibility, we consider the disjunctive truth-table reducibility: A set A is *disjunctively reducible* to a set B, if there is a polynomial-time computable function f mapping strings to sets of strings such that for all $x \in \Sigma^*$ it holds that $x \in A \iff f(x) \cap B \ne \emptyset$. Let SAT denote the set of satisfiable Boolean formulas. We next define promise problems.

Definition 1. [ESY84] *A promise problem is a pair of sets (Q,R). A set L is called a solution of the promise problem (Q,R) if for all $x \in Q$, $x \in L \Leftrightarrow x \in R$.*

Of particular interest to us is the promise problem $(1\text{SAT}, \text{SAT})$, where 1SAT contains precisely those Boolean formulas which have at most one satisfying assignment. Observe that any solution of the promise problem $(1\text{SAT}, \text{SAT})$ has to agree with SAT in the formulas having a unique satisfying assignment as well as in the unsatisfiable formulas. We next recall the definition of approximable sets.[3]

Definition 2. [BKS95] *A function g is an f-approximator for a set A if for every $x_1, x_2, \ldots, x_m$ with $m \ge f(max_i |x_i|)$,*

$$g(x_1, x_2, \ldots, x_m) \in \Sigma^m, \quad and$$

$$g(x_1, x_2, \ldots, x_m) \ne \chi_A(x_1, x_2, \ldots, x_m).$$

A set A is called f-approximable if it has an f-approximator.

---

[3] Approximability is called membership comparability in [Og95].

$FP_{||}^{NP}$ denotes the class of functions computable in polynomial time with parallel queries to an NP oracle and $FP^{NP}[log]$ denotes the class of functions computable in polynomial time with logarithmically many adaptive queries to an NP oracle.

Other notions from complexity theory used in this paper can be found in standard textbooks like [BDG88, Pa94].

3   Sparse Sets and Parallel Queries to NP

As preparation for the first result we prove the following lemma which is essentially implicit in [CNS96]. We are interested in solving the following decoding problem which we call the *hidden polynomial* problem: We are given as input a prime q, integers t, n and N such that $q \geq (n+1)t$, and $R_1, R_2, \ldots, R_N$, where $R_i \subseteq F_q \times F_q$ with the promise that there exists a polynomial P over F_q of degree at most n and a subset $I \subseteq \{1, 2, \ldots, N\}$ of size t such that $\cup_{i \in I} R_i = \mathrm{graph}(P)$. We want to compute a small set of polynomials over F_q containing P (with each polynomial output as a list of its coefficients).

Henceforth, we refer to the collection $R_1, R_2, \ldots, R_N$ as a *table* and the R_i's as its *rows*. Observe that the rows do not have to be all of the same size. We call a row R_i *correct* w. r. t. the polynomial P if $R_i \subseteq \mathrm{graph}(P)$.

The following lemma (based on [CNS96]) gives a precise answer to the above decoding problem.

Lemma 1. *There is an algorithm (that runs in time polynomial in n, N, and q) that takes as input q, t, n, N, and a table $T = \{R_1, R_2, \ldots, R_N\}$ of N rows as described above, and outputs a list of at most N polynomials, one of which is the hidden polynomial.*

Proof. Notice that there are exactly q^2 pairs (u, v), $u, v \in F_q$ of which exactly q pairs correctly define the graph of the hidden polynomial P. Since there is a set of t correct rows in the table T which completely specify the polynomial, by pigeon-hole principle there is one correct row which contains at least q/t pairs. Furthermore, notice that no correct row contains *inconsistent* pairs (u, v) and (u, w) where $v \neq w$.

Call a row of the table *long* if it has at least q/t pairs and does not contain any inconsistent pair. We know that there is at least one correct row which is long.

Writing the hidden polynomial $P(x)$ as $\sum_{i=0}^{n} a_i x^i$ we notice that each long row gives us a system of at least q/t linear equations in the $n+1$ unknowns $a_i, 0 \leq i \leq n$. Since $q/t \geq n+1$, we can pick any $n+1$ of the equations corresponding to a given long row which will have a unique solution in the a_i's since the coefficient matrix is a Vandermonde matrix which is invertible. Using Gaussian elimination we can efficiently compute this unique solution for each long row.

This yields a list of at most N polynomials, one for each long row in the table T, and we know that the hidden polynomial (corresponding to a correct long row of T) is in this list.

We prove now the first result of the paper.

Theorem 1. *If* SAT *is disjunctively reducible to a sparse set then* $\text{FP}^{\text{NP}}_{||} = \text{FP}^{\text{NP}}[\log]$.

Proof. We show that the function χ_{SAT}, which computes the characteristic sequence of a list of SAT queries, can be computed in $\text{FP}^{\text{NP}}[\log]$ under the assumption that SAT is disjunctively reducible to a sparse set. Since χ_{SAT} is complete for $\text{FP}^{\text{NP}}_{||}$, this clearly proves the equality $\text{FP}^{\text{NP}}_{||} = \text{FP}^{\text{NP}}[\log]$.

We now design an $\text{FP}^{\text{NP}}[\log]$ machine M for χ_{SAT}. Given a list of formulas $(x_1, x_2, \ldots, x_m)$ as input, the machine M first computes, with a binary search and queries to a suitable NP oracle, the cardinality k of $\{x_1, x_2, \ldots, x_m\} \cap \text{SAT}$.

Now consider the following set $Y = \{\langle q, u, v, k, x_1, x_2, \ldots, x_m \rangle \mid 0 \leq u, v \leq q - 1,$ and $\exists a \in \Sigma^m$ with k 1's such that if $a_i = 1$ then $x_i \in \text{SAT}$ and $\sum_{i=1}^{m} a_i u^{i-1} \equiv v \pmod{q}\}$.

Notice that $Y \in \text{NP}$. Also, observe now that if k is $||\{x_1, x_2, \ldots, x_m\} \cap \text{SAT}||$ then there is a unique vector $a \in \Sigma^m$ such that if $a_i = 1$ then $x_i \in \text{SAT}$. Thus, for a given triple q, u, v there is at most one vector $a \in \Sigma^m$ satisfying the above property.

Actually, we are interested only in those instances $\langle q, u, v, k, x_1, x_2, \ldots, x_m \rangle$ of Y where q is a small prime. More precisely, consider an instance $(x_1, x_2, \ldots, x_m)$ of χ_{SAT} of length n. Corresponding to this instance, we pick q to be a $c \log n$ bit prime number, where we will choose c later appropriately. Let F_q denote the finite field of size q. Notice that we can pick q and construct the field F_q efficiently (i.e. in time polynomial in n). Moreover, arithmetic in F_q can also be done efficiently.

Since $Y \in \text{NP}$ there is a disjunctive reduction f from Y to a sparse set S of density $||S^{\leq n}|| \leq p(n)$ for some polynomial p. I.e. f is an FP function that on input x produces a set of strings $f(x)$ such that $x \in Y$ iff $f(x) \cap S \neq \emptyset$. The length of $\langle q, u, v, k, x_1, x_2, \ldots, x_m \rangle$ using a standard pairing function can be bounded by $2n$ for large enough n since q, u, v and k can be encoded in $3c \log n + \log n$ bits. Now, since the reduction f from Y to S is polynomial-time computable there is a polynomial $r(n)$ which bounds both $||f(\langle q, u, v, k, x_1, x_2, \ldots, x_m \rangle)||$ and the length of each query in $f(\langle q, u, v, k, x_1, x_2, \ldots, x_m \rangle)$. Our aim is to apply Lemma 1. Let $Q = \bigcup_{u,v \in F_q} f(\langle q, u, v, k, x_1, x_2, \ldots, x_m \rangle)$. Write $Q = \{q_1, q_2, \ldots, q_N\}$. Build a table T with N rows R_i where (u, v) is in R_i if $q_i \in f(\langle q, u, v, k, x_1, x_2, \ldots, x_m \rangle)$. Note that $N \leq r(n)q^2$. We define row R_i to be *correct* if $q_i \in S$. Notice that there are at most $||S^{\leq r(n)}|| \leq r(n)p(r(n))$ correct rows in T. Now we choose the constant c (which determines the size of the field F_q) so that $q/r(n)p(r(n)) = n^c/r(n)p(r(n)) > n$, where we know that $n \geq m$.

Let $\chi_{\text{SAT}}(x_1, x_2, \ldots, x_m) = a_1 a_2 \ldots a_m$. Then we claim that a hidden polynomial specified by T is $\sum_{i=1}^{m} a_i x^{i-1}$. To see that the conditions of Lemma 1 are fulfilled, notice that each R_i is indeed correct w. r. t. this polynomial and $\bigcup_{q_i \in S} R_i$ contains precisely the q pairs (u, v) such that $\sum_{i=1}^{m} a_i u^{i-1} \equiv v \pmod{q}$. Applying the algorithm of Lemma 1 we can compute in time polynomial in n a list X of at most N polynomials of degree $m - 1$. Each of these polynomials gives us an m-bit vector of its coefficients. We discard from this list those m-bit vectors which have a number of 1's different from k. In the pruned list exactly one m-bit vector is $\chi_{\mathrm{SAT}}(x_1, x_2, \ldots, x_m)$ and every other m-bit vector has a 1 at a position where the corresponding formula in $(x_1, x_2, \ldots, x_m)$ is unsatisfiable.

The $\mathrm{FP}^{\mathrm{NP}}[\log]$ machine M can now find the *unique* correct m-bit vector in X by doing a standard binary search guided by at most $\log N = O(\log n)$ queries to a suitable NP oracle.

It is an open question if $P = NP$ can be derived from the assumption that SAT is disjunctively reducible to a sparse set. A first step here is to consider sparse sets of density lower than polynomial. It is known that if SAT is disjunctively reducible to a tally set then $P = NP$ [Uk83, Ya83]. However, the techniques of [Uk83, Ya83] do not work if we assume that SAT is disjunctively reducible to a set S of, say, polylog density. Reductions of SAT to sets of polylog density were considered by Buhrman and Hermo [BH95] where they showed that if SAT is Turing reducible to a set of polylog density then $NP(\log^k n) = NP$ for all k (where $NP(\log^k n)$ is the class of NP languages accepted by NP machines which make at most $\log^k n$ nondeterministic moves on inputs of length n). We recall here the result of Jenner and Torán [JT95] that if $\mathrm{FP}^{\mathrm{NP}}_{||} = \mathrm{FP}^{\mathrm{NP}}[\log]$ then $NP(\log^k n) = P$ for each $k > 0$. Combining the above-mentioned results of [BH95, JT95] with Theorem 1 immediately yields the following corollary.

Corollary 1. *If* SAT *is disjunctively reducible to a set of polylog density then* $P = NP$.

The question whether SAT is conjunctively reducible to a co-sparse set is, by complementation, equivalent to $\overline{\mathrm{SAT}}$ being disjunctively reducible to a sparse set. We show that we can apply again the technique of [CNS96] to derive $\mathrm{FP}^{\mathrm{NP}}_{||} = \mathrm{FP}^{\mathrm{NP}}[\log]$ as a consequence.

Theorem 2. *If* $\overline{\mathrm{SAT}}$ *is disjunctively reducible to a sparse set then* $\mathrm{FP}^{\mathrm{NP}}_{||} = \mathrm{FP}^{\mathrm{NP}}[\log]$.

Proof. Suppose $\overline{\mathrm{SAT}}$ is disjunctively reducible to a sparse set. It suffices to show that χ_{SAT} is in $\mathrm{FP}^{\mathrm{NP}}[\log]$. Let $(x_1, x_2, \ldots, x_m)$ be an instance of χ_{SAT}. We can assume w.l.o.g. that the variable sets of the formulas in $(x_1, x_2, \ldots, x_m)$ are pairwise disjoint. On input $(x_1, x_2, \ldots, x_m)$, $k = ||\{x_1, x_2, \ldots, x_m\} \cap \mathrm{SAT}||$ is first computed with an $\mathrm{FP}^{\mathrm{NP}}[\log]$ computation.

We introduce some notation. Let w denote an assignment to all variables in $\{x_1, x_2, \ldots, x_m\}$. Let $x_i(w)$ denote the value of formula x_i at w. We define a new predicate $U(\langle q, u, v, k, x_1, x_2, \ldots, x_m \rangle, a, w)$ as follows. It is true if and only if:

$$\left(\sum_{i=1}^{m} a_i = k \wedge \bigwedge_{a_i=1} x_i(w) = 1 \right) \implies \sum_{i=1}^{m} a_i u^{i-1} \equiv v \pmod{q}$$

Notice that U is polynomial-time computable. Consider now the following set

$Z = \{\langle q, u, v, k, x_1, x_2, \ldots, x_m \rangle \mid 0 \leq u, v \leq q-1, \text{and } \forall a \in \Sigma^m \ \forall \text{ assignments}$
$w\colon U(\langle q, u, v, k, x_1, x_2, \ldots, x_m \rangle, a, w)\}.$

Since U is a polynomial-time predicate it follows that $Z \in \text{coNP}$.

Observe that if $k = \|\{x_1, x_2, \ldots, x_m\} \cap \text{SAT}\|$ then $a_i = 1$ implies $x_i \in \text{SAT}$ for all i iff $a \in \Sigma^m$ is the characteristic vector of $x_1, x_2, \ldots, x_m$. We are interested in instances $\langle q, u, v, k, x_1, x_2, \ldots, x_m \rangle$ of Z for q picked to be a small prime. If $|\langle x_1, x_2, \ldots, x_m \rangle| = n$, we will pick q to be a $c \log n$ bit prime number, for an appropriate c.

Since $Z \in \text{coNP}$ there is a disjunctive reduction f from Z to a sparse set S of density $\|S^{\leq n}\| \leq p(n)$ for some polynomial p. I.e. f is an FP function that on input x produces a set of strings $f(x)$ such that $x \in Z$ iff $f(x) \cap S \neq \emptyset$. As in the proof of Theorem 1 $|\langle q, u, v, k, x_1, x_2, \ldots, x_m \rangle|$ can be bounded by $2n$. There is a polynomial r such that $r(n)$ bounds both $\|f(\langle u, v, k, x_1, x_2, \ldots, x_m \rangle)\|$ and the length of each query in $f(\langle q, u, v, k, x_1, x_2, \ldots, x_m \rangle)$ for $u, v \in F_q$.

The crucial property that we exploit is the following claim that is easy to check from the definition of Z.

Claim If $k = \|\{x_1, x_2, \ldots, x_m\} \cap \text{SAT}\|$ then $\langle q, u, v, k, x_1, x_2, \ldots, x_m \rangle \in Z$ iff $\sum_{i=1}^{m} a_i u^{i-1} \equiv v \pmod{q}$ holds for $a = \chi_{\text{SAT}}(x_1, x_2, \ldots, x_m)$.

Let $Q = \bigcup_{u, v \in F_q} f(\langle q, u, v, k, x_1, x_2, \ldots, x_m \rangle)$ and write $Q = \{q_1, q_2, \ldots, q_N\}$. In order to apply Lemma 1 we build a table T with N rows R_i and put (u, v) in row R_i if $q_i \in f(\langle q, u, v, k, x_1, x_2, \ldots, x_m \rangle)$. Note that $N \leq r(n)q^2$. We define row R_i to be *correct* if $q_i \in S$. Notice that there are at most $\|S^{\leq r(n)}\| \leq r(n)p(r(n))$ correct rows in T. Choose the constant c (which determines the size of the field F_q) so that $q/r(n)p(r(n)) = n^c/r(n)p(r(n)) > n$, where we know that $n \geq m$.

As in the proof of Theorem 1 we can find a list of at most N m-bit vectors one of which is $\chi_{\text{SAT}}(x_1, x_2, \ldots, x_m)$, which we can locate by doing a binary search with an $\text{FP}^{\text{NP}}[\log]$ computation.

Corollary 2. *If* $\overline{\text{SAT}}$ *is disjunctively reducible to a sparse set then there is a solution of* $(1\text{SAT}, \text{SAT})$ *in* P, *and consequently* NP = RP.

We next consider disjunctive reductions from Mod_kP to sparse sets. We first recall the definition of Mod_kP. For an NP machine N let $\text{acc}_N(x)$ denote the number of accepting computations of N on input $x \in \Sigma^*$. A language L is in the class Mod_kP if there is an NP machine N such that $x \in L \iff \text{acc}_N(x) \not\equiv 0 \pmod{k}$.

Theorem 3. *For prime* k, *if a* Mod_kP-*complete set is disjunctively reducible to a sparse set then there is a solution of* $(1\text{SAT}, \text{SAT})$ *in* P.

Proof. We sketch the proof. Details are as in the proof of Theorem 1. Let Y be the language consisting of tuples $\langle q, u, v, F \rangle$ satisfying:

- F is a Boolean formula and $0 \leq u, v \leq q - 1$, for nonnegative integers u and v, and positive q.
- If F is m-ary then $\|\{a \in \Sigma^m \mid F(a) = 1 \text{ and } \sum_{i=1}^{m} a_i u^{i-1} = v \pmod{q}\}\| \not\equiv 0 \pmod{k}$.

It is easily seen that Y is in Mod_kP. Furthermore, Y has the following useful property which is easy to check from its definition.

Claim: *If F is an instance of* $(1\text{SAT}, \text{SAT})$ *then* $\langle q, u, v, F \rangle \in Y$ *iff F is satisfiable and $\sum_{i=1}^{m} a_i u^{i-1} \equiv v \pmod{q}$ holds for the unique satisfying assignment a.*

Given an instance F of $(1\text{SAT}, \text{SAT})$ of length n, we pick a $c \log n$ bit prime number q (for appropriate c) and consider the disjunctive reduction f from Y to a sparse set S, as applied only to instances $\langle q, u, v, F \rangle$, for $u, v \in F_q$. As in the proof of Theorem 1, we can choose the constant c appropriately large (depending on density of S and the reduction f) so that Lemma 1 can be applied to yield a polynomially bounded set of vectors, one of which is the unique satisfying assignment of F assuming F is satisfiable. Thus, we can decide satisfiability for instances of $(1\text{SAT}, \text{SAT})$.

4   Approximability and Parallel Queries to NP

In this section we show that if SAT is $O(\log n)$-approximable then $\text{FP}_{||}^{\text{NP}} = \text{FP}^{\text{NP}}[\log]$. We prove this result by applying the main technical idea in [Si98] where he shows that if SAT is $O(\log n)$ approximable then the promise problem $(1\text{SAT}, \text{SAT})$ is in P.[4]

Theorem 4. *If SAT is $O(\log n)$ approximable then $\text{FP}_{||}^{\text{NP}} = \text{FP}^{\text{NP}}[\log]$.*

Proof. Assume SAT is $O(\log n)$ approximable. In other words, there are a constant c and an FP function f such that $f(\langle x_1, x_2, \ldots, x_k \rangle)$ is a k-bit vector different from $\chi_{\text{SAT}}(x_1, x_2, \ldots, x_k)$ for any k-tuple of formulas $\langle x_1, x_2, \ldots, x_k \rangle$ with $k \geq c \log(max_i |x_i|)$.

As before, we'll prove the result by giving an $\text{FP}^{\text{NP}}[\log]$ machine, call it M, that computes χ_{SAT}. Let $(x_1, x_2, \ldots, x_m)$ be an input instance for χ_{SAT}. The first step of M is to compute via a binary search guided by a suitable NP oracle the number k of the x_i's that are in SAT. As in the previous proof we will pick a suitable constant c' and efficiently construct the finite field F_q, where q is a $c' \log n$ bit prime number. For a binary vector $a = a_1 a_2 \ldots a_m \in \Sigma^m$ let P_a denote the univariate polynomial $\sum_{i=1}^{m} a_i x^{i-1}$ over F_q. We define the following new language: $Z = \{\langle u, j, k, x_1, x_2, \ldots, x_m \rangle \mid \exists a \in \Sigma^m$ with k 1's such that if $a_i = 1$ then $x_i \in \text{SAT}$ and the jth bit of $P_a(u)$ is 1$\}$.

Clearly, this language is in NP and is therefore $O(\log n)$ approximable by hypothesis. The next two technical steps are exactly as in [Si98].

It is not hard to see that we can apply [Si98, Corollary 3] to get in polynomial time for each $u \in F_q$ a set $S_u \subseteq F_q$ such that $||S_u|| \leq q^{1/3}$ and $P_a(u) \in S_u$. Next, applying [ALRS92] (as described in [Si98]) we can efficiently reconstruct a list of N polynomials of degree $m - 1$ that includes P_a and N is bounded by a polynomial in n, the length of $(x_1, x_2, \ldots, x_m)$.

---

[4] Both [Si98] and [BFT97] call the promise problem UniqueSAT which can be confused with USAT. In this paper we have used Selman's notation as in [ESY84]

We can recover from this list of N polynomials a list of N m-bit vectors. Afterwards we discard those vectors which contain a number of 1's different from k, we know that except $\chi_{SAT}(x_1, x_2, \ldots, x_m)$ which is in this list every other m-bit vector has a 1 in a position where the corresponding formula in $(x_1, x_2, \ldots, x_m)$ is unsatisfiable.

As described in the earlier proof we can find this vector by doing a binary search with at most $\log N$ queries to a suitable NP oracle.

In [BFT97] it is shown that if $\mathrm{FP}_{||}^{\mathrm{NP}} = \mathrm{FP}^{\mathrm{NP}}[\log]$ then SAT is $O(\log n)$ approximable. Combined with the above result and with Theorems 4, 1, and 2, we have the following corollaries.

Corollary 3. *SAT is $O(\log n)$ approximable iff* $\mathrm{FP}_{||}^{\mathrm{NP}} = \mathrm{FP}^{\mathrm{NP}}[\log]$.

Corollary 4. *If SAT or $\overline{\mathrm{SAT}}$ is disjunctively reducible to a sparse set then SAT is $O(\log n)$-approximable.*

5 Majority Reductions to Sparse Sets

We have a couple of observations regarding majority reductions to sparse sets. Majority reductions to sparse sets are interesting because they generalize both conjunctive and disjunctive reductions to sparse sets. A set A is *majority reducible* to a set B if there is an FP function f that on input x produces a set of strings $f(x)$ such that $x \in A$ iff $||f(x) \cap B|| > ||f(x)||/2$. In other words, the majority of strings in $f(x)$ are in B. By padding the list of queries generated by f with a suitable number of strings (we pad with copies of either a fixed string in B or a fixed string outside B), it follows that if A is conjunctively or disjunctively reducible to B, then A is also majority reducible to B. In [CNS96] randomized *bpp*-reductions are considered, and it is shown that if SAT is *bpp*-reducible to a sparse set then NP = RP. It is also easy to see that if A is majority reducible to B then, in fact, A is *bpp*-reducible to B, with the reduction having success probability $1/2 + 1/n^{O(1)}$. Combining this with the above stated result of [CNS96] we get the following corollary.

Corollary 5. *If SAT is majority reducible to a sparse set then NP = RP.*

We leave as an open question whether $\mathrm{FP}_{||}^{\mathrm{NP}} = \mathrm{FP}^{\mathrm{NP}}[\log]$ or the weaker consequence that $(1\mathrm{SAT}, \mathrm{SAT})$ has a solution in P, can be proved assuming that SAT is majority reducible to a sparse set. And finally, the main open problem is to show that P = NP if SAT is disjunctively reducible to a sparse set. We note that it is open if P=NP can be derived from the stronger assumption that the disjunctive reduction generates only a polylogarithmic number of queries.

Acknowledgments. We thank the referees for useful comments.

References

[ALRS92] S. AR, R. LIPTON, R. RUBINFELD, AND M. SUDAN. Reconstructing algebraic functions from erroneous data. In *Proc. 33rd Annual IEEE Symp. on Foundations of Computer Science*, 503–512, 1992.

[AKM96] V. ARVIND, J. KÖBLER AND M. MUNDHENK. Upper bounds for the complexity of sparse and tally descriptions. In *Mathematical Systems Theory*, 29:63–94, 1996.

[BDG88] J. BALCÁZAR, J. DÍAZ, AND J. GABARRÓ. *Structural Complexity I*. Springer-Verlag, 1988.

[BKS95] R. BEIGEL, M. KUMMER, AND F. STEPHAN. Approximable sets. *Information and Computation*, 120(2):304–314, 1995.

[BH77] L. BERMAN AND J. HARTMANIS. On isomorphisms and density of NP and other complete sets. *SIAM Journal on Computing*, 6(2):305–322, 1977.

[BFT97] H. BUHRMAN, L. FORTNOW, AND L. TORENVLIET. Six hypotheses in search of a theorem. In *Proc. 12th Annual IEEE Conference on Computational Complexity*, 2–12, IEEE Computer Society Press, 1997.

[BH95] H. BUHRMAN AND M. HERMO. On the sparse set conjecture for sets with low density. In *Proc. 12th Annual Symp. on Theoretical Aspects of Computer Science*, Lecture Notes in Computer Science 900, 609–618, Springer Verlag 1995.

[CNS96] J. Y. CAI, A. NAIK, AND D. SIVAKUMAR. On the existence of hard sparse sets under weak reductions. In *Proc. 13th Annual Symp. on Theoretical Aspects of Computer Science*, 307–318, 1996. See also the Technical Report 95-31 from the Dept. of Comp. Science and Engeneering, SUNY at Buffalo (1995).

[CO97] J. Y. CAI AND M. OGIHARA. Sparse sets versus complexity classes. Chapter in *Complexity Theory Retrospective II*, L. Hemaspaandra and A. Selman editors, Springer Verlag 1997.

[CS95] J. Y. CAI AND D. SIVAKUMAR. The resolution of a Hartmanis conjecture. In *Proc. 36th Foundations of Computer Science*, 362–373, 1995.

[ESY84] S. EVEN, A. SELMAN, AND Y. YACOBI. The complexity of promise problems with applications to public-key cryptography. *Information and Control*, 61:114-133, 1984.

[JT95] B. JENNER AND J. TORÁN. Computing functions with parallel queries to NP. *Theoretical Computer Science*, 141, 175–193, 1995.

[KL80] R. M. KARP AND R. J. LIPTON. Some connections between nonuniform and uniform complexity classes. In *Proc. 12th ACM Symposium on Theory of Computing*, 302–309. ACM Press, 1980.

[Og95] M. OGIHARA. Polynomial-time membership comparable sets. *SIAM Journal of Computing*, 24(5):1168–1181, 1995.

[OW91] M. OGIHARA AND O. WATANABE. On polynomial time bounded truth-table reducibility of NP sets to sparse sets. *SIAM Journal on Computing* 20(3):471–483 (1991).

[Pa94] C. PAPADIMITRIOU. *Computational Complexity*. Addison-Wesley, 1994.

[Si98] D. SIVAKUMAR. On membership comparable sets. In *Proc. 13th IEEE Computational Complexity Conference 1998*, 2–8, IEEE Computer Society Press, 1998.

[Uk83] E. UKKONEN. Two results on polynomial time truth-table reductions to sparse sets. *SIAM Journal on Computing*, 12(3):580–587, 1983.

[VV86] L. VALIANT AND V. VAZIRANI. NP is as easy as detecting unique solutions. *Theoretical Computer Science*, 47:85–93,1986.

[Ya83] C. YAP. Some consequences of non-uniform conditions on uniform classes. *Theoretical Computer Science*, 26:287–300, 1983.

External Selection

Jop F. Sibeyn

Max-Planck-Institut für Informatik, Im Stadtwald, 66123 Saarbrücken, Germany.
jopsi@mpi-sb.mpg.de. http://www.mpi-sb.mpg.de/~jopsi

Abstract. Sequential selection has been solved in linear time by Blum e.a. Running this algorithm on a problem of size N with $N > M$, the size of the main-memory, results in an algorithm that reads and writes $\mathcal{O}(N)$ elements, while the number of comparisons is also bounded by $\mathcal{O}(N)$. This is asymptotically optimal, but the constants are so large that in practice sorting is faster for most values of M and N.

This paper provides the first detailed study of the external selection problem. A randomized algorithm of a conventional type is close to optimal in all respects. Our deterministic algorithm is more or less the same, but first the algorithm builds an index structure of all the elements. This effort is not wasted: the index structure allows the retrieval of elements so that we do not need a second scan through all the data. This index structure can also be used for repeated selections, and can be extended over time. For a problem of size N, the deterministic algorithm reads $N + o(N)$ elements and writes only $o(N)$ elements and is thereby optimal to within lower-order terms.

1 Introduction

Selecting an element x with specified rank n from a set of size N is a problem that has attracted a substantial amount of interest. The first deterministic sequential algorithm running in $\mathcal{O}(N)$ time was presented in 1972 by Blum e.a. [2]. Later algorithms have tried to reduce the leading constant. Currently, the best results are the $2.95 \cdot N$ algorithms by Dor and Zwick [5] and Carlsson and Sundström [3].

What happens if N exceeds the size of the main-memory M? In other words: how about selection as an external-memory problem? The algorithm from [2] and most later algorithms work by repeatedly reducing the problem size by a constant factor, each of these reductions requiring a few passes over the data. So, summing over all reductions, this requires a constant number of passes over the data. The number of comparisons is the same as before. Thus, these sequential algorithms lead to external algorithms that read and write $\mathcal{O}(N)$ elements, while performing only $\mathcal{O}(N)$ comparisons. Asymptotically this is optimal. However, our estimate in Section 2.3 shows that the algorithm performs up to $8 \cdot N$ reads and $4 \cdot N$ writes. Considering that I/O operations are much more expensive than internal operations, this algorithm will be faster than a multiway-merge sorting algorithm (which requires $N \cdot (1 + \lceil \log N / \log M \rceil)$ reads and writes) only for extreme values of N/M.

The first question we wanted to address was whether in practice one can perform external selection faster than sorting. The answer is affirmative: a simple randomized algorithm requires only $N + o(N)$ reads and $o(N)$ writes with $N + n + o(N)$ comparisons. This algorithm, which is similar to earlier randomized (parallel) selection algorithms [6,8], almost hits the trivial lower bounds.

C. Meinel and S. Tison (Eds.): STACS'99, LNCS 1563, pp. 291–301, 1999.
© Springer-Verlag Berlin Heidelberg 1999

The second question was whether there is a deterministic algorithm with performance close to that of the randomized algorithm. Our basic algorithm performs selection either with $2 \cdot N + o(N)$ reads and $o(N)$ writes or with $N + o(N)$ reads and $N + o(N)$ writes. The value of the $o(N)$-terms depends on the number of comparisons we are willing to make: for $s \leq (M \cdot \log(N \cdot s/M^2))^{1/2}$, the lower-order terms are bounded to $\mathcal{O}(N \cdot \log(N \cdot s/M^2)/s)$ with $\mathcal{O}(\log s \cdot N)$ comparisons. The algorithm is based on the refined deterministic sampling technique from [9], which in turn is similar to the sampling technique in [4]. Methods of this type have also been applied in [1,7].

The variant with $N + o(N)$ reads and writes is most interesting, because this variant allows for several refinements. It can be made adaptive: as soon as it can be ruled out that an element is a candidate for the element to select, this element is not written away. On average-case inputs, this reduces the number of writes from N to at most $\ln 2 \cdot N$. Due to space limitations this algorithm is not described here. It will appear in the full version. A further refinement gives even better results, thus almost matching the performance of the randomized algorithm, and the lower bound, up to lower-order terms.

The deterministic algorithm has an additional advantage: in order to select an element with given rank, it first builds an index structure with which it can find the element with any specified rank in $o(N)$ time. This means that for repeated selections, it is more efficient than repeatedly running the randomized algorithm, which scans through all data. Furthermore, if the number of elements continues to grow, this index structure can be extended dynamically, so that the total amount of work for constructing it is only slightly more than if it had been constructed all at once.

All algorithms discussed in this paper can be parallelized a good deal, and on systems with several hard-discs, the full speed-up can be expected.

The remainder of this paper is organized as follows: after introducing some elementary facts in the following section, we analyze the randomized algorithm in Section 3. We then present the deterministic algorithm and its refinements. In Section 6 we consider several extensions.

2 Preliminaries

2.1 Problem and Goals

The *selection* problem is to select an element with a specified rank from a completely ordered set. Throughout this paper we will assume that the size of the set is N, and that the rank of the element to select is $n \leq N/2$. The goal of most sequential algorithms in this field has been to minimize the number of *comparisons* performed.

We are also particularly interested in selection problems for the case in which N exceeds the size of the main-memory M. This we call the *external selection* problem. As access to secondary storage is far more expensive than internal operations, a major concern of any external algorithm must be to minimize the number of these accesses.

Secondary storage is organized in *pages* and every time the program uses a datum that does not currently reside in the main-memory, a whole page of size B has to be brought into the main-memory. If the main-memory is full, a new page can be brought in only if another page is removed. The page to be removed can be selected in several ways, but whatever strategy is applied, there are two possibilities: if none of the data on this page has been changed, then it can be simply overwritten; otherwise, the page has to be written back to the secondary memory before it can be overwritten. The latter is more expensive, but the amount of additional expense depends on the underlying mechanism. In our analysis, we do not want to go into the technical details, but we do count the number of pages that has to be loaded into the main-memory, *load* operations, and the number of pages that has to be written away, *store* operations, separately.

The time required for performing a comparison is denoted by t_{comp}, that required for a load by t_{load} and that for a store by t_{store}. Our goal is to minimize the number of load and store operations, while keeping the number of comparisons as small as possible.

2.2 Sorting

In [1], it has been shown that sorting N numbers on a machine with M main-memory requires $\Omega(\log N/\log M) \cdot N/B$ paging operations. Sorting can be performed as follows: first all subsets of size M are sorted; then an $(M/B-1)$-way merge is performed as often as necessary. In this way sorting costs

$$T_{sort}(N) = \mathcal{O}(N \cdot \log N) \cdot t_{comp}$$
$$+ (1 + \lceil \log(N/M)/\log(M/B-1)\rceil) \cdot N \cdot (t_{load}/B + t_{store}/B).$$

In all practical cases (those with $M > (N/M+1) \cdot B$), all data are loaded and stored exactly twice.

2.3 Sequential Algorithm

We summarize the algorithm of [2] and analyze its quality as an external memory algorithm. The algorithm consists of the following steps:

Algorithm SEQUENTIAL_SELECTION(n, a)

1. Find the median in all groups of size a.
2. Find the median m of these medians.
3. Divide the input into two subsets; one whose members are smaller than m, $\mathcal{L}$, and one whose members are larger than m $\mathcal{R}$. Also determine the size L of $\mathcal{L}$.
4. If $n \leq L$, then recurse on $\mathcal{L}$, else set $n = n - L$ and recurse on $\mathcal{R}$.

If the size of the groups that are sorted together is denoted a, then the I/O-time $T(N)$ for a problem of size N can be expressed in terms of the costs of the subproblems:

$$T(N) \leq T(N/a) + T(N - \frac{a+1}{4 \cdot a} \cdot N)$$
$$+ 2 \cdot N \cdot t_{load}/B + (N/a + N) \cdot t_{store}/B.$$

In contrast to the sequential algorithm, the optimum is assumed for large a. For such a, solving gives

$$T(N) = (8 + o(1)) \cdot N \cdot t_{load}/B + (4 + o(1)) \cdot N \cdot t_{store}/B.$$

For practical values of M and N, sorting will be several times faster.

3   Randomized Algorithm

We give a randomized algorithm for external selection. Algorithms of this type have been given before for parallel selection. The most interesting point is the choice of the parameters so that the number of loads and stores are minimzed.

The algorithm is simple: a sample of sufficient size is selected at random; two of the elements of the sample are taken as bounds for the set of keys that have to be considered in a second round; all elements are traversed and those that lie between the bounds are singled out. From now on the algorithm is applied recursively on the reduced set. The essential difference with a deterministic algorithm is that a good sample can be selected at low costs. For selecting the element with rank n from a set $\mathcal{N}$ of cardinality N the algorithm proceeds as follows:

Algorithm RANDOMIZED_SELECTION(n, S, Δ)

1. Randomly and uniformly select a subset S of cardinality S out of the elements of $\mathcal{N}$.

2. Select from S the element with rank $\lfloor S/N \cdot n \rfloor - \Delta/2$ and assign it to x_{low}. Select from S the element with rank $\lfloor S/N \cdot n \rfloor + \Delta/2$ and assign it to x_{high}.

3. Traverse $\mathcal{N}$ and add all the elements l with $x_{\text{low}} \leq l \leq x_{\text{high}}$ to the subset $\mathcal{N}'$. Determine the number N' of selected elements, and the number R of elements l, with $l \geq x_{\text{high}}$.

4. Set $L = N - N' - R$. Select the element with rank $n' = n - L$ from $\mathcal{N}'$.

The complexity of the algorithm can be written as follows:

$$T(N) = 2 \cdot T(S) + T(N')$$
$$+ S \cdot t_{\text{load}} + S \cdot t_{\text{store}}/B + (2 \cdot N - R) \cdot T_{\text{comp}} + N \cdot T_{\text{load}}/B + N' \cdot t_{\text{store}}/B. \quad (1)$$

Lemma 1 *If Δ is taken $\Delta = (\log N \cdot S)^{1/2}$, then the element with rank n lies between x_{low} and x_{high}, with high probability.*

Proof: Denote the (unknown) element with rank n by x. Let X be the random variable giving the number of the elements selected in Step 1 that are smaller than x. For each element the probability that it is smaller than x equals n/N. Thus, X has an expected value of $n/N \cdot S$. Because all selections are independent, we can apply Chernoff bounds, which give

$$\Pr(X \geq n/N \cdot S + \Delta/2) \leq e^{-\Delta^2 \cdot N/(12 \cdot n \cdot S)}.$$

For $\Delta = (2 \cdot \log N \cdot n/N \cdot S)^{1/2} \leq (\log N \cdot S)^{1/2}$, the right-hand side is only $N^{-1/6}$. The estimate for a deviation below the expected value goes analogously.   □

Lemma 2 *If $\Delta = (\log N \cdot S)^{1/2}$, then $N' = (1 + o(1)) \cdot N \cdot (\log N/S)^{1/2}$, with high probability.*

Proof: The expected number of elements of $\mathcal{N}$ between any two elements of S is N/S. Thus, the expected number of elements of $\mathcal{N}$ between x_{low} and x_{high} is $\Delta \cdot N/S = N \cdot (\log N/S)^{1/2}$. Because of the process used in selecting x_{low} and x_{high}, the deviation from the expected value is smaller than if everything were independent. Thus, the estimate from the Chernoff bounds gives an upper-bound on the deviation.   □

Theorem 1 *With* $S = \log^{1/3} N \cdot (N/B)^{2/3}$ *and* $\Delta = (\log N \cdot S)^{1/2}$, *the given randomized algorithm solves external selection in*

$$T(N) = N \cdot t_{load}/B + (N + n) \cdot t_{comp}$$
$$+ \mathcal{O}(N^{2/3} \cdot (\log N \cdot B)^{1/3}) \cdot (t_{comp} + t_{load}/B + t_{store}/B),$$

with high probability.

4   Deterministic Algorithm

The algorithm of this section is in its structure similar to the randomized algorithm of Section 3, but deterministically, it is much harder to find a good sample at low costs.

By the *rank* of an element l in a set we mean the number of elements in the set that are smaller or equal than l. So, the smallest element has rank 1. Here we assume that all elements have different keys.

The algorithm first divides the input set $\mathcal{N}$ of size N into $2 \cdot N/M$ chunks of size $M/2$ each. For every such subset an index structure is constructed that later allows one to rather accurately estimate the ranks of the elements, and to retrieve the elements in a range of ranks with little effort:

Procedure FIRST_REDUCTION(s, M, N)
1. Out of the $M/2$ elements, select the elements with ranks $r \cdot M/(2 \cdot s), 1 \leq r \leq s$, as splitters.
2. Sort the elements into the s buckets, and store them away in this somewhat sorted order.

FIRST_REDUCTION is by far the most time consuming part of the whole algorithm: it dominates the I/O and the number of comparisons. Let $m_0 = 2 \cdot N/M$ denote the number of splitter sets after FIRST_REDUCTION. In $k \geq 0$ further reduction steps, the number of splitter sets is going to be repeatedly halved to finally become $m_k \leq M/(2 \cdot s)$. The reduction proceeds as follows. It is illustrated in Figure 1.

Procedure FURTHER_REDUCTION(s, m_0, k)
for $j = 1$ **to** k **do**
    for $i = 0$ **to** $\lfloor m_{j-1}/2 \rfloor$ **do**
        Load splitter set $2 \cdot i$ and $2 \cdot i + 1$, each consisting of s elements.
        Merge the sets and create a new set consisting of the s elements with even rank.
        Store the newly created set of splitters.
    if m_{j-1} is odd **then**
        Load splitter set $m_{j-1} - 1$ and store a copy along with the new splitter sets.
    $m_j = \lceil m_{j-1}/2 \rceil$.

Lemma 3 *The number k in* FURTHER_REDUCTION *should be taken* $k = \lceil \log(4 \cdot N \cdot s/M^2) \rceil$.

Proof: We must check that the k given is the smallest number that achieves $m_k \leq M/(2 \cdot s)$. Define the function f by $f(x) = \lceil x/2 \rceil$, and let $f^{(j)}$ denote the function f applied j times. The proof follows from the fact that $f^{(j)}(x) = \lceil x/2^j \rceil$, which can easily be proven by induction. $\qquad\square$

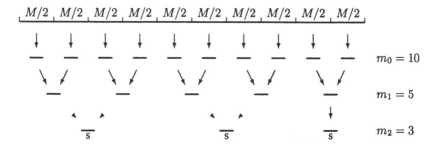

Fig. 1. The set $\mathcal{N}$ of size $N = 5 \cdot M$ is first reduced to ten subsets of size s each. Then these are merged and reduced until three subsets remain.

For each of the remaining splitters we now estimate the minimum and the maximum rank it may have in the set of size $2^k \cdot M/2$ from which it was extracted. Here, and in our further analysis, we assume that m_0 is a power of two. For other values of m_0, some subsets are less reduced. In these the ranks can be estimated more accurately. By the *subsets at Level j*, we mean the splitter sets that were obtained after j, $0 \leq j \leq k$, reduction rounds.

Lemma 4 *For any element l with rank r in its subset with s elements at Level j, $0 \leq j \leq k$, the rank R in the subset of size $2^j \cdot M/2$ from which it was extracted, satisfies*

$$r \cdot 2^j \cdot M/(2 \cdot s) \leq R \leq (r + j/2) \cdot 2^j \cdot M/(2 \cdot s).$$

Proof: The proof follows by induction on j. For $j = 0$, the lemma holds: by the selection in Step 1 of FIRST_REDUCTION, an element with rank r at Level 0 has exactly rank $r \cdot M/(2 \cdot s)$ in its mother set.

So, assume that the lemma holds for some $j - 1$. An element with rank r at Level j was obtained by merging sets $\mathcal{S}_0$ and $\mathcal{S}_1$ and extracting from the merged set the element with rank $2 \cdot r$. Let r_0, r_1 denote the rank of l in $\mathcal{S}_0$, $\mathcal{S}_1$, respectively. $r_0 + r_1 = 2 \cdot r$. Suppose, without loss of generality, that $l \in \mathcal{S}_0$, then

$$R \geq (r_0 \cdot 2^{j-1} + r_1 \cdot 2^{j-1}) \cdot M/(2 \cdot s)$$
$$= r \cdot 2^j \cdot M/(2 \cdot s).$$
$$R \leq ((r_0 + (j-1)/2) \cdot 2^{j-1} + (r_1 + (j-1)/2 + 1) \cdot 2^{j-1}) \cdot M/(2 \cdot s),$$
$$= (r + j/2) \cdot 2^j \cdot M/(2 \cdot s).$$

$\square$

In particular, Lemma 4 holds for $j = k$, the level at which all splitters can be loaded into the main-memory at the same time. For an arbitrary element l, we can now obtain an estimate of its rank in $\mathcal{N}$, *rank(l)*, by determining its rank in each of the m_k splitter sets at Level k. Denote the rank of l in splitter set i by $r_i(l)$. Then

$$\sum_{i=0}^{m_k-1} r_i(l) \cdot 2^k \cdot M/(2 \cdot s) \leq rank(l) \leq \sum_{i=0}^{m_k-1} (r_i(l) + k/2 + 1) \cdot 2^k \cdot M/(2 \cdot s). \quad (2)$$

Now we can determine elements x_{low} and x_{high} that are definitely smaller and larger, respectively, than the element x with *rank(x)* $= n$ that we are looking for.

Procedure DETERMINE_BOUNDS

1. Determine the largest number r_{low} so that $(r_{low}+(k/2+1)\cdot m_k)\cdot 2^k\cdot M/(2\cdot s) \leq n$. From the set of splitters at Level k, select the element with rank r_{low} and assign it to x_{low}.

2. Determine the smallest number r_{high} so that $r_{high} \cdot 2^k \cdot M/(2\cdot s) \geq n$. From the set of splitters at Level k, select the element with rank r_{high} and assign it to x_{high}.

$$r_{high} - r_{low} \leq (k/2+1) \cdot m_k + 1. \tag{3}$$

The following deterministic counterpart of Lemma 1 states that x_{low} and x_{high} are as desired:

Lemma 5 *The selected elements x_{low} and x_{high} satisfy $x_{low} \leq x \leq x_{high}$, where x is the element with $rank(x) = n$.*

Proof: Applying (2) to x_{low} and x_{high} yields $rank(x_{low}) \leq n \leq rank(x_{high})$. $\qquad\square$

The algorithm continues by tracing x_{low} and x_{high} in the $2 \cdot N/M$ subsets of size $M/2$ at the highest level, and writing away the elements that lie between them.

Procedure SINGLE_OUT
$L = N' = 0$.
for $i = 0$ **to** $2 \cdot N/M - 1$ **do**
   In splitter set i at Level 0, determine the largest splitter x'_{low} with
   $x'_{low} \leq x_{low}$ and the smallest splitter x'_{high} with $x'_{high} \geq x_{high}$.
   $L = L + M/(2 \cdot s) \cdot$ *rank-of-x'_{low}-in-subset-with-s-elements-at-Level 0* $- 1$.
   for each element l from a bucket between x'_{low} and x'_{high} **do**
      if $l < x_{low}$ **then** $L = L + 1$;
      else if $l \leq x_{high}$ **then** $N' = N' + 1$; Add l to $\mathcal{N}'$.

Lemma 6 $N' \leq ((\log(N \cdot s/M^2) + 5) \cdot M/(2 \cdot s) + 1) \cdot 4 \cdot N/M$.

Proof: From (3), it follows that in the set of splitters at Level k the ranks of x_{low} and x_{high} differ by at most $(k/2+1) \cdot m_k + 1$. Bounding $rank(x_{low})$ from below and $rank(x_{high})$ from above with (2) gives

$$N' \leq ((k+2) \cdot m_k + 1) \cdot 2^k \cdot M/(2 \cdot s).$$

Substituting the value of k from Lemma 3 gives the result. $\qquad\square$

The selection algorithm, which we call DETERMINISTIC_SELECTION, is completed recursively by selecting the element with rank $n - L$ from the set $\mathcal{N}'$ with N' elements.

Theorem 2 DETERMINISTIC_SELECTION *solves external selection in*

$$\begin{aligned}
T(N) = \ &N \cdot (1 + \mathcal{O}((s + B)/M + \log(N \cdot s/M^2)/s)) \cdot t_{load}/B \\
&+ N \cdot (1 + \mathcal{O}(s/M + \log(N \cdot s/M^2)/s)) \cdot t_{store}/B \\
&+ N \cdot \mathcal{O}(\log s) \cdot t_{comp}.
\end{aligned}$$

Proof: During FIRST_REDUCTION, all elements are loaded, and $N+2 \cdot N/M \cdot s$ elements are stored away. For each subset of size $M/2$, $\mathcal{O}(M \cdot \log s)$ comparisons are required, in total this gives $\mathcal{O}(N \cdot \log s)$ comparisons. The costs of FURTHER_REDUCTION can be estimated on twice the cost of its first round, because the problem size is halved in each round. Thus, $4 \cdot N/M \cdot s$ elements must be loaded and $2 \cdot N/M \cdot s$ elements stored. The number of comparisons is bounded by $\mathcal{O}(N/M \cdot s)$. The time-consumption of DETERMINE_BOUNDS is negligible in comparison to that of the previous procedures. In SINGLE_OUT all splitters at Level 0 must be loaded. Their number is $2 \cdot N/M \cdot s$. Thereafter, at most $N' + 2 \cdot (M/(2 \cdot s) + B) \cdot 2 \cdot N/M = N' + 2 \cdot N/s + 4 \cdot B \cdot N/M$ elements must be loaded, taking into account that for every splitter subset two almost useless pages may have to be loaded. Then the N' elements that are singled-out are stored away. The number of comparisons is small: $\mathcal{O}(N' + N/s)$. Summing over all procedures gives

$$T_{\text{load}}(N) = (N + 6 \cdot N/M \cdot s + N' + 2 \cdot N/s + 4 \cdot B \cdot N/M) \cdot t_{\text{load}}/B + T_{\text{load}}(N'),$$
$$T_{\text{store}}(N) = (N + 4 \cdot N/M \cdot s + N') \cdot t_{\text{store}}/B + T_{\text{store}}(N'),$$
$$T_{\text{comp}}(N) = \mathcal{O}(N \cdot \log s) \cdot t_{\text{comp}} + T_{\text{comp}}(N').$$

$\square$

The choice of s gives a trade-off between the computation time and the time for I/O. The I/O is minimized for $s = (M \cdot \log(N \cdot s/M^2))^{1/2}$. On a system where the amount of writable memory is small, most of the writing can be replaced by additional reading: if Step 2 of FIRST_REDUCTION is omitted, then the subset $\mathcal{N}'$ can be found by scanning through all the data, just as was done in Step 3 of RANDOMIZED_SELECTION.

5   Minimizing the Writing

In the presented deterministic algorithm, all elements that still had to be taken into account were stored away in bucket-sorted order. This has the advantage that all elements belonging to the same bucket are stored together, which is of great importance for accessing them at low cost. However, it also implies that the total amount of I/O (loads + stores) exceeds $2 \cdot N$. We show that large savings are possible.

At the lower levels the algorithm is the same as before. Only at the top level are there some differences. If a new batch of $M/2$ elements is loaded into the main-memory, they are processed as follows:

Procedure FIRST_REDUCTION'(s, M, N)

1. Select the elements with ranks $r \cdot M/(2 \cdot s)$, $1 \leq r \leq s$ out of the $M/2$ elements as splitters.

2. Traverse the elements. For each element l determine to which bucket b it belongs. Let i_l be the index of l within $\mathcal{N}$. Let l', with index $i_{l'}$, be the previous element allocated to bucket b (if l is the first element of bucket b, then $i_{l'} = 0$). Write $i_{l'} - i_l$ into bucket b. Store all buckets away.

Lemma 7 *For a given number of buckets s, the positions of the elements can be encoded in Step 2 in at most $2 \cdot N \cdot (1 + \log s)$ bits.*

Proof: We apply a simple encoding: each number $j = i_{l'} - i_l$ is written in binary, and between every two bits a 1 is inserted. After the last bit a 0 is added. In this way, j requires $2 \cdot (\lfloor \log j \rfloor + 1)$ bits. If in total z elements $j_0, \ldots, j_{z-1}$ are written whose sum equals $M/2$, then this requires at most $2 \cdot z \cdot (1 + \log(M/(2 \cdot z)))$ bits. In our case $z = M/(2 \cdot s)$, and there are a total of $2 \cdot s \cdot N/M$ buckets. $\square$

Later, during SINGLE_OUT, while searching the candidates together, it may happen that for each candidate a whole page has to be brought into the main-memory. From Lemma 6 we know that when s splitters are selected out of every $M/2$ elements, that this yields a total of $N' = \mathcal{O}(N/s \cdot \log N)$ candidates. Thus, we have to bring $\mathcal{O}(N/s \cdot \log N \cdot B)$ elements into the main-memory.

Theorem 3 *For a given word-length w and $s = \omega(\log N \cdot B)$, the algorithm runs in*

$$T(N) = (N + o(N)) \cdot t_{load}/B + \mathcal{O}(N \cdot \log s/w) \cdot t_{store}/B + \mathcal{O}(N \cdot \log M) \cdot t_{comp}.$$

For $w = \log N$ and $s = \log^2 N \cdot B$, the loading is close to optimal, while the storing is reduced to $2 \cdot N \cdot \log B / \log N + o(N)$. The encoding can easily be refined to reduce the factor 2 to less than 1.5. This means that even for current values of $\log B$ (between 10 and 14) and w (32 or 64), this approach may have practical importance.

6   Extensions

Multiple Selections. Consider the problem of selecting several elements with specified rank. Here we must distinguish the case in which all ranks are provided at the same time from the case in which several requests have to be served in turn.

If c elements must be selected that are all provided at the same time, then the randomized algorithm determines x_{low} and x_{high} for each of them. After sorting these values, the status of each scanned element can be determined with a binary search. Some extra care must be taken with overlapping intervals, but in principle the selection can be performed with N reading, $c \cdot N'$ writing and $\mathcal{O}(N \cdot \log c)$ comparisons. If the c elements are provided one-by-one, than the randomized algorithm given in Section 3 is not good: for every element all elements must be traversed, multiplying the time consumption by almost a factor of c.

Interestingly, for the deterministic algorithm it is no particular advantage to get all c elements specified at the start: in any case it will handle them one-by-one, using the same index structure again and again. Whether the refinement of Section 5 is advantageous depends on the value of c. Using the basic algorithm of Section 4 yields the following analogue of Theorem 2:

Theorem 4 DETERMINISTIC_SELECTION *solves c external selection problems on the same set $\mathcal{N}$ in*

$$\begin{aligned} T(N) = & \, N \cdot (1 + \mathcal{O}(c \cdot (s + B)/M + c \cdot \log(N \cdot s/M^2)/s)) \cdot t_{load}/B \\ & + N \cdot (1 + \mathcal{O}(s/M + c \cdot \log(N \cdot s/M^2)/s)) \cdot t_{store}/B \\ & + N \cdot \mathcal{O}(\log s + c \cdot \log(N \cdot s/M^2)/s) \cdot t_{comp}. \end{aligned}$$

The term $\mathcal{O}(N \cdot c \cdot s/M t_{\text{load}}/B)$ is due to the repeated loading of all splitters at Level 0. If all elements are specified at the start, then for large c, it becomes profitable to modify the algorithm so that these splitters have to be loaded only once. Then, s can be chosen close to M and c up to $c = o(\min\{M/B, M/\log(N/M)\})$.

Incremental Input. So far we have assumed that all N elements were provided at the start. However, one may imagine a set of elements that grows over time, while occasional selections must be made in between. For the deterministic algorithm, it is no great problem to build the index structure incrementally at little additional cost.

If we assume that new elements are added in multiples of $M/2$, then all reductions have to be performed once. Only the copying of the splitter sets at the far right side gives some waste, but this can be eliminated easily.

More Powerful Hardware. An important issue is whether an algorithm can be run on a system with $D > 1$ hard-disc units or on a system with $P > 1$ processors. For all algorithms presented this is easy: the time consumption is dominated by the first reduction round, and the last round in which the candidates are searched together. In the first round $2 \cdot N/M$ chunks of $M/2$ elements have to be loaded. This can easily be sped up by using up to $M/(2 \cdot B)$ hard-discs. All chunks could also be processed in parallel. In the final round, parallelism with P up to $2 \cdot N/M$ would again be no problem to organize. Applying the algorithm of Section 5, after decoding the indices, a considerable number of hard-discs can be used effectively for collecting all candidates which stand scattered over the whole input.

7 Conclusion

The algorithms presented in this paper solve the external selection problem almost completely. Only the lower-order terms might be further reduced. For the deterministic algorithm, one would also like to reduce the number of comparisons, even though on the extremely fast modern computers the time required to perform them is small in comparison to the time for scanning through the data.

Acknowledgements

This paper profited considerably by fruitful discussions with Ulrich Meyer and from helpful comments of an anonymous referee.

References

1. Aggarwal, A., J.S. Vitter, 'The Input/Output Complexity of Sorting and Related Problems,' *Communications of the ACM*, 31(9), pp. 1116–1127, 1988.
2. Blum, M., R.W. Floyd, V.R. Pratt, R.L. Rivest, R.E. Tarjan, 'Time Bounds for Selection,' *Journal of Computing System Sciences*, 7(4), pp. 448–461, 1972.
3. Carlsson, S., M. Sundström, 'Linear-Time In-Place Selection in Less than $3n$ Comparisons,' *Proc. 6th International Symposium on Algorithms and Computation*, LNCS 1004, pp. 245–253, Springer-Verlag, 1995.
4. Chaudhuri, S., T. Hagerup, R. Raman, 'Approximate and Exact Deterministic Parallel Selection,' *Proc. 18th Symposium on Mathematical Foundations of Computer Science*, LNCS 711, pp. 352–361, Springer-Verlag, 1993.

5. Dor, D., U. Zwick, 'Selecting the Median,' *Proc. 6th Symposium on Discrete Algorithms*, pp. 28–37, ACM-SIAM, 1995.
6. Floyd, R.W., R.L. Rivest, 'Expected Time Bounds for Selection,' *Communications of the ACM*, 18(3), pp. 165-172, 1975.
7. Nodine, M.H., J.S. Vitter, 'Deterministic Distribution Sort in Shared and Distributed Memory Multiprocessors,' *Proc. 5th Symposium on Parallel Algorithms and Architectures*, pp. 120–129, ACM, 1993.
8. Rajasekaran, S., 'Randomized Parallel Selection,' *Proc. Foundations of Software Technology and Theoretical Computer Science*, LNCS 472, pp. 215–223, Springer-Verlag, 1990.
9. Sibeyn, J.F., 'Sample Sort on Meshes,' *Proc. 3rd Euro-Par Conference*, LNCS 1300, pp. 389–398, Springer-Verlag, 1997.

Fast Computations of the Exponential Function

Timm Ahrendt

Institut für Informatik II, Universität Bonn,
Römerstr. 164, D - 53117 Bonn, Germany
ahrendt@informatik.uni-bonn.de
http://www.cs.uni-bonn.de/~ahrendt

Abstract. In this paper we present an algorithm which shows that the exponential function has algebraic complexity $O(\log^2 n)$, i.e., can be evaluated with relative error $O(2^{-n})$ using $O(\log^2 n)$ infinite-precision additions, subtractions, multiplications and divisions. This solves a question of J. M. Borwein and P. B. Borwein [9].

The best known lower bound for the algebraic complexity of the exponential function is $\Omega(\log n)$.

The best known upper and lower bounds for the bit complexity of the exponential function are $O(\mu(n) \log n)$ [10] and $\Omega(\nu(n))$ [4], respectively, where $\mu(n)$ denotes an upper bound and $\nu(n)$ denotes a lower bound for the bit complexity of n-bit integer multiplication.

The presented algorithm has bit complexity $O(\mu(n) \log n)$.

1 Introduction

This paper deals with fast algorithms for the computation of the exponential function. We consider the number of operations sufficient to evaluate $\exp(x)$ with relative error $< 2^{-n}$ for x in some (fixed nontrivial compact) interval $[p, q]$. For inputs $x \notin [p, q]$ we may apply range reduction techniques [13] (preserving the complexity bounds), thus we assume w.l.o.g. $[p, q] = [0, \ln(2)]$.

The algorithms are analyzed with respect to algebraic complexity [6], operational complexity [8], and bit complexity. (Some authors use these terms with different meanings.) These concepts of complexity are made precise in Sect. 2. "Fast" algorithms are meant to be asymptotically fast with respect to such a complexity measure.

The traditional way to compute the exponential function or the natural logarithm is to use a partial sum of the Taylor series or a related polynomial or rational approximation. When we wish to compute n digits of the natural logarithm using a Taylor series, then we employ a polynomial of degree n and perform $O(n)$ rational operations, while for the exponential function $O(n/\log n)$ rational operations are sufficient [9]. The improvement for the exponential function reflects the faster convergence of the Taylor series. Repeated use of $\ln(x^2) = 2\ln(x)$ or $\exp(2x) = (\exp(x))^2$, respectively, reduces the inputs x to values close to 1 or 0, respectively, leading to algorithms for evaluating $\ln(x)$ or $\exp(x)$ using Taylor series with $O(\sqrt{n})$ rational operations and bit complexity $O(\mu(n)\sqrt{n})$ [8].

C. Meinel and S. Tison (Eds.): STACS'99, LNCS 1563, pp. 302–312, 1999.

In the model of algebraic complexity, infinite-precision rational operations $(+, -, \times, /)$ with real numbers are accounted for with unit cost [6]. Using the arithmetic-geometric mean (AGM) iteration, it can be shown that the natural logarithm has algebraic complexity $O(\log^2 n)$. The exponential function may be evaluated using Newton's method for inverting the logarithm. This leads to an algorithm with algebraic complexity $O(\log^3 n)$. We refer to J. M. Borwein, P. B. Borwein [7], [8], [9], and R. P. Brent [10], [11] for these results.

The main result of this paper, established in Sect. 3.4, is an algorithm for the exponential function with algebraic complexity $O(\log^2 n)$. Our algorithm "couples" the $O(\log n)$ many AGM iterations performed for the computation of $\exp(x)$. By re-using internal results from prior iterations we speed up the internal square root operations. This saves a factor of $\log n$.

In the model of operational complexity, infinite-precision rational operations and infinite-precision extraction of mth roots of real numbers are accounted for with unit cost [8]. The AGM iteration leads to an algorithm with operational complexity $O(\log n)$ for the natural logarithm, and the use of Newton's method for inverting the natural logarithm gives an algorithm with operational complexity $O(\log^2 n)$ for the exponential function [9].

Both functions, natural logarithm and exponential function, have bit complexity $O(\mu(n) \log n)$, where $\mu(n)$ is an upper bound for the bit complexity of n-bit integer multiplication (assuming that $\mu(n)/n$ is monotonically nondecreasing). This bound follows from an analysis of the above algorithms (see again J. M. Borwein, P. B. Borwein [9] and R. P. Brent [11]).

The exponential function can be evaluated with Boolean circuits of depth $O(\log n)$ and size $n^{O(1)}$ (H. Alt [3]). This result has been improved by J. H. Reif [15] and others. In [14], Y. Okabe, N. Takagi, and S. Yajima describe Boolean circuits with depth $O(\log n)$ and size $O(n^4/\log^2 n)$. It is still unknown whether depth $O(\log n)$ and size $O(\mu(n) \log n)$ can be achieved simultaneously.

An overview of the results is shown in Table 1 (partially cited from [9]).

The idea of re-using internal results in AGM computations has been mentioned before by J. M. Borwein, P. B. Borwein [8], Y. Kanada [12], and T. Sasaki, Y. Kanada [16]. There, the idea was only used to reduce the constant factor in

Table 1. Overview of function and constant complexities.

Type of function/constant		O_{alg}	O_{op}	O_{bit}
(1)	Addition	1	1	n
(2)	Multiplication	1	1	$n \log n \log \log n$
(3)	Algebraic	$\log n$	1	$\mu(n)$
(4)	ln	$\log^2 n$	$\log n$	$\mu(n) \log n$
(5)	exp	$\log^3 n$	$\log^2 n$	$\mu(n) \log n$
		$\log^2 n$ ($\star new\star$)		
(6)	π, ln(2)	$\log^2 n$	$\log n$	$\mu(n) \log n$

the complexity bound, while it helps saving a factor of $\log n$ in the algorithm presented here. Particularly, the idea of using internal results from prior *nested iterations* seems to be new in this context.

The algorithm has bit complexity $O(\mu(n) \log n)$ which is the best known asymptotic bound. The result holds for computations performed on a multitape Turing machine.

In Sect. 2, we describe our models of computation. Section 3 starts with a short summary of Newton's method, AGM iteration, and their use for computing the exponential function. Finally, we introduce the idea of coupling iterations and present our new algorithm with its complexity estimates. Practical results are given in Sect. 4.

2   Model of Computation

For the description of the complexity measures we use definitions from [6] and [8]. The subscripts on the order symbols are for emphasis only.

Definition 1 (Algebraic Complexity).
A function f has algebraic complexity $O_{\mathrm{alg}}(s(n))$ *on a set A if there exists a sequence of rational functions R_n such that*
(i) $|R_n(x) - f(x)| < 2^{-n}$ for all $x \in A$;
(ii) R_n can be evaluated using at most $O(s(n))$ rational operations (i.e., infinite-precision additions, subtractions, multiplications, and divisions).

Many authors prefer the concept of *arithmetic circuits* [20] or *algebraic circuits* [6], where the *size* of the circuits coincides with what we call *algebraic complexity*.

Definition 2 (Operational Complexity).
A function f has operational complexity $O_{\mathrm{op}}(s(n))$ *on a set A if there exists a sequence of algebraic functions A_n such that*
(i) $|A_n(x) - f(x)| < 2^{-n}$ for all $x \in A$;
(ii) A_n can be evaluated using at most $O(s(n))$ rational operations or extractions of m-th roots.

This measure allows us, for example, to use square root extractions in the computation of approximations and to count them on an equal level with the operations $+$, $-$, $\times$, $/$. This is often appropriate because, from a bit complexity point of view, root extraction is equivalent to multiplication [2].

In practice, we have to take into account that low-precision operations are cheaper than high-precision operations.

Definition 3 (Bit Complexity).
A function f has bit complexity $O_{\mathrm{bit}}(s(n))$ *on a set A if there is a sequence of approximations B_n such that*
(i) $|B_n(x) - f(x)| < 2^{-n}$ for all $x \in A$;
(ii) there is a multitape Turing machine which (given input n and x) computes $B_n(x)$ with $O(s(n))$ steps.

Throughout this paper, we assume that $\mu(n)/n$ is monotonically non-decreasing. Our assumption is certainly valid if the Schönhage-Strassen method [19] is used to multiply n-bit integers (in the usual binary representation) in $O(n \log n \log \log n)$ single-digit operations.

Real numbers are assumed to be represented internally as *binary rationals* $x = u/2^n$ with $u \in \mathbb{Z}$, $n \in \mathbb{N}$. More precisely, every real number is approximated by binary rationals up to certain errors.

3   Algorithms

3.1   Newton Iteration

Given algorithms for addition and multiplication, algebraic functions can be calculated by applying Newton's method to solving equations of the form $f(y) - x = 0$. For example, Newton's method for $y^2 - x = 0$ leads to the iteration

$$y_{i+1} := y_i - (y_i^2 - x)/(2y_i) \tag{1}$$

which converges quadratically to $\sqrt{x}$. Thus $O(\log n)$ iteration steps give n digits of accuracy and we have an algorithm with algebraic complexity $O(\log n)$ for square root extraction.

Newton's method is *self-correcting* in the sense that a "small" perturbation in y_i does not change the limit. Therefore it is possible to start with a single-digit estimate and roughly double the precision with each iteration step. Thus the bit complexity of root extraction is $O(\mu(n) + \mu(\frac{n}{2}) + \mu(\frac{n}{4}) + \cdots + \mu(1)) = O(\mu(n))$, as the factors can be summed up in a geometric series.

Newton iteration for square root extraction is only a special case of the more general task to invert an arbitrary function. For any "reasonable" f, the iteration

$$y_{i+1} := y_i - (f(y_i) - x)/f'(y_i)$$

yields an algorithm for f^{-1} with same bit complexity as for f.

Using Newton's method without further considerations for inverting multiplies the algebraic and operational complexities by $\log n$. Thus the upper bound $O(\log^2 n)$ for the algebraic complexity of the natural logarithm yields $O(\log^3 n)$ as an upper bound for the algebraic complexity of the exponential function.

3.2   The Arithmetic-Geometric Mean (AGM)

Newton's method cannot suffice for the fast computation of elementary transcendental functions since, if f is algebraic in (2), then the limit f^{-1} is also algebraic.

The most familiar iteration that converges quadratically to a transcendental function is the arithmetic-geometric mean iteration of Gauß and Legendre for computing elliptic integrals.

Given two positive numbers $a = a_0$ and $b = b_0$ with $a_0 > b_0$, we define

$$a_{j+1} := (a_j + b_j)/2 , \quad b_{j+1} := \sqrt{a_j b_j}$$

for $j \in \mathbb{N}$. It can be shown that (a_j) and (b_j) converge quadratically to a common limit $\mathrm{AGM}(a,b)$ [8], which can be expressed in terms of a *complete elliptic integral of the first kind*,

$$I(a,b) = \int_0^{\pi/2} \frac{d\theta}{\sqrt{a^2 \cos^2 \theta + b^2 \sin^2 \theta}} ,$$

namely

$$\mathrm{AGM}(a,b) = \frac{\pi}{2 \cdot I(a,b)} .$$

The relation between the natural logarithm and the elliptic integral $I(\cdot, \cdot)$ can be stated as follows:

Proposition 1.

$$|\ln(4/k) - I(1,k)| < 4k^2(8 + |\ln k|) \quad \text{for } k \in (0,1] . \tag{2}$$

(For a proof see [7, Prop. 2].)

To get a low complexity algorithm for $\ln(y)$ we use estimate (2) and set $k := 4/(y \cdot 2^m)$ with *shift* m such that $k < 2^{-n/2}$, leading to

$$\ln(y) + m \cdot \ln(2) = \frac{\pi}{2 \cdot \mathrm{AGM}(1, 4/(y \cdot 2^m))} \cdot (1 + O(2^{-n})) .$$

The AGM iteration is not self-correcting. Therefore all iteration steps are performed with precision $O(n + \log n) = O(n)$, where the additional $O(\log n)$ bits are used to avoid cancellation of significant bits.

Up to computing π and $\ln(2)$, this allows for the derivation of algorithms with the complexity of entry (4) in Table 1. Algorithms for π and $\ln(2)$ can be derived from similar kinds of considerations (see [8] and others). Using Newton's method for inverting the natural logarithm, we get an algorithm with the complexity of entry (5) in Table 1.

3.3 An Example of Coupled Iteration

In this section we give an example of a coupled iteration which serves as motivation for this technique. Therefore we avoid a formalized description.

In a typical situation of Newton iteration, we perform $O(\log n)$ steps and the precision is doubled in each step. In Fig. 1(a) we see an iteration with a nested iteration. This is, from a bit complexity point of view, the situation in iteration (1), as the division by $2y_i$ involves Newton iteration for computing approximations for $1/(2y_i)$. Therefore a *coupled iteration* for $y_i \approx \sqrt{x}$ (━) and $z_i \approx 1/(2\sqrt{x})$

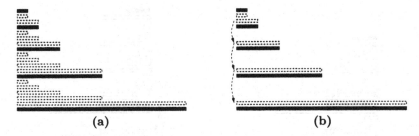

Fig. 1. Coupled iteration schemes: **(a)** typical scheme of a Newton iteration (━) with nested iteration (⋯⋯); **(b)** Newton iteration where the nested iteration uses internal results from prior iteration steps (⋯▸).

(⋯⋯) with re-use of internal results from prior iteration steps (Fig. 1(b)) seems to be considerably faster. The savings in comparison to case (a) are obvious.

We restate the main idea: Figure 1(b), applied to the Newton iteration for the computation of approximations to $\sqrt{x}$, corresponds to the coupled iteration

$$y_{i+1} := y_i - (y_i^2 - x) \cdot z_i , \quad z_{i+1} := z_i + (1 - 2y_i z_i) \cdot z_i .$$

By coupling these iterations we avoid the repeated computations of the beginnings of approximations to $1/(2y_i) \approx 1/(2\sqrt{x})$. A careful implementation of this idea leads to an algorithm which is superior to the Newton iteration without division which is preferred by many authors. Details on the coupled Newton iteration for square root extractions can be found in [1] and [17].

3.4   Coupling of Newton's Method and AGM Iterations

Remember that we assume that the input x of the algorithm for computing $\exp(x)$ lies in the interval $[p, q] = [0, \ln(2))$. This assumption is only for the purpose of a simplified description and done w.l.o.g., as we may use range reduction techniques [13] preserving the complexity bounds given in Table 1.

Simple **Re-use of Internal Results.** An AGM iteration can be split into two parts: In part I $a_j/b_j > 2$ holds, whereas in part II $a_j/b_j \leq 2$ is satisfied.

In part II we may easily re-use b_j (or a_j) as a starting value for the computation of an approximation for $b_{j+1} = \sqrt{a_j b_j}$, because the first significant bit(s) of b_j (or a_j, respectively) and b_{j+1} agree. This is what J. M. Borwein, P. B. Borwein [8], Y. Kanada [12], and T. Sasaki, Y. Kanada [16] suggest. This approach reduces the constant factor in the bit complexity $O(\mu(n) \log n)$, but the algebraic complexity is still $O(\log^2 n)$.

Applying Newton's method to equation $\ln(y) - x = 0$ leads to an iteration

$$y_{i+1} := y_i - (\ln(y_i) - x) \cdot y_i$$

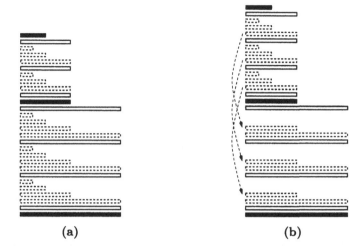

Fig. 2. Iteration schemes for $\exp(x)$: **(a)** traditional way with Newton iteration for the inversion of $\ln(y)$ (▬), AGM iteration for $\ln(y)$ (▭), and nested iteration for $\sqrt{a_i b_i}$ (⋯); **(b)** new algorithm with common shift N and re-use of internal results.

for approximations $y_i \approx \exp(x)$. The computation of y_{i+1} is performed with relative precision $O(2^{-2^{i+1}})$, i.e., $\ln(y_i)$ is computed with relative precision $O(2^{-2^{i+1}})$. This situation is shown in Fig. 2(a).

As $|y_{i+1} - y_i| = O(2^{-2^i})$, we have $|\ln(y_{i+1}) - \ln(y_i)| = O(2^{-2^i})$. At a first view, we cannot make use of this estimate, because the AGM for $\ln(y_{i+1})$ has to be computed with doubled precision (compared with $\ln(y_i)$) and with a new shift $m' \approx 2m$ which makes prior results unusable.

Coupling of *Two* AGM Iterations. Our idea is to get rid of these two drawbacks. Look at two "neighbouring" AGM iterations $(a_j), (b_j)$ and $(c_j), (d_j)$, where $a_0 = c_0 = 1$, $b_0 = 4/(y \cdot 2^N)$, $d_0 = 4/(\tilde{y} \cdot 2^N)$ with values y, $\tilde{y}$ related by $\tilde{y} = y \cdot (1 + O(2^{-\ell}))$. Note that both iterations use the same shift N. As y and $\tilde{y}$ are closely related, we call these AGM iterations "neighbouring". With a little effort it can be proved that the relations $c_j = a_j \cdot (1 + O(2^{-\ell}))$ and $d_j = b_j \cdot (1 + O(2^{-\ell}))$ hold for $j = 0, 1, 2, \ldots$. Thus, if b_{j+1} (or a sufficiently precise approximation to it) has been computed, it can serve as a starting value for the computation of an approximation to $d_{j+1} = \sqrt{c_j d_j}$. In two or three Newton steps we get a 4ℓ-digit approximation for d_{j+1}.

Consider the computation of $\ln(y)$ to relative precision $O(2^{-k})$ and $\ln(\tilde{y})$ to relative precision $O(2^{-2k})$, where $|\ln(\tilde{y}) - \ln(y)| = O(2^{-k/2})$. If the AGM iteration for $\ln(y)$ has been computed with relative precision $k' = k + O(\log N)$ (the additional $O(\log N)$ digits of precision are used again for avoiding cancellation) and if we have stored the approximations for the values $b_1, b_2, b_3, \ldots$, we can speed up the computation of $\ln(\tilde{y})$, because the approximations to the square roots $d_1 = \sqrt{c_0 d_0}$, $d_2 = \sqrt{c_1 d_1}, \ldots$ with relative error $O(2^{-2k'})$ can be computed in constant time. This precision is sufficient to get a $2k$-digit approximation for $\ln(\tilde{y})$.

Coupling of Newton Iteration and AGM Iterations. In the algorithm for computing an approximation to $\exp(x)$, we find again "neighbouring" AGM iterations. More precisely, the AGM iterations performed for $\ln(y_i)$ and $\ln(y_{i+1})$ are neighbouring for $i = 0, 1, 2, \ldots$.

To make use of the technique described above we *synchronize* the $O(\log n)$ many AGM iterations with a (sufficiently large) common shift $N = O(n)$. Once we have computed the AGM iteration for $\ln(y_0)$, the further procedure is straightforward.

This situation is shown in Fig. 2(b), where the additional iterations are caused by the common shift $N = O(n)$. Ideally, we are searching for an iteration, where we use a dynamically changed shift. We have not yet found such an algorithm which shares the same simultaneous upper bounds for the algebraic and bit complexity as our presented algorithm.

The initial computation of an approximation for $\ln(y_0)$ (with shift $O(n)$ for the AGM iteration) can be performed in $O(\log^2 n)$ rational operations, and as we can now evaluate the exponential function with $O(\log n)$ Newton iteration steps, each involving $O(\log n)$ AGM iteration steps which can be computed in constant time, we have proved the following theorem.

Theorem 1. *The exponential function has algebraic complexity $O(\log^2 n)$.*

The initial computation of an approximation for $\ln(y_0)$ (with shift $O(n)$ for the AGM iteration) can be performed with bit complexity $O(\mu(n) \log n)$. The relative precision is roughly doubled in each iteration step of the Newton iteration for the exponential function. Thus we can sum up the terms in a geometric series as $O(\mu(n) \log n) + O((\mu(n) + \mu(\frac{n}{2}) + \mu(\frac{n}{4}) + \cdots + \mu(\log n)) \log n) = O(\mu(n) \log n)$. We state this result as Theorem 2.

Theorem 2. *The described algorithm has bit complexity $O(\mu(n) \log n)$.*

Remark 1. For the purpose of a simplicity, we did not describe how to perform the computation of the initial AGM iteration with $O(\log n \log \log n)$ rational operations. We also neglected the possibility of coupling the Newton iterations for square roots in the sense of section 3.3, i.e., using prior approximations for the computation of approximations to the reciprocals $1/(2\sqrt{a_j b_j})$.

4   Practical Results

Our algorithms were implemented in the TP assembly language described in [18]. To achieve *fair* results we follow [18] in the concept of *typical time*: We compare the running times for inputs of (decimal) length n with relative error 10^{-n} in the outputs. The inputs are in $[\frac{1}{2}, 1]$ and were taken from random generators.

Routine *EXP* is an implementation of the traditional Newton iteration for the inversion of ln(y). Routine *Coupled_EXP* also computes this iteration, but in the new "coupled" way, performing three Newton steps in the square root computations. In practice, both routines have nearly the same running time. Our new algorithm can be improved additionally by using the ideas from Remark 1. This and revision of the square root routine, such that two Newton steps (instead of three) are sufficient, would save about 10–15% running time.

To make the measurements more informative and "system-independent", we made further comparisons with the packages *Mathematica 2.2* [21] and *MPFUN* [5]. The routine *MPEXPX* (*MPFUN*) uses a similar algorithm as routine *EXP*. The algorithm underlying the Exp-command of *Mathematica* is unknown to the author, but the factor of $\approx 6 = 2^{1+\log_2 3}$ in the running times when doubling the problem size makes the use of Taylor series with Karatsuba multiplication probable.

Table 2 summarizes our practical results.

Table 2. Running time comparison (CPU time in sec) *Mathematica – MPFUN – TP*. Timings do not include time for pre-computations of the constants π and ln(2). All timings are for a SPARC 20/512 with 50MHz.

n digits	Exp	MPEXPX	EXP	Coupled_EXP
1 000	1.16	1.68	0.17	0.18
2 000	6.6	4.9	0.5	0.5
5 000	65.2	16.1	2.9	2.9
10 000	367.	45.	11.	11.
20 000	2147.	119.	31.	32.
50 000	20730.	297.	117.	115.
100 000	121552.	747.	294.	285.
200 000	—	1541.	744.	735.
500 000	—	7742.	1972.	1946.
1 000 000	—	17239.	4594.	4412.

5   Conclusion and Future Work

A future version of our software package will include the improvements mentioned above, thus saving about 10–15% running time in the some thousand digit range and above.

The idea presented in this paper seems to be applicable for many iteration-like computations. It may lead to further progress if we find a general theory on the repeated use of intermediate results in the computation of (elementary) functions.

Acknowledgements

I would like to thank A. Schönhage for posing this problem and giving many useful hints. Helpful comments and critical remarks came from M. Bläser, P. Kirrinnis, and D. Reischert. Thanks go also to the referees for their suggestions.

References

1. AHRENDT, T. Fast high-precision computation of complex square roots. In *Proceedings of the 1996 International Symposium on Symbolic and Algebraic Computation: ISSAC '96*, LAKSHMAN, Y. N., Ed., ACM, New York, pp. 142–149.
2. ALT, H. Square rooting is as difficult as multiplication. *Computing 21* (1979), 221–232.
3. ALT, H. Comparison of arithmetic functions with respect to Boolean circuit depth. In *Proc. 16th Ann. ACM Symposium on Theory of Computing* (1984), pp. 466–470.
4. ALT, H. Multiplication is the easiest nontrivial arithmetic function. *Theoretical Comput. Sci. 36* (1985), 333–339.
5. BAILEY, D. H. A portable high performance multiprecision package. RNR Technical Report RNR-90-022, NAS Applied Research Branch, NASA Ames Research Center, Moffett Filed, CA 94035, May 1993.
6. BLUM, L., CUCKER, F., SHUB, M., AND SMALE, S. *Complexity and Real Computation*. Springer, New York, 1997.
7. BORWEIN, J. M., AND BORWEIN, P. B. The arithmetic-geometric mean and fast computation of elementary funtions. *SIAM Review 26*, 3 (1984), 351–366.
8. BORWEIN, J. M., AND BORWEIN, P. B. *Pi and the AGM*. John Wiley and Sons, New York, 1987.
9. BORWEIN, J. M., AND BORWEIN, P. B. On the complexity of familiar functions and numbers. *SIAM Review 30*, 4 (1988), 589–601.
10. BRENT, R. P. Fast multiple-precision evaluation of elementary functions. *J. ACM 23*, 2 (1976), 242–251.
11. BRENT, R. P. Multiple-precision zero-finding methods and the complexity of elementary function evaluation. In *Analytic Computational Complexity*, TRAUB, J. F., Ed. Academic Press, New York, 1976, pp. 151–176.
12. KANADA, Y. Vectorization of multiple-precision arithmetic program and $201,326,000$ decimal digits of π calculation. *Supercomputing 88: Volume II, Science and Applications* (1988), 117–128. In *Pi: A Source Book*, BERGGREN, L., BORWEIN, J. M., AND BORWEIN, P. B., Eds. Springer, New York, 1997.
13. MULLER, J.-M. *Elementary Functions: Algorithms and Implementation*. Birkhäuser, Boston, 1997.
14. OKABE, Y., TAKAGI, N., AND YAJIMA, S. Log-depth circuits for elementary functions using residue number system. *Electronics and Communications in Japan, Part 3 74*, 8 (1991), 31–38.
15. REIF, J. Logarithmic depth circuits for algebraic functions. In *Proc. 24th Ann. IEEE Symposium on Foundations of Computer Science* (1983), pp. 138–145.
16. SASAKI, T., AND KANADA, Y. Practically fast multiple-precision evaluation of $\log(x)$. *Journal of Information Processing 4*, 4 (1982), 247–250.
17. SCHÖNHAGE, A. Routines for square roots. Program documentation of the TP-routines SQRT, ISQRT, CSQRT, unpublished manuscript, July 1991.

18. SCHÖNHAGE, A., GROTEFELD, A. F. W., AND VETTER, E. *Fast Algorithms: a Multitape Turing Machine Implementation.* Bibliographisches Institut, Mannheim, 1994.
19. SCHÖNHAGE, A., AND STRASSEN, V. Schnelle Multiplikation großer Zahlen. *Computing 7* (1971), 281–292.
20. WEGENER, I. *Effiziente Algorithmen für grundlegende Funktionen.* Teubner, Stuttgart, 1989.
21. WOLFRAM, S. *Mathematica: A System for Doing Mathematics by Computer,* 2nd ed. Addison-Wesley, Redwood City, California, 1991.

A Model of Behaviour Abstraction for Communicating Processes

Maciej Koutny[1] and Giuseppe Pappalardo[2]

[1] Dept. of Comp. Sci., University of Newcastle, Newcastle upon Tyne NE1 7RU, U.K.
[2] Dipart. di Matematica, Università degli Studi di Catania, I-95125 Catania, Italy

Abstract. We investigate the notion that a system, or process, is an acceptable implementation of another base or target process, in the case that they have different interfaces. Base processes can be thought of as specifications, or ideal processes operating in an error-free environment, while implementations model their actual realisation, possibly employing a variety of fault-tolerant techniques. Using the CSP model, we relate implementations to targets in terms of their observable behaviours, through interface abstraction. We obtain two basic results: realisability and compositionality. The former ensures an implementation up to interface abstraction can be put to good use, in the sense that plugging it into an appropriate environment yields a conventional implementation. Compositionality requires that a target made up of subcomponents can be implemented by assembling their respective implementations.
Keywords: Theory of parallel and distributed computation, behaviour abstraction, communicating sequential processes.

1 Introduction

This work aims at formalising the notion that a system is an acceptable *implementation* of another *base* or *target* system, in the case that the two systems have different interfaces. The goal is to provide satisfactory answers to such questions as: In what sense does a serial link implement an abstract byte channel? Can handshake performed by a pair of asynchronous channels be seen as an implementation of a single synchronous channel? What does it mean that a replicated clock implements a single, precise clock?

To our knowledge the proposed problem has been paid scant attention so far. Yet, it has a clear practical interest extending e.g. to some fundamental fault tolerance techniques. Indeed, we have applied restricted versions of the treatment proposed [4,7] to the analysis of NMR (N-Modular Redundancy), whereby N possibly faulty copies of a base system mimic the behaviour of a non-faulty base. Another application we investigated [5] is the formalisation of CA actions, a structuring concept for distributed fault-tolerant systems [9].

Whatever the answer to the issue raised, it is clear that the implementation and the target should be compared *extensionally*, i.e. in terms of their observable behaviours, as determined by the communication taking place at their respective interfaces. We add that, when interfaces differ, communication performed by

C. Meinel and S. Tison (Eds.): STACS'99, LNCS 1563, pp. 313–322, 1999.

the implementation through its actual interface should first be interpreted as communication through the target's interface: the interpretation outcome is then suitable for comparison with the target's intended behaviour. Since it is natural to consider the target's communication as more abstract and understandable, we regard the noted interpretation as a form of abstraction or decoding.

Conventional formal methods can cope with interface difference between implementation and target only to a limited extent, by the two interpretation patterns represented by (partial) interface renaming and hiding. No special treatment is however needed in this case, because our notion of implementation oriented to interface interpretation is easily reduced to conventional ones. For in most formalisms hiding or renaming can be applied not only to *actions* (atomic instances of communication), but also as an operator, say f, to a system as a whole. Thus if P is a formal object representing the implementation system, $f(P)$ is also an object, whose communication is that of P interpreted through f. That P implements a target system rendered as K can be expressed as:

$$f(P) \, \mathcal{R} \, K \tag{1}$$

where relation $\mathcal{R}$ depends on the formalism. It could be, e.g., an observation preorder holding between process expressions $f(P)$ and K, or even satisfaction of assertion K by (the meaning of) $f(P)$ in a logic based framework.

However, the (rather obvious) reduction expressed by (1) is not applicable whenever interpretation cannot be lifted to the system level, as an operator f on systems. Indeed, it is often the case that interpretation can only be expressed at the level of communication *traces* (sequences of actions), as a mapping of traces at the implementation interface onto traces at the target interface; good examples are provided by NMR majority voting, or by a serial link viewed as an abstract byte channel. Furthermore, if correctness analysis is to be extended to non-deterministic aspects of communication, these are also to be interpreted across interface pairs. The following treatment will show precisely how useful interpretations need to be characterised by mathematical structures called *extraction patterns*. Anyway, it should be already clear that an extraction pattern, say again f with some abuse, is too complicated a device to be made into an operator on systems. The approach summarised by (1) is therefore not viable. Alternatively, we have to try and relate directly the implementation P and the target K, building straight into their relation the interpretation expressed by the extraction pattern f. Formally, we try to establish:

$$P \, \mathcal{R}_f \, K \tag{2}$$

What kind of device should the *implementation relation* (as we call it) $\mathcal{R}_f$ be?

For the problem at hand to be worth studying in general terms, f should play the role of an argument replacing a 'formal parameter' in a generic relation 'scheme' $\mathcal{R}_-$ (this is the case for those proposed later). But we saw no obvious or useful way of endowing standard relations, like e.g. observation preorders, with an extraction pattern parameter, so it turned out that we were treading on completely new ground here. Moreover, we did not it find easy to identify a single implementation relation, proving immediately acceptable to intuition. In fact, different relation schemes are likely to turn out to be useful for different applications, system classes or interconnection topologies.

On the other hand, candidate implementation relations can be tied firmly to intuition by placing two light but very natural requirements on them. The first, *accessibility* or *realisability*, ensures that the abstraction built into the implementation relation may be put to good use; in practice, this means that plugging an implementation into an appropriate environment[1] should yield a conventional implementation of the target. *Distributivity* or *compositionality*, the other constraint on the implementation relation, requires it to distribute over system composition; thus, a target composed of two connected systems may be implemented by connecting two of their respective implementations.

Looking at (2) reveals we must still make some choices to complete the formal framework of our investigation on the nature of the implementation relation $\mathcal{R}_f$. The implementation P must of course be rendered as a process expression (of which language is a secondary issue). The target K could be formalised as either another process expression or as a logical assertion. However, given the conceptual difficulties of relating P to K taking interpretation of communication into account, we see no reason to widen the semantic gap between P and K.

Finally, we need a semantic model for process expressions P and K. As noted above, the extraction pattern denoted by f in (2) normally interprets communication in terms of simple observable communication records, like traces, *refusals* (sets of actions refused) and *failures* (pairs combining a trace with a possible refusal after that trace). Thus, incorporating f smoothly into the implementation relation scheme requires that the meaning of a process should be firmly and directly based on its observable communication records. This is certainly true for the explicit FD (failure-divergence) model of CSP [1,3]; much less so for models that equate meaning to an observation equivalence class (see e.g. [6]), or for algebraic models that even depend on the set of process operators and axioms adopted [2]. Besides, the necessary choice of a specific equivalence or operator and axiom set would be unattractive and potentially misleading in the context of this investigation. In contrast, our treatment will not even require any combining device except process composition.

The FD model adopted is expressive, well-understood and provides a wealth of important theoretical results. A treatment could also be given within a pure trace model [7], but this would only allow partial (functional) correctness to be studied, whereas failures afford the ability to describe internal nondeterminism (hence deadlock properties). Divergences are traces after which a process might become uncontrollable, or diverge. While their presence can be regarded as a nuisance for applications like protocol verification [8], in the context at hand it provides a means of forbidding a network of implementations to diverge if the network of the respective targets does not. Altogether, the FD model provides a framework to reason about quite thorough a form of correctness.

---

[1] In [4] we build this environment with *disturbers*, feeding the implementation faulty but sufficiently redundant input, and *extractors* interpreting its output.

Finally, CSP is also well-suited to describing the intended family of target systems. Systems considered possess an interface that may be conceptually divided into ainput and output channels. Moreover, input channels should never refuse any input. These light restrictions do not imply a simpler asynchronous communication model could be employed. For we do not want to impose the same restrictions on implementations, both for the sake of generality and with a view to modelling controlled faults allowed within fault tolerant systems.

Proofs of all the theorems stated later and related results are in [5]. The reader ought to be familiar with the basic concepts of CSP [1,3].

2 Preliminaries

In the FD model of CSP [1,3] a process P is a triple $(\alpha P, \phi P, \delta P)$ where αP (alphabet) is a non-empty finite set of actions, ϕP (failures) is a subset of $\alpha P^* \times 2^{\alpha P}$, and δP (divergences) is a subset of αP^*. The set of traces of P is $\tau P = \{t \mid (t, R) \in \phi P\}$. We use structured actions of the form $b!v$, where v is a message and b a communication channel. For every channel b, the alphabet αb of b is the set of all valid actions $b!v$. For a set of channels B, $\alpha B = \bigcup_{b \in B} \alpha b$. We will associate with P a set of channels, $\operatorname{chan} P$, partitioned into input channels, $\operatorname{in} P$, and output channels, $\operatorname{out} P$, and stipulate $\alpha P = \alpha \operatorname{chan} P$. For a process P, let $\psi P = \{t \mid (t, \alpha \operatorname{out} P) \in \phi P\}$ (after $t \in \psi P$, P refuses any further output for the input within t). The following notation is similar to that of [3] (below t, u are traces; b, b', b'' channels; T and T' sets of traces; and B a set of channels):

- $t = \langle a_1, \ldots, a_n \rangle$ is the trace with i-th element a_i, and length $|t| = n$;
- $t \circ u$ is the trace obtained by appending u to t;
- $\leq$ is the prefix relation on traces;
- $t[b'/b]$ is a trace obtained from t replacing each action $b!v$ by $b'!v$;
- $t \lceil B$ is obtained by deleting from t all the actions that do not occur on channels in B; e.g., $\langle b''!3, b!1, b''!2, b!3, b'!3 \rangle \lceil \{b, b'\} = \langle b!1, b!3, b'!3 \rangle$;
- $f : T \to T'$ is: monotonic if $t \leq u$, $t, u \in T$ imply $f(t) \leq f(u)$; strict if $f(\langle \rangle) = \langle \rangle$; a homomorphism if $t, u, t \circ u \in T$ imply $f(t \circ u) = f(t) \circ f(u)$.

We use these CSP operators: parallel composition $P \| Q$; hiding $P \backslash B$ of communication at the channel set B; deterministic choice $P \,\square\, Q$; non-deterministic choice $P \sqcap Q$; and prefixing $a \to P$. Processes $P_1, \ldots, P_n$ form a network if $\operatorname{in} P_i \cap \operatorname{in} P_j = \emptyset = \operatorname{out} P_i \cap \operatorname{out} P_j$ for $i \neq j$; if so, $P = P_1 \otimes \cdots \otimes P_n$ denotes the parallel composition of processes $P_1 \cdots P_n$ with interprocess communication hidden, i.e. $P = (P_1 \| \cdots \| P_n) \backslash B$, where B is the set of channels shared by at least two different processes. We stipulate $\operatorname{in} P = \bigcup_{i=1}^{n} \operatorname{in} P_i - \bigcup_{i=1}^{n} \operatorname{out} P_i$ and $\operatorname{out} P = \bigcup_{i=1}^{n} \operatorname{out} P_i - \bigcup_{i=1}^{n} \operatorname{in} P_i$. Throughout the paper, we will often refer to the processes P, K and L shown in Figure 1.

Let U and V be two non-empty prefix-closed sets of traces over respectively $\operatorname{in} P$ and $\operatorname{out} P$. Then P is input-guarded w.r.t. U and V if, for every set of traces $T \subseteq \tau P$ satisfying $T \lceil \operatorname{in} P \subseteq U$: (IG1) $T \lceil \operatorname{out} P \subseteq V$; and (IG2) if T is infinite then so is $T \lceil \operatorname{in} P$. We denote this by $P \in \mathsf{IG}(U, V)$.

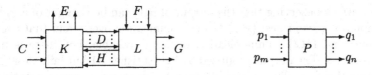

Fig. 1. Three generic processes (where $E \cup D \neq \emptyset$, $G \cup H \neq \emptyset$ and $n \geq 1$).

We consider as base processes the class GIO of *general input-output* processes. $P \in$ GIO iff it is input-guarded, i.e. $P \in$ IG(αin P^*, αout P^*), and never refuses any input, i.e. $R \cap \alpha$in $P = \emptyset$, for all $(t, R) \in \phi P$. An unbounded deterministic *buffer* with input channel p and output channel q, denoted by buffer$_{pq}$, belongs to GIO.

3   Extraction Patterns

Consider two processes, Sgl and Buf, as shown in Figure 2(a). The former generates a single binary pulse, belonging to $\mathbf{B} = \{0, 1\}$, on its output channel p, then terminates; the latter is a buffer process forwarding signals received on its input channel p. Sgl $= \Box_{v \in \mathbf{B}}\ p!v \rightarrow$ stop is a GIO process, and so is Buf $=$ buffer$_{pq}$. Suppose now Sgl and Buf are implemented using processes Sgl$_0$ and Buf$_0$ respectively, connected by two channels, r and s, as shown in Figure 2(b). Sgl$_0 = \Box_{v \in \mathbf{B}}\ (r!v \rightarrow$ stop $\|\ s!v \rightarrow$ stop), i.e. the signal is now duplicated by Sgl$_0$ and the two copies sent along r and s. However, Buf$_0$ (CSP definition omitted) only accepts one of the copies and passes it on, ignoring the other one. The scheme clearly works, since Sgl $\otimes$ Buf $=$ Sgl$_0 \otimes$ Buf$_0$. Suppose now communication is imperfect and two types of faulty behaviour can occur: Sgl$_1 =$ Sgl$_0 \sqcap$ stop and Sgl$_2 =$ Sgl$_0 \sqcap \Box_{v \in \mathbf{B}}\ r!v \rightarrow$ stop. I.e., Sgl$_1$ can break down completely, refusing to output any signals, while Sgl$_2$ can fail in a way that, although channel s is blocked, r can still transmit the signal. (This could model the following realistic situation: to improve performance, the 'slow' channel p is replaced by two channels, a high-speed unreliable channel s and a slow but reliable backup channel r.) Since Sgl $\otimes$ Buf $=$ Sgl$_2 \otimes$ Buf$_0$ but Sgl $\otimes$ Buf $\neq$ Sgl$_1 \otimes$ Buf$_0$ it follows that Sgl$_2$ is much 'better' an implementation than Sgl$_1$ of the Sgl process. We will now analyse the differences between Sgl$_1$ and Sgl$_2$ while introducing informally some basic concepts needed later.

Fig. 2. Two base processes and their implementations.

We start by observing that the output of Sgl_2 can be thought of as adhering to the following two rules: **(R1)** the communication over r and that over s are consistent; and **(R2)** communication over r is reliable, but there is no such guarantee for s. The output produced by Sgl_1 satisfies R1 but fails to satisfy R2. To express this formally, R1 and R2 must be set in some precise notation.

To capture the behavioural relationship between Sgl and Sgl_2 we will employ an (extraction) mapping extr which, given a trace over r and s, interprets it as a trace over p. E.g., $\langle \rangle \mapsto \langle \rangle$, $\langle r!0 \rangle \mapsto \langle p!0 \rangle$, $\langle s!0 \rangle \mapsto \langle p!0 \rangle$, $\langle s!1, r!1 \rangle \mapsto \langle p!1 \rangle$ and $\langle r!1, s!1 \rangle \mapsto \langle p!1 \rangle$. Note the extraction mapping need only be defined for traces satisfying R1. But the extraction mapping alone cannot suffice to identify the 'correct' implementation of Sgl in the presence of faults, since $\tau Sgl = \mathrm{extr}(\tau Sgl_1) = \mathrm{extr}(\tau Sgl_2)$. What one also needs is an ability to relate the refusals of Sgl_i with the refusals of the base process Sgl. This is much harder. For, simply applying extr also to refusals of Sgl_2 will not yield failures of Sgl. E.g. $(\langle \rangle, \{s!0\})$ is in ϕSgl_2, but the pair $(\mathrm{extr}(\langle \rangle), \{\mathrm{extr}(s!0)\}) = (\langle \rangle, \{p!0\})$ is *not* in ϕSgl: i.e., crude extraction of refusals is not going to work. A more sophisticated device is needed, in the form of another mapping, ref, constraining the possible refusals a process can exhibit after a given trace. This will help preventing, e.g., a sender process from blocking if its transmission is still incomplete. In the example at hand, this roughly amounts to requiring that communication on channel r should not be blocked before the sender, Sgl_2, has sent a signal over it. For example, we will stipulate that $\mathrm{ref}(\langle s!0 \rangle)$ must not comprise a refusal containing both $r!0$ and $r!1$. We will denote by dom the traces conveying information that can be regarded as complete. E.g. $\langle s!0 \rangle$ and $\langle s!1 \rangle$ should not belong to dom for, according to R2, we can expect them to be completed by some action over the reliable channel r.

The last notion we need in order to relate an implemantation and a target is a partial inverse inv of the extraction mapping. This will be used to ensure all the traces of a base process can be extracted from the traces of any implementation. Formally, an *extraction pattern* is defined for two non-empty sets of channels, B and B', respectively called *sources* and *targets*. It is a tuple $\mathrm{ep} = (\mathrm{dom}, \mathrm{extr}, \mathrm{ref}, \mathrm{inv})$ such that:

EP1   dom is a set of traces over sources; its prefix-closure is denoted Dom.

EP2   extr is a strict monotonic mapping for traces in Dom; for every t, $\mathrm{extr}(t)$ is a trace over targets.

EP3   ref is a mapping for traces in Dom; for every t, $\mathrm{ref}(t)$ is a non-empty family of proper subsets of αB such that $Y \subset X \in \mathrm{ref}(t)$ implies $Y \in \mathrm{ref}(t)$.

EP4   inv is a trace homomorphism from traces over targets to traces in Dom; for every trace w over targets, $\mathrm{extr}(\mathrm{inv}(w)) = w$.

The extraction mapping is monotonic, as receiving more information should not decrease the current knowledge about information exchange. $\alpha B \notin \mathrm{ref}(t)$ means that for the unfinished communication t we do not allow the sender to refuse all possible transmission. Since inv is a trace homomorphism, it suffices to define it for single actions over targets only. Note that ep is not decorated with source and target channels explicitly. We assume these can always be retrieved from the domains of extr and inv. By default, different extraction patterns have disjoint sources and disjoint targets, and the components of extraction patterns can be annotated to avoid ambiguity. A *basic* extraction pattern (see [4]) is one with a singleton target channel.

Let ep_1 and ep_2 be two extraction patterns with sources (targets) respectively B_1 and B_2 (B'_1 and B'_2). Then $ep = ep_1 \oplus ep_2$ is defined as the extraction pattern with sources $B = B_1 \cup B_2$ and targets $B' = B'_1 \cup B'_2$ such that:

$$\text{dom} \quad = \{t \in (\text{dom}_1 \cup \text{dom}_2)^* \mid t \lceil B_1 \cup B'_1 \in \text{dom}_1 \wedge t \lceil B_2 \cup B'_2 \in \text{dom}_2\}$$

$$\text{ref}(t) \quad = \{R \in \alpha B \mid R \cap \alpha B_1 \in \text{ref}_1(t \lceil B_1) \vee R \cap \alpha B_2 \in \text{ref}_2(t \lceil B_2)\}$$

$$\text{extr}(t \circ \langle a \rangle) = \begin{cases} \text{extr}(t) \circ u_1 & \text{if } a \in \alpha B_1 \wedge \text{extr}_1(w \lceil B_1 \circ \langle a \rangle) = \text{extr}_1(w \lceil B_1) \circ u_1 \\ \text{extr}(t) \circ u_2 & \text{if } a \in \alpha B_2 \wedge \text{extr}_2(w \lceil B_2 \circ \langle a \rangle) = \text{extr}_2(w \lceil B_2) \circ u_2 \end{cases}$$

$$\text{inv}(a) \quad = \text{inv}_i(a) \quad \text{if } a \in \alpha B'_i, \text{ for } i = 1, 2$$

The $\oplus$ operation is both associative and commutative and provides a way of building complex extraction patterns from simpler ones. An identity extraction pattern for a channel b, id_b, is one for which $B = B' = \{b\}$, $\text{dom} = \text{Dom} = \alpha b^*$, $\text{extr}(t) = \text{inv}(t) = t$ and $\text{ref}(t) = \{R \mid \alpha b \not\subseteq R\}$. For a set of channels $B'' = \{b_1, \ldots, b_k\}$, $id_{B''} = id_{b_1} \oplus \cdots \oplus id_{b_k}$.

For our example implementations of Sgl and Buf, the extraction pattern needed fs is defined by: $\text{dom} = \{t \in (\alpha r \cup \alpha s)^* \mid (t \lceil s)[p/s] \leq (t \lceil r)[p/r]\}$, $\text{ref}(t) = \{R \mid \alpha r \not\subseteq R\}$, $\text{extr}(t) = \max\{(t \lceil s)[p/s], (t \lceil r)[p/r]\}$ and $\text{inv}(p!v) = \langle r!v \rangle$.

4   Extraction in Acyclic Networks

Suppose that we implemented the base GIO process P using another process Q. The correctness of the implementation will be expressed in terms of two extraction patterns, ep and ep$'$. The former (with sources in Q and targets in P) will be used to relate the communication on the input channels of P and Q; ep$'$ (with sources out Q and targets out P) will serve a similar purpose for the output channels. There are four main properties Q has to satisfy: Firstly, if a trace t of Q projected on its input channels can be interpreted by ep, then it should be possible to interpret its projection on the output channels by ep$'$ (see WI1 below and IG1). Secondly, when connected to another process and supplied with valid input (i.e. belonging to Dom), Q should not introduce divergence (this rules out infinite uninterrupted communication on output channels — see WI1 and IG2), and should not refuse 'proper' input (this rules out refusals which might lead to a deadlock with another process that provides input to Q and whose refusals are constrained by ep — see WI2). Thirdly, if Q has received a complete input from process S and is to output to process W, then Q should not deadlock with W until it has sent W all the output produced (WI3). Finally, we ensure the functional behaviour of P (in terms of traces) can be realised by Q (WI4). Formally, Q is a *weak implementation* of P, denoted $Q \in \text{WI}(P, \text{ep}, \text{ep}')$, if:

WI1 $Q \in \text{IG}(\text{Dom}, \text{Dom}')$.

WI2   If $(t, R) \in \phi Q$ is such that $t \lceil \text{in } Q \in \text{Dom}$ then $\alpha \text{in } Q - R \not\subseteq \text{ref}(t \lceil \text{in } Q)$.

WI3   If $(t, R) \in \phi Q$ is such that $t \lceil \text{in } Q \in \text{dom}$ and $\alpha \text{out } Q \cap R \not\subseteq \text{ref}'(t \lceil \text{out } Q)$
then $t \lceil \text{out } Q \in \text{dom}'$ and $\text{extr}_{\text{ep} \oplus \text{ep}'}(t) \in \psi P$.

WI4 $\text{inv}_{\text{ep} \oplus \text{ep}'}(\tau P) \subseteq \tau Q$.

One can see that $P \in \text{WI}(P, \text{id}_{\text{in } P}, \text{id}_{\text{out } P}) \subseteq \text{GIO}$. For our example processes, we
have $\text{Sgl}_2 \in \text{WI}(\text{Sgl}, \text{fs})$ (we omitted ep since there are no input channels) and
$\text{Sgl}_1 \not\in \text{WI}(\text{Sgl}, \text{fs})$. Indeed, $\text{Sgl}_1 \not\in \text{WI}(\text{Sgl}, \text{ep}')$ for any extraction pattern ep' since
$(\langle \rangle, \alpha r \cup \alpha s) \in \text{Sgl}_1$ and $\alpha r \cup \alpha s \not\subseteq \text{ref}'(\langle \rangle)$ (see EP3).

To support modular treatment of networks of implementations we will now
establish two properties, namely compositionality and realisability. Referring to
Figure 1, the former is formulated thus.

Theorem 1. *Let $H = \emptyset$ and c, d, e, f, g be extraction patterns whose targets
are respectively the channel sets C, D, E, F, G. If $M \in \text{WI}(K, c, d \oplus e)$ and $N \in$
$\text{WI}(L, d \oplus f, g)$ then $M \otimes N \in \text{WI}(K \otimes L, c \oplus f, e \oplus g)$.*   □

Realisability is captured below where for GIO processes P and Q with the same
input and output channels, $Q \preceq P$ if $\tau Q = \tau P$ and $\psi Q \subseteq \psi P$.

Theorem 2. *If $Q \in \text{WI}(P, \text{id}_{\text{in } P}, \text{id}_{\text{out } P})$ then $Q \preceq P$.*   □

Why can this result be regarded as expressing an adequate notion of realisability?
First, note that given $Q' \in \text{WI}(P, \text{ep}, \text{ep}')$, where ep and ep' are general extraction
patterns, compositionality in many situations allows $Q \in \text{WI}(P, \text{id}_{\text{in } P}, \text{id}_{\text{out } P})$
to be constructed, e.g., using the extractors and disturbers discussed in [4,5].
As for the $\preceq$ relation, think of an environment for processes Q and P which
is represented by an arbitrary GIO process Rcv which is to receive the results
produced by Q and P, as shown in Figure 3. It turns out that $Q \preceq P$ implies:
$\tau(Q \otimes \text{Rcv}) = \tau(P \otimes \text{Rcv})$ and $\phi(Q \otimes \text{Rcv}) \subseteq \phi(P \otimes \text{Rcv})$. The former property
means that as far as functionality is concerned, $Q \otimes \text{Rcv}$ and $P \otimes \text{Rcv}$ are equivalent
processes. The latter property means that $Q \otimes \text{Rcv}$ is at least as deterministic a
process as $P \otimes \text{Rcv}$ in the sense of [3]. This makes $Q \otimes \text{Rcv}$ an as good as $P \otimes \text{Rcv}$
(and possibly much better) process to be used in the actual implementation. Note
that the underlying philosophy of [3] is to allow as non-deterministic processes
as possible for specification, but as deterministic as possible ones for actual
implementation; our approach is thus in line with that advocated there. Hence
Theorem 2 captures an adequate notion of realisability.

The above realisability result can be strengthened if we assume that $t \circ \langle a \rangle \not\in$
τP, for all $t \in \psi P$ and $a \in \alpha \text{out } P$ (this means P can refuse to generate any
output only if its functional specification in terms of traces rules this out). We
denote $P \in \text{GIO}'$ in such a case. GIO' is a subset of GIO, but is not closed under
composition $\otimes$. It is easily seen that if $P \in \text{GIO}'$ then $Q \otimes \text{Rcv} = P \otimes \text{Rcv}$.

The compositionality result of Theorem 1 is easily extended to process networks. Indeed, let $P_1, \ldots, P_k$ be a network of base GIO processes, and let Q_i
be a weak implementation of P_i, for all i. Using Theorem 1 and induction it
is straightforward to prove that the network $Q_1, \ldots, Q_k$ represents a weak implementation of the network $P_1, \ldots, P_k$. However, the possible topology of the
latter is restricted by the $H = \emptyset$ hypothesis of Theorem 1. This implies the base
network must be *acyclic* in the obvious sense.

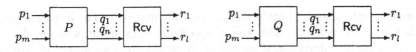

Fig. 3. Realisability of GIO processes

5 Extraction in Cyclic Networks

As demonstrated in [5], weak implementation is not sufficient to deal with cyclic networks of processes. We now take the pair of base processes K and L, as shown in Figure 1, without assuming $H = \emptyset$. The first problem we face is that $K \otimes L$ will not, in general, be a GIO process even though both K and L are. Our treatment is therefore restricted to well-behaved cases which do not lead to divergence. We say that the processes K and L are *compatible* if for every infinite set of traces $T \subseteq \tau(K \| L)$, the set $T \lceil \text{in}\, (K \otimes L)$ is also infinite. It follows that, if K and L are compatible GIO processes, then $K \otimes L \in$ GIO.

With the same assumptions as in the definition of WI, Q is a *strong implementation* of P, denoted $Q \in$ SI$(P, \text{ep}, \text{ep}')$, if the following hold:

SI1 If $T \subseteq \tau Q$ and $T \lceil \text{in}\, Q \subseteq$ Dom then: $T \lceil \text{out}\, Q \subseteq$ Dom$'$; if T is infinite then so is extr$(T \lceil \text{in}\, Q)$; and extr$_{\text{ep} \oplus \text{ep}'}(T) \subseteq \tau P$.

SI2 If $(t, R) \in \phi Q$ is such that $t \lceil \text{in}\, Q \in$ Dom then $\alpha \text{in}\, Q - R \notin$ ref$(t \lceil \text{in}\, Q)$.

SI3 If $(t, R) \in \phi Q$ is such that $t \lceil \text{in}\, Q \in$ Dom and $\alpha \text{out}\, Q \cap R \notin$ ref$'(t \lceil \text{out}\, Q)$ then: $t \lceil \text{out}\, Q \in$ dom$'$; and $t \in$ dom$_{\text{ep} \oplus \text{ep}'} \Rightarrow$ extr$_{\text{ep} \oplus \text{ep}'}(t) \in \psi P$.

SI4 inv$_{\text{ep} \oplus \text{ep}'}(\tau P) \subseteq \tau Q$.

The main difference between this and the previous definition is that IG2 in the definition of input-guardedness has been replaced by a stronger condition which can be interpreted as input-guardedness relative to extraction. Moreover, we have strengthened WI3 and added a third condition in SI1 to explicitly state that extracted valid traces of Q are traces of P. Being strong implementation implies being also weak implementation (but not vice versa); hence realisability follows from Theorem 2. Compositionality is as follows.

Theorem 3. *Let K and L be two compatible GIO processes, and let c, d, e, f, g and h be extraction patterns whose targets are respectively the channel sets C, D, E, F, G and H. If $M \in$ SI$(K, c \oplus h, d \oplus e)$ and $N \in$ SI$(L, d \oplus f, g \oplus h)$ then $M \otimes N \in$ SI$(K \otimes L, c \oplus f, e \oplus g)$.* □

Thus strong implementation is suitable for dealing with cyclic process networks, provided the base network can be decomposed down to single network components in such a way that at each stage compatibility is satisfied.

6 Concluding Remarks

We have presented a framework for relating a target process with its implementation, when each communicates with the environment through a possibly different interface, and/or pattern of information exchange. The framework supports a natural and expressive way of specifying fault assumptions as demonstrated in [4,5,7]. Moreover, it should be applicable to a variety of distributed architectures, provided the base processes employed as high level specifications adhere to the GIO properties (cf. Section 2). It should be noted that although the GIO process class is related to some of the existing models of communication with non-blocking input (such as IO-automata), this is not necessarily the case for processes implementing them. Therefore CSP, being more general a model, seems to provide a more appropriate framework.

This paper extends previous results in two ways, by introducing non-base extraction patterns and a more powerful notion of strong implementation for cyclic networks of processes. Future work will concentrate on extending the theory to cover non-GIO base processes, and on developing algorithms for verifying the implementation relations.

Acknowledgement We thank the referees for their useful comments.

References

1. S. D. Brookes and A. W. Roscoe. An improved failures model for communicating processes. Lecture Notes in Computer Science 197:281–305, 1985. Springer-Verlag.
2. J. C. M. Baeten and W. P. Weijland. *Process Algebra*. Camb. Univ. Press, 1990.
3. C. A. R. Hoare. *Communicating Sequential Processes*. Prentice Hall, 1985.
4. M. Koutny, L. Mancini, and G. Pappalardo. Two implementation relations and the correctness of communicating replicated processes. *Formal Aspects of Computing* 9:119-148, 1997.
5. M. Koutny and G. Pappalardo. The ERT model of fault-tolerant computing and its application to a formalisation of coordinated atomic actions. Technical Report 636, Dept. of Comp. Sci., University of Newcastle upon Tyne, 1998.
6. R. Milner. *Communication and Concurrency*. Prentice-Hall, 1989.
7. L. V. Mancini and G. Pappalardo. Towards a theory of replicated processing. Lecture Notes in Computer Science 331:175–192, 1988. Springer-Verlag.
8. G. Pappalardo. *Specification and Verification Issues in a Process Language*. PhD thesis, University of Newcastle upon Tyne, 1995.
9. J. Xu, B. Randell, A. Romanovsky, C. Rubira, R. Stroud, and Z. Wu. Fault tolerance in concurrent object-oriented software through coordinated error recovery. In *Proc. 25th Int. Symp. on Fault-Tolerant Comp.*, 450–457. IEEE Press, 1995.

Model Checking Lossy Vector Addition Systems

Ahmed Bouajjani[1] and Richard Mayr[2]

[1] VERIMAG, Centre Equation, 2 avenue de Vignate, 38610 Gières, France.
Ahmed.Bouajjani@imag.fr
[2] Institut für Informatik, TU-München, Arcisstr. 21, D-80290 München, Germany.
mayrri@informatik.tu-muenchen.de

Abstract. Lossy VASS (vector addition systems with states) are defined as a subclass of VASS in analogy to lossy FIFO-channel systems. They can be used to model concurrent systems with unreliable communication. We analyze the decidability of model checking problems for lossy systems and several branching-time and linear-time temporal logics. We present an almost complete picture of the decidability of model checking for normal VASS, lossy VASS and lossy VASS with test for zero.

1 Introduction

VASS's (vector addition systems with states) can model communicating systems through unbounded unordered buffers, and hence they can be seen as abstractions of fifo-channels systems, when the ordering between messages in the channels is not relevant but only their number. Communicating systems are often analyzed under the assumption that they communicate through unreliable channels. Hence, we consider *lossy* models of communicating systems, i.e. models where messages can be lost. Recent works are about lossy unbounded fifo-channels systems [AJ93, AJ96, CFI96]. The reachability problem is decidable for these models, which implies the decidability of the verification problem for safety properties. However, liveness properties cannot be checked for lossy fifo-channel systems, unless for very special ones like single eventualities. In particular, it is impossible to model check lossy channel systems under fairness conditions. Here we study verification problems for VASS and VASS with inhibitor arcs (counter machines) under the assumption of lossiness, i.e. the contents of a place/counter can spontaneously get lower at any time.

Using the approach introduced in [CFI96, ACJT96], it can be shown very easily that the set $pre^*(S)$ of predecessors of any set of configurations S is effectively constructible for lossy VASS even with inhibitor arcs, and that this set can be represented by simple linear constraints (SC for short), where integer variables can be compared only with constants. Moreover, for lossy VASS, the set $post^*(S)$ of successors is SC definable and effectively constructible, but interestingly, for lossy VASS with inhibitor arcs these sets are not constructible although they are SC definable.

Local model checking, or simply *model checking*, consists in deciding whether a given configuration of a system satisfies a given formula of a temporal logic, and

C. Meinel and S. Tison (Eds.): STACS'99, LNCS 1563, pp. 323–333, 1999.
© Springer-Verlag Berlin Heidelberg 1999

global model checking consists in constructing the set of all configurations that satisfy a given formula. We address these problems for a variety of linear-time and branching-time properties. We express these properties in a temporal logic, called AL (Automata Logic), which is based on automata on finite and infinite sequences to specify path properties (in the spirit of ETL), and the use of path quantifiers to express branching-time properties (like in ECTL* [Tho89]). The basic state predicates in this logic are SC constraints.

Our main positive result is that for lossy VASS, the global model checking is decidable for the logic ∃AL with only upward closed constraints, and dually for ∀AL with downward closed constraints (∀AL and ∃AL are the universal and existential positive fragments of AL. They subsume respectively the corresponding well-known fragments ∀CTL* and ∃CTL* [GL94] of the logic CTL*). When only infinite paths are considered our decidability result also holds for normal VASS. A corollary is that linear-time properties on finite and infinite paths (on infinite paths only) are decidable for lossy VASS (normal VASS). We can even construct the set of all the configurations satisfying these properties. This generalizes the result in [Esp97] where only model checking is considered. Notice also that ∀AL is strictly more expressive than all linear-time temporal logics.

These decidability results break down if we relax any of the restrictions: model checking becomes undecidable if we consider ∀AL or ∃AL formulae with both downward and upward closed constraints, or if we consider lossy VASS with inhibitor arcs. Also, even if we use only propositional constraints in the logic (i.e., only constraints on control locations) the use of negation must be restricted: model checking is undecidable for CTL and lossy VASS. However, it is decidable for the fragments EF and EG of CTL even for lossy VASS with inhibitor arcs, but surprisingly, global model checking is undecidable for EG and lossy VASS (while it is decidable for EF and lossy VASS with inhibitor arcs). As a side effect we obtain that normal VASS (Petri nets) and lossy VASS with inhibitor arcs (lossy counter machines) are incomparable.

The missing proofs can be found in the full version of the paper.

2   Vector Addition Systems with States

Definition 1. *A n-dim VASS S is a tuple $(\Sigma, \mathcal{X}, Q, \delta)$ where Σ is a set of action labels, $\mathcal{X}$ is a set of variables such that $|\mathcal{X}| = n$, Q is a finite set of control states, δ is a finite set of transitions of the form (q_1, a, Δ, q_2) where $a \in \Sigma$, $\Delta \in \mathbb{Z}^n$.*

A *configuration* of S is a pair $\langle q, u \rangle$ where $q \in Q$ and $u \in \mathbb{N}^n$. Let $\mathcal{C}(S)$ be the set of configurations of S. Given a configuration $s = \langle q, u \rangle$, we let $State(s) = q$ and $Val(s) = u$.

We define a *transition relation* $\longrightarrow$ on configurations as follows: $\langle q_1, u_1 \rangle \overset{a}{\longrightarrow} \langle q_2, u_2 \rangle$ iff $\exists \tau = (q_1, a, \Delta, q_2) \in \delta$, $u_2 = u_1 + \Delta$. Let $post_\tau(\langle q_1, u_1 \rangle)$ (resp. $pre_\tau(\langle q_2, u_2 \rangle)$) denote the configuration $\langle q_2, u_2 \rangle$ (resp. $\langle q_1, u_1 \rangle$), i.e., the immediate successor (resp. predecessor) of $\langle q_1, u_1 \rangle$ (resp. $\langle q_2, u_2 \rangle$) by the transition τ. Then, we let *post* (resp. *pre*) denote the union of the $post_\tau$'s (resp. pre_τ's)

for all the transitions $\tau \in \delta$. In other words, $post(\langle q, \boldsymbol{u} \rangle) = \{\langle q', \boldsymbol{u}' \rangle \; : \; \exists a \in \Sigma. \langle q, \boldsymbol{u} \rangle \xrightarrow{a} \langle q', \boldsymbol{u}' \rangle\}$, and $pre(\langle q, \boldsymbol{u} \rangle) = \{\langle q', \boldsymbol{u}' \rangle \; : \; \exists a \in \Sigma. \langle q', \boldsymbol{u}' \rangle \xrightarrow{a} \langle q, \boldsymbol{u} \rangle\}$.
Let $post^*$ and pre^* be the reflexive-transitive closures of $post$ and pre.

Given a configuration s, a *run* of the system S starting from s is a finite or infinite sequence $s_0 a_0 s_1 a_1 \ldots s_n$ such that $s = s_0$ and, for every $i \geq 0$, $s_i \xrightarrow{a_i} s_{i+1}$. We denote by $Run_f(s, S)$ (resp. $Run_\omega(s, S)$) the set of finite (resp. infinite) runs of S starting from s.

A *lossy VASS* is defined as a VASS with a *weak transition relation* $\Longrightarrow$ on configurations. We define the relation $\Longrightarrow$ as follows: $\langle q_1, \boldsymbol{u}_1 \rangle \xRightarrow{a} \langle q_2, \boldsymbol{u}_2 \rangle$ iff $\exists \boldsymbol{u}'_1, \boldsymbol{u}'_2 \in I\!\!N^n$, $\boldsymbol{u}_1 \geq \boldsymbol{u}'_1$, $\langle q_1, \boldsymbol{u}'_1 \rangle \xrightarrow{a} \langle q_2, \boldsymbol{u}'_2 \rangle$, and $\boldsymbol{u}'_2 \geq \boldsymbol{u}_2$.

The weak transition relation induces corresponding notions of runs, successor and predecessor functions defined by considering the weak transition relation $\Longrightarrow$ instead of $\longrightarrow$.

Definition 2. *We order vectors of natural numbers by* $(u_1, \ldots, u_n) \leq (v_1, \ldots, v_n)$ *iff* $\forall i \in \{1, \ldots, n\}. u_i \leq v_i$.
Given a set $S \subseteq I\!\!N^n$, *we denote by* $min(S)$ *the set of minimal elements of* S *w.r.t. the relation* $\leq$.
Let $S \subseteq I\!\!N^n$. *Then,* S *is* upward *(resp.* downward*) closed iff* $\forall \boldsymbol{u} \in I\!\!N^n. \boldsymbol{u} \in S \Rightarrow (\forall \boldsymbol{v} \in I\!\!N^n. \boldsymbol{v} \geq \boldsymbol{u}$ *(resp.* $\boldsymbol{v} \leq \boldsymbol{u}) \Rightarrow \boldsymbol{v} \in S)$. *Given a set* $S \subseteq I\!\!N^n$, *we denote by* $S\!\uparrow$ *(resp.* $S\!\downarrow$*) the* upward *(resp.* downward*) closure of* S, *i.e., the smallest upward (resp. downward) closed set which contains* S.

Lemma 3. *Every set* $S \subseteq I\!\!N^n$ *has a finite number of minimal elements. A set is upward closed if and only if* $S = min(S)\!\uparrow$. *The union and the intersection of two upward (resp. downward) closed sets is an upward (resp. downward) closed set. The complement of an upward closed set is downward closed and vice-versa.*

Definition 4 (Simple constraints, upward/downward closed constraints).
Let $\mathcal{X} = \{x_1, \ldots, x_n\}$ *be a set of variables ranging over* $I\!\!N$.

1. *A* simple constraint *over* $\mathcal{X}$, *SC for short, is any boolean combination of constraints of the form* $x \geq c$ *where* $x \in \mathcal{X}$ *and* $c \in I\!\!N \cup \{\infty\}$.
2. *An* upward closed *(resp.* downward closed*) constraint over* $\mathcal{X}$, *UC (resp. DC) for short, is any positive boolean combination of constraints of the form* $x \geq c$ *(resp.* $x < c$*) where* $x \in \mathcal{X}$ *and* $c \in I\!\!N \cup \{\infty\}$.

Constraints are interpreted in the standard way as a subset of $I\!\!N^n$ ($\leq$ is the usual ordering and $<$ is the strict inequality). Given a simple constraint ξ, we let $[\![\xi]\!]$ denote the set of vectors in $I\!\!N^n$ satisfying ξ. Notice that the constraints $x < 0$ and $x \geq \infty$ correspond to $\emptyset$ and that $x \geq 0$ and $x < \infty$ correspond to $I\!\!N$.

Definition 5. *A set* S *is SC (resp. UC, DC) definable if there exists an SC (resp. UC, DC)* ξ *such that* $S = [\![\xi]\!]$.

Definition 6 (Normal forms).
1. A canonical product *is a constraint of the form* $\ell \leq x \leq u$,
2. A canonical upward closed product *is a constraint of the form* $\ell \leq x$,
3. A canonical downward closed product *is a constraint of the form* $x \leq u$,

where $\ell \in I\!N^n$ *and* $u \in (I\!N \cup \{\infty\})^n$.
A SC (resp. UC, DC) in normal form *is either* $\emptyset$, *or a finite disjunction of canonical (resp. canonical upward closed, canonical downward closed) products.*

Lemma 7. *Every SC (resp. UC, DC) is equivalent to a SC (UC, DC) in normal form.*

Proposition 8. *SC definable sets are closed under boolean operations, and UC definable sets as well as DC definable sets are closed under union and intersection. The complement of a UC definable set is a DC definable set and vice-versa. A subset of* $I\!N^n$ *is UC definable (resp. DC definable) if and only if it is an upward (resp. downward) closed set. A set is SC definable if and only if it is a boolean combination of upward closed sets.*

Let $\mathcal{S} = (\Sigma, \mathcal{X}, Q, \delta)$ be a n-dim VASS with $Q = \{q_1, \ldots, q_m\}$. Then, every set of configurations of $\mathcal{S}$ is defined as a union $C = \{q_1\} \times S_1 \cup \cdots \cup \{q_m\} \times S_m$ where the S_i's are sets of n-dim vectors of natural numbers. The set of configurations C is SC (resp. UC, DC) definable if all the S_i's are SC (resp. UC, DC) definable. We represent SC definable sets by simple constraints in normal form coupled with control states. From now on, we consider a canonical product to be a pair of the form $\langle q, \ell \leq x \leq u \rangle$ where $q \in Q$. A simple constraint is either $\emptyset$ or a finite disjunction of canonical products. We use SC($Q, \mathcal{X}$) (resp. UC($Q, \mathcal{X}$), DC($Q, \mathcal{X}$)) to denote the set of simple constraints (resp. upward closed, downward closed constraints). We omit the parameters Q and $\mathcal{X}$ when they are known from the context.

3   Computing Successors and Predecessors

Lemma 9. *The class SC is effectively closed under the operations post and pre for any lossy VASS's.*

Proof. These operations are distributive w.r.t. union. Hence, it suffices to consider separately each transition $\tau = (q, a, \Delta, q')$ and perform them on canonical products:

1. $post_\tau(\langle q, \ell \leq x \leq u \rangle) = \langle q', x \leq u + \Delta \rangle$.
2. $pre_\tau(\langle q', \ell \leq x \leq u \rangle) = \langle q, (\ell - \Delta) \sqcap 0 \leq x \rangle$,

where $\forall u, v \in I\!N^n$, $u \sqcap v$ is the vector such that $\forall i \in \{1, \ldots, n\}$. $(u \sqcap v)_i = max(u_i, v_i)$. □

Notice that for lossy VASS's, the *pre* image of any set of configurations is upward closed and its *post* image is downward closed. This also holds for *pre** and *post**.

Theorem 10. *For every n-dim lossy VASS S, and every n-dim SC set S, the set $pre^*(S)$ is UC definable and effectively constructible.*

Proof. Since the set $pre^*(S)$ is upward closed, by Proposition 8 we deduce that it is UC definable. The construction of this set is similar to the one given in [CFI96, ACJT96] for lossy channel systems.   □

Theorem 11. *For every n-dim lossy VASS S, and every n-dim SC set S, the set $post^*(S)$ is DC definable and effectively constructible.*

Proof. Since $post^*(S)$ is downward closed, by Proposition 8 we deduce that $post^*(S)$ is DC definable. This set can be constructed using the Karp-Miller algorithm for the construction of the coverability graph [KM69].   □

4   Automata and Automata Logic

We use finite automata to express properties of computations. These automata are labeled on states and edges as well. State labels are associated with predicates on the configurations of a given system and edge labels are associated with the actions of the system.

Definition 12. *Let Λ and Σ be two finite alphabets. A labeled transition graph over (Λ, Σ) is a tuple $\mathcal{G} = (Q, q_{init}, \Pi, \delta)$ where Q is a finite set of states, q_{init} is the initial state, $\Pi : Q \to \Lambda$ is a state labeling function, $\delta \subseteq Q \times \Sigma \times Q$ is a finite set of labeled transitions. We write $q \xrightarrow{a} q'$ when $(q, a, q') \in \delta$.*
Given a state q, a run of $\mathcal{G}$ starting from q is a finite or infinite sequence $q_0 a_0 q_1 a_1 q_2 \ldots$ such that $q_0 = q$ and $\forall i \geq 0.\ q_i \xrightarrow{a_i} q_{i+1}$.

Definition 13 (Automata on finite sequences). *A finite-state automaton over (Λ, Σ) on finite sequences is a tuple $\mathcal{A}_f = (Q, q_{init}, \Pi, \delta, F)$ where $(Q, q_{init}, \Pi, \delta)$ is a labeled transition graph over (Λ, Σ), and $F \subseteq Q$ is a set of final states. A finite sequence $\lambda_0 a_0 \lambda_1 a_1 \ldots \lambda_n \in \Lambda(\Sigma\Lambda)^*$ is accepted by $\mathcal{A}_f$ if there is a run $q_0 a_0 q_1 a_1 \ldots q_n$ of $\mathcal{A}_f$ starting from q_{init} such that $\forall i \in \{0, \ldots, n\}.\ \Pi(q_i) = \lambda_i$, and $q_n \in F$. Let $L(\mathcal{A}_f)$ be the set of sequences in $\Lambda(\Sigma\Lambda)^*$ accepted by $\mathcal{A}_f$.*

Definition 14 (Büchi ω-automata). *A finite-state Büchi automaton over (Λ, Σ) is a tuple $\mathcal{A}_\omega = (Q, q_{init}, \Pi, \delta, F)$ where $(Q, q_{init}, \Pi, \delta)$ is a labeled transition graph over (Λ, Σ), and $F \subseteq Q$ is a set of repeating states. An infinite sequence $\lambda_0 a_0 \lambda_1 a_1 \ldots \lambda_n \in (\Lambda\Sigma)^\omega$ is accepted by $\mathcal{A}_\omega$ if there is a run $q_0 a_0 q_1 a_1 \ldots$ of $\mathcal{A}_\omega$ starting from q_{init} such that $\forall i \geq 0.\ \Pi(q_i) = \lambda_i$, and $\overset{\infty}{\exists} i \geq 0.\ q_i \in F$. We denote by $L(\mathcal{A}_\omega)$ the set of sequences in $(\Lambda\Sigma)^\omega$ accepted by $\mathcal{A}_\omega$.*

Definition 15 (Closed ω-automata). *A closed ω-automaton is a Büchi automaton $\mathcal{A}_{\omega c} = (Q, q_{init}, \Pi, \delta, F)$ such that $F = Q$.*

Remark. [Tho90] Büchi automata define ω-regular sets of infinite sequences. They are closed under boolean operations. Closed ω-automata define closed ω-regular sets in the Cantor topology (the class F in the Borel hierarchy). They correspond to the class of ω-regular safety properties. Closed ω-automata are closed under intersection and union, but not under complementation.

We introduce an automata-based branching-time temporal logic called AL (Automata Logic). This logic is defined in the spirit of the extended temporal logic ETL and is an extension of ECTL* [Tho89]. The logic AL is more expressive than CTL and CTL*, and allows to express all ∞-regular linear-time properties on finite and infinite computations.

Definition 16 (Automata Logic). *Given a set of control states Q and a set of variables $\mathcal{X}$, we let $\mathcal{F}$ denote a subset of $SC(Q, \mathcal{X})$, and we let π range over elements of $\mathcal{F}$. Then, the set of AL($\mathcal{F}$) formulae is defined by the following grammar:*

$$\varphi ::= \pi \mid \neg\varphi \mid \varphi \vee \varphi \mid \varphi \wedge \varphi \mid \exists\mathcal{A}_f(\varphi_1, \ldots, \varphi_m) \mid \forall\mathcal{A}_f(\varphi_1, \ldots, \varphi_m) \mid$$
$$\exists\mathcal{A}_\omega(\varphi_1, \ldots, \varphi_m) \mid \forall\mathcal{A}_\omega(\varphi_1, \ldots, \varphi_m)$$

where $\mathcal{A}_f$ (resp. $\mathcal{A}_\omega$) is a finite-state automaton on finite (resp. infinite) sequences over $(\Lambda = \{\lambda_1, \ldots, \lambda_m\}, \Sigma)$. We consider standard abbreviations like $\Rightarrow$.

Definition 17. *We use $\star$ to denote f or ω. Let $\mathcal{S} = (\Sigma, \mathcal{X}, Q, \delta)$ be a n-dim (lossy) VASS, We define a satisfaction relation between configurations of $\mathcal{S}$ and AL($\mathcal{F}$) as follows:*

$$s \models (q, \xi) \text{ iff } State(s) = q \text{ and } Val(s) \in [\![\xi]\!]$$
$$s \models \neg\varphi \text{ iff } s \not\models \varphi$$
$$s \models \varphi_1 \vee \varphi_2 \text{ iff } s \models \varphi_1 \text{ or } s \models \varphi_2$$
$$s \models \varphi_1 \wedge \varphi_2 \text{ iff } s \models \varphi_1 \text{ and } s \models \varphi_2$$
$$s \models \exists\mathcal{A}_\star(\varphi_1, \ldots, \varphi_m) \text{ iff } \exists\rho = s_0 a_0 \ldots \in Run_\star(s, \mathcal{S}). \exists\sigma = \lambda_{i_0} a_0 \ldots \in L(\mathcal{A}_\star).$$
$$|\sigma| = |\rho| \text{ and } \forall j. \, 0 \le j < |\rho|. \, s_j \models \varphi_{i_j}$$
$$s \models \forall\mathcal{A}_\star(\varphi_1, \ldots, \varphi_m) \text{ iff } \forall\rho = s_0 a_0 \ldots \in Run_\star(s, \mathcal{S}). \exists\sigma = \lambda_{i_0} a_0 \ldots \in L(\mathcal{A}_\star)$$
$$|\sigma| = |\rho| \text{ and } \forall j. \, 0 \le j < |\rho|. \, s_j \models \varphi_{i_j}$$

For every formula φ, let $[\![\varphi]\!]_{\mathcal{S}} := \{s \in \mathcal{S} \mid s \models \varphi\}$.

Definition 18 (Fragments of AL). *$\exists$AL($\mathcal{F}$) is the fragment of AL that uses only constraints from $\mathcal{F}$, conjunction, disjunction and existential path quantification. $\forall$AL($\mathcal{F}$) is the fragment of AL that uses only constraints from $\mathcal{F}$, conjunction, disjunction and universal path quantification. Let X be (some fragment of) the logic AL. Then X_f (resp. X_ω, $X_{\omega c}$) denote the fragment of X where only automata on finite sequences (resp. Büchi, closed ω-automata) are used.*

AL is a weaker logic than the modal μ-calculus, but many widely known temporal logics are fragments of AL. Every propositional linear-time property, in particular LTL properties, can be expressed in AL. CTL* is a fragment of AL since every path formula in CTL* corresponds to an LTL formula. Thus, CTL is also a fragment of AL. Clearly, $\forall$AL and $\exists$AL subsume the positive universal and existential fragments of CTL* denoted $\forall$CTL* and $\exists$CTL* (notice that LTL is a fragment of $\forall$CTL*).

We consider two fragments of CTL called EF and EG. The logic EF uses SC predicates, boolean operators, the one-step next operator and the operator EF which is defined by $[\![EF\varphi]\!] = pre^*([\![\varphi]\!])$, The logic EG is defined like EF, except that the operator EF is replaced by the operator EG, which is defined as follows: $s \models EG\varphi$ iff there exists a complete run that starts at s and always satisfies φ. By a complete run we mean either an infinite run or a finite run ending in a deadlock. We use the subscripts f or ω to denote the fragments of these logics obtained by interpreting their formulae on either finite or infinite paths only. Then, it can be seen that $EF = EF_f \subseteq CTL_f \subseteq CTL_f^* \subseteq AL_f$. It can also be seen that EG_ω is a fragment of $AL_{\omega c}$ but EG is not (due to the finite paths).

5   Model Checking

Definition 19 (Model checking and global model checking problems).

1. *The* model checking problem *is if $s \in [\![\varphi]\!]_S$ for configuration s and formula φ.*
2. *The* global model checking problem *is whether for any formula φ the set $[\![\varphi]\!]_S$ is effectively constructible.*

Lemma 20. *Let S be a lossy VASS. Then for every formula φ of the form $\exists\mathcal{A}_f(\pi_1, \ldots, \pi_m)$ where all the π_i are SC, the set $[\![\varphi]\!]_S$ is SC definable and effectively constructible.*

Proof. By a generalized pre* construction (see Theorem 10).   □

Theorem 21. *The global model checking problem for lossy VASS and the logic AL_f is decidable.*

Proof. By induction on the nesting-depth and Lemma 20.   □

The following results even hold for non-lossy VASS. The aim is to show decidability of the global model checking problem for VASS and the logic $\exists AL_\omega(UC)$. We define a generalized notion of configurations of VASS which includes the symbol ω. This symbol denotes arbitrarily high numbers of tokens on a place. It is used as an abbreviation in the following way: $\langle q, (\omega, \omega, \ldots, \omega, x_{k+1}, \ldots, x_n)\rangle \models \varphi$: $\iff \exists n_1, \ldots, n_k \in I\!N. \langle q, (n_1, n_2, \ldots, n_k, x_{k+1}, \ldots, x_n)\rangle \models \varphi$. (Of course the ω can occur at any position, e.g. $\langle q, (x_1, x_2, \omega, x_4, \omega, x_6)\rangle$.)

Lemma 22. *Let S be a VASS and φ a formula of the form $\exists A_\omega(\pi_1, \ldots, \pi_m)$ where all the π_i are in UC. Let s be a generalized configuration of S (i.e. it can contain ω). It is decidable if $s \models \varphi$.*

Proof. *(Sketch) First construct the Karp-Miller coverability graph [KM69]. Then check for the existence of cycles in this graph that have an overall positive effect of the fired transitions. These cycles may contain the same node several times. This check is done with the help of Parikh's Theorem. The property holds iff such a cycle with overall positive effect exists, because it can be repeated infinitely often.* □

Lemma 23. *Let S be a VASS and φ a formula of the form $\exists A_\omega(\pi_1, \ldots, \pi_m)$ where all the π_i are in UC. The set $[\![\varphi]\!]_S$ is UC definable and effectively constructible.*

Proof. $[\![\varphi]\!]_S$ *is upward closed, because all π_i are upward closed. Thus, it is characterized by the finite set of its minimal elements (see Lemma 3). To find the minimal elements, we use a construction that was described by Valk and Jantzen in [VJ85]. The important point here is that we can use Lemma 22 to check the existence of configurations that satisfy φ. For example, if $\langle q, (\omega, x_2, x_3) \rangle \models \varphi$ then we can check if $\langle q, (n_1, x_2, x_3) \rangle \models \varphi$ for $n_1 = 0$, $n_1 = 1$, $n_1 = 2$, ... until we find the minimal n_1 s.t. $\langle q, (n_1, x_2, x_3) \rangle \models \varphi$.* □

Theorem 24. *The global model checking problem is decidable for VASS and the logic $\exists AL_\omega(UC)$.*
Proof. *By induction on the nesting-depth of the formula and Lemma 23.* □

Theorem 25. *The global model checking problem is decidable for lossy VASS and the logic $\exists AL(UC)$.*
Proof. *By induction on the nesting depth and Theorems 21 and 24.* □

Theorem 26. *The model checking problem for lossy VASS and $AL_{\omega c}$ is decidable.*
Proof. *By induction on the nesting-depth of the formula and an analysis of all computations which is finite by Dickson's Lemma.* □

Theorem 27. *Model checking lossy VASS with the logic EG is decidable.*

Theorems 26 and 27 say that the model checking problem is decidable for a lossy VASS and an EG-formula/ $AL_{\omega c}$-formula φ. However, in both cases the set $[\![\varphi]\!]_S$ is not effectively constructible (although it is SC definable). If it were constructible then Lemma 20 could be used to decide model checking lossy VASS with formulae of the form $EFEG_\omega \pi$, where π is a constraint in SC. However, this problem has very recently been shown to be undecidable.

Proposition 28. *Model checking lossy VASS with formulae of the form $EFEG_\omega\pi$, where π is a constraint in SC is undecidable.*
Proof. This is a corollary of a more general undecidability result for lossy BPP (Basic Parallel Processes), which follows (not immediately) from the result on lossy counter machines in Proposition 31 (see [May98]).   □

Remark. This undecidability result also implies undecidability of model checking lossy VASS with the logic $\exists AL_\omega$. One can encode properties of the form $EFEG_\omega\pi$ in $\exists AL_\omega$ in the following way: Let $\mathcal{A}_\omega$ be an automaton with states q, q', and transitions $q \to q$, $q \to q'$ and $q' \to q'$ which are labeled with any action. The predicate *true* is assigned to q and the predicate π is assigned to q'. q is the initial state and q' is the only repeating state. Let $\mathcal{A}'_\omega$ be an automaton with only one state q which is the initial state and repeating and a transition $q \to q$ with any action. The predicate π is assigned to q. Then for any lossy VASS s we have $s \models EFEG_\omega\pi \iff s \models \mathcal{A}_\omega(true, \pi) \vee \mathcal{A}'_\omega(\pi)$.

Lossy VASS can be extended with inhibitor arcs. This means introducing transitions that can only fire if some defined places are empty (i.e. they can test for zero). Thus lossy VASS with inhibitor arcs are equivalent to lossy counter machines. Normal VASS with inhibitor arcs are Turing-powerful, but lossy VASS with inhibitor arcs are not.

Theorem 29. *For lossy VASS with inhibitor arcs*
 1. the global model checking problem is decidable for the logic AL_f.
 2. model checking is decidable for the logics $AL_{\omega c}$ and EG.

Inhibitor arcs can never keep a transition from firing, because one can just loose the tokens on the places that inhibit it. However, after such a transition has fired, the number of tokens on the inhibiting places is fixed and known exactly. Such a guarantee is impossible to achieve in lossy VASS without inhibitor arcs. Thus not all results for lossy VASS carry over to lossy VASS with inhibitor arcs.

Proposition 30. *Let S be a lossy VASS with inhibitor arcs. It is undecidable if there exists an initial configuration s s.t. there is an infinite run of (s, S).*
Proof. This is a corollary of a more general undecidability result for lossy counter machines in [May98]. The main idea is that one can enforce that lossiness occurs only finitely often in the infinite run.   □

Theorem 31. *Model checking lossy VASS with inhibitor arcs with the logic LTL is undecidable.*
Proof. We reduce the problem of Proposition 31 to the model checking problem. We construct a lossy VASS with inhibitor arcs S' that does the following: First it guesses an arbitrary configuration s of S doing only the atomic action a. Then it simulates S on s doing only the atomic action b. Let $\mathcal{A}_\omega$ be a Büchi-automaton with initial state q and repeating state q' and transitions $q \xrightarrow{a} q$, $q \xrightarrow{b} q'$ and $q' \xrightarrow{b} q'$. Let s' be the initial state of S'. We have reduced the question of Proposition 31 to the question if $(s', S') \models \exists \mathcal{A}_\omega(true, true)$. This question can be expressed in LTL.   □

It follows immediately that model checking lossy VASS with inhibitor arcs with $AL_\omega(UC)$ is undecidable. It is interesting to compare this result with Proposition 28. For undecidability it suffices to have either inhibitor arcs in the system or downward closed constraints in the logic. One can be encoded in the other and vice versa. The set $post^*(s)$ is DC definable since it is downward closed. However, it is not constructible for lossy VASS with inhibitor arcs (unlike for lossy VASS, see Theorem 11).

Theorem 32. $post^*(s)$ *is not constructible for lossy VASS with inhibitor arcs.*

Proof. Boundedness is undecidable for reset Petri nets [DFS98]. This result carries over to lossy reset Petri nets. Lossy VASS with inhibitor arcs can simulate lossy reset Petri nets. It follows that boundedness is undecidable for lossy VASS with inhibitor arcs and thus $post^*(s)$ is not constructible.   □

6   Conclusion

We have established results for normal VASS and lossy VASS with inhibitor arcs (lossy counter machines). Interestingly, it turns out that these two models are incomparable. Moreover, all the positive/negative results we obtained for lossy VASS with inhibitor arcs are the same as for lossy fifo-channel systems. Note that lossy fifo-channel systems can simulate lossy VASS with inhibitor arcs, but only with some additional deadlocks.

The following table summarizes the results on the decidability of model checking for VASS, lossy VASS with test for zero, lossy VASS and lossy fifo-channel systems. By '++' we denote the fact that for any formula φ the set $[\![\varphi]\!]$ is SC definable and effectively constructible (global model checking), while '+' means that only model checking is decidable. We denote by — that model checking is undecidable. The symbol '?' denotes an open problem.

Logic	VASS	Lossy VASS+0	Lossy VASS	Lossy FIFO
AL_f/EF	— [Esp97]	++	++ [AJ93]	++ [AJ93]
$\exists AL_\omega(UC)/LTL$	++/+[Esp97]	—	++	— [AJ96]
$\exists AL(UC)$	?	—	++	— [AJ96]
$AL_{\omega c}/EG$	— [EK95]	+	+ [AJ93]	+ [AJ93]
$\exists AL_\omega/CTL$	— [EK95]	—	—	— [AJ96]

The results in this table are new, except where references are given. For normal VASS and LTL, decidability of the model checking problem was known [Esp97], but the construction of the set $[\![\varphi]\!]$ is new. The results in [AJ93] are just about EF and EG formulae without nesting, not for the full logics AL_f and $AL_{\omega c}$.

Acknowledgment: We thank Peter Habermehl for interesting discussions.

References

[ACJT96] P. Abdulla, K. Cerans, B. Jonsson, and Y-K. Tsay. General Decidability Theorems for Infinite-state Systems. In *LICS'96*. IEEE, 1996.

[AJ93] P. Abdulla and B. Jonsson. Verifying Programs with Unreliable Channels. In *LICS'93*. IEEE, 1993.

[AJ96] P. Abdulla and B. Jonsson. Undecidable verification problems for programs with unreliable channels. *Information and Computation*, 130(1):71–90, 1996.

[CFI96] Gérard Cécé, Alain Finkel, and S. Purushothaman Iyer. Unreliable Channels Are Easier to Verify Than Perfect Channels. *Information and Computation*, 124(1):20–31, 1996.

[DFS98] C. Dufourd, A. Finkel, and Ph. Schnoebelen. Reset nets between decidability and undecidability. In *Proc. of ICALP'98*, volume 1443 of *LNCS*. Springer Verlag, 1998.

[EK95] J. Esparza and A. Kiehn. On the model checking problem for branching time logics and Basic Parallel Processes. In *CAV'95*, volume 939 of *LNCS*, pages 353–366. Springer Verlag, 1995.

[Esp97] J. Esparza. Decidability of model checking for infinite-state concurrent systems. *Acta Informatica*, 34:85–107, 1997.

[GL94] O. Grumberg and D. Long. Model Checking and Modular Verification. *ACM Transactions on Programming Languages and Systems*, 16, 1994.

[KM69] R. Karp and R. Miller. Parallel program schemata. *JCSS*, 3, 1969.

[May98] R. Mayr. Lossy counter machines. Technical Report TUM-I9827, TU-München, October 1998. wwwbrauer.informatik.tu-muenchen.de/~mayrri.

[Tho89] W. Thomas. Computation Tree Logic and Regular ω-Languages. LNCS 354, 1989.

[Tho90] W. Thomas. Automata on Infinite Objects. In *Handbook of Theo. Comp. Sci.* Elsevier Sci. Pub., 1990.

[VJ85] R. Valk and M. Jantzen. The Residue of Vector Sets with Applications to Decidability Problems in Petri Nets. *Acta Informatica*, 21, 1985.

Constructing Light Spanning Trees with Small Routing Cost

Bang Ye Wu[1], Kun–Mao Chao[2], and Chuan Yi Tang[1]

[1] Dept. of Computer Science, National Tsing Hua University, Hsinchu, Taiwan,
R.O.C. { dr838305, cytang}@cs.nthu.edu.tw
[2] Dept. of Computer Science and Information Management, Providence University.
Shalu, Taiwan, R.O.C. kmchao@csim.pu.edu.tw

Abstract. Let $G = (V, E, w)$ be an undirected graph with nonnegative
edge weight. For any spanning tree T of G, the weight of T is the total
weight of its tree edges and the routing cost of T is $\sum_{u,v \in V} d_T(u, v)$,
where $d_T(u, v)$ is the distance between u and v on T. In this paper,
we present an algorithm providing a trade off among tree weight, routing cost and time complexity. For any real number $\alpha > 1$ and an integer
$1 \leq k \leq 6\alpha - 3$, in $O(n^{k+1} + n^3)$ time, the algorithm finds a spanning tree
whose routing cost is at most $(1 + 2/(k + 1)) \alpha$ times the one of the minimum routing cost tree, and the tree weight is at most $(f(k) + 2/(\alpha - 1))$
times the one of the minimum spanning tree, where $f(k) = 1$ if $k = 1$
and $f(k) = 2$ if $k > 1$.
Keywords: approximation algorithms, network design, spanning trees.

1 Introduction

Constructing spanning trees of graphs is a classical network design problem.
Typically, we are given a graph $G = (V, E, w)$, where w is a nonnegative edge
weight function. The weight on each edge represents the distance and reflects
both the cost to install the link (building cost) and the cost to traverse the
link after the link is installed (routing cost). If we only consider the building
cost, we are looking for a spanning tree $T = (V, E_t)$ with minimum tree weight
$w(T) = \sum_{e \in E_t} w(e)$, i.e., the *minimum spanning tree* (MST) of the graph. If
only the routing cost is considered, the goal is to find the spanning tree T with
minimum $c(T) = \sum_{u,v \in V} d_T(u, v)$, where $d_T(u, v)$ is the distance (total weight
of the path) between u and v on T. Such a tree is called the *minimum routing
cost spanning tree* (MRCT) (or the *shortest total path length spanning tree*) of
the graph. Although both MST and MRCT tend to use light edges, a tree with
small weight may have a large routing cost and vice versa. For instance, we can
easily construct a graph such that the routing cost of its MST is $\Theta(n)$ times the
routing cost of its MRCT. Similarly, a spanning tree with a constant times the
minimum routing cost may have a tree weight as large as $\Theta(n)$ times the weight
of MST. Therefore, we often need to make a trade off between the two costs.

The minimum spanning tree is a fundamental problem and efficient polynomial time algorithms were developed (for example, see [2,3]). The MRCT problem is a special case of the *optimum communication spanning tree* problem [5].

C. Meinel and S. Tison (Eds.): STACS'99, LNCS 1563, pp. 334–344, 1999.

Unfortunately, the MRCT problem has been shown to be NP-hard [4,6], and a 2-approximation algorithm was presented in [8]. Recently, a *polynomial time approximation scheme* (PTAS) for the MRCT problem was presented in [9]. In this paper, we present an algorithm for finding a spanning tree that simultaneously approximates both the two costs with constant ratios.

Let $MRCT(G)$ and $MST(G)$ denote the MRCT and the MST of G respectively. We define the *light approximate routing cost spanning tree* (LART) of G as follows:

Definition 1: For $\alpha \geq 1$ and $\beta \geq 1$, an (α,β)-LART is a spanning tree T of G with $c(T) \leq \alpha \times c(MRCT(G))$ and $w(T) \leq \beta \times w(MST(G))$.

The main result of this paper is stated in the following theorem.

Theorem 1: Given a graph G, a $\left(\frac{k+3}{k+1}\alpha, \left(f(k) + \frac{2}{\alpha-1}\right)\right)$-LART can be constructed in $O(n^{k+1} + n^3)$ time for any real number $\alpha > 1$ and an integer $1 \leq k \leq 6\alpha - 3$, where $f(k) = 1$ if $k = 1$ and $f(k) = 2$ if $k > 1$.

Several results for trees realizing trade off between weight and distance requirements can be found in the literature. In [7], an algorithm for constructing an (α,β)-LAST (*light approximate shortest-path tree*) was presented. For any $v \in V(G)$, an (α,β)-LAST rooted at v is a spanning tree T of G with $d_T(i, v) \leq \alpha d_G(i, v)$ for any vertex $i \in V(G)$ and $w(T) \leq \beta w(MST(G))$. Khuller *et al.* showed that it is possible to construct an $(\alpha, 1 + 2/(\alpha - 1))$-LAST for any $\alpha > 1$. We used their algorithm as a kernel subroutine in our algorithm.

Considerable work has been done on the *spanner* of a graph recently. In general, a *t*-spanner of G is a low-weight subgraph of G such that, for any two vertices, the distance on the spanner is at most t times the distance on G. When the spanner is restricted to a spanning tree, it is called a tree *t*-spanner. Obviously, a tree *t*-spanner has a more strict distance requirement than the routing cost considered in this paper. However, because of the strict distance requirement, some graphs do not have a tree *t*-spanner for any constant t (e.g. a cycle with identical weight on each edge). Some methods for finding spanners of a weighted graph were presented in [1].

We briefly outline the basic ideas for finding an (α,β)-LART of a graph as follows.

1. Although the input graph of our problem is a general graph, we show that it can be reduced to the one with a metric input. In [9], an algorithm was developed for transforming a spanning tree of $\bar{G}$ to a spanning tree of G without increasing the routing cost, where $\bar{G}$ is the metric closure (defined later) of G. By observing that the algorithm does not increase the weight of the tree as well, we show that the problem of finding an (α,β)-LART of G can be reduced to that of finding an (α,β)-LART of $\bar{G}$.

2. A *k*-star is a spanning tree with at most k internal nodes. By showing that there exists a *k*-star with routing cost at most $\left(1 + \frac{2}{k+1}\right)$ times the one

of $MRCT(G)$, the PTAS in [9] finds the minimum routing cost k-star in $O(n^{2k})$ time, in which n is the number of vertices. However, the weight of the minimum k-star may be large. Consider the case when $k = 1$. The minimum 1-star is a shortest-path tree rooted at some vertex. The weight of the minimum 1-star may be $O(n)$ times the weight of the minimum spanning tree.

3. Consider the algorithm in [7] for constructing an $(\alpha, 1 + 2/(\alpha - 1))$-LAST rooted at a vertex. We show that the LAST with minimum routing cost is a $(2\alpha, 1 + 2/(\alpha - 1))$-LART. To find a LART achieving more general trade off, we use k-stars. Let R be a vertex set containing the k internal nodes of the k-star with approximate routing cost of the $MRCT(G)$. If R is given, we can construct a *light approximate shortest-path forest* with multiple roots R by the algorithm in [7]. Then we combine the forest into a tree T by adding the edges of the minimum spanning tree of R. We show that it is a $\left(\frac{k+3}{k+1}\alpha, \left(2 + \frac{2}{\alpha - 1}\right)\right)$-LART. Finally, since R is unknown, we try all possible subsets containing k vertices.

The remaining sections are organized as follows: In Section 2, some definitions are given. The reduction to metric input is shown in Section 3 and the algorithm is described in Section 4. The performance ratios are shown in Section 5.

2   Preliminaries

In this paper, a graph $G = (V, E, w)$ is a simple, connected, undirected graph, in which w is a nonnegative edge weight function. For any graph G, $V(G)$ denotes the vertex set and $E(G)$ denotes the edge set of G. We shall use n to denote the number of vertices in the input graph. We first give some definitions below:

Definition 2: A *metric graph* G is a complete graph, in which the edge weights satisfy the triangle inequality.

Definition 3: Let $G = (V, E, w)$ be a graph. we use $w(G) = \sum_{e \in E} w(e)$ to denote the graph weight. By $SP_G(u, v)$, we denote a shortest path between u and v on G, and $d_G(u, v) = w(SP_G(u, v))$ is the shortest path length. The *routing cost* of a tree T is defined by $c(T) = \sum_{u,v \in V(T)} d_T(u, v)$.

Definition 4: Let G be a graph. The *metric closure* of G, denoted by $\bar{G}$, is the complete graph with $V(\bar{G}) = V(G)$ and $\bar{w}(u, v) = d_G(u, v)$ for any u, v, where $\bar{w}$ is the edge weight function of $\bar{G}$.

Definition 5: Let $G = (V, E, w)$ be a graph and $r \in V$. A spanning tree T is a *shortest-path tree* of G rooted at r if $d_T(r, v) = d_G(r, v)$ for each $v \in V$.

Definition 6: Given a graph G, a *minimum routing spanning tree* of G, denoted by $MRCT(G)$, is a spanning tree T such that $c(T)$ is minimum among all possible spanning trees.

The following definition and lemma are given in [7].

Definition 7: Let $G = (V, E, w)$ be a graph and $r \in V$. For $\alpha \geq 1$ and $\beta \geq 1$, an (α,β)-LAST rooted at r is a spanning tree T of G with $d_T(r, v) \leq \alpha d_G(r, v)$ for each $v \in V$ and $w(T) \leq \beta \times w(MST(G))$.

Lemma 2 ([7]): Let G be a graph with nonnegative edge weights. For any $r \in V(G)$ and $\alpha > 1$, there exists an algorithm, named FIND-LAST, which can construct an $(\alpha, 1 + 2/(\alpha - 1))$-LAST rooted at r in $O(n)$ time if the minimum spanning tree and the shortest path tree of G are given.

3 Reduction to Metric Input

We now show that the problem of finding a LART of a graph can be reduced to that of finding a LART of its metric closure. The reduction is done by a transformation algorithm in [9]. It was developed for the MRCT problem, and can be shown that it also works for the LART problem.

Let $G = (V, E, w)$ and $\bar{G}$ be its metric closure with edge weight δ. Any edge (a, b) in $\bar{G}$ is called a *bad edge* if $(a, b) \notin E$ or $w(a, b) > \delta(a, b)$. For any bad edge $e = (a, b)$, there must exist a path $P = SP_G(a, b) \neq e$ such that $w(P) = \delta(a, b)$. Given any spanning tree T of $\bar{G}$, the algorithm can construct another spanning tree Y without any bad edge such that $c(Y) \leq c(T)$. Since Y has no bad edge, $\delta(e) = w(e)$ for each $e \in E(Y)$, and Y can be thought as a spanning tree of G with the same routing cost. The algorithm is listed in the following.

Algorithm Remove_bad
Compute all-pairs shortest paths of G.
while there exists a bad edge in T
    Pick a bad edge (a, b). Root T at a.
    /* assume $SP_G(a, b) = (a, x, ..., b)$ and y is the father of x*/
    if b is not an ancestor of x **then**
        $Y^* = T \cup (x, b) \setminus (a, b);\ Y^{**} = Y^* \cup (a, x) \setminus (x, y);$
    else
        $Y^* = T \cup (a, x) \setminus (a, b);\ Y^{**} = Y^* \cup (b, x) \setminus (x, y);$
    endif
    if $c(Y^*) < c(Y^{**})$ **then** $Y = Y^*$ **else** $Y = Y^{**}$ **endif**
    $T = Y$;
endwhile

The algorithm computes Y by iteratively replacing the bad edges until there is no bad edge left. It was shown in [9] that the cost is never increased at each iteration and it takes no more than $O(n^2)$ iterations. The result is stated in the following lemma.

Lemma 3 ([9]): For any spanning tree $\bar{T}$ of $\bar{G}$, it can be transformed into a spanning tree T of G in $O(n^3)$ time and $c(T) \leq c(\bar{T})$.

The main result of this section is summarized in the following lemma:

Lemma 4: If there is an algorithm for finding an (α,β)-LART of a metric graph with time complexity $t(n)$, then there is an algorithm for the (α,β)-LART of a general graph with time complexity $O(n^3 + t(n))$.

Proof: Given any graph G, we first construct its metric closure $\bar{G}$ in $O(n^3)$ time. We then find an (α,β)-LART $\bar{T}$ of $\bar{G}$ in $t(n)$ time. Finally, using the transformation algorithm, we obtain a spanning tree T of G.

By Lemma 3, $c(T) \leq c(\bar{T})$. Since Lemma 3 is true for any spanning tree, it also implies that $c(MRCT(G)) \leq c(MRCT(\bar{G}))$. Furthermore, it is easy to see that $c(MRCT(G)) \geq c(MRCT(\bar{G}))$. It follows that $c(MRCT(G)) = c(MRCT(\bar{G}))$, and we have

$$c(T) \leq c(\bar{T}) \leq \alpha c(MRCT(\bar{G})) = \alpha c(MRCT(G))$$

Since, in each iteration, the transformation algorithm does not increase the tree weight either, we have $w(T) \leq w(\bar{T})$. It is easy to show that $w(MST(\bar{G})) = w(MST(G))$. Thus,

$$w(T) \leq w(\bar{T}) \leq \beta w(MST(\bar{G})) = \beta w(MST(G))$$

Therefore, T is an (α,β)-LART of G.   □

By Lemma 4, we can focus on the problem with a metric input. In the rest of this paper, we shall assume the input graph G is a metric graph.

4   The Algorithm

In this section, we present the algorithm for finding a LART and analyze its time complexity.

Definition 8:   Let $G = (V, E, w)$ be a graph (not necessarily a metric), $R \subset V$, and $r \notin V$. We define $G^{+R} = (U, F, \bar{w})$, in which $U = V \cup \{r\}$, $F = E \cup \{(r, v) | \forall v \in R\}$, $\bar{w}(e) = w(e)\ \forall e \in E$ and $\bar{w}(e) = 0\ \forall e \in F \setminus E$.

The algorithm for finding a LART is listed below.

Algorithm FIND-LART

Input: a graph G, a real number $\alpha > 1$, and an integer $1 \leq k \leq 6\alpha - 3$.

Output: a $\left(\frac{k+3}{k+1}\alpha, \left(f(k) + \frac{2}{\alpha - 1} \right) \right)$-LART of G,

where $f(k) = 1$ if $k = 1$ and $f(k) = 2$ if $k > 1$.

Step 1:   Find $T_M = MST(G)$.

Step 2:   For each $R \subset V(G)$ and $|R| \leq k$, use the following method to construct a spanning tree, and keep T with minimum $c(T)$.

Step 2.0:   Assume $R = \{r_j | 1 \leq j \leq q\}$ and $E_R = \{(r, r_j) | 1 \leq j \leq q\}$.

Step 2.1:   Construct G^{+R}.

Step 2.2:   Find $MST(G^{+R})$.

Step 2.3:   Find the shortest path tree of G^{+R} rooted at r.

Step 2.4:   Call algorithm FIND-LAST to find an $(\alpha, 1 + 2/(\alpha - 1))$-LAST rooted at r.

/* Let the tree be T_1. We can assume $E_R \subset E(T_1)$ since $w(e) = 0 \ \forall e \in E_R$. */

Step 2.5:   Delete the edges E_R from T_1.

Step 2.6:   Find the $T_0 = MST(G|_R)$, where $G|_R$ is the induced subgraph with vertex set R.

Step 2.7:   Set $T = T_0 \cup T_1$.

Step 2.8:   Compute $c(T)$.

In the algorithm, **Step 1** is executed for the sake of time efficiency. With the results of **Step 1**, **Step 2.2** can be done in linear time (shown later). All the steps within **Step 2** can be easily done in $O(n^2)$ time by direct methods. Our goal is to show that they can be done in $O(n)$ time. To achieve this goal, we only need to show that **Step 2.2, 2.3** and **2.8** can be done in $O(n)$ time. The others are trivial.

We first focus on $MST(G^{+R})$. The following property of MST will be used: Let H be a graph and V_1, V_2 be any partition of $V(H)$. For any edge e crossing the cut (i.e. with one end point in V_1 and the other in V_2), e belongs to a MST of H if and only if e is the lightest edge crossing the cut [2].

Lemma 5:   Let $H = (V, E)$ be a graph and $Y = (V, E_Y) = MST(H)$. Assume $e \notin E$ and $H_1 = (V, E \cup \{e\})$, $Y_1 = (V, E_Y \cup \{e\})$. Then a MST of Y_1 is also a MST of H_1 and can be found in linear time.

Proof: For any edge (x, y) of $MST(H_1)$, by the property of MST, it is the lightest edge crossing some cut (V_x, V_y) of H_1. However, H_1 is obtained by inserting an edge e into H. Thus, either $(x, y) = e$ or (x, y) is the lightest edge crossing the cut (V_x, V_y) of H. This implies $(x, y) \in E(Y_1)$ and then $MST(H_1) \subset Y_1$. We have $w(MST(Y_1)) \leq w(MST(H_1))$. Furthermore, since Y_1 is a subgraph of H_1, $w(MST(H_1)) \leq w(MST(Y_1))$. It follows that $w(MST(H_1)) = w(MST(Y_1))$. Since there is only one cycle in Y_1, $MST(Y_1)$ can be found by deleting the heaviest edge in the cycle, which can be done in linear time.   □

Since $|R| \leq k$, the next lemma can be shown by iteratively applying Lemma 5 and we omit the proof.

Lemma 6: $MST(T_M^{+R})$ is a minimum spanning tree of G^{+R} and can be found in $O(kn)$ time.

The next two lemmas show the time complexities for constructing a shortest-path tree and for computing the routing cost of a tree respectively. The proofs are omitted here.

Lemma 7: The shortest path tree of G^{+R} rooted at r can be found in $O(kn)$ time.

Lemma 8: For any tree T, $c(T)$ can be computed in $O(n)$ time.

Now, the time complexity of the algorithm can be easily shown by Lemmas 2, 6, 7, and 8.

Lemma 9: The time complexity of **Algorithm FIND-LART** is $O(n^{k+1})$.

Lemma 9 and Lemma 4 prove the time complexity in Theorem 1. We prove the ratio bounds of the weight and routing cost in the following section.

5   The Performance Analysis

In this section, we analyze the weight and the routing cost of the tree T constructed by algorithm **FIND-LART**. The following lemma can be easily proved by Lemma 2 and observing that $w(MST(G|_R)) \leq w(MST(G))$. The proof is omitted.

Lemma 10: $w(T) \leq \left(f(k) + \frac{2}{\alpha-1}\right) w(MST(G))$, where $f(k) = 1$ if $k = 1$ and $f(k) = 2$ if $k > 1$.

To show the bound on the routing cost, we first introduce the δ-spine of a tree, which is defined in [9]. We shall use the δ-spine to derive a lower bound on $c(MRCT(G))$ and then show the approximation ratio of the routing cost.

5.1   The δ–Spine of a Tree

The δ-spine was introduced in [9]. For the completeness of this paper, we describe some necessary definitions and results here.

Definition 9: Let T be a spanning tree of G, and let S be a connected subgraph of T. For a positive number $\delta \leq 1/2$, S is a δ–*separator* of T if $|V(B)| \leq \delta n$ for every connected component B in the induced subgraph $T|_{V(T)\setminus V(S)}$. A δ–separator S is *minimal* if any proper subgraph of S is not a δ–separator of T.

Definition 10: Let T be a spanning tree of G, and let S be a connected subgraph of T. Deleting the edges in $E(S)$ from T will result in several subtrees. We use $VB(T, S, i)$ to denote the vertex set of the subtree containing i.

Definition 11: Let T be a tree and $P = SP_T(u, v)$ in which $|VB(T, P, u)| \geq |VB(T, P, v)|$. We define $P^a = |VB(T, P, u)|$, $P^b = |VB(T, P, v)|$, and $P^c = n - |VB(T, P, u)| - |VB(T, P, v)|$.

Definition 12: For a tree T and $0 < \delta \leq 0.5$, a δ–path of T is a path P such that $P^c \leq \delta n/2$.

Definition 13: Let $0 < \delta \leq 0.5$. A δ-spine $Y = \{P_1, P_2, ..., P_h\}$ of T is a set of pairwise edge-disjoint δ–paths in T such that $S = \bigcup_{1 \leq i \leq h} P_i$ is a minimal δ–separator of T. We define the *cut and leaf set* $CAL(Y)$ of a δ–spine Y to be the set of the endpoints of the paths in Y. In the case that Y is empty, $CAL(Y)$ is defined to be the vertex which is the minimal δ–separator.

It should be noted that for any pair of distinct paths P_i and P_j in the spine, they either do not intersect or, if they do, the intersection point is an endpoint of both paths. This can be easily shown by definition. The next lemma is given in [9], which shows the existence of the δ–spine with small cut and leaf set.

Lemma 11 ([9]): For any constant $0 < \delta \leq 0.5$, and spanning tree T of G, there exists a δ–spine Y of T such that $|CAL(Y)| \leq \lceil 2/\delta \rceil - 3$.

5.2 The Routing Cost

The following lemma is immediate and we omit the proof.

Lemma 12: For any spanning tree T of G, $c(T) = 2 \sum_{e \in T} e^a e^b w(e)$.

Let $d_G(v, U)$ denote $\min_{u \in U} \{d_G(v, u)\}$ for any $v \in V(G)$ and $U \subset V(G)$. We now derive a lower bound on $c(MRCT(G))$.

Lemma 13: Let Y be a δ-spine of a spanning tree T of G, and let $S = \bigcup_{P \in Y} P$ be a minimal δ–separator of T. Then $c(T) \geq 2(1-\delta)n \sum_{v \in V(G)} d_T(v, V(S)) + 2\delta(1 - \delta)n^2 w(S)$.

Proof: By Lemma 12, we have

$$c(T)/2 = \sum_{e \in E(T)} e^a e^b w(e) = \sum_{e \in E(T) \backslash E(S)} e^a e^b w(e) + \sum_{e \in E(S)} e^a e^b w(e) \quad (1)$$

By observing that

$$\sum_{v \in V(G)} d_T(v, V(S)) = \sum_{v \in V(G) \backslash V(S)} d_T(v, V(S)) = \sum_{e \in E(T) \backslash E(S)} e^b w(e)$$

and $e^a \geq (1 - \delta)n$ for each $e \in E(T) \setminus E(S)$, we have

$$\sum_{e \in E(T) \setminus E(S)} e^a e^b w(e) \geq (1 - \delta)n \sum_{v \in V(G)} d_T(v, V(S)) \tag{2}$$

For each $e \in E(S)$, since $e^a \geq e^b \geq \delta n$ and $e^a + e^b = n$, we have $e^a e^b \geq \delta(1-\delta)n^2$. Then

$$\sum_{e \in E(S)} e^a e^b w(e) \geq \sum_{e \in E(S)} \delta(1 - \delta)n^2 w(e) = \delta(1 - \delta)n^2 w(S) \tag{3}$$

By Equations (1), (2), and (3), we obtain

$$c(T)/2 \geq (1 - \delta)n \sum_{v \in V(G)} d_T(v, V(S)) + \delta(1 - \delta)n^2 w(S)$$

$$\square$$

Let $\hat{T}$ be the tree output by algorithm **FIND-LART**. The following lemma shows the ratio of the routing cost of $\hat{T}$ and completes the proof of Theorem 1.

Lemma 14: For any integer $k \leq 6\alpha - 3$, $c(\hat{T}) \leq \frac{k+3}{k+1}\alpha c(T^*)$, where $T^* = MRCT(G)$.

Proof: Let $\delta = 2/(k + 3)$. By Lemma 11, there must exist a δ-spine Y of T^* such that $|CAL(Y)| \leq k$. Let T be the tree constructed by algorithm **FIND-LART** with R selected in **Step 2**, in which $R = \{r_i | 1 \leq i \leq q\} = CAL(Y)$. Such a tree T always exists since we try all possible vertex subset with no more than k vertices. Since $c(\hat{T}) \leq c(T)$, we only need to prove the approximation ratio of T. In the following, T_0 and r are defined as in Section 4. Let $V = V(G) = V(T)$. By Lemma 12,

$$c(T)/2 = \sum_{e \in E(T)} e^a e^b w(e) = \sum_{e \in E(T) \setminus E(T_0)} e^a e^b w(e) + \sum_{e \in E(T_0)} e^a e^b w(e) \tag{4}$$

Similar to Equation (2) in Lemma 13 and observing that $e^a \leq n$, we have

$$\sum_{e \in E(T) \setminus E(T_0)} e^a e^b w(e) \leq n \sum_{v \in V} d_T(v, R) \tag{5}$$

For each $v \in V(G)$, by Lemma 2, $d_T(v, r) \leq \alpha d_G(v, r) = \alpha d_G(v, R)$. Since $d_T(v, R) = d_T(v, r)$, we have $d_T(v, R) \leq \alpha d_G(v, R)$. Thus,

$$n \sum_{v \in V} d_T(v, R) \leq \alpha n \sum_{v \in V} d_G(v, R) \tag{6}$$

Now turn to the second term in Equation (4). Since $e^a e^b \leq n^2/4$ for any edge e,

$$\sum_{e \in E(T_0)} e^a e^b w(e) \leq \left(n^2/4\right) \sum_{e \in E(T_0)} w(e) = \left(n^2/4\right) w(T_0) \tag{7}$$

By Equations (4), (5), (6), and (7), we have

$$c(T)/2 \le \alpha n \sum_{v \in V} d_G(v, R) + \left(n^2/4\right) w(T_0) \tag{8}$$

Now let us consider T^*. Let $Y = \{P_i | 1 \le i \le h\}$ and $S = \bigcup_{1 \le i \le h} P_i$ be the minimal δ-separator. Assume $P_i = SP_{T^*}(u_i, v_i)$, and let $V_i = V \setminus VB(T^*, P_i, u_i) \setminus VB(T^*, P_i, v_i)$ for any $i = 1, 2 \ldots h$, and V_0 denote $\bigcup_{v \in CAL(Y)} VB(T, S, v)$. We have $d_G(v, CAL(Y)) \le d_{T^*}(v, V(S))$ for any $v \in V_0$, and $d_G(v, CAL(Y)) \le d_{T^*}(v, V(S)) + w(P_i)/2$ for any $v \in V_i$ by triangle inequality. By the definition of δ-spine, $|V_i| = P_i^c \le \delta n/2$. Therefore, by Equation (8), we have

$$c(T)/2 \le \alpha n \sum_{v \in V} d_G(v, R) + \left(n^2/4\right) w(T_0)$$

$$= \alpha n \sum_{v \in V} d_G(v, CAL(Y)) + \left(n^2/4\right) w(T_0)$$

$$\le \alpha n \left(\sum_{v \in V} d_{T^*}(v, V(S)) + \sum_{1 \le i \le h} (\delta n/4) w(P_i) \right) + \left(n^2/4\right) w(T_0)$$

$$\le \alpha n \sum_{v \in V} d_{T^*}(v, V(S)) + \left(\alpha \delta n^2/4\right) w(S) + \left(n^2/4\right) w(T_0)$$

Since T_0 is a minimum spanning tree on $G|_R$ and S is a tree spanning $V(S) \supset R$, we have $w(T_0) \le w(S)$. Then,

$$c(T)/2 \le \alpha n \sum_{v \in V} d_{T^*}(v, V(S)) + \left(\alpha \delta n^2/4 + n^2/4\right) w(S) \tag{9}$$

By Lemma 13 and Equation (9), $c(T) \le \max\{\frac{\alpha}{1-\delta}, \frac{\alpha\delta+1}{4\delta(1-\delta)}\} c(T^*)$. Note that $\alpha > 1$ and $0 < \delta \le 1/2$. Let $g(\delta) = \max\{\frac{\alpha}{1-\delta}, \frac{\alpha\delta+1}{4\delta(1-\delta)}\}$. When $\delta \ge 1/(3\alpha)$, $\frac{\alpha}{1-\delta} \ge \frac{\alpha\delta+1}{4\delta(1-\delta)}$, and $g(\delta)$ decreases as δ decreases from $1/2$ to $1/(3\alpha)$. When $\delta \le 1/(3\alpha)$, $\frac{\alpha}{1-\delta} \le \frac{\alpha\delta+1}{4\delta(1-\delta)}$, and $g(\delta)$ increases as δ decreases from $1/(3\alpha)$. Therefore, $g(\delta)$ reaches its minimum when $\delta = 1/(3\alpha)$. Since $\delta = 2/(k + 3)$, we conclude that $c(T) \le \frac{k+3}{k+1} \alpha c(T^*)$, for any $k \le 6\alpha - 3$. $\qquad\square$

References

1. I. Althöfer, G. Das, D. Dobkin, D. Joseph, and J. Soares, On sparse spanners of weighted graphs, *Discrete and Computational Geometry*, 9(1):81-100,1993.
2. T.H. Cormen, C.E. Leiserson, and R.L. Rivest, *Introduction to Algorithms*, the MIT Press, 1994.
3. M. L. Fredman and R. E. Tarjan, Fibonacci heaps and their uses in improved network optimization algorithms, *J. of the ACM*, 34(3):596-615, 1987.
4. M.R. Garey and D.S. Johnson, *Computers and Intractability: A Guide to the Theory of NP-Completeness*, W.H. Freeman and Company, San Francisco, 1979.

5. T. C. Hu, Optimum communication spanning trees, *SIAM J. Computing*, 3(3):188-195, 1974.
6. D.S. Johnson, J.K. Lenstra, and A.H.G. Rinnooy Kan, The complexity of the network design problem, *Networks*, 8:279-285, 1978.
7. S. Khuller, B. Raghavachari, and N. Young, Balancing minimum spanning trees and shortest-path trees, *Algorithmica*, 14:305-321, 1995.
8. R. Wong, Worst-case analysis of network design problem heuristics. *SIAM J. Algebraic Discrete Mathematics*, 1:51-63, 1980.
9. B. Y. Wu, G. Lancia, V. Bafna, K. M. Chao, R. Ravi, and C. Y. Tang, A polynomial time approximation scheme for minimum routing cost spanning trees, *Proceedings of Ninth Annual ACM-SIAM Symposium on Discrete Algorithms (SODA '98)*, pp. 21-32, 1998.

Finding Paths with the Right Cost

Matti Nykänen* and Esko Ukkonen

Department of Computer Science
P.O. Box 26 (Teollisuuskatu 23)
FIN-00014 University of Helsinki, FINLAND.
matti.nykanen,esko.ukkonen@cs.helsinki.fi

Abstract. We study a problem related to finding shortest paths in weighted graphs. We ask whether or not there is a path between two nodes that is of a given cost. The edge weights of the graph can be both positive and negative integers, or even integer vectors. We show that most variants of this problem are **NP**-complete. We also develop a pseudo-polynomial algorithm for the case where the edge weights are integers. The running time of this algorithm is $O(M^2 N^3 + |w| \min(|w|, M)N^2)$ where N is the number of nodes in the graph, M is the largest absolute value of any edge weight, and w is the target cost. The algorithm is based on preprocessing the graph with a relaxation algorithm to eliminate the effects of weight sign alternations along a path.

1 Introduction

Finding shortest paths in weighted graphs is one of the most central problems in graph algorithms, with plenty of applications. The problem is now well-understood, with several polynomial-time solution algorithms [4, Chap. 25–26].

Here we study a related problem. Instead of finding a path with minimum cost we ask whether or not there is a path between two nodes that is of a given cost. We formulate the problem in a general setting, in which each edge has a weight that is a k-vector of (both positive and negative) integers as follows.

Let G be a weighted directed multi-graph with set of nodes $V(G)$ and set of edges $E(G)$. Every edge $e = p \xrightarrow{w} q \in E(G)$ carries a *cost vector* weight$(e) = w \in \mathbb{Z}^k$. The *cost of a path*

$$\mathcal{P} = p_0 \xrightarrow{w_1} p_1 \xrightarrow{w_2} p_2 \xrightarrow{w_3} \cdots \xrightarrow{w_m} p_m \qquad (1)$$

is cost$(\mathcal{P}) = \sum_{i=1}^{m} w_i$, the vector sum of all the edge costs along $\mathcal{P}$. We restrict attention to paths, whose *length* m is not zero. Our problem is as follows.

Definition 1. *Given two nodes $p, q \in V(G)$ from G, and a target cost vector w, the FIXED DISTANCE PROBLEM (FDP) is to determine if there is a path from p into q with cost exactly w in G.*

---

* Supported by Academy of Finland grant number 42977.

C. Meinel and S. Tison (Eds.): STACS'99, LNCS 1563, pp. 345–355, 1999.
© Springer-Verlag Berlin Heidelberg 1999

We will show FDP to be **NP**-complete. We also show it to remain **NP**-complete even for *scalar* weights, i.e., when the costs are integers. Our main result is a pseudo-polynomial time algorithm for the scalar version. The algorithm preprocesses G in time $O(M^2N^3)$ to account for sign alternations using a relaxation technique. Here N is the size of $V(G)$ and M is the largest absolute value of any edge weight in G. Obviously, a polynomial time algorithm is obtained if M is a constant. This happens for example when the weights are "trits" in $\mathbb{T} = \{-1, 0, +1\}$ (cf. [6]).

In addition, the same approach enables us to solve in pseudo-polynomial time the problem of finding a path with smallest *absolute* cost between given two nodes. This algorithm runs again in time $O(M^2N^3)$.

We would like to mention a couple of motivations for studying these problems. The first one comes from the analysis of multi-tape automata. In [9, 13] a query language for string databases has been developed such that string processing is carried out by two-way multi-tape finite state automata. It is important to be able to determine whether or not a given query has a finite answer [8, 14]. This requires determining whether or not a two-way multi-tape automaton can loop by moving back and forth on its input tapes while producing output. The answer to the query is finite if there are no such loops. A sufficient condition for this can be obtained using the transition graph of the automaton, but ignore the tape contents. For a k-tape automaton this graph is a weighted graph whose weights are in $\mathbb{T}^k$. The sufficient condition to be tested is that the graph has no loops with cost $\mathbf{0}$. Our second motivation stems from computational molecular biology, namely from the DNA sequence assembly problem. This leads in its classical form into the shortest common superstring problem widely studied by the algorithmic community [10, Chap. 16.17]. However, sometimes we know *a priori* the target length for the sequence to be constructed. This then reduces to the problem of finding a path of fixed length.

Finally note that if FDP is further restricted to only those paths $\mathcal{P}$ such that each prefix sum $s_i = \sum_{j=1}^{i} w_j$ is constrained to lie in the positive orthant (that is, no component of any s_i can be negative), then it becomes the reachability problem in vector addition systems or Petri nets [11, 12].

2   FDP Is NP-Complete

In this section we show our problem to be **NP**-complete.

Theorem 1. *FDP is **NP**-hard.*

Proof. Let us reduce the **NP**-complete INTEGER LINEAR PROGRAMMING (ILP) problem [15, Chap. 13] into FDP. An ILP instance (in standard form) consists of m linear equations $a_{i,1}x_1 + \cdots + a_{i,n}x_n = b_i$ with integer coefficients. The problem is then to determine whether or not this group of equations has a nonnegative integer solution.

Our reduction generates a graph G with two nodes p and q. Let A_j denote the column vector $\langle a_{1,j}, \ldots, a_{m,j} \rangle$ for each $j = 1, \ldots, n$. G consists of a loop

$p \xrightarrow{A_j} p$ for each A_j, and a bridge $p \xrightarrow{\langle -b_1,\dots,-b_m \rangle} q$. We ask whether this G has a path from p into q with weight 0. Clearly the equations have a solution $x_1 = c_1, \dots, x_n = c_n$ if and only if G contains a path from p into q with cost 0 such that the loop $p \xrightarrow{A_j} p$ is used exactly c_j times, for $j = 1, \dots, n$. $\square$

Let $\mathrm{sgn}(w) \in \mathbb{T}$ denote the *sign* of a given integer $w \in \mathbb{Z}$. It is shown next that restricting the weight vectors into $\mathbb{T}^k$ does not help in general. (However, Corollary 2 below shows that the case of scalar weights from $\mathbb{T}$ belongs to **P**.)

Corollary 1. *FDP remains **NP**-hard, even if all weight vectors belong to $\mathbb{T}^k$.*

Proof. Consider the proof of Theorem 1. Recall that ILP is **NP**-complete even in the *strong* sense [7, Problem MP1 and Chap. 4.2]. That is, there is a polynomial h so that already those ILP instances $\mathcal{I}$, whose numbers have absolute value at most $h(\text{length of } \mathcal{I})$, form an **NP**-complete language. Call this language SILP.

SILP can be reduced to our more restricted problem as follows. Perform first the reduction given in the proof of Theorem 1. Then repeatedly replace an edge $a \xrightarrow{\langle v_1,\dots,v_m \rangle} b$ having some $v_j \notin \mathbb{T}$ with a new node c and a path $a \xrightarrow{\langle \mathrm{sgn}(v_1),\dots,\mathrm{sgn}(v_m) \rangle} c \xrightarrow{\langle v_1-\mathrm{sgn}(v_1),\dots,v_m-\mathrm{sgn}(v_m) \rangle} b$ until no such edges remain. The existence of h ensures that this remains polynomial with respect to the size of the SILP instance. $\square$

Theorem 2. *FDP belongs to **NP**.*

Proof. The difficulty in this proof lies in the fact that the length of the path witnessing the desired cost need not be of polynomial length with respect to the size of the problem instance description: in for example the graph with edges $p \xrightarrow{2^n} p$ and $p \xrightarrow{-1} p$ even the shortest paths from p back into itself with cost 0 have lengths at least $2^n + 1$. We circumvent this difficulty by utilizing an *implicit* representation for these paths. Intuitively, this representation is the number of times each edge occurs on a path; technically, this is accomplished with an ILP.

We propose the following nondeterministic four-step algorithm:

1. Add into $E(G)$ a new edge $f = q \xrightarrow{0} p$ (even if $E(G)$ already contains such an edge).
2. Guess a subgraph H of G, which contains f and forms a (not necessarily maximal) strongly connected component.
3. Form the following ILP instance. There is one variable c_e for every $e \in E(H)$. There are no other variables. The constraints are as follows.
   (a) The variable c_f is constrained to be exactly 1.
       All other variables are constrained to be at least 1.
   (b) Add for every node $r \in V(H)$ the constraints $\sum_{e \in \mathrm{in}(r)} c_e = \sum_{g \in \mathrm{out}(r)} c_g$; here $\mathrm{in}(r)$ ($\mathrm{out}(r)$, respectively) consists of the edges entering (exiting, respectively) r in H.
   (c) Add finally the constraints $w = \sum_{e \in E(H)} c_e \cdot \mathrm{weight}(e)$.
4. Guess and verify a solution to the ILP instance constructed in Step 3.

Assume for specificity that G is given by listing $E(G)$ explicitly. Steps 1 to 3 can be performed in nondeterministic polynomial time with respect to the size of the problem instance description. Thus the generated ILP is also of polynomial size, and step 4 can also be performed in nondeterministic polynomial time.

Let us then verify the correctness of this algorithm. Assume first that G does have a path $\mathcal{P}$ from p into q with cost exactly $\boldsymbol{w}$. Denote as $\mathcal{Q}$ the closed loop induced by joining the last node q of $\mathcal{P}$ into its first node p with a new edge f having weight$(f) = \mathbf{0}$. Then we show that choosing each c_e to equal the number of times edge e occurs within $\mathcal{Q}$ provides a solution to the ILP constructed in step 3. Let H consist of the nodes and edges that occur in $\mathcal{Q}$. Constraints (3a) are satisfied by the choice of values for the c_e. Constraints (3b) are satisfied, because $\mathcal{Q}$ is a closed loop, and therefore every node r must be entered and exited an equal number of times. Finally, constraints (3c) reduce to the requirement that $\mathrm{cost}(\mathcal{P}) = \boldsymbol{w}$ by the choice of weight(f).

Assume conversely that the algorithm succeeds, and construct a path $\mathcal{P}$ from p into q with $\mathrm{cost}(\mathcal{P}) = \boldsymbol{w}$ as follows. By assumption, there is some subgraph H of G guessed in step 2, for which there exist values for the variables c_e that satisfy the constraints constructed in step 3. Form a multi-graph D by taking c_e copies of each edge $e \in E(H)$. This D does not omit any edges in H because of constraints (3a). On the other hand, H was chosen to be strongly connected in step 2. Therefore constraints (3b) ensure that D has an Eulerian tour $\mathcal{Q}$. Remove from $\mathcal{Q}$ the sole occurrence of f. This leaves an Eulerian path $\mathcal{P}$ from p into q in D with $\mathrm{cost}(\mathcal{P}) = \mathrm{cost}(\mathcal{Q}) = \boldsymbol{w}$ by constraints (3c), because weight$(f) = \mathbf{0}$ by step 1. Path $\mathcal{P}$ is by construction also a path in H, and therefore in G as well, if instead of many copies of the same edge we use the same edge many times. □

3   Pseudo-Polynomial Time Algorithm for Scalar FDP

Let us then consider FDP with edge costs restricted to scalars instead of vectors. At first sight, this does not seem to help:

Theorem 3. *FDP is* **NP**-*hard already for arbitrary scalar weights.*

Proof. Recall the **NP**-complete PARTITION problem [7, Problem SP12]: given a list $u_1, \ldots, u_m$ of nonzero natural numbers, determine whether the index set $I = \{1, \ldots, m\}$ contains a subset J such that $\sum_{i \in I \setminus J} u_i = \sum_{j \in J} u_j$. This is easily encoded as a graph G with nodes $E(G) = \{p_0, \ldots, p_m\}$ and edges $V(G) = \left\{ p_{i-1} \xrightarrow{+u_i} p_i, p_{i-1} \xrightarrow{-u_i} p_i : 1 \leq i \leq m \right\}$, and asking whether there exists a path $\mathcal{P}$ with cost 0 from p_0 into p_m. Then J corresponds to the numbers u_j for which the positive edge was chosen into $\mathcal{P}$ instead of the negative one. □

However, we develop a preprocessing stage, which yields a pseudo-polynomial solution to this sub-case. Preprocessing negative edge costs has already been suggested by Bertsekas [2, Sect. 4.2], but in a more restricted setting than ours.

3.1   Sign-Relaxation

Our preprocessing stage performs a certain relaxation operation on the given graph G with respect to the signs of its edge weights. The goal is to eliminate the need to consider paths, along which the edge signs change. Compared with other relaxation methods for path problems [4, Chap. 25.2–25.3], our method adds new "short-cut" edges instead of estimating path lengths.

Definition 2. *Graph H is a* sign-relaxation *of graph G if and only if the following holds for all $p, q \in V(G)$: G has a path $\mathcal{P}$ from p into q if and only if H has a path $\mathcal{Q}$ from p into q such that*

1. $\mathrm{cost}(\mathcal{Q}) = \mathrm{cost}(\mathcal{P})$, *and*
2. $\mathrm{sgn}(\mathrm{weight}(e)) = \mathrm{sgn}(\mathrm{cost}(\mathcal{Q}))$ *for all edges e that appear in $\mathcal{Q}$.*

Efficient computation of these sign-relaxations would imply $\mathbf{P} = \mathbf{NP}$ by the proof of Theorem 3. The problem is that computation on the edge weights is restricted to be polynomial with respect to the lengths of their representations, rather than their actual values. Thus we apply instead the following concept, which allows for adding these values, and leads therefore into the aforementioned pseudo-polynomiality.

Definition 3. *The* sign-closure $\mathrm{unsign}(G)$ *of a graph G is the closure of $E(G)$ with respect to the following edge addition rule: if $p \xrightarrow{w} q \xrightarrow{v} r \in E(\mathrm{unsign}(G))$ and $\mathrm{sgn}(w) \neq \mathrm{sgn}(v)$, then also $p \xrightarrow{v+w} r \in E(\mathrm{unsign}(G))$.*

Example 1. Consider the graph G_n^m with $E(G_n^m) = \left\{ p_{i-1} \xrightarrow{0} p_i : 0 < i < n \right\} \cup \left\{ p_{n-1} \xrightarrow{0} p_0, p_0 \xrightarrow{-1} p_0, p_0 \xrightarrow{m} p_0 \right\}$, where $m, n \in \mathbb{N}$.

1. The graph $\mathrm{unsign}(G_n^m)$ contains all edges $p_0 \xrightarrow{-1} p_i$ for $i = 1, 2, 3, \ldots, n-1$ by applying the edge addition rule of Definition 3 into $p_0 \xrightarrow{-1} p_{i-1} \xrightarrow{0} p_i$.
2. Then from $p_{n-1} \xrightarrow{0} p_0 \xrightarrow{-1} p_{n-1}$ the rule introduces edge $p_{n-1} \xrightarrow{-1} p_{n-1}$. Step 1 can then be repeated with p_{n-1} in the role of p_0. In this way the complete graph K_n^{-1} of n nodes and edges costing -1 emerges as a subgraph into $\mathrm{unsign}(G_n^m)$.
3. Step 2 can be repeated for the edge cost m instead of -1, producing K_n^m.
4. The rule application on $p_0 \xrightarrow{-1} p_0 \xrightarrow{m} p_0$ introduces the edge $p_0 \xrightarrow{m-1} p_0$, and then Step 3 can be applied for the edge cost $m-1$, producing K_n^{m-1}. This step can even be repeated for costs $m-2, m-3, m-4, \ldots, 0$.

Hence $\mathrm{unsign}(G_n^m)$ is the complete graph of n nodes and all edge costs in the range $-1, \ldots, m$. Thus $|E(\mathrm{unsign}(G_n^m))| = (m+2)n^2$ even though $|E(G_n^m)| = n + 2$.   ∎

```
1: H ← G; W ← E(G);
2: while W ≠ ∅ do
3:    Delete an arbitrary p --w--> q ∈ W;
4:    for all (q --v--> r ∈ E(H)) ∧ (sgn(v) ≠ sgn(w)) ∧ (e = p --v+w--> r ∉ E(H)) do
5:       E(H) ← E(H) ∪ {e}; W ← W ∪ {e}
6:    end for
7:    for all (r --v--> p ∈ E(H)) ∧ (sgn(v) ≠ sgn(w)) ∧ (e = r --v+w--> q ∉ E(H)) do
8:       E(H) ← E(H) ∪ {e}; W ← W ∪ {e}
9:    end for
10: end while
```

Fig. 1. A sign-closure algorithm.

The algorithm in Fig. 1 computes $H = \mathrm{unsign}(G)$. Assume for its analysis that $V(G) = \{1, \ldots, N\}$, and denote as $M = \max\left\{|v| : a \xrightarrow{v} b \in E(G)\right\}$ the magnitude of the edge weights. Assume further that these weights contain both positive and negative values. Observe finally that the edge weights in $\mathrm{unsign}(G)$ lie in the same range as in G.

Our computational model is a Random Access Machine (RAM), which can perform arithmetic on integers of magnitude $\max(M, N)$ in constant time. We now prove that our preprocessing stage is pseudo-polynomial and correct.

Theorem 4. *The algorithm in Fig. 1 runs in time* $O(M^2 N^3)$.

Proof. Let us consider the following implementation. Represent H as an $N \times N$ adjacency matrix A of integer lists, such that the list $A[p, q]$ contains w if and only if $p \xrightarrow{w} q \in H$. Moreover, these lists are kept in strictly ascending order. The maximum length of any such list is $2M + 1$ by our earlier observation. Then the first interior loop in steps 4–6 run in time $O(MN)$ by list merging:

> for $r \leftarrow 1, \ldots, N$ do
> > Let list L contain those elements of $A[q, r]$ with a different sign than w, in order;
> > Add w into every element in L;
> > Merge L into $A[p, r]$ so that whenever a new element u is to be added, add also the corresponding edge $p \xrightarrow{u} r$ into W
> end for

The second interior loop in steps 7–9 runs also in time $O(MN)$ with the same technique. On the other hand, an edge is added into W exactly when it is added into the graph under construction. Therefore W can be implemented with for example a list without the need to worry about duplicates, yielding $O(1)$ time insertion and deletion operations. The outer loop in steps 2–10 is executed at most $O(MN^2)$ times by the same reason. (It is executed $\Omega(MN^2)$ times in Example 1.) □

Theorem 5. *Graph* $\mathrm{unsign}(G)$ *is a sign-relaxation of graph* G.

Proof. We show $H = \text{unsign}(G)$ to satisfy the equivalence in Definition 2.

The forward direction is shown as follows. Consider any path $\mathcal{P}$ of G as in (1), and consider those pairs $p_{i-1} \xrightarrow{w_i} p_i \xrightarrow{w_{i+1}} p_{i+1}$ of consecutive edges along $\mathcal{P}$ such that $\text{sgn}(w_i) \neq \text{sgn}(w_{i+1})$; in other words, those nodes p_i, where the edge addition rule of Definition 3 could be applied.

If there are no such p_i, then choosing $\mathcal{Q} = \mathcal{P}$ suffices: edge weight signs do not change along $\mathcal{P}$, and $E(G) \subseteq E(\text{unsign}(G))$ by definition.

Otherwise select a *zenith* z from these p_i by further requiring that the absolute value $|s_i|$ of the corresponding prefix sum s_i mentioned in Sect. 1 must reach a maximum. Then divide $\mathcal{P}$ into the prefix $\mathcal{A}$ from p_0 into z, and the suffix $\mathcal{B}$ from z into p_m. $\mathcal{A}$ has now strictly fewer nodes, where the rule of Definition 3 could be applied, and therefore inductively $\text{unsign}(G)$ does have a path $\mathcal{A}'$ from p_0 into z such that $\text{cost}(\mathcal{A}') = \text{cost}(\mathcal{A})$ and $\text{sgn}(u) = \text{sgn}(\text{cost}(\mathcal{A}'))$ for all edge weights u along $\mathcal{A}'$. Similarly $\mathcal{B}$ has a corresponding path $\mathcal{B}'$ in $\text{unsign}(G)$ from z into p_m such that $\text{cost}(\mathcal{B}') = \text{cost}(\mathcal{B})$ and $\text{sgn}(v) = \text{sgn}(\text{cost}(\mathcal{B}'))$ for all edge weights v along $\mathcal{B}'$. Moreover, $\text{sgn}(\text{cost}(\mathcal{B}')) \neq \text{sgn}(\text{cost}(\mathcal{A}')) \neq 0$ by the choice of z. Therefore the edge addition rule of Definition 3 applies in the position $a \xrightarrow{u} z \xrightarrow{v} b$ of the combined path $\mathcal{A}'\mathcal{B}'$. Perform then the following rule applications:

- If $\text{sgn}(u + v) = \text{sgn}(\text{cost}(\mathcal{A}'))$, then omit z from the path by adding the edge $a \xrightarrow{u+v} b$. The rule applications continue with b as the next zenith, unless $b = p_m$, in which case rule applications cease.
- If $\text{sgn}(u + v) = -\text{sgn}(\text{cost}(\mathcal{A}'))$, then symmetrically to the previous case, z is omitted by $a \xrightarrow{u+v} b$, but now a becomes the next zenith, unless $a = p_0$.
- If $\text{sgn}(u + v) = 0$, then omit z as above with $a \xrightarrow{0} b$. Then a is a possible choice as the next zenith, unless $a = p_0$. Similarly, b is another possibility, unless $b = p_m$. If neither is possible, rule applications cease.

When these rule applications cease, the result is a path $\mathcal{Q}$ in $\text{unsign}(G)$, such that $\text{cost}(\mathcal{Q}) = \text{cost}(\mathcal{A}') + \text{cost}(\mathcal{B}') = \text{cost}(\mathcal{P})$ and $\text{sgn}(w) = \text{sgn}(\text{cost}(\mathcal{Q}))$ for all weights w along $\mathcal{Q}$, as required.

The converse direction of the equivalence is a consequence of the following claim: if $e = p \xrightarrow{w} q \in E(\text{unsign}(G))$, then G contains a path $\mathcal{P}$ from p into q with $\text{cost}(\mathcal{P}) = w$. This claim is shown induction on the genealogy of e.

If $e \in E(G)$, then $\mathcal{P} = e$ suffices. Otherwise e appeared into $\text{unsign}(G)$ as the result of applying the rule of Definition 3 into some edge pair $p \xrightarrow{v} r \xrightarrow{w-v} q$, which has already been added into $\text{unsign}(G)$, and $\text{sgn}(v) \neq \text{sgn}(w - v)$. Then inductively G has a path $\mathcal{A}$ from p into r with $\text{cost}(\mathcal{A}) = v$, and a path $\mathcal{B}$ from r into q with $\text{cost}(\mathcal{B}) = w - v$. Then path $\mathcal{A}\mathcal{B}$ is also in G and $\text{cost}(\mathcal{A}\mathcal{B}) = w$. $\square$

3.2 Solving the Scalar Version with Sign-Relaxation

Here we use the sign-relaxation concept introduced in Sect. 3.1 to provide algorithms for the scalar FDP. Our general method is to first use the algorithm in Fig. 1, and then post-process the result with known graph algorithms.

First we show that restricting the edge weights to be both scalars and in $\mathbb{T}$ at the same time yields a polynomial algorithm. Let $\mu(N)$ denote the time required to multiply together two $N \times N$ Boolean matrices (where addition is disjunction and multiplication conjunction); currently $\mu(N) = o(N^{2.376})$ [3].

Corollary 2. *FDP can be solved in $O(N^3 + \mu(N) \log_2 |w|)$ time for weights in $\mathbb{T}$, in which w is the target cost.*

Proof. Compute first unsign(G) with the algorithm in Fig. 1; this takes $O(N^3)$ time by Theorem 4. By our earlier observation, the resulting matrix A can be represented with three $N \times N$ Boolean matrices A_s, where $s \in \mathbb{T}$, so that $A_s[p, q] = 1$ if and only if $A[p, q]$ contains s.

By Theorem 5, the original graph G has a path from node p into node q with cost $w \neq 0$ if and only if the graph (whose adjacency matrix is) $A_{\mathrm{sgn}(w)}$ has a path from p into q with length $|w|$. Similarly for $w = 0$ we see that A_0 must have a path from p into q, and this can be checked within the stated time bound.

Whether A_s has a path from p into q of length $\ell > 0$ can in turn be decided by returning $A_s^\ell[p, q]$, the appropriate element of the ℓth power of A_s. This power can be computed with $O(\log_2 \ell)$ matrix multiplications by processing ℓ bit by bit, as is well known [4, Chap. 26.1]. $\square$

The algorithm in the proof of Corollary 2 provides also our first pseudo-polynomial solution to the scalar version of FDP, in which the edge weights are no longer constrained to $\mathbb{T}$: expand G into this unary form by adding at most $2M - 2$ new nodes for each original node in G, as in [1, Lemma 12].

However, when $|w|$ is moderate with respect to M and N (like $O(MN)$, say), we can improve the M^3 factor of this first algorithm into M^2 by post-processing the sign-closure of the graph with a breadth-first search instead of matrix multiplication of its unary representation.

Corollary 3. *FDP can be solved in $O(M^2 N^3 + |w| \min(|w|, M)N^2)$ time for general scalar weights.*

Proof. The adjacency matrix A for unsign(G) can be constructed in $O(M^2 N^3)$ time by Theorem 4. If $w = 0$, then the problem reduces again to checking whether or not unsign(G) has a path of edges with weight 0 from node p into node q. Assume therefore $w > 0$, and consider the post-processing algorithm in Fig. 2; the case $w < 0$ is obtained by switching signs.

Let F be the subgraph of unsign(G) consisting of the edges with sign $+1$. Its adjacency matrix, which is also denoted with F in the algorithm of Fig. 2, can be built from A in $O(MN^2)$ time; the algorithm in Fig. 1 can even maintain pointers to the first elements of the suffix $F[i, j]$ of each edge list $A[i, j]$ to build F during the construction of A at no extra cost. By Theorem 5, the original graph G has a path from node p into node q with cost w if and only if one can be found in F already.

The invariant of the main loop in steps 5–21 is as follows. List $C[i]$ contains u if and only if $u > 0$ and there exists a path $\mathcal{P}$ in F from node p into node i with cost $D + u \leq w$ such that the last edge in $\mathcal{P}$ has cost at least u.

```
1: D ← 0;
2: for all i ← 1, ... , N do
3:    Let list C[i] be the elements of list F[p, i] in the range 1, ... , w in the same order
4: end for
5: while D < w do
6:    Let B consist of those indices i such that the first element δ of list C[i] is smallest
      among the first elements of all the nonempty lists in the array C;
7:    if B = ∅ then
8:       D ← w
9:    else
10:      D ← D + δ;
11:      for all i ∈ B do
12:         Remove the first element δ from the list C[i]
13:      end for
14:      for all j ← 1, ... , N do
15:         Subtract δ from every element in the list C[j];
16:         for all i ∈ B do
17:            Merge into list C[j] all elements of list F[i, j] in range 1, ... , w − D
18:         end for
19:      end for
20:   end if
21: end while
22: Answer (D = w) ∧ (q ∈ B)
```

Fig. 2. A post-processing algorithm for FDP.

This invariant is first established by the initializations in steps 1-4. Assume then that this invariant holds for the current $D < w$. If $C[i] = \emptyset$, then F has no paths from node p into node i with cost in the range $I = \{D+1, \ldots, D+M\}$ by the invariant. If $B = \emptyset$, then accordingly F has no paths from node p with cost in I, and therefore F cannot have any paths from node p with cost more than $D + M$ either. Hence the algorithm can justifiably reply 'no' in this case.

Assume then $B \neq \emptyset$. By the invariant, $i \in B$ if and only if F contains some path $\mathcal{P}_i$ from node p into node i with cost $D + \delta$; moreover, there are no paths from node p with cost in the range $D+1, \ldots, D+\delta-1$ by the minimality of δ. The overall effect of the operations in steps 10-19 amounts to expanding each such path $\mathcal{P}_i$ with one additional edge having cost no greater than $w - D - \delta$, reinstating the invariant by moving δ from the elements of the lists $C[j]$ into D, and removing the leading zero elements that would have appeared whenever $j \in B$. In particular, if D is no longer smaller than w, the main loop terminates: a path from node p into node q with cost w was found if and only if a path $\mathcal{P}_q$ with cost w was found. The reply of the algorithm is again justified.

Each list $C[i]$ will contain a strictly ascending sequence of numbers within the range $1, \ldots, L = \min(w, M)$, extracted by using the aforementioned suffix pointers. Therefore the list processing steps 15 and 17 take time $O(L)$. The construction of the set B in step 6 can be performed in time $O(N)$. Then the interior loop in steps 14-19 takes time $O(LN^2)$ and dominates the time spent in the main loop in steps 5–21. This main loop is executed at most w times, leading to the stated time bound of $O(wLN^2)$ for the entire algorithm.   □

4   Minimum Absolute Cost Paths

Let us now turn our attention from FDP into the following minimization problem, which is readily solved with the sign-closure approach developed in Sect. 3.1.

Definition 4. *Given two nodes $p, q \in V(G)$ of a scalar graph G, what is the minimum absolute cost of any path from p into q in G, if any?*

This problem is similar to asking for the cost of the shortest path from p into q in graphs with both negative and positive edge weights [1, 5, 6, 16][4, Chap. 25.3]. However, finding a negative cost cycle warrants responding $-\infty$ in the shortest path problem, whereas our problem is "walking the tight-rope" by balancing the positive and negative contributions as well as possible.

Corollary 4. *The problem in Definition 4 can be solved in time $O(M^2 N^3)$.*

Proof. By Theorem 5, the original graph G has a path $\mathcal{P}$ from node p into node q with cost w if and only if $H = \text{unsign}(G)$ has a corresponding path $\mathcal{Q}$ from p into q with cost w via edges having the same sign as w. Assume furthermore that $|w|$ is minimal among all the paths $\mathcal{P}'$ from p into q with $\text{sgn}(\text{cost}(\mathcal{P}')) = \text{sgn}(w)$, and let $e = a \xrightarrow{u} b$ be an edge in $\mathcal{Q}$. Graph H cannot contain another edge $f = a \xrightarrow{v} b$ with $\text{sgn}(v) = \text{sgn}(w)$ but $|v| < |u|$, because path $\mathcal{Q}'$, which is as $\mathcal{Q}$, except that f is taken instead of e, still has $\text{sgn}(\text{cost}(\mathcal{Q}')) = \text{sgn}(w)$ but $|\text{cost}(\mathcal{Q}')| < |\text{cost}(\mathcal{Q})| = |w|$. Then G contains a path $\mathcal{P}'$ corresponding to $\mathcal{Q}'$, which contradicts the minimality of $|w|$. Hence a path of minimum absolute cost with edges of a given sign must always choose the edges of minimum absolute weight with the correct sign in H.

Compute therefore first H (that is, its adjacency matrix A) with the the implementation of the algorithm in Fig. 1 given in the proof of Theorem 4; this takes $O(M^2 N^3)$ time. Construct similarly to the proof of Corollary 2 three matrices A_s, $s \in \mathbb{T}$, where $A_s[p, q] = \min\{|u| : \text{sgn}(u) = s \wedge u \text{ is in list } A[p, q]\}$. (If there are no such u, then $A_s[p, q] = +\infty$.) By the reasoning above, it suffices to find shortest paths in the graphs (whose adjacency matrices are) A_s.

Therefore this problem reduces to first checking if there is a path from p into q in A_0, and answering 0 if this is the case; or otherwise determining (with for example Dijkstra's algorithm [4, Chap. 25.2]) the lengths $\ell_{\pm 1}$ of the shortest paths from p into q in $A_{\pm 1}$. Then the answer is $-\ell_{-1}$ if $\ell_{-1} < \ell_{+1}$, and ℓ_{+1} otherwise. (The answer $+\infty$ means there are no paths from p into q at all in G.)

This post-processing can be implemented to take $O(N^2)$ time. (Note also that using the adjacency matrix representation is natural, because the graph H is likely to be dense as the closure of the edge addition rule by Definition 3.)   □

5   Conclusions

We studied whether a directed graph has a path of exactly given total cost, where the edge weights can be negative as well as positive integers, or even vectors of such integers. Although the problem was shown to be **NP**-complete in most cases, we also gave pseudo-polynomial algorithms for some of these cases. Our algorithms were based on a relaxation algorithm to eliminate the effects of sign changes along the paths of the given graph. This relaxation algorithm was then used as a preprocessing stage, after which known graph algorithms were applicable, even though they do not normally cope with negative weights.

Acknowledgment. The authors thank E. Mayr for suggesting the use of ILP in Theorem 2.

References

[1] N. Alon, Z. Galil, and O. Margalit. On the exponent of the all pairs shortest path problem. *Journal of Computer and System Sciences*, 54:255–262, 1997.

[2] D.P. Bertsekas. An auction algorithm for shortest paths. *SIAM Journal on Optimization*, 1(4):425–447, 1991.

[3] D. Coppersmith and S. Winograd. Matrix multiplication via arithmetic progressions. *Journal of Symbolic Computation*, 9:251–280, 1990.

[4] T.H. Cormen, C.E. Leiserson, and R.L. Rivest. *Introduction to Algorithms*. McGraw-Hill, 1990.

[5] H.N. Gabow and R.E. Tarjan. Faster scaling algorithms for network problems. *SIAM Journal on Computing*, 18(5):1013–1036, 1989.

[6] Z. Galil and O. Margalit. All pairs shortest distances for graphs with small integer length edges. *Journal of Computer and System Sciences*, 54:243–254, 1997.

[7] M.R. Garey and D.S. Johnson. *Computers and Intractability: A Guide to the Theory of NP-Completeness*. Freeman, 1979.

[8] G. Grahne and M. Nykänen. Safety, translation and evaluation of Alignment Calculus. In *Advances in Databases and Information Systems*, Springer Electronic Workshops in Computing, pages 295–304, 1997.

[9] G. Grahne, M. Nykänen, and E. Ukkonen. Reasoning about strings in databases. In *ACM SIGACT-SIGMOD-SIGART Symposium on Principles of Database Systems*, pages 303–312, 1994.

[10] D. Gusfield. *Algorithms on Strings, Trees, and Sequences: Computer Science and Computational Biology*. Cambridge University Press, 1997.

[11] S.R. Kosaraju. Decidability of reachability in vector addition systems. In *ACM Symposium on Theory of Computing*, pages 267–281, 1982.

[12] E.W. Mayr. An algorithm for the general Petri net reachability problem. *SIAM Journal on Computing*, 13(3):441–460, 1984.

[13] M. Nykänen. *Querying String Databases with Modal Logic*. PhD thesis, Department of Computer Science, University of Helsinki, Finland, 1997.

[14] M. Nykänen. Using acceptors as transducers. In *Third International Workshop on Implementing Automata (WIA'98)*, 1998. In press.

[15] C.H. Papadimitriou and K. Steiglitz. *Combinatorial Optimization: Algorithms and Complexity*. Prentice-Hall, 1982.

[16] R. Seidel. On the all-pairs-shortest-path problem in unweighted undirected graphs. *Journal of Computer and System Sciences*, 51:400–403, 1995.

In How Many Steps the k Peg Version of the Towers of Hanoi Game Can Be Solved?

Mario Szegedy

ATT Shannon Labs, Florham Park NJ 07932, USA,
ms@research.att.com,
http://www.research.att.com/info/ms

Abstract. In this we paper we consider the version of the classical Towers of Hanoi games where the game-board contains more than three pegs. For k pegs we give a $2^{C_k n^{1/(k-2)}}$ lower bound on the number of steps necessary for transferring n disks from one peg to another. Apart from the value of the constants C_k this bound is tight.

Fig. 1. The board of the Towers of Hanoi game with four pegs

1 Introduction

"In an ancient city, so the legend goes, monks in a temple had to move a pile of 64 sacred disks from one location to another. The disks were fragile; only one could be carried at a time. A disk could not be placed on top of a smaller, less valuable disk. In addition, there was only one other location in the temple (besides the original and destination locations) sacred enough for a pile of disks to be placed there.

Using the intermediate location, the monks began to move disks back and forth from the original pile to the pile at the new location, always keeping the piles in order (largest on the bottom, smallest on the top). According to the legend, before the monks could make the final move to complete the new pile in the new location, the temple would turn to dust and the world would end." [12]

C. Meinel and S. Tison (Eds.): STACS'99, LNCS 1563, pp. 356–361, 1999.
© Springer-Verlag Berlin Heidelberg 1999

There is a single player game, called "The Towers of Hanoi," originating from the above legend with the following rules:

The game board has three pegs and n disks, which are originally arranged on the first peg, largest at the bottom and each disk sitting on a larger disk. The goal of the game is to transfer the n disks to another peg while observing the following conditions:

1. In each step the topmost disk on a peg is removed and placed on the top of the disks on another peg.
2. A disk cannot be placed on a disk smaller than itself.

The game has become one of the most popular examples for recursive algorithms (see e.g. [1], [6]). The unique shortest solution requires $2^n - 1$ steps. Since there is not much mathematical mystery left about the original game, its lovers developed various versions of it [3], [4], [9], [2]. In this article we consider a version where the number of pegs is some fixed $k > 3$ ([3], [4]). Define $h = k-2$, and s as the unique integer for which $\binom{h+s-1}{h} < n \le \binom{h+s}{h}$. In [3] and [7] different algorithms are presented that require exactly

$$a_k(n) = 2^s \left(n - \binom{h+s-1}{h} \right) + \sum_{t=0}^{s-1} 2^t \binom{h+t-1}{h} \qquad (1)$$

disk moves when transferring the pile of disks from the first peg to the second peg. To understand the above formula a little better observe that $a_k(n) - a_k(n-1) = 2^s$, where s is defined as above. This amount is perceived as the increase in the number of steps caused by the presence of the nth disk. (The formula appearing in [3] is equivalent to (1)) The order of magnitude of $a_k(n)$ is $2^{B_k n^{1/(k-2)}}$ where $B_k = (1 \pm o(1))(k-2)!^{1/(k-2)}$.

Our main result (Corollary 2) is a $2^{(1\pm o(1))C_k n^{1/(k-2)}}$ lower bound for the number of necessary moves to solve the k pegs version of the Towers of Hanoi game, which is optimal up to a constant factor in the exponent for fixed k.

Table 1. The values of $a_k(n)$ for $k \le 5$ and $n \le 10$.

$k\backslash n$	1	2	3	4	5	6	7	8	9	10
3	1	3	7	15	31	63	127	255	511	123
4	1	3	5	9	13	17	25	33	41	49
5	1	3	5	7	11	15	19	23	27	31

2 Motivation

The question about the optimal number of steps for the k peg variant of the Towers of Hanoi game was raised in the American Mathematical Monthly in

1939 (Problem 3918). The two solutions that it received in 1941 ([5], [10]) claim that the exact bound is $a_k(n)$, but the proofs are valid only for a very restricted class of algorithms. No unconditional lower bound has existed so far even though a few computer scientists have implemented algorithms for the problem (see e.g. [8], [13] for a nice web version). Since only a minimal background is required to understand the topic, the design of such algorithms is and ideal project for interested students.

The lower bound proof we present may also serve as a toy example for more involved complexity theoretic lower bound proofs, and its structure may be worthwhile to study for its own sake.

3   The Lower Bound Proof

Our lower bound will hold in a more general setup, namely when our only requirement is that between the initial and the final configuration each disk moves at least once. This generalization will allow us to use an induction. Getting the precise bound on the original problem might require a different kind of argument.

Definition 1. *For a game board with k pegs an arrangement of n disks is called a* configuration *if it obeys the "smallest disk on top of the larger" rule. For a configuration D as above let $g(D)$ be the minimal number of steps required to get every disk moved at least once, where all moves are taken according to rules 1. and 2. of the introduction. Let us define $g(n,k) = \min_D g(D)$ where D runs though over all possible configurations of n rings on a game board with k pegs.*

Remark 1. $g(D)$ *is finite for every configuration D.*

Proof. Since $g(D_0)$ is known to be finite, where D_0 is the the configuration where all disks pile up on the first peg, it is enough to show that this configuration can be reached from every other configuration. We can show this by using an induction on n. $\square$

Theorem 1. *For $k \geq 3$*

$$g(n,k) \geq 2^{(1 \pm o(1))C_k n^{1/(k-2)}}.$$

Here the constants C_k depend on k in the following way:

$$C_k = \frac{1}{2}\left(\frac{12}{k(k-1)}\right)^{1/(k-2)}.$$

Proof. We proceed by induction on k.

Remark 2. We shall make a little effort to get the optimal lower bound for $g(n,3)$, even though it would be enough for us to show that $g(n,3) = \Omega(2^n)$. We shall show that $g(n,3) \geq 2^{n-2} + 1$. From the configuration where the largest and the second largest disks are around the first peg, and all other disks are around the second peg we can see that this bound is sharp.

The case $k = 3$: We can suppose that $n \geq 2$. Consider an arbitrary initial arrangement of n disks. Let S be a sequence of the steps in which every disk moves at least once. Let j denote the first step that moves the largest disk. Let S_1 be the sequence of steps that proceed j, and let S_2 be the sequence of all steps following j. In the configurations before and after step j the disks other than the largest disk are piled up in a single tower. On the other hand, by our assumption, in either S_1 or in S_2 the second largest disk (which is at the bottom of the pile before and after step j) should move at least once. By symmetry we can assume that this happens during the moves of S_2.

It is easy to see that then S_2 must contain a solution to the three peg Towers of Hanoi problem on the $n - 2$ smallest disks, so S_2 must be at least $2^{n-2} - 1 + 1$ long. Here $2^{n-2} - 1$ comes from the lower bound for the classical game of Hanoi (see e.g. [6]) to which we can add one, because the second largest disk must also move. S contains at least one more step than S_2 (namely step j), so we obtain: $g(n, 3) \geq 2^{n-2} + 1$.

The case $k \geq 4$: First we prove a lemma which serves as the main lemma in our argument:

Lemma 1. *Suppose $k \geq 4$ and $0 < m < n/2k$. Then:*

$$g(n, k) \geq 2\min(g(n - 2km, k), g(m, k - 1)).$$

Proof. We say that a sequence S of steps *moves* a set H of disks if for every $h \in H$ there exists a step in S made by h. Let us call Z (small disks) the set of the smallest $n - 2km$ disks.

Consider an arbitrary configuration $\mathbf{D}$ of the disks around k pegs. There is a peg around which we have at least $2m$ disks from the largest $2km$ disks. Let us call X (extra large disks) the set of the largest m disks in this peg. Let us call L (large disks) the set of the next largest m disks in this peg . Note that X and L depend on $\mathbf{D}$ while Z does not. Clearly Z, L and X are disjoint, $\mid X \mid = \mid L \mid = m$, $\mid Z \mid = n - 2km$. Moreover every disk in X is larger than any disk in L and they are all larger than any disk in Z. Consider a sequence of steps that moves all the disks starting from the initial configuration $\mathbf{D}$. Define S_1 as the initial sequence of steps up to (but excluding) the first step by the topmost (i.e. the smallest) disk of X, and let S_2 be the sequence of all remaining steps. Obviously S_1 moves L and S_2 moves X. Moreover if S_1 does not move Z then the peg on which an idle member of Z is sitting is completely useless for the disks in L since they are all larger than any of the disks in Z.

This allows us in this case to estimate the number of steps made by the elements of L by $g(m, k - 1)$. The same argument shows that S_2 either moves Z or only $k - 1$ pegs were used when making the steps with the disks in X. Since according to the above argument both S_1 and S_2 contain at least $\min(g(n - 2km, k), g(m, k - 1))$ steps the proof of the lemma follows. $\square$

Let us denote $\log_2 g(n, k)$ by $\phi(n, k)$.

Corollary 1. *For $k \geq 4$ and $0 < m < n/2k$*

$$\phi(n, k) \geq 1 + \min(\phi(n - 2km, k), \phi(m, k - 1))$$

holds.

Lemma 2. $\phi(n, k)$ *is a monotone increasing function of n for any fixed $k \geq 3$.*

Lemma 2 follows from the fact that extra disks just make the task harder.

Lemma 3. *Suppose $k \geq 4$ and $\phi(n_i, k - 1) \geq i$ for $i = 1, .., s$. Then*

$$\phi(2k \sum_{i=1}^{s} n_i, k) \geq s.$$

Proof. We proceed by induction on s. The $s = 1$ case is straightforward. For $s \geq 2$:

$$\phi(2k \sum_{i=1}^{s} n_i, k) = \phi(2kn_s + 2k \sum_{i=1}^{s-1} n_i, k) \geq$$

$$1 + \min(\phi(2k \sum_{i=1}^{s-1} n_i, k), \phi(n_s, k - 1)) \geq s.$$

The first inequality comes from the corollary of Lemma 1, the second comes from the induction hypothesis on $s - 1$ and the assumption of the lemma on $\phi(n_s, k - 1)$.   □

Lemma 4. *Suppose the Theorem holds for $k - 1$. Define n_i as the smallest element of the set $\{n \mid \phi(n, k - 1) \geq i\}$. Then:*

$$\sum_{i=1}^{s} n_i \leq (1 \pm o(1)) \frac{s^{k-2}}{(k - 2) C_{k-1}^{k-3}}.$$

Proof. Since according to our assumption the theorem holds for $k - 1$ we have the $(1 \pm o(1))C_{k-1}n^{1/k-3}$ lower bound on $\phi(n, k - 1)$. Combining this with the monotonicity of $\phi(n, k - 1)$ in n we get the asymptotic upper bound on the integral of the inverse required by the lemma.   □

Now we are ready to prove the Theorem 1 for $k \geq 4$ assuming that it is true for $k - 1$. Let us denote $\sum_{i=1}^{s} n_i$ by N_s. From Lemma 4 we have $N_s \leq (1 \pm o(1)) \frac{2ks^{k-2}}{(k-2)C_{k-1}^{k-3}}$. On the other hand Lemma 3 says that $\phi(2kN_s, k) \geq s$. for every s. These two inequalities provide us with an asymptotic upper bound on the inverse of $\phi(n, k)$, which can be easily turned into the following asymptotic lower bound on $\phi(n, k)$ using the monotonicity in n:

$$\phi(k, n) \geq (1 \pm o(1)) C_{k-1}^{\frac{k-3}{k-2}} (\frac{k - 2}{2k})^{1/(k-2)} n^{k-2}.$$

Calculation shows that $C_k = C_{k-1}^{\frac{k-3}{k-2}} (\frac{k-2}{2k})^{1/(k-2)}$.   □

Corollary 2. *The k peg version of the Towers of Hanoi problem requires at least* $2^{(1-o(1))C_k n^{1/(k-2)}}$ *steps.*

Acknowledgment: The author wishes to thank L. Csirmaz for suggesting the more-than-three-pegs problem and for his useful comments and A. M. Hinz for making corrections in the manuscript.

References

1. A.V. Aho, J.E. Hopcroft, J.D. Ullman, Data Structures and Algorithms, Addison-Wesley, 1983
2. J.-P. Allouche, Note on the Cyclic Towers of Hanoi, Theoretical Comp. Sci. 123 (1994), 3-7.
3. S. Biswas and M.S. Krishnamoorthy, The generalized Towers of Hanoi, Unpublished manuscript, 1978
4. Br.A. Brousseau, Tower of Hanoi with More Pegs, Journal of Recreational Mathematics, 8:3, pp. 169-176, 1976
5. J.S. Frame, A Solution to AMM Problem 3918 (1939), American Mathematical Monthly, 48, pp.216-217, 1941
6. R. L. Graham, D. E. Knuth, O. Patashnik, Concrete mathematics. A foundation for computer science. Second edition. Addison-Wesley Publishing Company, Reading, MA, 1994. xiv+657 pp.
7. A. M. Hinz, An iterative algorithm for the tower of Hanoi with four pegs, Computing 42, pp. 133-140 (1989)
8. Xue-Miao Lu, A loopless approach to the multipeg Towers of Hanoi, International Journal of Computer Mathematics, vol. 33, no. 1/2, pp. 13-29, 1990.
9. S. Minsker, The Towers of Hanoi rainbow: coloring the rings, Journal of Algorithm, vol. 10, pp. 1-19, 1989.
10. B.M. Stewart, Solution to Problem 3918, American Mathematical Monthly, vol. 48, pp. 217-219, 1941.
11. D. Wood, Towers of Brahma and Hanoi Revisited, Journal of Recreational Mathematics, 14:1, pp.17-24, 1981-82
12. http://rialto.k12.ca.us/school/frisbie/mathfair/hanoilegend.html
13. The home page of Xue-Miao Lu at La Trobe: http://www.cs.latrobe.edu.au

Lower Bounds for Dynamic Algebraic Problems*

Gudmund Skovbjerg Frandsen[1]**, Johan P. Hansen[2], and
Peter Bro Miltersen[1]**

[1] BRICS, Centre of the Danish National Research Foundation, University of Aarhus,
Denmark. gudmund,bromille@brics.dk.
[2] Department of Mathematics, University of Aarhus, DK-8000 Aarhus C, Denmark.
matjph@imf.au.dk.

Abstract. We consider dynamic evaluation of algebraic functions (matrix multiplication, determinant, convolution, Fourier transform, etc.) in the model of Reif and Tate; i.e., if $f(x_1, \ldots, x_n) = (y_1, \ldots, y_m)$ is an algebraic problem, we consider serving on-line requests of the form "change input x_i to value v" or "what is the value of output y_i?". We present techniques for showing lower bounds on the worst case time complexity per operation for such problems. The first gives lower bounds in a wide range of rather powerful models (for instance history dependent algebraic computation trees over any infinite subset of a field, the integer RAM, and the generalized real RAM model of Ben-Amram and Galil). Using this technique, we show optimal $\Omega(n)$ bounds for dynamic matrix-vector product, dynamic matrix multiplication and dynamic discriminant and an $\Omega(\sqrt{n})$ lower bound for dynamic polynomial multiplication (convolution), providing a good match with Reif and Tate's $O(\sqrt{n} \log n)$ upper bound. We also show linear lower bounds for dynamic determinant, matrix adjoint and matrix inverse and an $\Omega(\sqrt{n})$ lower bound for the elementary symmetric functions. The second technique is the communication complexity technique of Miltersen, Nisan, Safra, and Wigderson which we apply to the setting of dynamic algebraic problems, obtaining similar lower bounds in the word RAM model. The third technique gives lower bounds in the weaker straight line program model. Using this technique, we show an $\Omega((\log n)^2 / \log \log n)$ lower bound for dynamic discrete Fourier transform. Technical ingredients of our techniques are the incompressibility technique of Ben-Amram and Galil and the lower bound for depth-two superconcentrators of Radhakrishnan and Ta-Shma. The incompressibility technique is extended to arithmetic computation in arbitrary fields.

Due to the space constraints imposed by these proceedings, in this version of the paper we only present the third technique, proving the lower bound for dynamic discrete Fourier transform and refer to the full version of the paper which is currently available as a BRICS technical report, for the rest of the proofs.

---

* Full version of this paper is available as BRICS technical report RS-98-11 from the web site www.brics.dk
** Supported by the ESPRIT Long Term Research Programme of the EU under project number 20244 (ALCOM-IT).

1 Introduction

Reif and Tate [RT97] considered the following setup of *dynamic algebraic algorithms*. Let $f_1, \ldots, f_m$ be a system of n-variate polynomials over a commutative ring or rational functions over a field. We seek an algorithm, that, when given an initial input vector $\mathbf{x} = (x_1, x_2, \ldots, x_n)$ to the system, does some preprocessing and then afterwards is able to efficiently handle on-line requests of two forms: "$\mathbf{change}_k(v)$: Change x_k to the new value v" and "$\mathbf{query}_k$: Return the value of output $f_k(\mathbf{x})$". Reif and Tate provided two general techniques for the design of efficient dynamic algebraic algorithms. They also presented lower bounds and time-space trade-offs for several problems. Apart from Reif and Tate's work, we also meet dynamic algebraic problems in the literature on the PREFIX SUM problem [Fre82, Fre81, Yao85, HF93, FS89, BAG91]; the specific case of $f_i(\mathbf{x}) = \sum_{j=1}^{i} x_i$ for $i = 1, \ldots, n$.

The aim of this paper is to present three techniques for showing lower bounds for dynamic algebraic problems. We use them to show lower bounds on the worst case time complexity per operation for several natural problems where Reif and Tate had no lower bounds or only lower bounds for the time-space trade-off.

1.1 Problems Considered

Given a commutative ring R, we look at the following systems of functions.

MATRIX-VECTOR MULTIPLICATION : $R^{n^2+n} \mapsto R^n$. The first n^2 components of the input are interpreted as an $n \times n$ matrix A, the last n components are interpreted as an n-vector $\mathbf{x}$, and $A\mathbf{x}$ is returned.

MATRIX MULTIPLICATION : $R^{2n^2} \mapsto R^{n^2}$. The input is interpreted as two $n \times n$ matrices which are multiplied.

CONVOLUTION : $R^{2n} \mapsto R^{2n}$: The input is interpreted as two n-vectors $\mathbf{x} = (x_0, \ldots, x_{n-1})$ and $\mathbf{y} = (y_0, \ldots, y_{n-1})$, whose convolution is returned. That is, the i'th component of the output is $z_i = \sum_{j+k=i} x_j y_k$.

DETERMINANT : $R^{n^2} \mapsto R$: The input is interpreted as a matrix, whose determinant is returned.

MATRIX ADJOINT : $R^{n^2} \mapsto R^{n^2}$ is the function that maps an $n \times n$ matrix A into the corresponding adjoint matrix given by MATRIX ADJOINT$(A)_{ij} = (-1)^{i+j} \det(A_{ji})$, where A_{ji} denotes the $(n-1) \times (n-1)$ matrix resulting when deleting the j'th row and the i'th column from A.

If k is a field, MATRIX INVERSE : $k^{n^2} \mapsto k^{n^2}$ is the partial function that maps a nonsingular $n \times n$ matrix A into the corresponding inverse matrix A^{-1}. Note that for a nonsingular matrix, MATRIX INVERSE$(A) = \frac{1}{\det A}$MATRIX ADJOINT(A).

DISCRIMINANT : $R^n \mapsto R$: The discriminant of the polynomial for which the n inputs are roots is returned, i.e.

$$\text{DISCRIMINANT}(x_1, \ldots, x_n) = \prod_{i \neq j}(x_i - x_j)$$

SYMMETRIC : $R^n \mapsto R^n$. All n elementary symmetric polynomials of the inputs are computed, i.e., the j'th component of the output is

$$y_j = \sum_{I \subseteq \{1,2,\ldots,n\}, |I|=j} \prod_{i \in I} x_i$$

POLYNOMIAL EVALUATION : $R^{n+2} \mapsto R$. A vector $(x, a_0, a_1, \ldots, a_n)$ is mapped to $a_0 + a_1 x + a_2 x^2 + \ldots + a_n x^n$.

Finally, the following problem is defined for any algebraically closed field k. Let ω be a primitive n'th root of unity k, and let F be the $n \times n$ matrix $F = (\omega^{ij})_{i,j}$. The Discrete Fourier Transform DFT : $k^n \mapsto k^n$, is the map $\mathbf{x} \to F\mathbf{x}$.

1.2   Models of Computation

A pivotal issue when considering lower bounds is the model of computation. For dynamic algebraic problems, this issue is quite subtle; models can vary according to *the algebraic domain* (reals, integers, finite fields, etc.), the *atomic operations allowed* (only arithmetic operations or more general operations), and the *possibility of influencing the control flow of the solution* (to what extent is the sequence of atomic operations performed allowed to depend on the previous history of the algorithm). We prove lower bounds in the following models of computation.

The straight line program model. This is the most basic model. Given the problem of dynamic evaluation of a function $f : k^n \mapsto k^m$, we assign a *straight line program* to each of the operations change$_1$, change$_2$, ..., change$_n$, query$_1$, query$_2$, ..., query$_m$. The programs corresponding to the change-operations take a single input x and have no output, while the programs corresponding to the query-operations have no input but one output. Each program is a sequence of instructions of the form $y_i \leftarrow y_j \circ y_k$, where $\circ \in \{+, -, *, /\}$, and y_j and y_k are either input variables, memory variables, or constants. We assume for convenience that we always initialize to some specific input vector and assign a corresponding initial value to each variable which appears somewhere in one of the programs. The complexity of a solution is the length of the longest program in the solution.

History dependent algebraic computation trees. In the straight line program model, it is not possible for the algorithm to modify the sequence of atomic operations performed. In the history dependent algebraic computation tree model, we allow the algorithm to control the sequence in a strong way. First, instead of assigning straight line programs to operations, we assign algebraic computation trees. As branching nodes, we do not just allow <-comparison (which only makes sense for certain fields), instead we allow branching according to *arbitrary* predicates of finite arity. Also, to each operation (such as change$_{12}$) we assign not one, but *several* (in fact infinitely many) algebraic computation trees: One for each *history*, where a history is every bit of discrete information the system has obtained so far; namely, the sequence of input variables that were changed and output variables that were queried, and the result of every branching test made so far during the execution of the operations performed. When we execute an operation, we find the tree corresponding to the current history and execute that. The complexity of a solution is the depth of its deepest tree. For an example of an algorithm where the added power of the history dependent computation tree over straight line programs seems necessary, see the algorithm for DISCRIMINANT in Section 1.4.

Random access machine models. A very general way of defining RAM models is outlined by Ben-Amram and Galil [BAG92]. Here, we will only give an informal discussion. A RAM has an infinite number of registers, indexed by the integers. It also has a finite number of CPU-registers with proper names. Each register contains an element of the domain of computation: if we consider computation over the reals, each register contains a real; if we consider computation over the integers, each register contains an integer. In any case, it is convenient if the integers (or at least a sufficiently large subset of the integers) is a subset of the domain of interest; this makes *indirect addressing* possible, an important feature of the RAM. The machine operates on the memory using a finite program containing the following kinds of instructions: direct and indirect reads and writes, conditional jumps and a finite number of atomic computational instructions operating on the CPU-registers. Each instruction is executed at unit cost. When the domain of the registers is the set of integers and the atomic operations are $+, -, *$, we get the *integer RAM*. Another model of interest is the *generalized real RAM* [BAG92]. Here, the registers contain arbitrary reals and as atomic operations we allow any set of functions $\mathbf{R}^c \mapsto \mathbf{R}$ for a constant c, with the property that for each function there is a countable closed set $C \subset \mathbf{R}^c$, so that the function is continuous in $\mathbf{R}^c \setminus C$.

The *word RAM* [FW93, FW94, Hag98] has a somewhat different flavor from the integer RAM and the real RAM. The integer RAM can be considered unreasonably powerful, since it can handle arbitrary integers with unit cost. Then again, the user can give it any sequence of n integers as input and measure the complexity of the computation as a function of n. The word RAM is the result of relaxing the power of both parties, the algorithm and the user. The word RAM does computation on words, i.e. integers in $\{0, 1, \ldots, 2^w - 1\}$ for some parameter w, intuitively determined at compile-time. The RAM has registers indexed by $\{0, 1, \ldots, 2^w - 1\}$; in particular, we assume $w \geq \log n$, so that the input can be given in registers and read. The RAM can operate on words using a number of unit cost operations including addition, subtraction, multiplication, integer division, bitwise Boolean operations, and left and right shifts. The algorithm should be correct for any value of $w \geq \log n$, but n, the number of words in the problem, should be the only variable appearing in the time bound. The word RAM has been extensively studied as a model for sorting and searching. The survey of Hagerup [Hag98] gives a good overview of this literature. When considered as a model for dynamic algebraic problems, the word RAM is appropriate when the function in question is a constant degree polynomial over the integers. This ensures that when the input is a vector of words, the output can be given in a constant number of words, i.e. we can at least *report* the output with unit cost.

For instance, dynamic matrix multiplication makes good sense in the word RAM model while we will not consider dynamic determinant in this model.

1.3   Our Results

We present three techniques for proving lower bounds for dynamic algebraic problems. The first technique is very robust. In particular, it holds under a wide range on assumptions about the algebraic domain and the operations allowed, and even if the algorithm is allowed to control the flow of computation in strong ways. The technique is closely related to the *incompressibility* technique of Ben-Amram and Galil [BAG92]. The second technique holds only for the word RAM model (where the first technique fails). It is a modest extension of communication complexity techniques of Miltersen *et al* [MNSW95]. With the first and second technique we show

Theorem 1. *Any solution to dynamic* MATRIX-VECTOR MULTIPLICATION, MA-TRIX MULTIPLICATION, MATRIX ADJOINT, MATRIX INVERSE, DETERMINANT, POLYNOMIAL EVALUATION *or* DISCRIMINANT *has worst case complexity* $\Omega(n)$ *per operation and any solution to dynamic* CONVOLUTION *or* SYMMETRIC *has worst case complexity* $\Omega(\sqrt{n})$ *per operation, in the following models of computation:*

- *Straight line programs over any fixed finite field (except for* POLYNOMIAL EVALUATION, DISCRIMINANT *and* SYMMETRIC*), with the allowed set of* change*-arguments being the field itself.*
- *History dependent algebraic computation trees over any infinite field, with the allowed set of* change*-arguments being any infinite subset of the field.*
- *The integer RAM (except for* MATRIX INVERSE*), with the allowed set of* change*-arguments being any infinite subset of the integers, and the generalized real RAM, with the allowed set of* change*-arguments being the reals.*
- *The word RAM (except for* MATRIX ADJOINT, MATRIX INVERSE, DETER-MINANT, POLYNOMIAL EVALUATION, DISCRIMINANT *and* SYMMETRIC*), with the allowed set of* change*-arguments being the set of words.*

We should note that the lower bound for dynamic POLYNOMIAL EVALUATION was also proved by Reif and Tate, though not for as wide a range of models as above. Reif and Tate present lower bounds for a number of other problems by reductions from POLYNOMIAL EVALUATION; we can apply the same reductions to get the lower bounds in the wider range of models.

We should also note that for certain models and certain of the above problems, there is an easier way of showing the same lower bound. For instance, we can show a lower bound for dynamic MATRIX-VECTOR MULTIPLICATION over the reals using arithmetic operations as follows: It is well known [Win67, Win70] that $n \times n$ matrices A over the reals exist so that computing $\mathbf{x} \to A\mathbf{x}$ requires $\Omega(n^2)$ arithmetic operations. Now, given an alleged dynamic algorithm for dynamic MATRIX-VECTOR MULTIPLICATION with complexity $o(n)$ per operation, we can initialize the matrix input to this matrix. Then, we can evaluate $A\mathbf{x}$ for any given **x** using n **change** and n **query** operations, i.e., a total of $o(n^2)$ arithmetic operations, a contradiction. The same technique was, in fact, used by Reif and Tate to show the lower bounds of their paper (using the fact that explicit hard polynomials exist, rather than the fact that explicit hard matrices exist). However, this argument does not seem to generalize to show, for instance, the linear lower bound for straight line programs over a finite field (where matrices requiring $\Omega(n^2)$ arithmetic operations do not exist [Sav74]), nor to show any lower bound for the generalized real RAM or the word RAM. Also, our technique applies to a wider variety of problems in a uniform way.

Our third technique is more fragile. It only works in the model of history *independent* straight line programs. A technical ingredient of the technique is the lower bound for depth-two superconcentrators by Radhakrishnan and Ta-Shma [RTS97]. With the third technique we show

Theorem 2. *Any solution to dynamic* DFT *in the straight line program model over an algebraically closed field of characteristic* 0, *with* **change**-*arguments restricted to any infinite subset of the field, has worst case complexity* $\Omega((\log n)^2 / \log \log n)$ *per operation.*

Due to the space constraints imposed by these proceedings, we shall in this paper only present our third technique, proving Theorem 2, and only for the case where the allowed set of input arguments is the entire field (and not some infinite subset). For a full account of the proof of Theorem 2, the proof of Theorem 1, and our first and second techniques, we refer the reader to the technical report BRICS RS-98-11, available at www.brics.dk.

1.4   Optimality (and Otherwise) of Results

The lower bounds for MATRIX-VECTOR MULTIPLICATION and MATRIX MULTI-PLICATION are tight, there are straightforward linear upper bounds. The lower bound for DISCRIMINANT is also tight, there is a linear upper bound for any infinite field (see Theorem 3), and a straightforward constant upper bound for any finite domain in the straight line program model. Interestingly, the linear upper bound does not seem to be implementable in the straight line program model. The lower bound for CONVOLUTION has a fairly good match in the $O(\sqrt{n \log n})$ upper bound of Reif and Tate [RT97] for the same problem. The upper and lower bounds for DETERMINANT, MATRIX ADJOINT, MATRIX INVERSE and SYMMETRIC are not tight, we don't know any solution for DETERMINANT, MATRIX ADJOINT and MATRIX INVERSE better than evaluating queries from scratch, and we don't know any better upper bound for dynamic SYMMETRIC than a (not quite obvious) $O(n)$ upper bound (see Theorem 4). Reif and Tate show an $O(\sqrt{n})$ upper bound for dynamic DFT which is valid in the straight line program model. This leaves a rather large gap between upper and lower bounds.

Theorem 3. *There is a computation tree solution of complexity* $O(n)$ *for dynamic evaluation of* DISCRIMINANT. *The solution works over any field.*

Proof. All the current inputs $x_1, \ldots, x_n$ are maintained, and so is the set of their (distinct) values together with the number of occurrences in a structure $L = \{[v_1, n_1], \ldots, [v_{|L|}, n_{|L|}]\}$, i.e. $n_i \geq 1$ and $\sum_i n_i = n$. Finally, we maintain the (nonzero) discriminant of the distinct values: $D = \prod_{i \neq j}(v_i - v_j)$. With this representation query is simple; if all n_i's are 1, we return D, otherwise we return 0. For change, we must update D and L, which is easily done in linear time (see Figure 1).

$\text{change}_i(v)$: assume $x_i = v_k$ for $[v_k, n_k] \in L$;   if $n_k > 1$ then $n_k := n_k - 1$
    else $D := D / \prod_{j \neq k}(-1)(v_j - v_k)^2$; $L := L \setminus \{[v_k, 1]\}$;
    if $v = v_l$ for some $[v_l, n_l] \in L$ then $n_l := n_l + 1$
    else $D := D \cdot \prod_j(-1)(v_j - v)^2$; $L := L \cup \{[v, 1]\}$;
    $x_i := v$;

Fig. 1. Computation tree solution for DISCRIMINANT.

Theorem 4. *There is a straight line program solution of complexity $O(n)$ for* SYMMETRIC. *The solution works over any commutative ring.*

Proof. All the current inputs $x_1, \ldots, x_n$ and corresponding outputs $y_1, \ldots, y_n$ are maintained. This makes the straight-line program for query_i trivial; it needs only return y_i. For the implementation of change, we observe that for any i, k, we have that $y_k = x_i z_{k-1, i} + z_{ki}$, where z_{ki} does not depend on x_i, which makes the solution in Figure 2 valid.

$\text{change}_i(v)$: $z_0 := 1$;
    for $k = 1 \ldots n$ do
        $z_k := y_k - x_i z_{k-1}$;
        $y_k := z_k + v z_{k-1}$;
    $x_i := v$;

Fig. 2. Straight line solution for SYMMETRIC.

2   Lower Bound for Dynamic DFT

Our technique is essentially based on the following *incompressibility* statement: If k is an algebraically closed field, a *rational* map $k^n \mapsto k^{n-1}$ can not be injective. Thus, it is closely related to the technique of Ben-Amram and Galil, who applied incompressibility in various domains to show a gap between the power of random access machines and pointer machines [BAG92].

First, a technical lemma stating a generalization of the above fact. Let k be an algebraically closed field. Recall that an algebraic subset $W \subset k^n$ is an intersection of sets of the form $\{\mathbf{x} \in k^n | p(\mathbf{x}) = 0\}$, where p is a non-trivial multivariate polynomial.

Lemma 5. *Let k be an algebraically closed field. Let W be an algebraic subset of k^m and let $\phi = (f_1/g_1, \ldots, f_n/g_n) : k^m \setminus W \mapsto k^n$ be a rational map where $f_i, g_i \in k[x_1, \ldots, x_m]$ for $i = 1, \ldots, n$. Assume that there exists $\mathbf{y} \in k^n$ such that $\phi^{-1}(\mathbf{y})$ is non-empty and finite. Then $m \leq n$.*

Using an approach similar to Valiant [Val76], we shall apply the notion of a superconcentrator to get a setting where we may use the incompressibility lemma. The equivalence of the following to the standard definition is due to Meshulam [Mes84]. An *n-superconcentrator* of depth 2 is a graph G with nodes $X \cup V \cup Y$, where X, V and Y are disjoint, $|X| = |Y| = n$, and with edges $E \subseteq (X \times V) \cup (V \times Y)$ such that for any l, for any $X_1 \subseteq X$ and for any $Y_1 \subseteq Y$ with $|X_1| = |Y_1| = l$, we have $|N(X_1) \cap N(Y_1)| \geq l$, where $N(X_1), N(Y_1) \subseteq V$ denote the neighbors to X_1, Y_1. Radhakrishnan and Ta-Shma [RTS97] proved that the number of edges in an n-superconcentrator of depth 2 is at least $\Omega(n \frac{\log^2 n}{\log \log n})$.

Let k be an algebraically closed field. Let $f : k^n \mapsto k^n$ be a function. Let $X = \{x_1, \ldots, x_n\}$ be the set of inputs, and let $Y = \{y_1, \ldots, y_n\}$ be the set of outputs. We say that f is *super-injective*, when for every l, for every $X_1 \subseteq X$ and for every $Y_1 \subseteq Y$ satisfying that $|X_1| = |Y_1| = l$ there is a $\mathbf{a} \in k^{n-l}$ such that $f_{\mathbf{a}} : k^l \mapsto k^l$ is injective, where $f_{\mathbf{a}}$ denotes the function arising from specializing f to the constants $\mathbf{a}$ on the inputs $X \setminus X_1$ and ignoring all outputs in $Y \setminus Y_1$.

Lemma 6. *Let k be an algebraically closed field. Let $f : k^n \mapsto k^n$ be a super-injective polynomial function. From any family of straight line programs for dynamic evaluation of f and of complexity d, we get an n-superconcentrator of depth 2 and with at most $3dn$ edges.*

Proof. From the dynamic solution for f, define a graph G as follows. The nodes of G is $X \cup V \cup Y$, where V is the variables used in the dynamic solution for f, i.e. we may assume that $V = \{v_1, \ldots, v_m\}$, where $m \leq 2dn$. The edges of G is $E \subseteq (X \times V) \cup (V \times Y)$ and $(x_i, v) \in E$, if the program for change$_i$ writes the variable v. Similarly, $(v, y_j) \in E$, if the program for query$_j$ reads the variable v. Clearly, $|E| \leq 3dn$. We shall argue that G is a superconcentrator.

Let l be given, and let $X_1 \subseteq X$, $Y_1 \subseteq Y$ be given such that $|X_1| = |Y_1| = l$. Let $V_1 = N(X_1) \cap N(Y_1)$. We need to argue that $|V_1| \geq l$. (After permutation of indices) we may assume that $X_1 = \{x_1, \ldots, x_l\}$ and $Y_1 = \{y_1, \ldots, y_l\}$. Use the super-injectivity of f to choose $\mathbf{a} \in k^{n-l}$ such that $f_{\mathbf{a}} : k^l \mapsto k^l$ is injective, where $f_{\mathbf{a}}$ denotes the function arising from specializing the inputs $(x_{l+1}, \ldots, x_n)$ to the constants $\mathbf{a} = (a_1, \ldots, a_{n-l})$ and ignoring all the outputs $(y_{l+1}, \ldots, y_n)$.

From the dynamic solution for f, construct an off-line solution $P = P_1; P_2; P_3$ for $f_{\mathbf{a}}$ as follows

$$P_1 : \text{change}_{l+1}(a_1); \cdots; \text{change}_n(a_{n-l})$$

$$P_2 : \text{change}_1(x_1); \cdots; \text{change}_l(x_l)$$

$$P_3 : y_1 := \text{query}_1; \cdots; y_l := \text{query}_l$$

Let $\tilde{X}_1$ denote the values of the input variables X_1 (before the execution of P_2), let $\tilde{V}_1$ denote the values of the variables V_1 after the execution of P_2 but before the execution of P_3, and let $\tilde{Y}_1$ denote the values of the output variables Y_1

(after the execution of P_3). Clearly, $\tilde{V}_1$ is a rational function of $\tilde{X}_1$. Denote this rational function by $g : k^l \mapsto k^{|V_1|}$. Similarly, $\tilde{Y}_1$ is a rational function of $\tilde{V}_1$, since the output does only depend on the input through the intermediate values $\tilde{V}_1$. Denote this rational function by $h : k^{|V_1|} \mapsto k^l$. We see that $f_\mathbf{a} = h \circ g$. Since $f_\mathbf{a}$ is injective, so must also g be injective, and by Lemma 5 this is only possible if $|V_1| \geq l$.

Lemma 6 and the lower bound for depth 2 superconcentrators of [RTS97] together implies the following lemma.

Lemma 7. *Let k be an algebraically closed field. Let $f : k^n \mapsto k^n$ be a super-injective polynomial function. Any family of straight line programs for dynamic evaluation of f has complexity $\Omega(\frac{\log^2 n}{\log \log n})$.*

It is obvious that a linear map is super-injective if and only if all minors of the corresponding matrix are non-zero. Thus, by Lemma 7, to show the lower bound for dynamic DFT claimed in Theorem 2 for the case where the allowed set of **change**-arguments is the entire field, we just need to show that this is the case for a large $(n^{\Omega(1)} \times n^{\Omega(1)})$ submatrix of the Fourier transform matrix. The following lemma accomplishes this and completes the proof of the special case of Theorem 2.

Lemma 8. *Let k be an algebraically closed field of characteristic 0, let $\omega \in k$ be a primitive n'th root of unity, and let $F = (a_{ij})$ be the $n \times n$ discrete Fourier transform matrix with $a_{ij} = \omega^{ij}$.*

Then F contains an $l \times l$ submatrix B for some $l = \Omega(\sqrt[3]{\frac{n}{\log \log n}})$ such that all minors of B are nonzero.

Proof. Let $l = \lfloor \sqrt[3]{\phi(n)} \rfloor$, where $\phi(n)$ denotes the Euler phi function, which is also the number of distinct primitive n'th roots of unity. It is known that $\lim \inf_{n \to \infty} \frac{\phi(n) \ln \ln n}{n} = e^{-\gamma} \approx 0.56$ (see Hardy and Wright [HW54] page 267, theorem 328), so $l = \Omega(\sqrt[3]{\frac{n}{\log \log n}})$ as required.

Let z be a variable and let $C(z)$ be the $l \times l$ matrix with the ij'th entry being $c_{ij} = z^{ij}$. Let $B = C(\omega)$ and note that B occurs as the $l \times l$ submatrix in the upper left corner of F.

We show that all minors of B are nonzero. Clearly, each minor of $C(z)$ is a polynomial in z with integer coefficients, and we will later show that no minor of $C(z)$ is the zero-polynomial. Therefore, each minor in $C(z)$ is a nonzero polynomial of degree strictly less than $l^3 \leq \phi(n)$ (assuming that $l \geq 2$). This implies that the minors of $B = C(\omega)$ are nonzero. To see this, observe that ω is a root of the nth cyclotomic polynomial which has degree $\phi(n)$ and is irreducible over the field $\mathbf{Q}$ (see Hungerford [Hun74], page 299, Proposition 8.3). Therefore ω is not root of any polynomial with integer coefficients and of degree strictly smaller than $\phi(n)$, as k has characteristic 0.

We now show that no minor in the matrix $C(z)$ is the zero-polynomial. Let an $m \times m$ minor D in $C(z)$ be given by row-indices $i_1 < \cdots < i_m$ and column indices $j_1 < \cdots < j_m$. By Lemma 9, $D = z^{i_1 j_1 + \cdots + i_m j_m} + p(z)$, where $p(z)$ is either the zero-polynomial or has degree strictly less than $i_1 j_1 + \cdots + i_m j_m$.

Lemma 9. *Let two sets of m positive integers each be given, namely I containing $i_1 < \cdots < i_m$ and J containing $j_1 < \cdots < j_m$. For any permutation σ of $\{1, \ldots, m\}$, let $S_\sigma = i_1 j_{\sigma(1)} + \cdots + i_m j_{\sigma(m)}$. Then $S_1 > S_\sigma$ for $\sigma \neq 1$, where 1 denotes the identity permutation.*

Proof. Let σ be a permutation on $\{1, \ldots, l\}$ such that $\sigma \neq 1$. We will argue that by changing σ slightly, we can get a new permutation τ (possibly with $\tau = 1$) such that $S_\tau > S_\sigma$, which suffices to prove the lemma.

Since $\sigma \neq 1$, we can find $a < b$ such that $\sigma(a) > \sigma(b)$. Define τ to be identical to σ except that $\tau(a) = \sigma(b)$ and $\tau(b) = \sigma(a)$. This implies that $S_\tau = S_\sigma - i_a j_{\sigma(a)} - i_b j_{\sigma(b)} + i_a j_{\tau(a)} + i_b j_{\tau(b)} = S_\sigma - i_a j_{\sigma(a)} - i_b j_{\sigma(b)} + i_a j_{\sigma(b)} + i_b j_{\sigma(a)} = S_\sigma + (i_b - i_a)(j_{\sigma(a)} - j_{\sigma(b)}) > S_\sigma$.

References

[BAG91]   Amir M. Ben-Amram and Zvi Galil. Lower bounds for data structure problems on RAMs (extended abstract). In *Proc. 32nd Annual Symposium on Foundations of Computer Science*, pages 622–631, 1991.

[BAG92]   Amir M. Ben-Amram and Zvi Galil. On pointers versus addresses. *J. Assoc. Comput. Mach*, 39:617–648, 1992.

[Fre81]   M.L. Fredman. Lower bounds on the complexity of some optimal data structures. *SIAM J. Comput.*, 10:1–10, 1981.

[Fre82]   M.L. Fredman. The complexity of maintaining an array and computing its partial sums. *J. Assoc. Comput. Mach.*, 29:250–260, 1982.

[FS89]   M.L. Fredman and M.E. Saks. The cell probe complexity of dynamic data structures. In *Proc. Twenty First Annual ACM Symposium on Theory of Computing*, pages 345–354, 1989.

[FW93]   M.L. Fredman and D.E. Willard. Surpassing the information-theoretic bound with fusion trees. *J. Comput. System Sci.*, 47:424–436, 1993.

[FW94]   M.L. Fredman and D.E. Willard. Trans-dichotomous algorithms for mimimum spanning trees and shortest paths. *J. Comput. System Sci.*, 48:533–551, 1994.

[Hag98]   Torben Hagerup. Sorting and searching on the Word RAM. In *Proc. 15th Annual Symposium on Theoretical Aspects of Computer Science, LNCS* vol. 1373, pages 366–398. Springer-Verlag, 1998.

[HF93]   H. Hampapuram and M.L. Fredman. Optimal bi-weighted binary trees and the complexity of maintaining partial sums. In *Proc. 34th Annual Symposium on Foundations of Computer Science*, pages 480–485, 1993.

[Hun74]   Thomas W. Hungerford. *Algebra*, volume 73 of *Graduate Texts in Mathematics*. Springer-Verlag, 1974.

[HW54]   G. H. Hardy and E. M. Wright. *An Introduction to the Theory of Numbers. (Third Edition).* Oxford University Press, 1954.

[Mes84]   Roy Meshulam. A geometric construction of a superconcentrator of depth 2. *Theoret. Comput. Sci.*, 32:215–219, 1984.

[MNSW95]   Peter Bro Miltersen, Noam Nisan, Shmuel Safra, and Avi Wigderson. On data structures and asymmetric communication complexity. In *Proc. 27th Annual ACM Symposium on the Theory of Computing*, pages 103–111, 1995.

[RT97]   John H. Reif and Stephen R. Tate. On dynamic algorithms for algebraic problems. *J. Algorithms*, 22:347–371, 1997.

[RTS97]   Jaikumar Radhakrishnan and Amnon Ta-Shma. Tight bounds for depth-two superconcentrators. In *Proc. 38th Annual Symposium on Foundations of Computer Science*, pages 585–594, 1997.

[Sav74]   J.E. Savage. An algorithm for the computation of linear forms. *SIAM J. Comput.*, 3:150–158, 1974.

[Val76]   Leslie G. Valiant. Graph-theoretic properties in computational complexity. *Journal of Computer and System Sciences*, 13(3):278–285, December 1976.

[Win67]   S. Winograd. On the number of multiplications required to compute certain functions. *Proc. Nat. Acad. Sci. U.S.A.*, 58:1840–1842, 1967.

[Win70]   S. Winograd. On the number of multiplications necessary to compute certain functions. *Comm. Pure Appl. Math.*, 23:165–179, 1970.

[Yao85]   A.C. Yao. On the complexity of maintaining partial sums. *SIAM J. Comput.*, 14:277–288, 1985.

An Explicit Lower Bound for
TSP with Distances One and Two

Lars Engebretsen

Dept. of Numerical Analysis and Computing Science
Royal Institute of Technology
SE-100 44 Stockholm, SWEDEN
enge@nada.kth.se

Abstract. We show that it is **NP**-hard to approximate the traveling salesman problem with distances one and two within $5381/5380 - \epsilon$, for any $\epsilon > 0$. Our proof is a reduction from systems of linear equations mod 2 with two unknowns in each equation and at most three occurrences of each variable.

1 Introduction

A common special case of the traveling salesman problem is the metric traveling salesman problem, where the distances between the cities satisfy the triangle inequality. In this paper, we study a further specialization: The traveling salesman problem with distances one and two between the cities. This problem was shown to be **NP**-complete by Karp [7]. Since this means that we have little hope of computing exact solutions, it is interesting to try to find an approximate solution, i.e., a tour with weight close to the optimum weight. Christofides [2] has constructed an elegant algorithm approximating the metric traveling salesman problem within $3/2$. This algorithm also applies to the traveling salesman problem with distances one and two, but it is possible to do better; Papadimitriou and Yannakakis [8] have shown that it is possible to approximate the latter problem within $7/6$. They also show a lower bound; that there exists some constant, which is never given explicitly in the paper, such that it is **NP**-hard to approximate the problem within that constant.

Recently, there has been a renewed interest in the hardness of approximating the traveling salesman problem with distances one and two. Fernandez de la Vega and Karpinski [3] and, independently, Fotakis and Spirakis [4] have shown that the hardness result of Papadimitriou and Yannakakis holds also for dense instances. We contribute to this line of research by showing an explicit lower bound on the approximability. More specifically, we construct a reduction from linear equations mod 2 with three occurrences of each variable to show that it is **NP**-hard to approximate the traveling salesman problem with distances one and two within $5381/5380 - \epsilon$.

C. Meinel and S. Tison (Eds.): STACS'99, LNCS 1563, pp. 373–382, 1999.
© Springer-Verlag Berlin Heidelberg 1999

2   Preliminaries

Definition 1. *We denote by* E2-Lin mod 2 *the following maximization problem: Given a system of linear equations mod 2 with exactly two variables in each equation, maximize the number of satisfied equations. We denote by* E2-Lin(3) mod 2 *the special case of E2-Lin mod 2 where there are exactly three occurrences of each variable.*

Definition 2. *We denote by* (1,2)-TSP *the traveling salesman problem where the distance matrix is symmetric and the off-diagonal entries are either one or two, and by* △-TSP *the traveling salesman problem where the distance matrix is symmetric and obeys the triangle inequality.*

We note in passing, that since (1,2)-TSP is a special case of △-TSP a lower bound on the approximability of (1,2)-TSP is also a lower bound on the approximability of △-TSP.

To describe a (1,2)-TSP instance, it is enough to specify the edges of weight one. We do this by constructing a graph G, and then let the (1,2)-TSP instance have the nodes of G as cities. The distance between two cities u and v is defined to be one if (u, v) is an edge in G and two otherwise. To compute the weight of a tour, it is enough to study the parts of the tour traversing edges of G.

Definition 3. *We call a node where the tour leaves or enters G an* endpoint. *A city with the property that the tour both enters and leaves G in that particular city is called a* double endpoint, *and counts as two endpoints.*

If c is the number of cities and $2e$ is the total number of endpoints, the weight of the tour is $c + e$, since every edge of weight two corresponds to two endpoints. When we analyze our reduction, we study an arbitrary tour restricted to certain subgraphs of G. Generally, such a restriction consists of several disjoint paths. To shorten the notation, we call these paths *partial tours*.

3   Our Construction

To obtain our hardness result we reduce from E2-Lin(3) mod 2. Previous reductions from integer programming [6] and satisfiability [8] to (1,2)-TSP make heavy use of the so called *xor gadget*. This gadget is used both to link variable gadgets with equation gadgets and to obtain a consistent assignment to the variables in the original instance. The xor gadget contains twelve nodes, which means that a gadget containing some twenty xor gadgets for each variable — which is actually the case in the previously known reductions — produces a very poor lower bound. To obtain a reasonable inapproximability result, we modify the previously used xor gadget to construct an *equation gadget*. A specific node in the equation gadget corresponds to one occurrence of a variable. Since each variable occurs three times, there are three nodes corresponding to each variable. These nodes are linked together in a *variable cluster*. The idea behind this is that

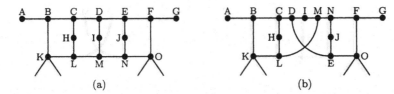

Fig. 1. The equation gadget is connected to other gadgets through the vertices A and G and through the edges shown above from the vertices K and O. Gadget (a) corresponds to an equation of the form $x + y = 1$ and gadget (b) to an equation of the form $x + y = 0$.

the extra edges in the cluster should force the nodes to represent the same value for all three occurrences of the variable. This construction contains 24 nodes for each variable, which is a vast improvement compared to earlier constructions.

We give our construction in greater detail below. In a sequel of lemmas, we show that an optimal tour can be assumed to have a certain structure. We do this by showing that we can transform, by local transformations which do not increase the length of the tour, any tour into a tour with the sought structure. This new tour, obtained after the local transformations, can then be used to construct an assignment to the variables in the original E2-Lin(3) mod 2 instance. Our main result follows from a recent hardness result of Berman and Karpinski [1] together with a correspondence between the length of the tour in the (1,2)-TSP instance and the number of unsatisfied equations in the E2-Lin(3) mod 2 instance.

3.1 The Equation Gadget

The equation gadget is shown in Fig. 1. It is connected to other gadgets in four places. The vertices A and G actually coincide with similar vertices at other gadgets to form a long chain. Thus, these vertices actually have degree two. The edges from the vertices K and O join the equation gadget with other equation gadgets. We study this closely in Sec. 3.2. No other vertex in the gadget is joined with vertices not belonging to the gadget.

Definition 4. *We from now on call the vertices K and O in Fig. 1 the* lambda-vertices *of the gadget and the boundary edges connected to these vertices* lambda-edges. *For short, we often refer to the pair of lambda-edges linked to a particular lambda-vertex as a* lambda. *(This name was chosen since the lambda-edges look like the Greek letter Λ.)*

Definition 5. *We say that a lambda is* traversed *if both lambda-edges are traversed by the tour,* untraversed *if none of the lambda-edges are traversed, and* semitraversed *otherwise.*

In the following lemmas, we study what an optimal tour can look like.

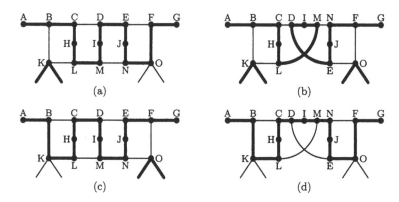

Fig. 2. Given that the lambda-edges are traversed as shown above, it is possible to construct a tour through the equation gadget such that there are no endpoints in the gadget.

Lemma 1. *Suppose that we have a tour traversing an equation gadget of the type shown in Fig. 1a in such a way that there are no semitraversed lambdas in it. If there is exactly one traversed lambda, it is possible to modify this tour, without increasing its length and without changing the tour on the lambdas, in such a way that there are no endpoints in the gadget. Otherwise, it is possible to construct a tour with two endpoints in the gadget and impossible to construct a tour with less than two endpoints in the gadget.*

Suppose that we have a tour traversing an equation gadget of the type shown in Fig. 1b in such a way that there are no semitraversed lambdas in it. If there are zero or two traversed lambdas, it is possible to modify this tour, without increasing its length and without changing the tour on the lambdas, in such a way that there are no endpoints in the gadget. Otherwise, it is possible to construct a tour with two endpoints in the gadget and impossible to construct a tour with less than two endpoints in the gadget.

Proof. Figures 2 and 3 show that there exists tours with the number of endpoints stated in the lemma. To complete the proof, we must show that it is impossible to construct better tours in the cases where the tour has two endpoints in the gadget. It is locally optimal to let the tour traverse the edge AB and the edge FG in Fig. 1. Thus we can assume that one partial tour enters the gadget through the vertex A and that another, or possibly the same, partial tour enters through the vertex G. Since there are no semitraversed lambdas in the gadget, the only way, other than through the above described vertices, a partial tour can leave the gadget is through an endpoint, which in turn implies that there is an even number of endpoints in the gadget.

If there is to be no endpoint in the gadget, all of the edges CHL, DIM and EJN must be traversed by the tour. Also, the edges CD and LM cannot be traversed simultaneously, neither can DE and MN. The only way we can avoid making the

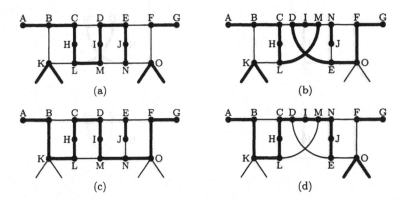

Fig. 3. Given that the lambda-edges are traversed as shown above, there must be at least two endpoints in the gadget. There are in fact many tours with this property, we show only a few above.

vertices D and M endpoints is to traverse either the edges CD and MN or the edges DE and LM.

Let us suppose that the lambdas in the gadget are traversed as shown in Fig. 3a. By our reasoning above and the symmetry of the gadget, we can assume that the edges AB, CHLMIDEJN, and FG are traversed by the tour. To avoid making the vertices C and N endpoints, the tour must traverse the edges BC and NO. But this is impossible, since the right lambda is already traversed. Thus, there is no tour with zero endpoints in the gadget, which implies that is impossible to construct a tour with less than two endpoints in the gadget. With a similar argument, we conclude that the same holds for the other cases shown in Fig. 3.

Lemma 2. *Suppose that we have a tour traversing an equation gadget in such a way that there is exactly one semitraversed lambda in it. Then it is possible to modify this tour, without increasing its length and without changing the tour on the lambdas, in such a way that there is one endpoint in the gadget, and it is impossible to construct a tour with less than one endpoint in the gadget.*

Proof. From Fig. 4, we see that we can always construct tours such that there is one endpoint in the gadget. We now show that it is impossible to construct a tour with fewer endpoints. As in the proof of Lemma 1, we can assume that one partial tour enters the gadget at A and that another, or the same, enters at G. Since there is one semitraversed lambda in the gadget, one partial tour enters the gadget at that lambda, which implies that there must be an odd number of endpoints in the gadget.

Lemma 3. *Suppose that we have a tour traversing an equation gadget in such a way that there are two semitraversed lambdas in it. Then it is possible to modify this tour, without increasing its length and without changing the tour on the lambdas, in such a way that there are two endpoints in the gadget, and it is impossible to construct a tour with less than two endpoints in the gadget.*

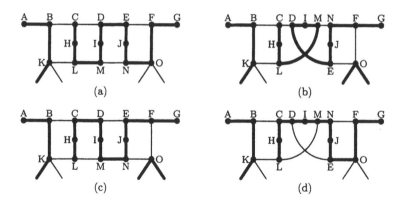

Fig. 4. If one lambda is semi-traversed, there must be at least one endpoint in the gadget.

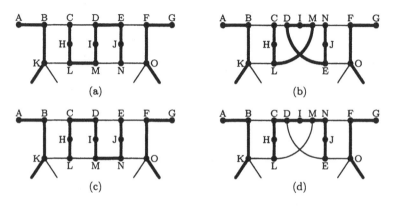

Fig. 5. If both lambdas in a gadget are semi-traversed, there must be at least two endpoints in the gadget.

Proof. From Fig. 5, we see that we can always construct tours such that there are two endpoints in the gadget. In order to prove the last part of the lemma we must argue that it is impossible to traverse the gadget in such a way that there are no endpoints in it. By an argument similar to that present in the proofs of Lemmas 1 and 2, a partial tour will enter (or leave) the gadget at four places, which implies that there is an even number of endpoints in the gadget. If there is to be no endpoints in the graph, there must be two partial tours in the gadget. Since the tours cannot cross each other and the gadget is planar we have two possible cases.

The first case is that the partial tour entering the gadget at A leaves it at G (for a gadget of the type shown in Fig. 1a) or at O (for a gadget of the type shown in Fig. 1b) and the partial tour entering at K leaves at O (for a gadget of the type shown in Fig. 1a) or at G (for a gadget of the type shown in Fig. 1b). These two partial tours cannot, however, traverse any of the edges CHL, DIM and EJN without crossing or touching each other. As noted in the proof of

Lemma 1, all three abovementioned edges must be traversed for the gadget to contain no endpoints. Thus, we can rule this case out.

The second case is that the partial tour entering the gadget at A leaves it at K and the partial tour entering at G leaves at O. Since these two partial tours cannot traverse all three edges CHL, DIM and EJN without crossing or touching each other, we conclude that at least two endpoints must occur within the equation gadget.

Lemma 4. *It is always possible to change a semitraversed lambda to either a traversed or an untraversed lambda without increasing the number of endpoints in the tour.*

Proof. First suppose that only one of the lambdas in the equation gadget is semitraversed. By Lemma 2 we can assume that the gadgets are traversed according to Fig. 4. Let us study the tour shown in Fig. 4a. By replacing it with the tour shown in Fig. 2a, we remove one endpoint from the equation gadget, but we may in that process introduce one endpoint somewhere else in the graph. In proof, let λ be the left lambda-vertex in Fig. 4a and v be the vertex adjacent to λ through the untraversed lambda edge. If v is an endpoint, we simply let the partial tour ending at v continue to λ, thereby saving one endpoint. If v is not an endpoint, we have to reroute the tour at v to λ. This introduces an endpoint at a neighbor of v, but that endpoint is set off against the endpoint removed from the equation gadget. To sum up, we have shown that it is possible to convert the tour in Fig. 4a to the one in Fig. 2a without increasing the total number of endpoints in the graph. In a similar way, we can convert the tour in Fig. 4b to the one in Fig 2b, the tour in Fig. 4c to the one in Fig 2c, and the tour in Fig. 4d to the one in Fig 2d, respectively.

Finally, suppose that both lambdas are semitraversed. By Lemma 3 we can assume that the gadgets are traversed according to Fig. 5. By the method described in the previous paragraph we can convert the tour in Fig. 5a to the one in Fig 4a, the tour in Fig. 5b to the one in Fig 4b, the tour in Fig. 5c to the one in Fig 4c, and the tour in Fig. 5d to the one in Fig 4d, respectively.

3.2   The Variable Cluster

The variable cluster is shown in Fig. 6. The vertices A and B coincide with similar vertices at other gadgets to form a long chain, as described in Sec. 3.3. Suppose that the variable cluster corresponds to some variable x. Then the upper three vertices in the cluster are lambda-vertices in the equation gadgets corresponding to equations where x occurs. The remaining two vertices in the cluster are not joined with vertices outside the cluster.

Lemma 5. *Suppose that we have a tour traversing a cluster in such a way that there are some semitraversed lambdas in it. Then, it is possible to modify the tour, without making it longer, in such way that there are no semitraversed lambdas in the cluster.*

Fig. 6. There is one lambda-vertex for each occurrence of each variable. The three lambda-vertices corresponding to one variable in the system of linear equations are joined together in a variable cluster. The three uppermost vertices in the figure are the lambda-vertices.

Proof. Suppose that there is one semitraversed lambda. If the semitraversed lambda is the middle lambda of the variable cluster it can, by Lemma 4, be transformed into either a traversed or an untraversed lambda. This moves the semitraversed lambda to the end of the cluster. By moving the endpoint in the variable cluster to the equation gadget corresponding to the semitraversed lambda, we can make the last semitraversed lambda traversed or untraversed without changing the number of endpoints.

Suppose now that there are two semitraversed lambdas. By Lemma 4, they can be transformed into either a traversed or an untraversed lambda without changing the number of endpoints in the tour. This implies that we can transform the tour in such a way that there is only one semitraversed lambda without changing the number of endpoints in the tour. Then we can use the method from the above paragraph to transform that tour in such a way that there are no semitraversed lambdas.

Finally, suppose that all three lambdas are semitraversed. By Lemma 4, the tour can be transformed in such a way that the two outer lambdas in the variable cluster are either traversed or untraversed without changing the weight of the tour. If the center lambda is not semitraversed after the transformation, the proof is complete. Otherwise we can apply the first paragraph of this proof.

3.3   The Entire (1,2)-TSP Instance

To produce the (1,2)-TSP instance, all equation gadgets are linked together in series, followed by all variable clusters. The first equation gadget is also linked to the last variable cluster. The construction is shown schematically in Fig. 7. The precise order of the individual gadgets in the chain is not important.

Our aim in the analysis of our construction is to show that we can construct, from a tour containing e edges of weight two, an assignment to the variables such that at most e equations are not satisfied. When we combine this with Berman's and Karpinski's recent hardness results for E2-Lin(3) mod 2 [1], we obtain Theorem 2, our main result.

Theorem 1. *Given a tour with $2e$ endpoints, we can construct an assignment leaving at most e equations unsatisfied.*

Proof. Given a tour, we can by Lemmas 1–5 construct a new tour, without increasing its length, such that for each variable cluster either all or no lambda-edges are traversed. Then we can construct an assignment as follows: If the

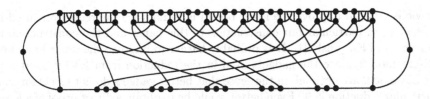

Fig. 7. All equation gadgets and all variable clusters are linked together in a circular chain as shown schematically above. The equation gadgets are at the top of the figure and the variable clusters at the bottom. The precise order of the gadgets is not important. For clarity, we have omitted nine vertices from each equation gadget.

lambda-edges in a cluster are traversed by the tour, the corresponding variable is assigned the value one; otherwise it is assigned zero. By Lemma 1, this assignment has the property that there are two endpoints in the equation gadgets corresponding to unsatisfied equations. Thus, the assignment leaves at most e equations unsatisfied if there are $2e$ endpoints.

Theorem 2. *It is **NP**-hard to decide whether an instance of the traveling salesman problem with distances one and two with $5376n$ nodes has an optimum tour with length above $(5381 - \epsilon_1)n$ or below $(5380 + \epsilon_2)n$.*

Corollary 1. *It is for any $\epsilon > 0$ **NP**-hard to approximate the traveling salesman problem with distances one and two within $5381/5380 - \epsilon$.*

Proof (of Theorem 2). The result of Berman and Karpinski [1] states that it is **NP**-hard to determine if an instance of E2-Lin(3) mod 2 with $336n$ equations has its optimum above $(332 - \epsilon_2)n$ or below $(331 + \epsilon_1)n$. If we construct from an instance of E2-Lin(3) mod 2 an instance of (1,2)-TSP as described above, the graph contains $48n$ nodes if the E2-Lin(3) mod 2 instance contains $2n$ variables and $3n$ equations. Thus, Theorem 1 and the above hardness result together imply that it is **NP**-hard to decide whether an instance of (1,2)-TSP with $5376n$ nodes has an optimum tour with length above $(5381 - \epsilon_1)n$ or below $(5380 + \epsilon_2)n$.

4   Concluding Remarks

We have shown in this paper that it is for any $\epsilon > 0$ **NP**-hard to approximate (1,2)-TSP within $5381/5380 - \epsilon$. Since the best known upper bound on the approximability is $7/6$, there is certainly room for improvements. Our lower bound follows from a sequence of reductions, which makes it unlikely to be optimal. The sequence starts with E3-Lin mod 2, systems of linear equations mod 2 with exactly three variables in each equation. Then follows reductions to, in turn, E2-Lin mod 2, E2-Lin(3) mod 2, and (1,2)-TSP. Thus, our hardness result ultimately follows from Håstad's optimal lower bound on E3-Lin mod 2 [5].

Obvious ways to improve the lower bound is to improve the reductions used in each step, in particular our construction in this paper and the construction of Berman and Karpinski [1]. It is probably harder to improve the lower bound on E2-Lin mod 2, since the gadgets used in the reduction from E3-Lin mod 2 to E2-Lin mod 2 are optimal, in the sense that better gadgets do not exists for that particular reduction [5,9]. Even better would be to obtain a direct proof of a lower bound on (1,2)-TSP. It would also be interesting to study the approximability of Δ-TSP in general, and try to determine if Δ-TSP is harder to approximate than (1,2)-TSP.

5   Acknowledgments

Viggo Kann contributed with fruitful discussions on this subject. Also, Gunnar Andersson and Viggo Kann gave valuable comments on early versions of the paper. One of the anonymous referees spotted a lapse in a previous version of this paper and helped improve the presentation of the results.

References

1. P. Berman and M. Karpinski. On some tighter inapproximability results. Technical Report 29, Electronic Colloquium on Computational Complexity, June 1998.
2. N. Christofides. Worst-case analysis of a new heuristic for the traveling salesman problem. Technical Report CS-93-13, Graduate School of Industrial Administration, Carnegie Mellon University, Pittsburgh, 1976.
3. W. Fernandez de la Vega and M. Karpinski. On approximation hardness of dense TSP and other path problems. Technical Report 24, Electronic Colloquium on Computational Complexity, Apr. 1998.
4. D. A. Fotakis and P. G. Spirakis. Graph properties that facilitate travelling. Technical Report 31, Electronic Colloquium on Computational Complexity, June 1998.
5. J. Håstad. Some optimal inapproximability results. In *Proc. Twenty-ninth Ann. ACM Symp. on Theory of Comp.*, pages 1–10. ACM, New York, 1997.
6. D. S. Johnson and C. H. Papadimitriou. Computational complexity. In E. L. Lawler, J. K. Lenstra, A. H. G. Rinnooy Kan, and D. B. Shmoys, editors, *The Traveling Salesman Problem*, chapter 3, pages 37–85. John Wiley & Sons, New York, 1985.
7. R. M. Karp. Reducibility among combinatorial problems. In R. E. Miller and J. W. Thatcher, editors, *Complexity of Computer Computations*, pages 85–103. Plenum Press, New York, 1972.
8. C. H. Papadimitriou and M. Yannakakis. The traveling salesman problem with distances one and two. *Math. of Oper. Res.*, 18(1):1–11, Feb. 1993.
9. L. Trevisan, G. B. Sorkin, M. Sudan, and D. P. Williamson. Gadgets, approximation, and linear programming. In *Proc. 37th Ann. IEEE Symp. on Foundations of Comput. Sci.*, pages 617–626. IEEE Computer Society, Los Alamitos, 1996.

Scheduling Dynamic Graphs*

Andreas Jakoby[1], Maciej Liśkiewicz[2]**, and Rüdiger Reischuk[1]

[1] Institut für Theoretische Informatik, Med. Universität zu Lübeck
Wallstr. 40, D-23560 Lübeck, Germany
{jakoby, reischuk}@informatik.mu-luebeck.de
[2] Wilhelm-Schickard Institut für Informatik, Universität Tübingen
Sand 13, D-72076 Tübingen, Germany
liskiewi@informatik.uni-tuebingen.de

Abstract. In parallel and distributed computing scheduling low level tasks on the available hardware is a fundamental problem. Traditionally, one has assumed that the set of tasks to be executed is known beforehand. Then the scheduling constraints are given by a *precedence graph*. Nodes represent the elementary tasks and edges the dependencies among tasks. This static approach is not appropriate in situations where the set of tasks is not known exactly in advance, for example, when different options how to continue a program may be granted.

In this paper a new model for parallel and distributed programs, the *dynamic process graph*, will be introduced, which represents all possible executions of a program in a compact way. The size of this representation is small – in many cases only logarithmically with respect to the size of any execution. An important feature of our model is that the encoded executions are directed acyclic graphs having a "regular" structure that is typical of parallel programs. Dynamic process graphs embed constructors for parallel programs, synchronization mechanisms as well as conditional branches. With respect to such a compact representation we investigate the complexity of different aspects of the scheduling problem: the question whether a legal schedule exists at all and how to find an optimal schedule. Our analysis takes into account communication delays between processors exchanging data. Precise characterization of the computational complexity of various variants of this compact scheduling problem will be given in this paper. The results range from easy, that is $\mathcal{NLOGSPACE}$-complete, to very hard, namely $\mathcal{NEXPTIME}$-complete.

1 Introduction

Scheduling tasks efficiently is crucial for fast executions of parallel and distributed programs. An intensive study of this scheduling problem has led to the development of a number of algorithms that cover a wide spectrum of strategies: from fully static, where the compiler completely precomputes the schedule, i.e. when and where each task will be executed, to fully dynamic, where tasks

---

* Supported by DFG Research Grant Re 672/2.
** On leave of Instytut Informatyki, Uniwersytet Wrocławski, Wrocław, Poland.

C. Meinel and S. Tison (Eds.): STACS'99, LNCS 1563, pp. 383–392, 1999.
© Springer-Verlag Berlin Heidelberg 1999

are scheduled at run-time only. Changing from static to dynamic strategies one gets the potential of reducing the total execution time of a program because the resources are better used, but in general there will be more effort necessary at run-time. Therefore, existing parallel systems fix most details of the schedule already at compile time (see e.g. [4]). For a more dynamic and also fault tolerant approach see for example the MAFT project [9,8].

In many cases, the set of tasks that have to be executed is not precisely known at compile time. El-Rewini and Ali have introduced a parallel program model that allows a suitable data representation for static scheduling algorithms [2]. The representation is based on two directed graphs: the *branch graph* and the *precedence graph*. This approach models conditional branchings quite well, but it is unsuited for parallel program constructors or synchronization mechanisms, for example the channel concept as implemented in the parallel programming language OCCAM.

1.1   Extension to a Dynamic Environment

In this paper we introduce a new model, the *dynamic process graph*, **DPG**, which allows a natural representation for parallel and distributed programs. In particular, it gives a concise description of static scheduling problems for highly concurrent programs. An important feature of DPGs is that they resemble the characteristics of executing typical parallel or distributed programs, and this in a space-efficient way. This representation can provide an exponential compaction compared to the length of execution sequences.

Our main technical contribution is to analyze the complexity of finding optimal schedules with respect to this compact program representation. We will concentrate on scheduling elementary tasks where each task has unit execution time. No bound will be put on the number of processors available.

1.2   Communication Delays

Papadimitriou and Yannakakis have argued in [12] that scheduling policies should take into account communication delays occurring when one processor sends a piece of data to another one. Thus, it will be faster to schedule dependent tasks on the same processor. For further results concerning scheduling with communication delays in the standard static setting see [7,6,13]. Here, we will extend the complexity analysis to the dynamic case.

Communication delays will be specified by a function $\delta : E \to \mathbb{N}$, which defines the time necessary to send the data from one processor to another one. For simplification we assume that this delay is independent of the particular pair of processors (alternatively that $\delta(e)$ gives an upper bound on the maximal delay). Scheduling with communication delays requires the following condition to be fulfilled: if a task v is executed on processor p at time t then for each direct predecessor u of v holds:

u has been finished either on p by time $t - 1$, or on some other processor p' by time $t - 1 - \delta(u, v)$.

1.3 Dynamic Dependencies

In many systems, dependencies among particular task-instances are determined by a scheduling policy, not by the program itself. Consider the following situation. A program contains a part P where several processes concurrently generate data. The program can continue as soon as one of these results is available. Such a situation can compactly be described using the ALT-constructor of OCCAM [1]. In the piece of code given in the left part of Figure 1 one process sends data down channel C1, while a second one behaves similarly using channel C2.

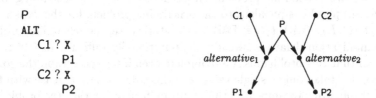

    P
    ALT
        C1 ? x
           P1
        C2 ? x
           P2

Fig. 1. *A dynamic precedence graph representing* P: *the output mode of* P *is* ALT *and the input mode of* $alternative_1$ *and* $alternative_2$ *is* PAR.

A scheduling policy has to select one (and only one) of the alternatives Pi. Hence, either P1 or P2 will be a successor of P. If both channels Ci get ready at the same time then a scheduler can choose arbitrarily. However, to minimize the total execution time it is helpful to know which process can be executed faster, P1 or P2? Even if the two channels do not get ready simultaneously executing a ready alternative immediately, may overall lead to a longer schedule than waiting until the other channel is ready.

We will consider different degrees of concurrency expressed by the ALT and the PAR constructor to create parallel processes, and analyze the complexity of the corresponding scheduling problems with respect to the amount of concurrency.

To represent such parallel programs in a natural way, we introduce dynamic process graphs, which are generalizations of standard precedence graphs. A dynamic process graph is an acyclic graph $G = (V, E)$ with two sets of labels $I(v), O(v) \in \{PAR, ALT\}$ attached to the nodes $v \in V$. Nodes represent tasks, edges dependencies between tasks. A complete formal definition will be given in the next section.

The label $I(v)$ describes the *input mode* of task v. If $I(v) = $ ALT then to execute v at least one of the predecessor tasks u with $(u, v) \in E$ has to be completed. $I(v) = $ PAR requires that executions of all predecessors of v have to be completed before v can start. If task v has been completed then according to the *output mode* $O(v)$ one of v's successors in case $O(v) = $ ALT (resp. all of them in case $O(v) = $ PAR) has to be initiated. Fig. 1 gives an example of such a representation.

Dynamic process graphs are a compact way to illustrate data dependencies of parallel programs written e.g. in a parallel programming language like OCCAM or Ada. Note that a standard precedence graph cannot represent such programs in a simple way. We should note that dynamic process graphs can also be modeled by a certain class of Petri nets and their reachability problem. However, this class does not seem to correspond to those subclasses that have been considered in more detail in the literature. We will therefore stick to the DPG model.

1.4   New Results

The scheduling problem for dynamic process graphs is a natural generalization of the static scheduling problem. In particular, the delay scheduling for a precedence graph G is equivalent to the scheduling problem for the dynamic process graph (G, I, O) with $I, O \equiv$ PAR. In the static case, the scheduling problem with communication delays is already computationally difficult. In [12] it has been shown that this problem is $\mathcal{NP}$-complete even if for each graph the communication delay takes only a single value, but this value has to increase with the size of the graph. We have improved this result in [6] showing that the problem remains $\mathcal{NP}$-complete even if we restrict the class of precedence graphs to (1,2)-trees. On the other hand, in [7] it has been shown that for fixed delay $\delta \equiv c$ independent of the precedence graphs the problem can be solved in polynomial time where the degree of the polynomial grows with c.

Due to the compact representation in our dynamic model it is no longer obvious that the dynamic scheduling problem can be solved in $\mathcal{NP}$ at all. In fact, one of our main results is that even restricted to constant communication delay the scheduling problem for dynamic process graphs is $\mathcal{NEXPTIME}$-complete. To prove this we construct a reduction of the SUCCINCT-3SAT problem. However, if we restrict the input mode I to ALT then the problem becomes $\mathcal{P}$-complete. Even more, also fixing the output mode the problem becomes $\mathcal{NLOGSPACE}$-complete. A similar complexity jump has been observed for classical graph problems in [3,11,10]. There it is shown that simple graph properties become $\mathcal{NP}$-complete when the graph is represented in a particular succinct way using generating circuits or a hierarchical decomposition. Under the same representation graph properties that are ordinarily $\mathcal{NP}$-complete, like HAMILTON CYCLE, 3-COLORABILITY, CLIQUE (of size $|V|/2$), etc., become $\mathcal{NEXPTIME}$-complete.

On the other hand, some restricted variants of this scheduling problem, which may seem to be easy at first glance, remain hard, namely $\mathcal{NP}$-complete. Fig. 2 summarizes our results about the complexity of the dynamic delay scheduling problem with respect to the input and output modes that may occur in the graphs.

We will also consider the question whether for a given dynamic process graph *there exists* a schedule for a program represented by the graph at all. It will be shown that the problem is $\mathcal{NP}$-complete even if the input mode is restricted to PAR and the output mode to ALT.

The remaining part of this paper is organized as follows. In Section 2 we give some examples and a formal definition of DPGs and the scheduling problem. Section 3 studies the complexity of the existence problem. The last section deals

input mode	output mode	complexity
ALT	ALT	$\mathcal{NLOGSPACE}$-complete
ALT	PAR	$\mathcal{NLOGSPACE}$-complete
ALT	ALT, PAR	$\mathcal{P}$-complete
PAR	ALT, PAR or PAR or ALT	$\mathcal{NP}$-complete
ALT, PAR	ALT	$\mathcal{NP}$-complete
ALT, PAR	PAR	$\mathcal{BH}_2$-hard
ALT, PAR	ALT, PAR	$\mathcal{NEXPTIME}$-complete

Fig. 2. *The complexity of dynamic scheduling with communication delays with respect to input and output modes.*

with the problem to find optimal schedules. Due to space limitations we will only present the technically most difficult result, which is scheduling unrestricted DPGs, and give a short sketch of the construction for the lower bound.

2   Dynamic Process Graphs and Runs

For illustration, consider the following parallel program P written in OCCAM (Fig. 3). This program contains branches, but the situation is not as bad as it could be since the branching does not depend on the current values of variables. Still there is the problem to determine the set of tasks that have to be executed at run time. It will turn out that even for such restricted programs the scheduling problem is quite hard. Depending on the chosen ALT branches the execution of this program is represented by one of the four possible runs shown in Fig. 4. The following definition tries to capture this dichotomy between parallel/distributed programs and their executions.

Definition 1. *A* **dynamic process graph** $\mathcal{G} = (G, I, O)$ *is a directed acyclic graph (DAG) $G = (V, E)$, with node labellings $I, O : V \to \{\text{ALT}, \text{PAR}\}$. $V = \{v_1, v_2, \ldots, v_n\}$ represents a set of* **processes** *and E dependencies among them. I and O describe* **input modes**, *(resp.* **output modes***) of the processes. A finite DAG $H_\mathcal{G} = (W, F)$ is a* **run** *of $\mathcal{G}$ iff the following conditions are fulfilled:*

1. *The set W is partitioned into subsets $W(v_1) \cup \ldots \cup W(v_n)$. The nodes in $W(v_i)$ are* **execution instances** *of the process v_i and will be called* **tasks**.

2. *Each source node of G has exactly one execution instance in $H_\mathcal{G}$.*

3. *For a process $v \in V$ let $pred(v) := \{u_1, u_2, \ldots, u_p\}$ denote the set of all predecessors of v and $succ(v) := \{w_1, w_2, \ldots, w_r\}$ its successors. For any execution instance x of v in $W(v)$ it has to hold*
   - *if $I(v) = \text{ALT}$ then x has a unique predecessor y belonging to $W(u_i)$ for some $i \in \{1, \ldots, p\}$;*
   - *if $I(v) = \text{PAR}$ then $pred(x) = \{y_1, y_2, \ldots, y_p\}$ with $y_i \in W(u_i)$ for each $i \in \{1, \ldots, p\}$;*

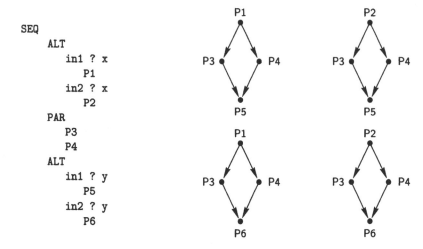

```
SEQ
  ALT
    in1 ? x
      P1
    in2 ? x
      P2
  PAR
    P3
    P4
  ALT
    in1 ? y
      P5
    in2 ? y
      P6
```

Fig. 3. *OCCAM program* P.   **Fig. 4.** *Possible runs of* P.

- *if $O(v) = $ ALT then x has a unique successor z belonging to $W(w_j)$ for some $j \in \{1, \ldots, r\}$;*
- *if $O(v) = $ PAR then $succ(x) = \{z_1, z_2, \ldots, z_r\}$ with $z_j \in W(w_j)$ for each $j \in \{1, \ldots, r\}$.*

Fig. 5 shows a node v of a DPG with input mode Q_1 and output mode Q_2. Through the paper we will illustrate the ALT-mode by a white box, the PAR-mode by a black box. For a source or a node with indegree 1 the input mode is obviously inessential. Hence we will ignore such a label, and similarly the output label in case of a sink or a node with outdegree 1. A dynamic process graph corresponding to the program of Fig. 3 is shown in Fig. 6 (a).

Observe that a run can be smaller than its defining dynamic process graph, e.g. the graph in Fig. 6 (a) has 10 nodes while its run (b) only 8. Hence, certain tasks are not executed at all. More typically, however, a run will be larger than the dynamic process graph itself since the PAR-constructor allows process duplications. If, for example, the output mode of vertex v_1 in Fig. 6 is changed from ALT into PAR then both processes P_1 and P_2 have to be executed. Hence, process v_2 with input mode ALT has to be duplicated in order to consume both processes. This in turn implies that P_3 and P_4 will have 2 execution instances each. Therefore, each run of the modified graph consists of 15 nodes. The following lemma gives an upper bound on this blow-up, resp. the possible compaction ratio of dynamic process graphs.

Lemma 1. *Let $\mathcal{G} = ((V, E), I, O)$ be a dynamic process graph and $H_{\mathcal{G}} = (W, F)$ be a corresponding run. Then it holds $|W| \leq 2^{|V|-1}$. Moreover, this general bound is tight.*

Hence, there are dynamic process graphs where processes have exponential many execution instances. Note that a similar effect occurs by using the repli-

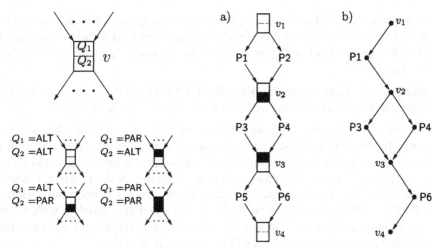

Fig. 5. *A node v with input label Q_1 and output label Q_2 and the schematic representation.*

Fig. 6. *(a) a dynamic process graph for program P, (b) a run of this graph.*

cated PAR- and ALT-constructor of OCCAM, which allows an exponential blow up of the number of active tasks.

Lemma 2. *Any dynamic process graph has at most double exponential many different runs and this bound can actually occur.*

Definition 2. *Let $\mathcal{G} = (G, I, O)$ be a dynamic process graph with communication delay $\delta : E \to I\!N$ between its processes. A schedule for $\mathcal{G}, \delta$ is a schedule of a run $H = (W, F)$, where the communication delay between each pair of execution instance $x \in W(u)$ and $y \in W(v)$ is given by $\delta(u, v)$.*

If S is a schedule of $\mathcal{G}, \delta$ then let $T(S)$ denote the duration of S, i.e. the amount of time necessary to complete all tasks in S. Define

$$T_{opt}(\mathcal{G}, \delta) := \min_{S \text{ for } \mathcal{G}, \delta} T(S).$$

This leads to the following decision problem:

Definition 3. DPG-SCHEDULING (DPGS): *Given a DPG (G, I, O), a communication delay δ, and a deadline T^*, does $T_{opt}(\mathcal{G}, \delta) \le T^*$ hold?*

3 The Execution Problem

The number of different runs of a DPG can be huge according to Lemma 2. On the other hand, it is not obvious that for any DPG an appropriate run exists at all. It is easy to see that dynamic process graphs with either only PAR labels, or with all input labels equal to ALT can always be executed. The first case corresponds to standard static precedence graphs. However, this is no longer true for arbitrary DPGs. For a simple example of a graph which has no run see Fig. 7. In this case the input mode of all nodes is PAR and their output mode ALT. This section studies the problem whether a given dynamic process graph has a legal run.

Definition 4. Execution Problem for DPGs (ExDPG): *Given a DPG $\mathcal{G}$, decide whether it can be executed, that means whether it has a run.*

As we have seen above for some restricted DPGs this question has a trivial answer. In general, a decision procedure may be complex. Of course, if a DPG has a run then it has also a schedule, and since we have estimated a bound on the size of the run, one can also compute an upper bound on the maximal schedule length given the maximal communication delay. Thus, the execution problem for DPGs could be solved by a reduction to the scheduling problem with a huge enough deadline. However, we want to capture the complexity of the execution problem more precisely and thus will investigate it directly.

The main negative result of this section says that for graphs with arbitrary input and output modes the execution problem is $\mathcal{NP}$-complete. We will prove even more, namely the problem remains $\mathcal{NP}$-complete even if the input mode of all nodes is PAR while the output mode is ALT. On the positive side, the complexity decreases drastically for DPGs with output labels restricted to PAR. Fig. 8 summarizes our results about the complexity of ExDPG Problem with respect to input and output modes that may occur in the graphs.

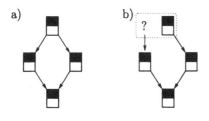

input mode	output mode	complexity
ALT	arbitrary	trivial
PAR	PAR	trivial
ALT, PAR	PAR	$\mathcal{NC}^2$
PAR	ALT	$\mathcal{NP}$-complete
ALT, PAR	ALT	$\mathcal{NP}$-complete
ALT, PAR	ALT, PAR	$\mathcal{NP}$-complete

Fig. 7. *Example of a dynamic process graph with $O \equiv$ ALT and $I \equiv$ PAR that has no run.*

Fig. 8. *The complexity of ExDPG-Problem with respect to input and output modes.*

4   Scheduling Dynamic Process Graphs

Let us now consider the problem to construct optimal schedules for dynamic process graphs. Since the execution problem is already hard in the less restricted cases one has to expect similar negative results for the scheduling problem. Our main result below, however, implies that the compaction provided by dynamic process graphs is quite efficient. The complexity of scheduling general DPGs increases to $\mathcal{NEXPTIME}$-complete. The hardness proof will be the topic of this section. We have also analysed the complexity for restricted classes of DPGs. They will be mentioned at the end of this section.

Theorem 1. *Scheduling dynamic process graphs is* $\mathcal{NEXPTIME}$-*complete,
even for constant communication delay.*

Sketch of the proof. For the reduction we will use the following problem

Definition 5. SUCCINCT-3SAT: *As input we are given a Boolean circuit
over the standard AND, OR, NOT-basis that succinctly codes a Boolean formula
in conjunctive normal form with the additional property that each clause has
exactly three literals and each literal appears exactly three times. Suppose that
the encoded formula consists of n variables and m clauses. On input $(0, i, k)$ with
$i \in \{0, \ldots, n-1\}$ and $k \in \{1, 2, 3\}$ (appropriately coded in binary), the coding
circuit returns the index of the clause where the literal $\neg x_i$ appears the k-th time.
On input $(1, i, k)$ it returns the index of the clause where x_i appears for the k-th
time. On input $(2, j, k)$ with $j \in \{0, \ldots, m-1\}$ and $k \in \{1, 2, 3\}$, it returns the
k-th literal of the j-th clause.
The problem is to decide whether the encoded formula is satisfiable.*

The $\mathcal{NEXPTIME}$-completeness for this problem has been proved by Papadimitriou and Yannakakis in [11].

For the reduction of SUCCINCT-3SAT to the scheduling problem, the first
crucial step is a transformation of a Boolean circuit B into a DPG $\mathcal{G}$. Our
encoding achieves the following property. Let $x_1, x_2, \ldots, x_r$ be the input gates
of B, and $v_1, v_2, \ldots, v_s$ be its output gates. In the coding graph $\mathcal{G}$ there will be
nodes x_{fi}, x_{ti} for $1 \le i \le r$, and nodes v_{fj}, v_{tj} for $1 \le j \le s$. The meaning of the
indices is as follows: x_{fi} codes the value *false* for x_i, while x_{ti} *true*; similarly v_{fj}
and v_{tj} code values *false* and *true* for v_j. To simplify notation, let $x_i(false) := x_{fi}$
and $x_i(true) := x_{ti}$, etc. Then the following holds: for any input $(b_1, b_2, \ldots, b_r)$,
B returns $(c_1, c_2, \ldots, c_s)$ iff there exists a run for $\mathcal{G}$ such that each of the process
nodes $x_1(b_1), \ldots, x_r(b_r)$ and $v_1(c_1), \ldots, v_s(c_s)$ has one execution instance, and
none of the complementary nodes $x_1(\neg b_1), \ldots, x_r(\neg b_r), v_1(\neg c_1), \ldots, v_s(\neg c_s)$ has
any execution instance.

The total construction consists of two disjoint graphs made up of the encoding
of the input circuit B and some specific auxiliary modules. The first graph will
be responsible for checking whether the circuit B generates a Boolean formula
according to the syntax of SUCCINCT-3SAT. The second graph checks whether
the encoded formula is satisfiable. The constructions are technically involved
and a complete description of them as well as $\mathcal{NEXPTIME}$ algorithm for the
scheduling problem can be found in [5].   □

If one restricts the DPGS problem to specific combinations of input or output
modes then in most cases the computational complexity decreases significantly.
In Figure 2 we list the results of our investigations. The proofs of these hardness
results and the algorithms to establish the upper bounds can be found in [5].

5   Conclusion

We have defined a dynamic model for scheduling process graphs. These dynamic
process graphs allow a compact representation of typical distributed programs

written in OCCAM or Ada style. We are not aware of another framework with a similar expressive power. Restricting the input and output mode of nodes different degrees of concurrency can be modelled. With respect to the degree of concurrency we have analysed how difficult it is to decide whether a dynamic process graph can be executed and to construct optimal schedules. An almost complete exact characterization could be given. Only for the scheduling problem with output mode restricted to PAR there remains a gap.

For the general case, we have shown an exponential complexity jump when scheduling process graphs in the dynamic setting. This implies that our compact dynamic representation is quite effective.

Finally, note that although constructing optimal schedules of standard graphs cannot be done efficiently, at least the problem can be approximated to a certain factor by simple algorithmic methods. This even holds when one takes communication delays into account. We do not have any nontrivial approximation results for scheduling dynamic process graphs.

References

1. A. Burns, *Programming in Occam 2*, Addison-Wesley Publishing Company, 1988.
2. H. El-Rewini and H. H. Ali, *Static Scheduling of Conditional Branches in Parallel Programs*, J. Par. Distrib. Comput. 24, 1995, 41-54.
3. H. Galperin, A. Wigderson, Succinct Representations of Graphs, Information and Control, 56, 1983, 183-198.
4. S. Ha, E. Lee, *Compile-time Scheduling and Assignment of Data-flow Program Graphs with Data-dependent Iteration*, IEEE Trans. Computers 40, 1991, 1225-1238.
5. A. Jakoby, M. Liśkiewicz, R. Reischuk, *Scheduling Dynamic Graphs*, Technischer Bericht, Institut für Theoretische Informatik, Med. Universität zu Lübeck, 1998.
6. A. Jakoby and R. Reischuk, *The Complexity of Scheduling Problems with Communication Delay for Trees*, Proc. 3. SWAT, 1992, 165-177.
7. H. Jung, L. Kirousis, P. Spirakis, *Lower Bounds and Efficient Algorithms for Multiprocessor Scheduling of DAGs with Communication Delays*, Proc. 1. SPAA, 1989, 254-264.
8. R. Kieckhafer, *Fault-Tolerant Real-Time Task Scheduling in the MAFT Distributed System*, Proc. 22. Hawaii Int. Conf. on System Science, 1989, 145-151.
9. R. Kieckhafer, C. Walter, A. Finn, P. Thambidurai, *The MAFT Architecture For Distributed Fault-Tolerance*, IEEE Trans. Computers, April 1988, 398-405.
10. T. Lengauer, K. Wagner, *The Correlation between the Complexities of the Non-hierarchical and Hierarchical Versions of Graph Problems*, J. CSS 44, 1992, 63-93.

11. C. Papadimitriou and M. Yannakakis, *A Note on Succinct Representations of Graphs*, Information and Control, 71, 1986, 181-185.
12. C. Papadimitriou and M. Yannakakis, *Towards an Architecture-Independent Analysis of Parallel Algorithms*, Proc. 20. STOC, 1988, 510-513, see also SIAM J. Comput. 19, 1990, 322-328.
13. B. Veltman, *Multiprocessor Scheduling with Communication Delays*, Ph.D. Thesis, University of Technology Eindhoven, Department of Computer Science, Eindhoven, The Netherlands, 1993.

Supporting Increment and Decrement
Operations in Balancing Networks*

William Aiello[1], Costas Busch[2], Maurice Herlihy[2], Marios Mavronicolas[3],
Nir Shavit[4], and Dan Touitou[5]

[1] AT&T Labs, 180 Park Avenue, Florham Park, NJ 07932, USA.
aiello@research.att.com
[2] Department of Computer Science, Brown University, Providence, RI 02912, USA.
{cb, mph}@cs.brown.edu
[3] Department of Computer Science, University of Cyprus, Nicosia CY-1678, Cyprus,
and Department of Computer Science and Engineering, University of Connecticut,
Storrs, CT 06269, USA.[§]
mavronic@turing.cs.ucy.ac.cy
[4] Department of Computer Science, School of Mathematical Sciences, Tel-Aviv
University, Tel-Aviv 69978, Israel.
shanir@math.tau.ac.il
[5] IDC Herzliya, Tel-Aviv, Israel.
danidin@math.tau.ac.il

Abstract. *Counting networks* are a class of distributed data structures
that support highly concurrent implementations of shared *Fetch&Incre-
ment* counters. Applications of these counters include shared pools and
stacks, load balancing, and software barriers [4, 12, 13, 18]. A limitation of counting networks is that the resulting shared counters can be
incremented, but not decremented.
A recent result by Shavit and Touitou [18] showed that the subclass of
tree-shaped counting networks can support, in addition, decrement operations. This paper generalizes their result, showing that *any* counting
network can be extended to support atomic decrements in a simple and
natural way. Moreover, it is shown that decrement operations can be supported in networks that provide weaker properties, such as *K-smoothing*.
In general, we identify a broad class of properties, which we call *boundedness properties*, that are preserved by the introduction of decrements:
if a balancing network satisfies a particular boundedness property for
increments alone, then it continues to satisfy that property for both increments and decrements.
Our proofs are purely combinatorial and rely on the novel concept of a
fooling pair of input vectors.

* This paper combines, unifies, and extends results appearing in preliminary form in [2]
and [6].
§ Partially supported by funds for the promotion of research at University of Cyprus.
Part of the work of this author was performed while at AT&T Labs – Research,
Florham Park, NJ, as a visitor to the Special Year on Networks, DIMACS Center
for Discrete Mathematics and Theoretical Computer Science, Piscataway, NJ.

C. Meinel and S. Tison (Eds.): STACS'99, LNCS 1563, pp. 393–403, 1999.

1   Introduction

Counting networks were originally introduced by Aspnes, Herlihy, and Shavit [4] and subsequently extended [1, 10, 11]. They support highly concurrent implementations of shared *Fetch&Increment* counters, shared pools and stacks, load balancing modules, and software barriers [4, 12, 13, 18].

Counting networks are constructed from basic elements called *balancers*. A balancer can be thought of as a routing switch for elements called *tokens*. It has a collection of input wires and a collection of output wires, respectively called the balancer's *fan-in* and *fan-out*. Tokens arrive asynchronously on arbitrary input wires, and are routed to successive output wires in a "round-robin" fashion. If one thinks of a balancer as having a state 'toggle" variable tracking which output wire the next token should exit on, then a token traversal amounts to a *Fetch&Toggle* operation, retrieving the value of the output wire and changing the toggle state to point to the next wire. The distribution of tokens on the output wires of a balancer thus satisfies the *step property* [4]: if y_i tokens exit on output wire i, then $0 \leq y_i - y_j \leq 1$ for any $j > i$.

A *balancing network* is a network of balancers, constructed by connecting balancers' output wires with other balancers' input wires in an acyclic fashion, in a way similar to the way *comparator networks* are constructed from *comparators* [9, Chapter 28]. The network itself has a number of input and output wires. A token enters the network on an input wire, traverses a sequence of balancers, and exits on an output wire. A balancing network is a *K-smoothing network* [1, 4] if, when all tokens have exited the network, the difference between the maximum and minimum number of tokens that exit on any output wire is bounded by K, regardless of the distribution of input tokens. Smoothing networks can be used for distributed load balancing.

A 1-smoothing network is a *counting network* if it satisfies the same step property as a balancer: when all tokens have traversed the network, if y_i tokens exit on output wire i, then $0 \leq y_i - y_j \leq 1$ for any $j > i$. Counting networks can be used to implement *Fetch&Increment* counters: the l-th token to exit on the j-th output wire returns the value $j + (l - 1)w_{out}$, where w_{out} is the network's fan-out.

A limitation of counting networks is that they support increments but not decrements. Many synchronization algorithms and tools require the ability to decrement shared objects.

Shavit and Touitou [18] devised the first counting network algorithm to support decrements for the class of networks that have the layout of a binary tree. They did so by introducing a new type of token for the decrement operation, which they named the *antitoken*.[1] Unlike a token, which traverses a balancer by fetching the toggle value and then advancing it, an antitoken sets the toggle back and then fetches it. Informally, an antitoken "cancels" the effect of the most recent token on the balancer's toggle state, and vice versa. They provide

---

[1] The name was actually suggested by Yehuda Afek (personal communication).

an operational proof that *counting trees* [19] count correctly when traversed by tokens and antitokens.

Shavit and Touitou [18] also introduced the notion of *elimination*. One can use a balancing network to implement a *pool*, a kind of concurrent stack. If a token representing an enqueue operation meets a token representing a dequeue operation in the network, then they can "cancel" one another immediately, without traversing the rest of the network.

It is natural to ask whether the same properties hold for *arbitrary* counting networks. More generally, what properties of balancing networks are preserved by the introduction of antitokens? In this paper, we give the first general answer to this question. We show the following results.

- If a balancing network is a counting network for tokens, then it is also a counting network for both tokens and antitokens. As a result *any* counting network can be extended to support a *Fetch&Decrement* operation.
- Any counting network, not just elimination trees, permits tokens and antitokens to eliminate one another.
- If a balancing network is a K-smoothing network when inputs are tokens, then it remains a K-smoothing network when inputs include both tokens and antitokens.
- We identify a broad class of properties, which we call *boundedness properties*, that are preserved by the introduction of antitokens: if a balancing network satisfies a particular boundedness property when inputs are tokens, then it continues to satisfy that property when inputs include both tokens and antitokens. The step property and the K-smoothing property are examples of boundedness properties.

Unlike earlier work [18], our proofs are combinatorial, not operational. They rely on the novel concept of a *fooling pair* of input vectors, which, we believe, is of independent interest.

We assign the value 1 to each token and -1 to each antitoken. We treat a balancer as an operator carrying an integer *input vector* to an integer *output vector*. The i-th entry in the input vector represents the *algebraic sum* of the tokens and antitokens received on the i-th input wire, and similarly for the output vector. For example, if this value is zero, then the same number of tokens and antitokens have arrived on that wire. We treat a balancing network in the same way, as an "operator" on integer vectors.

A *boundedness property* is a set of possible output vectors satisfying

- it is a subset of the K-smoothing property, for some $K \geq 1$, and
- it is closed under the addition of any *constant* vector.

Both the K-smoothing and the step property are examples of boundedness properties. Our principal result is that any balancing network that satisfies a boundedness property for non-negative integer input vectors also satisfies that property for arbitrary integer input vectors.

The *state* of a balancer is the "position" of its toggle. Two input vectors form a *fooling pair* to a balancer if, starting in the same state, each "drives" the balancer to the same state. Similarly, a balancing network state is given by its balancers' states. Two input vectors form a *fooling pair* for that network if, starting from the same state, each drives the network to the same state. For a specific initial state of a balancing network, its fooling pairs define equivalence classes of input vectors.

Roughly speaking, we prove our main equivalence result as follows. Consider any balancing network with some boundedness property; take any arbitrary integer input vector and the corresponding integer output vector. By adding to the input vector an appropriate vector that belongs to the equivalence class for some given initial state, we obtain a new input vector such that all of its entries are non-negative integers. We show that the output vector corresponding to the new input vector is, in fact, equal to the original output vector plus a constant vector. Hence, our main equivalence result follows from closure of the boundedness property under addition with a constant vector.

2   Framework

For any integer $g \geq 2$, $\mathbf{x}^{(g)}$ denotes the vector $\langle x_0, x_1, \ldots, x_{g-1} \rangle^{\mathrm{T}}$, while $\lceil \mathbf{x}^{(g)} \rceil$ denotes the integer vector $\langle \lceil x_0 \rceil, \lceil x_1 \rceil, \ldots, \lceil x_{g-1} \rceil \rangle^{\mathrm{T}}$. For any vector $\mathbf{x}^{(g)}$, denote $\|\mathbf{x}\|_1 = \sum_{i=0}^{g-1} x_i$. We use $\mathbf{0}^{(g)}$ to denote $\langle 0, 0, \ldots, 0 \rangle^{\mathrm{T}}$, a vector with g zero entries; similarly, we use $\mathbf{1}^{(g)}$ to denote $\langle 1, 1, \ldots, 1 \rangle^{\mathrm{T}}$, a vector with g unit entries. We use $\mathbf{r}^{(g)}$ to denote the *ramp vector* $\langle 0, 1, \ldots, g-1 \rangle^{\mathrm{T}}$. A *constant vector* is any vector of the form $c\, \mathbf{1}^{(g)}$, for any constant c.

Balancing networks are constructed from acyclically wired elements, called balancers, that route *tokens* and *antitokens* through the network, and *wires*. For generality, balancers may have arbitrary fan-in and fan-out, and they handle both tokens and antitokens.

For any pair of positive integers f_{in} and f_{out}, an $(f_{\mathrm{in}}, f_{\mathrm{out}})$-*balancer,* or *balancer* for short, is a routing element receiving tokens and antitokens on f_{in} input wires, numbered $0, 1, \ldots, f_{\mathrm{in}} - 1$, and sending out tokens and antitokens to f_{out} output wires, numbered $0, 1, \ldots, f_{\mathrm{out}} - 1$; f_{in} and f_{out} are called the balancer's *fan-in* and *fan-out,* respectively. Tokens and antitokens arrive on the balancer's input wires at arbitrary times, and they are output on its output wires. Roughly speaking, a balancer acts like a "generalized" *toggle*, which, on a stream of input tokens and antitokens, alternately forwards them to its output wires, going either down or up on each input token and antitoken, respectively. For clarity, we assume that all tokens and antitokens are distinct. Figure 1 depicts a balancer with three input wires and five output wires, stretched horizontally; the balancer is stretched vertically. In the left part, tokens and antitokens are denoted with full and empty circles, respectively; the numbering reflects the real-time order of tokens and antitokens in an execution where they traverse the balancer one by one (called a *sequential* execution).

For each input index i, $0 \leq i \leq f_{\mathrm{in}} - 1$, we denote by x_i the *balancer input state variable* that stands for the algebraic sum of the numbers of tokens and antitokens that have entered on input wire i; that is, x_i is the number of tokens

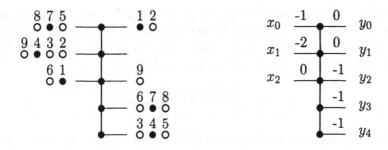

Fig. 1. A balancer

that have entered on input wire i *minus* the number of antitokens that have entered on input wire i. Denote $\mathbf{x}^{(f_{\text{in}})} = \langle x_0, x_1, \ldots, x_{f_{\text{in}}-1} \rangle^{\mathrm{T}}$; call $\mathbf{x}^{(w_{\text{in}})}$ an *input vector*. For each output index j, $0 \leq j \leq f_{\text{out}} - 1$, we denote by y_j the *balancer output state variable* that stands for the algebraic sum of the numbers of tokens and antitokens that have exited on output wire j; that is, y_j is the number of tokens that have exited on output wire j *minus* the number of antitokens that have exited on output wire j. The right part of Fig. 1 shows the corresponding input and output state variables. Denote $\mathbf{y}^{(f_{\text{out}})} = \langle y_0, y_1, \ldots, y_{f_{\text{out}}-1} \rangle^{\mathrm{T}}$; call $\mathbf{y}^{(f_{\text{out}})}$ an *output vector*.

The *configuration* of a balancer at any given time is the tuple $\langle \mathbf{x}^{(f_{\text{in}})}, \mathbf{y}^{(f_{\text{out}})} \rangle$; roughly speaking, the configuration is the collection of its input and output state variables. In the *initial configuration,* all input and output wires are empty; that is, in the initial configuration, $\mathbf{x}^{(f_{\text{in}})} = \mathbf{0}^{(f_{\text{in}})}$, and $\mathbf{y}^{(f_{\text{out}})} = \mathbf{0}^{(f_{\text{out}})}$. A configuration of a balancer is *quiescent* if there are no tokens or antitokens in the balancer. Note that the initial configuration is a quiescent one. The following formal properties are required for an $(f_{\text{in}}, f_{\text{out}})$-balancer.

1. *Safety property:* in any configuration, a balancer never creates either tokens or antitokens spontaneously.
2. *Liveness property:* for any finite number t of tokens and a of antitokens that enter the balancer, the balancer reaches within a finite amount of time a quiescent configuration where $t - e$ tokens and $a - e$ antitokens have exited the network, where e, $0 \leq e \leq \min\{t, a\}$, is the number of tokens and antitokens that are "eliminated" in the balancer.
3. *Step property:* in any quiescent configuration, for any pair of output indices j and k such that $0 \leq j < k \leq f_{\text{out}} - 1$, $0 \leq y_j - y_k \leq 1$.

From the safety and liveness properties it follows, for any quiescent configuration $\langle \mathbf{x}^{(f_{\text{in}})}, \mathbf{y}^{(f_{\text{out}})} \rangle$ of a balancer, that $\|\mathbf{x}^{(f_{\text{in}})}\|_1 = \|\mathbf{y}^{(f_{\text{out}})}\|_1$; that is, in a quiescent configuration, the algebraic sum of tokens and antitokens that exited the balancer is equal to the algebraic sum of tokens and antitokens that entered it. The equality of sums holds also for the case where some the tokens and antitokens are "eliminated" in the balancer.

For any input vector $\mathbf{x}^{(f_{in})}$, denote $\mathbf{y}^{(f_{out})} = b(\mathbf{x}^{(f_{in})})$ the output vector in the quiescent configuration that b will reach after all $\|\mathbf{x}^{(f_{in})}\|_1$ tokens and antitokens that entered b have exited; write also $b : \mathbf{x}^{(f_{in})} \to \mathbf{y}^{(f_{out})}$ to denote the balancer b. The output vector can also be written [1, 4, 8] as

$$\mathbf{y}^{(f_{out})} = \left\lceil \frac{\|\mathbf{x}^{(f_{in})}\|_1 \, \mathbf{1}^{(f_{out})} - \mathbf{r}^{(f_{out})}}{f_{out}} \right\rceil .$$

For any quiescent configuration $\langle \mathbf{x}^{(f_{in})}, \mathbf{y}^{(f_{out})} \rangle$ of a balancer $b : \mathbf{x}^{(f_{in})} \to \mathbf{y}^{(f_{out})}$, the *state* of the balancer b, denoted $state_b(\langle \mathbf{x}^{(f_{in})}, \mathbf{y}^{(f_{out})} \rangle)$, is defined to be

$$state_b(\langle \mathbf{x}^{(f_{in})}, \mathbf{y}^{(f_{out})} \rangle) = \|\mathbf{x}^{(f_{in})}\|_1 \bmod f_{out} ;$$

since the configuration is quiescent, it follows that

$$state_b(\langle \mathbf{x}^{(f_{in})}, \mathbf{y}^{(f_{out})} \rangle) = \|\mathbf{y}^{(f_{out})}\|_1 \bmod f_{out} .$$

Thus, for the sake of simplicity, we will denote

$$state_b(\mathbf{x}^{(f_{in})}) = state_b(\langle \mathbf{x}^{(f_{in})}, \mathbf{y}^{(f_{out})} \rangle) .$$

We remark that the state of an (f_{in}, f_{out})-balancer is some integer in the set $\{0, 1, \ldots, f_{out} - 1\}$, which captures the "position" to which it is set as a toggle mechanism. This integer is determined by either the balancer input state variables or the balancer output state variables in the quiescent configuration. Note that the state of the balancer in the initial configuration is 0. Moreover, the linearity of the modulus operation immediately implies linearity for the balancer state.

Lemma 1. *Consider a balancer* $b : \mathbf{x}^{(f_{in})} \to \mathbf{y}^{(f_{out})}$. *Then, for any input vectors* $\mathbf{x}_1^{(f_{in})}$ *and* $\mathbf{x}_2^{(f_{in})}$,

$$state_b(\mathbf{x}_1^{(f_{in})} + \mathbf{x}_2^{(f_{in})}) = (state_b(\mathbf{x}_1^{(f_{in})}) + state_b(\mathbf{x}_2^{(f_{in})})) \bmod f_{out} .$$

A (w_{in}, w_{out})-*balancing network* $\mathcal{B}$ is a collection of interwired balancers, where output wires are connected to input wires, having w_{in} designated input wires, numbered $0, 1, \ldots, w_{in} - 1$, which are not connected to output wires of balancers, having w_{out} designated output wires, numbered $0, 1, \ldots, w_{out} - 1$, similarly not connected to input wires of balancers, and containing no cycles. Tokens and antitokens arrive on the network's input wires at arbitrary times, and they traverse a sequence of balancers in the network in a completely asynchronous way till they exit on the output wires of the network.

For each input index i, $0 \le i \le w_{in} - 1$, we denote by x_i the *network input state variable* that stands for the algebraic sum of the numbers of tokens and antitokens that have entered on input wire i; that is, x_i is the difference of the number of tokens that have entered on input wire i *minus* the number of antitokens that have entered on input wire i. Denote $\mathbf{x}^{(w_{in})} = \langle x_0, x_1, \ldots, x_{w_{in}-1} \rangle^{\mathrm{T}}$;

call $\mathbf{x}^{(w_{in})}$ an *input vector*. For each output index j, $0 \leq j \leq w_{out} - 1$, we denote by y_j the *network output state variable* that stands for the algebraic sum of the numbers of tokens and antitokens that have exited on output wire j; that is, y_j is the number of tokens that have exited on output wire j *minus* the number of anti-tokens that have exited on output wire i. Denote $\mathbf{y}^{(w_{out})} = \langle y_0, y_1, \ldots, y_{w_{in}-1} \rangle^{\mathrm{T}}$; call $\mathbf{y}^{(w_{out})}$ an *output vector*.

The *configuration* of a network at any given time is the tuple of configurations of its individual balancers. In the *initial configuration*, all input and output wires of balancers are empty. The safety and liveness property for a balancing network follow naturally from those of its balancers. Thus, a balancing network eventually reaches a *quiescent configuration* in which all tokens and antitokens that entered the network have exited. In any quiescent configuration of $\mathcal{B}$ we have $\|\mathbf{x}^{(w_{in})}\|_1 = \|\mathbf{y}^{(w_{out})}\|_1$; that is, in a quiescent configuration, the algebraic sum of tokens and antitokens that exited the network is equal to the algebraic sum of tokens and antitokens that entered it.

Naturally, we are interested in quiescent configurations of a network. For any quiescent configuration of a network $\mathcal{B}$ with corresponding input and output vectors $\mathbf{x}^{(w_{in})}$ and $\mathbf{y}^{(w_{out})}$, respectively, the *state* of $\mathcal{B}$, denoted $state_{\mathcal{B}}(\mathbf{x}^{(w_{in})})$, is defined to be the collection of the states of its individual balancers. We remark that we have specified $\mathbf{x}^{(w_{in})}$ as the single argument of $state_{\mathcal{B}}$, since $\mathbf{x}^{(w_{in})}$ uniquely determines all input and output vectors of balancers of $\mathcal{B}$, which are used for defining the states of the individual balancers. Note that the state of the network in its initial configuration is a collection of 0's. For any input vector $\mathbf{x}^{(w_{in})}$, denote $\mathbf{y}^{(w_{out})} = \mathcal{B}(\mathbf{x}^{(w_{in})})$ the output vector in the quiescent configuration that $\mathcal{B}$ will reach after all $\|\mathbf{x}^{(w_{in})}\|_1$ tokens and antitokens that entered $\mathcal{B}$ have exited; write also $\mathcal{B} : \mathbf{x}^{(w_{in})} \to \mathbf{y}^{(w_{out})}$ to denote the network $\mathcal{B}$.

Not all balancing networks satisfy the step property. A (w_{in}, w_{out})-*counting network* is a (w_{in}, w_{out})-balancing network for which, in any quiescent configuration, for any pair of indices j and k such that $0 \leq j < k \leq w_{out} - 1$, $0 \leq y_j - y_k \leq 1$; that is, the output of a counting network has the step property.

The definition of a counting network can be weakened as follows [1, 4]. For any integer $K \geq 1$, a (w_{in}, w_{out})-K-*smoothing network* is a (w_{in}, w_{out})-balancing network for which, in any quiescent configuration, for any pair of indices j and k such that $0 \leq j, k \leq w_{out} - 1$, $0 \leq |y_j - y_k| \leq K$; that is, the output vector of a K-smoothing network has the K-*smoothing property*: all outputs are within K to each other.

For a balancing network $\mathcal{B}$, the *depth* of $\mathcal{B}$, denoted depth($\mathcal{B}$), is defined to be the maximum distance from any of its input wires to any of its output wires. In case depth($\mathcal{B}$) = 1, $\mathcal{B}$ will be called a *layer*. If depth($\mathcal{B}$) = d is greater than one, then $\mathcal{B}$ can be uniquely partitioned into layers $\mathcal{B}_1, \mathcal{B}_2, \ldots, \mathcal{B}_d$ from left to right in the obvious way.

Fix any integer $g \geq 2$. For any integer $K \geq 1$, the K-*smoothing property* [1] is defined to be the set of all vectors $\mathbf{y}^{(g)}$ such that for any entries y_j and y_k of $\mathbf{y}^{(g)}$, where $0 \leq j, k \leq g - 1$, $|y_j - y_k| \leq K$. A *boundedness property* is any subset of some K-smoothing property, for any integer $K \geq 1$, that is closed under addition with a constant vector. Clearly, the K-smoothing property is trivially a boundedness property; moreover, the set of all vectors $\mathbf{y}^{(g)}$ that have the *step property* [4] is a boundedness property, since any step vector is 1-smooth (but not vice versa). We remark that there are infinitely many boundedness properties.

A boundedness property captures precisely the two properties possessed by both K-smooth and step vectors upon which our later proofs will rely. Although we are unaware of any interesting property, other than the K-smoothing and step, that is a boundedness one, we chose to state our results for any general boundedness property in order to make explicit the two critical properties that are common to the classes of K-smooth vectors and step vectors; moreover, arguing in terms of a boundedness property will allow for a single proof of all claims found to hold for both the K-smoothing property and the step property.

Say that *a vector* $\mathbf{y}$ *has the boundedness property* Π if $\mathbf{y} \in \Pi$. Say that *a balancing network* $\mathcal{B} : \mathbf{x}^{(w_{\text{in}})} \to \mathbf{y}^{(w_{\text{out}})}$ *has the boundedness property* Π if for every input vector $\mathbf{x}^{(w_{\text{in}})}$, $\mathcal{B}(\mathbf{x}^{(w_{\text{in}})}) \in \Pi$.

3   Results

Input vectors $\mathbf{x}_1^{(f_{\text{in}})}$ and $\mathbf{x}_2^{(f_{\text{in}})}$ are a *fooling pair to balancer* $b : \mathbf{x}^{(f_{\text{in}})} \to \mathbf{y}^{(f_{\text{out}})}$ if

$$\text{state}_b(\mathbf{x}_1^{(f_{\text{in}})}) = \text{state}_b(\mathbf{x}_2^{(f_{\text{in}})}) \, ;$$

roughly speaking, inputs in a fooling pair drive the balancer to identical states.

Proposition 1. *Consider a balancer* $b : \mathbf{x}^{(f_{\text{in}})} \to \mathbf{y}^{(f_{\text{out}})}$. *Take any input vectors* $\mathbf{x}_1^{(f_{\text{in}})}$ *and* $\mathbf{x}_2^{(f_{\text{in}})}$ *that are a fooling pair to balancer* b. *Then, for any input vector* $\mathbf{x}^{(f_{\text{in}})}$,

(1) *the input vectors* $\mathbf{x}_1^{(f_{\text{in}})} + \mathbf{x}^{(f_{\text{in}})}$ *and* $\mathbf{x}_2^{(f_{\text{in}})} + \mathbf{x}^{(f_{\text{in}})}$ *are a fooling pair to balancer* b;

(2) $b(\mathbf{x}_1^{(f_{\text{in}})} + \mathbf{x}^{(f_{\text{in}})}) - b(\mathbf{x}_1^{(f_{\text{in}})}) = b(\mathbf{x}_2^{(f_{\text{in}})} + \mathbf{x}^{(f_{\text{in}})}) - b(\mathbf{x}_2^{(f_{\text{in}})})$.

Input vectors $\mathbf{x}_1^{(w_{\text{in}})}$ and $\mathbf{x}_2^{(w_{\text{in}})}$ are *a fooling pair to network* $\mathcal{B} : \mathbf{x}^{(w_{\text{in}})} \to \mathbf{y}^{(w_{\text{out}})}$ if for each balancer b of $\mathcal{B}$, the input vectors of b in quiescent configurations corresponding to $\mathbf{x}_1^{(w_{\text{in}})}$ and $\mathbf{x}_2^{(w_{\text{in}})}$, respectively, are a fooling pair to b; roughly speaking, a fooling pair "drives" all balancers of the network to identical states in the two corresponding quiescent configurations.

Proposition 2. *Consider a balancing network* $\mathcal{B} : \mathbf{x}^{(w_{\text{in}})} \to \mathbf{y}^{(w_{\text{out}})}$. *Take any input vectors* $\mathbf{x}_1^{(w_{\text{in}})}$ *and* $\mathbf{x}_2^{(w_{\text{in}})}$ *that are a fooling pair to network* $\mathcal{B}$. *Then, for any input vector* $\mathbf{x}^{(w_{\text{in}})}$,

(1) *the input vectors* $\mathbf{x}_1^{(w_{\text{in}})} + \mathbf{x}^{(w_{\text{in}})}$ *and* $\mathbf{x}_2^{(w_{\text{in}})} + \mathbf{x}^{(w_{\text{in}})}$ *are a fooling pair to network* $\mathcal{B}$;

(2) $\mathcal{B}(\mathbf{x}_1^{(w_{\text{in}})} + \mathbf{x}^{(w_{\text{in}})}) - \mathcal{B}(\mathbf{x}_1^{(w_{\text{in}})}) = \mathcal{B}(\mathbf{x}_2^{(w_{\text{in}})} + \mathbf{x}^{(w_{\text{in}})}) - \mathcal{B}(\mathbf{x}_2^{(w_{\text{in}})})$.

Say that $\mathbf{x}^{(w_{in})}$ is a *null vector* to network $\mathcal{B} : \mathbf{x}^{(w_{in})} \to \mathbf{y}^{(w_{out})}$ if the vectors $\mathbf{x}^{(w_{in})}$ and $\mathbf{0}^{(w_{in})}$ are a fooling pair to $\mathcal{B}$. Intuitively, a null vector "hides" itself in the sense that it does not alter the state of $\mathcal{B}$ by traversing it.

Proposition 3. *Consider a balancing network* $\mathcal{B} : \mathbf{x}^{(w_{in})} \to \mathbf{y}^{(w_{out})}$. *Take any input vectors* $\mathbf{x}_1^{(w_{in})}$ *and* $\mathbf{x}_2^{(w_{in})}$ *that are a fooling pair to network* $\mathcal{B}$. *If* $\mathbf{x}_2^{(w_{in})}$ *is a null vector to network* $\mathcal{B}$, *then,* $\mathbf{x}_1^{(w_{in})}$ *is also a null vector to network* $\mathcal{B}$.

Proposition 4. *Consider a balancing network* $\mathcal{B} : \mathbf{x}^{(w_{in})} \to \mathbf{y}^{(w_{out})}$. *Take any input vector* $\mathbf{x}^{(w_{in})}$ *that is null to* $\mathcal{B}$. *Then, for any integer* $k \geq 0$,

(1) $\mathcal{B}(k\mathbf{x}^{(w_{in})}) = k\mathcal{B}(\mathbf{x}^{(w_{in})})$;
(2) $k\mathbf{x}^{(w_{in})}$ *is a null vector to* $\mathcal{B}$.

For any balancing network $\mathcal{B}$, let $W_{out}(\mathcal{B})$ denote the product of the fan-outs of balancers of $\mathcal{B}$. For positive integer δ, say that δ *divides* $\mathbf{x}^{(g)}$ if δ divides each entry of $\mathbf{x}^{(g)}$.

Proposition 5. *Consider a balancing network* $\mathcal{B} : \mathbf{x}^{(w_{in})} \to \mathbf{y}^{(w_{out})}$. *If* $W_{out}(\mathcal{B})$ *divides* $\mathbf{x}^{(w_{in})}$, *then,* $\mathbf{x}^{(w_{in})}$ *is a null vector to* $\mathcal{B}$.

Proposition 6. *Consider any balancing network* $\mathcal{B} : \mathbf{x}^{(w_{in})} \to \mathbf{y}^{(w_{out})}$ *that has a boundedness property* Π. *If* $W_{out}(\mathcal{B})$ *divides* $\mathbf{x}^{(w_{in})}$, *then,* $\mathbf{y}^{(w_{out})}$ *is a constant vector.*

Here is our main result:

Theorem 1. *Fix any boundedness property* Π. *Consider any balancing network* $\mathcal{B} : \mathbf{x}^{(w_{in})} \to \mathbf{y}^{(w_{out})}$ *such that* $\mathbf{y}^{(w_{out})}$ *has the boundedness property* Π *whenever* $\mathbf{x}^{(w_{in})}$ *is a non-negative vector. Then,* $\mathcal{B}$ *has the boundedness property* Π.

4 Conclusion

We have shown that any balancing network that satisfies any boundedness property on all non-negative input vectors, continues to do so for any arbitrary input vector. Interesting examples of such properties are the step property and the K-smoothing property. A significant consequence of our result is that all known (deterministic) constructions of counting and smoothing networks [1, 3, 4, 5, 8, 10, 11, 14, 15, 19] will correctly handle both tokens and antitokens, and therefore support both increment and decrement operations. Another significant consequence is that the sufficient timing conditions for linearizability in counting networks established in [16, 17] immediately carry over to antitokens.

Aiello *et al.* [3] present a *randomized* counting network based on *randomized balancers* that toggle tokens according to some random permutation. We do not know whether such randomized networks can support antitokens.

A balancing network has the *threshold property* [4, 7] if $y_0 = \lceil \|\mathbf{x}^{(w_{\text{in}})}\|_1/w_{\text{out}} \rceil$, and the *weak threshold property* [7] if there is *some* output index j, possibly $j \neq 0$, such that $y_j = \lceil \|\mathbf{x}^{(w_{\text{in}})}\|_1/w_{\text{out}} \rceil$. Since we have not established that either of these properties is a boundedness property, our result does not necessarily apply, and it remains unknown whether these properties are preserved by the introduction of antitokens.

References

[1] E. Aharonson and H. Attiya. Counting networks with arbitrary fan-out. *Distributed Computing*, 8(4):163–169, 1995.

[2] W. Aiello, M. Herlihy, N. Shavit, and D. Touitou. Inc/dec counting networks. Manuscript, Dec. 1995.

[3] W. Aiello, R. Venkatesan, and M. Yung. Coins, weights and contention in balancing networks. In *Proceedings of the 13th Annual ACM Symposium on Principles of Distributed Computing (PODC'94)*, pages 193–205, Los Angeles, Aug. 1994.

[4] J. Aspnes, M. Herlihy, and N. Shavit. Counting networks. *Journal of the ACM*, 41(5):1020–1048, Sept. 1994.

[5] C. Busch, N. Hardavellas, and M. Mavronicolas. Contention in counting networks (abstract). In *Proceedings of the 13th annual ACM Symposium on Principles of Distributed Computing (PODC'94)*, page 404, Los Angeles, Aug. 1994.

[6] C. Busch and M. Mavronicolas. The strength of counting networks (abstract). In *Proceedings of the 15th Annual ACM Symposium on Principles of Distributed Computing (PODC'96)*, page 311, Philadelphia, May 1996.

[7] C. Busch and M. Mavronicolas. Impossibility results for weak threshold networks. *Information Processing Letters*, 63(2):85–90, July 1997.

[8] C. Busch and M. Mavronicolas. An efficient counting network. In *Proceedings of the 1st Merged International Parallel Processing Symposium and Symposium on Parallel and Distributed Processing (IPPS/SPDP'98)*, pages 380–385, Mar. 1998.

[9] T. H. Cormen, C. E. Leiserson, and R. L. Rivest. *Introduction to algorithms*. MIT Press and McGraw-Hill Book Company, Cambridge, MA, 1992.

[10] E. W. Felten, A. LaMarca, and R. Ladner. Building counting networks from larger balancers. Technical Report TR 93-04-09, University of Washington, Apr. 1993.

[11] N. Hardavellas, D. Karakos, and M. Mavronicolas. Notes on sorting and counting networks. In *Proceedings of the 7th International Workshop on Distributed Algorithms (WDAG'93)*, volume 725 of *Lecture Notes in Computer Science*, pages 234–248, Lausanne, Switzerland, Sept. 1993. Springer-Verlag.

[12] M. Herlihy, B.-H. Lim, and N. Shavit. Scalable concurrent counting. *ACM Transactions on Computer Systems*, 13(4):343–364, Nov. 1995.

[13] S. Kapidakis and M. Mavronicolas. Distributed, low contention task allocation. In *Proceedings of the 8th IEEE Symposium on Parallel and Distributed Processing (SPDP'96)*, pages 358–365, Washington, Oct. 1996.

[14] M. Klugerman. *Small-Depth Counting Networks and Related Topics*. PhD thesis, Department of Mathematics, Massachusetts Institute of Technology, Sept. 1994.

[15] M. Klugerman and C. G. Plaxton. Small-depth counting networks. In *Proceedings of the 24th Annual ACM Symposium on the Theory of Computing (STOC'92)*, pages 417–428, Victoria, B.C., Canada, May 1992.

[16] N. Lynch, N. Shavit, A. Shvartsman, and D. Touitou. Counting networks are practically linearizable. In *Proceedings of the 15th Annual ACM Symposium on Principles of Distributed Computing (PODC'96)*, pages 280–289, New York, May 1996.

[17] M. Mavronicolas, M. Papatriantafilou, and P. Tsigas. The impact of timing on linearizability in counting networks. In *Proceedings of the 11th International Parallel Processing Symposium (IPPS'97)*, pages 684–688, Los Alamitos, Apr. 1997.

[18] N. Shavit and D. Touitou. Elimination trees and the construction of pools and stacks. *Theory of Computing Systems*, 30(6):545–570, Nov./Dec. 1997.

[19] N. Shavit and A. Zemach. Diffracting trees. *ACM Transactions on Computer Systems*, 14(4):385–428, Nov. 1996.

Worst-Case Equilibria

Elias Koutsoupias[1] and Christos Papadimitriou[2]

[1] Univ of California, Los Angeles
elias@cs.ucla.edu
[2] Univ of California, Berkeley
christos@cs.berkeley.edu

Abstract. In a system in which noncooperative agents share a common resource, we propose the ratio between the worst possible Nash equilibrium and the social optimum as a measure of the effectiveness of the system. Deriving upper and lower bounds for this ratio in a model in which several agents share a very simple network leads to some interesting mathematics, results, and open problems.

1 Introduction

Internet users and service providers act selfishly and spontaneously, without an authority that monitors and regulates network operation in order to achieve some "social optimum" such as minimum total delay [1]. *How much performance is lost because of this?* This question appears to exemplify a novel and timely genre of algorithmic problems, in which we are investigating the cost of the lack of *coordination* —as opposed to the lack of *information* (on-line algorithms) or the lack of *unbounded computational resources* (approximation algorithms). As we show in this paper, this point of view leads to some interesting algorithmic and combinatorial questions and results.

It is nontrivial to arrive at a compelling mathematical formulation of this question. Independent, non-cooperative agents obviously evoke *game theory* [8], and its main concept of rational behavior, the *Nash equilibrium:* In an environment in which each agent is aware of the situation facing all other agents, a Nash equilibrium is a combination of choices (deterministic *or* randomized), one for each agent, from which no agent has an incentive to unilaterally move away. Nash equilibria are known not to always optimize overall performance, with the Prisoner's Dilemma [8,10] being the best-known example. Conditions under which Nash equilibria can achieve or approximate the overall optimum have been studied extensively ([10]; see also [5,7,11] for studies on networks). However, this line of previous work compares the overall optimum with the *best* Nash equilibrium, not the *worst*, as befits our line of reasoning. To put it otherwise, this previous research aims at achieving or approximating the social optimum by implicit acts of coordination, whereas we are interested in evaluating the loss to the system due to its deliberate lack of coordination.

Game-theoretic aspects of the Internet have also been considered by researchers associated with the Internet Society [1,12], with an eye towards designing variants of the Internet Protocols which are more resilient to video-like

C. Meinel and S. Tison (Eds.): STACS'99, LNCS 1563, pp. 404–413, 1999.
© Springer-Verlag Berlin Heidelberg 1999

traffic. Their point of view is also that of the mechanism design aspect of game theory, in that they try to design games (strategy spaces and reward tables) that encourage behaviors close to the social optimum. Understanding the worst-case distance of a Nash equilibrium from the social optimum in simple situations, which is the focus of the present paper, is a prerequisite for making rigorous progress in that project.

The Model

Let us make the general game-theoretic framework more precise. Consider a network in which each link has a law (curve) whereby traffic determines delay. Each of several agents wants to send a particular amount of traffic along a path from a fixed source to a fixed destination. This immediately defines a game-theoretic framework, in which each agent has as many pure strategies as there are paths from its origin to its destination, and the cost to an agent of a combination of strategies (one for each agent) is the negative of the total delay for each agent, as determined by the traffic on the links. There is also a well-defined optimization problem, in which we wish to minimize the *social* or *overall optimum*, the sum of all delays over all agents, say. The question we want to ask is, how far from the optimum total delay can be the total delay achieved by a Nash equilibrium? Numerical experiments reported in [6] imply that there are Nash equilibria which can be more than 20% off the overall optimum.

In this paper we address a very simple special case of this problem, in which the network is just a set of m parallel links from an origin to a destination, all with the same capacity (similar special cases are studied in other works in this field, e.g. [7]; we also briefly examine the case of two parallel links with unequal capacity). We model the delay of these links in a very simple way: Since the capacity is unit, we assume that the delay suffered by each agent using a link equals the total capacity of flow through this link. We assume that n agents have each an amount of traffic w_i, $i = 1, \ldots, n$ to send from the origin to the destination. Hence the resulting problem is essentially a scheduling problem with m links and n independent tasks with lengths w_i, $i = 1, \ldots, n$. The set of pure strategies for agent i is therefore $\{1, \ldots, m\}$, and a mixed strategy is a distribution on this set. Let $(j_1, \ldots, j_n) \in \{1, \ldots, m\}^n$ be a combination of pure strategies, one for each agent; its *cost* for agent i, denoted $C_i(j_1, \ldots, j_n)$, is simply

$$L^{j_i} + \sum_{j_k = j_i} w_k,$$

the finish time of the link chosen by i; here we assume that link j has in the beginning an initial task of length L^j scheduled, so it will be available for scheduling the agents' tasks only after L^j time units. This calculation assumes that, if agent i's task ends up in link j, it ends when all tasks on link j end; this is realistic if the tasks are broken in packets, which are then sent in a round-robin way. We also examine the alternative model, in which the tasks scheduled in link j are executed in a random batch order, and hence the cost to agent i is

$C_i(j_1, \ldots, j_n) = L^{j_i} + \frac{1}{2} \sum_{j_k=j_i} w_k$. We call this *the batch model*. Finally, the cost to agent i of a combination of mixed strategies is the expected cost of the corresponding experiment in which a pure strategy is chosen independently for each agent, with the probability assigned to it by the mixed strategy. The overall optimum in this situation, against which we propose to compare the Nash equilibria of the game just described, would be the optimum solution of the m-way load balancing (partition into m sets) problem for the n lengths $w_1, \ldots, w_n$.

The costs in our model are a simplification of the delays incurred in a network link when agents inject traffic into it. The actual delays are in fact not the sums of the individual delays, but nonlinear functions, as increased traffic causes increased loss rates and delays. We discuss briefly in the last section the open problems suggested by our work that are associated with more accurate modeling of network delays.

The Results of This Paper

In this paper we show upper and lower bounds on the ratio between the worst Nash equilibrium and the overall optimum solution.

- In a network with two parallel links, we show that the worst-case ratio is $\frac{3}{2}$ (both upper and lower bound), independent of the number n of agents (Theorems 1 and 2).
- The above result assumes that the two link speeds are the same. If the two links have different speeds, then the worst-case ratio increases to the golden ratio $\phi = 1.618 \ldots$ (lower bound, Theorem 3).
- Also, in the batch model of two links, the worst-case ratio is lower bounded by $\frac{29}{18} = 1.6111 \ldots$ which is also an upper bound if we have two agents (Theorem 4).
- We have not been able to determine the answers for three or more links. However, the worst-case ratio (in all of the above models) is bounded from below by the ratio suggested by the load-balancing aspect of the problem, that is to say, $\Omega(\frac{\log m}{\log \log m})$ (Theorem 6). Using the Azuma-Hoeffding inequality, we establish an $O(\sqrt{m \log m})$ upper bound (Theorem 8). A similar bound holds for links of different speeds (Theorem 9).

2   All Nash Equilibria

We consider the case of n agents sharing m identical links. Before describing all Nash equilibria, we need a few definitions. We usually use subscripts for agents and superscripts for links. For example, for a Nash equilibrium, we denote the probability that agents i selects link j with p_i^j. Let M^j denote the *expected traffic* on link j. If L^j is the initial load on link j, it is easy to see that

$$M^j = L^j + \sum_i p_i^j w_i. \tag{1}$$

From the point of view of agent i, its *finish time* when its own traffic w_i is assigned to link j is

$$c_i^j = w_i + L^j + \sum_{i \neq t} p_t^j w_t = M^j + (1 - p_i^j)w_i. \tag{2}$$

Probabilities p_i^j define a Nash equilibrium if there is no incentive for agent i to change its strategy. Thus, agent i will assign nonzero probabilities only to links j that minimize c_i^j. We will denote this minimum value by c_i, i.e.,

$$c_i = \min_j c_i^j,$$

and we will call the set of links $S_i = \{j : p_i^j > 0\}$ the *support* of agent i. More generally, let S_i^j be an indicator variable that takes value 1 when $p_i^j > 0$.

Conversely, a Nash equilibrium is completely defined by the supports $S_1, \ldots, S_n$ of all agents. More precisely, if we fix the S_i^j's, the strategies in a Nash equilibrium are given by

$$p_i^j = (M^j + w_i - c_i)/w_i \tag{3}$$

subject to

for all j: $M^j = L^j + \sum_i S_i^j (M^j + w_i - c_i)$

for all i: $\sum_j S_i^j (M^j + w_i - c_i) = w_i$

To see that these constraints indeed define an equilibrium, notice that the first set of equations is equivalent to (2). The constraints are equivalent to (1), and to the fact that the probabilities of agent i should sum up to exactly 1. Notice also that the set of constraints specify in general, a unique solution for c_i and M^j (there are $n + m$ constraints and $n + m$ unknowns). If the resulting probabilities p_i^j are in the interval $(0, 1]$, then the above equations define an equilibrium with support S_i^j. Thus, an equilibrium is completely defined by the supports of the agents (although not all supports give rise to a feasible equilibrium). As a result, the number of equilibria is, in general, exponential in n and m.

A natural quantity associated with an equilibrium is the *expected maximum traffic* over all links:

$$\text{cost} = \sum_{j_1=1}^{m} \cdots \sum_{j_n=1}^{m} \prod_{i=1}^{n} p_i^{j_i} \max_{j=1,\ldots,m} \{L^j + \sum_{t:j_t=j} w_t\}. \tag{4}$$

We call it the *social cost* and we wish to compare it with the social optimum opt. More precisely, we want to estimate the *coordination ratio* which is the worst-case ratio $R = \max \text{cost/opt}$ (the maximum is over all equilibria). Computing the social optimum opt is an NP-complete problem (partition problem), but for the purpose of upper bounding R here, it suffices to use two simple approximations of it: opt $\geq \max\{w_1, \sum_j M^j/m\} = \max\{w_1, (\sum_j L^j + \sum_i w_i)/m\}$ (we shall be assuming that $w_1 \geq w_2 \geq \cdots \geq w_n$).

3   Worst-Case Equilibria for 2 Links

We shall assume that there are no initial loads —that is, all L^j's are zero. This is no restriction at all for the standard model, because initial loads can be considered as jobs of m additional agents, each with a pure strategy. However, this may not be true for other models. In particular, in the batch model (the one with the $\frac{1}{2}$ factor in front of $\sum w_i$) it follows from our results that initial loads result in strictly worse ratio.

Our first theorem is trivial:

Theorem 1. *The coordination ratio for 2 links is at least* $3/2$.

Proof. Consider two agents with traffic $w_1 = w_2 = 1$. It is easy to check that probabilities $p_i^j = 1/2$ for $i, j = 1, 2$ give rise to a Nash equilibrium. The expected maximum load is cost $= 3/2$ and the social optimum is opt $= 1$ achieved by allocating each job to its own link.

Our main technical result of this section is a matching upper bound. To prove it, we find a way to upper bound the complicated expression (4) for the social cost. In fact, it is relatively easy to compute the strategies of a Nash equilibrium. There are 2 types of agents: *pure strategy agents* with support of size one and *stochastic agents* with support of size 2. Let d^j be the sum of all jobs of pure strategy agents assigned to link j. Also let $k > 1$ denote the number of stochastic agents. It is not difficult to verify that the system of equations (3) gives the following probabilities of a stochastic agent i:

$$p_i^j = \frac{1}{2} - \frac{d^1 + d^2 - 2d^j}{2(k-1)w_i}. \tag{5}$$

However, we don't see how to use this expression to upper bound (4).

Central to our proof of the upper bound is the notion *contribution probability*: The contribution probability q_i of agent i is equal to the probability that its job goes to the link of maximum load (if there are more than one maximum load links, we consider the lexicographically first such link, say). Clearly, the social cost is given by cost $= \sum_i q_i w_i$. The key idea in our proof is to consider the pairwise contribution to social cost. In particular, let t_{ik} be the *collision probability* of agents i and k, that is, the probability that the traffic of both agents goes to the same link. Observe then that both agents i and k can contribute to the social cost only if they collide, that is,

$$q_i + q_l \leq 1 + t_{ik}. \tag{6}$$

The following lemma provides a crucial property of collision probabilities. It holds for any number of links.

Lemma 1. *The collision probabilities of a Nash equilibrium of n agents and m links satisfy*

$$\sum_{k \neq i} t_{ik} w_k = c_i - w_i.$$

Proof. Observe first that $t_{ik} = \sum_j p_i^j p_k^j$. Therefore, we have

$$\sum_{k \neq i} t_{ik} w_k = \sum_j p_i^j \sum_{k \neq i} p_k^j w_k = \sum_j p_i^j (M^j - p_i^j w_i).$$

It follows from (3) that we can use $p_i^j w_i = M^j + w_i - c_i$. There is a minor technical point to be made here: the equality $p_i^j w_i = M^j + w_i - c_i$ holds only if link j is in the support of agent i ($p_i^j > 0$). However, observe that when $p_i^j = 0$ there is no harm in replacing $p_i^j w_i$ with any expression. We get

$$\sum_{k \neq i} t_{ik} w_k = \sum_j p_i^j (c_i - w_i) = c_i - w_i.$$

A final ingredient for the proof is the bound (which also holds for any number of agents and links):

$$c_i \leq \frac{\sum_i w_i}{m} + \frac{m-1}{m} w_i. \tag{7}$$

This follows from $c_i = \min_j c_i^j \geq \frac{1}{m} \sum_j (M^j + (1 - p_i^j) w_i) = \frac{\sum_j M_j}{m} + \frac{m-1}{m} w_i = \frac{\sum_k w_k}{m} + \frac{m-1}{m} w_i$.

Theorem 2. *The coordination ratio for any number of agents and $m = 2$ links is at most $3/2$.*

Proof. We have seen that pairwise the contribution probabilities satisfy $q_i + q_k \leq 1 + t_{ik}$. Therefore, $\sum_{k \neq i} (q_i + q_k) w_k \leq \sum_{k \neq i} (1 + t_{ik}) w_k$. Using Lemma 1 and bound (7), we get $\sum_{k \neq i} (q_i + q_k) w_k \leq \frac{3}{2} \sum_{k \neq i} w_k$. From this we can compute

$$\text{cost} = \sum_k q_k w_k = (\frac{3}{2} - q_i) \sum_k w_k + (2q_i - \frac{3}{2}) w_i.$$

Recall that opt $\geq \max\{\frac{1}{2} \sum_k w_k, w_i\}$. If for some agent i, $q_i \geq \frac{3}{4}$, then $(2q_i - \frac{3}{2}) w_i \leq (2q_i - \frac{3}{2})$opt and $cost \leq (\frac{3}{2} - q_i)2\text{opt} + (2q_i - \frac{3}{2})\text{opt} = \frac{3}{2}\text{opt}$. Otherwise, when all contribution probabilities are at most $\frac{3}{4}$, cost $= \sum_k q_k w_k \leq \frac{3}{4} \sum_k w_k \leq \frac{3}{2}$opt.

Links with Different Speeds

So far, we assumed that all links have the same speed or capacity. We now consider the general problem where links may have different speeds. Let s_j be the speed of link j. Without loss of generality, we shall assume $s_1 \leq \cdots \leq s_m$. We can estimate all Nash equilibria again. Equation (2) now becomes

$$c_i^j = (M^j + (1 - p_i^j) w_i)/s_j. \tag{8}$$

and the equilibria are given by:

$$p_i^j = (M^j + w_i - s_j c_i)/w_i \tag{9}$$

subject to

$$\text{for all } j: M^j = L^j + \sum_i S_i^j (M^j + w_i - s_j c_i)$$

$$\text{for all } i: \sum_j S_i^j (M^j + w_i - s_j c_i) = w_i$$

We can extend the lower bound Theorem 1 to this case:

Theorem 3. *The coordination ratio for two links with speeds $s_1 \le s_2$ is at least $R = 1 + s_2/(s_1 + s_2)$ when $s_2 \le \phi s_1$, where $\phi = (1 + \sqrt{5})/2$. The coordination ratio R achieves its maximum value ϕ when $s_2/s_1 = \phi$.*

Proof. We first describe the equilibria for any number of agents. Again let d^j be the sum of all traffic assigned to link j by pure agents. We give the probabilities p_i^1 of the stochastic agents ($p_i^2 = 1 - p_i^1$).

$$p_i^1 = \frac{s_2}{s_1 + s_2} - \frac{(s_2 - s_1)\sum_i w_i + (s_2 d^1 - s_1 d^2)}{(k-1)(s_1 + s_2)w_i}$$

It is not hard to verify that these probabilities indeed satisfy (9). To prove the theorem, we consider the case of no initial loads and two agents with jobs $w_1 = s_2$ and $w_2 = s_1$. The probabilities are $p_1^1 = \frac{s_1^2}{s_2(s_1+s_2)}$ and $p_2^1 = 1 - \frac{s_2^2}{s_1(s_1+s_2)}$. We can then compute cost $= (p_1^1 p_2^1/s_1 + p_1^2 p_2^2/s_2)(w_1 + w_2) + (p_1^1 p_2^2/s_1 + p_1^2 p_2^1/s_2)w_1 = (s_1 + 2s_2)/(s_1 + s_2)$ and opt $= 1$. The lower bound follows.

It is worth mentioning that when $s_2/s_1 > \phi$ the probabilities given above are outside the interval $[0, 1]$. Therefore, both agents have pure strategies and the coordination ratio is 1.

We believe that the proof of Theorem 2 can be appropriately generalized to the case of links of different speeds.

The Batch Model

For the batch model with two links we can prove the following bounds (proof omitted):

Theorem 4. *In the batch model with two identical links, the coordination ratio is between $\frac{29}{18} = 1.61\dots$ and 2. The lower bound $\frac{29}{18}$ is also an upper bound in the case of $n = 2$ agents.*

When the links have no initial load, the batch model and the standard model have the same equilibria and the same coordination ratio. However, in the general case, as the above theorem demonstrates, the batch model has higher coordination ratio. But it cannot be much higher:

Theorem 5. *For m links and any number of agents, the coordination ratios of the batch model and the standard model differ by at most a factor of 2.*

We omit the details of the proof, but we point out the main idea: We can consider the initial loads L^j of the batch model as pure strategy agents of weight $2L_j$. This preserves the equilibria and changes the social optimum by at most a factor of 2.

4 Worst-Case Equilibria for m Links

We now consider lower bounds for the coordination ratio for m links.

Theorem 6. *The coordination ratio for m identical links is $\Omega(\log m/\log\log m)$.*

Proof. Consider the case where there are m agents, each with a unit job, i.e., $w_i = 1$. If the links have no initial load, it is easy to see that the uniform strategies with $p_i^j = 1/m$ for $i, j = 1, \ldots, m$ is an equilibrium. This is identical to the problem of throwing m balls into m bins and asking for the expected maximum number of balls in a bin. The answer is well-known to be $\Theta(\log m/\log\log m)$.

We believe that this lower bound is tight: That is, if T_m denotes the expected maximum number of balls in a bin, we conjecture that the coordination ratio for any number of agents and m identical links is T_m (in the standard model). Theorem 2 shows that the conjecture holds for $m = 2$.

We believe that a proof of the conjecture can be obtained by appropriately generalizing the proof technique of Theorem 2; it seems however that a substantially deeper structural theorem about the Nash equilibria, similar to Lemma 1, is needed. Here, we give a weaker upper bound. But first we need the following theorem, which is interesting on its own.

Theorem 7. *For m identical links, the expected load M^j of any link j is at most $(2 - 1/m)$opt. For links with different speeds, M_j is at most $s_j(1 + \sqrt{m-1})$opt.*

Proof. For identical links the theorem follows directly from (7) by observing that $M^j \leq c_i \leq (\sum_i w_i)/m + (m-1)w_i/m \leq s_j(2 - 1/m)$opt.

The proof for links with different speeds has the same flavor with (7). This time we take a weighted average over the links (the weight for machine j is $s_j/\sum_r s_r$). Thus,

$$c_i^j \leq \frac{\sum_r M^r + (m-1)w_i}{\sum_r s_r}.$$

Also, $c_i^m \leq (M^m + w_i)/s_m \leq (\sum_r M^r + w_i)/s_m$. In summary,

$$c_i^j \leq \min\{\frac{\sum_r M^r + (m-1)w_i}{\sum_r s_r}, \frac{\sum_r M^r + w_i}{s_m}\}.$$

However, we can lower bound the social optimum by $\max\{w_i/s_m, \sum_r M^r/\sum_r s_r\}$. Thus, we get $c_i^j \leq \text{opt} + \min\{\frac{(m-1)w_i}{\sum_r s_r}, \frac{\sum_r M^r}{s_m}\}$. Using the obvious inequality $\min\{xa/b, c/d\} \leq \sqrt{x}\max\{a/d, c/b\}$, we get $c_i^j \leq (1 + \sqrt{m-1})$opt. We can then conclude that $M^j \leq s_j c_i^j \leq s_j(1 + \sqrt{m-1})$opt.

We can now prove an upper bound for the case of m identical links.

Theorem 8. *The coordination ratio of any number of agents and m identical links is at most $T = 3 + \sqrt{4m \ln m}$.*

Proof. Using a martingale concentration bound known as the Azuma-Hoeffding inequality [4], we will show that the load of a given link j exceeds $(T-1)$opt with probability at most $1/m^2$. Then, the probability that the maximum load on all links does not exceed $(T-1)$opt is at least $1 - 1/m$. It follows that the expected maximum load is bounded by $(1 - 1/m)(T - 1)$opt $+ 1/m(m$opt$) \leq T$opt.

It remains to show that indeed the probability that the load of a given link j exceeds $(T - 1)$opt is small (at most $1/m^2$). Let X_i be a random variable denoting the contribution of agent i to the load of link j. In particular, $Pr[X_i = w_1] = p_i^j$ and $Pr[X_i = 0] = 1 - p_i^j$. Clearly, the random variables $X_1, \ldots, X_n$ are independent. We are interested in estimating the probability $Pr[\sum_i X_i > (T - 1)$opt$]$. Since the weights w_i and the probabilities p_i^j may vary a lot, we don't expect the sum $\sum_i X_i$ to exhibit the good concentration bounds of sums of binomial variables. However, we can get a weaker bound using the Azuma-Hoeffding inequality. The inequality gives very good results for probabilities around $1/2$. Unfortunately, in our case the probabilities may be very close to 0 or 1.

Let $\mu_i = E[X_i]$ and consider the martingale $Y_t = X_1 + \cdots + X_t + \mu_{t+1} + \cdots + \mu_n$ (it is straightforward to verify $E[Y_{t+1}|Y_t] = Y_t$). Observe that $|Y_{t+1} - Y_t| = |X_{t+1} - \mu_{t+1}| \leq w_{t+1}$. We can then apply the Azuma-Hoeffding's inequality:

$$Pr[Y_n - Y_0 \geq x] \leq e^{-\frac{1}{2}x^2 / \sum_i w_i^2}.$$

Let $x = (T - 3)$opt. Since $Y_0 = \sum_i \mu_i = M^j \leq 2$opt (Theorem 7), we get that the load of link j exceeds $(T - 1)$opt with probability at most $e^{-\frac{1}{2}x^2 / \sum_i w_i^2}$. However, it is not hard to establish that

$$\sum_i w_i^2 \leq \max\{mw_1^2, m(\sum_i w_i/m)^2\} \leq m\text{opt}^2.$$

Thus the probability that the load of link j exceeds $(T - 1)$opt is at most $e^{-\frac{1}{2}(T-3)^2/m}$. For $T = 3 + \sqrt{4m \ln m}$, this probability becomes $1/m^2$ and the proof is complete.

It is worth noticing that the only structural property of Nash equilibria we needed in the proof of the above theorem is that the expected load of a link j is at most 2opt (and, of course, the independence of the agent strategies). We can use a similar proof to extend the theorem to the case of m links with different speeds:

Theorem 9. *The coordination ratio of any number of agents and m different links is $O(\sqrt{\frac{s_m}{s_1} \sum_j \frac{s_j}{s_1}} \sqrt{\log m})$.*

5   Discussion and Open Problems

We believe that the approach introduced in this paper, namely evaluating the worst-case ratio of Nash equilibria to the social optimum, may prove a useful calculation in many contexts. Although the Nash equilibrium is not trivial to reach without coordination, it does serve as an important indicator of the kinds of behaviors exhibited by noncooperative agents.

Besides bridging the gaps left open in our theorems, there are several extensions of this work that seem interesting, namely, investigating with the same point of view more complex and realistic cost models, for example, when the cost is given by $\frac{1}{C-\min\{C,\sum w_i\}}$ where C is the capacity of a link and $\sum w_i$ its load [7]. More important is the study of realistic Internet metrics, that result from the employed protocols such as the one related to TCP and the square root of the drop frequency [3]. Finally, it would be extremely interesting, once the relative quality of the Nash equilibria in such situations is better understood, to employ such understanding in the design of improved protocols [1].

References

1. B. Braden, D. Clark, J. Crowcroft, B. Davie, S. Deering, D. Estrin, S. Floyd, V. Jacobson, G. Minshall, C. Partridge, L. Peterson, K. Ramakrishnan, S. Shenker, J. Wroclawski, and L. Zhang. Recommendations on Queue Management and Congestion Avoidance in the Internet, April 1998.
   http://info.internet.isi.edu:80/in-notes/rfc/files/rfc2309.txt
2. Y. Cho and S. Sahni. Bounds for list schedules on uniform processors. *SIAM Journal on Computing*, 9(1):91–103, 1980.
3. S. Floyd and K. Fall. Router Mechanisms to Support End-to-End Congestion Control. Technical report, Lawrence Berkeley National Laboratory, February 1997.
4. G. R. Grimmet and D. R. Stirzaker. *Probability and Random Processes, 2nd ed.*. Oxford University Press, 1992.
5. Y. Korilis and A. Lazar. On the existence of equilibria in noncooperative optimal flow control. *Journal of the ACM* 42(3):584-613, 1995.
6. Y. Korilis, A. Lazar, A. Orda. Architecting noncooperative networks. *IEEE J. Selected Areas of Comm., 13,* 7, 1995.
7. R. La, V. Anantharam. Optimal routing control: Game theoretic approach. *Proc. 1997 CDC Conf.*
8. G. Owen. *Game Theory, 3rd ed.*. Academic Press, 1995.
9. K. Park, M. Sitharam, S. Chen. Quality of service provision in noncooperative network environments. Manuscript, Purdue Univ., 1998.
10. C. H. Papadimitriou, M. Yannakakis. On complexity as bounded rationality. In *Proceedings of the Twenty-Sixth Annual ACM Symposium on the Theory of Computing*, pages 726-733, Montreal, Quebec, Canada, 23-25 May 1994.
11. S. J. Shenker. Making greed work in networks: a game-theoretic analysis of switch service disciplines. *IEEE/ACM Transactions on Networking*, 3(6):819-831, Dec. 1995.
12. S. Shenker, D. Clark, D. Estrin, and S. Herzog. Pricing in Computer Network: Reshaping the Research Agenda. *Communications Policy*, 20(1), 1996.

A Complete and Tight Average-Case Analysis of Learning Monomials

Rüdiger Reischuk[1], * and Thomas Zeugmann[2], **

[1] Institut für Theoretische Informatik, Med. Universität zu Lübeck, Wallstraße 40,
23560 Lübeck, Germany
reischuk@informatik.mu-luebeck.de
[2] Department of Informatics, Kyushu University, Kasuga 816-8580, Japan
thomas@i.kyushu-u.ac.jp

Abstract. We advocate to analyze the average complexity of learning problems. An appropriate framework for this purpose is introduced. Based on it we consider the problem of *learning monomials* and the special case of learning *monotone* monomials *in the limit* and for *on-line predictions* in two variants: from positive data only, and from positive and negative examples. The well-known *Wholist algorithm* is completely analyzed, in particular its average-case behavior with respect to the class of *binomial distributions*. We consider different complexity measures: the *number of mind changes, the number of prediction errors, and the total learning time*. Tight bounds are obtained implying that worst case bounds are too pessimistic. On the average learning can be achieved *exponentially faster*.

Furthermore, we study a new learning model, *stochastic finite learning*, in which, in contrast to PAC learning, some information about the underlying distribution is given and the goal is to find a *correct* (not only approximatively correct) hypothesis. We develop techniques to obtain good bounds for stochastic finite learning from a precise average case analysis of strategies for learning in the limit and illustrate our approach for the case of learning monomials.

1. Introduction

Learning concepts efficiently has attracted considerable attention during the last decade. However, research following the traditional lines of inductive inference has mainly considered the update time, i.e., the effort to compute a *single* new hypothesis. Starting with Valiant's paper [18], the total amount of time needed to solve a given learning problem has been investigated as well. The

---

* Part of this work was performed while visiting the Department of Informatics at Kyushu University supported by the Japan Society for the Promotion of Science under Grant JSPS 29716102.
** Supported by the Grant-in-Aid for Scientific Research in Fundamental Areas from the Japanese Ministry of Education, Science, Sports, and Culture under grant no. 10558047.

C. Meinel and S. Tison (Eds.): STACS'99, LNCS 1563, pp. 414–423, 1999.
© Springer-Verlag Berlin Heidelberg 1999

complexity bounds proved within the PAC model are usually *worst-case* bounds. In experimental studies large gaps have often been observed between the time bounds obtained by a mathematical analysis and the actual runtime of a learner on typical data. This phenomenon can be explained easily. Data from running tests provide information about the *average-case* performance of a learner, rather than its worst-case behavior. Since algorithmic learning has a lot of practical applications it is of great interest to analyze the average-case performance, and to obtain tight bounds saying something about the typical behavior in practice.

Pazzani and Sarrett [14] have proposed a framework for analyzing the average-case behavior of learning algorithms. Several authors have followed their approach (cf., e.g., [12,13]). Their main goal is to predict the expected accuracy of the hypothesis produced with respect to the number of training examples. However, the results obtained so far are not satisfactory. Typically, the probability that a random example is misclassified by the current hypothesis is estimated by a complicated formula. The *evaluation* of this formula, and the computation of the corresponding expectation has been done by Monte-Carlo simulations. Clearly, such an approach does not provide general results about the average-case behavior for broader classes of distributions. Moreover, it is hard to compare these bounds with those proved for the PAC model.

We outline a new setting to study the average-case behavior of learning algorithms overcoming these drawbacks and illustrate it for learning monomials.

2. Preliminaries

Let $\mathbb{N} = \{0, 1, 2, \ldots\}$ be the set of all natural numbers, and let $\mathbb{N}^+ := \mathbb{N} \backslash \{0\}$. If M is a set, $|M|$ is used for its cardinality. For an infinite sequence d and $j \in \mathbb{N}^+$ let $d[j]$ denote the initial segment of d of length j. By $(0, 1)$ we denote the real interval from 0 to 1 excluding both endpoints. For $n \in \mathbb{N}^+$, let $\mathcal{X}_n = \{0, 1\}^n$ be the *learning domain* and $\wp(\mathcal{X}_n)$ the power set of $\mathcal{X}_n$. A subset c of $\mathcal{X}_n$ is called a *concept*, and a subset $\mathcal{C}$ of $\wp(\mathcal{X}_n)$ a *concept class*. The notation c is also used to denote the characteristic function of a subset, that is for $b \in \mathcal{X}_n$: $c(b) = 1$ iff $b \in c$. To define the classes of concepts we deal with in this paper let $\mathcal{L}_n = \{x_1, \bar{x}_1, x_2, \bar{x}_2 \ldots, x_n, \bar{x}_n\}$ be a set of literals. x_i is a *positive* literal and $\bar{x}_i$ a *negative* one. A conjunction of literals defines a *monomial*. For a monomial m let $\#(m)$ denote its length, that is the number of literals in it.

m describes a subset $L(m)$ of $\mathcal{X}_n$, in other words a concept, in the obvious way: the concept contains exactly those binary vectors for which the monomial evaluates to 1, that is $L(m) := \{b \in \mathcal{X}_n \mid m(b) = 1\}$. The collection of objects we are going to learn is the set $\mathcal{C}_n$ of all concepts that are describable by monomials over $\mathcal{X}_n$. There are two trivial concepts, the empty subset and $\mathcal{X}_n$ itself. $\mathcal{X}_n$, which will also be called "TRUE", can be represented by the empty monomial. The concept "FALSE" has several descriptions. To avoid ambiguity, we always represent "FALSE" by the monomial $x_1 \bar{x}_1 \ldots x_n \bar{x}_n$. Furthermore, we often identify the set of all monomials over $\mathcal{L}_n$ and the concept class $\mathcal{C}_n$. Note that $|\mathcal{C}_n| = 3^n + 1$. We also consider the subclass $\mathcal{MC}_n$ of $\mathcal{C}_n$ consisting of those concepts that can be described by *monotone* monomials, i.e., by monomials containing positive literals only. It holds $|\mathcal{MC}_n| = 2^n$.

3. Learning Models and Complexity Measures

The first learning model we are dealing with is the *on-line prediction* model going back to Barzdin, Freivald [1] and Littlestone [10]. In this setting the source of information is specified as follows. The learner is given a sequence of labeled examples $d = \langle d_j \rangle_{j \in \mathbb{N}^+} = \langle b_1, c(b_1), b_2, c(b_2), b_3, c(b_3), \ldots \rangle$ from the concept c, where the $b_j \in \mathcal{X}_n$, and $c(b_j) = 1$ if $b_j \in c$ and $c(b_j) = 0$ otherwise. The examples b_j are picked arbitrarily and the information provided is assumed to be without any errors. We refer to such sequences as *data sequences* and use $data(c)$ to denote the set of all data sequences for concept c.

A learner P must predict $c(b_j)$ after having seen $d[2j - 1] = \langle b_1, c(b_1), \ldots, b_{j-1}, c(b_{j-1}), b_j \rangle$. We denote this hypothesis by $P(d[2j-1])$. Then it receives the true value $c(b_j)$ and the next Boolean vector b_{j+1}. The learner has successfully learned if it eventually reaches a point beyond which it always predicts correctly.

Definition 1. *A concept class $\mathcal{C}$ is called **on-line predictable** if there is a learner P such that for all concepts $c \in \mathcal{C}$ and all data sequences $d = \langle d_j \rangle_{j \in \mathbb{N}^+} \in data(c)$ it holds: $P(d[2j - 1])$ is defined for all j, and $P(d[2j - 1]) = d_{2j}$ for all but finitely many j.*

For on-line prediction, the *complexity measure* considered is the *number of prediction errors* made. Note that the prediction goal can always be achieved trivially if the learning domain is finite. Therefore, we aim to minimize the number of prediction errors when learning monomials.

Next, let us define Gold-style [5] *learning in the limit*. One distinguishes between learning from positive and negative data, and learning from positive data only. For a concept c, let $info(c)$ be the set of those data sequences $\langle b_1, c(b_1), b_2, c(b_2), b_3, c(b_3), \ldots \rangle$ in $data(c)$ that contain each element b of the learning domain $\mathcal{X}_n$ at least once. Such a sequence is called *informant*. For ease of notation let us pair each element b_j with its classification $c(b_j)$. Then the j-th entry of an informant sequence d will be $d_j := (b_j, c(b_j))$.

A *positive presentation* of c is a data sequence that contains only elements of c and each one at least once. Thus all the values $c(b_j)$ are equal to 1 and thus could be omitted. In this case we will denote the sequence simply by $d = \langle d_j \rangle_{j \in \mathbb{N}^+} = \langle b_1, b_2, b_3, \ldots \rangle$. Let $d[j]^+ := \{b_i \mid 1 \le i \le j\}$ be the set of all examples contained in the prefix of d of length j, and let $pos(c)$ denote the set of all positive presentations of c. The elements of $pos(c)$ are also called a *text* for c.

A limit learner is an *inductive inference machine* (abbr. IIM). An IIM M works as follows. As inputs it gets incrementally growing segments of a positive presentation (resp. of an informant) d. After each new input, it outputs a hypothesis $M(d[j])$ from a predefined hypothesis space $\mathcal{H}$. Each hypothesis refers to a unique element of the concept class.

Definition 2. *Let $\mathcal{C}$ be a concept class and let $\mathcal{H}$ be a hypothesis space for it. $\mathcal{C}$ is called **learnable in the limit** from positive presentation (resp. from informant) if there is an IIM M such that for every $c \in \mathcal{C}$ and every $d \in pos(c)$ (resp. $d \in info(c)$): $M(d[j])$ is defined for all j, and $M(d[j]) = h$ for all but finitely many j, where $h \in \mathcal{H}$ is a hypothesis referring to c.*

For the concept class C_n we choose as hypothesis space the set of all monomials over $\mathcal{L}_n$, whereas for $\mathcal{MC}_n$ it is the set of all monotone monomials. Again, we are interested how efficiently C_n and $\mathcal{MC}_n$ can be learned in the limit.

The first complexity measure we consider is the *mind change* complexity. A mind change occurs iff $M(d[j]) \neq M(d[j+1])$. Clearly, this measure is closely related to the number of prediction errors. Both complexity measures say little about the total amount of data and time needed until a concept is guessed correctly. Thus, for learning in the limit we also measure the *time complexity*. As in [3] we define the total learning time as follows. Let M be any IIM learning a concept class C in the limit. Then, for $c \in C$ and a text or informant d for c, let $Con(M, d) =$ the least $i \in \mathbb{N}^+$ such that $M(d[j]) = M(d[i])$ for all $j \geq i$ denote the *stage of convergence* of M on d (cf. [5]). Moreover, by $T_M(d_j)$ we denote the number of steps to compute $M(d[j])$. We measure this quantity as a function of the length of the input and refer to it as the *update time*. Finally, the *total learning time* taken by the IIM M on a sequence d is defined as $TT(M, d) := \sum_{j=1}^{Con(M,d)} T_M(d[j])$. Given a probability distribution D on the data sequences d we evaluate the *expectation* of $TT(M, d)$ with respect to D, the *average total learning time*.

4. The Wholist Algorithm Learning Monomials

Next, we present Haussler's [6] Wholist algorithm for on-line prediction of monomials. For learning in the limit this algorithm can be modified straightforwardly. The limit learner computes a new hypothesis using only the most recent example received and his old hypothesis. Such learners are called *iterative* (cf. [8]). Let $c \in C_n$, let $d \in data(c)$, and let $b_i = b_i^1 b_i^2 \ldots b_i^n$ denote the i-th Boolean vector in d. Recall that "TRUE" is represented by the empty monomial.
Algorithm $\mathcal{P}$: On input sequence $\langle b_1, c(b_1), b_2, c(b_2), \ldots \rangle$ do the following:
  Initialize $h_0 := x_1 \bar{x}_1 \ldots x_n \bar{x}_n$.
  for $i = 1, 2, \ldots$ do
    let h_{i-1} denote $\mathcal{P}$'s internal hypothesis produced before receiving b_i;
    when receiving b_i predict $h_{i-1}(b_i)$; read $c(b_i)$;
    if $h_{i-1}(b_i) = c(b_i)$ then $h_i := h_{i-1}$
      else for $j := 1$ to n do
        if $b_i^j = 1$ then delete $\bar{x}_j$ in h_{i-1} else delete x_j in h_{i-1};
        let h_i be the resulting monomial
  end.
Note that the algorithm is monotone with respect to the sequence of its internal hypotheses: $h_i \geq h_{i-1}$ when considered as function on $\mathcal{X}_n$.
Theorem 1. *Algorithm* $\mathcal{P}$ *learns the set of all monomials within the prediction model. It makes at most $n + 1$ prediction errors.*

To learn the monotone concept class $\mathcal{MC}_n$, algorithm $\mathcal{P}$ can be easily modified by initializing $h_0 = x_1 x_2 \ldots x_n$ and by simplifying the loop appropriately. We refer to the modified algorithm as to $\mathcal{MP}$.

Theorem 1 can be directly reproved for $\mathcal{MP}$ with the only difference that now the worst-case bound for the number of prediction errors is n instead of $n + 1$.

5. Complexity Analysis: Best and Worst Case

For the learning models defined above we estimate the best-case complexity, the worst-case complexity, and the expectation of algorithm $\mathcal{P}$ and $\mathcal{MP}$. We start with the first two issues. Both algorithms do not make any prediction errors iff the initial hypothesis h_0 equals the target monomial. For $\mathcal{P}$ this means that the concept to be learned is "FALSE", while for $\mathcal{MP}$ the concept is the all-1 vector. These special concepts can be considered as *minimal* in their class. For them the best-case and the worst-case number of predictions errors coincide.

In the general case, we call the literals in a monomial m *relevant*. All other literals in $\mathcal{L}_n$ (resp. in $\{x_1, \ldots, x_n\}$ in the monotone case) are said to be *irrelevant* for m. There are $2n - \#(m)$ irrelevant literals in general, and $n - \#(m)$ in the monotone case. We call bit i **relevant** for m if x_i or $\bar{x}_i$ is relevant for m. By $k = k(m) = n - \#(m)$ we denote the number of irrelevant bits.

Theorem 2. *Let $c = L(m)$ be a non-minimal concept in $\mathcal{MC}_n$. Then algorithm $\mathcal{MP}$ makes 1 prediction error in the best case, and $k(m)$ prediction errors in the worst-case.*

If c is a non-minimal concept of $\mathcal{C}_n$ algorithm $\mathcal{P}$ makes 2 prediction errors in the best case and $1 + k(m)$ prediction errors in the worst-case.

As Theorem 2 shows, the gap between the best-case and worst-case behavior can be quite large. Thus, we ask what are the expected bounds for the number of prediction errors on randomly generated data sequences. Before answering this question we estimate the worst-case number of prediction errors averaged over the whole concept class $\mathcal{MC}_n$, resp. $\mathcal{C}_n$. Thus we get a complexity bound with respect to the parameter n, instead of $\#(m)$ as in Theorem 2. This averaging depends on the underlying probability distribution for selecting the target concepts (for the corresponding data sequences we consider the worst input). The average is shown to be linear in n if the literals are binomially distributed.

To generate the probability distributions we assume for $\mathcal{MC}_n$ the relevant positive literals to be drawn independently at random with probability p, $p \in (0, 1)$. Thus, with probability $1 - p$ a literal is irrelevant. The length of the monomials drawn by this distribution is binomially distributed with parameter p. Thus we call such a distribution on the concept class a *binomial distribution*.

Theorem 3. *Let the concepts in $\mathcal{MC}_n$ be binomially distributed with parameter p. Then the average number of prediction errors of $\mathcal{MP}$ for the worst data sequences is $n(1 - p)$.*

In case $p = 1/2$, the bound says that the maximal number of prediction errors when uniformly averaged over all concepts in $\mathcal{MC}_n$ is $n/2$.

Next, we deal with the class $\mathcal{C}_n$. For comparing it to the monotone case we have to clarify what does it mean for concepts in $\mathcal{C}_n$ to be binomially distributed. Since there are $3^n + 1$ many concepts in $\mathcal{C}_n$ for a uniform distribution each concept must have probability $1/(3^n + 1)$. For each position $i = 1, \ldots, n$ three options are possible, i.e., we may choose x_i, $\bar{x}_i$ or neither of them. This suggests the formula $\binom{n}{k_1, k_2, k_3} p_1^{k_1} p_2^{k_2} p_3^{k_3}$, where p_1 is the probability to take x_i, p_2 the probability to choose $\bar{x}_i$ and p_3 the probability to choose none, and $p_1 + p_2 + p_3 = 1$

and $k_1 + k_2 + k_3 = n$. $k_1 + k_2$ counts the number of relevant literals, resp. bits. However, this formula does not include the concept "FALSE." Thus, let us introduce $p_f \in (0, 1)$ for the probability to choose "FALSE." Then the formula becomes $(1 - p_f)\binom{n}{k_1, k_2, k_3} p_1^{k_1} p_2^{k_2} p_3^{k_3}$. We call such a probability distribution a *weighted multinomial distribution* with parameters (p_f, p_1, p_2, p_3).

Theorem 4. *Let the concepts in C_n occur according to a weighted multinomial distribution with parameters (p_f, p_1, p_2, p_3). Then the average number of prediction errors of $\mathcal{P}$ for the worst data sequences is $(1 - p_f)(1 + np_3)$.*

For the particular case that all concepts from C_n are equally likely, i.e., $p_1 = p_2 = p_3 = 1/3$ and $p_f = 1/(3^n + 1)$, we directly get that on the average less than $n/3 + 1$ errors are to be expected given the worst data sequences. Hence, in this case the class C_n seems to be easier to learn than $\mathcal{MC}_n$ with respect to the complexity measure *prediction errors*. However, this impression is a bit misleading, since the probabilities to generate an irrelevant literal are *different*, i.e., $1/3$ for C_n and $1/2$ for $\mathcal{MC}_n$. If we assume the probabilities to generate an irrelevant literal to be equal, say q, and make the meaningful assumption that "FALSE" has the same probability as "TRUE" then the average complexity is $\frac{1}{1+q^n}(1 + nq)$ for C_n and nq for $\mathcal{MC}_n$. Since for C_n it holds $q = 1 - (p_1 + p_2)$, and for $\mathcal{MC}_n$ $q = 1 - p$, under these assumptions $\mathcal{MC}_n$ is easier to learn than C_n. This insight is interesting, since it clearly shows the influence of the underlying distribution. In contrast, previous work has expressed these bounds in terms of the VC-dimension which is the same for both classes, i.e., n.

The results above directly translate to learning in the limit from informant or from positive presentations for the complexity measure *number of mind changes*.

What can be said about the total learning time? The best-case can be handled as above. Sine the update time is linear in n for both algorithms $\mathcal{MP}$ and $\mathcal{P}$, in the best case the total learning time is linear. The worst-case total learning time is *unbounded* for both algorithms, since every text and informant may contain as many repetitions of data not possessing enough information to learn the target.

Hence, as far as learning in the limit and the complexity measure *total learning time* are concerned, there is a huge gap between the best-case and the worst-case behavior. Since the worst-case is unbounded, it does not make sense to ask for an analogue to Theorem 3 and 4. Instead, we continue by studying the average-case behavior of the limit learner $\mathcal{P}$ and $\mathcal{MP}$.

6. Average-Case Analysis for Learning in the Limit from Text

For the following average case analysis we assume that the data sequences are generated at random with respect to some probability distribution D taken from a class of admissible distributions $\mathcal{D}$ specified below. We are interested in the *average number* of examples till an algorithm has converged to a correct hypothesis. **CON** denotes a random variable counting the number of examples till convergence. Let d be a text of the concept c to be learned that is generated at random according to D. If the concept to be learned is "FALSE" no examples are needed. Otherwise, if the target concept contains precisely n literals then one positive example suffices (note that this one is unique). Thus, for these two cases everything is clear and the probability distributions D on the set of positive examples for c are trivial.

For analyzing the nontrivial cases, let $c = L(m) \in \mathcal{C}_n$ be a concept with monomial $m = \bigwedge_{j=1}^{\#(m)} \ell_{i_j}$ such that $k = k(m) = n - \#(m) > 0$. There are 2^k positive examples for c. For the sake of presentation, we assume these examples to be *binomially distributed*. That is, in a random positive example all entries corresponding to irrelevant bits are selected independently of each other. With some probability p this will be a 1, and with probability $q := 1 - p$ a 0. We shall consider only nontrivial distributions where $0 < p < 1$. Note that otherwise the data sequence does not contain all positive examples. We aim to compute the expected number of examples taken by $\mathcal{P}$ until convergence.

The first example received forces $\mathcal{P}$ to delete precisely n of the $2n$ literals in h_0. Thus, this example always plays a *special* role. Note that the resulting hypothesis h_1 depends on b_1, but the number k of literals that remain to be deleted from h_1 until convergence is *independent* of b_1. Using tail bound techniques, we can show the following theorem.

Theorem 5. *Let $c = L(m)$ be a non-minimal concept in $\mathcal{C}_n$, and let the positive examples for c be binomially distributed with parameter p. Define $\psi := \min\{\frac{1}{1-p}, \frac{1}{p}\}$. Then the expected number of positive examples needed by algorithm $\mathcal{P}$ until convergence can be bounded by $E[\mathrm{CON}] \leq \lceil \log_\psi k(m) \rceil + 3$.*

A similar analysis can be given in the monotone setting for algorithm $\mathcal{MP}$.

Corollary 6. *For every binomially distributed text with parameter $0 < p < 1$ the average total learning time of algorithm $\mathcal{P}$ for concepts in $\mathcal{C}_n$ with μ literals is at most $O(n(\log(n - \mu + 2)))$.*

The expectation alone does not provide complete information about the average case behavior of an algorithm. We also like to deduce bounds on how often the algorithm exceeds the average considerably. The Wholist algorithm possesses two favorable properties that simplify this derivation considerably, i.e., it is *set-driven* and *conservative*. Set-driven means that for all $c \in \mathcal{C}_n$ all $d, h \in pos(c)$ and all $i, j \in \mathbb{N}^+$ the equality $d[i]^+ = h[j]^+$ implies $\mathcal{P}(d[i]) = \mathcal{P}(h[j])$. A learner is said to be conservative if every mind change is caused by an inconsistency with the data seen so far. Clearly, the Wholist algorithm satisfies this condition, too. Now, the following theorem establishes exponentially shrinking tail bounds for the expected number of examples needed in order to achieve convergence.

Theorem 7 ([16]). *Let CON be the sample complexity of a conservative and set-driven learning algorithm. Then $\Pr[\mathrm{CON} > 2t \cdot E[\mathrm{CON}]] \leq 2^{-t}$ for all $t \in \mathbb{N}$.*

A simple calculation shows that in case of exponentially shrinking tail bounds the variance is bounded by $O(E[\mathrm{CON}]^2)$.

7. Stochastic Finite Learning

Next we shall show how to convert the Wholist algorithm into a text learner that identifies all concepts in $\mathcal{C}_n$ *stochastically in a bounded number of rounds with high confidence*. A bit *additional knowledge* concerning the underlying class of probability distributions is required. Thus, in contrast to the PAC model, the resulting learning model is *not* distribution-free. But with respect to the quality of its hypotheses, it is stronger than the PAC model by requiring the output to be *probably exactly correct* rather than *probably approximately correct*. The *main* advantage is the usage of the additional knowledge to reduce the sample size, and hence the total learning time drastically. This contrasts to previous work in the area of PAC learning (cf., e.g., [2,4,7,9,11,17]). These papers have shown concepts classes to be PAC learnable from polynomially many examples given a known distribution or class of distributions, while the general PAC learnability of these concepts classes is not achievable or remains open. Note that our general approach, i.e., performing an average-case analysis and proving exponentially shrinking tail bounds for the expected total learning time, can also be applied to obtain results along this line (cf. [15,16]).

Definition 3. *Let $\mathcal{D}$ be a set of probability distributions on the learning domain, $\mathcal{C}$ a concept class, $\mathcal{H}$ a hypothesis space for $\mathcal{C}$, and $\delta \in (0,1)$. $(\mathcal{C}, \mathcal{D})$ is said to be* **stochastically finite learnable with δ-confidence** *with respect to $\mathcal{H}$ iff there is an IIM M that for every $c \in \mathcal{C}$ and every $D \in \mathcal{D}$ performs as follows. Given a random presentation d for c generated according to D, M stops after having seen a finite number of examples and outputs a single hypothesis $h \in \mathcal{H}$. With probability at least $1 - \delta$ (with respect to distribution D) h has to be correct, that is $L(h) = c$ in case of monomials. If stochastic finite learning can be achieved with δ-confidence for every $\delta > 0$ then we say that $(\mathcal{C}, \mathcal{D})$ can be learned stochastically finite* **with high confidence.**

We study the case that the positive examples are binomially distributed with parameter p. But we do not require precise knowledge about the underlying distribution. Instead, we reasonably assume that *prior knowledge* is provided by parameters p_{low} and p_{up} such that $p_{low} \leq p \leq p_{up}$ for the true parameter p. Binomial distributions fulfilling this requirement are called (p_{low}, p_{up})-*admissible distributions*. Let $\mathcal{D}_n[p_{low}, p_{up}]$ denote the set of such distributions on $\mathcal{X}_n$.

If bounds p_{low} and p_{up} are available, the Wholist algorithm can be transformed into a stochastic finite learner inferring all concepts with high confidence.

Theorem 8. *Let $0 < p_{low} \leq p_{up} < 1$ and $\psi := \min\{\frac{1}{1-p_{low}}, \frac{1}{p_{up}}\}$. Then $(\mathcal{C}_n, \mathcal{D}_n[p_{low}, p_{up}])$ is stochastically finitely learnable with high confidence from positive presentations. To achieve δ-confidence no more than $O(\log_2 1/\delta \cdot \log_\psi n)$ many examples are necessary.*

The latter example bound can even be improved to $\log_\psi n + \log_\psi 1/\delta + O(1)$ by performing a careful *error analysis*, i.e., for the Wholist algorithm, the confidence requirement increases the sample size by an additive term $\log_\psi 1/\delta$ only.

8. Average-Case Analysis for Learning in the Limit from Informant

Finally, we consider how the results obtained so far translate to the case of learning from informant. First, we investigate the uniform distribution over $\mathcal{X}_n$. Again, we have the trivial cases that the target is "FALSE" or m is a monomial without irrelevant bits. In the first case, no example is needed at all, while in the latter one, there is only one positive example having probability 2^{-n}. Thus the expected number of examples needed until successful learning is $2^n = 2^{\#(m)}$.

Theorem 9. *Let* $c = L(m) \in C_n$ *be a nontrivial concept. If an informant for c is generated from the uniform distribution by independent draws the expected number of examples needed by algorithm $\mathcal{P}$ until convergence is bounded by* $E[\mathrm{CON}] \leq 2^{\#(m)} \left(\lceil \log_2 k(m) \rceil + 3 \right)$.

Hence, as long as $k(m) = n - O(1)$, we still achieve an expected total learning time $O(n \log n)$. But if $\#(m) = \Omega(n)$ the expected total learning is exponential. However, if there are many relevant literals then even h_0 may be considered as a not too bad *approximation* for c. Thus, let $\varepsilon \in (0, 1)$ be an error parameter as in the PAC model. We ask if one can achieve an expected sample complexity for computing an ε-approximation that is polynomially bounded in $\log n$ and $1/\varepsilon$.

Let $err_m(h_j) := D(L(h_j) \triangle L(m))$ be the error made by hypothesis h_j with respect to monomial m. Here $L(h_j) \triangle L(m)$ is the symmetric difference of $L(h_j)$ and $L(m)$, and D the probability distribution with respect to which the examples are drawn. h_j is an ε-*approximation* for m if $err_m(h_j) \leq \varepsilon$. Finally, we redefine the stage of convergence. Let $d = (d_j)_{j \in \mathbb{N}^+} \in info(L(m))$, then

$$\mathrm{CON}_\varepsilon(d) = \text{the least number } j \text{ such that } err_m(P(d[i])) \leq \varepsilon \text{ for all } i \geq j.$$

Note that once the Wholist algorithm has reached an ε-approximate hypothesis all further hypotheses will also be at least that close to the target monomial. The following theorem gives an affirmative answer to the question posed above.

Theorem 10. *Let $c = L(m) \in C_n$ be a nontrivial concept. Assuming that examples are drawn independently from the uniform distribution, the expected number of examples needed by algorithm $\mathcal{P}$ until converging to an ε-approximation for c can be bounded by* $E[\mathrm{CON}_\varepsilon] \leq \frac{1}{\varepsilon} \cdot \left(\lceil \log_2 k(m) \rceil + 3 \right)$.

Thus, additional knowledge concerning the underlying probability distribution pays off again. Using Theorem 7 and modifying Section 7 *mutatis mutandis*, we achieve stochastic finite learning with high confidence for all concepts in C_n using $O(\frac{1}{\varepsilon} \cdot \log \frac{1}{\delta} \cdot \log n)$ many examples. However, the resulting learner now infers ε-approximations. Comparing this bound with the sample complexity given in the PAC model one notes an exponential reduction.

Finally, we generalize the last results to the case that the data sequences are binomially distributed for some parameter $p \in (0, 1)$. This means that any particular vector containing ν times a 1 and $n - \nu$ a 0 has probability $p^\nu (1 - p)^{n-\nu}$ since a 1 is drawn with probability p and a 0 with probability $1 - p$. First, Theorem 9 generalizes as follows.

Theorem 11. *Let $c = L(m) \in C_n$ be a nontrivial concept. Let m contain precisely π positive literals and τ negative literals. If the labeled examples for c are independently binomially distributed with parameter p and $\psi := \min\{\frac{1}{1-p}, \frac{1}{p}\}$ then the expected number of examples needed by algorithm $\mathcal{P}$ until convergence can be bounded by* $E[\mathrm{CON}] \leq \frac{1}{p^\pi (1-p)^\tau} \left(\lceil \log_\psi k(m) \rceil + 3 \right)$.

Theorem 10 directly translates into the setting of binomially distributed inputs.

Theorem 12. *Let $c = L(m) \in C_n$ be a nontrivial concept. Assume that the examples are drawn with respect to a binomial distribution with parameter p, and let $\psi = \min\{\frac{1}{1-p}, \frac{1}{p}\}$. Then the expected number of examples needed by algorithm $\mathcal{P}$ until converging to an ε-approximation for c can be bounded by* $E[\mathrm{CON}] \leq \frac{1}{\varepsilon} \cdot \left(\lceil \log_\psi k(m) \rceil + 3 \right)$.

Finally, one can also learn ε-approximations stochastically finite with high confidence from informant with an exponentially smaller sample complexity.
Theorem 13. *Let* $0 < p_{low} \leq p_{up} < 1$ *and* $\psi := \min\{\frac{1}{1-p_{low}}, \frac{1}{p_{up}}\}$. *For* $(\mathcal{C}_n, \mathcal{D}_n[p_{low}, p_{up}])$ ε*-approximations are stochastically finitely learnable with* δ*-confidence from informant for all* $\varepsilon, \delta \in (0,1)$. *Further,* $O\left(\frac{1}{\varepsilon} \cdot \log_2 1/\delta \cdot \log_\psi n\right)$, *resp.* $O\left(\frac{1}{\varepsilon} \cdot (\log_\psi 1/\delta + \log_\psi n)\right)$ *many examples suffice for this purpose.*

References

1. J.M. Barzdin, R.V. Freivald, On the prediction of general recursive functions. *Soviet Math. Doklady* 13:1224-1228, 1972.
2. G. Benedek and A. Itai. Learnability by fixed distributions. "Proc. 1988 Workshop on Computational Learning Theory," 81–90, Morgan Kaufmann, 1988.
3. R. Daley and C.H. Smith. On the complexity of inductive inference. *Inform. Control*, 69:12–40, 1986.
4. F. Denis and R. Gilleron. PAC learning under helpful distributions. "Proc. 8th International Workshop on Algorithmic Learning Theory," LNAI Vol. 1316, 132–145, Springer-Verlag, 1997.
5. E.M. Gold, Language identification in the limit. *Inform. Control* 10:447–474, 1967.
6. D. Haussler. Bias, version spaces and Valiant's learning framework. "Proc. 8th National Conference on Artificial Intelligence, 564–569, Morgan Kaufmann, 1987.
7. M. Kearns, M. Li, L. Pitt and L.G. Valiant. On the learnability of Boolean formula. "Proc. 19th Annual ACM Symposium on Theory of Computing," 285–295, ACM Press 1987.
8. S. Lange and T. Zeugmann. Incremental learning from positive data. *J. Comput. System Sci.* 53(1):88–103, 1996.
9. M. Li and P. Vitanyi. Learning simple concepts under simple distributions. *SIAM J. Comput.*, 20(5):911-935, 1991.
10. N. Littlestone. Learning quickly when irrelevant attributes are abound: A new linear threshold algorithm. *Machine Learning* 2:285–318, 1988.
11. B. Natarajan. On learning Boolean formula. "Proc. 19th Annual ACM Symposium on Theory of Computing," 295–304, ACM Press, 1987.
12. S. Okamoto and K. Satoh. An average-case analysis of k-nearest neighbor classifier. "Proc. 1st International Conference on Case-Based Reasoning Research and Development," LNCS Vol. 1010, 253–264, Springer-Verlag, 1995.
13. S. Okamoto and N. Yugami. Theoretical analysis of the nearest neighbor classifier in noisy domains. "Proc. 13th International Conference on Machine Learning, 355–363, Morgan Kaufmann 1996.
14. M.J. Pazzani and W. Sarrett, A framework for average case analysis of conjunctive learning algorithms. *Machine Learning* 9:349–372, 1992.
15. R. Reischuk and T. Zeugmann. Learning one-variable pattern languages in linear average time. "Proc. 11th Annual Conference on Computational Learning Theory," 198–208, ACM Press, 1998.
16. P. Rossmanith and T. Zeugmann. Learning k-variable pattern languages efficiently stochastically finite on average from positive data. "Proc. 4th International Colloquium on Grammatical Inference," LNAI Vol. 1433, 13–24, Springer-Verlag, 1998.
17. Y. Sakai, E. Takimoto and A. Maruoka. Proper learning algorithms for functions of k terms under smooth distributions. "Proc. 8th Annual ACM Conference on Computational Learning Theory," 206–213, ACM Press, 1995.
18. L.G. Valiant. A theory of the learnable. *Commun. ACM* 27:1134-1142, 1984.

Costs of General Purpose Learning*

John Case[1], Keh-Jiann Chen[2], and Sanjay Jain[3]

[1] Department of CIS
University of Delaware
Newark, DE 19716, USA
case@cis.udel.edu
[2] Institute for Information Sciences
Academica Sinica
Taipei, 15, Taiwan
Republic of China
[3] School of Computing
National University of Singapore
Singapore 119260
sanjay@comp.nus.edu.sg

Abstract. Leo Harrington surprisingly constructed a machine which can learn *any* computable function f according to the following criterion (called **Bc***-*identification*). His machine, on the successive graph points of f, outputs a corresponding infinite sequence of programs $p_0, p_1, p_2, \ldots$, and, for some i, the programs $p_i, p_{i+1}, p_{i+2}, \ldots$ each compute a variant of f which differs from f at only finitely many argument places. A machine with this property is called *general purpose*. The sequence $p_i, p_{i+1}, p_{i+2}, \ldots$ is called a *final sequence*. For Harrington's general purpose machine, for distinct m and n, the finitely many argument places where p_{i+m} fails to compute f can be very different from the finitely many argument places where p_{i+n} fails to compute f. One would hope though, that if Harrington's machine, or an improvement thereof, inferred the program p_{i+m} based on the data points $f(0), f(1), \ldots, f(k)$, then p_{i+m} would make very few mistakes computing f at the "near future" arguments $k+1, k+2, \ldots, k+\ell$, where ℓ is reasonably large. Ideally, p_{i+m}'s finitely many mistakes or anomalies would (mostly) occur at arguments $x \gg k$, i.e., ideally, its anomalies would be well placed beyond near future arguments. In the present paper, for general purpose learning machines, it is analyzed just how well or badly placed these anomalies may be with respect to near future arguments and what are the various tradeoffs. In particular, there is good news and bad. Bad news is that, for any learning machine $\mathbf{M}$ (including general purpose $\mathbf{M}$), for all m, there exist infinitely many computable functions f such that, infinitely often $\mathbf{M}$ incorrectly predicts f's next m near future values. Good news is that, for a suitably clever general purpose learning machine $\mathbf{M}$, for each computable f, for $\mathbf{M}$ on f, the *density* of any such associated bad prediction intervals of size m is vanishingly small.

---

* This paper considerably extends and improves the previously unpublished Chapter 5 of [Che81], in particular answering all its open questions.

C. Meinel and S. Tison (Eds.): STACS'99, LNCS 1563, pp. 424–433, 1999.
© Springer-Verlag Berlin Heidelberg 1999

Considered too is the possibility of providing a general purpose learner which *additionally* learns some interesting classes with respect to much stricter criteria than **Bc***-identification. Again there is good news and bad. The criterion of *finite identification* requires for success that a learner **M** on a function f output exactly one program which correctly computes f. **Bc**^{n}-*identification* is just like **Bc***-identification above except that the number of anomalies in each program of a final sequence is $\leq n$. Bad news is that there is a finitely identifiable class of computable functions $\mathcal{C}$ such that for *no* general purpose learner **M** and for no n, does **M** *additionally* **Bc**^{n}-identify $\mathcal{C}$. **Ex**-*identification* by **M** on f requires that **M** on f converges, after a few output programs, to a *single* final program which computes f. A *reliable* learner (by definition) never deceives by false convergence; more precisely: whenever it converges to a final program on a function f, it must **Ex**-identify f. Good news is that, for any class $\mathcal{C}$ that can be *reliably* **Ex**-identified, there is a general purpose machine which *additionally* **Ex**-identifies $\mathcal{C}$!

1 Introduction

The learning situation often studied in inductive inference [OSW86, JORS98] may be described as follows. A learner receives as input, one at a time, the successive graph points of a function f. As the learner is receiving its input, it conjectures a sequence of programs as hypotheses. To be able to learn the function f, the sequence of programs conjectured by the learner must have some desirable relation to the input function f. By appropriately choosing this desirable relation one gets different criteria of successful learning. One of the first such criteria studied is called **Ex**-*identification* ([Gol67, BB75, CS78, CS83]). The learner is said to **Ex**-*identify a function* f iff the sequence of programs output by it on f, after a few output programs, converges to a single final program which computes f.[1] A learner is said to **Ex**-*identify a class* iff it **Ex**-identifies each function in the class. A *class of functions is* **Ex**-*identifiable* iff some machine **Ex**-identifies the class.

Even though one cannot **Ex**-identify the class of all the computable functions [Gol67], there are large and useful classes of functions which can be **Ex**-identified. For example, any recursively enumerable class of computable functions such as the class of polynomials or the class of primitive recursive functions [Rog67a] is **Ex**-identifiable.

[Bär74, CS78, CS83] considered a generalization of **Ex**-identification called **Bc**-identification. In **Bc**-identification of a function f by a machine **M** one requires that the sequence of programs output by **M** on f either converges to a program for f, or the sequence of programs is infinite, with all but finitely many of them being (possibly different) programs for f. [CS78, CS83] also considered the variants of the **Ex** and **Bc**-identification criteria in which the final programs need not be perfect, but are allowed to have some anomalies or mistakes in their predictions of I/O behavior. For n a natural number, if the final programs are

---

[1] In general more formal definitions are in Section 2 below.

allowed to make at most n errors, then the criteria of inference are called $\mathbf{Ex}^n$ and $\mathbf{Bc}^n$ respectively. If the final programs are allowed to make at most finitely many errors, then the criteria of inference are called $\mathbf{Ex}^*$ and $\mathbf{Bc}^*$ respectively.

Harrington [CS83] constructed a machine which $\mathbf{Bc}^*$-identifies each computable function! In the present paper, we call machines which do this *general purpose*. However, on infinitely many computable functions, the final programs output by Harrington's machine become more and more degenerate, i.e., the finite sets of anomalies in successive final output programs, in general, grow in size without bound. We note that this is a property of any general purpose learner, and, in fact, the number of anomalies grows faster than any computable bound (Theorem 2 in Section 3 below).

Since the programs output by any general purpose learning machine make large numbers of mistakes (on infinitely many computable functions), it would be interesting to study how these errors are distributed. For example, in real life one probably cares more about "near future errors" than "distant future errors". Based on this motivation in Section 4 below we define new criteria of inference called $\mathbf{Bc}^n_m$. Informally, for a machine to $\mathbf{Bc}^n_m$-*identify* a function f, for its final programs, their predictions on the next m inputs should have at most n errors. In Section 4 we completely resolve the relationship between different $\mathbf{Bc}^n_m$ criteria of inference (Corollary 3 in Section 4). In particular, we show that for any learning machine $\mathbf{M}$, (including general purpose $\mathbf{M}$), for all m, there exist infinitely many computable functions f such that, infinitely often $\mathbf{M}$ incorrectly predicts f's next m near future values (Corollary 4)! Thus there is an ostensibly unpleasant cost to general purpose learning. As we will see, though, this can be assuaged at least in some interesting respects described below.

In contrast to the result mentioned above that any general purpose learning machine $\mathbf{M}$ predicts next m values wrongly infinitely often, we show that the *density* of such bad prediction intervals can be made very small (Theorem 6 in Section 5 below).

A *reliable* learner (by definition) never deceives by false convergence; more precisely: whenever it converges to a final program on a function f, it must $\mathbf{Ex}$-identify f [Min76, BB75, CJNM94]. For example, r.e. classes of computable functions (such as the class of polynomial functions and the class of primitive recursive functions [Rog67a]) as well as the class of total run time functions can be reliably $\mathbf{Ex}$-identified [BB75, CS83]. On a further positive note, we show that for every reliably $\mathbf{Ex}$-identifiable class of computable functions $\mathcal{S}$, there is a general purpose learning machine which $\mathbf{Ex}$-identifies $\mathcal{S}$ (Theorem 7 in Section 5 below)!

The criterion of *finite identification* requires for success that a learner $\mathbf{M}$ on a function f output exactly one program which correctly computes f. Learning by finite identification can be thought of as *one-shot learning*. We show, by contrast to the result in the immediately above paragraph (Theorem 7), that there is a class $\mathcal{S}$ which *is* finitely identifiable, yet for all n, no general purpose learner can additionally $\mathbf{Bc}^n$-identify $\mathcal{S}$ (Corollary 5 in Section 5 below).

We now proceed formally.

2   Notation and Preliminaries

Recursion-theoretic concepts not explained below are treated in [Rog67b]. N denotes the set of natural numbers. $*$ denotes a non-member of N and is assumed to satisfy $(\forall n)[n < * < \infty]$. Let $\in, \subseteq, \subset, \supseteq, \supset$, respectively denote membership, subset, proper subset, superset and proper superset relations for sets. $\emptyset$ denotes the emptyset. $\text{card}(S)$ denotes the cardinality of set S. So "$\text{card}(S) \leq *$" means that $\text{card}(S)$ is finite. $\min(S)$ and $\max(S)$, respectively, denote the minimum and maximum element in S. We take $\min(\emptyset)$ to be ∞ and $\max(\emptyset)$ to be 0.

$\langle \cdot, \cdot \rangle$ denotes a 1-1 computable mapping from pairs of natural numbers onto natural numbers. π_1, π_2 are the corresponding projection functions. $\langle \cdot, \cdot \rangle$ is extended to n-tuples in a natural way.

Λ denotes the empty function. η, with or without decorations, ranges over partial functions. $\eta(x){\downarrow}$ denotes that $\eta(x)$ is defined. $\eta(x){\uparrow}$ denotes that $\eta(x)$ is not defined. For $a \in N \cup \{*\}$, $\eta_1 =^a \eta_2$ means that $\text{card}(\{x \mid \eta_1(x) \neq \eta_2(x)\}) \leq a$. $\eta_1 \neq^a \eta_2$ means that $\neg[\eta_1 =^a \eta_2]$. (If η_1 and η_2 are both undefined on input x, then, as is standard, we take $\eta_1(x) = \eta_2(x)$.) If $\eta =^a f$, then we often call a program for η as an a-error program for f. $\text{domain}(\eta)$ and $\text{range}(\eta)$ respectively denote the domain and range of the partial function η.

f, g and h, with or without decorations, range over total functions. $\mathcal{R}$ denotes the class of all *computable* functions, i.e., total computable functions with arguments and values from N. $\mathcal{C}$ and $\mathcal{S}$, with or without decorations, range over subsets of $\mathcal{R}$. φ denotes a *fixed* standard or so-called acceptable programming system [Rog58, Rog67a, Ric80, Ric81, Roy87]. φ_i denotes the partial computable function computed by program i in the φ-system. Note that in this paper all programs are interpreted with respect to the φ-system.

2.1   Function Identification

We first describe inductive inference machines. We assume, without loss of generality, that the graph of a function is fed to a machine in canonical order. For any partial function η and $n \in N$ such that, for all $x < n$, $\eta(x){\downarrow}$, we let $\eta[n]$ denote the finite initial segment $\{(x, \eta(x)) \mid x < n\}$. Clearly, $\eta[0]$ denotes the empty segment. We let Λ denote the empty segment. SEG denotes the set of all finite initial segments, $\{f[n] \mid f \in \mathcal{R} \wedge n \in N\}$. We let σ and τ, with or without decorations, range over SEG. Let $|\sigma|$ denote the length of σ. We often identify (partial) functions with their graphs. Thus for example, for $\sigma = f[n]$ and for $x < n$, $\sigma(x)$ denotes $f(x)$. A learning machine (also called an *inductive inference machine* (IIM)) [Gol67] is an algorithmic device that computes a mapping from SEG into $N \cup \{?\}$. Intuitively, "?" above denotes the case when the machine may not wish to make a conjecture. Although it is not necessary to consider learners that issue "?" for identification in the limit, it becomes useful when the number of mind changes a learner can make is bounded. In this paper, we assume, without loss of generality, that once an IIM has issued a conjecture on some initial segment of a function, it outputs a conjecture on all extensions of that initial segment. This is without loss of generality because a machine wishing to emit "?" after making a conjecture can instead be thought of as repeating its previous conjecture. We let **M**, with or without decorations, range over learning machines. Since the set of all finite initial segments, SEG, can be coded onto N, we can view these machines as taking natural numbers as input and emitting natural numbers or ?'s as output. We say that $\mathbf{M}(f)$ converges to i (written: $\mathbf{M}(f){\downarrow} = i$) iff $(\forall^\infty n)[\mathbf{M}(f[n]) = i]$; $\mathbf{M}(f)$ is undefined if no such i exists. The next definitions describe several criteria of function identification.

Definition 1. [Gol67, BB75, CS83] Let $a \in N \cup \{*\}$. Let $f \in \mathcal{R}$.
(a) $\mathbf{M}$ $\mathbf{Ex}^a$-*identifies* f (written: $f \in \mathbf{Ex}^a(\mathbf{M})$) just in case, there exists an i such that $\mathbf{M}(f){\downarrow} = i$ and $\varphi_i =^a f$.
(b) $\mathbf{M}$ $\mathbf{Ex}^a$-*identifies* S iff $\mathbf{M}$ $\mathbf{Ex}^a$-identifies each $f \in S$.
(c) $\mathbf{Ex}^a = \{S \subseteq \mathcal{R} \mid (\exists \mathbf{M})[S \subseteq \mathbf{Ex}^a(\mathbf{M})]\}$.

We often write $\mathbf{Ex}$ for $\mathbf{Ex}^0$.
By definition of convergence, only finitely many data points from a function f had been observed by an IIM $\mathbf{M}$ at the (unknown) point of convergence. Hence, some form of learning must take place in order for $\mathbf{M}$ to learn f. For this reason, hereafter the terms *identify*, *learn* and *infer* are used interchangeably.

Definition 2. [Bār74, CS83] Let $a \in N \cup \{*\}$. Let $f \in \mathcal{R}$.
(a) $\mathbf{M}$ $\mathbf{Bc}^a$-*identifies* f (written: $f \in \mathbf{Bc}^a(\mathbf{M})$) iff, for all but finitely many $n \in N$, $\varphi_{\mathbf{M}(f[n])} =^a f$.
(b) $\mathbf{M}$ $\mathbf{Bc}^a$-*identifies* S iff $\mathbf{M}$ $\mathbf{Bc}^a$-identifies each $f \in S$.
(c) $\mathbf{Bc}^a = \{S \subseteq \mathcal{R} \mid (\exists \mathbf{M})[S \subseteq \mathbf{Bc}^a(\mathbf{M})]\}$.

We often write $\mathbf{Bc}$ for $\mathbf{Bc}^0$.
Some relationships between the above criteria are summarized in the following theorem.

Theorem 1. [CS83, BB75, Bār71]
$$\mathbf{Ex}^0 \subset \mathbf{Ex}^1 \subset \cdots \subset \mathbf{Ex}^* \subset \mathbf{Bc} \subset \mathbf{Bc}^1 \subset \cdots \subset \mathbf{Bc}^* = 2^{\mathcal{R}}.$$

Since $\mathcal{R} \in \mathbf{Bc}^*$, we often call a machine which $\mathbf{Bc}^*$-identifies $\mathcal{R}$ a general purpose learning machine.
We let **I** range over identification criteria defined in this paper. There exists an r.e. sequence $\mathbf{M}_0, \mathbf{M}_1, \mathbf{M}_2, \ldots$, of inductive inference machines such that, for all criteria **I** of inference considered in this paper, one can show that [OSW86, JORS98]:

for all $\mathcal{C} \in \mathbf{I}$, there exists an $i \in N$ such that $\mathcal{C} \subseteq \mathbf{I}(\mathbf{M}_i)$.

We assume $\mathbf{M}_0, \mathbf{M}_1, \mathbf{M}_2, \ldots$ to be one such sequence of machines.

3   General Purpose Machines and Their Mistakes

Unfortunately, the programs output by Harrington's machine become more and more degenerate, i.e. the finite set of anomalies in final programs output grows in size without bound. In fact the finite sets of anomalies cannot even be bounded by a computable function as the next theorem shows.

Theorem 2. *Suppose $\mathcal{R} \subseteq \mathbf{Bc}^*(\mathbf{M})$. Let g be a computable function. Then there exist infinitely many $f \in \mathcal{R}$ such that, for infinitely many n, $\varphi_{\mathbf{M}(f[n])} \neq^{g(n)} f$.*

4 Predicting Near Future Values

Based on Theorem 2, it would be interesting to study how the anomalies of the programs outputted by a machine are distributed. For example, in real life one probably cares more about "near future errors" than "distant future errors". This leads us to the following definition.

Definition 3. M $\mathbf{Bc}_m^n$-*identifies* f (written: $f \in \mathbf{Bc}_m^n(\mathbf{M})$), iff, for all but finitely many x, $\mathrm{card}(\{z < m \mid \varphi_{\mathbf{M}(f[x])}(x + z) \neq f(x + z)\}) \leq n$.
M $\mathbf{Bc}_m^n$-*identifies* $\mathcal{C}$, if it $\mathbf{Bc}_m^n$-identifies each $f \in \mathcal{C}$.
$\mathbf{Bc}_m^n = \{\mathcal{C} \mid \text{some } \mathbf{M} \; \mathbf{Bc}_m^n\text{-identifies } \mathcal{C}\}$.

Intuitively, one can view $\mathbf{Bc}_m^n$-identification of a function by a machine as follows. At any stage, the learning machine predicts the next m values. At all but finitely many stages, at least $m - n$ out of the m predictions are correct.

In this section we resolve the relationship between different $\mathbf{Bc}_m^n$-identification criteria.

Following four propositions follow directly from the definitions.

Proposition 1. *For $m \geq n$, $\mathbf{Bc}^n \subseteq \mathbf{Bc}_m^n$.*

Proposition 2. *Suppose $m \geq n$ and $k \in N$. Then $\mathbf{Bc}_m^n \subseteq \mathbf{Bc}_{m+k}^{n+k}$.*

Proposition 3. *Suppose $m \geq n \geq k$. Then $\mathbf{Bc}_m^k \subseteq \mathbf{Bc}_m^n$.*

Proposition 4. *Suppose $m \geq k \geq n$. Then $\mathbf{Bc}_m^n \subseteq \mathbf{Bc}_k^n$.*

$\mathbf{NV}''$ defined by Podnieks [Pod74], is same as $\mathbf{Bc}_1^0$. The following proposition can thus be proved by using the equality $\mathbf{Bc} = \mathbf{NV}''$.

Proposition 5. *For all $m > 0$, $\mathbf{Bc} = \mathbf{Bc}_m^0$.*

The following theorem shows some advantages of having to predict fewer correct values in the near future.

Theorem 3. *Suppose $m' > m$. Then, $\mathbf{Bc}_m^1 - \mathbf{Bc}_{m'}^{m'-m} \neq \emptyset$.*

Proof. Let
$$Z_k = \{x \mid k \cdot m' < x < k \cdot m' + m\},$$
$$E_k = \{k \cdot m'\} \cup \{x \mid k \cdot m' + m \leq x < (k+1) \cdot m'\},$$
and
$$U_k = Z_k \cup E_k = \{x \mid k \cdot m' \leq x < (k+1) \cdot m'\}.$$

We now consider the following two properties defined on total functions.

(PropA) f satisfies PropA iff, for all k, for all $x \in Z_k$, $f(x) = 0$.
(PropB) f satisfies PropB iff for all k, for all $x \in E_k$, $f(x) = f(k \cdot m')$.

Let $C = \{f \in \mathcal{R} \mid f \text{ satisfies PropA and PropB}\}$.

The above class is easily seen to be in $\mathbf{Bc}_m^1$. Proof of $C \notin \mathbf{Bc}_{m'}^{m'-m}$ uses a complicated diagonalization argument and is omitted due to lack of space. $\blacksquare$

The next theorem shows some advantages of being allowed to predict more wrong values in the near future.

Theorem 4. $\mathbf{Bc}^{n+1} - \mathbf{Bc}_{n+1}^n \neq \emptyset$.

Proof. Let $Z_f = \{x \mid f((n+2) \cdot x) \neq 0\}$.

Let $C = \{f \mid [\text{card}(Z_f) = \infty \wedge (\forall^\infty x \in Z_f)[\varphi_{f((n+2) \cdot x)} =^{n+1} f]] \vee [0 < \text{card}(Z_f) < \infty \wedge \varphi_{f((n+2) \cdot \max(Z_f))} =^{n+1} f]\}$.

It is easy to verify that $C \in \mathbf{Bc}^{n+1}$. A diagonalization argument can be used to show that $C \notin \mathbf{Bc}_{n+1}^n$. We omit the details due to lack of space. $\blacksquare$

As a corollary to the above theorem, if one looks at the errors committed by a general purpose machine on the next n inputs, then for infinitely many functions, at infinitely many positions, the machine commits n errors in predicting the next n inputs. Hence, there is an ostensibly unpleasant cost to general purpose learning. However, as we shall see, this can be assuaged at least in some interesting respects (Theorems 6 and 7 in Section 5 below).

Corollary 1. *Suppose $m > n$ and $m' > n'$. If $n > n'$, then $\mathbf{Bc}_m^n - \mathbf{Bc}_{m'}^{n'} \neq \emptyset$.*

Corollary 2. *Suppose $m' > n'$ and $m > n > 0$. If $m' - n' > m - n$, then $\mathbf{Bc}_m^n - \mathbf{Bc}_{m'}^{n'} \neq \emptyset$.*

Proof. If $n' < n$, then corollary follows from Corollary 1. So suppose $n \leq n'$. Theorem 3 shows that $\mathbf{Bc}_{m-n+1}^1 - \mathbf{Bc}_{m'}^{m'-m+n-1} \neq \emptyset$ (note that $m' > m-n+1$). Now, $\mathbf{Bc}_m^{1+n-1} \supseteq \mathbf{Bc}_{m-n+1}^1$ (by Proposition 2), and $\mathbf{Bc}_{m'}^{n'} \subseteq \mathbf{Bc}_{m'}^{m'-m+n-1}$ (since $n' \leq m' - m + n - 1$, and by Proposition 3). Corollary follows. $\blacksquare$

The following corollary resolves all relationships among the $\mathbf{Bc}_m^n$-criteria.

Corollary 3. *Suppose $m > n$ and $m' > n'$. Then: $\mathbf{Bc}_m^n \subseteq \mathbf{Bc}_{m'}^{n'}$ iff $[n = 0$ or $[n' \geq n$ and $m' - n' \leq m - n]]$.*

Proof. If $n > n'$ then Corollary 1 shows that $\mathbf{Bc}_m^n - \mathbf{Bc}_{m'}^{n'} \neq \emptyset$. If $m' - n' \leq m - n$ and $n > 0$, then Corollary 2 shows that $\mathbf{Bc}_m^n - \mathbf{Bc}_{m'}^{n'} \neq \emptyset$.

If $n = 0$, then $\mathbf{Bc}_m^n = \mathbf{Bc} \subseteq \mathbf{Bc}_{m'}^{n'}$.

If $n' \geq n$ and $m' - n' \leq m - n$, then $\mathbf{Bc}_m^n \subseteq \mathbf{Bc}_{m+n'-n}^{n'}$ (by Proposition 2) and $\mathbf{Bc}_{m+n'-n}^{n'} \subseteq \mathbf{Bc}_{m'}^{n'}$ (by Proposition 4). Corollary follows. $\blacksquare$

Corollary 4. *For all $m > n$, $\mathcal{R} \notin \mathbf{Bc}_m^n$.*

Thus, no general purpose learning machine can guarantee that anomalies are not concentrated in the near future.

5   Desirable Properties Achievable by General Purpose Learners

Since the errors committed by programs output by a general purpose learner can be arbitrarily bad, we look at how this may be assuaged for suitable general purpose learners, and we also determine some additional nice properties a general purpose learner can satisfy.

On can think of a program for a computable function as a *predictive explanation* for the function's I/O behavior [BB75, CS83]. Popper's Refutability Principle [Pop68] essentially says that explanations with mistakes should be refutable. As pointed out in [CS83] (see also [CJNM94]), an erroneous predictive explanation (program) for a computable function satisfies Popper's Principle if it computes a total function.[2] The following theorem says that one can construct a general purpose learner which, on computable function input, almost always outputs programs for total functions; hence, it almost always outputs predictive explanations which satisfy Popper's Principle.

Theorem 5. *There exists a machine $\mathbf{M}$, such that, for all $f \in \mathcal{R}$, (i) $\mathbf{M}$ $\mathbf{Bc}^*$-identifies f, and (ii) $(\forall^\infty n)[\varphi_{\mathbf{M}(f[n])} \in \mathcal{R}]$.*

The following shows that even though a general purpose machine may be locally bad for infinitely many positions, one can ensure that these bad positions have low density.

Theorem 6. *For all n, there exists a machine $\mathbf{M}$ such that,*
(a) $\mathbf{M}$ $\mathbf{Bc}^*$-identifies $\mathcal{R}$, and
(b) for all $f \in \mathcal{R}$, $\lim_{x \to \infty} \frac{\mathrm{card}(\{k \mid k \leq x \ \wedge \ (\forall z \leq n)[\varphi_{\mathbf{M}(f[k])}(k+z) = f(k+z)]\})}{x+1} = 1$.

Since general purpose learners are always quite erroneous (of course the density of erroneous, near future intervals can be made small), it is interesting to consider which classes a general purpose learner may additionally identify in a better or stricter sense.

Definition 4. [Min76, BB75, CJNM94] $\mathbf{M}$ is said to be *reliable* iff, for all f such that $\mathbf{M}(f)\downarrow$, $\mathbf{M}$ **Ex**-identifies f.

$\mathbf{M}$ is said to *reliably* **Ex**-*identify* $\mathcal{C}$, iff $\mathbf{M}$ is reliable and $\mathbf{M}$ **Ex**-identifies $\mathcal{C}$.
$\mathbf{RelEx} = \{\mathcal{C} \mid \text{some machine reliably } \mathbf{Ex}\text{-identifies } \mathcal{C}\}$.

---

[2] Then the halting problem [Rog67a] does not stand in the way of algorithmically locating the mistakes.

Intuitively, reliable machines do not deceive us by converging falsely. As noted above, r.e. classes of computable functions (such as the class of polynomial functions and the class of primitive recursive functions) as well as the class of total run time functions can be reliably **Ex**-identified.

The following theorem shows that for any class $\mathcal{S}$ in **RelEx**, one can create a general purpose learning machine which **Ex**-identifies $\mathcal{S}$!

Theorem 7. *Suppose* $\mathcal{S} \in$ **RelEx**. *Then there exists an* **M** *such that* **M** **Bc***-identifies* $\mathcal{R}$ *and* **Ex**-*identifies* $\mathcal{S}$.

On the other hand, the following corollary to the proof of Theorem 2 shows that **RelEx** cannot be replaced by finite identification [CS83].

Let $\mathcal{S}_0 = \{f \mid \varphi_{f(0)} = f\}$.

Proof of Theorem 2 can be used to show the following:

Corollary 5. *Suppose* $n \in N$ *and* **M** **Bc***-identifies* $\mathcal{R}$. *Then there exists an* $f \in \mathcal{S}_0$ *such that* **M** *does not* **Bc**^{n}-*identify* f.

It would be interesting to study what other useful properties a suitable general purpose learner can be made to satisfy.

6   Conclusions

Harrington [CS83] surprisingly constructed a general purpose learner, i.e., a machine which **Bc***-identifies all the computable functions. However, the programs output by Harrington's machine become more and more degenerate, i.e., in general, the finite set of anomalies in each final program grows without bound. In this paper we showed that this is unavoidable (Theorem 2 above).

Since the programs output by any general purpose learning machine make large number of errors on infinitely many functions, it is interesting to study how these errors are or can be distributed. Based on this motivation we defined new criteria of inference called **Bc**^n_m, and completely resolved the relationship between different **Bc**^n_m criteria of inference. Among other results, we showed that any general purpose learning machine is poor in predicting near future values. In particular any general purpose learning machine **M** predicts the next n values wrongly infinitely often. In contrast, though, we show that the density of such bad prediction points can be made vanishingly small (Theorem 6 above).

We constructed a general purpose learning machine **M** such that, on any computable function input, all but finitely many of the programs output by **M** are for total functions. Hence, almost all of its conjectures satisfy Popper's Refutability Principle.

We also showed that for every class of computable functions, $\mathcal{S}$, which can be **Ex**-identified by a reliable machine [Min76, BB75, CJNM94] (see definition in Section 5 above), some general purpose learning machine additionally **Ex**-identifies $\mathcal{S}$. We further show, though, that reliable identification in the just above statement cannot be replaced by finite identification.

It would be interesting to study which other useful properties a general purpose learner can or cannot have.

References

[Bär71]   J. Bārzdiņš. *Complexity and Frequency Solution of Some Algorithmically Unsolvable Problems.* PhD thesis, Novosibirsk State University, 1971. In Russian.

[Bär74]   J. Bārzdiņš. Two theorems on the limiting synthesis of functions. In *Theory of Algorithms and Programs, vol. 1*, pages 82–88. Latvian State University, 1974. In Russian.

[BB75]   L. Blum and M. Blum. Toward a mathematical theory of inductive inference. *Information and Control*, 28:125–155, 1975.

[Blu67]   M. Blum. A machine-independent theory of the complexity of recursive functions. *Journal of the ACM*, 14:322–336, 1967.

[Che81]   K. J. Chen. *Tradeoffs in Machine Inductive Inference.* PhD thesis, SUNY/Buffalo, 1981.

[CJNM94]   J. Case, S. Jain, and S. Ngo Manguelle. Refinements of inductive inference by Popperian and reliable machines. *Kybernetika*, 30:23–52, 1994.

[CS78]   J. Case and C. Smith. Anomaly hierarchies of mechanized inductive inference. In *Symposium on the Theory of Computation*, pages 314–319, 1978.

[CS83]   J. Case and C. Smith. Comparison of identification criteria for machine inductive inference. *Theoretical Computer Science*, 25:193–220, 1983.

[Gol67]   E. M. Gold. Language identification in the limit. *Information and Control*, 10:447–474, 1967.

[JORS98]   S. Jain, D. Osherson, J. Royer, and A. Sharma. *Systems that Learn: An Introduction to Learning Theory.* MIT Press, Cambridge, Mass., second edition edition, 1998. To appear.

[Min76]   E. Minicozzi. Some natural properties of strong identification in inductive inference. *Theoretical Computer Science*, pages 345–360, 1976.

[OSW86]   D. Osherson, M. Stob, and S. Weinstein. *Systems that Learn: An Introduction to Learning Theory for Cognitive and Computer Scientists.* MIT Press, 1986.

[Pod74]   K. Podnieks. Comparing various concepts of function prediction, Part I. In *Theory of Algorithms and Programs, vol. 1*, pages 68–81. Latvian State University, Riga, Latvia, 1974.

[Pop68]   K. Popper. *The Logic of Scientific Discovery.* Harper Torch Books, New York, second edition, 1968.

[Ric80]   G. Riccardi. *The Independence of Control Structures in Abstract Programming Systems.* PhD thesis, SUNY/Buffalo, 1980.

[Ric81]   G. Riccardi. The independence of control structures in abstract programming systems. *Journal of Computer and System Sciences*, 22:107–143, 1981.

[Rog58]   H. Rogers. Gödel numberings of partial recursive functions. *Journal of Symbolic Logic*, 23:331–341, 1958.

[Rog67a]   H. Rogers. *Theory of Recursive Functions and Effective Computability.* McGraw-Hill, 1967. Reprinted, MIT Press 1987.

[Rog67b]   H. Rogers. *Theory of Recursive Functions and Effective Computability.* McGraw-Hill, 1967. Reprinted by MIT Press in 1987.

[Roy87]   J. Royer. *A Connotational Theory of Program Structure*, volume 273 of *Lecture Notes in Computer Science*. Springer-Verlag, 1987.

Universal Distributions and Time-Bounded Kolmogorov Complexity

Rainer Schuler

Abt. Theoretische Informatik, Universität Ulm, D-89069 Ulm, Germany

Abstract. The equivalence of the universal distribution, the a priori probability and the negative exponential of Kolmogorov complexity is a well known result. The natural analogs of Kolmogorov complexity and of a priori probability in the time-bounded setting are not efficiently computable under reasonable assumptions. In contrast, it is known that for every polynomial p, distributions universal for the class of p-time computable distributions can be computed in polynomial time. We show that in the time-bounded setting the universal distribution gives rise to sensible notions of Kolmogorov complexity and of a priori probability.

1 Introduction

An universal distribution can be considered as representing the whole class of distributions, since for any distribution, the probability of an instance x is bounded (up to some constant factor) by the probability of x with respect to the universal distribution. To make this more precise consider the following two examples. In [7], Li and Vitányi show that any concept class that is PAC-learnable with respect to the universal distribution, is PAC-learnable with respect to any simple (i.e. r.e.) distribution provided the examples (in the learning phase) are given with respect to the universal distribution. A similar result holds if we consider universal distributions and PAC-learning with respect to the class of t-computable distributions.

As a second example observe that the average-case complexity (of any algorithm) with respect to the universal distribution gives an upper bound on the average-case complexity with respect to any distribution of the class. In fact [8] (see also [9]), in the unbounded case, worst-case complexity and average-case complexity with respect to the universal distribution coincide. (As an average-case complexity one might consider the expected value w.r.t. the conditional probability restricted to strings of length n).

In the unbounded case, an enumeration $\mu_1, \mu_2, \cdots$ of the r.e. semimeasures can be used to define an universal distribution (semimeasure) m (L.A.Levin cf. [7]). For all $x \in \Sigma^*$ let

$$m(x) = \sum_n \alpha(n) \cdot \mu_n(x), \qquad (1)$$

where α is some function such that $\sum_n \alpha(n) \leq 1$, e.g., $\alpha(n) = 1/n(n+1)$. However, an universal distribution can be defined equivalently using the algorithmic or the a priori probability on the strings [6].

C. Meinel and S. Tison (Eds.): STACS'99, LNCS 1563, pp. 434–443, 1999.
© Springer-Verlag Berlin Heidelberg 1999

Let U be a universal prefix Turing machine. (M is a *prefix* Turing machine if for all x and y, if $M(x)$ stops then M on input xy stops without reading any bit of y). The a priori probability Q_U is defined by $Q_U(x) = \sum_p 2^{-|p|}$, where p ranges over all strings such that $U(p) = x$ but $U(p')$ does not stop for any prefix p' of p. The algorithmic probability R of an instance x is defined by $R(x) = 2^{-K(x)}$, where $K(x)$ is the prefix Kolmogorov complexity of x, i.e., $K(x) = \min_l \{\exists p \in \Sigma^l : U(p) \text{ halts with output } x\}$.

Theorem 1. *(Levin cf. [7]) For some constant c and all strings x, $-\log m(x) = -\log Q_U(x) = -\log R(x) = K(x)$ up to an additive constant less than c.*

Recently, it has been shown that for some polynomial p and any nondecreasing time-constructible function t, there exists an enumeration $\mu_1, \mu_2, \cdots$ of $p(t)$-computable semimeasures containing all t-computable semimeasures [10]. Hence using Equation (1), a semimeasure m_t universal for the class of t-computable distributions is computable in time polynomial in t. In this paper we consider the question of whether equivalent definitions in terms of time-bounded a priori and time-bounded algorithmic probability exist.

First we note in Proposition 1 that the time-bounded prefix Kolmogorov complexity of a string x is not computable in time polynomial in the time-bound unless no polynomial-time computable pseudo-random generator exist (for definitions see next section). We then propose a restriction on Turing machines and show that for this model universal Turing machines exist. For a function t let Q_{U_t} and R_t denote the time bounded version of a priori and algorithmic probability (using this machine model). Then we can show the following theorem.

Theorem 2. *Let t be a nondecreasing time-constructible function. There is a constant c, polynomials p_1, p_2, and p_3 such that for all strings x, $-\log m_t(x) \leq -\log Q_{U_{p_1(t)}}(x) \leq -\log R_{p_2(t)}(x) \leq -\log m_{p_3(t)}(x)$ up to an additive constant less than c.*

In this paper we follow the notation and definitions given in [6] in particular Chapter 4 and Section 7.6.

2   Preliminaries

We consider distributions on the sample space Σ^*, where $\Sigma = \{0, 1\}$. More formally, a *semimeasure* is defined by a function μ from Σ^* to the real interval $[0, 1]$ such that $\sum_{x \in \Sigma^*} \mu(x) \leq 1$. If the sum converges to 1 then μ defines a *distribution* (or probability measure). The function μ is called density function, the distribution function μ^* is defined by $\mu^*(x) = \sum_{y \leq x} \mu(y)$ for all x.

A distribution (semimeasure) μ *dominates* a distribution ν, if $c \cdot \mu(x) \geq \nu(x)$ for some constant c and all $x \in \Sigma^*$. Let $\mathcal{F}$ be a class of distributions and h be a function from $\Sigma^* \to \mathbb{N}$. A distribution μ is *h-universal for $\mathcal{F}$* if $h \cdot \mu$ dominates every distribution ν in $\mathcal{F}$. If h is a constant then μ is *universal for $\mathcal{F}$*.

We will be considering two classes of distributions. A distribution μ is called *simple* [7] if it is dominated by a recursive enumerable semimeasure ν. (That is, the set $\{(x, p, q) \mid p/q \leq \nu(x)\}$ is r.e.). Let t be a nondecreasing time-constructible function and μ be a distribution. μ is called t-*computable* if the distribution function μ^* is computable in time t, i.e., $\mu^*(x) = f(x)/g(x)$ for some functions f, g computable in time t. μ is *polynomial-time computable* if it is p-time computable for some polynomial p. Note that, following [5,4,1], we require that the distribution μ^* function is t-computable (instead of the density function μ). In fact, if μ^* is polynomial-time computable then μ is polynomial-time computable, the converse is not true unless $\mathcal{P} = \mathcal{NP}$ [4].

We will be using three different notion of ordering of strings. A string x is lexicographically smaller than a string y, denoted by $x < y$, if $|x| < |y|$ or $|x| = |y|$ and $z0$ is a prefix of x and $z1$ is a prefix of y for some z. For a string $x = x_{k-1} \ldots x_0$, $x_i \in \{0, 1\}$, let $0.x$ denote the rational number $\sum_{i \leq k-1} x_i \cdot 2^i$. Then the value of x is smaller than the value of y if $0.x < 0.y$. For a string x let $\mathrm{Left}(x)$ be defined as follows.

$$\mathrm{Left}(x) = \{y \mid \exists z : z0 \text{ is a prefix of } x \text{ and } z1 \text{ is a prefix of } v\}$$

E.g., 010 is not in $\mathrm{Left}(0101)$ but 0100 is in $\mathrm{Left}(0101)$.

2.1   Time-Bounded Prefix Complexity

First let us note that a universal distribution defined using the time-bounded version of the prefix Kolmogorov complexity is not likely to be computable in time polynomial in the time bound. Let t be a nondecreasing time-constructible function. The t-time bounded prefix Kolmogorov complexity $\mathrm{K}^t(x)$ of a string x is defined as follows. Let U be a universal prefix Turing machine, then

$$\mathrm{K}^t(x) = \min_l \{\exists p \in \Sigma^l : U(p) \text{ halts with output } x \text{ within } \leq t(|x|) \text{ steps}\}.$$

The semimeasure $m_{\langle t \rangle}$ is defined as follows [7]. For all x, $m_{\langle t \rangle}(x) = 2^{-\mathrm{K}^{t'}(x)}$, where $t'(n) = t(n) \cdot \log t(n)$ for all n. Recall that for all p and p', if $U(p)$ stops and p is a prefix of p' then $U(p) = U(p')$. Hence it follows from König's Lemma that $\sum_x m_{\langle t \rangle}(x) \leq \sum_p 2^{-|p|} \leq 1$, where p ranges over all strings such that $U(p)$ stops in time $t(|U(p)|)$ but $U(p')$ does not stop for any prefix p' of p. Therefore, $m_{\langle t \rangle}$ defines indeed a semimeasure.

Theorem 3. *[7] Let t, $t(n) > n$, be a nondecreasing time-constructible function. The distribution $m_{\langle t \rangle}$ is universal for the class of t-computable distributions.*

From the definition of the time-bounded prefix Kolmogorov complexity it follows that $m_{\langle t \rangle}$ is computable in time $O(nt(n)2^n)$ and hence $m_{\langle t \rangle}^*$ is computable in time $O(nt(n)2^{2n})$. It is not difficult to see that under the assumption that $\mathcal{FP} = \#\mathcal{P}$, $m_{\langle t \rangle}^*$ is computable in time polynomial in n and t. However, without any (unproven) assumption, it is not known whether $m_{\langle t \rangle}$ is computable in time polynomial in n and t [7]. The next proposition shows that this is unlikely, since it is generally believed that polynomial-time computable pseudo random generators exist.

A function $G : \Sigma^n \to \Sigma^{2n}$ is a pseudo random generator [3], if for every set $A \in \mathcal{P}$ and infinitely many n $\|\mathrm{Pr}_{x \in \Sigma^{2n}}[A(x) = 1] - \mathrm{Pr}_{x \in \Sigma^n}[A(G(x)) = 1]\| \leq \frac{1}{n}$.

Proposition 1. *If for every polynomial t the distribution $m_{\langle t \rangle}$ is dominated by a polynomial-time computable distribution, then no polynomial-time computable pseudo random generator exists.*

Proof. Let $G : \Sigma^n \to \Sigma^{2n}$ be polynomial-time computable. Choose a constant k such that G is computable in time $O(n^k)$. Let $t(n) = n^{k+1}$ for all n, and assume that a polynomial-time computable distribution μ dominates $m_{\langle t \rangle}$. Choose a constant c such that $c \cdot \mu(x) \geq m_{\langle t \rangle}(x) = 2^{-K^{t'}(x)}$ for all x, where $t'(n) = t(n) \cdot \log t(n)$ for all n. This shows that for all x, $K^{t'}(x) \geq -\log \mu(x) - \log c$.

Let $A = \{x \mid \mu(x) \leq 2^{-(2/3)|x|}\}$. Observe that $A \in \mathcal{P}$ and for every strings x in A, $K^{t'}(x) \geq -\log \mu(x) - \log c \geq (2/3)|x| - \log c$. On the other hand, the strings generated by G have small $K^{t'}$ complexity which can be seen as follows. If x is the output of G on input y for some y, then x can be generated by U from a self-delimiting description of G followed by a self-delimiting description of y. Since U can be simulated in time $t(n) = n^{k+1}$ there exists a constant d independent of x, such that $K^{t'}(x) \leq |y| + 2\log|y| + d = |x|/2 + \log|x| + d$. Hence, there exists a constant n_0 such that $G(y) \notin A$ for all y, $|y| \geq n_0$.

Since $\sum_{x,|x|=n} \mu(x) \leq 1$, it follows that $\|A^{=n}\| \geq 2^n - 2^{2n/3}$. (Assume otherwise, then $\sum_{x,|x|=n} \mu(x) \geq \sum_{x \in \overline{A^{=n}}} \mu(x) > 2^{2n/3} \cdot 2^{-2n/3} = 1$). Hence for all $n > n_0$, $\|\mathrm{Pr}_{x \in \Sigma^n}[A(x) = 1] - \mathrm{Pr}_{x \in \Sigma^{n/2}}[A(G(x)) = 1]\| \geq \frac{2^n - 2^{2n/3}}{2^n} - 0 \geq 1 - 2^{-n/3}$. Therefore G is not a pseudo random generator.

2.2 Shannon-Fano Codes

A semimeasure μ defines a prefix code E, called Shannon-Fano code, on strings as follows [7]. Here and in the following we assume w.l.o.g. that $\mu(x) > 0$ for all x. For example, let μ be any distribution and let μ_{st} be defined by $\mu_{\mathrm{st}}(\lambda) = 1/2$ and $\mu_{\mathrm{st}}(x) = (2|x|(|x| + 1)2^{|x|})^{-1}$ for all $x \neq \lambda$. Then $\nu(x) = (\mu(x) + \mu_{\mathrm{st}}(x))/2$ dominates μ and $\nu(x) > 0$ for all $x \neq \lambda$.

Consider the interval $[0, 1)$ partitioned into disjoint half open intervals $I(x)$ of size $\mu(x)$ as follows. $I(x) = [\mu^*(x^-), \mu^*(x))$, where x^- is the predecessor of x. Recall that $\mu^*(x) = \sum_{y \leq x} \mu(x)$, i.e., $\mu(x) = \mu^*(x) - \mu^*(x^-)$. Let p be a string. Then $[0.p, 0.p + 2^{-|p|}]$ is the binary interval defined by p. Recall that we use $0.p$ to denote the dyadic rational number defined by $p_{k-1} \cdot 2^{k-1} + \cdots + p_0 \cdot 2^0$, where $p = p_{k-1} \cdots p_0$, $p_i \in \{0, 1\}$. Let I be any interval and l be the size of a largest binary interval contained in I. Then I is covered by at most 4 intervals of size l.

The *Shannon-Fano code* of x, $E(x)$, is the lexicographically first string p which defines a largest binary interval which is contained in $I(x)$. The code E has the following properties:

- E is a prefix code since for all p', if p is a prefix of p' then the binary interval defined by p' is contained in the binary interval defined by p.
- E is ordered, i.e., for all x and x', if $x \leq x'$ then $E(x) \in \mathrm{Left}(E(x'))$.
- for all x, $-\log(\mu(x)) \leq |E(x)| \leq -\log(\mu(x)) + 2$, since $\mu(x) \leq 4 \cdot 2^{-|E(x)|}$ and $2^{-|E(x)|} \leq \mu(x)$.

Let the inverse of the Shannon-Fano code E induced by μ, denoted by D, be defined as follows. For all p, $D(p) = x$, if $E(x)$ is a prefix of p or $(E(x^-) \in \text{Left}(p)$ and $p \in \text{Left}(E(x)))$, and $D(p)$ is undefined otherwise. That is, $D(p) = x$ for any string p such that the binary interval defined by p is contained in the interval $(E(x^-), E(x)]$.

Theorem 4. *[7] Assume that μ is t-computable. Then for some polynomial p, the Shannon-Fano code E and the inverse D of the Shannon-Fano code induced by μ are computable in time $p(t)$.*

Obviously, if μ is universal for the class of t-computable distributions then the Shannon-Fano code E_μ induced by μ is optimal in the following sense. For every t-computable distribution ν there exists a constant c such that for all x, $|E_\mu(x)| \leq |E_\nu(x)| + c$, where E_ν is the Shannon-Fano code induced by ν. In the following we discuss the properties of a Turing machine which, given $E(x)$, computes $x = D(E(x))$ and show that universal Turing machines exist.

3   Ordered Time-Bounded Prefix Complexity

In this section we propose a stronger version of time bounded Kolmogorov complexity and show that universal Turing machines exist in this model.

Recall the definition of the left set of a string u: $\text{Left}(u) = \{v \mid \exists z : z0$ is a prefix of v and $z1$ is a prefix of $u\}$. The relation (defined by) Left is a partial ordering on strings. Furthermore, for any set of prefix free strings, Left is a total ordering (i.e., if u is not a prefix of v and v is not a prefix of u, then $u \in \text{Left}(v)$ or $v \in \text{Left}(u)$).

Let t be a nondecreasing time-constructible function from $\Sigma^* \to \mathbb{N}$ and M be a prefix Turing machine. M is called t-*time nondecreasing* if for all u and v,

1. if $M(u)$ stops then the number of steps of $M(u)$ is bounded by $t(|M(u)|)$.
2. if $M(u)$ stops, $v \in \text{Left}(u)$, and $|v| \geq t(|u|)$ then $M(v)$ stops.
3. if $M(u)$ and $M(v)$ stop and $v \in \text{Left}(u)$ then $M(v) \leq M(u)$

Let M be a t-time nondecreasing prefix Turing machine. The nondecreasing prefix Kolmogorov complexity (Knd-complexity) of a string x w.r.t. M, in symbols $\text{Knd}_M(x)$, is defined as follows. For all x,
$$\text{Knd}_M(x) = \min_l \{\exists u \in \Sigma^l : M(u) \text{ halts with output } x\}.$$

Before we proceed to consider the existence of a universal Turing machine, we show that any string x has linear-time Knd-complexity of length $|x| + 2\log|x| + c$ for some constant c.

Proposition 2. *1. There exists a constant c and a linear-time nondecreasing prefix Turing machine M such that $\text{Knd}_M(x) \leq |x| + 2\log|x| + c$ for all $x \neq \lambda$.*

2. Let c be a constant and t, $t(n) \geq n$, be a nondecreasing time-constructible function. There exists a constant d, a polynomial p and a $p(t)$-time nondecreasing prefix Turing machine N such that for all x, if $K^t(x) \leq c \cdot \log|x|$ then $\text{Knd}_N(x) \leq d\log|x|$.

Proof. (1.) Consider the following coding of strings. For all $x \neq \lambda$ let $\tilde{x} = 1^{\log|x|}0s(|x|)x$, where $s(|x|) = z_{n-1}\cdots z_0$, $z_i \in \{0,1\}$, such that $z_{n-1} = 1$ and $|x| = z_{n-1}2^{n-1} + \cdots + z_0 2^0$. First we show that the coding is increasing. Given $y < x$, we distinguish three cases.

Assume that $\log|y| < \log|x|$. Then
$$\tilde{y} = 1^{\log|y|}0s(|y|)y \in \text{Left}(1^{\log|y|}1) \subseteq \text{Left}(1^{\log|x|}) \subseteq \text{Left}(\tilde{x}).$$
Assume that $\log|y| = \log|x|$ and $s(|y|) < s(|x|)$. Then
$$\tilde{y} \in \text{Left}(1^{\log|y|}0s(|y|+1)) \subseteq \text{Left}(1^{\log|x|}0s(|x|)) \subseteq \text{Left}(\tilde{x}).$$
Assume that $\log|y| = \log|x|$ and $s(|y|) = s(|x|)$ and $y < x$. Then
$$\tilde{y} \in \text{Left}(1^{\log|y|}0s(|y|)x) = \text{Left}(1^{\log|x|}0s(|x|)x) = \text{Left}(\tilde{x}).$$

It remains to show that there exists a linear-time nondecreasing Turing machine M computing x from $\tilde{x}$. M can be defined as follows.

On input y, if $y \in 1^*$ then output "?". Otherwise, choose k and z such that y is a prefix of $1^k 0z$. If $|z| \leq k + 2^{k-1}$ then output "?". Let $z = b_{k-1}\cdots b_0 x_0 \cdots x_m$. If $z_{k-1} = 0$ then output $0^{2^{k-1}}$. Otherwise, let $n = \sum_{i=0}^{k-1} b_i 2^i$. If $n \geq m$ output $x_0 \cdots x_n$ otherwise output "?".

(2.) Let $K[c \cdot \log n, t]$ denote the set of all strings x such that $K^t(x) \leq c \cdot \log|x|$. Given m, the elements in $K[c \cdot \log n, t] \cap \Sigma^{\leq m}$ can be enumerated in time polynomial in m and t. Therefore, there exists a polynomial-time nondecreasing prefix machine N which on input $\tilde{x}$ outputs the xth string in $K[c \cdot \log n, t]$.

In the following we show that for the Knd-complexity there is a reasonable notion of a universal Turing machine. Furthermore, (in contrast to what we know about time-bounded Kolmogorov complexity) given a string x it is possible to efficiently compute its Knd-complexity.

In general it is not decidable whether a given Turing machine is (t-time) nondecreasing, however it is possible to augment the computation of a prefix Turing machine to ensure that the machine is t time nondecreasing.

Lemma 1. *Let t, $t(n) \geq n$, be a nondecreasing time-constructible function. There exists a prefix Turing machine N such that*

1. *for every fixed string e, $N(e,p)$ is $O(n^2 t(n)^2 \log t(n))$ time nondecreasing,*
2. *for every t-time nondecreasing Turing machine M, there exists an index e such that for all p, if $M(p)$ stops then $N(e,p) = M(p)$.*

Proof. Let $M_1, M_2, \ldots$ be an effective enumeration of all prefix Turing machines. Consider a machine Minv defined as follows.

```
Minv(M, x)
    let n = |x|
    if M(0^{t(n)}) does not stop within t(n) steps then output λ
    v = λ
    for i = 1 to t(n) do
        if M(v10^{t(l)-i}) does not stop within t(n) steps then v = v0
        else if M(v10^{t(n)-i}) > x then v = v0 else v = v1.
    output v
```

Assume that M_e is t-time nondecreasing. In this case, $\mathrm{M}inv(M_e, x)$ computes (using binary search) a string v_x (of length $t(|x|)$) such that $M_e(v_x) \leq x$ and $M_e(u) > x$ for all u such that $v_x \in \mathrm{Left}(u)$ and $M_e(u)$ stops.

Now let M_e be an arbitrary prefix machine such that $M_e(0^k)$ stops for some k. Then for all x, $|x| \geq k$, $\mathrm{M}inv(M_e, x)$ is a string of length $t(|x|)$ such that $M_e(x)$ stops. Furthermore, for any y, if $|y| = |x|$ and $x < y$ then $\mathrm{M}inv(M_e, x)$ is equal to $\mathrm{M}inv(M_e, y)$ or $\mathrm{M}inv(M_e, x)$ is in $\mathrm{Left}(\mathrm{M}inv(M_e, y))$.

Using the procedure $\mathrm{M}inv$, we define M_e^{inv} a "inverse" of M_e in two steps as follows. In the first step, for $l = 1, 2, \cdots, |x|$ a string v_{1^l} is computed such that v_{1^l} is the maximal string v of length $t(l)$ such that $M_e(v)$ halts and $M_e(v) \leq 1^l$. If M_e is not t-time nondecreasing, then we make sure that for all l, v_{1^l} is a prefix of $v_{1^{l+1}}$ or $v_{1^l} \in \mathrm{Left}(v_{1^{l+1}})$.

input x, $|x| = n$
    for $l = 1$ to $l = n$ let $v_{1^l} = \mathrm{M}inv(M_e, 1^l)$
    for $l = 1$ to $l = n - 1$ if $v_l \notin \mathrm{Left}(v_{l+1})$ then $v_{l+1} = v_l 1^{t(l+1) - t(l)}$

In the second part, we use binary search (within $\Sigma^{|x|}$) to find a string v_x such that $M_e(v_x) \leq x$ and $x < M_e(u)$ for all u such that $v_x \in \mathrm{Left}(u)$ and $M_e(u)$ stops.

Let $v_{0^n} = \mathrm{M}inv(M_e, 0^n)$. If $v_{1^{n-1}} \notin \mathrm{Left}(v_{0^n})$ then $v_{0^n} = v_{1^{n-1}} 1^{t(n) - t(n-1)}$. For all $1 \leq i \leq n$ let

$$v_{x_1 \cdots x_i 0^{n-i}} = \begin{cases} v_{x_1 \cdots x_{i-1} 0^{n-i+1}}, & \text{if } \mathrm{M}inv(M_e, x_1 \cdots x_i 0^{n-i}) < v_{x_1 \cdots x_{i-1} 0^{n-i+1}} \\ v_{x_1 \cdots x_{i-1} 1^{n-i+1}}, & \text{if } \mathrm{M}inv(M_e, x_1 \cdots x_i 0^{n-i}) > v_{x_1 \cdots x_{i-1} 1^{n-i+1}} \\ \mathrm{M}inv(M_e, x_1 \cdots x_i 0^{n-i}), & \text{otherwise.} \end{cases}$$

$$v_{x_1 \cdots x_i 1^{n-i}} = \begin{cases} v_{x_1 \cdots x_i 0^{n-i}}, & \text{if } \mathrm{M}inv(M_e, x_1 \cdots x_i 1^{n-i}) < v_{x_1 \cdots x_i 0^{n-i}} \\ v_{x_1 \cdots x_{i-1} 1^{n-i+1}}, & \text{if } \mathrm{M}inv(M_e, x_1 \cdots x_i 1^{n-i}) > v_{x_1 \cdots x_{i-1} 1^{n-i+1}} \\ \mathrm{M}inv(M_e, x_1 \cdots x_i 1^{n-i}), & \text{otherwise.} \end{cases}$$

Note that by definition $v_{0^n} \leq M_e^{inv}(x) \leq v_{1^n}$. Furthermore, for all $k < n$, $v_{x_1 \cdots x_k 0^{n-k}}$ and $v_{x_1 \cdots x_k 1^{n-k}}$ is defined such that $v_{x_1 \cdots x_k 0^{n-k}}$ ($v_{x_1 \cdots x_k 1^{n-k}}$) is the maximal string v of length $t(n)$ such that $M_e(v)$ halts and $M_e(v) \leq x_1 \cdots x_k 0^{n-k}$ ($M_e(v) \leq x_1 \cdots x_k 1^{n-k}$). Furthermore, even if M_e is not t-time nondecreasing we make sure that $v_{x_1 \cdots x_k 0^{n-k}} \leq v_{x_1 \cdots x_k 1^{n-k}}$, for all $k < n$.

The Turing machine $N : \Sigma^* \times \Sigma^* \to \Sigma^*$ can now be defined as follows. On input (e, v) compute a string x such that $M_e^{inv}(x^-) \in \mathrm{Left}(v)$ but $M_e^{inv}(x) \notin \mathrm{Left}(v)$. If $v \in \mathrm{Left}(M_e^{inv}(x))$ or $M_e^{inv}(x)$ is a prefix of v then output x. Again, assume that M_e is t-time nondecreasing. In this case, if M_e stops on input u and if v is a prefix of u then $M_e(u) = x$.

input (e, v)
    $m = 0$
    while $M_e^{inv}(1^m) \in \mathrm{Left}(v)$ do $m = m + 1$
    if $M_e^{inv}(0^m) \notin \mathrm{Left}(v)$ then $y = 1^{m-1}$
    else

$y = \lambda$
for $l = 1$ to $m - 1$
  if $M_e^{inv}(y10^{m-l}) \notin \text{Left}(v)$ then $y = y0$ else $y = y1$
let x be the successor of y
if $v \in \text{Left}(M_e^{inv}(x))$ or $M_e^{inv}(x)$ is a prefix of v then output x

It remains to verify the time complexity of N. Observe that for every e, M_e^{inv} is computable in time $O(nt(n)^2)$. Since M_e^{inv} is simulated $O(n)$ times (where n is the length of the output) on strings of size at most n, the computation of N is $O(n^2 t(n)^2 \log t(n))$ time bounded.

A Turing machine U is called t-*time nondecreasing universal*, if U is nondecreasing t'-time bounded for a function t' (which should be not much larger than t) and for every nondecreasing t-time bounded Turing machine M there is a constant c such that $\text{Knd}_U(x) \leq \text{Knd}_M(x) + c$ for all x. We show that t-time nondecreasing universal Turing machines exist. Furthermore, the universal Turing machine is t'-time nondecreasing for some function t' polynomial in t.

Theorem 5. *Let* t, $t(n) \geq n$, *be a nondecreasing time-constructible function. There exists a* $n^4 \cdot t^2(n) \log t(n)$-*time nondecreasing Turing machine which is* t-*time nondecreasing universal.*

Proof. For all x and for all e let $q_e(x) = \max_{p,|p|=t(|x|)}\{N(e,p) \leq x\}$, i.e., $q_e(x) = M_e^{inv}(x)$, where $N(e,p)$ and M_e^{inv} are defined in Lemma 1. Recall that we identify a string x with a (dyadic) rational number denoted by $0.x$. For all x, $|x| = n$, let $q(x)$ be a string that satisfies the following equation:
$$0.q(x) = 1/2 \cdot 0.q_1(x) + 1/4 \cdot 0.q_2(x) + \cdots + 1/2^n \cdot 0.q_n(x).$$
Since $0.q_i(x) \leq 1$ for all i and $\sum_{i=1}^{n} 1/2^i < 1$, $q(x)$ is well defined. Let U be defined as follows.

input v
  $m = 0$
  while $q(1^m) \in \text{Left}(v)$ do $m = m + 1$
  if $q(0^m) \notin \text{Left}(v)$ then $y = 1^{m-1}$
  else
    $y = \lambda$
    for $l = 1$ to $m - 1$
      if $q(y10^{m-l}) \notin \text{Left}(v)$ then $y = y0$ else $y = y1$
    let x be the successor of y
    if $v \in \text{Left}(q(x))$ or $q(x)$ is a prefix of v then output x

First we observe that U is a nondecreasing t'-time bounded Turing machine. If U stops on input v and outputs a string x then either $q(x^-) \in \text{Left}(v)$ and $v \in \text{Left}(q(x))$ or $q(x)$ is a prefix of v. Hence U stops on input vz for any z and in this case $U(v) = U(vz)$. Now assume that $v \in \text{Left}(u)$ and U stops on input v and u. Let y_u and y_v denote the strings computed on input u and v, respectively. Since any string in $\text{Left}(v)$ is in $\text{Left}(u)$ it follows that $y_v \leq y_u$ and hence $U(v) \leq U(u)$.

Secondly we consider the time complexity of U. Since by Lemma 1 for any e, q_e is computable in time $O(n^2 t(n)^2)$, q is computable in time $O(n^3 t(n)^2) \log n$. Since q is evaluated $O(n)$ times (where n is the length of the output) on strings of size at most n, the computation of U is $O(n^4 t(n)^2 \log t(n))$ time bounded.

Now assume that M_e is t-time nondecreasing. We show that $\mathrm{Knd}_U(x) \leq \mathrm{Knd}_{M_e}(x) + e + 1$. Assume that for some x, $|x| > e$, $M_e(u) = x$. Since M_e is nondecreasing and prefix free $0.v \leq 0.u - 2^{-|u|}$ for all v such that $M_e(v) < x$. ($0.v \leq 0.u$ since M_e is t-time nondecreasing, and $|0.v - 0.u| \geq 2^{-|u|}$ since M_e is a prefix machine, i.e. if M_e stops on inputs u and v, then v is not a prefix of u and u is not a prefix of v). Since q_i is nondecreasing for all i, $0.q(x) \geq 0.q(x^-) + 2^{-(e+|u|)}$.

Let $q(x^-) = w_1 \ldots w_m$, where $w_i \in \{0.1\}$ for all $i \leq m$. Let p be a string of length $e + |u| + 1$ such that $0.p = 0.w_1 \cdots w_{e+|u|+1} + 2^{-(e+|u|+1)}$, where $w_i = 0$ for all $m < i \leq e + |u| + 1$. Then either $q(x^-) \in \mathrm{Left}(p)$ and $p \in \mathrm{Left}(q(x))$ or $q(x)$ is a prefix of p. Since $|p| = |u| + e + 1$ it follows that is $\mathrm{Knd}_U(x) \leq \mathrm{Knd}_{M_e}(x) + e + 1$.

In the following we fix for every t such a t-time nondecreasing universal Turing machine U_t and define $\mathrm{Knd}_t(x) = \mathrm{Knd}_{U_t}(x)$ for all x.

3.1   Time Bounded Universal Distributions

Let t be a time-constructible nondecreasing function and U_t a t-time nondecreasing universal Turing machine. Then the t-time bounded algorithmic probability R_t is defined as follows. For all x, $R_t(x) = 2^{-\mathrm{Knd}_t(x)}$. Then a t-time bounded priori probability Q_{U_t} is defined as follows. For all x, $Q_{U_t}(x) = \sum_p 2^{-|p|}$, where p ranges over all strings such that $U_t(p) = x$ but $U_t(p')$ does not stop for any prefix p' of p.

Let p be a polynomial and $\mu_1, \mu_2, \cdots$ be a enumeration of $p(t)$-computable semimeasures containing all t-computable semimeasures [10]. Then $m_t(x) = \sum_{n \leq |x|} \alpha(n) \cdot \mu_n(x)$ is universal for the class of t-computable semimeasure. Furthermore, m_t is computable in time polynomial in t.

Theorem 6. *Let t be a nondecreasing time-constructible function. There is a constant c and polynomials p_1, p_2, and p_3 such that for every x, $-\log m_t(x) \leq -\log Q_{U_{p_1(t)}}(x) \leq -\log R_{p_2(t)}(x) \leq -\log m_{p_3(t)}(x)$ up to an additive constant less than c.*

Proof. Recall from Section 2.2 the definition of a Shannon-Fano code E induced by m_t, and the inverse D of E. Then for all x, $-\log m_t(x) \leq |E(x)| + 2$. Since D is a nondecreasing prefix code and computable in time $p(t)$ for some polynomial p, it follows that for all x there exists a string v of length $|E(x)| + d$ such that $U_{p_1}(p) = x$ where d is some constant independent of x. Hence $2^{d+2} \cdot Q_{U_{p_1(t)}}(x) \geq m_t(x)$.

Since U is t-time nondecreasing, the probability of a string p such that $U_t(p) = x$ is at most $4 \cdot 2^{-|q|}$, where q is a shortest string such that $U_t(q) = x$. That is $4 \cdot \mathrm{Knd}_{U_t}(x) \geq \sum_p 2^{-|p|}$, where p ranges over all strings such that $U_t(p) = x$ but $U(p')$ does not stop for any prefix p' of p. That is $4 \cdot Q_{U_{p_1(t)}}(x) \leq R_{p_2}(x)$.

Finally note that $\mu(x) = 0.p - 0.q$, where $p = \mathrm{Knd}_t(x)$ and $q = \mathrm{Knd}_t(x^-)$ defines a semimeasure computable in time $p(t)$ for some polynomial p. Hence, for some constant d' and all x, $d' \cdot R_{p_2(t)}(x) \leq -\log m_{p_3(t)}(x)$.

Since by definition $R_{p_2(t)}(x) = \mathrm{Knd}_{p_2(t)}(x)$, theorem 6 can been seen as a justification of the definition of time-bounded nondecreasing Kolmogorov complexity. Finally we note that the concept classes shown to simple PAC-learnable in the t-time bounded setting [7] are simple PAC-learnable w.r.t. the universal distribution m_t.

A concept class is simple PAC-learnable (in the t-time bounded setting) if it is PAC-learnable [11] w.r.t. any simple (t-time bounded) distribution. In [6] Li and Vitányi show that any concept class that is PAC learnable under a t-universal distribution, is PAC learnable under all t-computable distributions, provided that (in the learning phase) the examples are drawn according to the universal distribution. In [7] Li and Vitányi (see also [2]) give several examples of concept classes that are PAC learnable w.r.t. the distribution $m_{(t)}$, e.g. the class of DNF of logarithmic Kolmogorov complexity.

We observe that the proofs of the learnability results depend on the fact that examples of low t-time bounded prefix Kolmogorov complexity have high probability w.r.t. $m_{(t)}$. By Proposition 2 and Theorem 6 this holds for any $p(t)$-universal distribution where p is some fixed polynomial.

References

1. S. Ben-David, B. Chor, O. Goldreich, and M. Luby. On the theory of average case complexity. *Journal of Computer and System Sciences*, 44:193–219, 1992.
2. J. Castro and J. Balcázar. Simple PAC learning of simple decision lists. In *Proceedings of the 6th Int'l Workshop on Algorithmic Learning Theory*, volume 997 of *Lecture Notes in Computer Science*, pages 239–248. Springer-Verlag, 1995.
3. Shafi Goldwasser and Silvio Micali. Probabilistic encryption. *Journal of Computer and System Sciences*, 28(2):270–299, April 1984.
4. Y. Gurevich. Average case completeness. *Journal of Computer and System Sciences*, 42(3):346–398, 1991.
5. L. Levin. Average case complete problems. *SIAM Journal on Computing*, 15:285–286, 1986.
6. M. Li and P. Vitányi. *An introduction to Kolmogorov complexity and its applications*. Springer-Verlag, second edition edition, 1997.
7. M. Li and P.M.B. Vitányi. Learning simple concepts under simple distributions. *SIAM Journal on Computing*, 20(5):911–935, 1991.
8. M. Li and P.M.B. Vitányi. Average case complexity under the universal distribution equals worst-case complexity. *Information Processing Letters*, 42:145–149, 1992.
9. P.B. Milterson. The complexity of malign measures. *SIAM Journal on Computing*, 22(1):147–156, 1993.
10. R. Schuler. A note on universal distributions for polynomial-time computable distributions. In *Proceedings of the 12th Annual IEEE Conference on Computational Complexity*. IEEE Computer Society Press, 1997.
11. L. Valiant. A theory of the learnable. *Communications of the ACM*, 27(11):1134–1142, 1984.

The Descriptive Complexity Approach to LOGCFL

Clemens Lautemann[1], Pierre McKenzie[2*], Thomas Schwentick[1], and
Heribert Vollmer[3]

[1] Institut für Informatik, Johannes-Gutenberg-Universität Mainz,
55099 Mainz, Germany.
[2] Informatique et recherche opérationnelle, Université de Montréal, C.P. 6128,
Succ. Centre-Ville, Montréal (Québec), H3C 3J7 Canada.
[3] Theoretische Informatik, Universität Würzburg, Am Exerzierplatz 3,
97072 Würzburg, Germany.

Abstract. Building upon the known generalized-quantifier-based first-order characterization of LOGCFL, we lay the groundwork for a deeper investigation. Specifically, we examine subclasses of LOGCFL arising from varying the arity and nesting of *groupoidal* quantifiers. Our work extends the elaborate theory relating *monoidal* quantifiers to NC^1 and its subclasses. In the absence of the BIT predicate, we resolve the main issues: we show in particular that no single outermost unary groupoidal quantifier with FO can capture all the context-free languages, and we obtain the surprising result that a variant of Greibach's "hardest context-free language" is LOGCFL-complete under quantifier-free BIT-free interpretations. We then prove that FO with unary groupoidal quantifiers is strictly more expressive with the BIT predicate than without. Considering a particular groupoidal quantifier, we prove that first-order logic with majority of pairs is strictly more expressive than first-order with majority of individuals. As a technical tool of independent interest, we define the notion of an aperiodic nondeterministic finite automaton and prove that FO translations are precisely the mappings computed by single-valued aperiodic nondeterministic finite transducers.

1 Introduction

In *Finite Automata, Formal Logic, and Circuit Complexity* [17], Howard Straubing surveys an elegant theory relating finite semigroup theory, first-order logic, and computational complexity. The gist of this theory is that questions about the structure of the complexity class NC^1, defined from logarithmic depth bounded fan-in Boolean circuits, can be translated back and forth into questions about the expressibility of first-order logic augmented with new predicates and quantifiers. Such a translation provides new insights, makes tools from one field available

---

* Research performed while on leave at the Universität Tübingen. Supported by the (German) DFG, the (Canadian) NSERC and the (Québec) FCAR.

C. Meinel and S. Tison (Eds.): STACS'99, LNCS 1563, pp. 444–454, 1999.
© Springer-Verlag Berlin Heidelberg 1999

in the other, suggests tractable refinements to the hard open questions in the separate fields, and puts the obstacles to further progress in a clear perspective.

In this way, although, for example, the unresolved strict containment in NC^1 of the class ACC^0, defined from bounded-depth polynomial-size unbounded fan-in circuits over {AND, OR, MOD}, remains a barrier since the work of Smolensky [15], significant progress was made in (1) understanding the power of the BIT predicate and the related circuit uniformity issues [3], (2) describing the regular languages within subclasses of NC^1 [2, 13], and (3) identifying the all-important role of the interplay between arbitrary and regular numerical predicates in the status of the ACC^0 versus NC^1 question [17, p. 169, Conjecture IX.3.4].

Barrington, Immerman and Straubing [3] introduced the notion of a *monoidal* quantifier and noted that, for any non-solvable group G, the class NC^1 can be described using first-order logic augmented with a monoidal quantifier for G. Loosely speaking, such a quantifier provides a constrained "oracle call" to the word problem for G (defined essentially as the problem of computing the product of a sequence of elements of G).

Bédard, Lemieux and McKenzie [4] later noted that there is a fixed finite *groupoid* whose word problem is complete for the class LOGCFL of languages reducible in logarithmic space to a context-free language [6, 18]. A groupoid G is a set with a binary operation on which no constraint—such as associativity or commutativity—is placed. The word problem for G is the set of all those sequences of elements of G that can be bracketed to evaluate to a fixed element of G. It is not hard to see that any context-free language is the word problem of some groupoid, and that any groupoid word problem is context-free (see [4, Lemma 3.1]).

It followed that LOGCFL, a well-studied class which contains nondeterministic logarithmic space [18] and is presumably much larger than NC^1, can be described by first-order logic augmented with *groupoidal* quantifiers. These quantifiers can be defined formally as Lindström quantifiers [10] for context-free languages.

In this paper, we take up this first-order characterization of LOGCFL, and initiate an investigation of LOGCFL from the viewpoint of descriptive complexity. The rationale for this study, which encompasses the study of NC^1, is that tools from logic might be of use in ultimately elucidating the structure of LOGCFL. We do not claim new separations of the major subclasses of LOGCFL here. But we make a first step, in effect settling necessary preliminary questions afforded by the first-order framework.

Our precise results concern the relative expressiveness of first-order formulas with ordering (written FO), interpreted over finite strings, and with: (1) nested versus unnested groupoidal quantifiers, (2) unary versus non-unary groupoidal quantifiers, (3) the presence versus the absence of the BIT predicate. Feature (3) was the focus of an important part of the work by Barrington, Immerman and Straubing [3] on uniformity within NC^1. Feature (2) was also considered, to a lesser extent, by the same authors, who left open the question of whether the "majority-of-pairs" quantifier could be simulated by a unary majority quantifier in the absence of the BIT predicate [3, p. 297]. Feature (1) is akin to comparing many-one reducibility with Turing reducibility in traditional complexity theory. Here we examine all combinations of features (1), (2) and (3). Our separation results are summarized on Fig. 1 in Sect. 5. In the absence of the BIT predicate, we are able to determine the following relationships:

- FO to which a single unary groupoidal quantifier is applied, written $Q_{\mathrm{Grp}}^{\mathrm{un}}\mathrm{FO}$, captures the CFLs, and is strictly less expressive than FO with nested unary quantifiers, written $\mathrm{FO}(Q_{\mathrm{Grp}}^{\mathrm{un}})$, which in its turn is strictly weaker than LOGCFL. A consequence of this result, as we will see, is an answer to the above mentioned open question from [3]: We show that first-order logic with the majority-of-pairs quantifier is strictly more expressive than first-order logic with majority of individuals, see Corollary 18.
- No single groupoid G captures all the CFLs as $Q_G^{\mathrm{un}}\mathrm{FO}$, i.e. as FO to which the single unary groupoidal quantifier Q_G^{un} is applied,
- FO to which a single *non-unary* groupoidal quantifier is applied, written $Q_{\mathrm{Grp}}\mathrm{FO}$, captures LOGCFL; our proof implies, remarkably, that adding a padding symbol to Greibach's hardest context-free language [7], see also [1], yields a language which is LOGCFL-complete under BIT-free quantifier-free interpretations.

When the BIT predicate is present, first-order with non-unary groupoidal quantifiers of course still describes LOGCFL. In the setting of monoidal quantifiers [3], FO with BIT is known to capture uniform circuit classes, notably uniform ACC^0, which have not yet been separated from NC^1. We face a similar situation here: the BIT predicate allows capturing classes (for example $\mathrm{FO}_{\mathrm{bit}}(Q_{\mathrm{Grp}}^{\mathrm{un}})$, verifying $\mathrm{TC}^0 \subseteq \mathrm{FO}_{\mathrm{bit}}(Q_{\mathrm{Grp}}^{\mathrm{un}}) \subseteq \mathrm{LOGCFL}$), which only a major breakthrough would seem to allow separating from each other. We are able to attest to the strength of the BIT predicate in the setting of unary quantifiers, proving that:

- $Q_{\mathrm{Grp}}^{\mathrm{un}}\mathrm{FO} \subsetneq Q_{\mathrm{Grp}}^{\mathrm{un}}\mathrm{FO}_{\mathrm{bit}}$, i.e. (trivially) some non-context-free languages are expressible using BIT and a single unary groupoidal quantifier,
- $\mathrm{FO}(Q_{\mathrm{Grp}}^{\mathrm{un}}) \subsetneq \mathrm{FO}_{\mathrm{bit}}(Q_{\mathrm{Grp}}^{\mathrm{un}})$, i.e. (more interestingly) BIT adds expressivity even when unary groupoidal quantifiers can be nested.

We also develop a technical tool of independent interest, in the form of an aperiodic (a. k. a. group-free, a. k. a. counter-free) nondeterministic finite automaton. Aperiodicity has been studied intensively, most notably in connection with the star-free regular languages [14], but, to the best of our knowledge, always in a deterministic context. Here we define a NFA A to be aperiodic if the DFA resulting from applying the subset construction to A is aperiodic. The usefulness of this notion lies in the fact, proved here, that first-order translations are precisely those mappings which are computable by single-valued aperiodic nondeterministic finite transducers.

Due to lack of space most proofs are omitted in this abstract. A full paper including complete proofs of all our claims can be obtained as ECCC Report 98-59 from http://www.eccc.uni-trier.de.

2 Preliminaries

2.1 Complexity Theory

REG and CFL refer to the regular and to the ϵ-free context-free languages respectively. The CFL results in this paper could be adapted to treat the empty string ϵ in standard ways. We will make scant reference to the inclusion chain $AC^0 \subsetneq ACC^0 \subseteq TC^0 \subseteq NC^1 \subseteq NL \subseteq LOGCFL = SAC^1 \subseteq P$. We refer the reader to [17] for definitions of these classes.

2.2 The First-Order Framework

We consider first-order logic with linear order. We restrict our attention to *string signatures*, i. e. signatures of the form $\langle P_{a_1}, \ldots, P_{a_s} \rangle$, where all the predicates P_{a_i} are unary, and in every structure $\mathcal{A}$, $\mathcal{A} \models P_{a_i}(j)$ iff the jth symbol in the input is the letter a_i. Such structures are thus words over the alphabet $(a_1, \ldots, a_s)$, and first-order variables range over positions within such a word, i. e. from 1 to the word length n. For technical reasons that will become apparent shortly, we assume here, as in the rest of the paper, a linear order on each alphabet and we write alphabets as sequences of symbols to indicate that order.

Our basic formulas are built from variables in the usual way, using the Boolean connectives $\{\wedge, \vee, \neg\}$, the relevant predicates P_{a_i} together with $\{=, <\}$, the constants min and max, the quantifiers $\{\exists, \forall\}$, and parentheses. We will occasionally use the binary predicate $\mathrm{BIT}(x, y)$, defined to be true iff the xth bit in the binary representation of y is 1. We write $\mathrm{BC}(\mathcal{L})$ to denote the Boolean closure of the set $\mathcal{L}$ of languages (i. e. closure under intersection, union, and complement) and $\mathrm{BC}^+(\mathcal{L})$ to denote the closure under union and intersection only.

Definition 1. *Lindström quantifier.* Consider a language L over an alphabet $\Sigma = (a_1, a_2, \ldots, a_s)$, $s > 1$. Let $\bar{x}$ be a k-tuple of variables (each of which ranges from 1 to the "input length" n, as we have seen). In the following, we assume the lexical ordering on $\{1, 2, \ldots, n\}^k$, and we write $X_1, X_2, \ldots, X_{n^k}$ for the sequence of potential values taken on by $\bar{x}$. The k-ary groupoidal quantifier Q_L binding $\bar{x}$ takes a meaning if $s - 1$ formulas, each having as free variables the variables in $\bar{x}$ (and possibly others), are available. Let $\phi_1(\bar{x})$, $\phi_2(\bar{x})$, $\ldots$, $\phi_{s-1}(\bar{x})$ be these $s - 1$ formulas. Then $Q_L \bar{x}[\phi_1(\bar{x}), \phi_2(\bar{x}), \ldots, \phi_{s-1}(\bar{x})]$ holds on a string $w = w_1 \cdots w_n$ (where $w_i \in \Sigma$ for all i), iff the word of length n^k whose ith letter, $1 \leq i \leq n^k$, is

$$
\begin{cases}
a_1 \text{ if } w \models \phi_1(X_i), \\
a_2 \text{ if } w \models \neg\phi_1(X_i) \wedge \phi_2(X_i), \\
\quad \ldots \\
a_s \text{ if } w \models \neg\phi_1(X_i) \wedge \neg\phi_2(X_i) \wedge \ldots \wedge \neg\phi_{s-1}(X_i),
\end{cases}
$$

belongs to L. Thus the formulas $[\phi_1(\bar{x}), \phi_2(\bar{x}), \ldots, \phi_{s-1}(\bar{x})]$ fix a function mapping an input word/structure w of length n to a word of length n^k. This function is called the *reduction* or *transformation* defined by $[\phi_1(\overline{x}), \phi_2(\overline{x}), \ldots, \phi_{s-1}(\overline{x})]$. In case we deal with the binary alphabet ($s = 2$) we omit the braces and write $Q_L \overline{x} \phi(\overline{x})$ for short.

Definition 2. A *groupoidal quantifier* is a Lindström quantifier Q_L where L is a context-free language.

The Lindström quantifiers of Definition 1 are more precisely what has been refered to as "Lindström quantifiers on strings" [5]. The original more general definition [10] uses transformations to arbitrary structures, not necessarily of string signature. However, in the context of this paper reductions to CFLs play a role of utmost importance, and hence the above definition seems to be the most natural. The terminology *groupoidal quantifier* stems from the fact that any context-free language is a word problem over some groupoid [4, Lemma 3.1], and vice-versa every word problem of a finite groupoid is context-free. Thus a Lindström quantifier on strings defined by a context-free language is nothing else than a Lindström quantifier (in the classical sense) defined by a structure that is a finite groupoid multiplication table.

Barrington, Immerman, and Straubing, defining *monoidal quantifiers* in [3], in fact proceed along the same avenue: they first show how monoid word problems can be seen as languages, and then define generalized quantifiers given by such languages (see [3, pp. 284f.]).

2.3 Unary Quantifiers and Homomorphisms

We will encounter unary groupoidal quantifiers repeatedly. Here we show how these relate to standard formal language operations. In a different context, a result very similar to the next theorem is known as *Nivat's Theorem* [11, Theorem 3.8, p. 207].

Theorem 3. *Let B be an arbitrary language, and let A be describable in $Q_B^{un}FO$, that is, by a first order formula preceded by one unary Lindström quantifier (i. e. binding exactly one variable). Then there are length-preserving homomorphisms g, h and a regular language D such that $A = h(D \cap g^{-1}(B))$.*

2.4 Groupoid-Based Language Classes

Fix a finite groupoid G. Each $S \subseteq G$ defines a language $\mathcal{W}(S, G)$ composed of all words w, over the alphabet G, which "multiply out" to an element of S when an appropriate legal bracketing of w is chosen.

Definition 4. $Q_G FO$ is the set of languages describable by applying a single groupoidal quantifier Q_L to an appropriate tuple of FO formulas, where $L = \mathcal{W}(S, G)$ for some $S \subseteq G$.
$Q_{Grp}FO$ is the union, over all finite groupoids G, of $Q_G FO$.

FO(Q_G) and FO(Q_{Grp}) are defined analogously, but allowing groupoidal quantifiers to be used as any other quantifier would (i. e. allowing arbitrary nesting). Q_G^{un}FO and FO$_{\text{bit}}(Q_{\text{Grp}}^{\text{un}})$, etc, are defined analogously, but possibly allowing the BIT predicate (signaled by subscripting FO with bit) and/or restricting to *unary* groupoidal quantifiers (signaled by the superscript "un").

3   An Automaton Characterization of FO-Translations

As a technical tool, it will be convenient to have an automata-theoretic characterization of *first-order translations*, i. e. of reductions defined by FO-formulas with one free variable. Since FO precisely describes the (regular) languages accepted by aperiodic deterministic finite automata [12], one might expect aperiodic deterministic finite transducers to capture FO-translations. This is not the case however because, e.g. the FO-translation which maps every string $w_1 \cdots w_n$ to w_n^n cannot be computed by such a device.

We show in this section that the appropriate automaton model to use is that of a single-valued aperiodic nondeterministic finite transducer, which we define and associate with FO-translations in this section. But first, we discuss the notion of an aperiodic NFA.

Definition 5. A deterministic or nondeterministic FA M is *aperiodic* (or group-free) iff there is an $n \in \mathbb{N}$ such that for all states s and all words w,

$$\delta(s, w^n) = \delta(s, w^{n+1}).$$

Here δ is the extension of M's transition function from symbols to words. Observe that if M is nondeterministic then $\delta(t, v)$ is a set of states, i. e. locally here we abuse notation by not distinguishing between M's extended transition function δ and the function δ^* as defined in the context of a nondeterministic transducer below.

Remark 6. This definition of aperiodicity for a DFA is the usual one (see [16]). For a NFA, a statement obviously equivalent to Definition 5 would be that A is aperiodic iff applying the subset construction to A yields an aperiodic DFA. Hence [14] a language L is star-free iff some aperiodic (deterministic or nondeterministic) finite automaton accepts L.

Definition 7. A *finite transducer* M is given by a set Q of *states*, an *input alphabet* Σ, an *output alphabet* Γ, an *initial state* q_0, a *transition relation* $\delta \subseteq Q \times \Sigma \times \Gamma \times Q$ and a set $F \subseteq Q$ of *final states*. For a string $w = w_1 \cdots w_n \in \Sigma^*$ we define the *set $O_M(w)$ of outputs* of M on input w as follows. A string $v \in \Gamma^*$ of length n is in $O_M(w)$, if there is a sequence $s_0 = q_0, s_1, \ldots, s_n$ of states, such that $s_n \in F$ and, for every i, $1 \leq i \leq n$, we have $(s_{i-1}, w_i, v_i, s_i) \in \delta$.

We say that M is *single-valued* if, for every $w \in \Sigma^*$, $|O_M(w)| = 1$. If M is single-valued it naturally defines a function $f_M : \Sigma^* \to \Gamma^*$.

For every string $u \in \Sigma^*$ and every state $s \in Q$ we write $\delta^*(s, u)$ for the set of states s' that are reachable from s on input u (i. e., there are $s_1, \ldots, s_{|u|} = s'$ and $v_1 \cdots v_{|u|}$ such that, for every i, $1 \le i \le |u|$, we have $(s_{i-1}, u_i, v_i, s_i) \in \delta$).

As per Definition 5, M is *aperiodic* if there is an $n \in \mathbb{N}$ such that for all states q and all strings w, $\delta^*(q, w^n) = \delta^*(q, w^{n+1})$.

Theorem 8. *A function* $f \colon \Sigma^* \to \Gamma^*$ *is defined by an* FO *translation if and only if it is defined by a single-valued aperiodic finite transducer.*

4   First-Order with Groupoidal Quantifiers

4.1   Non-unary Groupoidal Quantifiers

Clearly all classes from our framework fall within LOGCFL. But even one groupoidal quantifier to the left is sufficient:

Theorem 9. *There is a fixed groupoid* G *such that* $Q_G\mathrm{FO}_{\mathrm{bit}} = \mathrm{FO}_{\mathrm{bit}}(Q_{\mathrm{Grp}}) =$ LOGCFL.

As we show next, the largest attainable class LOGCFL is even characterizable without bit, as long as we allow arbitrary arities of the quantifiers:

Theorem 10. *There is a fixed groupoid* G *such that* LOGCFL $\subseteq Q_G\mathrm{FO}$*, hence* $Q_{\mathrm{Grp}}\mathrm{FO} = \mathrm{FO}(Q_{\mathrm{Grp}}) =$ LOGCFL.

Corollary 11. *Greibach's hardest context-free language with a neutral symbol is complete for* LOGCFL *under quantifier-free interpretations without* BIT.

4.2   Unary Groupoidal Quantifiers without BIT

In the previous subsection, we have shown that the situation with non-unary groupoidal quantifiers is clearcut, since a single such quantifier, even without the BIT predicate, captures all of LOGCFL. Here we examine the case of unary quantifiers. Let us first turn to the case of unary groupoidal quantifiers without BIT.

Theorem 12. $Q_{\mathrm{Grp}}^{\mathrm{un}}\mathrm{FO} = \mathrm{CFL}$.

Proof. The direction from right to left is obvious. The converse direction is proved by appealing to Theorem 3 and observing that the context-free languages have the required closure properties.

It follows immediately that nesting unary groupoidal quantifiers (in fact merely taking the Boolean closure of $Q_{\mathrm{Grp}}^{\mathrm{un}}\mathrm{FO}$) adds expressiveness:

Corollary 13. $Q_{\mathrm{Grp}}^{\mathrm{un}}\mathrm{FO} = \mathrm{CFL} \subsetneq \mathrm{BC}^+(Q_{\mathrm{Grp}}^{\mathrm{un}}\mathrm{FO}) = \mathrm{BC}^+(\mathrm{CFL})$
$$\subsetneq \mathrm{BC}(Q_{\mathrm{Grp}}^{\mathrm{un}}\mathrm{FO}) = \mathrm{BC}(\mathrm{CFL})$$
$$\subseteq \mathrm{FO}(Q_{\mathrm{Grp}}^{\mathrm{un}}).$$

Can we refine Theorem 12 and find a universal finite groupoid G which captures all the context-free languages as $Q_G^{\mathrm{un}}\mathrm{FO}$? Intuition from the world of monoids [3, p. 303] suggests that the answer is no. Proving that this is indeed the case is the content of Theorem 15 below. We first make a definition and state a lemma.

Let D_t be the context-free one-sided Dyck language over $2t$ symbols, i. e. D_t consists of the well-bracketed words over an alphabet of t distinct types of parentheses. Recall that a PDA is a nondeterministic automaton which reads its input from left to right and has access to a pushdown store with a fixed pushdown alphabet. We say that a PDA A is k-pushdown-limited, for k a positive integer, iff the pushdown alphabet of A has size k, and A pushes no more than k symbols on its stack between any two successive input head motions.

Lemma 14. No k-pushdown-limited PDA accepts D_t when $t \geq (k+1)^k + 1$.

Theorem 15. Any finite groupoid G verifies $Q_G^{\mathrm{un}}\mathrm{FO} \subsetneq \mathrm{CFL}$.

Proof. Suppose to the contrary that G is a finite groupoid such that $Q_G^{\mathrm{un}}\mathrm{FO} = \mathrm{CFL}$. Then there is a FO-translation from each context-free language to a word problem for G. This means that a finite set of PDAs (one for each word problem $\mathcal{W}(\cdot, G)$) can take care of answering each "oracle question" resulting from such a FO-translation. By Theorem 8, each FO-translation is computed by a single-valued NFA. Although the NFAs differ for different context-free languages (and this holds in particular when language alphabets differ), the NFAs do not bolster the "pushdown-limits" of the PDAs which answer all oracle questions. Hence if k is a fixed integer such that all word problems $\mathcal{W}(\cdot, G)$ for G are accepted by a k-pushdown-limited PDA, then for any positive integer t, D_t is accepted by a k-limited-pushdown PDA. This contradicts Lemma 14 when $t = (k+1)^k + 1$. $\square$

In the next subsection we will see that the BIT-predicate provably adds expressive power to the logic $Q_{\mathrm{Grp}}^{\mathrm{un}}\mathrm{FO}$. Since it is known that BIT can be expressed either by plus and times [9] (cf., [8]) or by the majority of pairs quantifier [3], the following two simple observations about the power of $Q_{\mathrm{Grp}}^{\mathrm{un}}\mathrm{FO}$ are of particular interest.

Theorem 16. *1. The majority quantifier is definable in $Q_{\mathrm{Grp}}^{\mathrm{un}}\mathrm{FO}$.*
2. Addition is definable in $Q_{\mathrm{Grp}}^{\mathrm{un}}\mathrm{FO}$.

4.3 Unary Groupoidal Quantifiers with BIT

What are $Q_{\mathrm{Grp}}^{\mathrm{un}}\mathrm{FO}_{\mathrm{bit}}$ and $\mathrm{FO}_{\mathrm{bit}}(Q_{\mathrm{Grp}}^{\mathrm{un}})$? It seems plausible that $Q_{\mathrm{Grp}}^{\mathrm{un}}\mathrm{FO}_{\mathrm{bit}} \subsetneq \mathrm{FO}_{\mathrm{bit}}(Q_{\mathrm{Grp}}^{\mathrm{un}}) \subsetneq \mathrm{LOGCFL}$, but we are unable to prove these separations. Since TC^0 is captured by first-order logic with bit and majority quantifiers (definable by a context-free language), we conclude $\mathrm{TC}^0 \subseteq \mathrm{FO}_{\mathrm{bit}}(Q_{\mathrm{Grp}}^{\mathrm{un}})$, hence proving the latter separation would prove $\mathrm{TC}^0 \neq \mathrm{LOGCFL}$, settling a major open question in complexity theory.

Easily $\mathrm{CFL} = Q_{\mathrm{Grp}}^{\mathrm{un}}\mathrm{FO} \subsetneq Q_{\mathrm{Grp}}^{\mathrm{un}}\mathrm{FO}_{\mathrm{bit}}$, since $\left\{ 0^{2^n} \mid n \in \mathbb{N} \right\}$ is in the difference of the two classes. The remainder of this subsection is devoted to documenting a more complicated setting in which the BIT predicate provably adds expressiveness.

Theorem 17. $\mathrm{FO}_{\mathrm{bit}}(Q_{\mathrm{Grp}}^{\mathrm{un}})$ *is not contained in* $\mathrm{FO}(Q_{\mathrm{Grp}}^{\mathrm{un}})$.

As a particular case we can now solve an open question of [3], addressing the power of different arity for majority quantifiers.

Corollary 18. *Majority of pairs can not be expressed in first-order logic with unary majority quantifiers.*

Proof. In Theorem 16 it was observed that the unary majority quantifier can be simulated in $\mathrm{FO}(Q_{\mathrm{Grp}}^{\mathrm{un}})$. On the other hand in [3] it is shown that majority of pairs is sufficient to simulate the BIT predicate. But as $\mathrm{FO}_{\mathrm{bit}}(Q_{\mathrm{Grp}}^{\mathrm{un}})$ is not contained in $\mathrm{FO}(Q_{\mathrm{Grp}}^{\mathrm{un}})$ the BIT predicate and hence the majority of pairs is not definable in $\mathrm{FO}(Q_{\mathrm{Grp}}^{\mathrm{un}})$, hence it can not be simulated by unary majority quantifiers.

Corollary 19. *Multiplication is not definable in* $\mathrm{FO}(Q_{\mathrm{Grp}}^{\mathrm{un}})$.

5   Conclusion

Figure 1 depicts the first-order groupoidal-quantifier-based classes studied in this paper. Together with the new characterization of FO-translations by means of aperiodic finite transducers, the relationships shown on Fig. 1 summarize our contribution.

A number of open questions are apparent from Figure 1. Clearly, it would be nice to separate the $\mathrm{FO}_{\mathrm{bit}}$-based classes, in particular $\mathrm{FO}_{\mathrm{bit}}(Q_{\mathrm{Grp}}^{\mathrm{un}})$ from $\mathrm{FO}_{\mathrm{bit}}(Q_{\mathrm{Grp}})$, but this is a daunting task. A sensible approach then is to begin with $Q_{\mathrm{Grp}}^{\mathrm{un}}\mathrm{FO}_{\mathrm{bit}}$. How does this compare with TC^0 for example? Can we at least separate $Q_{\mathrm{Grp}}^{\mathrm{un}}\mathrm{FO}_{\mathrm{bit}}$ from LOGCFL? We know that $Q_{\mathrm{Grp}}^{\mathrm{un}}\mathrm{FO}_{\mathrm{bit}} \not\subseteq \mathrm{FO}(Q_{\mathrm{Grp}}^{\mathrm{un}})$; a witness for this is the set $\left\{ 0^{n^2} \mid n \in \mathbb{N} \right\}$, cf. the proof of Theorem 17.

An important fundamental question of course is if we can hope for an algebraic theory of groupoids to explain the detailed structure of CFL, much in the way that an elaborate theory of monoids is used in the extensive first-order parameterization of REG.

Acknowledgments. We thank Dave Barrington, Gerhard Buntrock, Volker Diekert, Klaus-Jörn Lange, Ken Regan, Heinz Schmitz, Denis Thérien, Wolfgang Thomas, Klaus Wagner, and Detlef Wotschke for useful discussions in the course of this work.

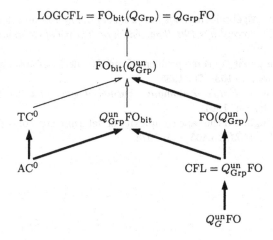

Fig. 1. The new landscape. Here G stands for any fixed groupoid, and a thick line indicates strict inclusion.

References

[1] J.-M. Autebert, J. Berstel, and L. Boasson. Context-free languages and pushdown automata. In R. Rozenberg and A. Salomaa, editors, *Handbook of Formal Languages*, volume I, chapter 3. Springer Verlag, Berlin Heidelberg, 1997.

[2] D. A. Mix Barrington, K. Compton, H. Straubing, and D. Thérien. Regular languages in NC^1. *Journal of Computer and System Sciences*, 44:478–499, 1992.

[3] D. A. Mix Barrington, N. Immerman, and H. Straubing. On uniformity within NC^1. *Journal of Computer and System Sciences*, 41:274–306, 1990.

[4] F. Bédard, F. Lemieux, and P. McKenzie. Extensions to Barrington's M-program model. *Theoretical Computer Science*, 107:31–61, 1993.

[5] H.-J. Burtschick and H. Vollmer. Lindström quantifiers and leaf language definability. *International Journal of Foundations of Computer Science*, 9:277–294, 1998.

[6] S. A. Cook. Characterizations of pushdown machines in terms of time-bounded computers. *Journal of the ACM*, 18:4–18, 1971.

[7] S. Greibach. The hardest context-free language. *SIAM Journal on Computing*, 2:304–310, 1973.

[8] N. Immerman. *Descriptive Complexity*. Springer Verlag, New York, 1998.

[9] S. Lindell. manuscript, 1994. e-mail communication by Kenneth W. Regan.

[10] P. Lindström. First order predicate logic with generalized quantifiers. *Theoria*, 32:186–195, 1966.

[11] A. Mateescu and A. Salomaa. Aspects of classical language theory. In R. Rozenberg and A. Salomaa, editors, *Handbook of Formal Languages*, volume I, chapter 4. Springer Verlag, Berlin Heidelberg, 1997.

[12] R. McNaughton and S. Papert. *Counter-Free Automata*. MIT Press, 1971.

[13] P. Péladeau P. McKenzie and D. Thérien. NC^1: The automata-theoretic viewpoint. *Computational Complexity*, 1:330–359, 1991.

[14] M. P. Schützenberger. On finite monoids having only trivial subgroups. *Information & Control*, 8:190–194, 1965.

[15] R. Smolensky. Algebraic methods in the theory of lower bounds for Boolean circuit complexity. In *Proceedings 19th Symposium on Theory of Computing*, pages 77–82. ACM Press, 1987.

[16] J. Stern. Complexity of some problems from the theory of automata. *Information & Computation*, 66:163–176, 1985.

[17] H. Straubing. *Finite Automata, Formal Logic, and Circuit Complexity*. Birkhäuser, Boston, 1994.

[18] I. H. Sudborough. On the tape complexity of deterministic context-free languages. *Journal of the ACM*, 25:405–414, 1978.

The Weakness of Self-Complementation

Orna Kupferman[1*] and Moshe Y. Vardi[2**]

[1] Hebrew University, The institute of Computer Science, Jerusalem 91904, Israel
orna@cs.huji.ac.il, http://www.cs.huji.ac.il/~orna
[2] Rice University, Department of Computer Science, Houston, TX 77251-1892, U.S.A.
vardi@cs.rice.edu, http://www.cs.rice.edu/~vardi

Abstract. Model checking is a method for the verification of systems with respect to their specifications. Symbolic model-checking, which enables the verification of large systems, proceeds by evaluating fixed-point expressions over the system's set of states. Such evaluation is particularly simple and efficient when the expressions do not contain alternation between least and greatest fixed-point operators; namely, when they belong to the *alternation-free μ-calculus* (AFMC). Not all specifications, however, can be translated to AFMC, which is exactly as expressive as *weak monadic second-order logic* (WS2S). Rabin showed that a set $\mathcal{T}$ of trees can be expressed in WS2S if and only if both $\mathcal{T}$ and its complement can be recognized by nondeterministic Büchi tree automata. For the "only if" direction, Rabin constructed, given two nondeterministic Büchi tree automata $\mathcal{U}$ and $\mathcal{U}'$ that recognize $\mathcal{T}$ and its complement, a WS2S formula that is satisfied by exactly all trees in $\mathcal{T}$. Since the translation of WS2S to AFMC is nonelementary, this construction is not practical. Arnold and Niwiński improved Rabin's construction by a direct translation of $\mathcal{U}$ and $\mathcal{U}'$ to AFMC, which involves a doubly-exponential blow-up and is therefore still impractical. In this paper we describe an alternative and quadratic translation of $\mathcal{U}$ and $\mathcal{U}'$ to AFMC. Our translation goes through *weak alternating tree automata*, and constitutes a step towards efficient symbolic model checking of highly expressive specification formalisms.

1 Introduction

In *model checking*, we verify the correctness of a system with respect to a desired behavior by checking whether a structure that models the system satisfies a formula that specifies this behavior. Commercial model-checking tools need to cope with the exceedingly large state-spaces that are present in real-life designs, making the so-called *state-explosion problem* one of the most challenging areas in computer-aided verification. One of the most important developments in this

* Part of this work was done when this author was visiting Cadence Berkeley Laboratories.
* Supported in part by the NSF grants CCR-9628400 and CCR-9700061, and by a grant from the Intel Corporation. Part of this work was done when this author was a Varon Visiting Professor at the Weizmann Institute of Science.

C. Meinel and S. Tison (Eds.): STACS'99, LNCS 1563, pp. 455–466, 1999.
© Springer-Verlag Berlin Heidelberg 1999

area is the discovery of *symbolic* model-checking methods [BCM+92, McM93]. In particular, the use of BDDs [Bry86] for model representation has yielded model-checking tools that can handle systems with 10^{120} states and beyond [CGL93].

Typically, symbolic model-checking tools proceed by computing fixed-point expressions over the model's set of states. For example, to find the set of states from which a state satisfying some predicate p is reachable, the model checker starts with the set S of states in which p holds, and repeatedly add to S the set $\exists \bigcirc S$ of states that have a successor in S. Formally, the model checker calculates the least fixed-point of the expression $S = p \vee \exists \bigcirc S$. The evaluation of such expressions is particularly simple when they contain no alternation between least and greatest fixed-point operators. Formally, the evaluation of expressions in *alternation-free μ-calculus* (AFMC) [Koz83, EL86] can be solved in time that is linear in both the size of the model and the length of the formula [AC88, CS91]. In contrast, the evaluation of expressions in which there is a single alternation takes time that is quadratic in the size of the model. Since the models are very large, the difference with the linear complexity of AFMC is very significant [HKSV97]. Hence, it is desired to translate specification to AFMC. Not all specifications, however, can be translated to AFMC [KV98a], and known translations to AFMC involve a blow-up that makes them impractical. In this paper we describe an alternative translation of specifications to AFMC.

Second-order logic is a powerful formalism for expressing properties of sequences and trees. We can view all common program logics as fragments of second-order logic. Second-order logic also serves as the specification language in the model-checking tool MONA [EKM98, Kla98]. While in first-order logic one can only quantify individual variables, second-order logic enables also the quantification of sets[1]. For example, the formula

$$\exists X. \epsilon \in X \wedge \forall z(z \in X \leftrightarrow \neg(succ(z) \in X)) \wedge \forall z(z \in X \rightarrow P(z))$$

specifies sequences in which P holds at all even positions. We distinguish between two types of logic, linear and branching. In second-order logic with one successor (S1S), the formulas describe sequences and contain, as the example above, the successor operator *succ*. In second-order logic with two successors (S2S), formulas describe trees and contain both left-successor and right-successor operators. For example, the S2S formula

$$\exists X. \epsilon \in X \wedge \forall z(z \in X \rightarrow (P(z) \wedge (succ_l(z) \in X \vee succ_r(z) \in X)))$$

specifies trees in which P holds along at least one path.

Second-order logic motivated the introduction and study of *finite automata on infinite objects*. Like automata on finite objects, automata on infinite objects either accept or reject their input. Since a run on an infinite object does not have a final state, acceptance is determined with respect to the set of states visited infinitely often during the run. For example, in *Büchi* automata, some of

---

[1] Thus, we consider *monadic* second-order logic, where quantification is over unary relations

the states are designated as accepting states, and a run is accepting iff it visits states from the accepting set infinitely often [Büc62] (when the run is a tree, it is required to visit infinitely many accepting states along each path). More general are *Rabin* automata, whose acceptance conditions involve a set of pairs of sets of states. The tight relation between automata on infinite objects and second-order logic was first established for the linear paradigm. In [Büc62], Büchi translated S1S formulas to nondeterministic Büchi word automata. Then, in [Rab69], Rabin translated S2S formulas to nondeterministic Rabin tree automata. These fundamental works led to the solution of the decision problem for S1S and S2S, and were the key to the solution of many more problems in mathematical logic [Tho90].

Recall that we are looking for a fragment of S2S that can be translated to AFMC. Known results about the expressive power of different types of automata enabled the study of definability of properties within fragments of second-order logic. In [Rab70], Rabin showed that nondeterministic Büchi tree automata are strictly less expressive than nondeterministic Rabin tree automata, and that they are not closed under complementation. Rabin also showed that for every set T of trees, both T and its complement can be recognized by nondeterministic Büchi tree automata iff T can be specified in a fragment of S2S, called *weak second-order logic* (WS2S), in which set quantification is restricted to finite sets. For the "only if" direction, Rabin constructed, given two nondeterministic Büchi tree automata U and U' that recognize T and its complement, a WS2S formula that is satisfied by exactly all trees in T.

It turned out that WS2S is exactly the fragment of S2S we are looking for, thus WS2S=AFMC. In [AN92], Arnold and Niwiński showed that every AFMC formula can be translated to an equivalent WS2S formula. For the other direction, they constructed, given U and U' as above, an AFMC formula that is satisfied by exactly all trees accepted by U. The translation in [AN92] is doubly exponential. Thus, if U and U' has n and m states, respectively, the AFMC formula is of length $2^{2^{nm}}$. While this improves the nonelementary translation of Rabin's WS2S formula to AFMC, it is still not useful in practice. In this paper we present a quadratic translation of U and U' to an AFMC formula that is satisfied by exactly all trees accepted by U. Our translation goes through *weak alternating automata* [MSS86]. Thus, while the characterizations of WS2S in [Rab70, AN92] go from logic to automata and then back to logic, our construction provides a clean, purely automata-theoretic, characterization of WS2S.

In an *alternating* automaton [BL80, CKS81], the transition function can induce both existential and universal requirements on the automaton. For example, a transition $\delta(q, \sigma) = q_1 \vee (q_2 \wedge q_3)$ of an alternating word automaton means that the automaton in state q accepts a word $\sigma \cdot \tau$ iff it accepts the suffix τ either from state q_1 or from both states q_2 and q_3. In a *weak automaton*, the automaton's set of states is partitioned into partially ordered sets. Each set is classified as accepting or rejecting. The transition function is restricted so that in each transition, the automaton either stays at the same set or moves to a set smaller in the partial order. Thus, each run of a weak automaton eventually gets trapped in some set in the partition. Acceptance is then determined according to the classification of this set. It is shown in [MSS86] that formulas of WS2S can be translated to weak alternating tree automata. Moreover, it is shown in [KV98a] that weak alternating automata can be linearly translated to AFMC.

Given two nondeterministic Büchi tree automata $\mathcal{U}$ and $\mathcal{U}'$ that recognize a language and its complement, we construct a weak alternating tree automaton $\mathcal{A}$ equivalent to $\mathcal{U}$. The number of states in $\mathcal{A}$ is quadratic in the number of states of $\mathcal{U}$ and $\mathcal{U}'$. Precisely, if $\mathcal{U}$ and $\mathcal{U}'$ has n and m states, respectively, the automaton $\mathcal{A}$ has $(nm)^2$ states. The linear translation of weak alternating tree automata to AFMC then completes a translation to AFMC of the same complexity. Our translation can be viewed as a step towards efficient symbolic model checking of highly expressive specification formalisms such as the fragment of the branching temporal logic CTL* that can be translated to WS2S. A step that is still missing in order to complete this goal is a translation of CTL* formulas to nondeterministic Büchi tree automata, when such a translation exists. From a theoretical point of view, our translation completes the picture of "quadratic weakening" in both the linear and the branching paradigm. The equivalence in expressive power of nondeterministic Büchi and Rabin word automata [McN66] implied that WS1S is as expressive as S1S [Tho90]. The latter equivalence is supported by an automaton construction: given a nondeterministic Büchi word automaton, one can construct an equivalent weak alternating word automaton of quadratic size [KV97]. In the branching paradigm, WS2S is strictly less expressive than S2S, and a nondeterministic Büchi tree automaton can be translated to a weak alternating tree automaton only if its complement can also be recognized by a nondeterministic Büchi tree automaton. It follows from our construction that the size of the equivalent weak alternating automaton is then quadratic in the sizes of the two automata.

2   Tree Automata

A *full infinite binary tree* (tree) is the set $T = \{l, r\}^*$. The elements of T are called *nodes*, and the empty word ε is the *root* of T. For every $x \in T$, the nodes $x \cdot l$ and $x \cdot r$ are the *successors* of x. A *path* π of a tree T is a set $\pi \subseteq T$ such that $\varepsilon \in \pi$ and for every $x \in \pi$, exactly one successor of x is in π. For two nodes x_1 and x_2 of T, we say that $x_1 \leq x_2$ iff x_1 is a prefix of x_2; i.e., there exists $z \in \{l, r\}^*$ such that $x_2 = x_1 \cdot z$. We say that $x_1 < x_2$ iff $x_1 \leq x_2$ and $x_1 \neq x_2$. A *frontier* of an infinite tree is a set $E \subset T$ of nodes such that for every path $\pi \subseteq T$, we have $|\pi \cap E| = 1$. For example, the set $E = \{l, rll, rlr, rr\}$ is a frontier. For two frontiers E_1 and E_2, we say that $E_1 \leq E_2$ iff for every node $x_2 \in E_2$, there exists a node $x_1 \in E_1$ such that $x_1 \leq x_2$. We say that $E_1 < E_2$ iff for every node $x_2 \in E_2$, there exists a node $x_1 \in E_1$ such that $x_1 < x_2$. Note that while $E_1 < E_2$ implies that $E_1 \leq E_2$ and $E_1 \neq E_2$, the other direction does not necessarily hold. Given an alphabet Σ, a Σ-*labeled tree* is a pair $\langle T, V \rangle$ where T is a tree and $V : T \to \Sigma$ maps each node of T to a letter in Σ. We denote by V_Σ the set of all Σ-labeled trees. For a Σ-labeled tree $\langle T, V \rangle$ and a set $A \subseteq \Sigma$, we say that E is an A-frontier iff E is a frontier and for every node $x \in E$, we have $V(x) \in A$.

Automata on infinite trees (tree automata) run on infinite Σ-labeled trees. We first define *nondeterministic Büchi tree automata* (NBT). An NBT is $\mathcal{U} = \langle \Sigma, Q, \delta, q_0, F \rangle$ where Σ is the input alphabet, Q is a finite set of states, $\delta : Q \times \Sigma \to 2^{Q \times Q}$ is a transition function, $q_0 \in Q$ is an initial state, and $F \subseteq Q$ is a Büchi acceptance condition. Intuitively, each pair in $\delta(q, \sigma)$ suggests a nondeterministic choice for the automaton's next configuration. When the automaton is in a state q as it reads a node x labeled by a letter σ, it proceeds by first choosing a pair $\langle q_l, q_r \rangle \in \delta(q, \sigma)$, and then splitting into two copies. One copy enters the state q_l and proceeds to the node $x \cdot l$ (the left successor of x), and the other copy enters the state q_r and proceeds to the node $x \cdot r$ (the right successor of x). Formally, a *run* of $\mathcal{U}$ on an input tree $\langle T, V \rangle$ is a Q-labeled tree $\langle T, r \rangle$, such that the following hold:

- $r(\varepsilon) = q_0$.
- Let $x \in T$ with $r(x) = q$. There exists $\langle q_l, q_r \rangle \in \delta(q, V(x))$ such that $r(x \cdot l) = q_l$ and $r(x \cdot r) = q_r$.

Note that each node of the input tree corresponds to exactly one node in the run tree.

Given a run $\langle T, r \rangle$ and a path $\pi \subseteq T$, let $inf(\pi) \subseteq Q$ be such that $q \in inf(\pi)$ if and only if there are infinitely many $x \in \pi$ for which $r(x) = q$. That is, $inf(\pi)$ contains exactly all the states that are visited infinitely often in π. A path π satisfies a Büchi acceptance condition $F \subseteq Q$ if and only if $inf(\pi) \cap F \neq \emptyset$. A run $\langle T, r \rangle$ is *accepting* iff all its paths satisfy the acceptance condition. Equivalently, $\langle T, r \rangle$ is accepting iff $\langle T, r \rangle$ contains infinitely many F-frontiers $G_0 < G_1 < \dots$. A tree $\langle T, V \rangle$ is accepted by $\mathcal{U}$ iff there exists an accepting run of $\mathcal{U}$ on $\langle T, V \rangle$, in which case $\langle T, V \rangle$ belongs to the language, $\mathcal{L}(\mathcal{U})$, of $\mathcal{U}$. We say that a set $\mathcal{T}$ of trees is in NBT iff there exists an NBT $\mathcal{U}$ such that $\mathcal{L}(\mathcal{U}) = \mathcal{T}$. We say that $\mathcal{T}$ is in co-NBT iff the complement of $\mathcal{T}$ is in NBT; i.e., there exists an NBT $\mathcal{U}$ such that $\mathcal{L}(\mathcal{U}) = V_\Sigma \setminus \mathcal{T}$.

Alternating tree automata generalize nondeterministic tree automata and were first introduced in [MS87]. In order to define alternating tree automata, we first need some notations. For a given set X, let $\mathcal{B}^+(X)$ be the set of positive Boolean formulas over X (i.e., Boolean formulas built from elements in X using $\land$ and $\lor$), where we also allow the formulas **true** and **false** and, as usual, $\land$ has precedence over $\lor$. For a set $Y \subseteq X$ and a formula $\theta \in \mathcal{B}^+(X)$, we say that Y *satisfies* θ iff assigning **true** to elements in Y and assigning **false** to elements in $X \setminus Y$ satisfies θ.

A finite alternating automaton over infinite binary trees is $\mathcal{A} = \langle \Sigma, Q, \delta, q_0, F \rangle$ where Σ, Q, q_0, and F are as in NBT, and $\delta : Q \times \Sigma \to \mathcal{B}^+(\{l, r\} \times Q)$ is a transition function. A run of an alternating automaton $\mathcal{A}$ over a tree $\langle T, V \rangle$ is a $(T \times Q)$-labeled tree $\langle T_r, r \rangle$. The tree T_r is not necessarily binary and it may have states with no successors. Thus, $T_r \subseteq \mathbb{N}^*$ is such that if $x \cdot c \in T_r$ where $x \in \mathbb{N}^*$ and $c \in \mathbb{N}$, then also $x \in T_r$. For every $x \in T_r$, the nodes $x \cdot c$, with $c \in \mathbb{N}$, are the *successors* of x. Each node of T_r corresponds to a node of T. A node in T_r, labeled by (x, q), describes a copy of the automaton that reads the node x of T and visits the state q. Note that many nodes of T_r can correspond to the same node of T; in contrast, in a run of a nondeterministic automaton over $\langle T, V \rangle$ there is a one-to-one correspondence between the nodes of the run and the nodes of the tree. The labels of a node and its successors have to satisfy the transition function. Formally, $\langle T_r, r \rangle$ satisfies the following:

1. $\varepsilon \in T_r$ and $r(\varepsilon) = (\varepsilon, q_0)$.
2. Let $y \in T_r$ with $r(y) = (x, q)$ and $\delta(q, V(x)) = \theta$. Then there is a possibly empty set $S = \{(c_0, q_0), (c_1, q_1), \ldots, (c_n, q_n)\} \subseteq \{\mathbf{l}, \mathbf{r}\} \times Q$, such that the following hold:
   - S satisfies θ, and
   - for all $0 \leq i \leq n$, we have $y \cdot i \in T_r$ and $r(y \cdot i) = (x \cdot c_i, q_i)$.

For a run $\langle T_r, r \rangle$ and an infinite path $\pi \subseteq T_r$, we define $inf(\pi)$ to be the set of states that are visited infinitely often in π, thus $q \in inf(\pi)$ if and only if there are infinitely many $y \in \pi$ for which $r(y) \in T \times \{q\}$. A run $\langle T_r, r \rangle$ is accepting if all its infinite paths satisfy the Büchi acceptance condition. As with NBT, a tree $\langle T, V \rangle$ is accepted by $\mathcal{A}$ iff there exists an accepting run of $\mathcal{A}$ on $\langle T, V \rangle$, in which case $\langle T, V \rangle$ belongs to $\mathcal{L}(\mathcal{A})$.

Example 1. We define an alternating Büchi tree automaton $\mathcal{A}$ that accepts exactly all $\{a, b, c\}$-labeled binary trees in which all paths have a node labeled a and there exists a path with two successive b labels. Let $\mathcal{A} = \langle \{a, b, c\}, \{q_0, q_1, q_2, q_3\}, \delta, q_0, \emptyset \rangle$, where δ is defined in the table below.

q	$\delta(q, a)$	$\delta(q, b)$	$\delta(q, c)$
q_0	$(0, q_1) \vee (1, q_1)$	$(0, q_3) \wedge (1, q_3) \wedge$ $((0, q_2) \vee (1, q_2))$	$(0, q_3) \wedge (1, q_3) \wedge$ $((0, q_1) \vee (1, q_1))$
q_1	$(0, q_1) \vee (1, q_1)$	$(0, q_2) \vee (1, q_2)$	$(0, q_1) \vee (1, q_1)$
q_2	$(0, q_1) \vee (1, q_1)$	true	$(0, q_1) \vee (1, q_1)$
q_3	true	$(0, q_3) \wedge (1, q_3)$	$(0, q_3) \wedge (1, q_3)$

In the state q_0, the automaton checks both requirements. If a is true, only the second requirement is left to be checked. This is done by sending a copy in state q_1, which searches for two successive b's in some branch, to either the left or the right child. If b is true, $\mathcal{A}$ needs to send more copies. First, it needs to check that all paths in the left and right subtrees have a node labeled a. This is done by sending copies in state q_3 to both the left and the right children. Second, it needs to check that one of these subtrees contains two successive b's. This is done (keeping in mind that the just read b may be the first b in a sequence of two b's) by sending a copy in state q_2 to one of the children. Similarly, if c is true, $\mathcal{A}$ sends copies that check both requirements. As before, a requirement about a is sent universally and a requirement about the b's is sent existentially.   ⊏

In [MSS86], Muller et al. introduce *weak alternating tree automata* (AWT). In an AWT, we have a Büchi acceptance condition $F \subseteq Q$ and there exists a partition of Q into disjoint sets, Q_i, such that for each set Q_i, either $Q_i \subseteq F$, in which case Q_i is an *accepting set*, or $Q_i \cap F = \emptyset$, in which case Q_i is a *rejecting set*. In addition, there exists a partial order $\leq$ on the collection of the Q_i's such that for every $q \in Q_i$ and $q' \in Q_j$ for which q' occurs in $\delta(q, \sigma)$, for some $\sigma \in \Sigma$, we have $Q_j \leq Q_i$. Thus, transitions from a state in Q_i lead to states in either the same Q_i or a lower one. It follows that every infinite path of a run of an AWT ultimately gets "trapped" within some Q_i. The path then satisfies the acceptance condition if and only if Q_i is an accepting set. Indeed, a run visits infinitely many states in F if and only if it gets trapped in an accepting set.

2.1   Traps

Let $\mathcal{U} = \langle \Sigma, S, s_0, M, F \rangle$ and $\mathcal{U}' = \langle \Sigma, S', s'_0, M', F' \rangle$ be two NBT, and let $|S| \cdot |S'| = m$. In [Rab70], Rabin studies the composition of a run of $\mathcal{U}$ with a run of $\mathcal{U}'$. Recall that an accepting run of $\mathcal{U}$ contains infinitely many F-frontiers $G_0 < G_1 < \dots$, and an accepting run of $\mathcal{U}'$ contains infinitely many F'-frontiers $G'_0 < G'_1 < \dots$. It follows that for every tree $\langle T, V \rangle \in \mathcal{L}(\mathcal{U}) \cap \mathcal{L}(\mathcal{U}')$ and accepting runs $\langle T, r \rangle$ and $\langle T, r' \rangle$ of $\mathcal{U}$ and $\mathcal{U}'$ on $\langle T, V \rangle$, the composition of $\langle T, r \rangle$ and $\langle T, r' \rangle$ contains infinitely many frontiers $E_i \subset T$, with $E_i < E_{i+1}$, such that $\langle T, r \rangle$ reaches an F-frontier and $\langle T, r' \rangle$ reaches an F'-frontier between E_i and E_{i+1}. Rabin shows that the existence of m such frontiers, in the composition of some runs of $\mathcal{U}$ and $\mathcal{U}'$, is sufficient to imply that the intersection $\mathcal{L}(\mathcal{U}) \cap \mathcal{L}(\mathcal{U}')$ is not empty. Below we repeat Rabin's result, with some different notations.

Let $\mathcal{U}$ and $\mathcal{U}'$ be as above. We say that a sequence $E_0, \dots, E_m$ of frontiers of T is a *trap for $\mathcal{U}$ and $\mathcal{U}'$* iff $E_0 = \varepsilon$ and there exists a tree $\langle T, V \rangle$ and runs $\langle T, r \rangle$ and $\langle T, r' \rangle$ of $\mathcal{U}$ and $\mathcal{U}'$ on $\langle T, V \rangle$, such that for every $0 \leq i \leq m - 1$, the following hold.

- $\langle T, r \rangle$ contains an F-frontier G_i such that $E_i \leq G_i < E_{i+1}$, and
- $\langle T, r' \rangle$ contains an F'-frontier G'_i such that $E_i \leq G'_i < E_{i+1}$.

We say that $\langle T, r \rangle$ and $\langle T, r' \rangle$ *witness* the trap for $\mathcal{U}$ and $\mathcal{U}'$.

Theorem 1. [Rab70] *Consider two nondeterministic Büchi tree automata $\mathcal{U}$ and $\mathcal{U}'$. If there exists a trap for $\mathcal{U}$ and $\mathcal{U}'$, then $\mathcal{L}(\mathcal{U}) \cap \mathcal{L}(\mathcal{U}')$ is not empty.*

3   From NBT and Co-NBT to AWT

Theorem 2. *Let $\mathcal{U}$ and $\mathcal{U}'$ be two NBT with $\mathcal{L}(\mathcal{U}') = V_\Sigma \setminus \mathcal{L}(\mathcal{U})$. There exists an AWT $\mathcal{A}$ such that $\mathcal{L}(\mathcal{A}) = \mathcal{L}(\mathcal{U})$ and the size of $\mathcal{A}$ is quadratic in the sizes of $\mathcal{U}$ and $\mathcal{U}'$.*

Proof. Let $\mathcal{U} = \langle \Sigma, S, s_0, M, F \rangle$ and $\mathcal{U}' = \langle \Sigma, S', s'_0, M', F' \rangle$, and let $|S| \cdot |S'| = m$. We define the AWT $\mathcal{A} = \langle \Sigma, Q, q_0, \delta, \alpha \rangle$ as follows.

- $Q = ((S \times \{\bot, \top\}) \times S' \times \{0, \ldots, m\}) \setminus (S \times \{\top\}) \times S' \times \{m\}$. Intuitively, a copy of $\mathcal{A}$ that visits the state $\langle (s, \gamma), s', i \rangle$ as it reads the node x of the input tree corresponds to runs r and r' of $\mathcal{U}$ and $\mathcal{U}'$ that visit the states s and s', respectively, as they read the node x of the input tree. Let $\rho = y_0, y_1, \ldots, y_{|x|}$ be the path from ε to x. Consider the joint behavior of r and r' on ρ. We can represent this behavior by a sequence $\tau_\rho = \langle t_0, t_0' \rangle, \langle t_1, t_1' \rangle, \ldots, \langle t_{|x|}, t_{|x|}' \rangle$ of pairs in $S \times S'$ where $t_j = r(y_j)$ and $t_j' = r'(y_j')$. We say that a pair $\langle t, t' \rangle \in S \times S'$ is an F-pair iff $t \in F$ and is an F'-pair iff $t' \in F'$. We can partition the sequence τ_ρ to blocks $\beta_0, \beta_1, \ldots, \beta_i$ such that we close block β_k and open block β_{k+1} whenever we reach the first F'-pair that is preceded by an F-pair in β_k. In other words, whenever we open a block, we first look for an F-pair, ignoring F'-pairness. Once an F-pair is detected, we look for an F'-pair, ignoring F-pairness. Once an F'-pair is detected, we close the current block and we open a new block. Note that a block may contain a single pair that is both an F-pair and an F'-pair. The number i in $\langle (s, \gamma), s', i \rangle$ is the index of the last block in τ_ρ. The status $\gamma \in \{\bot, \top\}$ indicates whether the block β_i already contains an F-pair, in which case $\gamma = \top$, or β_i does not contain an F-pair, in which case $\gamma = \bot$.
  For a status $\gamma \in \{\bot, \top\}$ and an index $i \in \{0, \ldots, m\}$, let $Q_{\gamma, i} = (S \times \{\gamma\}) \times S' \times \{i\}$.
- $q_0 = \langle (s_0, \bot), s_0', 0 \rangle$.
- In order to define the transition function δ, we first define two functions, $new\gamma : Q \setminus Q_{\bot, m} \to \{\bot, \top\}$ and $newi : Q \setminus Q_{\bot, m} \to \{0, \ldots, m\}$, as follows.

$$new\gamma(\langle (s, \gamma), s', i \rangle) = \begin{bmatrix} \top & \text{If } s' \notin F' \text{ and } (\gamma = \top \text{ or } s \in F). \\ \bot & \text{Otherwise.} \end{bmatrix}$$

$$newi(\langle (s, \gamma), s', i \rangle) = \begin{bmatrix} i + 1 & \text{If } s' \in F' \text{ and } (\gamma = \top \text{ or } s \in F). \\ i & \text{Otherwise.} \end{bmatrix}$$

Intuitively, $new\gamma$ and $newi$ are responsible for the recording and tracking of blocks. Recall that the status γ indicates whether an F-pair in the current block has already been detected. As such, the new status is $\top$ whenever s is in F or γ is $\top$, unless s' is in F', in which case $\langle s, s' \rangle$ is the last pair in the current block, and the new status is $\bot$. Similarly, the index i is increased to $i + 1$ whenever we detect an F'-pair that is either also an F-pair or, as indicated by γ, preceded by an F-pair in the same block.
The automaton $\mathcal{A}$ proceeds as follows. Essentially, for every run $\langle T, r' \rangle$ of $\mathcal{U}'$, the automaton $\mathcal{A}$ guesses a run $\langle T, r \rangle$ of $\mathcal{U}$ such that for every path ρ of T, the run $\langle T, r \rangle$ visits F along ρ at least as many times as $\langle T, r' \rangle$ visits F' along ρ. Since $\mathcal{L}(\mathcal{U}) \cap \mathcal{L}(\mathcal{U}') = \emptyset$, no run $\langle T, r \rangle$ can witness with $\langle T, r' \rangle$ a trap for $\mathcal{U}$ and $\mathcal{U}'$. Consequently, recording of visits to F and F' along ρ can be completed once $\mathcal{A}$ detects that τ_ρ contains m blocks as above.
Formally, let $q = \langle (s, \gamma), s', i \rangle$ be such that $M(s, \sigma) = \{\langle u_1, v_1 \rangle, \ldots, \langle u_n, v_n \rangle\}$ and $M'(s', \sigma) = \{\langle u_1', v_1' \rangle, \ldots, \langle u_{n'}', v_{n'}' \rangle\}$. We distinguish between two cases

- If $q \notin Q_{\perp,m}$, then $\delta(q,\sigma)$ is

$$\bigwedge_{1 \le k \le n'} \bigvee_{1 \le p \le n} (\mathbf{l}, \langle (u_p, new\gamma(q)), u'_k, newi(q) \rangle) \wedge (\mathbf{r}, \langle (v_p, new\gamma(q)), v'_k, newi(q) \rangle).$$

- If $q \in Q_{\perp,m}$, then $\delta(q,\sigma)$ is
  * $\bigwedge_{1 \le k \le n'} \bigvee_{1 \le p \le n} (\mathbf{l}, \langle (u_p, \perp), u'_k, m \rangle) \wedge (\mathbf{r}, \langle (v_p, \perp), v'_k, m \rangle)$, if $s \notin F$.
  * **true**, otherwise.

- $\alpha = (S \times \{\top\}) \times S' \times \{0 \dots, m-1\}$. Thus, α makes sure that infinite paths of the run visits infinitely many states in which the status is $\top$.

The automaton $\mathcal{A}$ is indeed an AWT. Clearly, each set $Q_{\gamma,i}$ is either contained in α or is disjoint from α. The partial order on the sets is defined by $Q_{\gamma',i'} \le Q_{\gamma,i}$ iff either $i < i'$, or $i = i'$ and $\gamma' = \top$. Note that, by the definition of α, a run is accepting iff no path of it gets trapped in a set of the form $Q_{\perp,i}$, namely a set in which $\mathcal{A}$ is waiting for a visit of $\mathcal{U}$ in a state in F. The size of $\mathcal{A}$ is $O(m^2)$.

We prove that $\mathcal{L}(\mathcal{A}) = \mathcal{L}(\mathcal{U})$. We first prove that $\mathcal{L}(\mathcal{U}) \subseteq \mathcal{L}(\mathcal{A})$. Consider a tree $\langle T, V \rangle$. With every run $\langle T, r \rangle$ of $\mathcal{U}$ on $\langle T, V \rangle$ we can associate a single run $\langle T_R, R \rangle$ of $\mathcal{A}$ on $\langle T, V \rangle$. Intuitively, the run $\langle T, r \rangle$ directs $\langle T_R, R \rangle$ in the only nondeterminism in δ. Formally, recall that a run of $\mathcal{A}$ on a tree $\langle T, V \rangle$ is a $(T \times Q)$-labeled tree $\langle T_R, R \rangle$, where a node $y \in T_R$ with $R(y) = \langle x, q \rangle$ corresponds to a copy of $\mathcal{A}$ that reads the node $x \in T$ and visits the state q. We define $\langle T_R, R \rangle$ as follows.

- $\varepsilon \in T_R$ and $R(\varepsilon) = (\varepsilon, \langle (s_0, \perp), s'_0, 0 \rangle)$.
- Consider a node $y \in T_R$ with $R(y) = (x, \langle (s, \gamma), s', i \rangle)$. By the definition of $\langle T_R, R \rangle$ so far, we have $r(x) = s$. Let $r(x \cdot \mathbf{l}) = u$ and $r(x \cdot \mathbf{r}) = v$. Also, let $M'(s', \sigma) = \{\langle u'_1, v'_1 \rangle, \dots, \langle u'_{n'}, v'_{n'} \rangle\}$, $\gamma' = new\gamma(\langle (s, \gamma), s', i \rangle)$, and $i' = newi(\langle (s, \gamma), s', i \rangle)$. We define $S = \{(\mathbf{l}, \langle (u, \gamma'), u'_1, i' \rangle), (\mathbf{r}, \langle (v, \gamma'), v'_1, i' \rangle), \dots, (\mathbf{l}, \langle (u, \gamma'), u'_{n'}, i' \rangle), (\mathbf{r}, \langle (v, \gamma'), v'_{n'}, i' \rangle)\}$. By the definition of δ, the set S satisfies $\delta(\langle (s, \gamma), s', i \rangle, V(x))$. For all $0 \le j \le n' - 1$, we have $y \cdot 2j \in T_R$ with $R(y \cdot 2j) = (x \cdot \mathbf{l}, \langle (u, \gamma'), u'_j, i' \rangle)$, and $y \cdot 2j + 1 \in T_R$ with $R(y \cdot 2j + 1) = (x \cdot \mathbf{r}, \langle (v, \gamma'), v'_j, i' \rangle)$.

Consider a tree $\langle T, V \rangle \in \mathcal{L}(\mathcal{U})$. Let $\langle T, r \rangle$ be an accepting run of $\mathcal{U}$ on $\langle T, V \rangle$, and let $\langle T_R, R \rangle$ be the run of $\mathcal{A}$ on $\langle T, V \rangle$ induced by $\langle T, r \rangle$. It is easy to see that $\langle T_R, R \rangle$ is accepting. Indeed, as $\langle T, r \rangle$ contains infinitely many F-frontiers, no infinite paths of $\langle T_R, R \rangle$ can get trapped in a set $Q_{\perp,i}$.

It is left to prove that $\mathcal{L}(\mathcal{A}) \subseteq \mathcal{L}(\mathcal{U})$. For that, we prove that $\mathcal{L}(\mathcal{A}) \cap \mathcal{L}(\mathcal{U}') = \emptyset$. Since $\mathcal{L}(\mathcal{U}) = V_\Sigma \setminus \mathcal{L}(\mathcal{U}')$, it follows that every tree that is accepted by $\mathcal{A}$ is also accepted by $\mathcal{U}$. Consider a tree $\langle T, V \rangle$. With each run $\langle T_R, R \rangle$ of $\mathcal{A}$ on $\langle T, V \rangle$ and run $\langle T, r' \rangle$ of $\mathcal{U}'$ on $\langle T, V \rangle$, we can associate a run $\langle T, r \rangle$ of $\mathcal{U}$ on $\langle T, V \rangle$. Intuitively, $\langle T, r \rangle$ makes the choices that $\langle T_R, R \rangle$ has made in its copies that correspond to the run $\langle T, r' \rangle$. Formally, we define $\langle T, r \rangle$ as follows.

- $r(\varepsilon) = s_0$.
- Consider a node $x \in T$ with $r(x) = s$. Let $r'(x) = s'$. The run $\langle T, r' \rangle$ fixes a pair $\langle u', v' \rangle \in M'(s', V(x))$ that $\mathcal{U}'$ proceeds with when it reads the node x. Formally, let $\langle u', v' \rangle$ be such that $r'(x \cdot \mathbf{l}) = u'$ and $r'(x \cdot \mathbf{r}) = v'$. By the definition of $r(x)$ so far, the run $\langle T_R, R \rangle$ contains a node $y \in T_R$ with $R(y) = \langle x, \langle (s, \gamma), s', i \rangle \rangle$ for some γ and i. If $\delta(\langle (s, \gamma), s', i \rangle, V(x)) = \mathbf{true}$, we define the reminder of $\langle T, r \rangle$ arbitrarily. Otherwise, by the definition of δ, the successors of y in T_R fix the pair in $M(s, V(x))$ that $\mathcal{A}$ proceeds with per each pair in $M'(s', V(x))$. In particular, T_R contains at least two nodes $y \cdot c_1$ and $y \cdot c_2$ such that $R(y \cdot c_1) = \langle x \cdot \mathbf{l}, \langle (u, \gamma'), u', i' \rangle \rangle$ and $R(y \cdot c_2) = \langle x \cdot \mathbf{r}, \langle (v, \gamma'), v', i' \rangle \rangle$, for some γ' and i'. We then define $r(x \cdot \mathbf{l}) = u$ and $r(x \cdot \mathbf{r}) = v$.

We can now prove that $\mathcal{L}(\mathcal{A}) \cap \mathcal{L}(\mathcal{U}') = \emptyset$. Assume, by way of contradiction, that there exists a tree $\langle T, V \rangle$ such that $\langle T, V \rangle$ is accepted by both $\mathcal{A}$ and $\mathcal{U}'$. Let $\langle T_R, R \rangle$ and $\langle T, r' \rangle$ be the accepting runs of $\mathcal{A}$ and $\mathcal{U}'$ on $\langle T, V \rangle$, respectively, and let $\langle T, r \rangle$ be the run of $\mathcal{U}$ on $\langle T, V \rangle$ induced by $\langle T_R, R \rangle$ and $\langle T, r' \rangle$. We claim that then, $\langle T, r \rangle$ and $\langle T, r' \rangle$ witness a trap for $\mathcal{U}$ and $\mathcal{U}'$. Since, however, $\mathcal{L}(\mathcal{U}) \cap \mathcal{L}(U') = \emptyset$, it follows from Theorem 1, that no such trap exists, and we reach a contradiction. To see that $\langle T, r \rangle$ and $\langle T, r' \rangle$ indeed witness a trap, define $E_0 = \varepsilon$, and define, for $0 \leq i \leq m - 1$, the set E_{i+1} to contain exactly all nodes x for which there exists $y \in T_R$ with $R(y) = \langle x, \langle (r(x), \gamma), r'(x), i \rangle \rangle$ and $newi(\langle (r(x), \gamma), r'(x), i \rangle) = i + 1$. That is, for every path ρ of T, the set E_{i+1} consists of the nodes in which the i'th block is closed in τ_ρ. By the definition of δ, for all $0 \leq i \leq m - 1$, the run $\langle T, r \rangle$ contains an F-frontier G_i such that $E_i \leq G_i < E_{i+1}$ and the run $\langle T, r' \rangle$ contains an F'-frontier G_i' such that $E_i \leq G_i' < E_{i+1}$. Hence, $E_0, \ldots, E_m$ is a trap for $\mathcal{U}$ and $\mathcal{U}'$.   □

4   Discussion

Today, automata on infinite objects are used for specification and verification of nonterminating programs. By translating specifications to automata, we reduce questions about programs and their specifications to questions about automata. More specifically, questions such as satisfiability of specifications and correctness of programs with respect to their specifications are reduced to questions such as nonemptiness and language containment [VW86, BVW94, Kur94, VW94]. The automata-theoretic approach separates the logical and the combinatorial aspects of verification. The translation of specifications to automata handles the logic and shifts all the combinatorial difficulties to automata-theoretic problems. There are many types of automata, and choosing the appropriate type for the application is important.

We believe that weak alternating automata are often a good choice. The special structure of weak alternating automata is reflected in their attractive computational properties. For example, while the best known complexity for solving the 1-letter emptiness problem for Büchi alternating automata is quadratic

;ime, we know how to solve the problem for weak alternating automata in linear time [BVW94]. In addition, weak alternating automata can be very eas-
ly complemented. In the linear paradigm, where WS1S=S1S, weak alternating
word automata (AWW) can recognize all the ω-regular languages. In particular,
the translation of LTL formulas to AWW is linear, and follows the syntax of
the formula [Var96]. Moreover, it is known how to translate other types of automata to AWW efficiently [KV97, KV98b]. In the branching paradigm, where
WS2S<S2S, AWT can recognize exactly all specifications that can be efficiently
checked symbolically. The translation of CTL and AFMC formulas to AWT is
linear and simple [BVW94]. As we have seen in this paper, the translation of
two NBT for a specification and its complementation to AWT involves only a
quadratic blow up. In particular, we believe that model-checking tools like MONA
EKM98, Kla98], which have WS1S and WS2S as their specification languages,
may benefit from employing weak alternating automata.

References

[AC88]   A. Arnold and P. Crubille. A linear algorithm to solve fixed-point equations. *Information Processing Letters*, 29(2):57–66, September 1988.

[AN92]   A. Arnold and D. Niwiński. Fixed point characterization of weak monadic logic definable sets of trees. In *Tree Automata and Languages*, pp. 159–188, 1992. Elsevier.

[BCM+92] J.R. Burch, E.M. Clarke, K.L. McMillan, D.L. Dill, and L.J. Hwang. Symbolic model checking: 10^{20} states and beyond. *Information and Computation*, 98(2):142–170, 1992.

[BL80]   J.A. Brzozowski and E. Leiss. Finite automata and sequential networks. *Theoretical Computer Science*, 10:19–35, 1980.

[Bry86] R.E. Bryant. Graph-based algorithms for boolean-function manipulation. *IEEE Trans. on Computers*, C-35(8), 1986.

[Büc62] J.R. Büchi. On a decision method in restricted second order arithmetic. In *Proc. Internat. Congr. Logic, Method and Philos. Sci. 1960*, pages 1–12, 1962.

[BVW94] O. Bernholtz, M.Y. Vardi, and P. Wolper. An automata-theoretic approach to branching-time model checking. In *Proc. 6th CAV*, LNCS 818, pages 142–155, 1994.

[CGL93] E.M. Clarke, O. Grumberg, and D. Long. Verification tools for finite-state concurrent systems. In *Decade of Concurrency – Reflections and Perspectives (Proceedings of REX School)*, LNCS 803, pages 124–175, 1993.

[CKS81] A.K. Chandra, D.C. Kozen, and L.J. Stockmeyer. Alternation. *Journal of the Association for Computing Machinery*, 28(1):114–133, January 1981.

[CS91]   R. Cleaveland and B. Steffen. A linear-time model-checking algorithm for the alternation-free modal μ-calculus. In *Proc. 3rd CAV*, LNCS 575, pages 48–58, 1991.

[EKM98] J. Elgaard, N. Klarlund, and A. Möller. Mona 1.x: new techniques for WS1S and WS2S. In *Proc 10th CAV*, LNCS 1427, pages 516–520, 1998.

[EL86]   E.A. Emerson and C.-L. Lei. Efficient model checking in fragments of the propositional μ-calculus. In *Proc. 1st LICS*, pages 267–278, 1986.

[HKSV97] R.H. Hardin, R.P. Kurshan, S.K. Shukla, and M.Y. Vardi. A new heuristic for bad cycle detection using BDDs. In *Proc. 9th CAV*, LNCS 1254, pages 268–278, 1997.

[Kla98] N. Klarlund. Mona & Fido: The logic-automaton connection in practice. In *Computer Science Logic, '97*, Lecture Notes in Computer Science, 1998.

[Koz83] D. Kozen. Results on the propositional μ-calculus. *Theoretical Computer Science*, 27:333–354, 1983.

[Kur94] R.P. Kurshan. *Computer Aided Verification of Coordinating Processes*. Princeton Univ. Press, 1994.

[KV97] O. Kupferman and M.Y. Vardi. Weak alternating automata are not that weak. In *Proc. 5th ISTCS*, pages 147–158. IEEE Computer Society Press, 1997.

[KV98a] O. Kupferman and M.Y. Vardi. Freedom, weakness, and determinism: from linear-time to branching-time. In *Proc. 13th LICS*, pages 81-92, 1998.

[KV98b] O. Kupferman and M.Y. Vardi. Weak alternating automata and tree automata emptiness. In *Proc. 30th STOC*, pages 224–233, 1998.

[McM93] K.L. McMillan. *Symbolic Model Checking*. Kluwer Academic Publishers 1993.

[McN66] R. McNaughton. Testing and generating infinite sequences by a finite automaton. *Information and Control*, 9:521–530, 1966.

[MS87] D.E. Muller and P.E. Schupp. Alternating automata on infinite trees. *Theoretical Computer Science*, 54,:267–276, 1987.

[MSS86] D.E. Muller, A. Saoudi, and P.E. Schupp. Alternating automata, the weak monadic theory of the tree and its complexity. In *Proc. 13th ICALP*, 1986.

[Rab69] M.O. Rabin. Decidability of second order theories and automata on infinite trees. *Transaction of the AMS*, 141:1–35, 1969.

[Rab70] M.O. Rabin. Weakly definable relations and special automata. In *Proc. Symp. Math. Logic and Foundations of Set Theory*, pages 1–23. North Holland, 1970.

[Tho90] W. Thomas. Automata on infinite objects. *Handbook of Theoretical Computer Science*, pages 165–191, 1990.

[Var96] M.Y. Vardi. An automata-theoretic approach to linear temporal logic In F. Moller and G. Birtwistle, editors, *Logics for Concurrency: Structure versus Automata*, LNCS 1043, pages 238–266, 1996.

[VW86] M.Y. Vardi and P. Wolper. An automata-theoretic approach to automatic program verification. In *Proc. 1st LICS*, pages 322–331, 1986.

[VW94] M.Y. Vardi and P. Wolper. Reasoning about infinite computations. *Information and Computation*, 115(1):1–37, November 1994.

On the Difference of Horn Theories

(Extended Abstract)

Thomas Eiter[1], Toshihide Ibaraki[2], and Kazuhisa Makino[3]

[1] Institut und Ludwig Wittgenstein Labor für Informationssysteme, Technische Universität Wien, Treitlstraße 3, A-1040 Wien, Austria. (eiter@kr.tuwien.ac.at)
[2] Department of Applied Mathematics and Physics, Graduate School of Informatics, Kyoto University, Kyoto 606, Japan. (ibaraki@kuamp.kyoto-u.ac.jp)
[3] Department of Systems and Human Science, Graduate School of Engineering Science, Osaka University, Toyonaka, Osaka, 560, Japan. (makino@sys.es.osaka-u.ac.jp)

Abstract. In this paper, we consider computing the difference between two Horn theories. This problem may arise, for example, if we take care of a theory change in a knowledge base. In general, the difference of Horn theories is not Horn. Therefore, we consider Horn approximations of the difference in terms of Horn cores (i.e., weakest Horn theories included in the difference). We study the problem under the familiar representation of Horn theories by Horn CNFs, as well as under the recently proposed model-based representation in terms of the characteristic models. For all problems and representations, polynomial time algorithms or proofs of intractability for the propositional case are provided.

Keywords: computational issues in AI, knowledge compilation, difference of Horn theories, Horn approximation, Horn core

1 Introduction

Among the basic operations for combining logical theories are the Boolean operations, i.e., conjunction $\wedge$, disjunction $\vee$, and complement $\neg$. In this paper, we consider another basic operation, the difference $\setminus$, which can represent the complement $\neg\Sigma$ of a theory $\Sigma \subseteq \{0,1\}^n$ (i.e., a set of models) by $\{0,1\}^n \setminus \Sigma$. In principle, we are interested in a particular fragment of theories (such as Horn theories), and would like to know whether the result of such operations also belong to this fragment.

Along this line, we study the problem of computing the Boolean difference between two Horn theories Σ_1 and Σ_2, i.e., $\Sigma = \Sigma_1 \setminus \Sigma_2$. In general, the resulting theory Σ is not Horn. Therefore, we consider approximating Σ by a Horn theory, in order to maintain the desired closure property. Different such approximations are possible, among which Horn cores [11,12] are quite natural.

A Horn theory $\Pi \subseteq \{0,1\}^n$ is a *Horn core* (or a *Horn greatest lower bound*) of a theory $\Sigma \subseteq \{0,1\}^n$, if $\Pi \subseteq \Sigma$, i.e., Π logically implies Σ, and there is no

C. Meinel and S. Tison (Eds.): STACS'99, LNCS 1563, pp. 467–477, 1999.

weaker Π of this property, i.e., no Horn Π' exists such that $\Pi \subset \Pi' \subseteq \Sigma$. Observe that, in general, a theory Σ has more than one Horn core; e.g., $\Sigma = \{(110), (101)\}$ has two Horn cores $\Sigma_1 = \{(110)\}$ and $\Sigma_2 = \{(101)\}$, respectively. Approximating a propositional logic theory by Horn theories (or Horn cores) is used for *knowledge compilation* in [11]. Because of its theoretical and practical importance, semantical and computational issues on Horn cores have been studied extensively, and a number of results have been obtained, cf. [11,2,12,7,1].

The main contributions of the present paper can be summarized as follows.

• We present characterizations of the Horn cores of a Horn difference $\Sigma_1 \setminus \Sigma_2$, which will form a basis of the algorithms discussed in this paper.
• We either present a polynomial time algorithm or prove intractability (unless P=NP) for each of the problems mentioned above.
• Besides the familiar representation in terms of Horn CNFs, we also consider the model-based representation of Horn theories through their sets of characteristic models [10]. This alternative has also been studied repeatedly, since it offers advantages to formula-based representation in certain cases; see [13,10,4] for more details.

Our results on the complexity of these issues are summarized in Table 1, which gives a complete picture of the tractability/intractability frontier of these problems. The table also shows results on the *Horn envelope* [11,12] of the difference, i.e., the (unique) least Horn theory E such that $\Sigma_1 \setminus \Sigma_2 \subseteq E$, which we do not discuss here. The interested reader is referred to [6].

The results on formula-based representation will complement those results known from [2,11,1], which focus on theories represented by CNFs or restricted classes of CNFs, i.e., the problem input is a (possibly restricted) CNF. In contrast, we also deal with non-CNF formulas. The tractability of computing one Horn core of a Horn difference is a positive result, since this problem is intractable for a general theory.

For the case in which a theory is represented by a CNF formula φ, [2] contains a polynomial time algorithm for computing a Horn core of φ by consulting an NP oracle, and shows that the problem is $P^{NP}[O(\log n)]$-hard. Most recently, in [1], another algorithm for computing a Horn core from a given CNF is represented, which is based on the classical Davis-Putnam procedure for the satisfiability problem. Observe, however, that these algorithms are not efficiently applicable to our problem, since they require a CNF formula for input; by rewriting the difference of Horn CNFs φ_1 and φ_2 to a CNF, the size of the formula representing $\varphi_1 \setminus \varphi_2$ might exponentially increase. Our result that the Horn cores of a Horn difference $\varphi_1 \setminus \varphi_2$ can be enumerated with polynomial delay parallels the analogous result if all models of a theory Σ are given as input [12].

Our results find an application in the area of theory change, which has grown into an important research area within AI during the last decade. Taking the difference between Horn theories is, for instance, meaningful in the following scenario. Assume that we have a Horn theory Σ, which is a description of all the possible worlds of a state of affairs; i.e., the real "world" amounts to one of

Table 1. Complexity of problems on the Horn difference $\Sigma_1 \setminus \Sigma_2$ (Σ_1, Σ_2 and Π are Horn)

	representation	
Problem	formula-based (Horn CNF)	model-based (characteristic set)
Is $\Sigma_1 \setminus \Sigma_2$ Horn?	P	P
Compute $\Sigma_1 \setminus \Sigma_2$, if it is Horn	P	P
Is Π a Horn core of $\Sigma_1 \setminus \Sigma_2$?	P	P
Compute one Horn core of $\Sigma_1 \setminus \Sigma_2$	P	P
Compute all Horn cores of $\Sigma_1 \setminus \Sigma_2$	polynomial delay	not polynomial total time unless P = NP
Is Π the Horn envelope of $\Sigma_1 \setminus \Sigma_2$?	co-NP-complete	P
Compute the Horn envelope of $\Sigma_1 \setminus \Sigma_2$	not polynomial total time unless P = NP	P

the models in Σ, say w, but is unknown to us. Suppose that we now obtain the information that a formula φ is false in w. Since w is a categorical description of the world, either ψ or $\neg\psi$ is true in it (but not both), for any formula ψ. Hence, in this case, we conclude that $\neg\varphi$ is true in w, and thus all models v that satisfy φ can be discarded from Σ. This amounts to updating Σ to $\Sigma_{new} = \Sigma \setminus \Sigma'$, where Σ' is the set of models satisfying φ; if φ is Horn, this gives an instance of our problem. If Σ_{new} is not Horn, the Horn cores and the Horn envelope provide sound and complete approximations of the set Σ_{new} in terms of Horn theories. Our algorithms can be applied for computing Σ_{new} or its Horn approximations. Observe that computing a Horn core and the Horn envelope in the context of theory change was studied in [7], in which the theory $\varphi \circ \psi$, given by two Horn CNFs φ and ψ, is studied for some revision operators "$\circ$" (see [9]). However, their work is only weakly related to ours, since taking the difference of φ and ψ was not considered, and moreover, the formula ψ is restricted to be a single Horn clause.

For space reasons, most proofs are omitted. They are given in [6].

2 Preliminaries

We assume a supply of propositional variables $x_1, x_2, \ldots, x_n$, where each x_i evaluates to either 1 (true) or 0 (false). Negated variables are denoted by $\bar{x}_i$. The x_i and $\bar{x}_i$ are called literals. A clause is a disjunction $c = \ell_1 \vee \cdots \vee \ell_k$ of literals, while a term is a conjunction $t = \ell_1 \wedge \cdots \wedge \ell_k$ of literals. By $P(c)$ and $N(c)$ (resp., $P(t)$

and $N(t)$) we denote the sets of variables occurring positively and negatively in c (resp., t); $\perp$ (resp., $\top$) denotes the empty clause (resp., empty term) representing falsity (resp., truth). A *conjunctive normal form* (CNF) (resp., *disjunctive normal form* (DNF)) is a conjunction of clauses $\varphi = \bigwedge_i c_i$ (resp., a disjunction of terms $\varphi = \bigvee_i t_i$).

A *model* is a vector $v \in \{0,1\}^n$, whose i-th component is denoted by v_i. The models $(0,0,\ldots,0)$ and $(1,1,\ldots,1)$ are denoted by $\mathbf{0}$ and $\mathbf{1}$, respectively. We use $v \le w$ for the usual bitwise ordering of models, i.e., $v_i \le w_i$ for all $i = 1,\ldots,n$, where $0 \le 1$. A *theory* is any set $\Sigma \subseteq \{0,1\}^n$ of models. A model $v \in \Sigma$ is *minimal* in Σ, if no $w \in \Sigma$ exists such that $w < v$.

For any formula φ, let $T(\varphi) = \{v \in \{0,1\}^n \mid \varphi(v) = 1\}$ be the set of its models. A formula φ represents a theory Σ if $T(\varphi) = \Sigma$. If unambiguous, we do not distinguish a formula from the theory it represents. We write $\varphi \le \psi$, if $T(\varphi) \subseteq T(\psi)$ holds. A nontautological clause c (resp., noncontradictory term t) is an *implicate* (resp., *implicant*) of a theory Σ if $c(v) = 1$ for all $v \in \Sigma$, i.e., $\{0,1\}^n \supset T(c) \supseteq \Sigma$ (resp., $t(v) = 0$ for all $v \notin \Sigma$, i.e., $\emptyset \subset T(t) \subseteq \Sigma$); it is *prime*, if no proper subclause (resp., subterm) is an implicate (resp., implicant) of Σ.

A theory Σ is *Horn* if $\Sigma = Cl_\wedge(\Sigma)$ holds, where $Cl_\wedge(S)$ is the closure of $S \subseteq \{0,1\}^n$ under bitwise AND (i.e., intersection) of models v and w, denoted by $v \wedge w$. Observe that any Horn theory Σ has the least (unique smallest) model, which is given by $\bigwedge_{v \in \Sigma} v$. For a Horn theory Σ, a model $v \in \Sigma$ is called *characteristic* [10], if $v \notin Cl_\wedge(\Sigma \setminus \{v\})$ holds. The set of all characteristic models of Σ, the *characteristic set of* Σ, is denoted by $C^*(\Sigma)$. Note that every Horn theory Σ has the unique characteristic set $C^*(\Sigma)$. For example, the theory $\Sigma = \{(0101), (1001), (1000), (0001), (0000)\}$ is Horn, and has $C^*(\Sigma) = \{(0101), (1001), (1000)\}$.

A clause c is *Horn* (resp., *negative, positive*) if $|P(c)| \le 1$ (resp., $|P(c)| = 0$, $|N(c)| = 0$). A CNF is *Horn* (resp., *negative, positive*) if it contains only Horn (resp., negative, positive) clauses. It is well-known that a theory Σ is Horn if and only if it is represented by some Horn CNF, and that all prime implicates of a Horn theory are Horn. A theory is *negative* (resp., *positive*), if it is represented by some negative (resp., positive) CNF. If φ represents a theory Σ, and ψ is Horn CNF representing a Horn core of Σ, then ψ is also called a *Horn core* of φ.

3   Horn Property of the Difference

3.1   Formula-Based Representation

Let $\varphi_1 = \bigwedge_{j=1}^{m_1} c_{1,j}$ and $\varphi_2 = \bigwedge_{j=1}^{m_2} c_{2,j}$ be Horn CNFs. Then

$$\varphi_1 \setminus \varphi_2 \equiv \bigvee_{j=1}^{m_2} (\varphi_1 \wedge \neg c_{2,j}) = \bigvee_{j=1}^{m_2} (\varphi_1 \wedge t_{2,j}), \qquad (3.1)$$

where $t_{2,j}$ is the term equivalent to $\neg c_{2,j}$; e.g., if $c_{2,j} = (\overline{x}_1 \vee \overline{x}_2 \vee x_3)$, then $t_{2,j} = x_1 x_2 \overline{x}_3$. Note that each $\psi_j = \varphi_1 \wedge t_{2,j}$ is a Horn CNF. Hence $\varphi_1 \setminus \varphi_2$ can be represented by the disjunction of m_2 Horn theories ψ_j. Observe that in

general, deciding whether a disjunction of Horn CNFs is Horn is co-NP-complete
[5]. However, we sketch a proof that checking the Horn property of the difference
of Horn theories is polynomial.

If $\varphi_2 \equiv \bot$, then

$$\varphi_1 \setminus \varphi_2 \equiv \varphi_1 \tag{3.2}$$

clearly holds, and hence the difference is Horn. Otherwise, we compute the least
model v in φ_2, and separately consider the two cases: (1) $\varphi_1(v) = 0$ and (2)
$\varphi_1(v) = 1$.

Case (1). Since φ_1 is Horn, there is a Horn clause $c^* \in \varphi_1$ such that $c^* \geq \varphi_1$
and $c^*(v) = 0$. For a model u, let $HC(u)$ denote the set of all maximal Horn
clauses c such that $c(u) = 0$. E.g., if $u = (111000)$, then $HC(u) = \{(\overline{x}_1 \vee \overline{x}_2 \vee
\overline{x}_3 \vee x_4), (\overline{x}_1 \vee \overline{x}_2 \vee \overline{x}_3 \vee x_5), (\overline{x}_1 \vee \overline{x}_2 \vee \overline{x}_3 \vee x_6)\}$. Now, some c in $HC(v)$ satisfies
$c \geq c^* (\geq \varphi_1)$.

It can be shown that if $v \neq \mathbf{1}$, then

$$\varphi_1 \setminus \varphi_2 \equiv \varphi_1 \setminus (\varphi_2 \wedge x_i) \tag{3.3}$$

holds for the x_i such that $P(c) = \{x_i\}$. On the other hand, if $v = \mathbf{1}$, then
$T(\varphi_2) = \{\mathbf{1}\}$ and it follows

$$\varphi_1 \setminus \varphi_2 \equiv \varphi_1. \tag{3.4}$$

Case (2). Clearly $(\varphi_1 \setminus \varphi_2)(v) = 0$. Thus, if $\varphi_1 \setminus \varphi_2$ represents a Horn theory,
there exists a Horn clause c^* such that $c^* \geq \varphi_1 \setminus \varphi_2$ and $c^*(v) = 0$. Some c in
$HC(v)$ then satisfies $c \geq c^* (\geq \varphi_1 \setminus \varphi_2)$. It can be shown that if $v \neq \mathbf{1}$, then

$$\varphi_1 \setminus \varphi_2 \equiv (\varphi_1 \wedge c) \setminus (\varphi_2 \wedge x_i) \tag{3.5}$$

holds, where the x_i appears in c (i.e., $P(c) = \{x_i\}$).

On the other hand, if $v = \mathbf{1}$, then in a manner similar to (3.4), we prove

$$\varphi_1 \setminus \varphi_2 \equiv \varphi_1 \wedge c. \tag{3.6}$$

Now we iterate as long as possible. As soon as $\varphi_2 = \bot$ or $v = \mathbf{1}$ holds, we can
conclude that $\varphi_1 \setminus \varphi_2$ is Horn by (3.2), (3.4) or (3.6). If there is no $c \in HC(v)$
such that $c \geq \varphi_1 \setminus \varphi_2$, then $\varphi_1 \setminus \varphi_2$ is not Horn. In the remaining cases, we apply
(3.3) or (3.5) (i.e., φ_1 is modified to $\varphi_1 \wedge c$ in case of (3.5), and φ_2 to $\varphi_2 \wedge x_i$ in
both cases).

It can be shown that this procedure halts in finitely many steps. Formally,
the algorithm can be written as follows.

Algorithm CHECK-HORN
Input: Horn CNFs φ_1 and φ_2.
Output: If $\varphi_1 \setminus \varphi_2$ is a Horn theory Σ, then output a Horn CNF for Σ; otherwise,
"No".

Step 0. $\phi_1 := \varphi_1$; $\phi_2 := \varphi_2$;
Step 1. **if** $\phi_2 \equiv \bot$ **then** output ϕ_1 and halt
    else begin compute the least model v of ϕ_2
        if $\phi_1(v) = 0$ **then goto** Step 2
        else (i.e., $\phi_1(v) = 1$) **goto** Step 3
    end;
Step 2. **if** $v = 1$ **then** output ϕ_1 and halt
    else begin
        find a Horn clause c in $HC(v)$ such that $c \geq \phi_1$;
        $\phi_2 := \phi_2 \wedge x_i$, where $P(c) = \{x_i\}$;
        goto Step 1
    end;
Step 3. Find a Horn clause c in $HC(v)$ such that $c \geq \phi_1 \setminus \phi_2$;
    if no such c exists **then** output "No" and halt
    else begin $\phi_1 := \phi_1 \wedge c$;
        if $v = 1$ **then** output ϕ_1 and halt
        else $\phi_2 := \phi_2 \wedge x_i$, where $P(c) = \{x_i\}$;
        goto Step 1
    end.   ⊏

Example 1. Let us apply CHECK-HORN to $\varphi_1 = (\overline{x}_1 \vee \overline{x}_2) \wedge (\overline{x}_1 \vee x_3) \wedge (\overline{x}_2 \vee x_3) \wedge (\overline{x}_2 \vee x_4)$ and $\varphi_2 = \overline{x}_4 \wedge (\overline{x}_1 \vee \overline{x}_3) \wedge (\overline{x}_2 \vee \overline{x}_3)$. As we can check, the theory represented by $\varphi_1 \setminus \varphi_2$ is $\Sigma = \{(0111), (1011), (1010), (0011), (0001)\}$. Let $\phi_1 := \varphi_1$ and $\phi_2 := \varphi_2$. Since $\phi_2 \not\equiv \bot$ in Step 1, the least model $v = (0000)$ of ϕ_2 is computed. Since $\phi_1(v) = 1$, we branch to Step 3. Then $HC(v) = \{x_i \mid i = 1, 2, \ldots, 4\}$, and we can check that $\phi_1 \setminus \phi_2 \not\leq x_i$ holds for all $i = 1, 2, \ldots, 4$. Thus the algorithm outputs "No" in Step 3. Observe that this is correct, since (1010), $(0001) \in \Sigma$ but $(0000) = (1010) \wedge (0001) \notin \Sigma$, which means that Σ is not Horn.

On the other hand, let φ_1 as above and $\varphi_2' = \overline{x}_4 \wedge (\overline{x}_2 \vee \overline{x}_3)$. Then, $\varphi_1 \setminus \varphi_2'$ represents $\Sigma' = \{(0111), (1011), (0011), (0001)\}$. Let $\phi_1 := \varphi_1$ and $\phi_2 := \varphi_2'$. Then, in Step 1 the least model is $v = (0000)$, and we again branch to Step 3 where $HC(v)$ is as previous. Now, $\varphi_1 \setminus \varphi_2' \leq x_4$ holds. Hence, we update ϕ_1 to $\phi_1 = \phi_1 \wedge x_4$ and ϕ_2 to $\phi_2 \wedge x_4$. Returning to Step 1, we find that $\phi_2 \equiv \bot$ holds. Hence,

$$\phi_1 = (\overline{x}_1 \vee \overline{x}_2) \wedge (\overline{x}_1 \vee x_3) \wedge (\overline{x}_2 \vee x_3) \wedge (\overline{x}_2 \vee x_4) \wedge x_4$$
$$= (\overline{x}_1 \vee \overline{x}_2) \wedge (\overline{x}_1 \vee x_3) \wedge (\overline{x}_2 \vee x_3) \wedge x_4$$

is output. Observe that Σ' is represented by this ϕ_1, and thus the output is correct.   ⊏

An analysis of the time complexity yields the first result. Note $\varphi \equiv \bot$ for Horn φ is decidable in linear time [3], and that by (3.1), a Horn difference $\varphi_1 \setminus \varphi_2$ is representable as Horn disjunction $\psi_1 \vee \cdots \vee \psi_{m_2}$. Thus, $c \geq \varphi_1 \setminus \varphi_2$ is equivalent to Horn $\psi_i \wedge \neg c \equiv \bot$ for all $i = 1, 2, \ldots, m_2$.

Theorem 1. *Let φ_1 and φ_2 be Horn CNFs. Then, algorithm* CHECK-HORN *checks whether $\varphi_1 \setminus \varphi_2$ is Horn in $O(n^4 + n^3|\varphi_2| + n^2|\varphi_1| + n|\varphi_1||\varphi_2|)$ time.*

3.2   Model-Based Representation

Let Σ_1 and Σ_2 be Horn theories. Let S be the set of models defined by

$$S = C^*(\Sigma_1) \cup \{v \bigwedge w \mid v, w \in C^*(\Sigma_1)\}. \tag{3.7}$$

That is, S augments $C^*(\Sigma_1)$ by one step of the closure operator $Cl_\wedge(\cdot)$. We split S into S_1 and S_2 as follows:

$$S_1 \cap \Sigma_2 = \emptyset \quad \text{and} \quad S_2 \subseteq \Sigma_2. \tag{3.8}$$

It is easy to see that $S_1 \subseteq \Sigma_1 \setminus \Sigma_2$ and $Cl_\wedge(S_2) \subseteq \Sigma_2$.

Lemma 1. *For a Horn theory Σ, let S as in (3.7), and let S_1 and S_2 be sets of models such that $S_1 \cup S_2 = S$. Then $Cl_\wedge(S_1) \cup Cl_\wedge(S_2) = \Sigma$.*

By this lemma, it holds for the S_1 and S_2 of (3.8) that

$$Cl_\wedge(S_1) \cup Cl_\wedge(S_2) = \Sigma_1. \tag{3.9}$$

The next lemma leads then to a polynomial time algorithm for our problem.

Lemma 2. *Let Σ_1 and Σ_2 be Horn theories, and let S_1 as in (3.8). Then $Cl_\wedge(S_1) \cap \Sigma_2 = \emptyset$ holds if and only if $\Sigma_1 \setminus \Sigma_2$ is a Horn theory. Furthermore, if $\Sigma_1 \setminus \Sigma_2$ is a Horn theory, then $Cl_\wedge(S_1) = \Sigma_1 \setminus \Sigma_2$ (i.e., $C^*(S_1) = C^*(\Sigma_1 \setminus \Sigma_2)$) holds.*

It is known [4] that given $Q_1, Q_2 \subseteq \{0,1\}^n$, deciding $Cl_\wedge(Q_1) \cap Cl_\wedge(Q_2) = \emptyset$ is possible in $O(n(|Q_1| + |Q_1|))$ time. Since $\Sigma_2 = Cl_\wedge(C^*(\Sigma_2))$, we thus obtain from Lemma 2, turned into a straightforward algorithm, the following result.

Theorem 2. *Let Σ_1, Σ_2 be Horn theories. Given $C^*(\Sigma_1)$ and $C^*(\Sigma_2)$, we can check whether $\Sigma_1 \setminus \Sigma_2$ is Horn in $O(n|C^*(\Sigma_1)|^2|C^*(\Sigma_2)|)$ time. Furthermore, if $\Sigma_1 \setminus \Sigma_2$ is Horn, $C^*(\Sigma_1 \setminus \Sigma_2)$ can be computed in $O(n|C^*(\Sigma_1)|^2(|C^*(\Sigma_1)|^2 + |C^*(\Sigma_2)|))$ time.*

4   Horn Cores

4.1   Formula-Based Representation

We first consider the problem of computing one Horn core of the difference $\varphi_1 \setminus \varphi_2$, where φ_1 and φ_2 are Horn CNFs. The algorithm is a modification of algorithm CHECK-HORN, which checks whether the difference $\varphi_1 \setminus \varphi_2$ is Horn.

In Step 3 of algorithm CHECK-HORN, if no Horn clause c is in $HC(v)$ such that $c \geq \phi_1 \setminus \phi_2$, we conclude that the difference is not Horn and halt. To compute a Horn core, however, we update ϕ_1 and ϕ_2 as $\phi_1 := \phi_1 \wedge c$ and $\phi_2 := \phi_2 \wedge x_i$, respectively, for an *appropriate* $c \in HC(v)$. This is because no Horn core contains v.

Denote for a formula ψ and a model w, by $\min_{\geq w}(\psi)$ the set of minimal models u such that $\psi(u) = 1$ and $u \geq w$. Then, it holds that any clause $c \in HC(v)$ such that $c(u) = 1$ for some $u \in \min_{\geq v}(\phi_1 \setminus \phi_2)$ is appropriate for our purpose.

For models v, u with $v \leq u$, let $HC(v; u)$ be the set of Horn clauses $c \in HC(v)$ whose positive literal x_j satisfies $u_j = 1$. E.g., if $v = (111000)$ and $u = (111110)$, then $HC(v; u) = \{(\overline{x}_1 \vee \overline{x}_2 \vee \overline{x}_3 \vee x_4), (\overline{x}_1 \vee \overline{x}_2 \vee \overline{x}_3 \vee x_5)\}$. To make the discussion clear and speed up the algorithm, based on a model $u \in \min_{\geq v}(\phi_1 \setminus \phi_2)$, we update ϕ_1 and ϕ_2 respectively by

$$\phi_1 := \phi_1 \wedge \bigwedge_{c \in HC(v;u)} c \quad \text{and} \quad \phi_2 := \phi_2 \wedge \bigwedge_{u_j=1 \wedge v_j=0} x_j. \tag{4.10}$$

Thus, we have the following algorithm.

Algorithm HORN-CORE1
Input: Horn CNFs φ_1 and φ_2.
Output: A Horn core of $\varphi_1 \setminus \varphi_2$.

    Steps 0.–2. as in CHECK-HORN
    Step 3. if there exists a Horn clause c in $HC(v)$ such that $c \geq \phi_1 \setminus \phi_2$
        then begin $\phi_1 := \phi_1 \wedge c$;
            if $v = 1$ then output ϕ_1 and halt
            else $\phi_2 := \phi_2 \wedge x_i$, where $P(c) = \{x_i\}$
        end
        else begin find a model $u \in \min_{\geq v}(\phi_1 \setminus \phi_2)$;
            $\phi_1 := \phi_1 \wedge \bigwedge_{c \in HC(v;u)} c$;
            $\phi_2 := \phi_2 \wedge \bigwedge_{u_j=1 \wedge v_j=0} x_j$
        end
        goto Step 1.   ⊏

Let $\phi_1^{(k)}$ and $\phi_2^{(k)}$ respectively denote the ϕ_1 and ϕ_2 in Step 1 of the k-th iteration, and let $v^{(k)}$ be the least model in $\phi_2^{(k)}$. Then, it can be shown that

$$v^{(k)} < v^{(k+1)}, \quad \text{for } k = 1, 2, \ldots. \tag{4.11}$$

This implies that the number of iterations is at most $n + 1$. The only nonobvious polynomial operation in HORN-CORE1 is finding some $u \in \min_{\geq v}(\phi_1 \setminus \phi_2$ in Step 3. The models $\{w \geq v \mid (\phi_1 \setminus \phi_2)(w) = 1\}$ can be described by a disjunction of m Horn CNFs $\psi_1, \ldots, \psi_m$, where m is the number of clauses in ϕ_2. Consequently, the models in $\min_{\geq v}(\phi_1 \setminus \phi_2)$ are the minimal models among the least models of $\psi_1, \ldots, \psi_m$. Each ψ_i results by fixing the value of some literals in ϕ_1, and its least model can be computed in $O(|\phi_1| + n)$ time exploiting [3]. Looking at the number of 1-bits in each model, we can thus find some $u \in \min_{\geq v}(\phi_1 \setminus \phi_2)$ in $O(m(|\phi_1| + n) + mn) = O(m(|\phi_1| + n))$ time. An analysis of the time complexity gives us then the following result.

Theorem 3. *Let φ_1 and φ_2 be Horn CNFs. Then algorithm HORN-CORE1 outputs a Horn core of $\varphi_1 \setminus \varphi_2$ in $O(n^4 + n^3|\varphi_2| + n^2|\varphi_1| + n|\varphi_1||\varphi_2|)$ time.*

As regards computing all Horn cores of $\varphi_1 \setminus \varphi_2$, we note that this problem is provably not solvable in polynomial time, since in general, an exponential number of Horn cores might exist [6]. However, we have a *polynomial total time* algorithm (called *output-polynomial* in [8]), i.e., it runs in polynomial time in the combined size of the input and the output. In fact, it enumerates all Horn cores with polynomial delay [8].

Note that in HORN-CORE1, the choice of $u \in \min_{\geq v}(\phi_1 \setminus \phi_2)$ in the else-statement of Step 3 results in a Horn core ψ of $\varphi_1 \setminus \varphi_2$ such that $\psi(u) = 1$ and $\psi(u') = 0$ for all other models $u' \in \min_{\geq v}(\phi_1 \setminus \phi_2)$. We can also show that every Horn core ψ satisfies $\psi(u) = 1$ for some u. This means that every Horn core can be constructed by algorithm HORN-CORE1, if it properly chooses a model u in the else-statement of Step 3.

Thus, all Horn cores can be generated as follows.

Algorithm ALL-HORN-CORES(φ_1, φ_2)
Input: Horn CNFs φ_1 and φ_2.
Output: All Horn cores of $\varphi_1 \setminus \varphi_2$.

> **Steps 0.–2.** as in HORN-CORE1 and CHECK-HORN.
> **Step 3.** **if** there exists a Horn clause c in $HC(v)$ such that $c \geq \phi_1 \setminus \phi_2$
> > **then begin** $\phi_1 := \phi_1 \wedge c$;
> > > **if** $v = 1$ **then** output ϕ_1 and exit;
> > > **else begin** $\phi_2 := \phi_2 \wedge x_i$, where $P(c) = \{x_i\}$; **goto** Step 1 **end**
> > **end**
> > **else for** each a model $u \in \min_{\geq v}(\phi_1 \setminus \phi_2)$ **do**
> > > **begin** $\phi_1 := \phi_1 \wedge \bigwedge_{c \in HC(v;u)} c$;
> > >
> > > $$\phi_2 := \phi_2 \wedge \bigwedge_{u_j=1 \wedge v_j=0} x_j;$$
> > >
> > > call ALL-HORN-CORES(ϕ_1, ϕ_2)
> > **end**{for}. □

An analysis of the time complexity leads us to the following result.

Theorem 4. *Algorithm ALL-HORN-CORES correctly generates all Horn cores of $\varphi_1 \setminus \varphi_2$ with polynomial delay, where the delay is bounded by the time of computing one Horn core, i.e., $O(n^4 + n^3|\varphi_2| + n^2|\varphi_1| + n|\varphi_1||\varphi_2|)$.*

4.2   Model-Based Representation

In Subsection 3.2, we have seen that S_1 of (3.8) characterizes the difference $\Sigma_1 \setminus \Sigma_2$ of two Horn theories Σ_1 and Σ_2. In this subsection, we show that any maximal set $Q \subseteq S_1$ such that $Cl_\wedge(Q) \cap \Sigma_2 = \emptyset$ gives a Horn core of $\Sigma_1 \setminus \Sigma_2$, and conversely any Horn core of $\Sigma_1 \setminus \Sigma_2$ can be generated from such a maximal set $Q \subseteq S_1$.

Note that, by the definition of S_1, any maximal set $Q \subseteq S$ with $Cl_\wedge(Q) \cap \Sigma_2 = \emptyset$ is contained in S_1. Thus we prove the above statements by using a set S instead of S_1.

Theorem 5. *Let Σ_1, Σ_2 be Horn theories and let S as in (3.7). Then, Π is a Horn core of $\Sigma_1 \setminus \Sigma_2$ iff $\Pi = Cl_\wedge(Q)$ for some maximal $Q \subseteq S$ such that $Cl_\wedge(Q) \cap \Sigma_2 = \emptyset$.* □

Based on Theorem 5, we obtain the following polynomial time algorithm.

Algorithm HORN-CORE2
Input: Characteristic sets $C^*(\Sigma_1)$ and $C^*(\Sigma_2)$ of Horn theories Σ_1 and Σ_2.
Output: The characteristic set $C^*(\Pi)$ of a Horn core Π of $\Sigma_1 \setminus \Sigma_2$.

Step 0. $Q := \emptyset$;
    compute the set of models S given by (3.7);
Step 1. for each $w \in S$ **do begin**
    if $Cl_\wedge(Q \cup \{w\}) \cap \Sigma_2 = \emptyset$ **then** $Q := Q \cup \{w\}$
    end{for};
Step 2. output $C^*(Q)$ and halt. □

Theorem 6. *Let Σ_1 and Σ_2 be Horn theories. Then, HORN-CORE2 outputs the characteristic set of a Horn core of $\Sigma_1 \setminus \Sigma_2$ in $O(n|C^*(\Sigma_1)|^2(|C^*(\Sigma_1)|^2 + |\Sigma_2|))$ time.*

Is also possible to decide whether a given theory Π is a Horn core of the difference $\Sigma_1 \setminus \Sigma_2$ in polynomial time. A suitable algorithm is straightforward by proper modification of algorithm HORN-CORE2. We omit the details (see [6]).

As for the efficient computation of all Horn cores, we have a negative result. There is no polynomial total time algorithm for generating all Horn cores, unless $P = NP$.

Theorem 7. *There is no algorithm which, given $C^*(\Sigma_1)$ and $C^*(\Sigma_2)$ of Horn theories Σ_1 and Σ_2, computes the characteristic sets $C^*(\Pi)$ of all Horn cores Π of $\Sigma_1 \setminus \Sigma_2$ in polynomial total time (i.e., polynomial in the combined size of the input and the output), unless $P = NP$.*

5   Conclusion and Further Work

The results of the present and companion papers [4,5] establish operations towards a Boolean "calculus" on Horn theories, in which Horn theories are combined using the operations of conjunction ($\wedge$), disjunction ($\vee$), and difference ($\setminus$). These operations are useful, for example, in the context of changing theories that are possible world representations of a state of affairs.

Several issues remain for our ongoing work. One is giving a precise account to computing Horn cores and the Horn envelope of the complement $\overline{\Sigma} = \{0, 1\}^n \setminus \Sigma$

of a Horn theory. Our results imply polynomial time algorithms in some cases, but a complete picture remains to be drawn. Another issue is a more accurate account of the complexity of the polynomial cases in Table 1. Under formula-based representation, all these problems are complete for P under logspace reductions; this is an easy consequence of the fact that deciding the satisfiability of a Horn CNF is complete for P under logspace reductions. Therefore, parallelization of these problems is most likely not possible. Another interesting issue is a study of the effect of Horn renamings (cf. [1]).

Acknowledgments. The authors appreciate the comments of the anonymous reviewers.

References

1. Y. Boufkhad. Algorithms for Propositional KB Approximation. In *Proc. National Conference on AI (AAAI '98)*, Madison, Wisconsin, pp. 280–285, July 26–30 1998.
2. M. Cadoli. Semantical and Computational Aspects of Horn Approximations. In *Proc. IJCAI-93*, pp. 39–44, 1993.
3. W. Dowling and J. H. Gallier. Linear-time Algorithms for Testing the Satisfiability of Propositional Horn Theories. *J. Logic Programming*, 3:267–284, 1984.
4. T. Eiter, T. Ibaraki, and K. Makino. Computing Intersections of Horn Theories for Reasoning with Models. In: *Proc. AAAI '98*, pp. 292–297, 1998.
5. T. Eiter, T. Ibaraki, and K. Makino. Disjunctions of Horn Theories and their Cores. In *Proc. 9th Int'l Symp. on Algorithms and Computation (ISAAC '98)*, Dec. 1998.
6. T. Eiter, T. Ibaraki, and K. Makino. On the Difference of Horn Theories. RUTCOR Research Report RRR 25-98, Rutgers University 1998.
7. G. Gogic, C. Papadimitriou, and M. Sideri. Incremental Recompilation of Knowledge. *J. Artificial Intelligence Research*, 8:23–37, 1998.
8. D. S. Johnson, M. Yannakakis, and C. H. Papadimitriou. On Generating All Maximal Independent Sets. *Information Processing Letters*, 27:119–123, 1988.
9. H. Katsuno and A. O. Mendelzon. Propositional Knowledge Base Revision and Minimal Change. *Artificial Intelligence*, 52:253–294, 1991.
10. H. Kautz, M. Kearns, and B. Selman. Reasoning With Characteristic Models. In *Proc. AAAI '93*, 1993.
11. H. Kautz and B. Selman. Knowledge Compilation and Theory Approximation. *JACM*, 43(2):193–224, 1996.
12. D. Kavvadias, C. Papadimitriou, and M. Sideri. On Horn Envelopes and Hypergraph Transversals. In *Proc. ISAAC '93*, LNCS 762, pp. 399–405, 1993.
13. R. Khardon and D. Roth. Reasoning with Models. *Art. Intelligence*, 87(1/2):187–213, 1996.

On Quantum Algorithms for
Noncommutative Hidden Subgroups

Mark Ettinger[1] and Peter Høyer[2,1,*]

[1] Los Alamos National Laboratory, Mail Stop B-230, Los Alamos, NM 87545
ettinger@lanl.gov
[2] Odense University, Department of Mathematics and Computer Science,
Campusvej 55, DK-5230 Odense M, Denmark
u2pi@imada.ou.dk

Abstract. Quantum algorithms for factoring and finding discrete logarithms have previously been generalized to finding hidden subgroups of finite Abelian groups. This paper explores the possibility of extending this general viewpoint to finding hidden subgroups of noncommutative groups. We present a quantum algorithm for the special case of dihedral groups which determines the hidden subgroup in a linear number of calls to the input function. We also explore the difficulties of developing an algorithm to process the data to explicitly calculate a generating set for the subgroup. A general framework for the noncommutative hidden subgroup problem is discussed and we indicate future research directions.

1 Introduction

All known quantum algorithms which run super-polynomially faster than the most efficient probabilistic classical algorithm solve special cases of what is called the Abelian Hidden Subgroup Problem. This general formulation includes Shor's celebrated algorithms for factoring and finding discrete logarithms [15]. A very natural question to ask is if quantum computers can efficiently solve the Hidden Subgroup Problem in *noncommutative* groups. This question has been raised regularly [1,10,11,12], and seems important for at least three reasons.

The first reason is that determining if two graphs are isomorphic reduces to finding hidden subgroups of symmetric groups. The second reason is that the noncommutative hidden subgroup problem arguably represents a most natural line of research in the area of quantum algorithmics. The third reason is that an efficient quantum algorithm for a hidden subgroup problem could potentially be used to show an exponential gap between quantum and classical two-party probabilistic communication complexity models [7,6].

The heart of the idea behind the quantum solution to the Abelian hidden subgroup problem is Fourier analysis on Abelian groups. The difficulties of Fourier analysis on noncommutative groups makes the noncommutative version of the problem very challenging.

* Supported in part by BRICS—Basic Research in Computer Science, Centre of the Danish National Research Foundation.

C. Meinel and S. Tison (Eds.): STACS'99, LNCS 1563, pp. 478–487, 1999.
© Springer-Verlag Berlin Heidelberg 1999

In this paper, we present the first known quantum algorithm for a noncommutative subgroup problem. We focus on dihedral groups because they are well-structured noncommutative groups, and because they contain an exponentially large number of different subgroups of small order, making classical guessing infeasible. Our main result is that there exists a quantum algorithm that solves the dihedral subgroup problem using only a linear number of evaluations of the function which is given as input. This is the first time such a result has been obtained for a noncommutative group.

However, we hasten to add that our algorithm does *not* run in polynomial time, even though it only uses few evaluations of the given function. The reason for this is as follows: Our algorithm first applies a certain polynomial-time quantum subroutine a linear number of times, each time producing some output data, and each time using just one application of the given input function. The collection of all the output data determines the hidden subgroup with high probability. We know how to find the subgroup from those data in exponential time, but we do not know if this task can be done efficiently.

Three important questions are left open. The first question is if there exists a polynomial-time algorithm (classical or quantum) to postprocess the output data from our quantum subroutine. The second is whether our algorithm can be used to show an exponential gap between quantum and classical probabilistic communication complexity models, as mentioned above. Currently, the state-of-the-art is an exponential separation between error-free models, and a quadratic separation between probabilistic models [6]. The third open question is for what other noncommutative groups similar results can be obtained.

2 Algorithm for Dihedral Groups

The *Hidden Subgroup Problem* is defined as follows:

Given: A function $\gamma : G \to R$, where G is a finite group and R an arbitrary finite range.

Promise: There exists a subgroup $H \leqslant G$ such that γ is *constant* and *distinct* on the left cosets of H.

Problem: Find a generating set for H.

We say of such a function γ that it *fulfills the subgroup promise* with respect to H. We also say of γ that it *has hidden subgroup H*. Note that we are not given the order of H. Without loss of generality we assume γ is constant and distinct on *left* cosets because we may formally rename group elements and convert multiplication on the right to multiplication on the left. We assume throughout this paper that function γ is given as a black box, so that it is not possible to obtain knowledge about it by any other means than evaluating it on points in its domain.

If G is Abelian, then we refer to this problem as the Abelian Subgroup Problem. Similarly, if the given group is dihedral, then we refer to it as the Dihedral Subgroup Problem. Classically, if γ is given as a black box, then the

Abelian subgroup problem is intractable: If $G = \mathbb{Z}_2^n$, then just to determine if H is non-trivial or not takes time exponential in n [16]. Here, $\mathbb{Z}_2$ denotes the cyclic group of order 2. In contrast, the Abelian subgroup problem can be solved efficiently on a quantum computer [3,5,8,12,15,16].

Theorem 1 (Abelian case). *Let* $\gamma : G \to R$ *be a function that fulfills the Abelian subgroup promise with respect to H. There exists a quantum algorithm that outputs a subset $X \subseteq H$ such that X is a generating set for H with probability at least $1 - 1/|G|$, where $|G|$ denotes the order of G. The algorithm uses $O(\log |G|)$ evaluations of γ, and runs in time polynomial in $\log |G|$ and in the time required to compute γ.*

We review the quantum solution to the Abelian subgroup problem in terms of group representation theory in Section 4 below. For other reviews, see for example [4,11].

The dihedral group of order $2N$ is the symmetry group of an N–sided polygon. It is isomorphic to a semidirect product of the two cyclic groups $\mathbb{Z}_N$ and $\mathbb{Z}_2$ of order N and 2, respectively,

$$D_N = \mathbb{Z}_N \rtimes_\phi \mathbb{Z}_2 \qquad (1)$$

with multiplication defined by

$$(a_1, b_1)(a_2, b_2) = \big(a_1 + \phi(b_1)(a_2), b_1 + b_2\big).$$

The homomorphism $\phi : \mathbb{Z}_2 \to \mathrm{Aut}(\mathbb{Z}_N)$ is given by $1 \mapsto \phi(1)(a) = -a$. An element $(a, b) \in D_N$ is a *rotation* if $b = 0$, and a *reflection* if $b = 1$. The group D_N contains N rotations and N reflections, and the N rotations comprise the cyclic subgroup $\mathbb{Z}_N \times \{0\} \leqslant D_N$ of index 2.

Theorem 2 (Main theorem). *Let* $\gamma : D_N \to R$ *be a function that fulfills the dihedral subgroup promise with respect to H. There exists a quantum algorithm that given γ, uses $\Theta(\log N)$ evaluations of γ and outputs a subset $X \subseteq H$ such that X is a generating set for H with probability at least $1 - \frac{2}{N}$.*

Theorem 2 constitutes our main result that the dihedral subgroup problem can be solved with few applications of the given function γ. The essential part of the proof is that it is possible to find a hidden reflection.

Theorem 3 (Finding a reflection). *Let* $\gamma : D_N \to R$ *be a function that fulfills the dihedral subgroup promise with respect to H. Suppose we are promised that $H = \{0\}$ is either trivial, or $H = \{0, r\}$ is generated by a reflection $r \in D_N$. Then there exists a quantum algorithm that given γ, outputs either "trivial" or the reflection r. If H is trivial then the output is always "trivial", otherwise the algorithm outputs r with probability at least $1 - \frac{1}{2N}$. The algorithm uses at most $89 \log(N) + 7$ evaluations of γ, but it runs in time $O(\sqrt{N})$.*

In Theorem 3, suppose N is even and consider the decision problem of determining, for a hidden reflection $r = (k_0, 1)$, whether integer k_0 is even or odd. By that theorem, we can solve this decision problem with vanishing small error probability using a number of evaluations of γ linear in $\log N$. In contrast, this decision problem is infeasible classically: just to obtain success probability $1/2 + 2^{-n/3}$ requires more than $2^{n/3}$ evaluations of γ, where $n = \log(2N)$.[1] The reduction of the general problem given in Theorem 2 to the special case of order-2 subgroups in Theorem 3 can be found in the appendix, so from now on, we consider only hidden subgroups of order 2 and of order 1.

We assume that the reader is familiar with the basic notions of quantum computation [2]. The quantum algorithm we shall use to prove Theorem 3 is

$$\mathcal{V}_\gamma = \left(\mathbf{F}_N \otimes \mathbf{W} \otimes \mathbf{I}\right) \circ \mathbf{U}_\gamma \circ \left(\mathbf{F}_N^{-1} \otimes \mathbf{W} \otimes \mathbf{I}\right). \tag{2}$$

Here, $\mathbf{I}$ is the identity operator and $\mathbf{U}_\gamma$ is any unitary operator that satisfies that

$$\mathbf{U}_\gamma |a\rangle|b\rangle|0\rangle = |a\rangle|b\rangle|\gamma(a,b)\rangle \tag{3}$$

for all elements $(a, b) \in D_N$. The operator $\mathbf{F}_N = \frac{1}{\sqrt{N}} \sum_{i,j=0}^{N-1} \omega_N^{ij} |j\rangle\langle i|$ is the quantum Fourier transform for the cyclic group $\mathbb{Z}_N$, where $\omega_N = \exp(2\pi\sqrt{-1}/N)$ is the Nth principal root of unity. When $N = 2$, then the Fourier transform $\mathbf{F}_2$ is equal to the Walsh–Hadamard transform $\mathbf{W}$ which maps a qubit in state $|b\rangle$ to the superposition $\frac{1}{\sqrt{2}}(|0\rangle + (-1)^b|1\rangle)$.

Suppose for a moment that we were not given a function defined on the dihedral group $D_N = \mathbb{Z}_N \rtimes_\phi \mathbb{Z}_2$, but instead a function defined on the Abelian group $\mathbb{Z}_N \times \mathbb{Z}_2$. Or equivalently, suppose for the moment that $\phi : \mathbb{Z}_2 \to \operatorname{Aut}(\mathbb{Z}_N)$ is the trivial homomorphism. Then by Theorem 1, we can find any hidden subgroup with probability exponentially close to 1 by applying the experiment

$$(a, b) = \mathcal{M}_{1,2} \circ \mathcal{V}_\gamma |0\rangle|0\rangle|0\rangle \tag{4}$$

a number of $O(\log N)$ times. Here, $\mathcal{M}_{1,2}$ denotes a measurement of the first two registers with outcome (a, b). A natural question to ask is, how much information, if any, would we gain by performing the experiment given in (4) when γ is defined on D_N and not on $\mathbb{Z}_N \times \mathbb{Z}_2$. Rewriting the state $\mathcal{V}_\gamma|0\rangle|0\rangle|0\rangle$ as a superposition over the basis states shows that we indeed learn something, as quantified in the following lemma.

Lemma 4. *Let $\gamma : D_N \to R$ fulfill the subgroup promise with respect to $H = \{0, r\}$, where $r = (k_0, 1)$ is a reflection. Then, if we apply quantum algorithm $\mathcal{V}_\gamma$ on the initial state $|0\rangle|0\rangle|0\rangle$, the probability that a measurement of the first two registers yields $(a, 0)$, is*

$$\frac{1}{2N}\left(1 + \cos(2\pi k_0 a/N)\right) = \frac{1}{N}\cos^2(\pi k_0 a/N). \tag{5}$$

Furthermore, the probability that the outcome is $(a, 1)$, is $\frac{1}{N}\sin^2(\pi k_0 a/N)$.

---

[1] The classical algorithm fails since there are N possible hidden reflections, and the algorithm can rule out at most T^2 of them by using T queries to γ. This argument is similar to the one used in [16,5] for the Abelian group $\mathbb{Z}_2^n$.

Let $\mathbf{Z}$ denote the discrete random variable defined by the probability mass function

$$\text{Prob}[\mathbf{Z} = z] = \alpha \cos^2(\pi k_0 z/N) \qquad (0 \leq z < N), \qquad (6)$$

where $\alpha = 1/N$ if $k_0 = 0$ or $2k_0 = N$, and $\alpha = 2/N$ otherwise. Lemma 4 provides us with a quantum algorithm for sampling from $\mathbf{Z}$. Intuitively, since the random variable $\mathbf{Z}$ depends on k_0, the more samples we draw from $\mathbf{Z}$, the more knowledge we gather about k_0 and the hidden reflection $r = (k_0, 1)$. The crucial question therefore becomes, how many samples from $\mathbf{Z}$ do we need, to be able to identify k_0 correctly with high probability. Theorem 5 below states that we only need a logarithmic number of samples. We postpone its proof till the next section.

Theorem 5. *Let* $m \geq \lceil 64 \ln N \rceil$, *and let* $z_1, \ldots, z_m$ *be* m *independent samples from* $\mathbf{Z}$. *Let* $\kappa \in \{1, \ldots, \lfloor N/2 \rfloor\}$ *be such that the sum* $\sum_{i=1}^{m} \cos(2\pi\kappa z_i/N)$ *is maximal. Then* $\kappa = \min\{k_0, N - k_0\}$ *with probability at least* $1 - \frac{1}{2N}$.

Proof (of Theorem 3). The algorithm starts by disposing the possibility that $r = (0, 1)$ by evaluating $\gamma(0, 0)$ and $\gamma(0, 1)$. If the two values are equal, then the algorithm outputs the reflection $(0, 1)$ and stops. If N is even, then the algorithm proceeds by disposing the possibility that $r = (N/2, 1)$, too.

Now, the algorithm applies the quantum experiment given in (4) a number of $m' = 2\lceil 64 \ln N \rceil$ times. Let m denote the number of times it measures zero in the second register. Let $\{a_1, \ldots, a_m\}$ denote the outcomes in the first register conditioned to that the measurement of the second register yields a zero.[2]

Suppose $m \geq m'/2$. The algorithm continues with classical post-processing. It finds $1 \leq \kappa \leq \lfloor N/2 \rfloor$ such that the sum $\sum_{i=1}^{m} \cos(2\pi\kappa a_i/N)$ is maximized. It then computes $\gamma(\kappa, 1)$ and compares it with $\gamma(0, 0)$. If they are equal, it outputs the reflection $(\kappa, 1)$ and stops. Otherwise, it performs the same test for $\gamma(N - \kappa, 1)$. If that one also fails, it outputs "trivial".

If $m < m'/2$, then the algorithm performs the same classical post-processing except that it uses the $m' - m$ measurements for which the output in the second register is 1, and except that it now seeks to maximize $\sum_{i=1}^{m} \sin(2\pi\kappa a_i/N)$.

If H is trivial, then the algorithm returns "trivial" with certainty. If $H = \{0, r\}$, then it outputs $r = (k_0, 1)$ with probability at least $1 - 1/2N$ by Theorem 5. The total number of evaluations of γ is at most $m' + 5 < 89 \log N + 5$. Unfortunately, we do not know how to find κ any faster than in time $O(\sqrt{N})$. □

3   Proof of Theorem 5

The proof of Theorem 5 requires two lemmas, the first of them being a result by Hoeffding [9] on the sum of bounded random variables. Hoeffding's lemma says that the probability that the sum of m independent samples are off from its expected value by a constant fraction in m drops exponentially in m.

---

[2] Alternatively, we could apply amplitude amplification [5] to ensure that we will always measure 0 in the second register, instead of as here, only with probability 1/2

Lemma 6 (Hoeffding). *Let* $X_1, \ldots, X_m$ *be independent identically distributed random variables with* $\ell \leq X_1 \leq u$. *Then, for all* $\alpha > 0$,

$$\mathrm{Prob}[S - E[S] \geq \alpha m] \leq e^{-2\alpha^2 m/(u-\ell)^2},$$

where $S = \sum_{i=1}^{m} X_i$.

Let $0 < k < N$, and suppose we want to test if $k \overset{?}{=} k_0$ or $k \overset{?}{=} N - k_0$, where k_0 is given as in Lemma 4. Clearly, we can answer that question just by testing if $\gamma(0,0) \overset{?}{=} \gamma(k,1)$ or $\gamma(0,0) \overset{?}{=} \gamma(N - k, 1)$. Lemma 7 provides us with another probabilistic method: First draw m samples $\{z_i\}_{i=1}^{m}$ from Z, and then compute the sum $\sum_{i=1}^{m} \cos(2\pi k z_i/N)$. Conclude that $k \neq k_0$ and $k \neq N - k_0$ if and only if that sum is at most $m/4$.

Lemma 7. *Fix an integer* k *with* $0 < k < N$. *Let* $z_1, \ldots, z_m$ *be* m *independent samples from* Z. *Then with probability at most* $e^{-m/32}$, *we have*

$$\sum_{i=1}^{m} \cos(2\pi k z_i/N) \leq m/4$$

if $k = k_0$ *or* $k = N - k_0$, *and*

$$\sum_{i=1}^{m} \cos(2\pi k z_i/N) \geq m/4$$

otherwise.

Proof. Let f denote the function of Z defined by $f(z) = \cos(2\pi k z/N)$, and let $X = f(Z)$ denote the random variable defined by f. Then $-1 \leq X \leq 1$ and the expected value of X is

$$E[X] = \begin{cases} 1 & \text{if } 2k = 2k_0 = N \\ \frac{1}{2} & \text{if either } k = k_0 \text{ or } k = N - k_0 \\ 0 & \text{otherwise.} \end{cases}$$

If $k \neq k_0$ and $k \neq N - k_0$, then apply Hoeffding's lemma on m independent random variables all having the same probability distribution as X. If $k = k_0$ or $k = N - k_0$, then apply Hoeffding's lemma on m independent random variables all having the same probability distribution as the random variable $E[X] - X$. □

We are not only concerned about testing for a specific $0 < k \leq N/2$ if $k \overset{?}{=} k_0$ or $k \overset{?}{=} N - k_0$, but in testing every one of them. Fortunately, the probability $e^{-m/32}$ is diminutive, so we can reuse the same m samples in all $N/2$ tests, and still it is very likely that the sum $\sum_{i=1}^{m} \cos(2\pi k z_i/N)$ is larger than $m/4$ if and only if $k = k_0$ or $k = N - k_0$.

Proof (of Theorem 5). This is a simple consequence of Lemma 7. Let $k_0' = \min\{k_0, N - k_0\}$. The probability that $\sum_{i=1}^{m} \cos(2\pi k_0' z_i/N) \leq m/4$ is at most $e^{-m/32} \leq \frac{1}{N^2}$. Furthermore, for every integer $0 < k \leq N/2$ not equal to k_0', the probability that $\sum_{i=1}^{m} \cos(2\pi k z_i/N) \geq m/4$ is also at most $\frac{1}{N^2}$. If $\kappa \neq k_0'$, then one of these $\lfloor N/2 \rfloor$ events must have happened, and the probability for that is upper bounded by $\lfloor \frac{N}{2} \rfloor \frac{1}{N^2} \leq \frac{1}{2N}$. □

4   Abelian Hidden Subgroups

Theorem 1 in Section 2 states that the Abelian subgroup problem can be solved efficiently on a quantum computer. The algorithm which accomplishes this is most easily understood using some basic representation theory for finite Abelian groups which we now briefly review. For more details see the excellent references [13,14]. For any Abelian group G the group algebra $\mathbb{C}[G]$ is the Hilbert space of all complex-valued functions on G equipped with the standard inner product. A *character* of G is a homomorphism from G to $\mathbb{C}$. The set of characters admits a natural group structure via pointwise multiplication and is a basis for the group algebra. The *Fourier transform* is the linear transformation from the point mass basis of the group algebra to the basis of characters. Finally, for any subgroup $H \leqslant G$, there exists a subgroup of the character group called the orthogonal subgroup $H^{\perp}$ which consists of all characters χ such that $\chi(h) = 1$ for all $h \in H$.

We now sketch the quantum algorithm for solving the Abelian hidden subgroup problem. In the interest of clarity we omit all normalization factors in our description. The state of the computer is initialized in the superposition

$$\sum_{g \in G} |g\rangle |\gamma(g)\rangle.$$

We then observe the second register with outcome, say, $q \in R$. This action serves to place the first register into a superposition of all elements that map to q under γ. Because γ is constant and distinct on left cosets of H we may write the new state of the computer as

$$\sum_{h \in H} |s + h\rangle |q\rangle$$

for some coset $s + H$ chosen by the observation of the second register. We then apply the quantum Fourier transform on the first register, producing the state

$$\sum_{\chi \in H^{\perp}} \chi^{*}(s) |\chi\rangle |q\rangle,$$

where $\chi^{*}(s)$ denotes the complex conjugate of $\chi(s)$. Finally, we observe the first register. Notice that this results in a uniformly random sample from $H^{\perp}$.

It can easily be shown that by repeating this experiment of order $\log |H^{\perp}|$ times, we find a subset $X \subseteq H^{\perp}$ that generates $H^{\perp}$ with probability exponentially close to 1. The hidden subgroup $H \leqslant G$ can then be calculated efficiently from $H^{\perp}$ on a classical computer, essentially by linear algebra. In summary, the sole purpose of the quantum machine in the above algorithm is to sample uniformly from $H^{\perp}$. It is known that an arbitrary good approximation to the quantum Fourier transform can be performed efficiently [12], so, assuming the given function γ can be computed in polynomial time, then so does the complete algorithm.

5   A Generalized $H^\perp$

We now briefly discuss the main ideas of harmonic analysis on groups, stating as facts the main results that we require. For more detailed information see for example [13,14]. Let G be a (possibly noncommutative) finite group. A representation of G is a homomorphism $\rho : G \to \mathrm{GL}(V_\rho)$ where V_ρ is called the *representation space* of the representation. The dimension of V_ρ, denoted d_ρ, is called the dimension of the representation.

The representation ρ is *irreducible* if the only invariant subspaces of V_ρ are 0 and V_ρ itself. Two representations ρ_1 and ρ_2 are *equivalent* if there exists an invertible linear map $S : V_{\rho_1} \to V_{\rho_2}$ such that $\rho_1(g) = S^{-1}\rho_2(g)S$ for all $g \in G$. Let $\Gamma = \{\rho_1, \rho_2, \ldots, \rho_r\}$ be a complete set of inequivalent, irreducible representations of G. Then the identity $\sum_{i=1}^{r} d_{\rho_i}^2 = |G|$ holds. Furthermore, we may assume that the representations are unitary, i.e., that $\rho(g)$ is a unitary matrix for all $g \in G$ and all $\rho \in \Gamma$. The functions defined by $\rho_{ij} = \rho(g)_{ij}$ for $1 \le i, j \le d_\rho$ are called *matrix coefficients*, and by the previous identity it follows that there are $|G|$ matrix coefficients. It is a fundamental fact that the set of all *normalized* matrix coefficients obtained from any fixed Γ is an orthonormal basis of the group algebra $\mathbb{C}[G]$. The *Fourier transform* with respect to a chosen Γ, is a change of basis transformation of the group algebra from the basis of point masses to the basis of matrix coefficients.

If G is commutative, then these definitions reduce to those discussed in Section 4, since in that case, all representations are 1-dimensional and each matrix coefficient is just a character. If G is noncommutative, then there exists at least 1 irreducible representation of G with higher dimension, and in this case the Fourier transform depends on the choice of bases for the irreducible representations. It seems as though this is what complicates the extension of the quantum algorithm for commutative groups to the noncommutative scenario.

It turns out that for our present application it is most useful to use an equivalent notion of the Fourier transform. One may also think of the matrix coefficients as collected together in matrices. In this view the Fourier transform is a matrix-valued function on Γ. For each $f \in \mathbb{C}[G]$, we define the value of the Fourier transform at an irreducible representation $\rho \in \Gamma$ to be

$$\hat{f}(\rho) = \sqrt{\frac{d_\rho}{|G|}} \sum_{g \in G} f(g)\rho(g).$$

If we take individual entries of these matrices, then we recover the coefficients in the basis of matrix coefficients. There is a Fourier inversion formula and therefore f is determined by the matrices $\left\{\hat{f}(\rho)\right\}_{\rho \in \Gamma}$.

We may now describe the noncommutative version of $H^\perp$. Let V_ρ^H be the elements of V_ρ that are *pointwise* fixed by H,

$$V_\rho^H = \{v \in V_\rho \mid \rho(h)v = v \text{ for all } h \in H\}.$$

Let P_ρ^H be the projection operator onto V_ρ^H. Then define

$$H^\perp = \left\{ P_\rho^H \right\}_{\rho \, \in \, \Gamma}.$$

The significance of this definition follows from the following elementary result.

Theorem 8. *Let I_H be the indicator function on the subgroup $H \leqslant G$. Then, for all $\rho \in \Gamma$, we have that $\widehat{I_H}(\rho) = P_\rho^H$.*

Corollary 9. *Let sH be any coset of $H \leqslant G$. Then Theorem 8 immediately yields, for all $\rho \in \Gamma$, we have $\widehat{I_{sH}}(\rho) = \rho(s) P_\rho^H$.*

Let us briefly summarize the role of this result in the quantum algorithm. If we straight-forwardly apply the quantum algorithm described in the previous section to the case where G is noncommutative, then we must determine the resulting probability amplitudes and the information gained by sampling according to these amplitudes.

Recall that the state of the quantum system after the first observation is a superposition of states corresponding to the members of one coset. Thus the state may be described by the indicator function of a coset I_{sH}. The final observation results in observing the name of a matrix coefficient $|\rho, i, j\rangle$. The probability of observing $|\rho, i, j\rangle$ is given by $|c_{\rho,i,j}|^2$ where $c_{\rho,i,j}$ is the coefficient of ρ_{ij} in the expansion of I_{sH} in the basis of matrix coefficients. The corollary above allows us, in theory, to compute these probability amplitudes.

The algorithm described in the first part of this paper may be derived from these general methods. For a general noncommutative group it seems that these methods are necessary for an analysis of the resulting probability amplitudes.

Acknowledgements

We would like to thank Dan Rockmore, David Maslen and Hans J. Munkholm from whom we learned noncommutative Fourier analysis, and Richard Hughes, Robert Beals and Joan Boyar for helpful conversations on this problem.

References

1. Beals, R.: Quantum computation of Fourier transforms over symmetric groups. Proc. 29th Annual ACM Symposium on Theory of Computing (1997) 48–53
2. Berthiaume, A.: Quantum computation. In: Hemaspaandra, L. A., Selman, A. L. (eds.): Complexity Theory Retrospective II. Springer-Verlag (1997) 23–51
3. Boneh, D., Lipton R.: Quantum cryptoanalysis of hidden linear functions (Extended abstract). Advances in Cryptology—CRYPTO '95. Lecture Notes of Computer Science, Vol. 963. Springer-Verlag (1995) 424–437
4. Brassard, G., Høyer, P.: On the power of exact quantum polynomial time. Unpublished (1996). Available on Los Alamos e-Print archive (http://xxx.lanl.gov) as quant-ph/9612017
5. Brassard, G., Høyer, P.: An exact quantum polynomial-time algorithm for Simon's problem. Proc. Fifth Israeli Symposium on Theory of Computing and Systems. IEEE Computer Society Press (1997) 12–23

6. Buhrman, H., Cleve, R., Wigderson, A.: Quantum vs. classical communication and computation. Proc. 30th ACM Symposium on Theory of Computing (1998) 63–68
7. Cleve, R., Buhrman, H.: Substituting quantum entanglement for communication. Physical Review A **56** (1997) 1201–1204
8. Grigoriev, D.: Testing shift-equivalence of polynomials by deterministic, probabilistic and quantum machines. Theoretical Computer Science **180** (1997) 217–228
9. Hoeffding, W.: Probability inequalities for sums of bounded random variables. Journal of the American Statistical Association **58** (1963) 13–30
10. Høyer, P.: Efficient quantum transforms. Unpublished (1997). Available on Los Alamos e-Print archive (http://xxx.lanl.gov) as quant-ph/9702028
11. Jozsa, R.: Quantum algorithms and the Fourier transform. Proceedings of the Royal Society, London **A454** (1998) 323–337
12. Kitaev, A.: Quantum measurements and the Abelian stabilizer problem. Unpublished (1995). Available on Los Alamos e-Print archive (http://xxx.lanl.gov) as quant-ph/9511026
13. Maslen, D., Rockmore, D.: Generalized FFTs — A survey of some recent results. Proc. 1996 DIMACS Workshop in Groups and Computation. American Mathematical Society (1997) 183–238
14. Rockmore, D.: Some applications of generalized FFTs. Proc. 1996 DIMACS Workshop in Groups and Computation. American Mathematical Society (1997) 329–370
15. Shor, P.: Polynomial-time algorithms for prime factorization and discrete logarithms on a quantum computer. SIAM Journal on Computing **26** (1997) 1484–1509
16. Simon, D.: On the power of quantum computation. SIAM Journal on Computing **26** (1997) 1474–1483

Appendix: Proof of Theorem 2

Proof (of Theorem 2). The following commutative diagram illustrates our approach:

$$H_1 \hookrightarrow H \longrightarrow H/H_1$$
$$\mathbb{Z}_N \times \{0\} \hookrightarrow \mathbb{Z}_N \rtimes_\phi \mathbb{Z}_2 = D_N \longrightarrow D_N/H_1$$

First, apply Theorem 1 to produce a subset $X_1 \subseteq H_1 = H \cap (\mathbb{Z}_N \times \{0\})$ such that X_1 generates H_1 with probability at least $1 - 1/N$ by using $O(\log N)$ queries to γ. Let $x_1 \in X_1$ generate $\langle X_1 \rangle$.

The subgroup $\langle x_1 \rangle$ is normal in D_N, and the quotient group $D_N/\langle x_1 \rangle$ is isomorphic to D_M with $M = [\mathbb{Z}_N \times \{0\} : \langle x_1 \rangle]$. Define $\gamma_2 : D_N/\langle x_1 \rangle \to R$ by $\gamma_2(g + \langle x_1 \rangle) = \gamma(g)$. Then γ_2 has hidden subgroup $H/\langle x_1 \rangle$.

Suppose $\langle x_1 \rangle = H_1$. Then $H/\langle x_1 \rangle$ is either trivial or generated by a reflection $r_2 + \langle x_1 \rangle$. Apply the algorithm in Theorem 3 with γ_2 a number of $= \lceil \log(2N)/\log(2M) \rceil$ times, ensuring we find $r_2 + \langle x_1 \rangle$ with probability at least $1 - 1/2N$, provided it exists.

Finally, output x_1, and output also the coset representative $r_2 \in D_N$ if it exists. The overall success probability is at least $(1 - 1/N)(1 - 1/2N) > 1 - 2/N$. The total number of evaluations of γ is at most $O(\log N) + t(89 \log M + 7)$, as each evaluation of γ_2 requires just one evaluation of γ. □

On the Size of Randomized Obdds and Read-Once Branching Programs for k-Stable Functions

Martin Sauerhoff*

FB Informatik, LS 2, Univ. Dortmund, 44221 Dortmund, Germany
sauerhoff@ls2.cs.uni-dortmund.de

Abstract. In this paper, a simple technique which unifies the known approaches for proving lower bound results on the size of deterministic, nondeterministic, and randomized Obdds and kObdds is described.
This technique is applied to establish a generic lower bound on the size of randomized Obdds with bounded error for the so-called "k-stable" functions which have been studied in the literature on read-once branching programs and Obdds for a long time. It follows by our result that several standard functions are not contained in the analog of the class BPP for Obdds.
It is well-known that k-stable functions are hard for deterministic read-once branching programs. Nevertheless, there is no generic lower bound on the size of randomized read-once branching programs for these functions as for Obdds. This is proven by presenting a randomized read-once branching program of polynomial size, even with zero error, for a certain k-stable function. As a consequence, we obtain that $P \subsetneq ZPP \cap NP \cap coNP$ for the analogs of these classes defined in terms of the size of read-once branching programs.

1 Introduction

Branching programs (BPs) are established as a standard model for the study of space-bounded computations. Basic definitions are given in the next section.

Obdds (ordered binary decision diagrams) are a restricted type of branching programs which have been introduced by Bryant [7] as a data structure for Boolean functions and turned out to be extremely useful in various fields of application. Jain, Bitner, Abadir, and Fussell [14] have extended Obdds to kIBDDs in order to have succinct representations for a larger class of functions. Roughly speaking, a kIBDD is a branching program which can be decomposed into at most k layers such that each layer is an OBDD (possibly with a different variable ordering for each layer). A kOBDD is a kIBDD where the variable orderings have to be identical for all layers.

Apart from practical issues, these restricted types of branching programs are also interesting as objects of theory. The first exponential lower bounds for Obdds are due to Bryant [8]. Jukna [16], Krause [21] and Gergov [11] have proven exponential lower bounds even for kObdds. Bollig, Sauerhoff, Sieling, and Wegener [5] have shown that the classes of sequences of functions representable in polynomial size by kIBDDs and by kObdds form a proper hierarchy with respect to k.

---

* This work has been supported by DFG grant We 1066/8-1.

C. Meinel and S. Tison (Eds.): STACS'99, LNCS 1563, pp. 488–499, 1999.
© Springer-Verlag Berlin Heidelberg 1999

Here we are concerned with the nondeterministic and probabilistic modes of computation for branching programs. Although nondeterministic variants of branching programs are well-known (even the first model of Shannon has in fact been nondeterministic), the probabilistic mode of computation has only recently gained attention. The complexity theoretical analysis of randomized OBDDs has been launched by Ablayev and Karpinski [2]. They have presented a function which is representable by randomized OBDDs of polynomial size with small one-sided error, but which has exponential size for deterministic OBDDs and even deterministic kOBDDs. Later on, they have extended the lower bound also to nondeterministic kOBDDs [3]. Lower bounds for randomized OBDDs have been proven independently by Ablayev [1] and the author [24]. Using these results, the relations between the complexity classes P, NP, RP, and BPP defined in terms of the size of OBDDs could be completely characterized (see [18]). Recently, Karpinski and Mubarakzjanov [19] have also resolved the relation between the classes P and ZPP for OBDDs by using a similar result for one-way communication complexity [10]. They have shown that, surprisingly, these classes coincide for OBDDs. A lower bound technique for the more general case of randomized read-once and randomized read-k-times BPs has been described in [25].

A large part of the known lower bound proofs on the size of branching programs either explicitly or implicitly rely on results from communication complexity theory. These communication complexity theoretical proof techniques have been applied in different disguises by several people. Hromkovič [13] has already described a "unifying" approach to the proof techniques for kOBDDs (which can be generalized to the case of read-k-times BPs) using a new measure for communication complexity, so-called "overlapping" communication complexity.

In this paper, we show that all the known proofs of lower bounds on the size of deterministic, nondeterministic, and randomized variants of OBDDs and kOBDDs boil down to rectangular reductions in the sense of communication complexity theory. Such reductions allow to apply known results on communication complexity to prove results on the size of OBDDs and kOBDDs in a simple way.

We apply this "reduction technique" to the class of so-called "k-stable" functions. The definition of k-stable functions goes back to Dunne [9]; the name itself has been introduced by Jukna [17]. Several authors have observed that such a function has size $2^k - 1$ for deterministic read-once BPs. To put it intuitively, a k-stable function is a function which has to compute a "pointer function" as a subproblem. By a pointer function, we mean a function which first computes the index (address) of a variable from its input and then outputs the value of this variable as the result. We show here that an arbitrary k-stable function has size $2^{\Omega(k)}$ for randomized OBDDs with bounded error. It immediately follows for many standard functions from the literature on read-once BPs that they are not contained in the analog of the class BPP for OBDDs.

Since randomization does not seem to help very much for functions which have to output a single bit of its input as the result, and since k-stable functions are known to be hard for deterministic read-once BPs, it is tempting to conjecture that they are also hard for randomized read-once BPs. Jukna, Razborov, Savický, and Wegener [15] have used a certain k-stable function to show that the analogs of the classes P and NP ∩ coNP for read-once BPs are distinct and have asked whether their function can even be shown to have exponential size for randomized read-once BPs. Here we answer this question in the negative sense: It turns out that this function can even be computed by randomized read-once BPs of polynomial size with zero error. As a consequence of this result, we obtain that the classes P and ZPP ∩ NP ∩ coNP for read-once BPs are different.

The rest of the paper is organized as follows. In Section 2, we introduce some basic notions concerning branching programs. In Section 3, we describe the general proof technique. After this, we apply this technique to prove the generic lower bound on the size of randomized OBDDs for k-stable functions (Section 4). Section 5 is devoted to the upper bound result for the function of Jukna, Razborov, Savický, and Wegener.

2 Definitions

We start with a review of basic definitions concerning branching programs and OBDDs

Definition 1. *A* branching program (BP) *on the variable set* $\{x_1, \ldots, x_n\}$ *is a directed acyclic graph with one source and two sinks, the latter labeled by the constants 0 and 1 Each non-sink node is labeled by a variable* x_i *and has two outgoing edges labeled by 0 or 1. This graph represents a Boolean function* $f: \{0,1\}^n \to \{0,1\}$ *in the obvious way: Call an edge labeled by* $c \in \{0,1\}$ *leaving a node labeled by* x_i activated by a a *if* $a_i = c$. *Then each input* $a \in \{0,1\}^n$ *corresponds to exactly one path of activated edges from the source to one of the sinks (called the* computation path *for a). The value of the sink reached by this path determines* $f(a)$. *The* size *of a branching program G is the number of its nodes and is denoted by* $|G|$.

A read-once branching program *is a branching program where on each path from the source to a sink, each variable may appear at most once.*

An OBDD *is a branching program with a* variable ordering, *given by a permutation* π *on the set* $\{1, \ldots, n\}$. *On each path from the source to the sinks, the variables at the nodes have to appear in the order prescribed by* π *(where some variables may be left out). A* π-OBDD *is an OBDD ordered with respect to* π.

A kOBDD *is a branching program with a variable ordering* π *whose set of nodes can be partitioned into k parts (called* layers) $\mathcal{L}_1, \ldots, \mathcal{L}_k$ *such that (i) edges starting in* $\mathcal{L}_i$ *end in* $\mathcal{L}_j$ *with* $j \geq i$; *and (ii) all edges within each* $\mathcal{L}_i$ *fulfill the ordering restriction for OBDDs with respect to* π.

Ablayev and Karpinski [2] have introduced randomized OBDDs defined analogously to probabilistic circuits. In the following, we give a definition for randomized general branching programs.

Definition 2. *Let a branching program G with the following special properties be given: (i) G has three types of sinks, labeled by 0, 1 or "?"; (ii) G is defined on two disjoint sets of variables* $X = \{x_1, \ldots, x_n\}$ *and* $Z = \{z_1, \ldots, z_r\}$; *and (iii) on each path from a source to a sink, each variable from Z appears at most once. The variables from Z are called* probabilistic variables, *and nodes labeled by such variables are called* probabilistic nodes.

By an obvious extension of the usual semantics for deterministic branching programs (see above), G represents a function $g: \{0,1\}^n \times \{0,1\}^r \to \{0,1,?\}$.

We say that G as a randomized branching program represents a function $f : \{0, 1\}^n \to \{0, 1\}$ with

- unbounded error if for all $x \in \{0, 1\}^n$ it holds that $\mathrm{Pr}_z\{g(x, z) = f(x)\} > 1/2$;
- two-sided error (bounded error) at most ε, where $0 \leq \varepsilon < 1/2$, if for all $x \in \{0, 1\}^n$ it holds that $\mathrm{Pr}_z\{g(x, z) = f(x)\} \geq 1 - \varepsilon$;
- one-sided error at most ε, where $0 \leq \varepsilon < 1$, if for all $x \in \{0, 1\}^n$ it holds that
  $$\mathrm{Pr}_z\{g(x, z) = 0\} = 1, \qquad \text{if } f(x) = 0;$$
  $$\mathrm{Pr}_z\{g(x, z) = 1\} \geq 1 - \varepsilon, \quad \text{if } f(x) = 1;$$
- zero error and failure probability at most ε, $0 \leq \varepsilon < 1$, if for all $x \in \{0, 1\}^n$ it holds that
  $$\mathrm{Pr}_z\{g(x, z) = 1\} = 0 \ \wedge \ \mathrm{Pr}_z\{g(x, z) = ?\} \leq \varepsilon, \quad \text{if } f(x) = 0;$$
  $$\mathrm{Pr}_z\{g(x, z) = 0\} = 0 \ \wedge \ \mathrm{Pr}_z\{g(x, z) = ?\} \leq \varepsilon, \quad \text{if } f(x) = 1.$$

In these expressions, z is an assignment to the probabilistic variables which is chosen according to the uniform distribution from $\{0, 1\}^r$.

Definitions for randomized variants of restricted branching programs are derived from this in a straightforward way by requiring that the non-probabilistic variables fulfill the restriction. Randomized OBDDs are thus randomized branching programs with a variable ordering π on the non-probabilistic variables, and the tests of the non-probabilistic variables on each path from the source to a sink have to be consistent with π (analogously for kOBDDs). As for Turing machines, nondeterministic branching programs occur as a special case of randomized branching programs with one-sided error.

For a type $\mathcal{R} \in \{\mathrm{BP1}, \mathrm{OBDD}, k\mathrm{OBDD}\}$ of restricted branching programs (where "BP1" stands for read-once BPs), we denote the classes of sequences of functions with polynomial size deterministic, nondeterministic, and randomized branching programs with zero, one-sided, bounded or unbounded error of the respective type by P-$\mathcal{R}$, NP-$\mathcal{R}$, ZPP-$\mathcal{R}$, RP-$\mathcal{R}$, BPP-$\mathcal{R}$, and PP-$\mathcal{R}$, resp. For a class $\mathcal{C}$ of sequences of functions, co-$\mathcal{C}$ denotes the class of sequences of functions $(f_n)_{n \in \mathbb{N}}$ with $(\neg f_n)_{n \in \mathbb{N}} \in \mathcal{C}$.

In the following, we will also apply well-known concepts from communication complexity theory. For the definition of the respective notions and a thorough introduction, we have to refer to the monographs of Hromkovič [12] and Kushilevitz and Nisan [23].

The Reduction Technique

In this section, we describe the known techniques for proving lower bounds on deterministic, nondeterministic, and randomized OBDDs and kOBDDs in a unified way. This general approach is called "reduction technique" here.

To put it intuitively, the known proofs of lower bounds on the size of OBDDs are all based on the fact that a large amount of information has to be exchanged across a suitably chosen cut in the graph in order to evaluate the represented function. Results from communication complexity theory are then explicitly or implicitly used to get lower bounds on the necessary amount of information. A similar approach works for kOBDDs.

Our goal is to clearly separate the communication complexity theoretical part of these proofs from the conclusions on the size of the OBDD or kOBDD. We will directly handle the more general case of kOBDDs. The following definition will be used to establish the connection between the size of kOBDDs and communication complexity with respect to a fixed partition.

Definition 3. *Let* $f: \{0,1\}^n \to \{0,1\}$ *be a function defined on the variable set* $X = \{x_1, \ldots, x_n\}$. *Let a variable ordering on* X *be given by* $\pi: \{1, \ldots, n\} \to \{1, \ldots, n\}$ *Let* $1 \leq p \leq n-1$ *and* $L := \{x_{\pi(1)}, \ldots, x_{\pi(p)}\}$, $R := \{x_{\pi(p+1)}, \ldots, x_{\pi(n)}\}$. *Define the function* $f': \{0,1\}^p \times \{0,1\}^{n-p} \to \{0,1\}$ *on assignments* x *to* L *and* y *to* R *by* $f'(x, y) := f(x + y)$, *where* $x + y$ *denotes the joint assignment to* X *obtained from* x *and* y. *Then we call* f' *the* partitioned version *of* f *with respect to* π *and* p *and denote this function by* $f^{\pi,p}$.

Since we usually cannot directly analyze the communication complexity of the function represented by a kOBDD, it is important to be able to identify hard subproblems which can be handled by the available tools. As for Turing machines, we use a reduction to show that the whole function is at least as hard as the considered subproblem.

Several notions of reducibility defined analogously to the well-known notions for Turing machines have been introduced in communication complexity theory. The most common type is the *rectangular reduction*, which is the analog of many-one reducibility for Turing machines.

Definition 4 (Rectangular reduction). *Let* X_f, Y_f *and* X_g, Y_g *be finite sets. Let* $f: X_f \times Y_f \to \{0,1\}$ *and* $g: X_g \times Y_g \to \{0,1\}$ *be arbitrary functions. Then we call a pair* (φ_1, φ_2) *of functions* $\varphi_1: X_f \to X_g$ *and* $\varphi_2: Y_f \to Y_g$ *a* rectangular reduction *from* f *to* g *(or simply "reduction" for short) if* $g(\varphi_1(x), \varphi_2(y)) = f(x, y)$ *for all* $(x, y) \in X_f \times Y_f$. *If such a pair of functions exists for* f *and* g, *we say that* f *is* reducible *to* g.

Now we are ready to formally describe the connection between the size of kOBDDs and communication complexity.

Lemma 1. *Let* $g: \{0,1\}^n \to \{0,1\}$ *be defined on the variable set* $X = \{x_1, \ldots, x_n\}$ *Let* π *be a variable ordering on* X. *Assume that there is a function* $f: U \times V \to \{0,1\}$ *where* U *and* V *are finite sets, and a parameter* p *with* $1 \leq p \leq n-1$ *such that* f *is reducible to the partitioned version* $g^{\pi,p}$ *of* g. *Let* G *be a randomized* kOBDD *ordered according to* π *which represents* g *with two-sided error at most* ε. *Then it holds that*

$$\lceil \log |G| \rceil \geq R_\varepsilon^{2k-1}(f)/(2k-1),$$

where $R_\varepsilon^{2k-1}(f)$ *denotes the minimal number of bits exchanged by a randomized* $(2k-1)$*-round communication protocol for* f *with two-sided error at most* ε. *Analogous assertions hold for deterministic, nondeterministic, and randomized OBDDs with zero, one-sided or unbounded error and the corresponding measures for* $(2k-1)$*-round communication complexity.*

Proof. Since f is reducible to $g^{\pi,p}$, it follows that $R_\varepsilon^{2k-1}(g^{\pi,p}) \geq R_\varepsilon^{2k-1}(f)$ (any r-round protocol for $g^{\pi,p}$ can be used to define an r-round protocol for f with the same amount of communication). Hence, it is sufficient to show that $(2k-1)\lceil\log|G|\rceil \geq R_\varepsilon^{2k-1}(g^{\pi,p})$. To prove this inequality, we construct a randomized $(2k-1)$-round protocol for $g^{\pi,p}$ from G. The basic ideas behind this construction go back to Jukna [16] and Krause [21].

A *cut* in G is a set C of nodes with the property that each path from the source to the sinks runs through exactly one node in C. The set which only contains the source of G and the set containing the sinks are obviously cuts. Call these cuts C_0 and C_{2k}, resp. Furthermore, using the fact that G is a kOBDD we can choose cuts $C_1, \ldots, C_{2k-1}$ in G such that the following holds: (i) The subgraph consisting of the paths between the nodes in C_{2i} and C_{2i+2}, the *ith layer* of G, is a randomized π-OBDD with several sources and sinks). (ii) The cut C_{2i+1} decomposes this π-OBDD into an *upper part* of paths where the non-probabilistic variables are labeled by variables from $L := \{x_{\pi(1)}, \ldots, x_{\pi(p)}\}$, and a *lower part* where the non-probabilistic variables are labeled by variables from $R := \{x_{\pi(p+1)}, \ldots, x_{\pi(n)}\}$,

Now we are ready to sketch a randomized protocol P by which two players, called Alice and Bob, can evaluate $g^{\pi,p}$ in $(2k-1)$ rounds. Player Alice obtains an assignment x to the variables in L and player Bob an assignment y to R as inputs. Both use the graph G as an "oracle." Player Alice starts the communication. It is easy to see how Alice and Bob can jointly follow the computation path for the combined assignment $x + y$ in G by exchanging the numbers of nodes on the cuts $C_1, \ldots, C_{2k-1}$ lying on such a path. When a player encounters a node labeled by a probabilistic variable in her (his) part of the graph, she (he) locally chooses a value for this variable at random and precedes to the corresponding successor. The last player (who is always Bob) outputs the value of the reached sink as the result of the protocol.

Since each probabilistic variable can appear at most once on each computation path in G, both players can choose the values of the probabilistic variables independently. Because of the error guarantee of G, it follows that the above protocol P computes $g^{\pi,p}$ with error at most ε. Furthermore, the number of exchanged bits of communication is at most $\sum_{i=1}^{2k-1} \lceil\log|C_i|\rceil \leq (2k-1)\lceil\log|G|\rceil$.   □

Lower Bounds for k-Stable Functions

Now we apply the lower bound technique presented in the last section to the class of k-stable functions.

Definition 5. *Let* $k \in \{1, \ldots, n-1\}$. *A function* $f\colon \{0,1\}^n \to \{0,1\}$ *defined on the variable set* X *($|X| = n$), is called* k-stable *if the following holds. For an arbitrary set* $X_1 \subseteq X$, $|X_1| = k$, *and each variable* $x \in X_1$ *there is an assignment* b *to the variables in* $X\backslash X_1$ *such that either* $f(a+b) = a(x)$ *for all assignments* a *to* X_1 *or* $f(a+b) = \neg a(x)$ *for all assignments* a *to* X_1.

It is a well-known fact that k-stable functions have size at least $2^k - 1$ for deterministic read-once BPs. Lower bounds of this type have been proven by several authors, e. g., by Dunne [9], Jukna [17], Krause [20] and Jukna, Razborov, Savický, and Wegener [15]. We list some examples from these papers.

Examples.

(1) Let $N := \binom{n}{2}$ and $1 \le k \le n$. Define the function $\text{cl}_{n,k} \colon \{0,1\}^N \to \{0,1\}$, on the Boolean variables $X := (x_{ij})_{1 \le i < j \le n}$. Let $G(X)$ be the undirected graph on the nodes from $\{1, \ldots, n\}$ described by X, i. e., edge $\{i,j\}$ exists in $G(X)$ iff $x_{ij} = 1$. Let $\text{cl}_{n,k}(X) = 1$ iff the graph $G(X)$ contains a k-clique.

It holds that $\text{cl}_{n,k}$ is s-stable for $s := \min\{\binom{k}{2} - 1, (n - k + 2)/2\}$. (This can be proven easily by using the ideas contained in [17] and [27]. Jukna has proven a similar result for the directed version of the clique-function.)

(2) By $\text{PM}_n, \text{DET}_n \colon \{0,1\}^{n^2} \to \{0,1\}$, which are both functions defined on an $n \times n$-matrix of Boolean variables $X := (x_{ij})_{1 \le i,j \le n}$ as an input, denote the *Boolean permanent* and the *Boolean determinant*, resp. The functions PM_n and DET_n are both $(n - 1)$-stable [20].

(3) Let $n = q^2 + q + 1$, where $q = p^m$, p is a prime and m an arbitrary natural number. Let $P = \{1, \ldots, n\}$ be the set of "points" of a projective plane of order q and let $L_1, \ldots, L_n \subseteq P$ be the "lines." A set $A \subseteq P$ is called a *blocking set* if $A \cap L_i \ne \emptyset$ for $i = 1, \ldots, n$. Define $B_n \colon \{0,1\}^n \to \{0,1\}$ by $B_n(x_1, \ldots, x_n) = 1$ iff $\{i \mid x_i = 1\}$ is a blocking set of size at most $q + k$, where where $k := (q+1)/2$ if q is prime, $k := \lceil \sqrt{q} \rceil$ otherwise.

The proof of the lower bound on the size of deterministic read-once BPs for B_n from [15] shows that B_n is k-stable.

(4) Let $n = 2^l$, and define $m := \lfloor n/l \rfloor$. Let $|u|_2$ denote the value of a Boolean vector u interpreted as a binary number. Define $\lambda \colon \{0,1\}^m \to \{0,1\}$ as follows. Chop the input vector from $\{0,1\}^m$ into $s := \lfloor \sqrt{m} \rfloor$ blocks of size s each. Then λ is the disjunction of the conjunctions of all variables in each of these blocks.

Finally, define $\text{ADDR}_n \colon \{0,1\}^n \to \{0,1\}$ by $\text{ADDR}_n(x_0, \ldots, x_{n-1}) := x_a$, $a := |(\lambda(x^{l-1}), \ldots, \lambda(x^0))|_2$, where $x^i := (x_{im}, \ldots, x_{(i+1)m-1}), i = 0, \ldots, l-1$. It is easy to verify that ADDR_n is $(s - 1)$-stable (see [15] or [17]).

The following technical lemma describes a large class of functions which are hard for randomized OBDDs with bounded error. We will show that the functions defined above belong to this class.

Lemma 2. *Let $\text{INDEX}_m \colon U \times V \to \{0,1\}$, where $U := \{0,1\}^m$, $V := \{1, \ldots, m\}$ be defined by $\text{INDEX}_m(u, v) := u_v$ for $u = (u_1, \ldots, u_m) \in U$ and $v \in V$.*
Let $g \colon \{0,1\}^n \to \{0,1\}$ be defined on the variable set X. Let k with $1 \le k \le n - 1$ be fixed. Assume that for each variable ordering π on X, there is a parameter p with $1 \le p \le n - 1$ such that INDEX_k is reducible to the partitioned version $g^{\pi,p}$ of g. Let G be a randomized OBDD for g with arbitrary two-sided error ε, $\varepsilon < 1/2$. Then it holds that $|G| = 2^{\Omega(k)}$.

Proof. Kremer, Nisan, and Ron [22] have shown that each randomized one-way protocol which computes INDEX_k with two-sided error smaller than $1/8$ needs $\Omega(k)$ bits of communication. If the error of the randomized OBDD for g is smaller than $1/8$, then the lower bound of order $2^{\Omega(k)}$ immediately follows from Lemma 1.

To obtain the claimed lower bound for an arbitrary error probability $\varepsilon < 1/2$, we use the fact that the error probability of randomized OBDDs can be decreased below a arbitrary small constant while maintaining polynomial size by "probability amplification," as shown in [4] and [25].   □

The above lemma can be applied to all k-stable functions, which yields the desired generic lower bound.

Lemma 3. Let $g: \{0,1\}^n \to \{0,1\}$ be a k-stable function. Let G be a randomized OBDD for g with arbitrary two-sided error $\varepsilon < 1/2$. Then it holds that $|G| = 2^{\Omega(k)}$.

Proof. We are going to construct a rectangular reduction from INDEX_k to a suitable partitioned version of g.

Let an arbitrary variable ordering π on the variable set X of g be given. For the ease of notation, we assume here that π maps indices to its respective variables, i.e., π is a function of the form $\pi: \{1,\ldots,n\} \to X$. Define $L := \{\pi(1),\ldots,\pi(k)\}$ and $R := \{\pi(k+1),\ldots,\pi(n)\}$. We observe that, since f is k-stable, for each variable $x \in L$, there is an assignment b_x to R such that either $f(a + b_x) = a(x)$ for all assignments a to L or $f(a + b_x) = \neg a(x)$ for all assignments a to L. Let us first assume that always the former case occurs. In the following, we define a rectangular reduction (φ_1, φ_2) from INDEX_k to $g^{\pi,k}$.

The function $\varphi_1: U \to \{0,1\}^k$ is only a permutation of the bits of its input vector. For an arbitrary input $u = (u_1,\ldots,u_k) \in U = \{0,1\}^k$, define the assignment a to the variables in L by $a(x) := u_{\pi^{-1}(x)}$ for $x \in L$. Set $\varphi_1(u) := a$. The function $\varphi_2: V \to \{0,1\}^{n-k}$ is defined by $\varphi_2(v) := b_{\pi(v)}$, where $v \in V = \{1,\ldots,k\}$. For arbitrary $(u,v) \in U \times V$, we have $g^{\pi,k}(\varphi_1(u), \varphi_2(v)) = \text{INDEX}_k(u,v)$, hence, (φ_1, φ_2) is a rectangular reduction from INDEX_k to $g^{\pi,k}$.

We still have to handle the case that for some variables $x \in L$, it holds that $f(a + b_x) = \neg a(x)$ for all assignments a to L. For this case, we slightly extend our reduction concept. Additional to the transformation of the input by the pair of functions (φ_1, φ_2), we allow to negate the result for the "target problem," $g(\varphi_1(u), \varphi_2(v))$, dependently on the input $v \in V$. More precisely, such a reduction consists of φ_1, φ_2 and an additional function $\nu: V \times \{0,1\} \to \{0,1\}$ for which $\nu(v, g(\varphi_1(u), \varphi_2(v))) = f(u,v)$ for all $(u,v) \in U \times V$.

It is easy to see that an analogous versions of Lemma 1 from the last section holds for this extended type of reductions. Here we choose $\nu(v,c) = c$ for $c \in \{0,1\}$ if $f(a + b_{\pi(v)}) = a(\pi(v))$ for all assignments a to L, and $\nu(v,c) = \neg c$ for $c \in \{0,1\}$ if $f(a + b_{\pi(v)}) = \neg a(\pi(v))$ for all assignments a to L. One easily verifies that for this choice of ν and φ_1, φ_2 it holds that $\nu(v, g(\varphi_1(u), \varphi_2(v))) = \text{INDEX}_k(u,v)$ for all $(u,v) \in U \times V$.   □

From this lemma, we immediately obtain that the examples of k-stable functions already mentioned are all hard for randomized OBDDs with bounded error:

Theorem 1. $\text{cl}_{n,n/2}, \text{PM}_n, \text{DET}_n, B_n, \text{ADDR}_n \notin \text{BPP-OBDD}$.

Apart from these functions, there are some "pointer functions" in the sense of the informal definition from the introduction which cannot be k-stable for large k because they are contained in the class P-BP1. For these functions, we can still apply Lemma 2. As examples we consider the following standard functions from the literature on OBDDs.

Definition 6.
(1) The function HWB_n *("hidden weighted bit") is defined on* $x = (x_1, \ldots, x_n)$. *Define* $\mathrm{sum}(x) := \sum_{i=1}^{n} x_i$ *and let* $x_0 := 0$. *Then* $\mathrm{HWB}_n(x) := x_{\mathrm{sum}(x)}$.
(2) The function ISA_n *("indirect storage access") is defined on* $n = 2^r + r$ *variables* $x_0, \ldots, x_{2^r-1}$ *and* $y_0, \ldots, y_{r-1}$. *Let* $s := \lfloor 2^r/r \rfloor$. *Let* $\mathrm{ISA}_n(x, y) = x_j$, *where* $j := |(x_{ir}, \ldots, x_{(i+1)r-1})|_2$ *and* $i := |(y_{r-1}, \ldots, y_0)|_2$ *(if* $i \geq s$, *let* $\mathrm{ISA}(x, y) = 0$).

The functions HWB_n and ISA_n have been introduced by Bryant [8] and Breitbart, Hunt, and Rosenkrantz [6], resp., who have also shown that these functions have exponential size for deterministic OBDDs. Sieling and Wegener [26] have proven that both functions are contained in the class P-BP1. We complement this by the following result.

Theorem 2. $\mathrm{HWB}_n, \mathrm{ISA}_n \notin$ BPP-OBDD.

Sketch of Proof. This follows immediately by Lemma 2 and the known lower bound proofs for these functions (from [6] and [8], resp.). These proofs can be seen as rectangular reductions from the function INDEX to appropriate partitioned versions of ISA_n and HWB_n, resp. (An explicit construction for the function ISA_n can be found in the ECCC report [24].)   □

5   A k-Stable Function with Small Randomized Read-Once BPs

In this section, we show that there is no generic lower bound on the size of randomized read-once BPs for k-stable functions as it is the case for randomized OBDDs. We prove that the function ADDR_n from the paper of Jukna, Razborov, Savický, and Wegener [15] can even be computed by a randomized read-once BP with zero error.

For notational convenience, we consider the function ADDR_n only for input sizes where we can do without floors or ceilings.

Theorem 3. *Let* $n = 2^l$ *and* $l = 2^{\tilde{l}}$. *The function* ADDR_n *can be represented by a randomized read-once BP of polynomial size with zero error and failure probability at most* $1/2$.

Proof. Define $m := n/l = 2^{l-\tilde{l}}$. Let λ, x^i $(i = 0, \ldots, l - 1)$, and x_a be as in the definition of ADDR_n. Call the bits $\lambda(x^0), \ldots, \lambda(x^{l-1})$ "address bits" and the bit x_a "output bit." Imagine the input variables of ADDR_n to be arranged as an $l \times m$-matrix with rows $x^0, \ldots, x^{l-1}$. The algorithm implemented by the randomized read-once BP for ADDR_n will consist of two phases. In the first phase, we read some rows of the input matrix and compute the respective address bits. After that, only a small set A of possible output bits will be left. The second phase consists of evaluating all remaining address bits and "storing" the values of all variables in A in the branching program. Finally, we have determined the complete address. With probability at least $1/2$, the addressed bit will belong to the stored values.

By $v = (v_{l-1}, \ldots, v_0) \in \{0, 1, *\}^l$ we describe the address bits computed so far in the algorithm, let $v_i = *$ if the ith bit is not yet known. The bits $v_0, \ldots, v_{l-\tilde{l}-1}$, called "column address bits" in the following, determine the *column* where the output bit is

ound. Likewise, the "row address bits" $v_{l-\tilde{l}}, \ldots, v_{l-1}$ determine the *row* of the output
it.

For an arbitrary vector v, let $C(v) \subseteq \{0, \ldots, m-1\}$ be the set of columns which
re addressed by vectors v' which are obtained from v by assigning constant values to
he $*$-bits, and let $R(v) \subseteq \{0, \ldots, l-1\}$ the set of rows addressed in this way. Define
$A(v) := \{im + j \mid i \in R(v), j \in C(v)\}$ as the set of indices of addressed output bits.
Now we describe our randomized algorithm for the computation of ADDR_n.

Algorithm 1.

0) Initialize v: For $i = 0, \ldots, l-1$, let $v_i := *$.
1) Choose $z \in \{0, 1\}$ uniformly at random.
2) **Case $z = 0$:**

   Phase 1: For $i \in \{l - \tilde{l}, \ldots, l-1\}$, read x^i and compute $v_i := \lambda(x^i)$. Let $r :=$
   $|(v_{l-1}, \ldots, v_{l-\tilde{l}})|_2 \in \{0, \ldots, l-1\}$, i.e., r is the row within which the output bit
   lies. Then $R(v) = \{r\}$. If $r \geq l - \tilde{l}$, we have "lost" and output "?".
   Now assume that $r \in \{0, \ldots, l - \tilde{l} - 1\}$. For $i \in \{0, \ldots, l - \tilde{l} - 1\} \setminus \{r\}$ read the
   row x^i and compute $v_i = \lambda(x^i)$. After this we have also determined all bits of the
   column address except one. Hence, $|C(v)| = 2$ and thus also $|A(v)| = 2$.
   Phase 2: As the final step, we evaluate the last missing address bit $v_r = \lambda(x^r)$.
   While we compute v_r, we store the values of the two variables x_j with $j \in A(v)$
   (these variables lie within row r). Afterwards, we know the complete address of the
   output bit, $a = |(v_{l-1}, \ldots, v_0)|_2$. Since we have stored both possible output bits,
   we can output the correct value.

3) **Case $z = 1$:**

   Phase 1: For $i \in \{0, \ldots, l - \tilde{l} - 1\}$ read the row x^i of the input matrix and compute
   $v_i := \lambda(x^i)$. After this, we have $C(v) = \{c\}$, where $c = |(v_{l-\tilde{l}-1}, \ldots, v_0)|_2$, and
   hence, $A(v) = \{im + c \mid l - \tilde{l} \leq i \leq l-1\}$. Notice that $|A(v)| = \tilde{l} = \log \log n$.
   Phase 2: Now read all remaining rows x^i with $i \in \{l - \tilde{l}, \ldots, l-1\}$, but again store
   all values of variables x_j with $j \in A(v)$ (i.e., the variables in column c). Finally,
   we know the complete address $a = |(v_{l-1}, \ldots, v_0)|_2$ of the output bit. If it holds
   that $\lfloor a/m \rfloor \leq l - \tilde{l} - 1$, i.e., the row where the output bit is found has already been
   read in Phase 1, output "?". Otherwise, we can output the stored value of x_a.

t is easy to verify that the above algorithm in fact has zero error and outputs "?" with
probability at most $1/2$. The algorithm can be coded into a randomized read-once BP by
he standard construction techniques for branching programs. We have ensured already
n the description of the algorithm that each variable is only read once. For the evalu-
tion of the bits v_i, we use polynomial size branching programs for λ as sub-modules.
We can at any time store the parts of the vector v computed so far since the whole vec-
or only has length l. The second phases can be represented in polynomial size since
lways $|A(v)| \leq \log \log n$ and hence, we need only to enlarge the width of the branch-
ng program by a logarithmic factor in order to store all the needed values.   □

We have thus obtained an exponential gap between "Las Vegas" and deterministic al-
orithms for the read-once BP model. Together with the result of Jukna, Razborov,
avický, and Wegener that ADDR_n is representable in polynomial size by nondeter-
inistic and co-nondeterministic read-once BPs, we obtain:

Corollary 1.   P-BP1 $\subsetneq$ ZPP-BP1 $\cap$ NP-BP1 $\cap$ coNP-BP1.

Acknowledgement. I would like to thank Ingo Wegener for proofreading and many useful hints, Matthias Krause for pointing out the connection between the size of branching programs and communication complexity to me some time ago, and Martin Dietzfelbinger and Detlef Sieling for helpful discussions.

References

1. F. Ablayev. Randomization and nondeterminism are incomparable for polynomial ordered binary decision diagrams. In *Proc. of 24th ICALP, LNCS 1256*, 195–202. Springer, 1997.
2. F. Ablayev and M. Karpinski. On the power of randomized branching programs. In *Proc. of 23rd ICALP, LNCS 1099*, 348–356. Springer, 1996.
3. F. Ablayev and M. Karpinski. On the power of randomized ordered branching programs. TR98-004, Electr. Coll. on Computational Complexity, 1998.
4. M. Agrawal and T. Thierauf. The satisfiability problem for probabilistic ordered branching programs. In *Proc. of the 13th IEEE Int. Conf. on Computational Complexity*, 81–90, 1998.
5. B. Bollig, M. Sauerhoff, D. Sieling, and I. Wegener. Hierarchy theorems for *k*OBDDs and *k*IBDDs. *Theoretical Computer Science*, 205(1):45–60, 1998.
6. Y. Breitbart, H. Hunt III, and D. Rosenkrantz. On the size of binary decision diagrams representing Boolean functions. *Theoretical Computer Science*, 145:45 – 69, 1995.
7. R. E. Bryant. Graph-based algorithms for Boolean function manipulation. *IEEE Trans. Computers*, C-35(8):677–691, Aug. 1986.
8. R. E. Bryant. On the complexity of VLSI implementations and graph representations of Boolean functions with application to integer multiplication. *IEEE Trans. Computers*, C-40(2):205–213, Feb. 1991.
9. P. E. Dunne. Lower bounds on the complexity of 1-time only branching programs. In *Proc. of FCT, LNCS 199*, 90–99. Springer, 1984.
10. P. Ďuriš, J. Hromkovič, J. D. P. Rolim, and G. Schnitger. Las Vegas versus determinism for one-way communication complexity, finite automata, and polynomial-time computations. In *Proc. of 14th STACS, LNCS 1200*, 117–128. Springer, 1997. To appear in *Information and Computation*.
11. J. Gergov. Time-space tradeoffs for integer multiplication on various types of input oblivious sequential machines. *Information Processing Letters*, 51:265 – 269, 1994.
12. J. Hromkovič. *Communication Complexity and Parallel Computing*. Springer, Berlin, 1997.
13. J. Hromkovič. Communication complexity and lower bounds on multilective computations. In *Proc. of 23rd MFCS, LNCS 1450*, 789 – 797. Springer, 1998.
14. J. Jain, J. Bitner, M. S. Abadir, J. A. Abraham, and D. S. Fussell. Indexed BDDs: Algorithmic advances in techniques to represent and verify Boolean functions. *IEEE Trans. Computers*, 46:1230–1245, 1997.
15. S. Jukna, A. Razborov, P. Savický, and I. Wegener. On P versus NP $\cap$ co-NP for decision trees and read-once branching programs. In *Proc. of 22nd MFCS, LNCS 1295*, 319–326. Springer, 1997. To appear in *Computational Complexity*.
16. S. P. Jukna. Lower bounds on communication complexity. *Mathematical Logic and Its Applications*, 5:22–30, 1987.
17. S. P. Jukna. Entropy of contact circuits and lower bounds on their complexity. *Theoretical Computer Science*, 57:113 – 129, 1988.
18. M. Karpinski. On the computation power of randomized branching programs. In *Randomized Algorithms, Proc. of the International Workshop*, 1–12, Brno, 1998.

9. M. Karpinski and R. Mubarakzjanov. A note on Las Vegas OBDDs. *Manuscript*, Nov. 1998.

20. M. Krause. Exponential lower bounds on the complexity of local and real-time branching programs. *Journal of Information Processing and Cybernetics, EIK*, 24(3):99–110, 1988.

21. M. Krause. Lower bounds for depth-restricted branching programs. *Information and Computation*, 91(1):1–14, Mar. 1991.

22. I. Kremer, N. Nisan, and D. Ron. On randomized one-round communication complexity. In *Proc. of 27th STOC*, 596 – 605, 1995.

23. E. Kushilevitz and N. Nisan. *Communication Complexity*. Cambridge University Press, Cambridge, 1997.

24. M. Sauerhoff. A lower bound for randomized read-k-times branching programs. TR97-019, Electr. Coll. on Computational Complexity, 1997.

25. M. Sauerhoff. Lower bounds for randomized read-k-times branching programs. In *Proc. of 15th STACS, LNCS 1373*, 105 – 115. Springer, 1998.

26. D. Sieling and I. Wegener. Graph driven BDDs—a new data structure for Boolean functions. *Theoretical Computer Science*, 141:283 – 310, 1995.

27. I. Wegener. On the complexity of branching programs and decision trees for clique functions. *Journal of the ACM*, 35(2):461–471, Apr. 1988.

How To Forget a Secret

(Extended Abstract)

Giovanni Di Crescenzo[1*], Niels Ferguson[2**], Russell Impagliazzo[1], and
Markus Jakobsson[3* * *]

[1] Computer Science Department, University of Califonia San Diego,
La Jolla, CA, 92093-0114. giovanni.russell@cs.ucsd.edu
[2] Counterpane Systems. niels@counterpane.com
[3] Information Sciences Research Center, Bell Labs, Murray Hill, NJ 07974.
markusj@research.bell-labs.com

Abstract. We uncover a new class of attacks that can potentially affect
any cryptographic protocol. The attack is performed by an adversary
that at some point has access to the physical memory of a participant,
including all its previous states.
In order to protect protocols from such attacks, we introduce a cryptographic primitive that we call *erasable memory*. Using this primitive,
it is possible to implement the essential cryptographic action of *forgetting* a secret. We show how to use a small erasable memory in order to
transform a large non-erasable memory into a large and erasable memory. In practice, this shows how to turn any type of storage device into a
storage device that can selectively forget. Moreover, the transformation
can be performed using the minimal assumption of the existence of any
one-way function, and can be implemented using any block cipher, in
which case it is quite efficient. We conclude by suggesting some concrete
implementations of small amounts of erasable memory.

1 Introduction

All of cryptography is based on the assumption that some information can be
kept private, accessible only by certain parties. At some level, physical control
over the storage device containing the information must be maintained. While
it is reasonable to assume that such control can be maintained for some finite
amount of time, as the duration of a protocol, it is not realistic to assume it
continues indefinitely. Thus, it is important to be able to *forget* information.

Practical considerations. Current computer systems treat erasing memory
only from the point of view of efficiency, not security. Erasing a memory location
enables the system to use that location for other purposes; it does not guarantee
that the information is destroyed. In fact, in many practical systems the old

---

* Part of Giovanni's work done while at Bellcore
** Part of this work done while visiting UCSD
* * * Part of this work done while at UCSD

C. Meinel and S. Tison (Eds.): STACS'99, LNCS 1563, pp. 500–509, 1999.
© Springer-Verlag Berlin Heidelberg 1999

value might still be retrievable, making the assumption that data is forgotten, if not specifically remembered, a *fallacy*. For example, it has been observed that if a message is stored for a long time in some types of RAM, the chip will slowly 'learn' this value. Overwriting the data or powering down the chip will not erase it. When the chip is powered up again there is a good chance that the data will again appear in the RAM [4]. Some of these effects are due to aging effects in the silicon that depend on the bits stored in the RAM. Another example is the case of magnetic media. Here, the problem is more severe, since these devices are notoriously hard to wipe. It has long been known that simply overwriting the data is not sufficient; US government specifications call for overwriting the data *three times* for non-classified information, in order to minimize the chances of their retrieval by an adversary. However, the magnetic fields used to store the data have a tendency to migrate away from the read/write head and bury themselves deep down in the magnetic material, so even if these fields are no longer readable by the original read/write head, other techniques might well retrieve them. The problem gets even more complicated if the operating system supports virtual memory. A typical virtual memory system will write a piece of RAM to disk in a swap-file without telling the program using the RAM. When the program tries to access this RAM, the operating system reads the data from the swap file back into RAM for the program to use. Suppose the program overwrites the RAM in an attempt to erase the data and then exits. The old copy in the swap file is not erased by the operating system, but just marked as 'available'. A direct inspection of the swap file will reveal the old values of the RAM.

Cryptographic protocols. In the security and cryptography literature, many protocols make the silent assumption that we can 'forget' information. For example, the randomly chosen 'temporary secret key' in DSS [15] and similar schemes [7,18] should be forgotten by the signer after it has been used. If it can be found by an attacker, he will be able to reconstruct the secret key of the signer from the signature and the temporary secret key. More generally, participants to cryptographic protocols should forget all partial results, randomness values and temporarily stored information they use while executing such protocols. Otherwise this information could later fall in the hands of some adversary, who will be able to use it to attack the scheme. Clearly, this is an issue that can potentially arise in just any cryptographic protocol (apart possibly from very few exceptions). Known cryptographic techniques, such as 'proactivization' (e.g., [8,13,12,17]), or 'forward security' (e.g., [6]) do not avoid the problem. The former technique maintains secrecy with respect to intruders that occasionally gain access to some of the storage devices, by changing the way secrets are shared between such devices. This, however, is of no help if the old shares cannot be forgotten. The latter techniques guarantee security with respect to adversaries that obtain the current state of the memory, but still knows nothing about the previous states (which we consider here).

Our results. We investigate methods for storing data in such a way that information can be securely forgotten. To this purpose, we pt forward the new notion of *erasable memory*, a memory which allows to definitely and reliably erase values, by keeping only the most recently stored ones. We then formalize the type of attack that an adversary can mount on any cryptographic protocol which involves storage of (secret) values on any non-erasable memory. This leads to the formal definition of a secure erasable memory implementation as a method using a small piece of erasable memory in order to transform a large piece of non-erasable memory into large *and* erasable memory. Then, our main result consists in exhibiting such a secure erasable memory implementation under the assumption of existence of any pseudo-random permutation, which is known to exist under the existence of any one-way function (using [11,9,14]). We can show that such assumption is minimal too. In practice, pseudo-random permutations can be implemented using a block cipher such as some composed version of DES [1] (e.g., Triple-DES). In this case, our method also seems practical and efficient: the amount of erasable memory needed is only the amount to store the key for and compute one Triple-DES application; the storage overhead of our solution is only linear, and the computational overhead per memory access is a logarithmic number (in the size of the memory) of encryptions (this can be amortized to constant in some typical cases). While one would probably not want to use our method for all data, it would be quite reasonable for the small amount of data that is important for security purposes.

Outline of the paper: We introduce our terminology and model in Section 2, we present our construction of large erasable memory in Section 3, and discuss three possible concrete realization of small erasable memory in Section 4.

2   Definitions and Model

In this section we recall the notion of pseudo-random permutations [9,14], we introduce the model for secure erasable memory implementation and briefly discuss a first (inefficient) solution to our problem.

2.1   Pseudo-random Permutations.

Pseudo-random functions have been introduced in [9]. In this paper we will use the formalization of finite pseudorandom functions given in [2] (a concrete version of the formalization in [9,14]).

Let $\mathcal{P}$ be the family of all permutations, and let $G = \{E(k, \cdot), D(k, \cdot)\}_k \subseteq \mathcal{P}$ be a family of permutations, where $E(k, \cdot)$ denotes a finite permutation and $D(k, \cdot)$ its inverse. Each element from G is specified by a K-bit key k; therefore, uniformly choosing an element from G is equivalent to uniformly choosing $k \in \{0, 1\}^K$ and returning $(E(k, \cdot), D(k, \cdot))$. Let A be an algorithm with oracle access to an element from G. We let $\mathbf{E}[A^G] = \Pr[(E(k, \cdot), D(k, \cdot)) \xleftarrow{R} G : A^{E(k, \cdot)} = 1]$, i.e. the probability that A outputs 1 when A's oracle is selected to be a random permutation from G. If G, G' are two families of finite permutations, we let $\mathrm{Adv}_A(G, G') = \mathbf{E}[A^G] - \mathbf{E}[A^{G'}]$ denote the *advantage* of A in distinguishing G from G'. Here, following [9], we are considering the following game, or statistical test. Algorithm A is given as oracle a permutation g chosen at random from either G or G', the choice being made randomly according to a bit b. Then A will try to predict b. The advantage is $\Pr[A^g = b] - 1/2$, i.e., the amount that the probability that A's output is correct minus the probability that a random guess for b is correct. We say that the family of permutations G is (t, q, ϵ)-pseudorandom if there is no A which, running in time t and making q oracle queries, is able to obtain that $\mathrm{Adv}_A(G, \mathcal{P}) \geq \epsilon$.

2.2 Erasable Memory Implementation.

By the term *memory* we will denote any type of storage device. We will then consider two main types of memory: *persistent memory* and *erasable memory*. Persistent memory doesn't reliably forget former values (i.e., it allows the retrieval of formerly stored values). In fact, in order to prove the strongest result, we make the pessimistic assumption that all values that were ever written to the persistent memory can be retrieved by an attacker. In contrast, erasable memory reliably forgets old values (i.e., at any location, only the most recently stored value can be retrieved). Due to the practical considerations previously discussed, we should think of a typical computer storage device as being entirely persistent and of the erasable memory as a different and small piece of memory having (typically) some kind of physical implementation (we discuss three possible ones later, in Section 4).

We will consider an *erasable memory implementation* as a probabilistic data structure algorithm that translates read and write operations for a logical array, called the *virtual memory*, into read and write operations to a physically implemented array, called the *physical memory*. At the beginning of the system, the physical memory is preprocessed into two parts: the erasable memory and the persistent memory, where we should think of the erasable memory as being much smaller than the persistent one. The goal of the erasable memory implementation is to transform the physical memory into a single erasable memory.

Informally speaking, an attack on an erasable memory implementation proceeds in the following way. An adversary picks two sequences of read and write operations which are not trivially distinguishable. The data structure is then simulated on both sequences, computing the final state of the erasable memory and the entire history of writes to the persistent memory. Then the two pairs of histories and final erasable memory states are sent to the adversary in a random order. The adversary tries to predict which physical memory history corresponds to which sequence of operations. The implementation is considered secure if no adversary can succeed significantly better than by a random guess.

Definition 1. Let m, e, p be integers, and let M, EM be arrays denoting, respectively, the memory and the erasable memory, and such that $|M| = m$, $|EM| = e$. Let PM be an array denoting the persistent memory, and made of p lists $pm_1, \ldots, pm_p$, such that $pm_i = (v_{i,1}, \ldots, v_{i,s_i})$ denotes the sequence of values ever written into location i of the persistent memory (i.e., $v_{i,1}$ is the less recent and v_{i,s_i} the most recent, where s_i is the number of values ever stored into location i). Let $OP = \{\text{READ}, \text{WRITE}\}$ be the set of *memory operations*. An *instance* of a memory operation $op \in OP$ is the tuple $ins = (op; i; inp_1, \ldots, inp_c;$

$out_1, \ldots, out_d$), where argument i points to the input location, arguments inp_j are additional inputs for op and arguments out_j are outputs of op when executed on input location i and additional inputs $inp_1, \ldots, inp_c$. We will denote as *valid* all instances of the form

1. (READ; i; EM, PM; $cont$);
2. (WRITE; i; $EM, PM, cont$; EM, PM);

where $i \in \{1, \ldots, p\}$, $cont \in \Sigma$, for some alphabet Σ. Here, the read operation transfers into $cont$ the most recent value v_{i,s_i} previously stored at location i of array PM; the operation possibly uses EM, and returns $cont$. The write operation, possibly using array EM, inserts $cont$ into the last position of the list associated with location i in array PM. Namely, it increments s_i by 1 and sets $v_{i,s_i} = cont$. We say that two sequences of length T of valid instances of memory operations are *T-equivalent* if they contain the same subsequence of pairs (op, i), for $op =$ WRITE (but possibly different additional inputs or outputs, e.g. $cont$, or different subsequences with READ operations).[1]

We are now ready to define a secure erasable memory implementation.

Definition 2. An *erasable memory implementation* (EMI) is a pair of probabilistic algorithms EMI=(PREPROCESS,UPDATE). On input a memory array M, algorithm PREPROCESS returns an erasable memory array EM and a persistent memory array PM. On input arrays EM, PM, and an instance of a memory operation ins, algorithm UPDATE checks if the operation ins is valid; if so, it returns updated arrays EM, PM; otherwise it returns the unchanged input arrays EM, PM. Now, let A be an algorithm and let DISTINGUISH$_A(M, OP)$ be the following probabilistic experiment:

$\{$ $(EM, PM) \leftarrow$ PREPROCESS(M);
    $((ins_{0,1}, \ldots, ins_{0,T}), (ins_{1,1}, \ldots, ins_{1,T})) \leftarrow A(EM, PM)$;
    $EM_0 \leftarrow EM$; $PM_0 \leftarrow PM$; $EM_1 \leftarrow EM$; $PM_1 \leftarrow PM$;
    $(EM_0, PM_0) \leftarrow$ UPDATE$(ins_{0,i}, EM_0, PM_0)$, for $i = 1, \ldots, T$;
    $(EM_1, PM_1) \leftarrow$ UPDATE$(ins_{1,i}, EM_1, PM_1)$, for $i = 1, \ldots, T$;
    $b \xleftarrow{R} \{0,1\}$; $d \leftarrow A(EM_b, PM_b, EM_{1-b}, PM_{1-b})$:
       if $((ins_{0,1}, \ldots, ins_{0,T}), (ins_{1,1}, \ldots, ins_{1,T}))$ are T-equivalent and $b = d$ then
          return: 1 else return: 0; $\}$

We say that the erasable memory implementation is (T, ϵ)-*secure* if for any adversary A making T memory operations, the probability that the experiment DISTINGUISH$_A(M, OP)$ returns 1 is at most $1/2 + \epsilon$.

We note that in the above definition we are asking the adversary to perform only T-equivalent sequences of valid virtual memory operations. The reason for this is to avoid the adversary to have trivial ways of distinguishing the two sequences (e.g., as for two sequences writing to different memory locations).

---

[1] We note that we can also handle different definitions of T-equivalent sequences, in different scenarios.

Related notions. The notion of *crypto-paging* [19,20] was introduced to study the problem of hiding information from secondary storage devices, and can be used give a preliminary solution to our problem (see Section 2.3). A problem similar to ours is considered in [3], where a method similar to crypto-paging is suggested, but without a formal model or proofs. In [10] the authors study software protection in a model where one wants to protect any information about the communication between CPU and memory, including any access patterns. In [16] the authors study the problem of storing information in memory in such a way that the location of the stored item is hidden to the observer of the communication. The fact that in our model an adversary is allowed at some time to look at the content of both the erasable memory and the memory (i.e., reading private keys and opening any encryption) makes these last two models incomparable to ours.

2.3 An Inefficient Solution

A simple (but inefficient) solution to the problem of constructing a secure erasable memory implementation can be obtained as an immediate adaptation of the crypto-paging concept [20], as we now show. The preprocessing operation consists in uniformly choosing a key k for a pseudo-random permutation, and encrypting all the data using k; then the encrypted data is stored in the persistent memory, and the key k is stored in the erasable memory. In order to read a value, using the current key k in the erasable memory, the encryption of the value is decrypted and the data is recovered. In order to write a value, a new key k' for the pseudo-random permutation is uniformly chosen, the value is encrypted under key k' and the encryption is stored in the persistent memory. In order to guarantee security, the new key k' needs to be stored into the erasable memory, replacing the old key k; moreover, all items in the memory need to be re-encrypted using the new key k'. Notice that the number of encryptions/decryptions at each memory operation is quite large (*linear* in the size of the memory); this motivates the search for more efficient solutions.

3 Our Method for Obtaining Large Erasable Memory

In this section we present our construction for obtaining a secure erasable memory implementation. The main cryptographic tool we use is a family of pseudo-random permutations $\{E(k, \cdot), D(k, \cdot)\}_k$ (which, in practice, can be implemented by any block cipher). Formally, we achieve the following:

Theorem 3. Let m be the size of the memory. Given a family of pseudo-random permutations F$= \{E(k, \cdot), D(k, \cdot)\}_k$, there exists a secure erasable memory implementation EMI$=$(PREPROCESS,UPDATE), such that, for any integer $l > 1$, the following holds:

1. *Security.* If F is (T, q, ϵ)-pseudorandom, then EMI is (T', ϵ')-secure, for
    $T' = T - O(lT \log_l m)$, and $\epsilon' = \epsilon/(T \log_l m)$;

2. *Space Complexity.* If F has block size a and key size K, and is computable in space s, then EMI uses a persistent memory of size $\Theta(m)$ and an erasable memory of size $\Theta(a + s + K)$;

3. *Time Complexity.* Let $\mathcal{D}$ be the distribution of the location accessed to the memory; if $E(k, \cdot), D(k, \cdot)$ can be computed in time t then, in order to process T memory operations, EMI takes time $\Theta(Ttl \log_l m)$ for an arbitrary $\mathcal{D}$ or time $\Theta(t(T + l \log_l m))$ if $\mathcal{D}$ returns consecutive values.

We observe that the size of the persistent memory can be any polynomial in the size of the erasable memory input to EMI. Moreover, our erasable memory implementation requires only a logarithmic (in the size of the memory) number of encryption/decryption operations in the worst case distribution over the possible memory operations, and it achieves an amortized constant number of encryption/decryption operations in the case the memory operations are made on consecutive locations. In the following, we first present an overview of our construction and then a formal description. Further issues on the efficiency of our scheme, discussions about dividing the write operation into a write and an erase stage, and a proof that our construction satisfies Theorem 3 are in [5].

An overview of our construction. Let $G = \{E(k, \cdot), D(k, \cdot)\}_k$ denote a family of pseudo-random permutations (or, secure block ciphers). We arrange the persistent memory in a complete l-ary tree H, for $l \geq 2$. The root of the tree is contained in the erasable memory, the internal nodes correspond to the persistent memory addresses, and the leaves contain the data from the virtual memory. Therefore, we need more persistent memory than virtual memory, mainly for the interior nodes, whose number is that of the virtual locations divided by $l - 1$ (thus, increasing l decreases the memory overhead). At each interior node x there is an associated key k_x and the list of values ever stored at this node's location. The key associated to a leaf is equal to the content of the corresponding location in the persistent memory. At each physical location x, we store $E(k_{p(x)}, k_x \circ j)$ where x is the j'th child of its parent $p(x)$. To perform either a read or a write operation on a certain position of the persistent memory, we need to access the content of the corresponding leaf in the tree; therefore we follow its path starting from the root, and decrypting each physical location's contents with its parent's key to get its key. Then, in the case of a read operation, we just return the most recently stored value at that leaf. In the case of a write operation, we first insert the value to be written into the list associated to that leaf. Then, we follow the same path now starting from the leaf, and we pick new keys for each node along the path. For each such node, we decrypt its children's most recently stored value and re-encrypt their keys with the parent's new key. This encryption is then inserted into the list associated to that node. We note that before and while computing the encryption/decryption function, we must transfer to the erasable memory all input values, keys and intermediate values used. However, we can do the above so that only a constant number constant number of keys are in the erasable memory at any one time.

A formal description of our construction. Our formal description considers the most general case of arbitrary distribution over the memory operations (the

:ase of consecutive locations can be directly derived from it and is omitted). In ▸rder to make the description cleaner, we assume that the intermediate values n the computation of algorithms E and D which need to be stored into the nemory will always be stored into the erasable memory. By $p(x)$ and $c_j(x)$ we lenote the parent of node x, and the j-th child of node x, respectively.

The algorithm PREPROCESS

nput: an array $M[1\ldots m]$.

nstructions:

1. Initialize empty arrays $PM[1\ldots p]$ and $EM[1\ldots e]$.
2. Arrange PM as a complete l-ary tree H of height $\lceil\log_l m\rceil$, where $2 \le l \le 2^{a-K}$.
3. Denote by l_i the i-th leaf of H and let $k_{l_i} = M[i]$, for $i = 1,\ldots,m$.
4. For each node x of H,
   if x is not a leaf then uniformly choose a key $k_x \in \{0,1\}^K$;
   let j be such that x is the j-th child of node $p(x)$;
   store $k_x \circ j$ into $EM[2]$ and $k_{p(x)}$ into $EM[3]$ and set $z_x = E(k_{p(x)}, k_x \circ j)$;
   if x is the root then store k_x into $EM[1]$;
       else insert z_x into list pm_i,
           where i is the location in PM associated with node x;
5. return: (EM, PM) and halt.

The algorithm UPDATE

nput:

1. an instance ins of an operation $op \in OP = \{\text{READ}, \text{WRITE}\}$;
2. arrays $PM[1\ldots p]$ (storing tree H) and $EM[1\ldots e]$.

nstructions:

1. If ins is not valid then return: $\bot$ and halt.
2. Let q be the path in tree H starting from the root and
   finishing at node associated with location loc to be read or written.
3. For all nodes x in path q (in descending order),
   if $p(x)$ is the root of H then set $k_{p(x)} = EM[1]$;
       else if x is not the root then
           store z_x into $EM[2]$ and $k_{p(x)}$ into $EM[3]$ and set $y_x = D(k_{p(x)}, z_x)$;
           let $k_x \in \{0,1\}^K$ and $j \in \{1,\ldots,l\}$ be such that $y_x = k_x \circ j$;
4. If $ins = (\text{READ}; loc; EM, PM; cont)$ then
       set $cont = k_x$, return: $cont$ and halt.
5. If $ins = (\text{WRITE}; loc; EM, PM, cont; EM, PM)$ then
       for all nodes x in path q (in ascending order),
           if x is not a leaf then
               uniformly choose $k'_x \in \{0,1\}^K$;
               for $j = 1,\ldots,l$,
                   store $z_{c_j(x)}$ into $EM[2]$ and k_x into $EM[3]$ and set $y_{c_j(x)} = D(k_x, z_{c_j(x)})$;
                   let $k_{c_j(x)} \in \{0,1\}^K$ and $j \in \{1,\ldots,l\}$ be s.t. $y_{c_j(x)} = k_{c_j(x)} \circ j$;
                   set $z_{c_j(x)} = E(k'_x, y_{c_j(x)})$;
               if x is the root then store k_x into $EM[1]$;

else insert z_x into list pm_i,
   where i is the location in PM associated with node x;
set $k_x = k_x'$;
return: (EM, PM) and halt.

4   Concrete Realizations of Small Erasable Memory

We describe different concrete ways of constructing physically erasable memory. All of these ways seem prohibitive for large amounts of memory, but much more reasonable for the small amount such as what we require. Discussions about comparing and combining the above methods can be found in [5].

Trusted erasable memory. The simplest example of an implementation for an erasable memory is to use a *trusted* or *guardable* piece of memory. In this solution, the erasable memory is not so much a construction as a trust relationship. The system trusts the erasable memory *not to reveal any information*, which is at least as strong as *forgetting* former information. In a network system, the erasable memory might be a separate computer in a locked room. Due to the better physical security of this machine, it is less likely to be corrupted, and thus more likely to forget, in the sense that the information is not revealed to other parties. In the example of a multi-user personal computer (e.g., a device used by many people *one at a time*), each user might maintain physical control of one piece of memory, say a smart card or disk, and use that piece of memory as their personal erasable memory.

Disposable erasable memory. Another construction for erasable memory is based on *limited physical destruction*. More specifically, we can associate the *physical* erasable memory with a disposable memory device, such as an inexpensive smart card or disk. Each time we need to forget a value, a new device is obtained, and the old one physically destroyed, e.g., burnt or melted. (This frequency can be limited by a combination of methods, e.g., that of a *temporarily* trusted memory that is replaced at regular intervals.)

Randomly updated erasable memory. The main problem with erasing memory in physical sources is that imprints are left if the memory has the same value for a long period. Updating with the same value will not prevent this. However it is reasonable to assume that updating a memory device frequently with uncorrelated random values will not leave traces. Heuristically, our reasoning is as follows: The memory device has a maximum amount of retrievable data that can be stored on it. If there is no distinction between old and new data in terms of time the data has been stored, or correlations with other data, then it seems reasonable to assume that the newer data will be the easiest to recover. If at most a finite amount of data is possible to recover, and newer data is easier to recover than older data, it follows that sufficiently old data is impossible to recover. We can use this to make a fixed item erasable even if held for a long period of time. We distribute the item m by sharing it over two or more memory devices. At any time, one device holds a random string r, and the other $m \oplus r$. The value of r

s updated regularly, and much more frequently than m, by $\oplus$'ing both memory
devices with the same random string. (Note that the random strings do not have
to be kept secret, so they could probably be generated pseudo-randomly even if
the seed becomes known later. The randomness is intended to 'fool' the memory
device, not the adversary.) This is reminiscent of the *zero-sharing* method in
proactive secret sharing schemes [8,13,17].

References

1. B. Aiello, M. Bellare, G. Di Crescenzo, and R. Venkatesan, *Security amplification by composition: the case of doubly-iterated, ideal ciphers*, Proc. of CRYPTO 98.
2. M. Bellare, J. Kilian and P. Rogaway, *The security of cipher block chaining*, Proc. of CRYPTO 94.
3. D. Boneh, and R. Lipton, *A revocable backup system*, Proc. of USENIX 97.
4. J. Bos, *Booting problems with the JEC Computer*, personal communication, 1983.
5. G. Di Crescenzo, N. Ferguson, R. Impagliazzo, and M. Jakobsson, *How to forget a secret*, full version of this paper, available from authors.
6. W. Diffie, P. Van Oorschot, and M. Wiener, *Authentication and authenticated key exchanges*, Design, Codes and Cryptography, vol. 2, 1992.
7. T. ElGamal, *A public-key cryptosystem and a signature scheme based on discrete logarithms*, Proc. of CRYPTO 84.
8. Y. Frankel, P. Gemmell, P. MacKenzie, M. Yung, *Proactive RSA*, Proc. of CRYPTO 97.
9. O. Goldreich, S. Goldwasser and S. Micali, *How to construct random functions*, Journal of the ACM, Vol. 33, No. 4, 210–217, (1986).
10. O. Goldreich and R. Ostrovsky, *Software protection and simulation by oblivious RAMs*, Journal of the ACM, 1996.
11. J. Hastad, R. Impagliazzo, L. Levin, and M. Luby, *Construction of a pseudorandom generator from any one-way function*, SIAM Journal on Computing, to appear (previous versions: FOCS 89, and STOC 90).
12. A. Herzberg, M. Jakobsson, S. Jarecki, H. Krawczyk, M. Yung, *Proactive public key and signature systems*, Proc. of ACM CCS 97.
13. A. Herzberg, S. Jarecki, H. Krawczyk, M. Yung, *Proactive secret sharing, or how to cope with perpetual leakage*, Proc. of CRYPTO '95.
14. M. Luby and C. Rackoff, *How to construct pseudorandom permutations from pseudorandom functions*, SIAM Journal on Computing, Vol. 17, No. 2, April 1988.
15. National Institute for Standards and Technology, *Digital signature standard (DSS)*, Federal Register Vol. 56 (169), Aug 30, 1991.
16. R. Ostrovsky and V. Shoup, *Private information storage*, Proc. of STOC 1997.
17. R. Ostrovsky and M. Yung, *How to withstand mobile virus attacks*, Proc. of PODC 91.
18. C. P. Schnorr, *Efficient signature generation for smart cards*, Proc. CRYPTO 89.
19. B. Yee, D. Tygar, *Secure coprocessors in electronic commerce applications*, Proc. of USENIX 95.
20. B. Yee, *Using secure coprocessors*, Ph.D. Thesis, CMU-CS-94-149, 1994.

A Modal Fixpoint Logic with Chop

Markus Müller-Olm

Department of Computer Science
University of Dortmund
44221 Dortmund, Germany
mmo@ls5.informatik.uni-dortmund.de

Abstract. We study a logic called FLC (Fixpoint Logic with Chop) that extends the modal mu-calculus by a chop-operator and termination formulae. For this purpose formulae are interpreted by predicate transformers instead of predicates. We show that any context-free process can be characterized by an FLC-formula up to bisimulation or simulation. Moreover, we establish the following results: FLC is strictly more expressive than the modal mu-calculus; it is decidable for finite-state processes but undecidable for context-free processes; satisfiability and validity are undecidable; FLC does not have the finite-model property.

1 Introduction

Imperative programming languages typically offer a sequential composition operator which allows the straightforward specification of behavior proceeding in successive phases. Similar operators are provided by interval temporal logics, where they are called *chop*-operators. Important examples are Moszkowski's Interval Temporal Logic ITL [13] and the Duration Calculus DC [17]. As far as we know, however, no point-based temporal logic and, in particular, no branching-time logic with a chop operator has been proposed up to now. Indeed, at first glance there seems to be no natural way for explaining the meaning of sequentially composed formulae ϕ_1 ; ϕ_2 in the setting of point-based temporal or modal logic, as there is no natural notion of where interpretation of ϕ_1 stops and interpretation of ϕ_2 starts.

In this paper we present a logic called FLC (Fixpoint Logic with Chop) that extends the modal mu-calculus [8], a popular point-based branching-time fixpoint logic, by a chop operator ; and *termination formulae* term. For this purpose we utilize a 'second-order' interpretation of formulae. While (closed) formulae of usual temporal logics are interpreted by sets of states, i.e. represent *predicates*, we interpret formulae by mappings from states to states, i.e. by *predicate transformers*. A similar idea has been used by Burkart and Steffen [3] in a model checking procedure for modal mu-calculus formulae and context-free processes. However, while we use a second-order interpretation of formulae, they rely on a second-order interpretation of states as property transformers.

It turns out that FLC is strictly more expressive than the modal mu-calculus but is still decidable for finite-state processes. Consequently, FLC-based mode

C. Meinel and S. Tison (Eds.): STACS'99, LNCS 1563, pp. 510–520, 1999.
© Springer-Verlag Berlin Heidelberg 1999

checking is, in our opinion, an interesting alternative to modal mu-calculus based model checking as it enables to verify non-regular properties. The chop-operator also enables a straightforward specification of phased behavior. Other results shown in this paper are that FLC is undecidable for context-free processes, that satisfiability (and thus also validity) is undecidable, and that the logic does not have the finite model property. These results are inferred from the existence of formulae characterizing context-free processes up to bisimulation and simulation.

The remainder of this paper is structured as follows. The next section recalls bisimulation, simulation and context-free processes. In Section 3 we introduce the logic FLC and show that it conservatively extends the modal mu-calculus. Section 4 shows that context-free processes can be characterized up to bisimulation and simulation by single FLC-formulae. These facts, besides being of interest in their own, provide the main means for establishing the results on expressiveness and decidability, which are presented in Section 5. The paper finishes with a discussion of the practical utility of FLC.

2   Preliminaries

Processes, Bisimulation, and Simulation. A commonly used basic operational model of processes is that of rooted labeled transition systems. Assume for the remainder of this paper given a finite set Act of actions. Then a *labeled transition system* (over Act) is a structure $T = (S, Act, \rightarrow)$, where S is a set of *states*, and $\rightarrow \subseteq S \times Act \times S$ is a *transition relation*. We write $s \xrightarrow{a} s'$ for $(s, a, s') \in \rightarrow$. A *process* is a pair $P = (T, s_0)$ consisting of a labeled transition system $T = (S, Act, \rightarrow)$ and an *initial state* (or *root*) $s_0 \in S$. A process is called *finite-state* if the underlying state set S is finite.

Transition systems provide a rather fine-grained model of processes. Therefore, various equivalences and preorders have been studied in the literature that identify or order processes on the basis of their behavior. Classic examples are strong bisimulation [15,11] denoted by $\sim$ and simulation denoted by $\preceq$.

For two given processes $P = ((S, Act, \rightarrow_P), s_0)$ and $Q = ((T, Act, \rightarrow_Q), t_0)$ both bisimulation $\sim$ and simulation $\preceq$ are first defined as relations between the state sets S and T. These definitions are then lifted to the processes themselves. As relations between S and T they can be characterized as the greatest fixpoints $\nu F_\sim$ and $\nu F_\preceq$ of certain monotonic functionals $F_\sim$ and $F_\preceq$. These functionals operate on the complete lattice of relations $R \subseteq S \times T$ ordered by set inclusion and are defined by

$$F_\sim(R) \stackrel{\text{def}}{=} \{(s,t) \mid \forall a, s' : s \xrightarrow{a}_P s' \Rightarrow \exists t' : t \xrightarrow{a}_Q t' \wedge (s', t') \in R$$
$$\wedge \forall a, t' : t \xrightarrow{a}_Q t' \Rightarrow \exists s' : s \xrightarrow{a}_P s' \wedge (s', t') \in R\}$$

and $F_\preceq(R) \stackrel{\text{def}}{=} \{(s,t) \mid \forall a, s' : s \xrightarrow{a}_P s' \Rightarrow \exists t' : t \xrightarrow{a}_Q t' \wedge (s', t') \in R\}$. The processes P and Q are called *bisimilar* if $s_0 \sim t_0$. Similarly, Q is said to *simulate* P if $s_0 \preceq t_0$. By abuse of notation we denote these relationships by $P \sim Q$ and $P \preceq Q$ and view $\sim$ and $\preceq$ also as relations between processes.

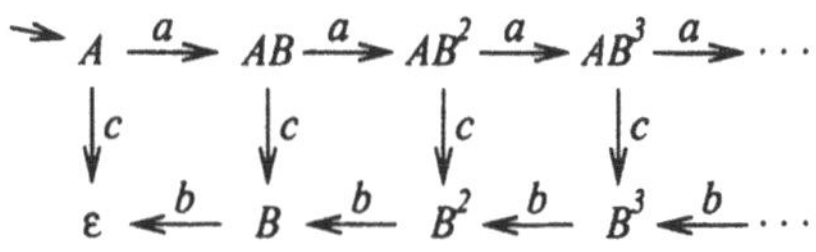

$$\begin{array}{ccccc}
\twoheadrightarrow A & \xrightarrow{a} AB & \xrightarrow{a} AB^2 & \xrightarrow{a} AB^3 & \xrightarrow{a} \cdots \\
\downarrow c & \downarrow c & \downarrow c & \downarrow c & \\
\varepsilon & \xleftarrow{b} B & \xleftarrow{b} B^2 & \xleftarrow{b} B^3 & \xleftarrow{b} \cdots
\end{array}$$

Fig. 1. A context-free process

Context-Free Processes. Context-free processes, also called BPA (basic process algebra) processes [1], are a certain type of finitely generated infinite-state processes. Their name derives from the fact that they are induced by leftmost derivations of context-free grammars in Greibach normal form, where the terminal symbols are interpreted as actions and the non-terminals induce the state. Greibach normal form means that all rules have the form $A ::= a\alpha$, where A is a non-terminal symbol, a a terminal symbol, and α a string of non-terminals. Formally, context-free processes can be defined as an instance of Rewrite Transition Systems as introduced by Caucal [4].

A *context-free process rewrite system* (over *Act*) is a triple $R = (V, Act, \Delta)$ consisting of a finite set V of *process variables*, the assumed finite set *Act* of *actions*, and a finite set $\Delta \subseteq V \times Act \times V^*$ of *rules*. The labeled transition system induced by $R = (V, Act, \Delta)$, called a *context-free transition system*, is $T_R = (V^*, Act, \rightarrow)$, where $\rightarrow \subseteq V^* \times A \times V^*$ is the smallest relation obeying the prefix rewrite rule

$$\text{PRE} \quad \frac{(A, a, \alpha) \in \Delta}{A\beta \xrightarrow{a} \alpha\beta} \quad .$$

Note that the states in a context-free transition system are words of process variables of the underlying context-free process rewrite system. A *context-free process* is a pair (T_R, α_0), consisting of a context-free transition system T_R and an initial state $\alpha_0 \in V^*$. As an example we picture in Fig. 1 the context-free process (T_R, A) where $R = (V, Act, \Delta)$, $V = \{A, B\}$, $Act = \{a, b, c\}$, and $\Delta = \{(A, a, AB), (A, c, \varepsilon), (B, b, \varepsilon)\}$.

The following two results are crucial for the remainder of this paper: firstly, there are context-free processes that are not bisimilar to any finite-state process (the process in Fig. 1 is an example) and, secondly, simulation between context-free processes is undecidable [6]. The reader interested in learning more about context-free processes and other classes of Rewrite Transition Systems is pointed to the surveys [12] and [2] and the many references there.

3   The Logic FLC

In the remainder of this paper the letter X ranges over an infinite set *Var* of *variables*, a over the assumed finite action set *Act* and p over an assumed finite set *Prop* of *atomic propositions*. We assume that *Prop* contains the propositions true and false.

The modal mu-calculus [8] is a small, yet expressive process logic that has been used as underlying logic in a number of model checkers. Modal mu-calculus formulae in positive normal form are constructed according to the grammar

$$\phi ::= p \mid [a]\phi \mid \langle a \rangle \phi \mid \phi_1 \wedge \phi_2 \mid \phi_1 \vee \phi_2 \mid X \mid \mu X . \phi \mid \nu X . \phi .$$

In the modal mu-calculus the modal operators $[a]$ and $\langle a \rangle$ do not have the status of formulae but can only be used in combination with already constructed formulae ϕ to form composed formulae $[a]\phi$ and $\langle a \rangle \phi$. We now define FLC (*Fixpoint Logic with Chop*), an extension of the modal mu-calculus that gives the modal operators the status of formulae. More importantly, FLC provides a *chop operator* ;, which intuitively represents sequential composition of behavior, and a *termination formula* term, which intuitively requires the behavior of the sequential successor formula.

We consider again formulae in positive form which are now constructed according to the following grammar:

$$\phi ::= p \mid [a] \mid \langle a \rangle \mid \phi_1 \wedge \phi_2 \mid \phi_1 \vee \phi_2 \mid X \mid \mu X . \phi \mid \nu X . \phi \mid \mathsf{term} \mid \phi_1 ; \phi_2 .$$

As in the modal mu-calculus, the two *fixpoint operators* μX and νX bind the respective variable X and we will apply the usual terminology of free and bound variables in a formula, closed formula etc. Moreover, we write for a finite set M of formulae $\bigwedge M$ and $\bigvee M$ for the conjunction and disjunction of the formulae in M. As usual, we agree that $\bigwedge \emptyset = \mathsf{true}$ and $\bigvee \emptyset = \mathsf{false}$.

Both the modal mu-calculus as well as FLC are basically interpreted over a given labeled transition system $T = (S, Act, \rightarrow)$. Furthermore, an *interpretation* $I \in (Prop \rightarrow 2^S)$ is assumed, which assigns to each atomic proposition the set of states for which it is valid. We assume that interpretations always interpret true and false in the standard way, i.e. such that $I(\mathsf{true}) = S$ and $I(\mathsf{false}) = \emptyset$.

In the modal mu-calculus the meaning of a closed formula essentially is a subset of the state set S, i.e. a *predicate* on states. In order to explain the meaning of the new types of formulae, we interpret FLC-formulae by monotonic predicate transformers. A *(monotonic) predicate transformer* is simply a mapping $f : 2^S \rightarrow 2^S$ which is monotonic w.r.t. the inclusion ordering on 2^S. It follows from well-known results of lattice theory that the set of monotonic predicate transformers, which we denote by $MTrans_T$, together with the pointwise extension $\sqsubseteq$ of the inclusion ordering on 2^S defined by

$$f \sqsubseteq f' \quad \text{iff} \quad f(x) \subseteq f'(x) \text{ for all } x \subseteq S$$

is a complete lattice. We denote the join and meet operations by $\sqcup$ and $\sqcap$.

It is customary to refer to *environments*, in order to explain the meaning of open formulae. In the modal mu-calculus, environments are partial mappings of type $\rho : Var \xrightarrow{\text{part.}} 2^S$; they interpret (at least) the free variables of the formula in question by a set of states. In FLC we interpret free variables by predicate transformers. Thus, we use environments of type $\delta : Var \xrightarrow{\text{part.}} MTrans_T$. The predicate transformer assigned to an FLC-formula ϕ, denoted by $\mathcal{C}_T^I(\phi)(\delta)$, is

$$C_T^I(p)(\delta)(x) = I(p)$$

$$C_T^I([a])(\delta)(x) = \{s \mid \forall s' : s \xrightarrow{a} s' \Rightarrow s' \in x\}$$

$$C_T^I(\langle a\rangle)(\delta)(x) = \{s \mid \exists s' : s \xrightarrow{a} s' \wedge s' \in x\}$$

$$C_T^I(\phi_1 \wedge \phi_2)(\delta)(x) = C_T^I(\phi_1)(\delta)(x) \cap C_T^I(\phi_2)(\delta)(x)$$

$$C_T^I(\phi_1 \vee \phi_2)(\delta)(x) = C_T^I(\phi_1)(\delta)(x) \cup C_T^I(\phi_2)(\delta)(x)$$

$$C_T^I(X)(\delta) = \delta(X)$$

$$C_T^I(\mu X \,.\, \phi)(\delta) = \sqcap\{f \in MTrans_T \mid C_T^I(\phi)(\delta[X \mapsto f]) \sqsubseteq f\}$$

$$C_T^I(\nu X \,.\, \phi)(\delta) = \sqcup\{f \in MTrans_T \mid C_T^I(\phi)(\delta[X \mapsto f]) \sqsupseteq f\}$$

$$C_T^I(\mathsf{term})(\delta)(x) = x$$

$$C_T^I(\phi_1 \,;\, \phi_2)(\delta) = C_T^I(\phi_1)(\delta) \circ C_T^I(\phi_2)(\delta)$$

Fig. 2. Semantics of FLC

inductively defined in Fig. 2. The similar definition of the predicate $\mathcal{M}_T^I(\phi)(\rho)$ assigned to a modal mu-calculus formulae ϕ is omitted due to lack of space. It can be found in many papers on the modal mu-calculus.

Note that the fixpoint formulae of FLC are interpreted by the corresponding fixpoints in the set of predicate transformers and not in the set of predicates as in the modal mu-calculus. Also note that the chop operator is interpreted by functional composition and that **term** denotes the identity predicate transformer. Thus, **term** is the neutral element of ;.

As the meaning of a closed formula ϕ does not depend on the environment, we sometimes write just $C_T^I(\phi)$ ($\mathcal{M}_T^I(\phi)$) for $C_T^I(\phi)(\delta)$ ($\mathcal{M}_T^I(\phi)(\rho)$), where δ (ρ) is an arbitrary environment. We also omit the indices T and I if they are clear from the context.

The set of states *satisfying* a given closed formula ϕ is $C(\phi)(S)$. A process $P = (T, s_0)$ is said to satisfy ϕ if its initial state s_0 satisfies ϕ. It might appear somewhat arbitrary that the predicate transformer $C(\phi) : 2^S \to 2^S$ is applied to the full state set S in the definition of satisfaction. As far as expressiveness is concerned, however, the choice of a specific set x to which $C(\phi)$ is applied is largely arbitrary, as long as x can be described by a closed FLC formula ϕ_x: assume $x = C(\phi_x)(S)$; then $C(\phi)(x)$ equals $C(\phi \,;\, \phi_x)(S)$. As Lemma 1 below shows, sets x expressible in this way include at least all state sets that can be described by a modal mu-calculus formula (i.e. all modal mu-calculus definable properties). The formula $\phi_{DL} \stackrel{\text{def}}{=} \bigwedge_{a \in Act}[a] \,;\, \mathsf{false}$, for instance, characterizes the set of deadlocked states.

Any modal mu-calculus formula ϕ can straightforwardly be translated to FLC: just replace all sub-formulas of the form $[a]\psi$ or $\langle a\rangle\psi$ by $[a] \,;\, \psi$ or $\langle a\rangle \,;\, \psi$, respectively. We call the resulting FLC-formula $T(\phi)$. A rather straightforward structural induction shows that the interpretation of $T(\phi)$ is just the constant predicate transformer mapping any state set to the interpretation of the original modal mu-calculus formula.

Lemma 1. *Let ϕ be a modal mu-calculus formula and $\rho : Var \xrightarrow{part.} 2^S$ a modal mu-calculus environment. Let δ be the environment defined by $\text{dom}\,\delta = \text{dom}\,\rho$ and $\delta(X) = \lambda y\,.\,\rho(X)$ for $X \in \text{dom}\,\rho$. (Note that δ assigns constant predicate transformers to the variables.) Then $C(T(\phi))(\delta) = \lambda y\,.\,M(\phi)(\rho)$.*

As a consequence, FLC is at least as expressive as the modal mu-calculus.

Corollary 1. *Suppose ϕ is a closed modal mu-calculus formula and P is a process. Then P satisfies ϕ (in the sense of the mu-calculus) iff P satisfies $T(\phi)$ (in the sense of FLC).*

Equation systems. A (closed) *equation systems* of FLC-formula is a set $E = \{X_i = \phi_i \mid 1 \leq i \leq n\}$ consisting of $n \geq 0$ equations $X_i = \phi_i$, where $X_1, \ldots, X_n$ are mutually distinct variables and $\phi_1, \ldots, \phi_n$ are FLC-formulae having at most $X_1, \ldots, X_n$ as free variables. An environment $\delta : \{X_1, \ldots, X_n\} \to MTrans$ is a *solution* of equation system E, if $\delta(X_i) = C(\phi_i)(\delta)$ for $i = 1, \ldots, n$. By the Knaster-Tarski fixpoint theorem every equation system has a largest solution as the corresponding functional on environments is easily seen to be monotonic. We denote the largest solution of E by νE.

While it proves convenient to refer to equation systems, they do not increase the expressive power. Any predicate transformer that can be obtained as a component of the largest solution of an equation system E can just as well be characterized by a single formula. In order to show this, Gauß elimination [10] can be applied to the equation system (see e.g. [14]).

Proposition 1. *Let E be a closed equation system and X a variable bound in E. Then there is a closed FLC-formula ϕ such that $C(\phi) = (\nu E)(X)$.*

4   Characteristic Formulae for Context-Free Processes

The goal of this section is to show that any context-free process can be characterized up to bisimulation or simulation by an FLC-formula.[1] As a stepping stone, we construct equation systems that capture the contribution of the single process variables to bisimulation and simulation. Characteristic formulae for various other (bi-)simulation-like relations, in particular the weak versions, can be constructed along this line too.

In the following, we assume given a context-free process rewrite system $R = (V, Act, \Delta)$ and agree on the following variable conventions: the letters A and $B, B_1, B_2, \ldots$ range over V, a ranges over Act, and α and β range over V^*. For notational convenience we use the process variables $A \in V$ also as variables of the logic.

We consider the three equation systems $E_\sim = \{A = \phi_{\sim A} \mid A \in V\}$, $E_{\preceq} = \{A = \phi_{\preceq A} \mid A \in V\}$, and $E_{\succeq} = \{A = \phi_{\succeq A} \mid A \in V\}$. Analogously to the

---

[1] For the simulation case we shall actually construct two formulae. One of them characterizes the set of processes that are simulated by the process in question and the other the set of processes that simulate it.

finite-state case [16,14], the formulae $\phi_{\sim A}$, $\phi_{\preceq A}$, and $\phi_{\succeq A}$ mirror the conditions in the definition of bisimulation and simulation and are defined by

$$\phi_{\sim A} \overset{\text{def}}{=} \phi_{\succeq A} \wedge \phi_{\preceq A} \ ,$$

$$\phi_{\preceq A} \overset{\text{def}}{=} \bigwedge_{a \in Act} [a] ; \bigvee_{(A,a,B_1 \cdots B_l) \in \Delta} B_1 ; \ldots ; B_l \ , \text{ and}$$

$$\phi_{\succeq A} \overset{\text{def}}{=} \bigwedge_{a \in Act} \bigwedge_{(A,a,B_1 \cdots B_l) \in \Delta} \langle a \rangle ; B_1 ; \ldots ; B_l \ .$$

Now, suppose given an arbitrary transition system $T = (S, Act, \rightarrow)$ and an arbitrary interpretation $I : Prop \rightarrow 2^S$. (The specific interpretation does not matter as only the atomic propositions true and false appear in the characteristic equation systems.) Let $\delta_{\sim} = \nu E_{\sim} : V \rightarrow (2^S \rightarrow 2^S)$ be the largest solution of $E_{\sim}$ on T. The following lemma intuitively shows that the A-component of this solution represents the contribution of the process variable A to bisimulation.

Lemma 2. $\delta_{\sim}(A)(\{s \in S \mid s \sim \beta\}) = \{s \in S \mid s \sim A\beta\}$ for all $A \in V$, $\beta \in V^*$.

The '$\supseteq$'-direction can be proved by a fixpoint induction for $\delta_{\sim}$ and the '$\subseteq$'-direction by a fixpoint induction for $\sim = \nu F_{\sim}$. Combined with Proposition 1, Lemma 2 shows that there is a closed formula $\varphi_{\sim A}$ for each $A \in V$ such that for all $\beta \in V^*$:

$$\mathcal{C}_T(\varphi_{\sim A})(\{s \in S \mid s \sim \beta\}) = \{s \in S \mid s \sim A\beta\} \ . \tag{1}$$

These formulae $\varphi_{\sim A}$ can now be used to construct characteristic formulae for context-free processes with underlying process rewrite system R.

Theorem 1 (Characteristic formulae). *For each context-free process P there is a (closed) FLC-formula $\psi_{\sim P}$ such that, for any process Q, Q satisfies $\psi_{\sim P}$ iff $Q \sim P$.*

Proof. Let $P = (T_R, B_1 \cdots B_l)$ and let $\psi_{\sim P}$ be the formula $\varphi_{\sim B_1} ; \ldots ; \varphi_{\sim B_l} ; \phi_{DL}$, where ϕ_{DL} is the formula characterizing the set of deadlocked states from Section 3.

Suppose $Q = ((S, Act, \rightarrow), s_0)$ is an arbitrary process. Clearly, a state $s \in S$ is bisimilar to the state ε in T_R if and only if it satisfies ϕ_{DL}. It follows by repeated application of (1) that, for $i = 1, \ldots, l$, a state $s \in S$ satisfies $\varphi_{\sim B_i} ; \ldots ; \varphi_{\sim B_l} ; \phi_{DL}$ if and only if $s \sim B_i \cdots B_l$. Thus, Q is bisimilar to P if and only if it satisfies $\psi_{\sim P}$. □

An analogue of Lemma 2 for $E_{\preceq}$ and $E_{\succeq}$ ensures the existence of closed formulae $\varphi_{\preceq A}$ and $\varphi_{\succeq A}$ such that $\mathcal{C}_T(\varphi_{\preceq A})(\{s \in S \mid s \preceq \beta\}) = \{s \in S \mid s \preceq A\beta\}$ and $\mathcal{C}_T(\varphi_{\sim A})(\{s \in S \mid s \succeq \beta\}) = \{s \in S \mid s \succeq A\beta\}$ for arbitrary A and β. These formulae are used to establish the final theorem of this section.

Theorem 2 (Characteristic formulae for simulation). *For each context-free process P there are (closed) FLC-formulae $\psi_{\preceq P}$ and $\psi_{\succeq P}$ such that, for any process Q, Q satisfies $\psi_{\preceq P}$ iff $Q \preceq P$, and Q satisfies $\psi_{\succeq P}$ iff $Q \succeq P$.*

Proof. Let $P = (T_R, B_1 \cdots B_l)$ and $Q = ((S, Act, \rightarrow), s_0)$.

A state $s \in S$ is simulated by ε if and only if it satisfies ϕ_{DL}. Thus, $\psi_{\preceq P}$ can be chosen as the formulae $\varphi_{\preceq B_1} ; \ldots ; \varphi_{\preceq B_l} ; \phi_{DL}$.

On the other hand, every state $s \in S$ simulates ε. In other words, a state simulates ε if and only if it satisfies the formulae true. Thus, $\psi_{\succeq P}$ can be chosen as the formulae $\varphi_{\succeq B_1} ; \ldots ; \varphi_{\succeq B_l} ; \text{true}.^2$ □

5 Decidability and Expressiveness Issues

Clearly, FLC is decidable for finite-state processes: given a finite-state process $P = (T, s_0)$, an interpretation I, and an FLC-formula ϕ, $\mathcal{C}_T^I(\phi)$ can effectively be computed inductively over ϕ. The usual approximation of fixpoints terminates as $MTrans_T$ is finite.

Theorem 3. *FLC is decidable for finite-state processes.*

However, FLC is not decidable for context-free processes. This is a consequence of the existence of characteristic formulae for simulation. A decision procedure for FLC could namely be used to decide simulation between context-free processes, which is – as mentioned in Section 2 – undecidable: given two context-free processes P and Q one would just have to check, whether Q satisfies $\psi_{\succeq P}$ in order to decide, whether $P \preceq Q$.

Theorem 4. *FLC is undecidable for context-free processes.*

There is an interesting duality between the decidability of FLC for finite-state processes and the decidability of the modal mu-calculus for context-free (and even push-down) processes [3]. Both scenarios relate an inherently 'regular' structure with a structure of at least 'context-free strength'. While the former is concerned with the at least 'context-free' logic FLC and 'regular' finite-state processes, the latter relates the 'regular' modal mu-calculus (recall that the mu-calculus can be translated to monadic second order logic, which closely corresponds to finite automata) with context-free processes.

The existence of characteristic formulae for simulation also implies that satisfiability (and hence validity) of FLC is undecidable: assume given two context-free processes P and Q. It is easy to see that Q simulates P if and only if the formula $\psi_{\succeq P} \wedge \psi_{\preceq Q}$ is satisfiable. Thus, decidability of satisfiability would again imply decidability of simulation between context-free processes.

Theorem 5. *Satisfiability and validity of FLC are undecidable.*

An interesting consequence of the existence of characteristic formulae for bisimulation is that FLC does not enjoy the finite-model property:[3] choose a

---

[2] The final true could be omitted due to our definition of satisfaction.

[3] A modal logic has the finite-model property, if any satisfiable formula has a finite model.

$$0 \xrightarrow{a} 1 \xrightarrow{b} 2 \xrightarrow{a} 3 \xrightarrow{b} 4 \xrightarrow{b} 5 \xrightarrow{c} 6$$

Fig. 3. A linear, finite process

context-free process P that is not bisimilar to any finite-state process and take its characteristic formulae $\psi_{\sim P}$. Then this formulae is satisfiable (namely by P itself). But it cannot be satisfied by a finite-state process, as this finite-state process would then be bisimilar to the context-free process which yields a contradiction.

Theorem 6. *FLC does not enjoy the finite-model property.*

The modal mu-calculus on the other hand does enjoy the finite-model property [9]. Hence, in general, context-free processes cannot have characteristic modal mu-calculus formulae: let P again be a context-free process that is not bisimilar to a finite-state process and assume there would be a modal mu-calculus formula ϕ characterizing P up to bisimulation. Then – by the finite-model property – there would be a finite-state process Q satisfying ϕ. But this would mean that P and Q are bisimilar, which contradicts the choice of P.

As a consequence, FLC is strictly more expressive than the modal mu-calculus. However, if this increase of expressiveness would not show through on finite-state processes, it would be useless, as far as automatic model checking is concerned.

Fortunately, we can show that FLC already is more expressive on the class of finite-state processes, and even on a small subclass, that of *finite linear processes*. These are processes corresponding to finite words over Act. Formally, the process corresponding to a word $w = w_0 \cdots w_k \in Act^*$ is $P_w = ((\{0, \ldots, k\}, Act, \rightarrow), 0)$ where $\rightarrow = \{(i, w_i, i+1) \mid 0 \leq i < k\}$. As an example, the process corresponding to the word $ababbc$ is pictured in Fig. 3. The class of finite linear processes is $\{P_w \mid w \in Act^*\}$ and subclasses of it can straightforwardly be identified with sets of words over Act. The modal mu-calculus can be translated to monadic second-order logic. Therefore, the class of finite linear models of a modal mu-calculus formula ϕ corresponds to a regular set of words. The class of finite linear models of the FLC-formula $(\mu X . (\text{term} \vee \langle a \rangle X \langle b \rangle))$; ϕ_{DL}, however, correspond to the set $\{a^n b^n \mid n = 0, 1, \ldots\}$, which is well-known not to be regular [7].

Theorem 7. *FLC is strictly more expressive than the modal mu-calculus, even on finite linear processes, and therefore also on finite-state processes.*

It is interesting to note that FLC can even characterize certain non-context-free sets of finite linear processes due to the presence of conjunction: let, for arbitrary actions a and b, ϕ_{ab} be the formula $\mu X . (\text{term} \vee (\langle a \rangle \; ; \; X \; ; \; \langle b \rangle))$ and ψ_a be the formula $\mu X . (\text{term} \vee \langle a \rangle \; ; \; X)$. Then the finite linear models of

$$(\phi_{ab} \; ; \; \phi_c \; ; \; \phi_{DL}) \wedge (\phi_a \; ; \; \phi_{bc} \; ; \; \phi_{DL})$$

correspond to the set $\{a^n b^n c^n \mid n = 0, 1, \ldots\}$, which is context-sensitive but not context-free. A more thorough study of the expressiveness of FLC is left for future research.

5 Conclusion

We have proposed a modal logic FLC with fixpoints, a chop-operator, and termination formulae. The basic idea has been to interpret formulae by predicate transformers instead of predicates and to take fixpoint construction over predicate transformers as well. As a stepping stone in the technical development we have shown that FLC allows to characterize context-free processes up to bisimulation and simulation. FLC is strictly more expressive than the modal mu-calculus but is still decidable for finite-state processes.

Like the modal mu-calculus, FLC is perhaps not so much suited as a direct vehicle for specification. Rather it provides an expressive core logic, into which various other logics can be translated. An FLC-based model checking system could handle non-regular specification formalisms that are beyond the reach of modal mu-calculus based model checkers. An interesting example of such a formalism from a practical point of view are the timing diagrams studied by K. Fisler in [5]. It is a topic of future research whether they can actually be embedded into FLC.

A simple global model checking algorithm for FLC and finite-state processes can straightforwardly be constructed from the usual iterative computation of fixpoints. This procedure in general has an exponentially larger storage requirement compared to a straightforward global modal mu-calculus model checker: we have to store a mapping $2^S \to \mathbb{B}$ per state and formula (where $\mathbb{B}$ denotes the set of the Boolean values true and false) instead of just a single Boolean value.[4] Also the time complexity of FLC seems to be much higher than that of modal mu-calculus as $(2^S \to 2^S)$ has exponentially longer chains than 2^S such that fixpoint computation can require exponentially more iterations. Thus, at first glance model checking FLC seems to be impractical. (We currently know that model checking with a fixed formula is at least PSPACE-hard.)

However, the exponential blow up can be avoided for FLC-formulae corresponding to modal mu-calculus formulae. The idea is to represent the above mentioned mappings of type $2^S \to \mathbb{B}$ by binary decision diagrams (BDDs). As a consequence of Lemma 1 these mappings are constant for all FLC-formulae corresponding to modal mu-calculus formulae. It is, moreover, easy to see that the intermediate functions occurring during fixpoint iteration are constant too and correspond to the Boolean values that would be observed in a mu-calculus model checking procedure. Therefore, only a linear penalty arises for both space and time when model checking FLC-formula corresponding to modal mu-calculus formulae because constant BDDs can be represented in constant space. In this sense the increased expressiveness of FLC is obtained for free: the exponential blow-up can only occur in cases that cannot be handled by a modal mu-calculus model checker at all!

The above comparison applies to straight-forward global model checking. If and how more elaborate global and local mu-calculus model checking procedures

---

[4] A collection consisting of one of those mappings for each state in S represents a mapping $2^S \to 2^S$, i.e. a predicate transformer, which is the meaning of a formula.

can be adapted to FLC remains to be seen. Other topics for future research are a more thorough study of the complexity and expressiveness of FLC, in particular its relationship to context-free and context-sensitive languages and, last not least, the implementation and empirical evaluation of an FLC-based model checker. It is, moreover, interesting to study, whether the idea of a 'second-order' interpretation of formulae by predicate transformers can advantageously be applied to other logics.

References

1. J. A. Bergstra and J. W. Klop. Algebra of communicating processes with abstraction. *Theoretical Computer Science*, 37:77–121, 1985.
2. O. Burkart and J. Esparza. More infinite results. *ENTCS*, 6, 1997. URL: http://www.elsevier.nl/locate/entcs/volume6.html.
3. O. Burkart and B. Steffen. Model checking the full modal mu-calculus for infinite sequential processes. In *ICALP '97*, LNCS 1256, 419–429. Springer-Verlag, 1997.
4. D. Caucal. On the regular structure of prefix rewriting. *Theoretical Computer Science*, 106:61–86, 1992.
5. K. Fisler. Containment of regular languages in non-regular timing diagram languages is decidable. In *CAV'97*, LNCS 1254. Springer-Verlag, 1997.
6. J. F. Groote and H. Hüttel. Undecidable equivalences for basic process algebra. *Information and Computation*, 115(2):354–371, 1994.
7. J. E. Hopcroft and J. D. Ullman. *Introduction to Automata Theory, Languages and Computation*. Addison-Wesley, 1979.
8. D. Kozen. Results on the propositional mu-calculus. *Theoretical Computer Science*, 27:333–354, 1983.
9. D. Kozen. A finite model theorem for the propositional mu-calculus. *Studia Logica*, 47:233–241, 1988.
10. A. Mader. Modal mu-calculus, model checking and Gauss elimination. In *TACAS'95*, LNCS 1019, 72–88. Springer-Verlag, 1995.
11. R. Milner. *Communication and Concurrency*. Prentice Hall, 1989.
12. F. Moller. Infinite results. In *CONCUR'96*, LNCS 1119, 195–216. Springer-Verlag, 1996.
13. B. Moszkowski. A temporal logic for multi-level reasoning about hardware. *IEEE Computer*, 18(2):10–19, 1985.
14. M. Müller-Olm. Derivation of characteristic formulae. *ENTCS*, 18, 1998. URL: http://www.elsevier.nl/locate/entcs/volume18.html.
15. D. M. R. Park. Concurrency and automata on infinite sequences. In LNCS 154, 561–572. Springer-Verlag, 1981.
16. B. Steffen and A. Ingólfsdóttir. Characteristic formulae for processes with divergence. *Information and Computation*, 110(1):149–163, 1994.
17. Zhou Chaochen, C. A. R. Hoare, and A. P. Ravn. A calculus of durations. *Information Processing Letters*, 40(5):269–276, 1991.

Completeness of Neighbourhood Logic

Rana Barua[1] *, Suman Roy[2] **, and Zhou Chaochen[3] * * *

[1] Stat-Math. unit, I.S.I., 203 B.T. Road, Calcutta 700 035, India.
rana@isical.ac.in
[2] Computer Science and Automation, I.I.Sc., Bangalore 560 012, India.
suman@csa.iisc.ernet.in, corresponding author
[3] International Institute for Software Technology, UNU/IIST, P.O. Box 3057, Macau
zcc@iist.unu.edu

Abstract. This paper presents a completeness result for a first-order interval temporal logic, called Neighbourhood Logic (NL) which has two neighbourhood modalities. NL can support the specification of liveness and fairness properties of computing systems as well as formalisation of many concepts of real analysis. These two modalities are also adequate in the sense that they can derive other important unary and binary modalities of interval temporal logic. We prove the completeness result for NL by giving a Kripke model semantics and then mapping the Kripke models to the interval models for NL.

1 Introduction

In many applications, digital systems reacting with environment and events have to produce an output before a certain delay has elapsed. Time requirements - both qualitative as well as quantitative - have to be considered to reason about such systems. Thus, for such purposes one has to consider a real-time logic. Various such logics have been proposed. Some of these formalisms interpret formulas over intervals of time [5,11,16,17,19]; notably among them are Interval Temporal Logic (ITL) [11] and Duration Calculus (DC) [6,19]. ITL is a first-order interval modal logic which uses a binary modal operator "⌢" which is interpreted as the operation of "chopping" an interval into two parts. DC is an extension of ITL in the sense that temporal variables are written in the form of the integrals of "states".

Since chop "⌢" is a *contracting* modality, ITL-based logics can succintly express properties of the real-time systems, such as; "for all time intervals of a given length, ϕ must be true", or "if ϕ holds for a time interval, then there is a sub-interval where ψ holds". However, these logics cannot express *liveness* properties, which depend on intervals lying outside the reference interval, like;

---

* The work was done when the author visited UNU/IIST as a fellow during May-Aug.'97
* The work was done when the author visited UNU/IIST as a fellow Jun.'96–Jan.'97
* On leave of absence from Software Institute, Chinese Academy of Sciences

C. Meinel and S. Tison (Eds.): STACS'99, LNCS 1563, pp. 521–530, 1999.

"eventually there is an interval where ϕ is true, and "ϕ will hold infinitely often in the future".

Another limitation of these logics is that when they are used in the specification of hybrid systems the notions of real analysis such as limit, continuity and differentiability cannot be suitably formalised in them. These notions are neighbourhood properties of a point which cannot be defined in those logics. Although an informal mathematical theory of calculus can be assumed as in extended duration calculus [20], Hybrid Statecharts [10], Hybrid Automata [1] and TLA$^+$ [9], a formalization of real analysis may help in developing theorem provers for supporting the design of hybrid systems.

In order to improve the expressiveness of ITL, *expanding* modalities have been used. Venema [17] gives a complete axiomatization of a propositional calculus with three binary modalities; in addition to chop (designated as C) it has modalities T and D, which can represent properties outside the interval. Some of the axioms and rules in it are quite complicated. Other expanding modalities which are *unary* have been considered in Halpern and Shoham [5]. But many notions of real analysis cannot be formalised without first-order quantifiers.

In [18], Zhou and Hansen proposed a first-order interval logic called Neighbourhood Logic (NL) which has provisions for specifying liveness and fairness properties as well as formalising some notions in real analysis. This logic has two expanding modalities $\Diamond_l$ and $\Diamond_r$, called the *left* and *right* neighbourhood modality respectively. These modalities refer to some *past* and *future* intervals of time respectively with respect to the original interval of time being observed.

Although, it is not very hard to see that the Propositional NL is complete with respect to Kripke models, it seems to be quite inadequate to derive the modalities considered in [5,17] (but not conversely). Moreover, Propositional NL forms a fragment of the complete logic proposed by Venema in [16]. Nevertheless, the adequacy of the neighbourhood modalities can be established by deriving the other unary and binary modalities of [5] in a first-order logic of the neighbourhood modalities and the interval length (cf.[18]). Thus *first-order* Neighbourhood Logic seems to have more expressive power than those of [4,5,17] with a minimum number of modalities.

This paper presents the syntax and semantics of first-order Neighbourhood Logic and then establishes a *completeness* result. First, we establish a completeness theorem for the NL formulas in the Kripke model (or possible world model) Then we map the Kripke model to the interval model and prove the completeness of NL in the interval model. Dutertre [4] has proved a similar completeness result for ITL with chop modality. Both follow the approach suggested by [8,17].

2 Neighbourhood Logic (NL)

2.1 Syntax of NL

A language $\mathcal{L}$ for NL consists of an infinite collection of *global variables*, $\mathbf{V} \triangleq \{x, y$ $z, \ldots\}$ and also an infinite collection of *temporal variables*, $\mathbf{T} \triangleq \{\ell, v_1, v_2, \ldots\}$

where ℓ is a special symbol which will denote the length of an interval. "ℓ" will depict the "natural" properties of length in the axioms of the logic to be introduced later. In addition, the language contains an infinite set of *global function* symbols **F** and global *predicate* symbols **H**. These symbols are called global because their meaning will be independent of time. With each of the function and predicate symbols is associated an arity $n \geq 0$. Function symbols of arity 0 will be called *constants*. Predicate symbols of arity 0 are *propositions* which include two Boolean symbols **true** and **false**. **F** includes the symbols $+, -$ and **H** includes $=, \geq$ etc. There is also an infinite set of *temporal propositional letters* $\mathbf{P} \cong \{X, Y, \ldots\}$ which will be interpreted as Boolean-valued functions on intervals. The vocabulary also consists of propositional connectives $\neg$ and $\vee$, the existential quantifier $\exists$ and the *left neighbourhood modality* $\Diamond_l$ and the *right neighbourhood modality* $\Diamond_r$. The other usual connectives $\wedge, \Rightarrow, \Leftrightarrow$ as well as the universal quantifier $\forall$ are introduced as abbreviations.

The *terms* denoted as θ, θ_i, are defined by the following abstract syntax:
$$\theta ::= x \mid \ell \mid v \mid f^n(\theta_1, \ldots, \theta_n); \quad x \in \mathbf{V}, \quad v \in \mathbf{T}, \quad f \in \mathbf{F}.$$
The *formulas*, denoted as ϕ, ψ, are defined by the following abstract syntax:
$$\phi ::= X \mid G^n(\theta_1, \ldots, \theta_n) \mid \neg\phi \mid \phi \vee \psi \mid (\exists x)\phi \mid \Diamond_l\phi \mid \Diamond_r\phi; \quad x \in \mathbf{V}, \quad X \in \mathbf{P}, \quad G \in \mathbf{H}.$$

A term is *global* or *rigid*, if it does not contain any temporal variables. A formula is *global* or *rigid*, if it does not contain any temporal variables, any temporal propositional letters, or any neighbourhood modalities.

2.2   Semantics of NL

We fix our domain to be a non-empty set $I\!D$ (containing the constant symbol 0) which will be the underlying representation of time as well as lengths of intervals. Traditional semantics, however, distinguishes between *temporal domain* T (which is generally a totally ordered set) and *duration domain* $I\!D$ which represents durations of time intervals. The duration domain satisfies certain constraints since their elements are supposed to measure "lengths" of time intervals(cf.[4]). Here, for simplicity, we take the time domain to be the same as the duration domain. As in [4], we want $I\!D$ to have certain properties which are specified by the following axioms.

D 1 Axioms for $=$:
The standard axioms for $=$ are assumed (*cf.* [15]).

D 2 Axiom for $+$:

1. $x + 0 = x$.   3. $x + (y + z) = (x + y) + z$.
2. $x + y = y + x$.   4. $(x + y = x + z) \Rightarrow y = z$.

D 3 Axioms for $\geq$:

1. $0 \geq 0$.   3. $x \geq y \Leftrightarrow \exists z \geq 0.(x = y + z)$.
2. $x \geq 0 \wedge y \geq 0 \Rightarrow x + y \geq 0$.   4. $\neg(x \geq y) \Leftrightarrow (y > x)$,

where we write $y > x$ if $(y \geq x) \wedge (y \neq x)$.

D 4 Axioms for $-$:

1. $x - y = z \Leftrightarrow x = y + z$.

Clearly, $\mathbb{R}$ (the reals), $\mathbb{Q}$ (the rationals) and $\mathbb{Z}$ (the integers) are examples o domains which satisfy the above axioms. From the above axioms it can be show that $(\mathbb{D}, +)$ is a commutative group with 0 as the additive identity and $-y$ as the additive inverse of y.

The time domain is $\mathbb{D}$ and the set of all intervals $\mathbb{I}$ is given by:
$$\mathbb{I} \,\hat{=}\, \{[a,b] : a,b \in \mathbb{D} \text{ and } (b \geq a)\},$$
where the *interval* $[a, b]$ is defined as
$$[a,b] \,\hat{=}\, \{x \in \mathbb{D} : b \geq x \geq a\}.$$
The global variables are assigned meaning through a *valuation* or *value assign* ment $\nu : \mathbf{V} \to \mathbb{D}$. Given an interval $[a, b] \in \mathbb{I}$ the meaning of the tempora variables, propositional letters, function and predicate symbols is given by a *interpretation function* $\mathcal{I}$ such that,

1. $\mathcal{I}(0, [a,b]) = 0$,
2. $\mathcal{I}(\ell, [a,b]) = b - a$,
3. $\mathcal{I}(v, [a,b]) \in \mathbb{D}$; for $v \in \mathbf{T}$,
4. $\mathcal{I}(X, [a,b]) \in \{\text{tt,ff}\}$ for $X \in \mathbf{P}$,
5. $\mathcal{I}(f, [a,b]) = \underline{f}$, for an n-ary function $f \in \mathbf{F}$,
   where $\underline{f} : \mathbb{D}^n \to \mathbb{D}$ is any standard interpretation of f,
6. $\mathcal{I}(G, [a,b]) = \underline{G}$, for an n-ary predicate symbol $G \in \mathbf{H}$,
   where $\underline{G} : \mathbb{D}^n \to \{\text{tt,ff}\}$ is any standard interpretation of G

Note that "$+$" and "$-$" are interpreted as the associated binary operations o $\mathbb{D}$.

Given a valuation ν, the terms are interpreted in the usual way by induction o the length of terms [15].

We shall call the pair $\mathcal{M} \,\hat{=}\, \langle \mathbb{D}, \mathcal{I} \rangle$ an *interval model*. Let $\mathcal{M}, \nu, [a, b] \models A$ denot that the formula A is satisfied in the interval $[a, b]$ (also called the *referenc interval*) with respect to the model $\mathcal{M}$ and valuation ν. Satisfiability can the be defined by induction on the formulas in a standard way [15,8]. We only stat the cases for formulas with modalities $\Diamond_l$ and $\Diamond_r$.

1. $\mathcal{M}, \nu, [a, b] \models \Diamond_l A$ iff there exists c, $a \geq c$ such that $\mathcal{M}, \nu, [c, a] \models A$.
2. $\mathcal{M}, \nu, [a, b] \models \Diamond_r A$ iff there exists d, $d \geq b$ such that $\mathcal{M}, \nu, [b, d] \models A$.

We say that A is *valid*, written as $\models A$, iff for any model $\mathcal{M}$, any valuation ν and interval $[a, b]$, $\mathcal{M}, \nu, [a, b] \models A$. Also A is *satisfiable* iff for some model $\mathcal{M}$ valuation ν, and some interval $[a, b]$, $\mathcal{M}, \nu, [a, b] \models A$.

3   The Proof System for NL

In the following set of axioms and rules (as well as elsewhere), $\Diamond(\square)$ can b instantiated by either $\Diamond_l$ or $\Diamond_r$ ($\square_l$ or $\square_r$ respectively). The following abbreviatio will be adopted.

$$\bar{\Diamond} \,\hat{=}\, \begin{cases} \Diamond_r, \text{ if } \Diamond = \Diamond_l \\ \Diamond_l, \text{ if } \Diamond = \Diamond_r \end{cases}$$

$$\square \,\hat{=}\, \neg \Diamond \neg$$

$$\bar{\square} \,\hat{=}\, \neg \, \bar{\Diamond} \, \neg$$

$$\Diamond^\circ \,\hat{=}\, \bar{\Diamond} \, \Diamond$$

Axioms

A1 Global formulas are not connected to intervals.

$\Diamond A \Rightarrow A$, provided A is a global formula.

A2 Interval length is non-negative.

$\ell \geq 0$

A3 Neighbourhood can be of arbitrary length.

$x \geq 0 \Rightarrow \Diamond(\ell = x)$

A4 Neighbourhood modalities can be distributed over disjunction and existential quantifier.

$$\Diamond(A \vee B) \Rightarrow \Diamond A \vee \Diamond B$$
$$\Diamond\exists x.A \quad \Rightarrow \exists x.\Diamond A$$

(The second part of **A4** implies that the analogue of Barcan Formula is true.)

A5 A left (right) neighbourhood coincides with any other left (right) neighbourhood provided they have the same length. In other words, neighbourhood is determined by its length.

$\Diamond((\ell = x) \wedge A) \Rightarrow \Box((\ell = x) \Rightarrow A)$

A6 Left (right) neighbourhoods of an interval always start at the same point.

$\Diamond\bar{\Diamond}A \Rightarrow \Box\bar{\Diamond}A$

A7 Left (right) neighbourhood of the ending (beginning) point of an interval is the interval itself, if it has the same length as the interval.

$(\ell = x) \Rightarrow (A \Leftrightarrow \Diamond^c((\ell = x) \wedge A))$

A8 Two consecutive left (right) expansions can be replaced by a single left (right) expansion, if the third expansion has a length of the sum of the first two.

$((x \geq 0)\wedge(y \geq 0)) \Rightarrow (\Diamond((\ell = x)\wedge\Diamond((\ell = y)\wedge\Diamond A)) \Leftrightarrow \Diamond((\ell = x+y)\wedge\Diamond A))$

Rule schemas
M (Monotonicity)   If $\phi \Rightarrow \psi$ then $\Diamond\phi \Rightarrow \Diamond\psi$.
N (Necessity)   If ϕ then $\Box\phi$.
MP (Modus Ponens) If ϕ and $\phi \Rightarrow \psi$ then ψ.
G (Generalization)   If ϕ then $(\forall x)\phi$.
The proof system also contains axioms **D1–D4** and axioms of propositional logic and first-order predicate logic. They can be taken as any complete system for first-order logic except for some restrictions on the instantiation of quantified formulas. A term θ is called *free for x* in ϕ if x does not occur freely in ϕ within a scope of $\exists y$ or $\forall y$, where y is any variable occurring in θ. We also adopt the following axioms:

$\forall x.\phi(x) \Rightarrow \phi(\theta)$ $\left(\begin{array}{l} \text{if either } \theta \text{ is free for } x \text{ in } \phi(x) \text{ and } \theta \text{ is rigid} \\ \text{or } \theta \text{ is free for } x \text{ in } \phi(x) \text{ and } \phi(x) \text{ is modality free.} \end{array} \right)$

$\phi(\theta) \Rightarrow \exists x.\phi(x)$

A *proof* of an NL formula A is a finite sequence of NL formulas $A_1, \ldots, A_n$, where A_n is A, and each A_i is either an instance of one of the axiom schemas mentioned above or obtained by applying one of the inference rules, also mentioned above to the previous members of the sequence. We write ⊢ A to mean that there exists a proof of A in NL and we say that A is a *theorem* in NL (or A is *provable in* NL).

The following is easy to check by induction on the length of proof.

Theorem 1 ((Soundness in NL)). *An NL formula which can be proved in the calculus must be valid (in any interval model).*

4 Kripke Completeness

Kripke Model   A Kripke model $\mathcal{K}$ for NL is a quintuple $\langle W, R_l, R_r, I\!\!D, \mathcal{I} \rangle$ where

- W is a non-empty set of *possible worlds*,
- R_l and R_r are binary relations on W, called *accessibility relations*,
- $I\!\!D$ is a non-empty set, called the *domain*,
- $\mathcal{I}$ is an *interpretation* function which assigns to each symbol s and world w an interpretation $\mathcal{I}(s, w)$ satisfying the following,
  1. If s is an n-ary function symbol, then $\mathcal{I}(s, w)$ is a function $I\!\!D^n \to I\!\!D$.
  2. If s is an n-ary predicate symbol, then $\mathcal{I}(s, w)$ is a function $I\!\!D^n \to \{\text{tt,ff}\}$

  3. If s is a constant or a temporal variable, then $\mathcal{I}(s, w) \in I\!\!D$.
  4. If s is a temporal propositional letter, then $\mathcal{I}(s, w) \in \{\text{tt,ff}\}$.
  5. If s is a global symbol, then $\mathcal{I}(s, w_1) = \mathcal{I}(s, w_2)$, for all worlds $w_1, w_2 \in W$ i.e., its interpretation is the same in all worlds.

Semantics   Given a Kripke model $\mathcal{K}$, each term t is assigned a meaning on $I\!\!D$ in each world of W. Given an interpretation $\mathcal{I}$, a valuation ν and a world w the semantics of a term is defined by induction [2] on its length in a standard way and is written as $\mathcal{I}_\nu(t, w)$. For a rigid term, the interpretation of the term is the same in all worlds.

Now we describe the semantics of the formulas. We shall write $\mathcal{K}, \nu, w \models A$ to denote that a formula A is satisfied in the world w under the Kripke model $\mathcal{K}$ and valuation ν. It can be defined by induction on formulas in a standard way with R_l and R_r playing the role similar to binary accessibility relation in ordinary modal logic [8]. We illustrate the cases for modal operators.

1. $\mathcal{K}, \nu, w \models \Diamond_l A$ iff there exists $w' \in W$ such that $R_l(w, w')$ and $\mathcal{K}, \nu, w' \models A$
2. $\mathcal{K}, \nu, w \models \Diamond_r A$ iff there exists $w' \in W$ such that $R_r(w, w')$ and $\mathcal{K}, \nu, w' \models A$

We say that $\mathcal{K}$ *satisfies* a formula A (or A has a *Kripke model* $\mathcal{K}$), if there are a world w and a valuation ν such that $\mathcal{K}, \nu, w \models A$. An NL formula A is *valid* in a Kripke model $\mathcal{K}$ if for any valuation ν and world w, $\mathcal{K}, \nu, w \models A$. An NL formula A is *valid* if A is valid in every Kripke model.

A set Γ of sentences is *consistent* [8] if there does not exist any finite subset $\{A_1, \ldots, A_n\}$ of Γ such that ⊢ $\neg(A_1 \wedge \ldots \wedge A_n)$. If, in addition, there does not exist any consistent set Γ' such that $\Gamma' \supset \Gamma$, then Γ is called a *maximal consistent set (mcs)*.

Let $\mathbb{B}$ be a countably infinite set of symbols not occurring in the language $\mathcal{L}$. Let $\mathcal{L}^+$ be the language obtained by adding to $\mathcal{L}$ all the symbols in $\mathbb{B}$ as rigid constants. Denote the extended proof system by NL^+.

A set Γ of sentences is said to have *witnesses* in $\mathbb{B}$ if for every sentence in Γ of the form $\exists x.\phi(x)$, there exists a constant $b \in \mathbb{B}$ such that $\phi(b)$ is in Γ.

Let Q be a sentence not provable in NL. Suppose $\Gamma = \{\neg Q\}$. It is easy to show that Γ is consistent. Enumerating the sentences of $\mathcal{L}^+$ and adding appropriate sentences to Γ in stages, one can obtain a *mcs* $\Gamma^* \supseteq \Gamma$ in $\mathcal{L}^+$ such that Γ^* has a witness in $\mathbb{B}$ (cf. [4]). Let Σ be the set of rigid formulas of Γ^*. We shall now construct the desired Kripke model $\mathcal{K}_\Gamma = \langle W, R_l, R_r, \mathbb{D}, \mathcal{I} \rangle$. Let

$$W = \{\Delta : \Delta \text{ is a } mcs \text{ with witnesses in } \mathbb{B} \text{ and } \Delta \supseteq \Sigma\}.$$

W is non-empty since $\Gamma^* \in W$. The accessibility relations R_l, R_r are defined as follows.

$$R_l(\Delta_1, \Delta_2) \stackrel{\text{def}}{\Leftrightarrow} \Diamond_l \Delta_2 \subseteq \Delta_1 \text{ and } R_r(\Delta_1, \Delta_2) \stackrel{\text{def}}{\Leftrightarrow} \Diamond_r \Delta_2 \subseteq \Delta_1.$$

The domain $\mathbb{D}$ is defined as follows. In $\mathbb{B}$ define a relation $\equiv$ by

$$a \equiv b \text{ iff } a = b \in \Sigma.$$

The axioms **D 1** for equality show that $\equiv$ is an equivalence relation on $\mathbb{B}$. Let

$$\mathbb{D} = \{[b] : b \in \mathbb{B}\}$$

be the set of equivalence classes, where $[a]$ denotes the equivalence class containing a.

The interpretation function $\mathcal{I}$ is defined as follows.

1. If v is a temporal variable, then $\mathcal{I}(v, \Delta) = [a]$ iff $v = a \in \Delta$
2. If a is a constant, then $\mathcal{I}(a, \Delta) = [c]$ iff $a = c \in \Delta$
3. If f is an n-ary function symbol, then
   $$\mathcal{I}(f, \Delta)([b_1], \ldots, [b_n]) = [c] \text{ iff } f(b_1, \ldots, b_n) = c \in \Delta.$$
4. If G is an n-ary predicate symbol, then
   $$\mathcal{I}(G, \Delta)([b_1], \ldots, [b_n]) = tt \text{ iff } G(b_1, \ldots, b_n) \in \Delta$$
5. If X is a propositional letter, then $\mathcal{I}(X, \Delta) = tt$ iff $X \in \Delta$.

Lemma 1 ((Truth Lemma)). *For any formula $A(x_1, \ldots, x_n)$, where the free variables in A are among $x_1, \ldots, x_n$, for any world $\Delta \in W$ and valuation ν,*

$$\mathcal{K}_\Gamma, \nu, \Delta \models A(x_1, \ldots, x_n) \text{ iff } A(b_1, \ldots, b_n) \in \Delta, \text{ where } \nu(x_i) = [b_i]; 1 \leq i \leq n.$$

Since $\neg Q \in \Gamma^*$; by Lemma 1, $\mathcal{K}_\Gamma, \nu, \Gamma^* \models \neg Q$ for any valuation ν. Moreover, if a sentence A is a theorem of NL then it is in Γ^* and so $\mathcal{K}_\Gamma, \nu, \Gamma^* \models A$. Actually it is not required to use all the axioms in the proof of Kripke completeness [2].

5 Completeness in Interval Models

We now translate the Kripke world to the interval models and prove a completeness result in the interval models.

Consider the Kripke Model $\mathcal{K}_\Gamma = \langle W, R_l, R_r, D, \mathcal{I} \rangle$ such that $\mathcal{K}_\Gamma, \nu, \Gamma^* \models \neg Q$, as build earlier. From this Kripke model through a sequence of steps, we shall construct an interval model $\mathcal{M} = \langle D^*, \mathcal{I}^* \rangle$ and an interval $[a, b]$ such that $\mathcal{M}, \nu, [a, b] \models \neg Q$, for any valuation ν.

Define $D^* = D$. It is quite straightforward to check that D^* satisfies all the axioms **D 1 - D 4**. Let $\Delta_0 \in W$ such that $\ell = 0 \in \Delta_0$ and $\diamondsuit_l \Delta_0 \subseteq \Gamma^*$. Such a Δ_0 exists (See [2]). Recall that D is a set of equivalence classes (of rigid constants added to $\mathcal{L}$). From now on we shall not distinguish between a and the equivalence class $[a]$ containing a. Given an interval $[a, b], a, b \in D$, we shall construct a world $\Delta_{[a,b]}$ as described below.

Construction of $\Delta_{[a,b]}$ We think of the world Δ_0 as representing 0. We consider the following cases.

Case 1 $a \geq 0$.

Let Δ_1 be a world in W such that $\ell = a \in \Delta_1$ and $\diamondsuit_r(\Delta_1) \subseteq \Delta_0$. Then $\Delta_{[a,b]}$ is a world such that $(\ell = b - a) \in \Delta_{[a,b]}$ and $\diamondsuit_r(\Delta_{[a,b]}) \subseteq \Delta_1$.

The existence and uniqueness of such worlds can be established [2]. (Think of Δ_1 as representing the interval $[0, a]$ which is to the *right* of 0 represented by Δ_0. Then $\Delta_{[a,b]}$ represents the interval of length $(b - a)$ to the right of Δ_1; see Figure 1.)

Case 2 $a < 0$.

Let $\Delta_2, \Delta_3 \in W$ such that $\ell = -a \in \Delta_2$ and $\diamondsuit_l(\Delta_2) \subseteq \Delta_0$. Also, $\ell = 0 \in \Delta_3$ and $\diamondsuit_l(\Delta_3) \subseteq \Delta_2$.

Then, $\Delta_{[a,b]}$ is a world such that $(\ell = b - a) \in \Delta_{[a,b]}$ and $\diamondsuit_r(\Delta_{[a,b]}) \subseteq \Delta_3$. Such a world $\Delta_{[a,b]}$ can be uniquely found [2]. (Think of Δ_2 as representing the interval $[-a, 0]$ which is to the left of 0 represented by Δ_0. Also Δ_3 represents the point interval $[-a, -a]$. Then $\Delta_{[a,b]}$ represents the interval of length $(b - a)$ to the right of Δ_3; see Figure 1).

Now, define the function $\mathcal{I}^*$ as, $\mathcal{I}^*(s, [a, b]) = \mathcal{I}(s, \Delta_{[a,b]})$, for any symbol s. From the definition of $\mathcal{I}$ it follows that $\mathcal{I}^*(\ell, [a, b]) = \mathcal{I}(\ell, \Delta_{[a,b]}) = b - a$.

We now need the following lemma which can be proved by taking induction on formulas (*cf.* [2]).

Lemma 2. *For any interval $[a, b]$, valuation ν and formula A*
$$\mathcal{M}, \nu, [a, b] \models A \quad \text{iff} \quad \mathcal{K}, \nu, \Delta_{[a,b]} \models A.$$

We have $\mathcal{K}, \nu, \Gamma^* \models \neg Q$. Now it can be shown that (see [2]) for some $c \geq 0$, $\Gamma^* = \Delta_{[0,c]}$, for any valuation ν. Thus we have $\mathcal{K}, \nu, \Delta_{[0,c]} \models \neg Q$, for all valuation ν. Hence by Lemma 2, $\mathcal{M}, \nu, [0, c] \models \neg Q$, for any valuation ν. Thus Q is not valid. Hence we have,

Theorem 2 ((Completeness of NL)). *If a sentence Q is valid in interval models, then Q is provable in NL.*

6   Discussion

A complete axiomatic system for a first-order interval logic with two neighbour-hood modalities has been presented in this paper. Barua and Zhou [2] have

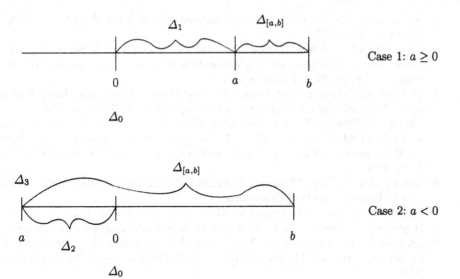

Fig. 1. Construction of $\Delta_{[a,b]}$

extended NL by introducing two more modalities in the upward and downward directions and have proposed a two-dimensional neighbourhood logic NL^2. They have proved a completeness result in NL^2 using the same construction. Their work suggests that it may be possible to obtain a proof system of Neighbourhood Logic in any dimension using the same technique. The logic of NL^2 can be used to specify the behaviour of the real-time systems where timeless computation is taken into account [3].

In [14] NL has been extended to obtain a Duration Calculus (DC) where temporal variables are expressed in the form of integrals (durations) of state variables. It is interesting to note that the proof system of DC is relatively complete, *i.e.* it is complete provided all valid NL formulas (with time domain and valuation domain taken to be reals) are considered as axioms in the proof system of DC (cf. [14]).

Applications of NL (and NL^2) are being investigated. In [13] NL is combined with a linear temporal logic to give a real-time semantics for an OCCAM-like language, where timeless computation was assumed. Further NL is applied for Interval Algebra in the area of Artificial Intelligence [12].

References

1. Alur R., Courcoubetis C., Henzinger T., Ho P-H.: *Hybrid Automata: An algorithmic approach to the specification and verification of Hybrid systems*, in Hybrid Systems, R. L. Grossman, A. Nerode, A. P. Ravn and H. Rischel (Eds.), LNCS **736**, pp. 209–229, Springer-Verlag, 1993.

2. Barua Rana, Zhou Chaochen: Neighbourhood Logics: NL and NL^2, *UNU/IIST Report no. 120*, 1997. [1]

3. Berry Gérard, Gonthier Georges: The Esterel Synchronous Programming Language: Design, Semantics and Implementation, in *Science of Computer Programming*, vol. **19**, pp. 87–152, Elsevier, 1992.

4. Dutertre B.: *Complete Proof Systems for First Order Interval Logic*, Tenth Annual IEEE Symp. on Logic in Computer Science, pp. 36-43, IEEE Press, 1995.

5. Halpern J., Shoham Y.: *A Propositional Modal Logic of Time Intervals*, Journal of the ACM **38** (4) pp. 935–962, 1991. Also appeared in Proceedings of the First IEEE Symposium on Logic in Computer Science, pp. 279-292, Computer Society Press, 1986.

6. Hansen Michael, Zhou Chaochen: *Duration Calculus: Logical Foundations*, To appear in Formal Aspects of Computing.

7. Humberstone: *Interval semantics for Tense Logic*, Jour of Phil. Logic, **8**, 1979.

8. Hughes G. E., Creswell M. J.: An introduction to Modal Logic, Routledge, 1990.

9. Lamport L.: *Hybrid systems in TLA^+*, in in Hybrid Systems, R. L. Grossman, A Nerode, A. P. Ravn and H. Rischel (Eds.), LNCS **736**, pp. 77–102, Springer-Verlag 1993.

10. Manna Z., Pnueli A.: *Verifying hybrid systems*, in Hybrid Systems, R. L. Grossman, A. Nerode, A. P. Ravn and H. Rischel (Eds.), LNCS **736**, pp. 4–35, Springer-Verlag 1993.

11. Moszkowski B.: *A Temporal Logic for Multilevel Reasoning about Hardware*, IEEE Computer **18** (2) pp. 10-19, 1985.

12. Pujari Arun K: Neighbourhood Logic & Interval Algebra. *UNU/IIST Report no 116*, 1997.

13. Qiu Zongyan, Zhou Chaochen: A Combination of Interval Logic and Linear Temporal Logic. *UNU/IIST Report no. 123*, 1997 (accepted by PROCOMET'98).

14. Roy Suman, Zhou Chaochen: Notes in Neighbourhood Logic, *UNU/IIST Report no. 97*, 1997.

15. Shoenfield J., Mathematical Logic, Addison-Wesley, Reading, Mass., 1967.

16. Venema Y.: *Expressiveness and Completeness of an Interval Tense Logic*, Notre Dame Journal of Formal Logic, Vol. **31**, No. 4, pp. 529-547, 1990.

17. Venema Y.: *A Modal Logic for Chopping Intervals*, Journal of Logic and Computation, Vol. **1**, pp. 453-476, Oxford University Press, 1991.

18. Zhou Chaochen, Hansen Michael R.: An Adequate First Order Interval Logic *UNU/IIST Report No. 91*, Revised report, December 1996.

19. Zhou Chaochen, Hoare C. A. R., Ravn A. P.: *A Calculus of Durations*, Information Processing Letters, Vol. **40**, No. 5, pp. 269-276, 1991.

20. Zhou Chaochen, Ravn A. P., Hansen Michael R.: *An extended duration calculus for hybrid systems*, in Hybrid Systems, R. L. Grossman, A. Nerode, A. P. Ravn and H. Rischel (Eds.), LNCS **736**, pp. 36–59, Springer-Verlag, 1993.

---

[1] All the research reports of UNU/IIST can be accessed at the URL
http://www.iist.unu.edu

Eliminating Recursion in the μ-Calculus[*]

Martin Otto

RWTH Aachen

Abstract. Consider the following problem: given a formula of the modal μ-calculus, decide whether this formula is equivalently expressible in basic modal logic. It is shown that this problem is decidable, in fact in deterministic exponential time. The decidability result can be obtained through a model theoretic reduction to the monadic second-order theory of the complete binary tree, which by Rabin's classical result is decidable, albeit of non-elementary complexity. An improved analysis based on tree automata yields an exponential time decision procedure.

1 Introduction

The propositional μ-calculus L_μ has, since its introduction in its present form in [12], emerged as one of the major logical formalism that can deal with interesting aspects of the dynamic and temporal behaviour of processes or programs. As such, it comprises the expressive power of several other well developed logical formalism for reasoning about transition systems, among them computation tree logic CTL and propositional dynamic logic PDL. Conceptually and model theoretically the μ-calculus is a modal logic. It extends propositional modal logic ML by a least fixed point operation, and it shares with basic modal logic the crucial semantic property of being invariant under bisimulation. The least fixed point construct, which is essentially second-order in nature, adds to modal logic a powerful, yet tractable form of recursion. It is this aspect of recursion that boosts the expressiveness of the μ-calculus in allowing it to express truly dynamic features of transition systems that go far beyond the more static and local properties expressible in basic modal logic. Liveness, safety or termination conditions are typical examples of L_μ-definable properties. Given the broad applicability of the μ-calculus and its fragments for specification and model checking uses, it is natural to consider the issue whether a formalization of some supposedly interesting condition on transition systems requires the use of the recursive features of the μ-calculus in an essential way. It can be that although some given L_μ-specification syntactically involves μ-constructs, it is logically equivalent to a much simpler, static and local assertion in basic modal logic.

[*] This is an extension of the original submission; the EXPTIME result, based on an automata theoretic analysis, is new here. Moreover, the original model theoretic approach has been simplified. I am very grateful to Moshe Vardi for having, with his comments on an earlier version, inspired these improvements.

C. Meinel and S. Tison (Eds.): STACS'99, LNCS 1563, pp. 531–540, 1999.
© Springer-Verlag Berlin Heidelberg 1999

Similar questions about the possibility to eliminate recursion have of course been asked and investigated in other contexts. The so-called *boundedness problem* for Datalog programs, which arises in connection with the issue of database query optimization, is a case in point. In the context of classical model theory, boundedness was originally introduced and studied by Barwise and Moschovakis [1]. Several results in the first-order context and for the applications to Datalog query optimization show the boundedness problem to be undecidable even in very restricted settings [7,9,11]. Probably the strongest known exception concerns monadic Datalog (or the boundedness of simple monadic fixed points over existential first-order formulae without equality or negation), shown to be decidable in [3].

In the present paper we propose a proof that the eliminability of recursion from arbitrarily nested L_μ-formulae does indeed constitute a decidable problem.

Main Theorem *The following problem is decidable, in fact even in* EXPTIME. *given a formula in the modal μ-calculus, decide whether this formula can equivalently be expressed in plain modal logic.*

By way of interpreting this result in a somewhat wider context, we recall how ML and L_μ are characterized as exactly the bisimulation-invariant fragments of first-order logic and monadic second-order logic, respectively, see [2] and [10].

Theorem (van Benthem) *A first-order formula $\varphi(x)$ is equivalent to a formula of* ML *if and only if the class of its models, $\mathrm{Mod}(\varphi)$, is closed under bisimulation.*

Theorem (Janin, Walukiewicz) *A monadic second-order formula $\varphi(x)$ is equivalent to a formula of L_μ if and only if $\mathrm{Mod}(\varphi)$ is closed under bisimulation.*

In view of these characterizations one may rephrase our main result as follows: given bisimulation-invariance, the distinction between first-order and true monadic second-order becomes decidable. This provides a nice analogy between the bisimulation-invariant scenario and the much more limited scenario of word structures. For word structures the corresponding distinction is known to be decidable due to the classical results of Büchi, Elgot, Trakhtenbrot and Schützenberger, McNaughton, Papert, since it coincides with star-freeness of regular languages, see [15].

Turning back to the related issue of boundedness, we may look at the boundedness problem for modal logic as a special case of our decision problem: given a formula $\varphi(X)$ of modal logic in which X occurs only positively, decide whether there is some $n \in \mathbb{N}$ such that the least fixed point $\mu_X \varphi(X)$ associated with $\varphi(X)$ is always reached within n iterations. By a straightforward variation of a theorem of Barwise and Moschovakis [1], $\varphi(X)$ is bounded if and only if $\mu_X \varphi(X)$ is ML-definable. Thus, our main theorem implies in particular the decidability of the boundedness problem for modal formula. This contrasts sharply with the undecidability of the boundedness problem for two-variable first-order logic as established in [11], and adds to the comparative study of modal versus two-variable logics which has emerged in related research, see e.g. [17,8].

Following preparations in Section 2, we shall complete in Section 3 the proof of the decidability claim of the main theorem by a reduction to S2S based on a bounded branching property for the issue of ML-expressibility. In the final Section 4 we then present an alternative automata theoretic analysis which furthermore yields the EXPTIME result.

2 Basic Definitions and Preliminaries

Tree structures. Since the μ-calculus and modal logic satisfy the tree model property, it will throughout suffice to consider tree structures rather than arbitrary Kripke structures. We shall also restrict attention to the notationally simpler case in which only a single binary transition relation (accessibility relation) E is present. For us, therefore, a tree structure of type $\tau = \{P_1, \ldots, P_l\}$ is a structure $\mathfrak{A} = (A, E^{\mathfrak{A}}, P_1^{\mathfrak{A}}, \ldots, P_l^{\mathfrak{A}}, 0^{\mathfrak{A}})$ where $(A, E^{\mathfrak{A}}, 0^{\mathfrak{A}})$ is a tree with root $0^{\mathfrak{A}}$ and the $P_i^{\mathfrak{A}}$ are subsets A. Here (A, E) being a tree with root a means that a is the unique element of zero in-degree w.r.t. E and that every element is reachable from a on a unique directed E-path (whose length is the *height* of that element). The following classes of tree structures will be important:

(i) $\mathcal{T}[\tau]$ consisting of all tree structures of type τ;

(ii) $\mathcal{T}_n[\tau] \subseteq \mathcal{T}[\tau]$ consisting of those tree structures whose branching is bounded by n;

(iii) $\mathcal{T}_{n;m}[\tau] \subseteq \mathcal{T}[\tau]$ consisting of those tree structures whose branching is bounded by n in all nodes of height less than m.

For an element a of a tree structure $\mathfrak{A}$ we denote by $\langle a \rangle$ the elements of the subtree rooted at a, by $\mathfrak{A} \restriction \langle a \rangle$ that subtree itself. By $\langle a \rangle^m$ we denote the set of elements whose height in $\mathfrak{A} \restriction \langle a \rangle$ is at most m, by $\mathfrak{A} \restriction \langle a \rangle^m$ the induced subtree rooted a. We denote by $E^{\mathfrak{A}}[a]$ the set of immediate E-successors of a in $\mathfrak{A}$.

Prunings and end extensions. If $\mathfrak{A} \subseteq \mathfrak{B}$ and both $\mathfrak{A}$ and $\mathfrak{B}$ are tree structures we call $\mathfrak{A}$ a *pruning* of $\mathfrak{B}$ to stress the view that $\mathfrak{A}$ is obtained from $\mathfrak{B}$ through cutting away subtrees. Note that $\mathfrak{A} \subseteq \mathfrak{B}$ for tree structures implies that $A \subseteq B$ is an *initial* subset in the tree $\mathfrak{B}$, i.e. $0^{\mathfrak{B}} \in A$ and A is $E^{\mathfrak{B}}$-connected.

We say that $\mathfrak{A}$ is a *finite pruning* of $\mathfrak{B}$ if there is some $n \in \mathbb{N}$ such that for all $a \in A$ whose distance from the root is at least n, $E^{\mathfrak{A}}[a] = E^{\mathfrak{B}}[a]$.

$\mathfrak{B}$ is an *end extension* of the tree $\mathfrak{A}$, $\mathfrak{A} \subseteq_{\text{end}} \mathfrak{B}$, if $\mathfrak{A} \subseteq \mathfrak{B}$ and if $E^{\mathfrak{B}}[a] = E^{\mathfrak{A}}[a]$ for all interior nodes (non-leaves) a of $\mathfrak{A}$.

Propositional modal logic. We write ML for *propositional modal logic*. ML$[\tau]$ for $\tau = \{P_1, \ldots, P_l\}$ has the following formulae: each P_i is a formula; ML$[\tau]$ is closed under Boolean connectives $\wedge, \neg$ (and $\vee$, which, however, we regard as defined); and if φ is a formula, then so are $\Diamond\varphi$ (and dually $\Box\varphi$, which again we regard as defined). The semantics is defined over Kripke structures in the natural way. $(\mathfrak{A}, a) \models P_i$ if $a \in P_i^{\mathfrak{A}}$; the Boolean connectives behave as usual; and $(\mathfrak{A}, a) \models \Diamond\varphi$ if there is some $a' \in E^{\mathfrak{A}}[a]$ for which $(\mathfrak{A}, a') \models \varphi$. If $\mathfrak{A}$ is a tree structure of the appropriate type, with root $0^{\mathfrak{A}}$, we simply write $\mathfrak{A} \models \varphi$ for

$(\mathfrak{A}, 0^{\mathfrak{A}}) \models \varphi$. I.e. we always regard the root as the distinguished element of a tree structure unless otherwise specified.

The propositional μ-calculus. L_μ augments the syntax and semantics of ML by means of a monadic least fixed point constructor. A formula $\varphi(X)$ is *positive* in the monadic second-order variable X, if X only occurs in the scope of an even number of negations. In this case $\varphi(X)$ induces a monotone operator on subsets P of τ-structures $\mathfrak{A}$ according to $P \mapsto \{a \in A \mid (\mathfrak{A}, a) \models \varphi(P)\}$. Owing to its monotonicity, this operator has a *least fixed point* $[\mu_X \varphi(X)]^{\mathfrak{A}}$, which may also be obtained as the limit of the monotone sequence of its *stages* $P_\alpha = \{a \in A \mid (\mathfrak{A}, a) \models \varphi(\bigcup_{\beta < \alpha} P_\beta)\}$. Now $\mu_X \varphi(X)$ is itself a formula of L_μ with semantics according to $(\mathfrak{A}, a) \models \mu_X \varphi(X)$ if $a \in [\mu_X \varphi(X)]^{\mathfrak{A}}$.

The following observation will be useful in the analysis of L_μ-formulae. It follows from monotonicity considerations.

Observation 1 *If $\varphi(X)$ is positive in X and if $P \subseteq A$ is any stage of $\mu_X \varphi(X)$ over $\mathfrak{A}$, then $(\mathfrak{A}, a) \models \varphi(P) \Leftrightarrow (\mathfrak{A}, a) \models \varphi(P \setminus \{a\})$.*

Relativization. It is useful to associate with classes $\mathcal{C}$ of tree structures of type τ, and with a unary predicate $U \notin \tau$, the class of all those tree structures of type $\tau \,\dot\cup\, \{U\}$ for which the root is an element of the U-part, and for which the tree structure of type τ induced on the largest initial subset contained in the U-part is a member of $\mathcal{C}$. We call this derived class the relativization of $\mathcal{C}$ to U and denote it $\mathcal{C}^U$. It is easy to see that $\mathcal{C}^U$ is ML-definable or L_μ-definable respectively, if $\mathcal{C}$ is so definable. In fact the following inductively defined U-relativization of L_μ-formulae provides the desired formulae: $\varphi^U = U \wedge \varphi$ for atomic φ; $(\neg \varphi)^U = U \wedge \neg \varphi^U$; $(\varphi_1 \wedge \varphi_2)^U = \varphi_1^U \wedge \varphi_2^U$; $(\Diamond \varphi)^U = U \wedge \Diamond \varphi^U$; $(\mu_X \varphi(X))^U = U \wedge \mu_X \varphi^U(X)$. Note that the translation $\varphi \mapsto \varphi^U$ increases the length only linearly.

Bisimulation. Two tree structures $\mathfrak{A}$ and $\mathfrak{B}$ with roots $0^{\mathfrak{A}}$ and $0^{\mathfrak{B}}$ are *bisimulation equivalent*, $\mathfrak{A} \sim \mathfrak{B}$, if there is an $R \subseteq A \times B$, such that $(0^{\mathfrak{A}}, 0^{\mathfrak{B}}) \in R$ and for all $(c, d) \in R$:

$$c \in P_j^{\mathfrak{A}} \Leftrightarrow d \in P_j^{\mathfrak{B}} \text{ for all } P_j, \quad \text{and} \quad \begin{array}{l} \forall c' \in E^{\mathfrak{A}}[c] \,\exists d' \in E^{\mathfrak{B}}[d] : (c', d') \in R, \\ \forall d' \in E^{\mathfrak{B}}[d] \,\exists c' \in E^{\mathfrak{A}}[c] : (c', d') \in R. \end{array}$$

We shall also deal with finite approximations of bisimulation equivalence in the form of *n-bisimulation equivalence* $\sim_n$, which for tree structures can be characterized as follows: $\mathfrak{A} \sim_n \mathfrak{B}$ if and only if $\mathfrak{A} \restriction \langle 0^{\mathfrak{A}} \rangle^n \sim \mathfrak{B} \restriction \langle 0^{\mathfrak{B}} \rangle^n$.

The model theoretic proof of the decidability result in our main theorem relies on a restriction of the issue to some subclass $\mathcal{T}_{n;m}$ of initially n-branching trees. Let us say that some model theoretic condition *holds in restriction to* $\mathcal{T}_{n;*}$ if this condition is true in restriction to $\mathcal{T}_{n;m}$ for some (and hence for all sufficiently large) m.

Lemma 2 *For a bisimulation-closed class $\mathcal{C} \subseteq \mathcal{T}[\tau]$ and for any n the following are equivalent:*

(i) C is ML-definable in restriction to $T_{n;*}[\tau]$.
(ii) there is some $m \in \mathbb{N}$ such that for any two tree structures $\mathfrak{A}$ and $\mathfrak{A}'$ in $T_{n;m}$:
if $\mathfrak{A} \upharpoonright \langle 0^{\mathfrak{A}} \rangle^m \simeq \mathfrak{A}' \upharpoonright \langle 0^{\mathfrak{A}'} \rangle^m$, then $\mathfrak{A} \in C \Leftrightarrow \mathfrak{A}' \in C$.

Sketch of proof. (i) $\Rightarrow$ (ii) follows directly from the fact that a modal formula of quantifier rank m is insensitive to parts of the structure whose distance from the distinguished node (the root) is greater than m. For (ii) $\Rightarrow$ (i) we observe first that, for any given m, there are only finitely many m-bisimulation classes of τ-trees, each of which is characterized by a single ML-formula of quantifier rank m. So C is ML-definable over $T_{n;m}$ as a finite union of ML-definable q-bisimulation types, provided we can show that (ii) and closure of C under bisimulation together imply that C is actually m-bisimulation closed over $T_{n;m}$. Let to this end $\mathfrak{A}, \mathfrak{B} \in T_{n;m}$, $\mathfrak{B} \sim_m \mathfrak{A}$, $\mathfrak{A} \in C$; we show that $\mathfrak{B} \in C$. From $\mathfrak{A} \sim_m \mathfrak{B}$ we may obtain an $\mathfrak{A}' \in T_{n;m}$, through duplication of subtrees rooted within $\langle 0 \rangle^m$ in $\mathfrak{A}$, with the following properties: $\mathfrak{A}' \sim \mathfrak{A}$ and $\mathfrak{A}' \sim_m \mathfrak{B}$ via a bisimulation R between $\mathfrak{A}' \upharpoonright \langle 0 \rangle^m$ and $\mathfrak{B} \upharpoonright \langle 0 \rangle^m$ that is the graph of a function f from $\mathfrak{A}' \upharpoonright \langle 0 \rangle^m$ onto $\mathfrak{B} \upharpoonright \langle 0 \rangle^m$. Let now $\mathfrak{A}''$ be the result of replacing, for all $c \in A'$ at height n, each $\mathfrak{A}' \upharpoonright \langle c \rangle$ by the corresponding $\mathfrak{B} \upharpoonright \langle f(c) \rangle$. It follows that $\mathfrak{A}'' \sim \mathfrak{B}$, $\mathfrak{A}'' \upharpoonright \langle 0 \rangle^m \simeq \mathfrak{A}' \upharpoonright \langle 0 \rangle^m$, $\mathfrak{A}'' \sim \mathfrak{A}$. Now $\mathfrak{A} \in C$ by assumption, $\mathfrak{A}' \in C$ by $\sim$-closure, $\mathfrak{A}'' \in C$ by (ii), and therefore finally $\mathfrak{B} \in C$ by $\sim$-closure again. $\square$

It is easy to see that condition (ii) of the lemma is further equivalent to the following condition on expansions of the complete n-ary tree: *There is a finite initial subset V of the complete n-branching tree T_n such that for all initial $W \subseteq V$ and all $P_1, \ldots, P_l \subseteq W$: either all end extensions of $(T_n, P_1, \ldots, P_l) \upharpoonright W$ are in C, or none is.* Omitting the straightforward application of well-known interpretation techniques, we note that this condition is expressible in monadic second-order over the complete n-branching tree, provided C itself is monadic second-order definable (as is clearly the case for L_μ-definable classes). But by Rabin's famous theorem [14], the monadic second-order theories of the complete n-branching trees are all decidable — actually uniformly in n, since all these theories are uniformly interpretable in that of the complete binary tree, S2S. Thus we have the following for the restriction of the ML-expressibility issue to classes $T_{n;*}$.

Proposition 3 *The following decision problem is decidable via reduction to S2S, uniformly in n and in the vocabulary of φ: given $\varphi \in L_\mu$ and $n \in \mathbb{N}$, decide whether there is a formula $\psi \in ML$ that is equivalent to φ in restriction to $T_{n;*}$.*

3   Prunings, Preservation, and Decidability via S2S

Prunings offer a canonical means to govern the branching degree of tree structures. The idea is to associate with each node a set of properties which are relevant for its direct E-successors, and to consider those prunings, which – at each node a – retain sufficiently many immediate successors so as to still realize the same relevant properties in the remaining successors of a.

Definition 4 Let Γ be a set of classes of tree structures of type τ. A pruning $\mathfrak{B} \subseteq \mathfrak{A}$ is called Γ-*elementary*, if for all $b \in B$ and for all $\mathcal{C} \in \Gamma$: if there is some $a \in E^{\mathfrak{A}}[b]$ such that $\mathfrak{A} \restriction \langle a \rangle \in \mathcal{C}$, then there also is an $a \in E^{\mathfrak{B}}[b]$ such that $\mathfrak{A} \restriction \langle a \rangle \in \mathcal{C}$.

Observation 5 *Given Γ and $\mathfrak{A}$, there is a Γ-elementary pruning of $\mathfrak{A}$, whose branching degree is bounded by $|\Gamma|$.*

Definition 6 Let Γ be a set of classes of tree structures of type τ. A class $\mathcal{C}_0 \subseteq T[\tau]$ is *finitely preserved with respect to Γ* if for all Γ-elementary finite prunings $\mathfrak{A} \subseteq \mathfrak{B}$, $\mathfrak{B} \in \mathcal{C}_0$ iff $\mathfrak{A} \in \mathcal{C}_0$. Γ is a *preservation set* if each $\mathcal{C} \in \Gamma$ is finitely preserved w.r.t. Γ.

These notions apply in the context of the well-known small branching property for L_μ which plays a role in satisfiability considerations, see e.g. [13,5] and compare the so-called Fischer-Ladner closure of φ from [5,6] and others.

Proposition 7 *Any L_μ-definable class $\mathcal{C} = \text{Mod}(\varphi)$ is a member of some preservation set Γ_φ of L_μ-definable classes, whose size is bounded by the length of φ.*

Sketch of proof. By induction with respect to the structure of the defining formula $\varphi \in L_\mu$. We first argue that $\text{Mod}(\varphi)$ is finitely preserved w.r.t. some Γ_φ consisting of $\text{Mod}(\varphi)$ and fewer than $|\varphi|$ other L_μ-definable classes; identifying these classes with their defining L_μ-formulae, we regard Γ_φ as a subset of L_μ, with $\varphi \in \Gamma_\varphi$. The atomic case is obvious with $\Gamma_\varphi = \{\varphi\}$; negation is dealt with by putting $\Gamma_{\neg\varphi} = \Gamma_\varphi \cup \{\neg\varphi\}$ (the complement of $\mathcal{C}$ is finitely preserved w.r.t. Γ if $\mathcal{C}$ itself is). Similarly we may put $\Gamma_{\varphi_1 \wedge \varphi_2} = \Gamma_{\varphi_1} \cup \Gamma_{\varphi_2} \cup \{\varphi_1 \wedge \varphi_2\}$. Modal quantification is covered in $\Gamma_{\Diamond\varphi} = \Gamma_\varphi \wedge \{\Diamond\varphi\}$.

Finally, let $\varphi = \mu_X \psi(X)$. Note that $\Gamma_{\psi(X)}$ is a set of formulae in a vocabulary involving X as a basic proposition. Let $\Gamma[\mu_X\psi/X]$ be the result of substituting $\mu_X\psi(X)$ for every occurrence of X in all formulae in Γ. We claim that $\Gamma_\varphi = \Gamma_{\psi(X)}[\mu_X\psi/X] \cup \{\varphi\}$ is good for φ. We have to show that φ itself as well as any other member of Γ_φ is finitely preserved w.r.t. Γ_φ. We first argue for formulae other than φ, i.e. for $\chi[\mu_X\psi/X]$ for $\chi(X) \in \Gamma_{\psi(X)}$. Under the assumption that $\mu_X\psi(X)$ itself is preserved, preservation for $\chi[\mu_X\psi/X]$ is inherited from the corresponding preservation property of Γ_ψ: it corresponds to the special case of the latter in which X happens to be interpreted as $\mu_X\psi(X)$. It remains to show that $\mu_X\psi(X)$ is finitely preserved w.r.t. Γ_φ. Since we are claiming preservation only w.r.t. finite prunings, we may prove the preservation claim by induction over subtrees and may consider a pruning in just one single node. Assume that $\mathfrak{B} \subseteq \mathfrak{A}$ is a Γ_φ-elementary pruning obtained from $\mathfrak{A}$ through deletion of subtrees rooted in elements $a' \in E^{\mathfrak{A}}[a]$. Using the assumption that $(\mathfrak{B}, b) \models \mu_X\psi(X)$ iff $(\mathfrak{A}, b) \models \mu_X\psi(X)$, for all $b \in \langle a \rangle^{\mathfrak{B}} \setminus \{a\}$, we find that the pruning $(\mathfrak{B}, [\mu_X\psi]^{\mathfrak{A}} \setminus \{a\}, a) \subseteq (\mathfrak{A}, [\mu_X\psi]^{\mathfrak{A}} \setminus \{a\}, a)$ is $\Gamma_{\psi(X)}$-elementary. This gives the desired preservation of φ at a, through an application of Observation 1. $\square$

Lemma 8 *Let $C \subseteq T[\tau]$ be a member of a preservation set Γ. If C is ML-definable in restriction to $T_{n;*}$ for $n = 2 \cdot |\Gamma|$, then C is ML-definable over $T[\tau]$.*

Proof. Let $n = 2 \cdot |\Gamma|$ and assume that C is ML-definable in restriction to $T_{n;m}[\tau]$. Suppose, towards a contradiction, that C were not ML-definable over $T[\tau]$. By Lemma 2, we may therefore find tree structures $\mathfrak{A}$ and $\mathfrak{A}'$ such that $\mathfrak{A} \in C$, $\mathfrak{A}' \notin C$, and $\mathfrak{A} \upharpoonright \langle a \rangle^m \simeq \mathfrak{A}' \upharpoonright \langle a' \rangle^m$. W.l.o.g. assume that $\mathfrak{A} \upharpoonright \langle a \rangle^m = \mathfrak{A}' \upharpoonright \langle a' \rangle^m$, and that $\mathfrak{A}$ and $\mathfrak{A}'$ are disjoint beyond height m. Let U and V be monadic predicates not in τ and put $\hat{\tau} := \tau \cup \{U, V\}$. Let $\hat{\Gamma} = \{C^U \mid C \in \Gamma\} \cup \{C^V \mid C \in \Gamma\}$ the set of classes obtained by relativizing those in Γ to U or V, respectively. Note that $|\hat{\Gamma}| = n$. Let $\mathfrak{B}$ be the tree structure of type $\hat{\tau}$ obtained from $\mathfrak{A} \cup \mathfrak{A}'$ by putting $U^{\mathfrak{B}} := A$ and $V^{\mathfrak{B}} := A'$. Let $\mathfrak{B}_0 \subseteq \mathfrak{B}$ be a $\hat{\Gamma}$-elementary finite pruning of $\mathfrak{B}$ such that $\mathfrak{B}_0 \in T_{n;m}[\hat{\tau}]$ (cf. Observation 5). Note that in $\mathfrak{B}_0$, U and V still are initial subsets. Let $\mathfrak{A}_0$ be the tree structure of type τ obtained as the restriction of $\mathfrak{B}_0$ to $U^{\mathfrak{B}_0}$, $\mathfrak{A}_0'$ similarly induced by the restriction to $V^{\mathfrak{B}_0}$. Now, $\mathfrak{A}_0$ and $\mathfrak{A}_0'$ are in particular Γ-elementary finite prunings of $\mathfrak{A}$ and $\mathfrak{A}'$, respectively, whence $\mathfrak{A}_0 \in C$ and $\mathfrak{A}_0' \notin C$. Obviously still $\mathfrak{A}_0 \upharpoonright \langle a \rangle^m = \mathfrak{A}_0' \upharpoonright \langle a' \rangle^m$. But this contradicts the assumption that C was ML-definable in restriction to $T_{n;m}[\tau]$, by Lemma 2. □

Corollary 9 *For $\varphi \in L_\mu$ there is an equivalent formula in ML if and only if φ is equivalent to some formula of ML in restriction to $T_{n;*}$ for $n = 2|\varphi|$.*

With Proposition 3, this yields the decidability claim of the main theorem.

4 Tree Automata and Exponential Time Complexity

The following is a variant of Lemma 2, which is in fact easier inasmuch as branching degrees are disregarded. Let $C \subseteq T[\tau]$, $U \notin \tau$, and C^U the corresponding relativization. Let ΔC be the class of those $\mathfrak{A} \in T_{\text{fin}}[\tau \cup \{U\}]$ that have two different end extensions $\mathfrak{A} \subseteq_{\text{end}} \mathfrak{B}_i$ such that $\mathfrak{B}_1 \in C^U$, $\mathfrak{B}_2 \notin C^U$.

The *tallness* of a tree structure is the minimum over the heights of the leaves. A class of (finite) tree structures is of *bounded tallness* if there is a uniform finite bound on the tallness of its members.

Lemma 10 *For any bisimulation-closed $C \subseteq T[\tau]$, and with the induced classes C^U and ΔC as above, the following are equivalent:*

i) C is ML-definable.

*ii) there is some $m \in \mathbb{N}$ such that for any two tree structures $\mathfrak{A}$ and $\mathfrak{A}'$ in $T[\tau]$:
if $\mathfrak{A} \upharpoonright \langle 0^{\mathfrak{A}} \rangle^m \simeq \mathfrak{A}' \upharpoonright \langle 0^{\mathfrak{A}'} \rangle^m$, then $\mathfrak{A} \in C \Leftrightarrow \mathfrak{A}' \in C$.*

iii) ΔC is of bounded tallness.

For L_μ-definable C we want to view condition (iii) in an automata theoretic setting, having ΔC accepted by some suitable tree automaton. Firstly, however, we need to present finite tree structures and some information about their end extensions in a way that fits automata.

A Λ-*labelling* of a (naked) tree $\mathfrak{A} \in T_{\mathrm{fin}}[\emptyset]$ is a mapping $\lambda \colon A \to \Lambda$. Clearly, any tree structure in $T_{\mathrm{fin}}[\tau]$ may be coded as a Λ-labelled tree for $\Lambda = \mathcal{P}(\tau)$, by putting $\lambda(a) = \{P_i \in \tau \mid \mathfrak{A}, a \models P_i\}$. A *leaf labelling* on $\mathfrak{A}$ is a labelling defined on the set of leaves of $\mathfrak{A}$ rather than the entire universe. We may regard a finite tree structure of type τ together with a leaf labelling π in alphabet Σ as a (naked) finite tree with a labelling in the alphabet $\Lambda = \mathcal{P}(\tau) \times (\Sigma \,\dot\cup\, \{*\})$, where $\pi(a) \in \mathcal{P}(\tau) \times \{*\}$ for all interior nodes a. It will therefore suffice to deal with the format given by

$$T_{\mathrm{fin}}^\ell[\Lambda] = \big\{T = (\mathfrak{A}, \lambda) \mid \mathfrak{A} \in T_{\mathrm{fin}}[\emptyset], \lambda \colon A \to \Lambda \text{ a labelling }\big\}.$$

A *(deterministic leaves-to-root) tree automaton* over $T_{\mathrm{fin}}^\ell[\Lambda]$ is given as $\mathcal{A} = (Q, \delta)$ where Q is the finite set of states, and δ a transition function of the form $\delta \colon \mathcal{P}(Q) \times \Lambda \to Q$. Its run on $T \in T_{\mathrm{fin}}^\ell[\Lambda]$ is described by an induced labelling $\rho \colon A \to Q$ defined inductively according to $\rho(a) = \delta(\{\rho(a') \mid a' \in E^{\mathfrak{A}}[a]\}, \lambda(a))$. The tree language accepted by $\mathcal{A}$ w.r.t. some $F \subseteq Q$ is $L(\mathcal{A}, F) = \bigcup_{q \in F} L(\mathcal{A}, q)$ where

$$L(\mathcal{A}, q) = \big\{T \in T_{\mathrm{fin}}^\ell[\Lambda] \mid \rho(0) = q \text{ for the run } \rho \text{ of } \mathcal{A} \text{ on } T \big\}.$$

For an application as an acceptor of trees whose branching degree is bounded by some n (which is the standard format used in the literature), we may of course transcribe δ into a function $\delta^n \colon \bigcup_{m \leqslant n} Q^m \times \Sigma \to Q$. Let $\mathcal{A}^n = (Q, \delta^n)$ be this specialization of $\mathcal{A}$ to trees of n-bounded branching.

Observation 11 *The size of* $\mathcal{A} = (Q, \delta)$ *is* $|\delta| \leqslant |\Lambda| \cdot 2^{|Q|}$. *Its restriction* $\mathcal{A}^n$ *to trees of n-bounded branching is of size* $|\delta^n| \leqslant |\Lambda| \cdot |Q|^n$.

It is one of the main points in our application, though, that the branching should not a priori be bounded. The exponential blow-up between $\mathcal{A}^n$ for fixed n and $\mathcal{A}$ itself is the reason that we shall want to deal with a special subspecies of the above kind of automata, for which the transition function can be given in a more compact format. Assume that $\mathcal{A} = (Q, \delta)$ where Q is of the form $Q \subseteq \mathcal{P}(\Gamma)$ for some finite set Γ, Q closed under union. We say that $\mathcal{A} = (Q, \delta)$ is *of $\cup$-type* w.r.t. Γ if, for all $q \subseteq Q$ and for all $r \in \Lambda$, the value of $\delta(q, r)$ only depends on $\bigcup q$. The following is then straigtforward.

Lemma 12 *Let* $\mathcal{A} = (Q, \delta)$ *be of $\cup$-type w.r.t.* Γ *and let* $|\Gamma| = n$. *Then, for any* $F \subseteq Q$, $L(\mathcal{A}, F)$ *is of bounded tallness iff* $L(\mathcal{A}^n, F)$ *is of bounded tallness.*

An inspection of typical results for standard (fixed branching) tree automata shows that these carry over to our slightly more general notion of tree automata with the above notion of size. We refer in particular to M. Vardi's discussion of tree automata and their applications in [16], and to the handbook article [15] by W. Thomas for background.

Theorem 13 *Bounded tallness of $L(\mathcal{A}, F)$ is decidable in time polynomial in the size of $\mathcal{A}$.*

For a fixed finite set $\Gamma \subseteq L_\mu[\tau]$ let $\text{tp}^\Gamma(\mathfrak{A}) = \{\psi \in \Gamma \mid \mathfrak{A} \models \psi\}$, and $\text{Tp}^\Gamma(\mathfrak{A}, a) = \bigcup_{a' \in E^{\mathfrak{A}}[a]} \text{tp}^\Gamma(\mathfrak{A} \restriction \langle a' \rangle)$. Consider an end extension $\mathfrak{B}$ of a finite tree structure $\mathfrak{A}$ of type τ, $\mathfrak{A} \subseteq_{\text{end}} \mathfrak{B}$. With $\mathfrak{B}$ associate the leaf labelling that maps a leaf a of $\mathfrak{A}$ to $\pi(a) = \text{Tp}^\Gamma(\mathfrak{B}, a)$. If Γ is a preservation set, then the leaf labelling induced by an end extension $\mathfrak{B}$ of $\mathfrak{A}$ fully determines $\text{tp}^\Gamma(\mathfrak{B} \restriction \langle a \rangle)$ for all $a \in A$. Indeed, a tree automaton can compute these types from the leaf labelling over $\mathfrak{A}$. Note that $(\mathfrak{A}, \pi)$ is a Σ_Γ-leaf-labelled tree structure where $\Sigma_\Gamma = \{\text{Tp}^\Gamma(\mathfrak{B}, b) \mid \mathfrak{B} \in T[\tau]\} \subseteq \mathcal{P}(\Gamma)$. If Γ is a set of at most n L_μ-formulae whose length is at most n, then this alphabet itself is recognizable in time exponential in n: $p \in \Sigma_\Gamma$ if and only if the L_μ-formula $\bigwedge_{\psi \in p} \Diamond \psi \wedge \bigwedge_{\psi \in \Gamma \setminus p} \neg \Diamond \psi$ is satisfiable. But L_μ-satisfiability is in ExpTime due to [4]. Note that the necessity to make this labelling alphabet explicit eventually turns our automata theoretic decision procedure into a reduction to L_μ-satisfiability. We code $(\mathfrak{A}, \pi)$ as a tree $T \in T_{\text{fin}}^\ell[\Lambda_\Gamma]$, where $\Lambda_\Gamma \subseteq \mathcal{P}(\tau) \times (\mathcal{P}(\Gamma) \,\dot{\cup}\, \{*\})$. Let in this sense $[\mathfrak{A}; \mathfrak{B}]^\Gamma$ stand for the $T \in T_{\text{fin}}^\ell[\Lambda_\Gamma]$ associated with an end extension $\mathfrak{A} \subseteq_{\text{end}} \mathfrak{B}$. The following extends Proposition 7. The inductive proof, which is an elaboration of that given for Proposition 7 above, is omitted here.

Proposition 14 *For every $\varphi \in L_\mu[\tau]$ there is a preservation set $\Gamma = \Gamma_\varphi$ of size $|\Gamma_\varphi| \leqslant |\varphi|$ and with $\varphi \in \Gamma_\varphi$, and an automaton $\mathcal{A}_\Gamma$ with state set $Q_\Gamma = \mathcal{P}(\Gamma)$, of $\dot{\cup}$-type w.r.t. Γ, such that for all $\mathfrak{A} \subseteq_{\text{end}} \mathfrak{B}$: $[\mathfrak{A}; \mathfrak{B}]^\Gamma \in L(\mathcal{A}_\Gamma, q) \Leftrightarrow \text{tp}^\Gamma(\mathfrak{B}) = q$.*

We turn to criterion (iii) from Lemma 10. Let $\varphi \in L_\mu[\tau]$ and consider two end extensions $\mathfrak{B}_1$ and $\mathfrak{B}_2$ of $\mathfrak{A} \in T_{\text{fin}}[\tau \cup \{U\}]$. For $\Delta\mathcal{C}$ we are interested in the case that $\mathfrak{B}_i \models \varphi^U$ and $\mathfrak{B}_2 \not\models \varphi^U$. The $\mathfrak{B}_i$ induce two leaf labellings on $\mathfrak{A}$ w.r.t. $\Gamma = \Gamma_{\varphi^U}$, which we may code into one with a labelling alphabet consisting of the product of the original Σ_Γ with itself. This turns the triple $(\mathfrak{A}, \mathfrak{B}_1, \mathfrak{B}_2)$ into a labelled tree

$$[\mathfrak{A}; \mathfrak{B}_1; \mathfrak{B}_2] \in T_{\text{fin}}^\ell[\Lambda^*], \quad \text{where } \Lambda^* = \mathcal{P}(\tau) \times ((\Sigma_\Gamma \times \Sigma_\Gamma) \cup \{*\}).$$

Consider now an automaton $\mathcal{A}$ which simulates in parallel two copies of the automaton $\mathcal{A}_\Gamma$ of Proposition 14, one working with the first component of the leaf labelling, the other with the second component. The natural way of performing this parallel simulation uses a state set $Q_\Gamma^* := Q_\Gamma \times Q_\Gamma$, so that $\mathcal{A}_\Gamma^*$ operates like $\mathcal{A}_\Gamma$ in both components. Identifying $\mathcal{P}(\Gamma) \times \mathcal{P}(\Gamma)$ with $\mathcal{P}(\Gamma \times \{1, 2\})$, and writing Γ^* for $\Gamma \times \{1, 2\}$, we may regard $\mathcal{A}_\Gamma^*$ as of $\dot{\cup}$-type w.r.t. Γ^*.

Now $\Delta\mathcal{C}$ consists of those $\mathfrak{A}$ for which some $[\mathfrak{A}; \mathfrak{B}_1; \mathfrak{B}_2]$ is accepted by $\mathcal{A}^* = \mathcal{A}_\Gamma^*$ in some state (q, q') where $\varphi^U \in q$ and $\varphi^U \notin q'$. Therefore, the tallness problem for $\mathcal{A}^*$ turns out to settle ML-expressibility of φ, by Lemma 10:

$$\varphi \text{ is expressible in ML} \quad \Leftrightarrow \quad \begin{array}{l} L(\mathcal{A}^*, F) \text{ is of bounded tallness,} \\ \text{where } F = \{(q, q') \mid \varphi^U \in q, \varphi^U \notin q'\}. \end{array}$$

Note that the size of $\mathcal{A}^*$ is doubly exponential in $|\varphi|$. But by Lemma 12 we may equivalently consider the tallness problem for the automaton $(\mathcal{A}^*)^n$ for $n = |\Gamma^*| = 2 \cdot |\Gamma|$, because $\mathcal{A}^*$ is of $\sqcup$-type. By Observation 11 and Theorem 13 bounded tallness of $(\mathcal{A}^*)^n$ is decidable in simply exponential time in $|\Gamma|$, and hence in $|\varphi|$. This proves the EXPTIME bound in the main theorem.

This bound is essentially optimal, since there is a straightforward reduction of L_μ-satisfiability to ML-expressibility: if $\xi \in L_\mu$ is *not* ML-expressible, and if U and V are not in ξ or φ, then φ is *unsatisfiable* if and only if the formula $\Diamond\varphi^U \wedge \xi^V$ *is* equivalent to a formula in ML (namely to $\bot$).

References

1. J. BARWISE AND Y.N. MOSCHOVAKIS, *Global inductive definability*, Journal of Symbolic Logic, 43(3), 1978, pp. 521–534.
2. J.F.A.K. VAN BENTHEM, Modal Logic and Classical Logic, Bibliopolis, 1985.
3. S.S. COSMADAKIS, H. GAIFMAN, P.C. KANELLAKIS, AND M.Y. VARDI, *Decidable optimization problems for database logic programs*, Proc. 20th ACM Symp. on Theory of Computing, 1988, pp. 477–490.
4. E.A. EMERSON AND C. JUTLA, *The complexity of tree automata and logics of programs*, Proc. 29thSymp. on Foundations of Computer Science, 1988, pp. 328–337.
5. E.A. EMERSON AND R. STREETT, *An automata theoretic decision procedure for the propositional mu-calculus*, Information and Computation, 81, 1989, pp. 249–264.
6. M.J. FISCHER AND R.E. LADNER, *Propositional dynamic logic of regular programs*, Journal of Computer and System Sciences, 18, 1979, pp. 194–211.
7. H. GAIFMAN, H. MAIRSON, Y. SAGIV, AND M.Y. VARDI, *Undecidable optimization problems for database logic problems*, Journal of the Association for Computing Machinery, 40, 1993, pp. 683–713.
8. E. GRÄDEL AND M. OTTO, *On logics with two variables*, to appear in Theoretical Computer Science.
9. G.G. HILLEBRAND, P.C. KANELLAKIS, H.G. MAIRSON, AND M.Y. VARDI, *Undecidable boundedness problems for Datalog programs*, Journal of Logic Programming, 25, 1995, pp. 163–190.
10. D. JANIN AND I. WALUKIEWICZ, *On the expressive completeness of the propositional mu-calculus with respect to monadic second order logic*, Proc. of 7th Int. Conf. on Concurrency Theory, 1996, pp. 263–277.
11. P.G. KOLAITIS AND M. OTTO, *On the boundedness problem for two-variable first-order logic*, Proc. 13th Symp. on Logic in Computer Science, LICS'98, pp. 513-524.
12. D. KOZEN, *Results on the propositional μ-calculus*, Theoretical computer Science, 27, 1983, pp. 333–354.
13. D. KOZEN, *A finite model theorem for the propositional μ-calculus*, Studia Logica, 47, 1987, pp. 233–241.
14. M.O. RABIN, *Decidability of second-order theories and automata on infinite trees*, Transactions of the American Mathematical Society, 141, 1969, pp. 1–35.
15. W. THOMAS, *Languages, Automata, and Logic*, in Handbook of Formal Languages, vol. 3, G. Rozenberg and A. Salomaa, ed., Springer, 1997, pp. 389–456.
16. M.Y. VARDI, *Automata theory for database theoreticians*, in Theoretical Studies in Computer Science, J.D. Ullman, ed., Academic Press, 1992, pp. 153–180.
17. M.Y. VARDI, *Why is modal logic so robustly decidable*, DIMACS Series in Discrete Mathematics and Theoretical Computer Science 31, AMS, 1997, 149–184.

On Optimal Algorithms and Optimal Proof Systems

Jochen Messner

Abteilung Theoretische Informatik
Universität Ulm, 89069 Ulm, Germany
messner@informatik.uni-ulm.de

Abstract. A deterministic algorithm O accepting a language L is called (polynomially) optimal if for any algorithm A accepting L there is a polynomial p such that $\mathrm{time}_O(x) \leq p(|x| + \mathrm{time}_A(x))$ for every $x \in L$. It is shown that an optimal acceptor for a language L exists if there is a p-optimal proof system for L. If L is a p-cylinder also the inverse implication holds. This result widely generalizes work from Krajíček and Pudlák who showed the result for $L = \mathrm{TAUT}$. It is further shown how to construct an optimal acceptor for a p-cylinder L, given an acceptor for L which runs fast on every easy subset of L. Then we investigate the relationship of this notion of an 'optimal acceptor' to a more general notion of optimality. Here, instead of considering time-complexity on each individual string x, worst-case time-bounds are considered. It is observed that every set complete for exponential time under linearly length-bounded polynomial-time many-one reducibility has an acceptor with an optimal time-bound whereas on the other hand no set hard for exponential time under polynomial-time many-one reducibility has a p-optimal proof system. Finally we show how these results can be translated to nondeterministic algorithms and optimal proof systems.

1 Introduction

The major aim in the development of algorithms for hard sets is to decrease the runtime. A related line of research is to design heuristics which have a good performance on important instances or to identify efficiently decidable subsets (see, e.g., [5]). It seems to be ambitious to ask for an algorithm which has in some sense the fastest possible runtime on every input, an algorithm which runs fast on easy instances, and in some sense includes all possible heuristics for the problem, even those which are not known yet. However, Levin [8] proved that such an optimal algorithm exists for the functional task to find witnesses for elements of a given set in $\mathcal{NP}$ (cf., Theorem 1). For example, using the random access machine (RAM) model of computation, one can construct an algorithm O which finds satisfying assignments for formulas $\varphi \in \mathrm{SAT}$ such that for any other algorithm A solving the same task, there is a constant c with $\mathrm{time}_O(x) \leq c(\mathrm{time}_A(x) + |x|)$ for every $x \in \mathrm{SAT}$ (O may not halt on other inputs). One can rephrase Levin's result in terms of inverting polynomial time computable functions as follows (again we state the result using the RAM model).

C. Meinel and S. Tison (Eds.): STACS'99, LNCS 1563, pp. 541–550, 1999.
© Springer-Verlag Berlin Heidelberg 1999

Theorem 1 ([8]). *For each (partial) function h computable by a RAM in polynomial time p there is a RAM M inverting h such that for every RAM M' inverting h there is a constant $c > 0$ with* $\text{time}_M(y) \leq c \cdot (\text{time}_{M'}(y) + p(|M'(y)|))$ *for every y in the range of h.*

It is noted in [12] that one can transfer the result to the Turing machine model if one replaces the term $c \cdot (\ldots)$ above by $c' \cdot ((\ldots) + \log(\ldots))$.

To study the existence of optimal algorithms in a more machine independent fashion it is suitable to use the following definition of optimality. Intuitively, for some task to solve, let us call an algorithm O *optimal for this task on instances from* $S \subseteq \Sigma^*$ if for any other algorithm A solving the same task there is a polynomial p such that $\text{time}_O(x) \leq p(\text{time}_A(x) + |x|)$ for every $x \in S$. Using the Turing machine model of computation (as we will do in the rest of the paper) one can formalize this intuition as follows.

Definition 1. *Let $\mathcal{A}$ be a collection of Turing machines, let $S \subseteq \Sigma^*$. A Turing machine $M \in \mathcal{A}$ is called* polynomially time-optimal *(short: optimal) for $\mathcal{A}$ on S if for any $M' \in \mathcal{A}$ there is a polynomial p such that $\text{time}_M(x) \leq p(\text{time}_{M'}(x) + |x|)$ for each $x \in S$.*

The definition implies that any task which can be solved in polynomial time has an optimal algorithm. As a consequence of Theorem 1 one obtains

Corollary 1. *For each (partial) function $f \in \mathcal{FP}$ there is a Turing transducer which is optimal on the range of f for the deterministic transducers inverting f.*

Contrasting with this functional task, in this paper we primarily investigate the existence of an algorithm which is optimal on L for the deterministic algorithms accepting the language L. For short, such an algorithm is called an *optimal acceptor for L*. We show that the existence of an optimal acceptor for L is closely related to the existence of a p-optimal proof system for L. In [4] Cook and Reckhow considered a function $h \in \mathcal{FP}$ with range L as an *(abstract) proof system* for L. A proof system h for L is called *p-optimal* if every proof system f for L is p-simulated by h which means that there is a function $g \in \mathcal{FP}$ such that $h(g(x)) = f(x)$ for any x in the domain of f. Connections between the existence of p-optimal proof systems and other complexity theoretical notions have also been studied in [7, 11, 9, 6, 3, 10]. Krajíček and Pudlák showed in [7] that an optimal acceptor for TAUT exists if, and only if, there is a p-optimal proof system for TAUT. Recently, using an idea from [6], Sadowski [10] showed that the result holds also for SAT instead of TAUT. A main objective of this paper is to generalize the result to further languages L. In fact, using further ideas from [6], we prove that for any language L, an optimal acceptor for L exists if there is a p-optimal proof system for L. The reverse implication is proved under the assumption that L is a p-cylinder.

Schnorr shows in [12] that for self-reducible sets L, the complexity of the functional problem which is the problem to find witnesses for membership in L is closely related to the complexity of the decision problem. Using the result of Levin he is able to construct 'optimal' acceptors for self-reducible problems like, e.g., SAT. However, the notion of optimality used here is a more general one than the notion of optimality considered above. Instead of considering the time-complexity of the algorithms on each individual string, time-bounds are considered. A function $t : I\!N \to I\!N$ is called a *time-bound for an algorithm A* on $S \subseteq \Sigma^*$ if $\text{time}_A(x) \leq t(n)$ for every $x \in S$ with $|x| \leq n$. Let us call t a *time-bound for the set L* if t is a time-bound on L for a deterministic Turing machine accepting L. If, in addition, for any time bound s for L there is a polynomial p such that $t(n) \leq p(s(n))$, t is called an *optimal time-bound for L*. Rephrasing the result in [12], one can construct an acceptor for SAT with an optimal time-bound. Note however that such an acceptor may have exponential run-time on every instance from SAT even on those instances which can be solved efficiently by some known algorithm. On the other hand, under the assumption $\mathcal{P} = \mathcal{NP}$ this algorithm has a polynomial time-bound on instances from SAT (the algorithm may not halt on other inputs).

The relation between both notions of optimality is examined in Section 5. There we show that any deterministic time class $\text{DTIME}(t(n))$ determined by a time-constructible function t with $\mathcal{P} \subseteq \text{DTIME}(t(n))$ contains a set that has no optimal acceptor but an optimal time-bound. This implies for example that no set $\leq_m^p$-hard for exponential time has a p-optimal proof system. On the other hand it is shown that any set complete for exponential time under linearly length-bounded many-one reducibility has an optimal time-bound. In Section 3 we show a relationship between the performance of acceptors or proof systems on easy instances from L, and the existence of optimal acceptors or p-optimal proof systems for L. Some implications stated in the main theorem in Section 4 are already proved there. Finally in Section 6 we briefly discuss how the results can be transferred to nondeterministic algorithms and optimal proof systems.

Due to the limited space several proofs are shortened or omitted in this version of the paper.

2   Preliminaries

We assume some familiarity with standard notions of computational complexity theory, and refer the reader to books like [2] for notions not defined in this paper. Let Σ be some fixed finite alphabet containing 0 and 1. The output of a Turing transducer M on input $x \in \Sigma^*$ is denoted by $M(x)$; we write $M(x) = \bot$ if M does not accept or runs forever on input x. Similarly for a partial function $f : \Sigma^* \to \Sigma^*$ we write $f(x) = \bot$ if f is undefined on x; a transducer M computes f if $f(x) = M(x)$ for every $x \in \Sigma^*$. Let $\mathcal{FP}$ denote the class of partial functions computed by transducers which have a polynomial time bound on Σ^*. Let h be a function with range $R \subseteq \Sigma^*$; For a function f (and also for a transducer M computing f) we say that f *(resp. M) inverts h* on $S \subseteq R$ if $h(f(y)) = y$ for $y \in S$. If additionally $f \in \mathcal{FP}$ we say that h *is $\mathcal{FP}$-invertible* on S. Following [2] a function h is called *1-invertible* (in polynomial time) if there is a function, denoted h^{-1}, in $\mathcal{FP}$ such that $h^{-1}(h(x)) = x$ for $x \in \Sigma^*$; h is called *length-increasing* if $|h(x)| > |x|$ for any x; we call h *linearly length-bounded* if

$|h(x)| \leq c \cdot |x|$ for some constant c and any x. Given a subset S of the domain of f, $f(S)$ denotes the set $\{f(x) \mid x \in S\}$. A set $A \subseteq \Sigma^*$ *many-one reduces to* B via a total function $f \in \mathcal{FP}$ (in symbols: $A \leq_m^p B$) if for all $x \in \Sigma^*$, $x \in A$ if, and only if, $f(x) \in B$. If, additionally, f is 1-invertible with range Σ^*, which means $B \leq_m^p A$ via f^{-1}, A and B are called *p-isomorphic*. A set A which is p-isomorphic to $A \times \Sigma^*$ is called *p-cylinder*. The following Lemma shows the property of p-cylinders that makes the notion useful for this paper.

Lemma 1. *L is a p-cylinder if, and only if, any set which is $\leq_m^p$-reducible to L, is $\leq_m^p$-reducible to A via a 1-invertible, length increasing function.*

A proof of the lemma is found in [2]. It is further shown there that L is a p-cylinder if, and only if, L is 1-invertible paddable (L is called *1-invertible paddable* if there is a 1-invertible function $g \in \mathcal{FP}$ such that $g(\langle x, y \rangle) \in L$ if, and only if, $x \in L$ for all $x, y \in \Sigma^*$).

Let $\mathcal{E}$ denote the class $\text{DTIME}(2^{O(n)})$ and let $\mathcal{NE}$ be its nondeterministic counterpart.

3   Optimality and Easy Tasks

In this section we show several constructions that relate the performance of acceptors or proof systems on easy instances from L to the existence of optimal acceptors or p-optimal proof systems. So it is easy to see that (intuitively stated) an optimal acceptor runs fast on easy instances from L (see Proposition 1); similarly when considering a p-optimal proof system we obtain that proofs for easy instances are easy to compute (see Proposition 2). More surprising is probably that also an inverse version of these statements holds when L is a p-cylinder. So given a proof system for L which allows one to compute proofs for easy instances easily, one obtains a p-optimal proof system for L (see Proposition 3). Similarly given an acceptor for L which runs fast on every easy instance one obtains an optimal acceptor for L. The proof of the latter statement is the most involved of these constructions and is delayed to the proof of the main theorem in the next section.

Proposition 1. *Let M be an optimal deterministic acceptor for L, and let S be a subset of L with $S \in \mathcal{P}$. Then there is a polynomial p such that $\text{time}_M(x) \leq p(|x|)$ for $x \in S$.*

Proposition 2. *A p-optimal proof system for L is $\mathcal{FP}$-invertible on any subset S of L with $S \in \mathcal{P}$.*

Proof. Let h be a p-optimal proof system for L and let $S \subseteq L$, $S \in \mathcal{P}$. Let g be a proof system defined as follows

$$g(w) = \begin{cases} h(v) & \text{if } w = 0v, \\ v & \text{if } w = 1v \text{ and } v \in S, \\ \bot & \text{otherwise} \end{cases}$$

As h is p-optimal there is a function $f \in \mathcal{FP}$ such that $h(f(w)) = g(w)$ for any $w \in \Sigma^*$. Now, let r be a polynomial time computable function with $r(x) = f(1x)$ for $x \in S$. One easily checks that $h(r(x)) = h(f(1x)) = g(1x) = x$ for any $x \in S$.

For any set L let $T(L)$ be defined by

$$T(L) = \{\langle M, x, 0^s\rangle \mid x \in \Sigma^*, s \geq 0, \ M \text{ is a det. Turing transducer,}$$
$$\text{and time}_M(x) \leq s \text{ implies } M(x) \in L\},$$

and fixing some transducer M let

$$T(L)_M = \{\langle M, x, 0^s\rangle \in T(L) \mid x \in \Sigma^*, s \geq 0\}.$$

Notice, if L is the range of the function computed by the transducer M then $T(L)_M$ is the trivial language $\{\langle M, x, 0^s\rangle \mid x \in \Sigma^*, s \geq 0\}$.

In [6] it had been observed that $T(L)$ is polynomial time many-one equivalent to L for $L \neq \Sigma^*$. We state this observation in Lemma 2. It shows that in some sense $T(L)$ is the hardest set which is $\leq_m^p$-reducible to L: any set $\leq_m^p$-reducible to L is reducible to $T(L)$ via a very simple many-one reduction.

Lemma 2. *For any set $L \subseteq \Sigma^*$ the following holds*

- $T(L) \leq_m^p L$ *if* $L \neq \Sigma^*$.
- $A \leq_m^p L$ *via f implies that A is reducible to $T(L)$ via $g : x \mapsto \langle M, x, 0^{p(|x|)}\rangle$ where M is a transducer computing f in polynomial time p.*

It is interesting to note that the Lemmas 1 and 2 imply that $T(L)$ is a p-cylinder.

We will now see that an inverse version of Proposition 2 holds. In the proof of Proposition 3 we show how to construct a p-optimal proof system for a p-cylinder L given a proof system for L which is $\mathcal{FP}$-invertible on every easy subset of L. Let us first state the following lemma which holds for any language L.

Lemma 3. *Let g be a proof system for $T(L)$ such that for any transducer M computing a proof system for L, g is $\mathcal{FP}$-invertible on $T(L)_M$. Then there is a p-optimal proof system for L.*

Proof. We will observe that the following algorithm computes a p-optimal proof system h.

   input $\langle M_1, M_2, x, 0^s, 0^n\rangle$
   if time$_{M_1}(\langle M_2, x, 0^s\rangle) \leq n$ **then**
      let w be the output of M_1 on input $\langle M_2, x, 0^s\rangle$
      if $g(w) = \langle M_2, x, 0^s\rangle$ and time$_{M_2}(x) \leq s$ **then**
         output $M_2(x)$ and halt;
   otherwise reject.

Notice that $g(w) = \langle M_2, x, 0^s\rangle$ implies $\langle M_2, x, 0^s\rangle \in T(L)$, then $M_2(x) \in L$ if time$_{M_2}(x) \leq s$. This shows that the range of h is a subset of L. We now show that h p-simulates any proof system f for L. By assumption there is a transducer M_2 computing f in polynomial time p and a function $r \in \mathcal{FP}$ such that $g(r(\langle M_2, x, 0^s\rangle)) = \langle M_2, x, 0^s\rangle$ for all x and s. Let M_1 be a transducer computing r in polynomial time q. Now f is p-simulated by h via the translation $x \mapsto \langle M_1, M_2, x, 0^s, 0^n\rangle$ where $s = p(|x|)$ and $n = q(|\langle M_2, x, 0^s\rangle|)$.

One may also replace the proof system g in the proof above by a suitable acceptor for L which yields the following Lemma.

Lemma 4. *Let R be a Turing acceptor for $T(L)$ which has a polynomial time bound on $T(L)_M$ for any transducer M computing a proof system for L. Then there is a p-optimal proof system for L.*

When L is a p-cylinder one can replace the set $T(L)$ in Lemma 3 above by L itself which yields the Proposition 3.

Proposition 3. *Let L be a p-cylinder, and let h be a proof system for L which is $\mathcal{FP}$-invertible on any $S \in \mathcal{P}$ with $S \subseteq L$. Then there is a p-optimal proof system for L.*

Proof. Let f be a 1-invertible reduction from $T(L)$ to L. We will show that a function $g \in \mathcal{FP}$ with $g(\langle x, w \rangle) = x$ if $f(x) = h(w)$ is a proof system for $T(L)$ which fulfills the conditions of Lemma 3. If a transducer M computes a proof system for L we have $T(L)_M \in \mathcal{P}$. Then $S = f(T(L)_M)$ is also in $\mathcal{P}$ as f is 1-invertible. As h is $\mathcal{FP}$-invertible on S there is a function $r \in \mathcal{FP}$ such that $h(r(y)) = y$ for $y \in S$. Observe $g(\langle x, r(f(x)) \rangle) = x$ for $x \in T(L)_M$.

Again a similar result holds using an acceptor.

Proposition 4. *Let L be a p-cylinder, and let R be an acceptor for L which has a polynomial time-bound on every $S \in \mathcal{P}$ with $S \subseteq L$. Then a p-optimal proof system for L exists.*

Notice that the implications in the Propositions 3 and 4 can be generalized to further languages L. For the proofs to go through it suffices that there is a $\leq_m^p$-reduction from $T(L)$ to L such that $f(T(L)_M) \in \mathcal{P}$ for every machine M computing a proof system for L. On the other hand the author does not believe that any recursively enumerable language which has a maximal subset in $\mathcal{P}$, has an optimal acceptor. This would show that the implications in Propositions 3 and 4 do not hold for every non-p-levelable set (see [1] for definitions).

4   Optimal Algorithms and Optimal Proof Systems

We now complete the proof of the main theorem. The theorem generalizes the result of Krajíček and Pudlák [7] for TAUT to any p-cylinder.

Theorem 2. *For a p-cylinder L, the following statements are equivalent.*

1. *There is a p-optimal proof system for L.*
2. *There is a proof system for L which is $\mathcal{FP}$-invertible on any $S \in \mathcal{P}$ with $S \subseteq L$.*
3. *There is an optimal acceptor for L.*
4. *There is an acceptor for L which has a polynomial time-bound on every $S \in \mathcal{P}$ with $S \subseteq L$.*

In addition, the implications $1 \Rightarrow 2$, $3 \Rightarrow 4$, and $1 \Rightarrow 3$ hold for any language L.

The proof for the theorem is obtained by several constructions partially already given in the previous section. More detailed, the implications $1 \Rightarrow 2$, $3 \Rightarrow 4$, $2 \Rightarrow 1$, $4 \Rightarrow 1$, and $1 \Rightarrow 3$ are proved by the Propositions 2, 1, 3, 4, and 5. The most involved of these constructions is given by the implication $1 \Rightarrow 3$ which we prove now.

We need the following Lemma from [6].

Lemma 5. *If A has a p-optimal proof system then any set $\leq_m^p$-reducible to A has a p-optimal proof system, too.*

In an intuitive sense we can use a p-optimal proof system for L to certificate efficiently and uniformly the fact that a machine M_i accepts only strings from L. This intuition is reflected in the proof of the following lemma.

Lemma 6. *Let L be a set possessing a p-optimal proof system. Then there is a recursive function mapping any transducer M_i to a transducer M_i' such that*

$$L(M_i') = L(M_i) \cap L$$

and, if $L(M_i) \subseteq L$, there is a polynomial p such that for any $x \in \Sigma^*$

$$\text{time}_{M_i'}(x) \leq p(\text{time}_{M_i}(x) + |x|).$$

Proof. It suffices to prove the result for $L \neq \Sigma^*$. Let

$$A(L) = \{\langle M, x, 0^s \rangle \mid x \in L \text{ if the DTM } M \text{ accepts } x \text{ in } \leq s \text{ steps}\}.$$

Observe $A(L) \leq_m^p L$. Therefore, assuming the existence of a p-optimal proof system for L, there is a p-optimal proof system $h \in \mathcal{FP}$ for $A(L)$ by Lemma 5. Let I_h be an optimal transducer inverting h which is given by Corollary 1.

On input x the machine M_i' first proceeds like M_i. If M_i rejects x, M_i' rejects. If M_i accepts x in s steps then M_i' runs I_h on input $\langle M_i, x, 0^s \rangle$ and accepts x iff I_h produces some output. Because I_h inverts h, I_h produces an output on input $\langle M_i, x, 0^s \rangle$ if, and only if, $\langle M_i, x, 0^s \rangle$ is in the range $A(L)$ of h. Therefore $L(M_i') = L(M_i) \cap L$. Now assume $L(M_i) \subseteq L$. Observe that $\langle M_i, x, 0^s \rangle \in A(L)$ for any $x \in \Sigma^*$, $s > 0$. By Proposition 2 there is a function $r \in \mathcal{FP}$ such that $h(r(\langle M_i, x, 0^s \rangle)) = \langle M_i, x, 0^s \rangle$ for $x \in \Sigma^*$, $s > 0$. Due to the optimality of I_h this implies that $\text{time}_{I_h}(\langle M_i, x, 0^s \rangle) \leq q(|\langle M_i, x, 0^s \rangle|)$ for all $x \in \Sigma^*$, $s > 0$, and some polynomial q. As $\text{time}_{M_i'}(x) = t + \text{time}_{I_h}(\langle M_i, x, 0^t \rangle)$ where $t = \text{time}_{M_i}(x)$ we obtain $\text{time}_{M_i'}(x) \leq p(\text{time}_{M_i}(x) + |x|)$ for some polynomial p and any $x \in \Sigma^*$.

Proposition 5. *If L has a p-optimal proof system then an optimal acceptor for L exists.*

Proof. Let $M_1', M_2', \ldots$ be the recursive enumeration of acceptors for subsets of L that can be obtained by Lemma 6 from a standard enumeration $M_1, M_2, \ldots$ of deterministic Turing machines. Let U be some universal Turing machine which on input $\langle i, x \rangle$ can simulate s steps of M_i' on input x in time $c_i s^2 + c_i$ (U first constructs a suitable encoding of M_i' in less than c_i steps).

On input x the optimal acceptor Opt proceeds in stages, each second stage Opt spends for simulating one step of U on input $\langle 1, x \rangle$, each 4th stage is spent for simulating one step of U on $\langle 2, x \rangle$ each 8th stage is spent for $\langle 3, x \rangle$, and so on. Generally said, in stage $2^{i-1} + 2^i(j-1)$, $j \geq 1$, the jth step of U on $\langle i, x \rangle$ is simulated. If in some simulation a machine accepts, Opt also accepts and halts (rejects are discarded). Note that stage n can be performed in time $O(n)$.

As each machine in $M_1', M_2', \ldots$ accepts only strings from L, it is clear that $L(\text{Opt}) \subseteq L$. Let now M_i be some acceptor for L. By Lemma 6 M_i' is also an acceptor for L with $\text{time}_{M_i'}(x) \leq p(\text{time}_{M_i}(x) + |x|)$. Now Opt will accept x in stage $2^{i-1} + 2^i(c_i p(\text{time}_{M_i}(x) + |x|)^2 + c_i - 1)$ due to M_i' (or earlier due to some other machine). The time needed to reach this stage is bounded by a polynomial in $\text{time}_{M_i}(x) + |x|$.

5   Sets without an Optimal Acceptor

In this section we show that one encounters sets which have no optimal acceptor immediately when one leaves $\mathcal{P}$ in the deterministic world. We construct a set which has an optimal time-bound but has no optimal acceptor. The result implies that no p-cylinder $\leq_m^p$-hard for (e.g.) exponential time has an optimal acceptor nor a p-optimal proof system. On the other hand we show that any set complete for $\mathcal{E}$ under linearly length-bounded many-one reducibility has an optimal time bound.

A function $t : \mathbb{N} \to \mathbb{N}$ is called *time-constructible* if there is Turing transducer M and a constant $c > 0$ such that M on input 0^n outputs $0^{t(n)}$ in not more than $c \cdot t(n)$ steps.

Theorem 3. *Let $t : \mathbb{N} \to \mathbb{N}$ be a time-constructible function such that for every polynomial p there is a number n with $p(n) \leq t(n)$. Then there is a language $L \in DTIME(t(n))$ for which the following holds*

- *there is no optimal acceptor for L.*
- *$t(n)$ is an optimal time bound for L.*

Proof. Let $M_1, M_2, \ldots$ be a standard enumeration of deterministic Turing acceptors, and let U be a universal machine which on input of $0^i 1x$ can simulate s steps of M_i on input x in time $\leq c_i \cdot s^2 + c_i$. For any $i > 0$ let L_i be the regular language described by the expression $0^i 10^*$. Define

$$L_i' = \{x \in L_i \mid U \text{ does not accept } x \text{ in less than } t(|x|) \text{ steps}\},$$

and let $L = \bigcup_{i>0} L_i'$. Clearly $L \in DTIME(t(n))$. This construction guarantees that for any machine M_i accepting L it holds $L_i' = L_i$ and $c_i \text{time}_{M_i}(x)^2 + c_i \geq t(|x|)$ for $x \in L_i$. Using this one obtains in a straightforward way that L has the stated properties. We omit the details due to the limited space.

Observe that any time-constructible function t with $\mathcal{P} \subseteq DTIME(t(n))$ fulfills the condition of Theorem 3.

Together with Theorem 2 and Lemma 5 one obtains the following two corollaries.

Corollary 2. *No set $\leq_m^p$-hard for $\mathcal{E}$ has a p-optimal proof system.*

Corollary 3. *No p-cylinder $\leq_m^p$-hard for $\mathcal{E}$ has an optimal acceptor.*

The proof of Theorem 3 does not rely on the monotonicity of time-bounds implied by the definition of the notion 'time-bound' in this article. Using the monotonicity of time-bounds one obtains the following theorem. Due to the limited space we state it without a proof.

Theorem 4. *Every set hard for $\mathcal{E}$ under linearly length-bounded polynomial time many-one reducibility has an optimal time-bound.*

5 Optimal Nondeterministic Acceptors and Optimal Proof Systems

In this section we briefly sketch how the results in the previous sections can be translated to the nondeterministic case. Clearly the notions of an optimal acceptor has a straightforward nondeterministic correspondence. A proof system h for L is called *optimal* if every proof system f for L is simulated by h which means that there is a polynomial p such that for any x in the domain of f there is $y, |y| \leq p(|x|)$, with $h(y) = f(x)$. Basically, a proof system h can be associated with the nondeterministic acceptor N which on input x guesses w and accepts if $h(w) = x$. Symmetrically a nondeterministic acceptor N can be transformed to a proof system h with $h(\langle x, \alpha \rangle) = x$ if α denotes an accepting path of N on input x. Therefore it is a trivial observation that there is an optimal nondeterministic acceptor for L if, and only if, there is an optimal proof system for L. Nonetheless, the result corresponding to the remaining equivalence of Theorem 2 is of some interest. Again this generalizes a result from [7] for TAUT. Due to the limited space the proof is omitted.

Theorem 5. *For a p-cylinder L, the following statements are equivalent.*

1. *There is an optimal proof system for L.*
2. *There is an optimal nondeterministic acceptor for L.*
3. *There is a nondeterministic acceptor for L which has a polynomial time-bound on every $S \in \mathcal{NP}$ with $S \subseteq L$.*
4. *There is a nondeterministic acceptor for L which has a polynomial time-bound on every $S \in \mathcal{P}$ with $S \subseteq L$.*

With very few modifications the proof of Theorem 3 can be adjusted to the nondeterministic case. This yields

Theorem 6. *Let $t : I\!N \to I\!N$ be a time-constructible function such that for every polynomial p there is a number n with $p(n) \leq t(n)$. Then there is a language $L \in co\text{-}NTIME(t(n))$ for which no optimal proof system exists.*

It is shown in [6] that any set $\leq_m^p$-reducible to a set possessing an optimal proof system has an optimal proof system, too. Together with Theorem 6 this yields the following corollary.

Corollary 4. *No set $\leq_m^p$-hard for $co\text{-}\mathcal{NE}$ has an optimal proof system.*

7   Acknowledgments

I whish to thank Johannes Köbler for his permanent advice, and for the motivating discussions about the topic.

References

[1] José Luis Balcázar, Josep Díaz, and Joaquim Gabarró. *Structural Complexity II* Springer-Verlag, 1990.

[2] José Luis Balcázar, Josep Díaz, and Joaquim Gabarró. *Structural Complexity I.* Springer-Verlag, 2 edition, 1995.

[3] Shai Ben-David and Anna Gringauze. On the existence of optimal propositional proof system and oracle-relativized propositional logic. Technical Report TR98-021, Electronic Colloquium on Computational Complexity, 1998.

[4] Stephen A. Cook and Robert A. Reckhow. The relative efficiency of propositional proof systems. *The Journal of Symbolic Logic*, 44(1):36–50, 1979.

[5] Jun Gu, Paul Purdom, John Franco, and Benjamin Wah. *Algorithms for the Satisfiability Problem*. Cambridge University Press, to appear.

[6] Johannes Köbler and Jochen Messner. Complete problems for promise classes by optimal proof systems for test sets. In *Proceedings of the 13th Annual IEEE Conference on Computational Complexity, CC 98*, pages 132–140, 1998.

[7] Jan Krajíček and Pavel Pudlák. Propositional proof systems, the consistency of first order theories and the complexity of computations. *The Journal of Symbolic Logic*, 54(3):1063–1079, 1989.

[8] Leonid A. Levin. Universal search problems (in russian). *Problemy Peredach Informatsii*, 9(3):115–116, 1973. English translation in *Problems of Information Transmission* 9(3):265–266, 1973. Revised translation as an appendix to [13].

[9] Jochen Messner and Jacobo Torán. Optimal proof systems for propositional logic and complete sets. In *Proceedings of the 15th Symposium on Theoretical Aspects of Computer Science (STACS'98)*, number 1373 in Lecture Notes in Computer Science, pages 477–487. Springer-Verlag, 1998.

[10] Zenon Sadowski. On an optimal deterministic algorithm for SAT. Presented on the Annual Conference of the European Association for Computer Science Logic CSL '98. To be published in the LNCS-series of Springer-Verlag.

[11] Zenon Sadowski. On an optimal quantified propositional proof system and a complete language for $\mathcal{NP} \cap$ co-$\mathcal{NP}$. In *Proceedings of the 11-th International Symposium on Fundamentals of Computing Theory, FCT'97*, number 1279 in Lecture Notes in Computer Science, pages 423–428. Springer-Verlag, 1997.

[12] Claus-Peter Schnorr. Optimal algorithms for self-reducible problems. In *Proceedings of the third International Colloquium on Automata, Languages, and Programming (ICALP'76)*, pages 322–337. Edinburgh University Press, 1976.

[13] Boris A. Trakhtenbrot. A survey of russian approaches to perebor (brute-force search) algorithms. *Annals of the History of Computing*, 6(4):384–400, 1984.

Space Bounds for Resolution

Juan Luis Esteban[1] * and Jacobo Torán[2] **

[1] Dept. Llenguatges i sistemes informàtics, Universitat Politècnica de Catalunya
c/ Jordi Girona Salgado 1-3, 28023 Barcelona, Spain
esteban@lsi.upc.es
[2] Abt. Teoretische Informatik, Universität Ulm
Oberer Eselsberg, 89069 Ulm, Germany
toran@informatik.uni-ulm.de

Abstract. We introduce a new way to measure the space needed in a resolution refutation of a CNF formula in propositional logic. With the former definition [6] the space required for the resolution of any unsatisfiable formula in CNF is linear in the number of clauses. The new definition allows a much finer analysis of the space in the refutation, ranging from constant to linear space. Moreover, the new definition allows to relate the space needed in a resolution proof of a formula to other well studied complexity measures. It coincides with the complexity of a pebble game in the resolution graphs of a formula, and as we show, has strong relationships to the size of the refutation. We also give upper and lower bounds on the space needed for the resolution of unsatisfiable formulas.

1 Introduction and Definitions

In this paper we deal exclusively with propositional logic, and the only refutation system considered is resolution. Due to its simplicity and to its importance in automatic theorem proving and logic programming systems, resolution is one of the best studied refutation systems. Resolution contains only one inference rule: if $A \vee x$ and $B \vee \bar{x}$ are clauses, then the clause $A \vee B$ may be inferred by the resolution rule resolving the variable x. A resolution refutation of a conjunctive normal form (CNF) formula φ is a sequence of clauses $C_1 \ldots C_s$ where each C_i is either a clause of φ or is inferred from earlier clauses in the refutation by the resolution rule, and C_s is the empty clause, $\square$. One way to measure the complexity of resolution applied to a specific formula, is to measure the minimum *size* of a refutation for it. This is defined as the number of clauses in the refutation. More than a decade ago, Haken [5] gave the first proof of an exponential lower bound on the number of clauses needed in any resolution refutation of a family of formulas expressing the pigeonhole principle. In following years, the original proof has been greatly simplified and extended to other classes of formulas [10,3,2,9].

* Supported by ESPRIT LTR Project 20244, ALCOM-IT and CICYT Project TIC97-1475-CE, KOALA: DGES PB95-0787, SGR: CIRIT 1997SGR-00366
* Partially supported by the DFG

A less studied measure for the complexity of a resolution refutation is the amount of *space* it needs. This measure was defined in [6] in the following way:

Definition 1. *[6] Let $k \in \mathbb{N}$, we say that an unsatisfiable CNF formula φ has resolution refutation bounded by space k if there is a series of CNF formulas $\varphi_1, \ldots, \varphi_s$, such that $\varphi = \varphi_1$, $\square \in \varphi_s$, in any φ_i there are at most k clauses, and for each $i < s$, φ_{i+1} is obtained from φ_i by deleting (if wished) some of its clauses and adding the resolvent of two clauses of φ_i.*

Intuitively this expresses the idea of keeping a set of *active* clauses in the refutation, and producing from this set a new one by copying clauses from the previous set and resolving one pair of clauses, until the empty clause is included in the set. Initially the set of active clauses consists of all the clauses of φ, and the space needed is the maximum number of clauses that are simultaneously active in the refutation.

In [6] it is proven that any unsatisfiable CNF formula φ with n variables and m clauses can be refuted in space $m + n$, and in [4] it is observed that the space upper bound $2m$ can also be obtained.

Although natural, the above definition has the drawback that the space needed in a refutation can never be less than the number of clauses in the formula being refuted. This is so because this formula is the first one in the sequence used to derive the empty clause. Making an analogy with a more familiar computation model, like the Turing machine, this is the same as saying that the space needed cannot be less than the size of the input being processed. To be able to study problems in which the *working space* is smaller than the size of the input, the space needed in the input tape is usually not taken into consideration. We do the same for the case of resolution and introduce the following alternative definition for the space needed in a refutation.

Definition 2. *Let $k \in \mathbb{N}$, we say that an unsatisfiable CNF formula φ has resolution refutation bounded by space k if there is a series of CNF formulas $\varphi_1, \ldots, \varphi_s$, such that $\varphi_1 \subseteq \varphi$, $\square \in \varphi_s$, in any φ_i there are at most k clauses, and for each $i < s$, φ_{i+1} is obtained from φ_i by deleting (if wished) some of its clauses, adding the resolvent of two clauses of φ_i, and adding (if wished) some of the clauses of φ (initial clauses).*

The space needed for the resolution of an unsatisfiable formula is the minimum k for which the formula has a refutation bounded by space k.

In the new definition it is allowed to add initial clauses to the set of active clauses at any stage in the refutation. Therefore this clauses do not need to be stored and do not consume much space since in any moment at most two of them are needed simultaneously. The only clauses that consume space are the ones derived at intermediate stages. As we will see in Section 2, there are natural classes of formulas that can be refuted using only logarithmic space (in the number of initial clauses), or even constant space.

There is another natural way to look at this definition using pebble games on graphs, a traditional model used for space measures in complexity theory and for register allocation problems (see [8]). Resolution refutations can be represented as directed acyclic graphs of in-degree two, in which the nodes are the clauses used in the refutation, and a vertex (clause) has outgoing edges to the resolvents obtained using this clause. In this graph the sources are the initial clauses, all the other nodes have in-degree two, and the unique sink is the empty clause. In case that in the refutation no derived clauses are reused, that is, when all the nodes (except maybe the sources) have out-degree one, the proof is called tree-like.

The space required for the resolution refutation of a CNF formula φ (as expressed in Definition 2) corresponds to the minimum number of pebbles needed in the following game played on the graph of a refutation of φ.

Definition 3. *Given a connected directed acyclic graph with one sink the aim of the pebble game is to put a pebble on the sink of the graph (the only node with no outgoing edges) following this set of rules:*

1) *A pebble can be placed in any initial node, that is, a node with no predecessors.*

2) *Any pebble can be removed from any node at any time.*

3) *A node can be pebbled provided all its parent nodes are pebbled.*

3') *If all the parent nodes of node are pebbled, instead of placing a new pebble on it, one can shift a pebble from a parent node.*

There are different variations of this simple pebble game in the literature. In fact, in [11] it is shown that the inclusion of rule 3' in the game can at most decrease by one the number of pebble needed to pebble a graph, but in the worst case the saving is obtained at the price of squaring the number of moves needed in the game. We include rule 3' so that the number of pebbles coincides exactly with the space in Definition 2. This fact is stated in the following Lemma.

Lemma 1. *Let φ be an unsatisfiable CNF formula. The space needed in a resolution refutation of φ coincides with the number of pebbles needed for the pebble game played on the graph of a resolution refutation of φ.*

This second characterization of space in resolution proofs allow us to use techniques introduced for the estimation of the number of pebbles required for pebbling certain graphs, for computing the space needed in resolution refutations. However the estimation of the number of pebbles needed in the refutation of a formula is harder than the estimation of the number of pebbles needed for a graph, since in the first case one has to consider all the possible refutation graphs for the formula. From now on we will refer indistinctly to the space needed for the refutation of a formula or to the number of pebbles needed for the game on its refutation graphs.

In Section 3 we give upper and lower bounds for the amount of space needed for resolution. When measuring the space relative to the number of variables in the initial formula we show that any unsatisfiable CNF formula with n variables has a resolution proof that uses space $n + 1$, and we also obtain a matching lower bound, that is, we show that there are formulas on n variables whose refutation needs space $n + 1$. If we are interested to measure the space relative to the number of initial clauses, we can prove that any unsatisfiable CNF formula with m clauses can be resolved in space m, but the best lower bound we get is $\log m$ for the space needed in the refutation of certain formulas. Later, in Section 4 this lower bound is improved for the cases of tree-like and regular resolution, two restrictions of the resolution procedure. These results are obtained from an upper bound on the size of a refutation of a formula in terms of the space needed for its resolution and the depth of the refutation. (The depth of a refutation is the size of the longest path from the empty clause to an initial clause in the refutation graph). We prove in Theorem 6 that if a formula φ has a resolution refutation of depth d that uses space s, then it has a resolution refutation of size $\leq 2^s d s^3$. For types of resolution in which the depth of the proofs is bounded (like in the case of regular resolution), this provides an exponential upper bound for the resolution size in terms of the resolution space.

We include at the end of the paper a section of conclusions and open problems. Due to space reasons some of the proofs are not included in this version of the paper.

2   Some Examples

In this section we give two examples of families of unsatisfiable formulas that can be refuted within less space than its number of clauses. The first example are the formulas whose clauses are all possible combinations of literals in such a way that every variable appears once in every clause. We will see that the space needed to refute these formulas is bounded by the number of different variables in it. In fact we will prove a more general result about the space needed in a tree-like resolution.

Definition 4. *We say that a graph G_1 is embedded in a graph G_2 if a graph isomorphic to G_2 can be obtained from G_1 by adding nodes and edges or inserting nodes in the middle of edges of G_1.*

Observe that the number of pebbles needed for pebbling any graph is greater than or equal to the number of pebbles needed for pebbling any embedded subgraph in it. This is true since any pebbling strategy for the graph, also pebbles the embedded subgraph.

Theorem 1. *Let φ be an unsatisfiable CNF formula with a treelike resolution of size s, then φ has a resolution refutation of space $\lceil \log s \rceil + 1$.*

Proof. We will show that the resolution tree in the refutation of φ can be pebbled with $d + 1$ pebbles, where d is the depth of the biggest complete binary tree embedded in the refutation graph. As the biggest possible complete binary tree embedded in a tree of size s has depth $\lceil \log s \rceil$, the theorem holds. It is a well known fact (see for example [8]) that $d + 1$ pebbles suffice to pebble a complete binary tree of depth d (with the directed edges pointing to the root). In fact $d+1$ pebbles suffice to pebble any binary tree whose biggest embedded complete binary tree has depth d. In order to see this we use induction on the size of the tree. The base case is obvious. Let T be refutation tree, and T_1 and T_2 be the two subtrees from the root. Let us call $d_c(T)$ the depth of the biggest embedded subtree in T. So

$$d_c(T) = \begin{cases} \max(d_c(T_1), d_c(T_2)) & \text{if } d_c(T_1) \neq d_c(T_2) \\ d_c(T_1) + 1 & \text{if } d_c(T_1) = d_c(T_2) \end{cases}$$

By induction hypothesis one can pebble T_1 with $d_c(T_1) + 1$ pebbles and T_2 with $d_c(T_2)+1$ pebbles. Let us suppose that $d_c(T_1) < d_c(T_2)$, then $d_c(T) = d_c(T_2)$ and one can pebble first T_2 with $d_c(T_2) + 1$ pebbles, leave a pebble in the root of T_2 and then pebble T_1 with $d_c(T_1) + 1$. For this second part of the pebbling one needs $d_c(T_1) + 2 \leq d_c(T_2) + 1$. The other case is similar. ∎

We can apply the above lemma to compute the space needed in the refutation of the following formula.

Definition 5. *Let $n \in \mathbb{N}$, COMPLETE-TREE$_n$ is the CNF formula on the set of variables $\{x_1, \ldots, x_n\}$, whose clauses are all possible combinations of literals with the restriction that each variable appears once in each clause.*

$$\text{COMPLETE-TREE}_n = (x_1 x_2 \ldots x_n), (\bar{x}_1 x_2 \ldots x_n), \ldots, (\bar{x}_1 \bar{x}_2 \ldots \bar{x}_n).$$

Observe that this formula has 2^n clauses. It is not hard to see that COMPLETE-TREE$_n$ can be refuted using space $n+1$. This is so since a straightforward tree-like resolution of the formula that resolves the variables in different stages, has size $2^{n+1} - 1$. The previous lemma assures that this refutation can be pebbled with $n + 1$ pebbles. In the next section we will see that this amount of space is also necessary.

As second example, consider the class of unsatisfiable formulas in CNF with at most two literals per clause.

Theorem 2. *Any unsatisfiable CNF formula with at most two literals in each clause can be resolved within constant space.*

3 Upper and Lower Bounds

For the results in this section the following concept will be very useful.

Definition 6. *We say that a CNF unsatisfiable formula is minimally unsatisfiable if removing any clause the formula becomes satisfiable.*

The next result attributed to [1] has been proved independently many times.

Lemma 2. *Any minimally unsatisfiable CNF formula must have more clauses than variables.*

We start by giving bounds with respect to the number of variables.

Theorem 3. *Every unsatisfiable formula with n variables can be resolved using resolution in space at most $n + 1$.*

Proof. As mentioned in the proof of Theorem 1, for pebbling a tree of depth d, $d+1$ pebbles suffice. If we consider regular tree-like resolution, which is complete, we have refutation trees whose depth is at most the number of variables in the formula being refuted.   ∎

There is a matching lower bound, since there are formulas of n variables whose refutation graphs can only be pebbled with $n + 1$ pebbles. This is a consequence of the following result:

Theorem 4. *Let φ an unsatisfiable CNF formula and k the smallest number of literals of a clause of φ. Any resolution refutation of φ needs at least space $k+1$*

Proof. It is a well known fact that in a resolution refutation, for every truth assignment of the variables there is a unique path that goes from the empty clause to an initial clause, and the assignment gives value false to all the clauses in the path. Let us call these paths *truth assignment paths*.

For any pebbling strategy, there is a first step, let us call it i, in which the set of pebbled clauses becomes unsatisfiable. This step must exist because the first pebbling step consists of pebbling an initial clause, which is always satisfiable and the last step pebbles the empty clause. In step $i - 1$, there was a path in the resolution graph that goes from the empty clause to a initial clause and does not contain any pebbles. Otherwise the set of pebbled clauses in step $i - 1$ would be unsatisfiable since every truth assignment will make false at least one of the pebbles (the ones in the truth assignment path).

In step i, an initial clause has to be pebbled since according to the pebbling rules the only other possibility would be to pebble a clause with both parents pebbled, and this step would not transform the set of pebbled clauses into an unsatisfiable set. Therefore the set of pebbled clauses at step i contains at least k variables (the ones of the initial clause).

Let us suppose than the set of pebbled clauses at step i is minimally unsatisfiable, then, by Lemma 2, it has at least $k + 1$ clauses because it has at least k variables. On the other hand, if this set is not minimally unsatisfiable, we can throw aside clauses until the remaining set becomes minimally unsatisfiable Notice that we cannot delete the initial clause last added to the set, otherwise the set of clauses would be a subset of the clauses at stage $i - 1$ and becomes therefore satisfiable. So, $k + 1$ clauses are still needed because the initial clause is contained in the set and has at least k variables.   ∎

Since all the clauses in COMPLETE-TREE$_n$ have n variables, we obtain:

Corollary 1. *For all $n \in I\!N$ any resolution refutation of COMPLETE-TREE$_n$ requires at least space $n + 1$.*

Theorem 4 can be strengthen to allow to prove lower bounds for the space needed in the refutation of a more general class of formulas. This is done in Theorem 5 using the following lemma.

Let φ a CNF-formula, and α a (partial) truth assignment to the variables in φ. φ_α is a modification of φ according to α. For every variable v in α if its truth value is 1, all the clauses in φ containing the positive literal v are deleted and all occurrences of $\bar{v}$ are deleted. If the truth value of v is 0, then all clauses in φ containing $\bar{v}$ are deleted and all occurrences of the literal v are deleted.

The proof of the next lemma is an easy adaptation of [7, Theorem 1].

Lemma 3. *Let R be a resolution refutation of the CNF-formula $\varphi(p, q)$, where $p \cup q$ are the variables of φ, and let α any truth assignment of p. Then there is a resolution refutation of $\varphi(\alpha, q)$ whose resolution graph is embedded in R.*

Theorem 5. *Let φ be a unsatisfiable CNF formula, and let k be the maximum over all partial assignments α of the minimum number of literals of a clause in φ_α. The space needed in a resolution refutation of φ is at least k.*

The upper and lower space bounds measured with respect to the number of variables coincide. We have not been able to prove a matching result when measuring the space with respect to the number of initial clauses. Observe that from Lemma 2 and Theorem 3 we immediately obtain:

Corollary 2. *Every unsatisfiable formula with m clauses can be resolved using resolution in space at most m.*

Proof. Just consider a minimally unsatisfiable subset of the initial clauses. This subset contains at most $m - 1$ variables. ∎

However Theorems 4 and 5 can only provide a lower bound of $\log m$ for the space needed in the refutation of any formula. In the next section we improve this lower bound for some restrictions of resolution.

4 Relationships between Space and Size

The main result of this section provides an upper bound on the size of resolution refutations of a formula, in terms of the space and the depth needed in a refutation. Recall that the depth of a resolution refutation is the size of the longest path from the empty clause to an initial clause in the graph of the refutation.

Theorem 6. *If an unsatisfiable CNF formula φ has a resolution refutation of depth d and space s, then φ has a resolution refutation of size at most $2^s \cdot d \cdot s^3$.*

Proof. Let R be the resolution refutation proof of depth d that can be pebbled with s pebbles. As in Theorem 4 one can follow the pebbling strategy, placing pebbles in R until the set of pebbled clauses becomes unsatisfiable for the first time. Let us call this set this set of clauses φ_1. One can then start the pebbling strategy again from the beginning until all the clauses in the set φ_1 (except its initial clauses) can be inferred using resolution from the new set of pebbled clauses. Define this set to be φ_2. φ_2 always exists since the clauses in φ_1 are part of the refutation and therefore have been inferred using resolution. In the same way one can define the set of clauses φ_3, φ_4, and so on. Observe that for any i, φ_i contains at most s clauses, since all of them are simultaneously pebbled at a certain stage.

The last set in this series is the set φ_z formed by a set of clauses that can be inferred from the first s initial clauses being pebbled. In order to see how large z can be, define d_i to be the sum of the depths in R of all the clauses in φ_i, and p_i to be the number of clauses in φ_i. Clearly there cannot be two sets with the same pair (d_i, p_i) since one of the sets is inferred from the other one (and from some initial clauses) and therefore all the clauses in both sets cannot be at the same depth. Since for every i d_i is smaller than $d \cdot s$, and $p_i \leq s$, we get $z < d \cdot s^2$.

We show now how to build a new refutation of bounded size. The idea is to infer sequentially from φ_{i+1} the clauses in φ_i. Since these sets have bounded size, the number of resolution steps in between can also be bounded. This is clear for the last step since φ_1 is an unsatisfiable set with at most s clauses, and by Lemma 2 it contains an unsatisfiable set of clauses with at most $s - 1$ variables. From such a set the empty clause can be derived with a refutation of size bounded by 2^s (using for example regular tree-like resolution which is refutation complete). In order to prove this bound for a derivation of φ_i from φ_{i+1} for all i, we will start by the initial clauses, eventually inferring new sets of clauses φ_i' from which the clauses in φ_{i+1} can still be derived. The clauses in φ_i' have the property that for every clause C in φ_i there is a clause C' in φ_i' whose literals are a subset of those of C. Because of this property, from φ_1' the empty clause can still be derived.

We start inferring the clauses in φ_z. Let C be any clause in this set. We consider the derivation of C from (some of) the initial clauses. Let us call these clauses φ^C and this proof R^C. We apply Lemma 3 to R^C defining a partial truth assignment on the variables of C. The partial truth assignment used is:

$$\alpha(v) = \begin{cases} 1 \text{ if literal } \bar{v} \in C \\ 0 \text{ if literal } v \in C \end{cases}$$

We get a proof of the empty clause $\square$ from φ_α^C. φ_α^C is an unsatisfiable set and has at most s clauses. We get rid of all useless clauses in order to transform it into a minimally unsatisfiable formula denoted by φ'^C_α. By Lemma 2 φ'^C_α has at most $s - 1$ variables, so there is a refutation of F'^φ_α with at most 2^s clauses. This proof has none of the literals in C. Adding them again to the new proof we get a derivation of a clause $C' \subseteq C$. If this clause is different from C we modify φ_z by substituting C by C'. This is done for all the clauses in φ_z obtaining the set φ_z'. For this derivation at most $2^s \cdot s$ clauses are needed.

For the next step we have to infer a (possibly modified) set φ_{z-1} from the set φ_z'. For every clause C in φ_{z-1} we derive a clause $C' \subseteq C$. Let us call φ_z^C the set of clauses from the original φ_z that were used to derive C. Now we possibly do not have this set, but there is a set φ'^C_z containing a subclause of each of the clauses in φ^C_z. With this and the argument used in the previous step we can derive the clause C'. Finally from φ'_1 (if not before), the empty clause is derived.

We have shown that in order to infer φ'_i from φ'_{i+1} at most $s \cdot 2^s$ clauses are needed. Since there are at most $z < d \cdot s^2$ intermediate sets of clauses, the total size of the new refutation is bounded by $2^s \cdot d \cdot s^3$.   ■

We get different consequences from this result, for example:

Corollary 3. *The set of unsatisfiable CNF formulas with resolution refutations of polynomial depth and logarithmic space, have resolution refutations of polynomial size.*

In some types of resolution, the depth of the proof is automatically bounded. For example in regular resolution it is required that in every path from the empty clause to an initial clause in the refutation graph, every variable appears at most once. Clearly in this case the number of variables is a bound on the depth of the proof.

Corollary 4. *If an unsatisfiable CNF-formula on n variables has a regular resolution refutation of space s, then it has a resolution refutation of size at most $2^s \cdot n \cdot s^3$.*

For the case of regular resolution this upper bound allows to improve the lower bound on the space of a refutation measured with respect to the number of initial clauses stated in Section 3.

Beame and Pitassi show in [2] that for sufficiently large n, any resolution refutation of the formula expressing the pigeonhole principle for n pigeons, PHP^n_{n-1}, requires size at least $2^{n/20}$. As a direct consequence of the above theorem we get:

Corollary 5. *For sufficiently large n, any regular resolution refutation of the formula PHP^n_{n-1} requires space $\Omega(n)$.*

Since the formula PHP^n_{n-1} has $\Theta(n^3)$ clauses, measured in terms of the number m of initial clauses of this formula this means a lower bound of $\Omega(m^{1/3})$ for the space needed for its regular refutation.

For the case of tree-like resolution, Theorem 1 and the mentioned bound for PHP^n_{n-1}, provide the following result:

Corollary 6. *For sufficiently large n, any tree-like resolution refutation of the formula PHP^n_{n-1} requires space $n/20$.*

An interesting question is whether the depth of the refutation can be taken out of the bound given by Theorem 6. A way to do this would be by showing that a refutation of a formula can be transformed into another one that uses the same amount of space, but has bounded depth. It is not clear at all that this result holds, but as we see in our next result, it does hold for the case of tree-like resolution.

Theorem 7. *If φ is a CNF unsatisfiable formula with a tree-like resolution refutation of space s, then φ has a tree-like regular resolution refutation that uses the same amount of space.*

5 Conclusions and Open Problems

We have introduced a new definition to measure the space needed in the resolution of an unsatisfiable formula. This definition is more natural that the existing one since it is closer to space measures in other complexity models and can be characterized in terms of a well studied pebble game. We have obtained upper and, in some cases, matching lower bounds for the space needed, as well as relationships between the space and the size of a refutation. These results bring new insight in the structure of resolution and hopefully will be useful in the analysis of refutations.

There are however several interesting problems that remain open. One of them is to match the upper and lower bounds for the space needed for general resolution, measured in terms of the number of initial clauses. Recall from Section 3 that the bounds we have a respectively m and $\log m$. Other important question are whether Theorem 6 can be modified so that the depth is not a parameter in the right side of the upper bound for the size, or whether it is true that every unsatisfiable formula that can be resolved in logarithmic space, has a resolution refutation of polynomial size (an improvement of Corollary 3).

References

1. Aharoni, R., Linial, N.: Minimal non-two-colorable hypergraphs and minimal unsatisfiable formulas. Journal of Combinatorial Theory, **43** (1986) 196–204.
2. Beame, P., Pitassi, T.: Simplified and Improved Resolution Lower Bounds. Proc 37th IEEE Symp. on Foundations of Computer Science, **37** (1996) 274–282.
3. Chvátal, V., Szemerédi, E.: Many hard examples for resolution. Journal of the ACM **35** (1988) 759–768.
4. Esteban, J.L.: Complejidad de la resolución en espacio y tiempo. Masters thesis Facultad de Informática de Barcelona. 1995.
5. Haken, A.: The intractability of resolution. Theoretical Computer Science. **39, 2-3** (1985) 297–308
6. Kleine-Büning, H., Lettmann, T.: Aussagenlogik: Deduktion und Algorithmen. B.G Teubner. Stuttgart. 1994.
7. Pudlák, P.: Lower Bounds for Resolution and Cutting Plane Proofs and Monotone Computations. The Journal of Symbolic Logic. **62, 3** (1997) 981–998.
8. Savage, J.: Models of Computation. Addison-Wesley. 1998.
9. Schöning, U.: Resolution proofs, exponential bounds and Kolmogorov complexity In Proc. 22nd MFCS Conference, Springer Verlag LNCS. **1295** (1997) 110–116.
10. Urquhart, A.: Hard examples for resolution. Journal of the ACM. **34** (1987) 209–219.
11. van Emde Boas, P., van Leeuwen, J.: Move rules and trade-offs in the pebble game In Proc. 4th GI Conference, Springer Verlag LNCS. **67** (1979) 101–112.

Upper Bounds for Vertex Cover
Further Improved

Rolf Niedermeier[*1] and Peter Rossmanith[2]

[1] Wilhelm-Schickard-Institut für Informatik, Universität Tübingen, Sand 13,
D-72076 Tübingen, Fed. Rep. of Germany,
niedermr@informatik.uni-tuebingen.de
[2] Institut für Informatik, Technische Universität München, Arcisstr. 21,
D-80290 München, Fed. Rep. of Germany,
rossmani@in.tum.de

Abstract. The problem instance of Vertex Cover consists of an undirected graph $G = (V, E)$ and a positive integer k, the question is whether there exists a subset $C \subseteq V$ of vertices such that each edge in E has at least one of its endpoints in C with $|C| \leq k$. We improve two recent worst case upper bounds for Vertex Cover. First, Balasubramanian *et al.* showed that Vertex Cover can be solved in time $O(kn + 1.32472^k k^2)$, where n is the number of vertices in G. Afterwards, Downey *et al.* improved this to $O(kn + 1.31951^k k^2)$. Bringing the exponential base significantly below 1.3, we present the new upper bound $O(kn + 1.29175^k k^2)$.

1 Introduction

Vertex Cover is a problem of central importance in computer science:

- It was among the first NP-complete problems [7].
- There have been numerous efforts to design efficient approximation algorithms [3], but it is also known to be hard to approximate [1].
- It is of central importance in parameterized complexity theory and has one of the most efficient *fixed parameter algorithms* [4], which is also subject of [2, 5] and this paper.
- It has important applications, e.g., in computational biochemistry, where it is used to resolve conflicts between sequences by excluding some of them from a sample and, for this reason, the algorithm of Balasubramanian *et al.* [2] has been implemented as part of the DARWIN project at ETH Zürich [5, 8]. In particular, exact algorithms are important here.

An instance of Vertex Cover is an undirected graph $G = (V, E)$ and a positive integer k. The question is whether there exists a *vertex cover set* $C \subseteq V$ with $|C| \leq k$ such that for all edges (u, v) in E, it holds that $u \in C$ or $v \in C$. A

---

* Supported by a Feodor Lynen fellowship of the Alexander von Humboldt-Stiftung, Bonn, and the Center for Discrete Mathematics, Theoretical Computer Science and Applications (DIMATIA), Prague, Czech Republic.

C. Meinel and S. Tison (Eds.): STACS'99, LNCS 1563, pp. 561–570, 1999.
© Springer-Verlag Berlin Heidelberg 1999

straightforward greedy algorithm shows that Vertex Cover is approximable to a ratio 1 (cf. [13]). However, unless $P = NP$, Vertex Cover has no polynomial time approximation scheme [1] and it is known to be *not* approximable to a ratio 0.1666 [9].

Vertex Cover has seen quite some history of progress with respect to fixed parameter algorithms ("fixed parameter" refers to k, see [4, 11] for details). Recently, Balasubramanian *et al.* [2] came up with a greatly improved fixed parameter algorithm for Vertex Cover, running in time $O(kn + 1.324718^k k^2)$. They employ an intricate, improved search tree algorithm. Very recently, this result was slightly improved to $O(kn + 1.31951^k k^2)$ by Downey *et al.* [5]. Note that according to the authors this "tiny difference amounts to a 21% improvement in the running time for $k = 60$." In the following we prove a better upper bound of $O(kn + 1.29175^k k^2)$, thus breaking the 1.3 barrier in the base of the exponential term. Adopting the above example for $k = 60$, our new result means an improvement of 78% to the result of Balasubramanian *et al.* A technical report [12] contains all details that had to be omitted due to lack of space.

2   Preliminaries and Basic Notation

Let $G = (V, E)$ be an undirected graph. A set $C \subseteq V$ is a *vertex cover* of G, if for every edge $(i, j) \in E$, either $i \in C$ or $j \in C$ or both $i, j \in C$. A vertex cover is *minimal* or *optimal* if it has minimum size, i.e., if there is no vertex cover that has less vertices. By $N(x)$ we denote the set of neighbors, i.e., adjacent vertices, of a vertex x. For the ease of notation, we often write $\{x, N(y)\}$ instead of $\{x\} \cup N(y)$ or $N(\{x, y, \})$ instead of $N(x) \cup N(y)$ to denote sets of vertices. A graph is called *r-regular* if every vertex has degree r; it is called *regular* if it is r-regular for some r. A graph is *connected* if there is a path between each pair of vertices. A *component* of a graph is a maximal connected subgraph. Three vertices a, b, c are a *bridge* of a vertex x if x, a, b, c form a cycle (a closed path). We say b is a *bridge vertex* of x. A cycle of length 3 is a *triangle*.

Our algorithm works recursively. The number of recursions is the number of nodes in the according tree. This number is governed by homogeneous, linear recurrences with constant coefficients. It is well known, how to solve them and the asymptotic solution is determined by the roots of the characteristic polynomial. We use the same notation as Kullmann and Luckhardt [10]. If the algorithm solves a problem of size n and calls itself recursively for problems of sizes $n - d_1, \ldots, n - d_k$, then $(d_1, \ldots, d_k)$ is called the *branching vector* of this recursion. It corresponds to the recurrence $t_n = t_{n-d_1} + \cdots + t_{n-d_k}$ with the characteristic polynomial $z^d = z^{d-d_1} + \cdots + z^{d-d_k}$, where $d = \max\{d_1, \ldots, d_k\}$. If α is a root of the characteristic polynomial with maximum absolute value, then t_n is α^n up to a polynomial factor. We call $|\alpha|$ the *branching number* that correspond to the branching vector $(d_1, \ldots, d_2)$. Moreover, if α is a single root, then even $t_n = O(\alpha^n)$ and all branching numbers that will occur in this paper are single roots.

In this paper, the size of the search tree is therefore $O(\alpha^k)$, where k is the parameter and α is the biggest branching number that will occur; it is about 1.291742754 and belongs to the branching vector $(3, 5, 8, 8)$ which occurs in Section 5 (Case 5.2.1).

3   General Outline of the Algorithm

Our algorithm works, in essence, as all previous algorithms for Vertex Cover. The main part is to build a *bounded search tree*: To cover an edge, we have to put at least one of its two endpoints into the (optimal) vertex cover set. Thus, starting with an arbitrary edge, we can make a binary decision between its two endpoints. In each subcase, we delete the corresponding vertex chosen and its incident edges and repeat this until we have built a search tree of size 2^k. Altogether, it is easy to see that this leads to an algorithm running in time $O(2^k n)$ [4], where n denotes the number of vertices in the graph. All results (including ours) to get more efficient algorithms are based on efforts to make the search tree smaller. So, Balasubramanian *et al.* [2] presented an algorithm with search tree size 1.32472^k and this was improved to 1.31951^k by Downey *et al.* [5]. We further improve this size to 1.29175^k.

Before we give an overview of our approach we still have to explain briefly a technique called *reduction to problem kernel*, which is a kind of preprocessing. The main idea is that vertices of degree $> k$ *must* be part of a vertex cover, if its size is at most k. Deleting all those edges leaves a graph, which can still be very big. If, however, it is connected and bigger than $2k^2$ then there cannot exist a vertex cover of size k since there are more than k^2 edges. Hence, after reduction to problem kernel we can assume that the size of the graph is at most $2k^2$.

In parameterized complexity theory the resulting algorithm is known as *Buss' algorithm* [4], but basically the same approach can be traced back to older ideas from VLSI–theory, e.g., Evans [6]. It is not difficult to see, using appropriate subalgorithms, that Buss' algorithm has running time $O(kn + (2k^2)^k k^2)$. Combining reduction to problem kernel with the search tree algorithm described before, we get easily an $O(kn + 2^k k^2)$ algorithm for Vertex Cover. All subsequent improvements concentrated on replacing the exponential term 2^k by a smaller one.

The algorithm we describe is closer in spirit to the one of Balasubramanian *et al.* [2] than to that of Downey *et al* [5]. The main difference between both approaches is that Downey *et al.* employ a different reduction to problem kernel, which not only works as preprocessing, but is also applied during the search tree construction. We refer to Downey *et al.* [5] for details. By way of contrast, Balasubramanian *et al.* [2] and we use the "more classical approach" where the search tree deals also with vertices of degree 2 and 3 and reduction to problem kernel is only applied once as a preprocessing phase. In the rest of the paper, we concentrate on shrinking the search tree size.

3.1   Overall Structure of the New Search Tree Algorithm

The algorithms finds recursively an optimal vertex cover as follows. Given a graph G, we choose several subgraphs $G_1,\ldots,G_k$ and compute optimal vertex covers for all of them. From them we can construct an optimal vertex cover for G For example, let x be some vertex of G and let G_1 be the subgraph that results from G by deleting x and all incident edges. A vertex cover of G_1, together with x, is then a vertex cover of G. Moreover, if there is an optimal vertex cover for G that contains x, then we can construct an optimal vertex cover from an optimal vertex cover of G_1. Otherwise, if no optimal vertex cover of G contains x, they must contain all neighbors of x. Hence, let G_2 be the graph that results from G by deleting all neighbors of x. Again, we can construct an vertex cover of G by taking a vertex cover of G_2 and adding all neighbors of x. If we start from optimal vertex covers for G_1 and G_2, then one of the resulting covers for G must be optimal, since either x or its neighbors must be part of any vertex cover. We say we *branch according to x and $N(x)$*, where $N(x)$ denotes the neighbors of x In the first branch, x will be part of the vertex cover and in the second branch it will be $N(x)$. The vertex cover constructed grows in size with each step. Since its size cannot exceed k, the goal, the algorithm terminates.

In principle that is the way our algorithm works, but we choose the subgraphs $G_1,\ldots,G_k$ in a more complicated way and branch according to much more com plicated sets. The rules how to choose those branching sets are as follows, if the graph is connected:

1. If there is a vertex x with degree 1, then branch according to $N(x)$ (and nothing else). There is no other branch, since there is always an optimal vertex cover that contains $N(x)$ and does not contain x.
2. If there is a vertex x with degree 6 or more, then branch according to x and $N(x)$.
3. If there are no vertices with degree 1 or at least 6, but there is a vertex with degree 2, then proceed as shown in Section 4.
4. If 1.–3. do not apply and if the graph is regular, then choose some vertex x with maximum degree and branch according to x and $N(x)$. (This can happen at most three times in each path of the search tree and increases its size at most by a small constant factor.)
5. If 1.–4. do not apply and if there is a vertex with degree 3 then proceed as shown in Section 5.
6. Otherwise, there must be a vertex with degree 4 and all other vertices have degrees between 4 and 5. Proceed as shown in Section 6.

If the graph is not connected, then the algorithm chooses some component G' and tests recursively if G' has a vertex cover of size k or less and, if it has finds out the optimal size k' of a vertex cover for G'. Then it proceeds to test if $G - G'$, the other components, have a vertex cover of size $k - k'$. In this way the algorithm finds out whether the whole graph has a vertex cover of size k.

4   Degree-2-Vertices

If the graph is 2-regular (and connected), all vertices constitute a cycle and it is very easy to construct an optimal cover in linear time. Otherwise let x be a vertex with degree 2 and a, b its neighbors, where a has degree ≥ 3. The algorithm chooses the first of the following four cases that applies.

Case 1. There is an edge between a and b or x has a bridge whose bridge vector has degree 2. Then include $\{a, b\}$ into the vertex cover, which is optimal. No branching is necessary.

Case 2. Assume that $|N(a) \cup N(b)| \geq 4$. Then branch according to $\{a, b\}$ and $N(a) \cup N(b)$, whose branching vector is at least $(2, 4)$.

Case 3. Assume x has exactly one bridge. Then a's degree must be 3 and b's degree must be 2. Otherwise $|N(a) \cup N(b)| \geq 4$. Then there is an optimal cover that does not contain both a and y, the bridge vertex: If a and y are part of an optimal vertex cover then we can assume that x is also in the cover (but b is not). Replacing a by $N(a)$ produces another vertex cover that is not bigger. Hence, we branch according to $N(y)$ and $N(a)$. The branching vector is at least $(3, 3)$.

Case 4. Finally, let x have two bridges. Then the degrees of both a and b must be 3 since otherwise $|N(a) \cup N(b)| \geq 4$. Let y and z be the bridge vertices. We can branch according to y and $N(y)$. If y is in an optimal cover, including y and z, but not a or b is optimal, since two further vertices are necessary anyways to cover all incident edges of a and b. Hence, we can branch according to $N(y)$ and $\{x, y, z\}$ with a branching vector at least $(3, 3)$.

5   Degree-3-Vertices

In this section, the graph can contain vertices with degrees between 3 and 5. Particularly there must be at least one vertex with degree 3. Due to the lack of space, many details and proofs of correctness had to be omitted.

For Cases 1, 2, 3, and 4 let x be such a vertex and let a, b, and c be its neighbors. The first four cases distinguish on the structure of the subgraph around x, in particular on the degree of its neighbors and whether x has triangles or bridges. Case 5 is different, it rather assumes that no vertices exist in the whole graph, for which one of the first four cases applies.

Case 1. Assume that x is part of a triangle, e.g., let $\{x, a, b\}$ be the triangle (but there can be more triangles). Then we can branch according to $N(x)$ and $N(c)$. If x is not part of the cover, $N(x)$ is. If x is part of the cover, then is a or b. If c is also in the cover, then two neighbors of x are and we can replace x by $N(x)$. The branching vector is at least $(3, 3)$.

Case 2. Assume that x has at least two bridges (separate ones or a double bridge). Let y and z be the middle vertices on the bridges. We can branch according to $N(x)$ and $\{x, y, z\}$. The branching vector is at least $(3, 3)$.

Case 3. Next, assume that x has exactly one bridge, let us say between a and b. Call the center vertex on the bridge again y. Let us further assume that a or b has degree 3, without loss of generality a. Then we branch according to $N(x)$ and $N(a)$. The branching vector is at least $(3, 3)$.

Case 4. Now assume again that x has exactly one bridge as in the case above, but both a and b have degrees of at least 4. Then we can branch according to $N(x)$, $N(a)$, and $\{a, x, N(b), N(c)\}$. Since we can assume that x is not part of a triangle and there is exactly one bridge, we get the branching vector $(3, 4, 7)$.

Case 5. Finally, we can assume that there is no vertex with degree 3 that has a bridge or a triangle.

Case 5.1. Assume that there is a vertex x with degree 3 and neighbors a, b, c, two of which have degree at least 4, say, a and b. We pick either $N(x)$ or $\{x, N(a), N(b)\}$ or $\{x, a, N(b), N(c)\}$ or $\{x, b, N(a), N(c)\}$, using that, in order to get an optimal vertex cover, at most one of the neighbors can be chosen together with x. This yields the branching vector $(3, 7, 7, 7)$.

Case 5.2. Otherwise, we can assume that each degree 3 vertex has at most one neighbor with degree ≥ 4. We assumed further in this section that the graph is not regular and has at least one vertex with degree 3. Since the graph is connected there must be some vertex with degree 3 that has exactly one neighbor with degree 4 or 5.

Case 5.2.1 Let us assume that there is no cycle of length 5 with the following two properties: (1) each vertex on the cycle has degree 3 and (2) there is a vertex on the cycle that has a neighbor with degree at least 4. We choose some vertex with degree 3 that has a neighbor with degree 4 or 5. Call this vertex a_3 and the neighbor b_3. The other two neighbor of a_3 must have degree 3. From each vertex with degree 3 we can inductively follow some path that consists solely of degree–3 vertices: Just choose a neighbor with degree 3, but not that one you came from. Start such a path from a_3 and call the vertices a_2, a_1, a_0. Start another path and call the vertices a_4, a_5, a_6. Each of the a_i has at least two neighbors with degree 3, i.e., a_{i-1} and a_{i+1}. The third neighbor is called b_i and might have degree 3, 4, or 5.

Figure 1 shows the resulting part of the graph. This picture does not necessarily denote a subgraph of G: Firstly, all vertices b_i are shown with degree 4 but some of them might also have degree 3 or 5. On the other hand, we know that all a_i have *exactly* degree 3. Secondly, not all vertices shown in picture mus be necessarily distinct. For example, we could have $b_1 = a_4$ (then b_1 would have degree 3), since it does not violate any assumptions we made for this subcase. The picture in Figure 1 is therefore merely a sceletal structure leaving open many details. The freedom of these details lays in variation of degree of the b_i's and pairs of vertices being identical.

Our algorithm's behavior must depend on these details, but mostly we branch according to

1. $\{a_2, b_3, a_4\}$,
2. $\{a_1, b_2, a_3, b_4, a_5\}$,
3. $\{a_1, b_2, N(b_3), N(b_4), b_5, a_6\}$, and
4. $\{a_0, b_1, N(b_2), N(b_3), b_4, a_5\}$,

which can be found marked in Figure 1. The correctness is seen as follows: The first branch handles the case that a_3 is not in the cover, the remaining branches that it is. The second branch assumes that a_2 and a_4 are not in the cover. The third branch assumes that a_2 is not, but a_4 is in the cover. We can then further assume that b_2 is not in the cover, otherwise there is another optimal cover that contains $N(a_3)$ instead of a_3. Moreover, we can assume that also b_4 and a_5 are not part of the cover, since otherwise we can replace a_4 by $N(a_4)$, which is then handled by the second branch. Hence, the neighbors of b_3, b_4, and a_5 are in the cover and we get altogether $\{a_1, b_2, N(b_3), N(b_4), b_5, a_6\}$. The third and fourth branches are symmetric.

The resulting branching vector is $(3, 5, n_1, n_2)$. Clearly a_2, b_3, a_4 are pairwise distinct. Furthermore $b_2 \neq b_4$, $a_1 \neq b_4$, $b_2 \neq a_5$ (otherwise a_3 has a bridge) making also a_1, b_2, a_3, b_4, a_5 pairwise distinct and yielding the first two components of the branching vector. ($a_1, \ldots, a_5$ are pairwise distinct, since they do not constitute a cycle.) We concentrate now on the third branch, since the fourth one is quite similar to it and the same reasoning applies.

In $\{a_1, b_2, N(b_3), N(b_4), b_5, a_6\}$ we count 11 or 12 vertices, but some of them might be identical. If we could prove that the size of this set is always at least 8, we got the branching vector $(3, 5, 8, 8)$, which is good enough. Unfortunately, this is not possible. We proceed as follows: First we find out under what circumstances the size of the set can be smaller than 8. It will turn out that there is only one pathological possibility. Then we provide a different type of branching suited exactly for this exception. If neither the third nor the symmetrical fourth branch are pathological we get a branching vector of $(3, 5, 8, 8)$ using the above branching scheme. For the pathological cases we can even prove a branching vector $(3, 5, 7)$. We omit any details.

Case 5.2.2 Finally, we assume that there is a cycle of length 5 that consists of vertices with degree 3 and at least one of them has a neighbor with degree at least 4. This cycle shall consist of $a_0, \ldots, a_4$ with neighbors $b_0, \ldots, b_4$ outside the cycle, where b_2 is the neighbor with degree at least 4. We branch by either picking $\{a_1, b_2, a_3\}$ or $\{a_0, b_1, a_2, b_3, a_4\}$ or $\{a_0, b_1, N(b_2), N(b_3), b_4\}$ or $\{b_0, N(b_1), N(b_2), b_3, a_4\}$. Similar considerations as in Case 5.2.1 show that we

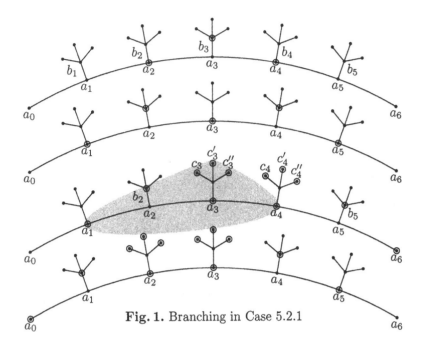

Fig. 1. Branching in Case 5.2.1

can get branching vectors $(3, 5, 8, 8)$ or $(3, 5, 7)$, in the latter case omitting the fourth branching set.

6   Degree-4-Vertices

In this section, we can assume that all vertices have either degree 4 or 5 and that there is at least one vertex with degree 4 and at least one vertex with degree 5. Again, several details had to be omitted.

Case 1. Assume that there is a vertex x of degree 4 that is part of a triangle and has a neighbor y with degree 5. Let a and b be the neighbors of x such that $\{a, b, x\}$ is a triangle (a or b can but need not coincide with y).

First, we assume that $a, b \neq y$. Let $c \notin \{a, b, y\}$ be another neighbor of x. We can branch according to $N(x)$, $N(y)$, and $\{x, y, N(c)\}$: The resulting branching vector is at least $(4, 5, 4)$.

Now let us assume that $a = y$. Then we can further assume that the remaining two neighbors c and d of x, i.e., not a or b, are *not* connected by an edge (Otherwise we can choose *them* to play the role of a and b above.) Branch according to $N(x)$, $N(c)$, and $\{x, c, N(d)\}$, which is sufficient for the same reason as above. The branching vector is again at least $(4, 4, 5)$, because $c \notin N(d)$.

Case 2. We assume in this case that there is no vertex x with degree 4 that has the following properties simultaneously: (1) x has a neighbor y with degree 5

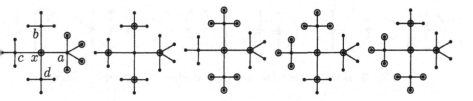

Fig. 2. All marked vertices in each of the five branches are distinct if there are no bridges that do not involve a. If there is such a bridge, we can assume it matches two marked vertices in the last branch because of symmetry.

(2) x has at least two bridges of whom y is not part of (this might be independent bridges or a double bridge). We can further assume that there are no triangles that contain a vertex with degree 4 that has a neighbor with degree 5, i.e., that Case 1 does not apply. Choose some vertex x with degree 4 that has a neighbor a with degree 5 (such vertex exists, otherwise the graph would be regular or not connected). Let b, c, d be the other neighbors of x (which can have degrees between 4 and 5). We can branch according to $N(a)$, $N(x)$, and $\{a, x\}$, but we split the last branch according to whether b and/or d are part of the optimal vertex cover. Altogether, we branch according to $N(a)$, $N(x)$, $\{x, a, N(b), N(d)\}$, $\{x, a, d, N(b), N(c)\}$, and $\{x, a, b, N(c), N(d)\}$ (see Figure 2). We get the branching vector $(5, 4, 8, 9, 9)$ if x has no bridge that a is no part of.

There might be *one* bridge that a is no part of. Without loss of generality we can assume that this bridge goes over c and d (b, c, d are symmetric and can be mutually exchanged in the above branching scheme). Then all vertices in the third and fourth branch must be still mutually distinct, but c and d now share another neighbor besides x and therefore two of the marked vertices in the last branch coincide. This leaves a branching vector of $(5, 4, 8, 9, 8)$.

Case 3. Now we assume that there is a vertex x that has exactly the properties that were forbidden in Case 2: It has degree 4 and two bridges. There is a neighbor y with degree 5 that is not part of either of the two bridges. There are two possibilities, which are depicted in Figure 3: Either x has two separate bridges or a double bridge. The algorithm can branch according to $N(y)$, $N(x)$, and $\{x, y\}$. In the last branch, if $\{x, y\}$ is part of the cover, then we can assume that the vertices on the bridge are members of the cover, too. Otherwise, their neighbors would be part of the cover and that implies that at least three of x's neighbors are in the vertex cover. Then, however, we can take $N(x)$ instead of x. That is, this cover is already handled by the first branch. Altogether, this implies a branching vector $(5, 4, 4)$.

7   Conclusion

Improving previous work [2, 5], we presented the so far best known algorithm for the NP-complete Vertex Cover problem, running in time $O(kn + 1.29175^k k^2)$.

Fig. 3. How to treat two separate bridges or a double bridge.

Besides the theoretical interest, this result may also be relevant for practical applications. So, for example, the Vertex Cover algorithm of Balasubramanian et al. [2] has been implemented for applications in computational biology [5, 8], our algorithm now being a natural candidate to replace or complement it. Note that our as well as previous algorithms that are good in the worst case have the clear potential to perform much better on average.

References

[1] S. Arora, C. Lund, R. Motwani, M. Sudan, and M. Szegedy. Proof verification and hardness of approximation problems. In *Proceedings of the 33d IEEE Conference on Foundations of Computer Science*, pages 14–23, 1992.

[2] R. Balasubramanian, M. R. Fellows, and V. Raman. An improved fixed parameter algorithm for vertex cover. *Information Processing Letters*, 65(3):163–168, 1998.

[3] P. Crescenzi and V. Kann. A compendium of NP optimization problems. Available at http://www.nada.kth.se/theory/problemlist.html, April 1997.

[4] R. G. Downey and M. R. Fellows. *Parameterized Complexity*. Springer-Verlag, 1998.

[5] R. G. Downey, M. R. Fellows, and U. Stege. Parameterized complexity: A framework for systematically confronting computational intractability. In F. Roberts, J. Kratochvíl, and J. Nešetřil, editors, *The Future of Discrete Mathematics: Proceedings of the First DIMATIA Symposium, June 1997*, AMS-DIMACS Proceedings Series. AMS, 1998. To appear. Available through http://www.inf.ethz.ch/personal/stege.

[6] R. C. Evans. Testing repairable RAMs and mostly good memories. In *Proceedings of the IEEE Int. Test Conf.*, pages 49–55, 1981.

[7] M. Garey and D. Johnson. *Computers and Intractability: A Guide to the Theory of NP-completeness*. Freeman, San Francisco, 1979.

[8] M. Hallett, G. Gonnet, and U. Stege. Vertex cover revisited: A hybrid algorithm of theory and heuristic. Manuscript, 1998.

[9] J. Håstad. Some optimal inapproximability results. In *Proceedings of the 29th ACM Symposium on Theory of Computing*, pages 1–10, 1997.

[10] O. Kullmann and H. Luckhardt. Deciding propositional tautologies: Algorithms and their complexity. 1997. Submitted to *Information and Computation*.

[11] R. Niedermeier. Some prospects for efficent fixed parameter algorithms (invited paper). In B. Rovan, editor, *Proceedings of the 25th Conference on Current Trends in Theory and Practice of Informatics (SOFSEM)*, number 1521 in Lecture Notes in Computer Science, pages 168–185. Springer-Verlag, 1998.

[12] R. Niedermeier and P. Rossmanith. Upper bounds for vertex cover further improved. Technical Report KAM-DIMATIA Series 98-411, Faculty of Mathematics and Physics, Charles University, Prague, November 1998.

[13] C. H. Papadimitriou. *Computational Complexity*. Addison-Wesley, 1994.

Online Matching for Scheduling Problems

Marco Riedel*

Heinz Nixdorf Institute, University of Paderborn, Germany
barcom@uni-paderborn.de

Abstract. In this work an alternative online variant of the matching problem in bipartite graphs is presented. It is motivated by a scheduling problem in an online environment. In such an environment, a task is unknown up to its disclosure. However, in that moment it is not necessary to take a decision on the service of that particular task. In reality, an online scheduler has to decide on how to use the current resources. Therefore, our problem is called *online request server matching* (ORSM). It differs substantially from the *online bipartite matching* problem of Karp *et al.* [KVV90]. Hence, the analysis of an optimal, deterministic online algorithm for the ORSM problem results in a smaller competitive ratio of 1.5.
We also introduce an extension to a weighted bipartite matching problem. A lower bound of $\frac{\sqrt{5}+1}{2} \approx 1.618$ and an upper bound of 2 is given for the competitive ratio.

1 Introduction

Motivation. The problem, which is investigated here, is motivated by online scheduling problems with deadlines. Such problems can be found in server systems for continuous data streams (e. g. video transmissions). Such a server consists of a set of memory modules (e. g. hard disks) to store a huge amount of data and to provide a high bandwidth. The latter is necessary to serve a large number of customers simultaneously. The memory modules are connected to a set of I/O-ports via a communication network inside the server system. Customers are connected to one of the I/O-ports in order to receive a demanded continuous data stream. It is clear that such a server is a real-time system and that the requests for data have hard deadlines. Data which violate their deadlines are not of interest anymore and can be omitted. Fortunately, the situation is relaxed a little bit because it seems to be acceptable when a small amount of the required data will not be delivered. A first question, which needs to be solved concerns the mapping of the data to the memory modules. To increase the usability of the system and to reduce the number of conflicts, the data or parts of it can be stored in more than one copy in different memory modules. Secondly, the access to memory modules and network resources has to be controlled. The difficulty of

* Supported by DFG-Graduiertenkolleg "Parallele Rechnernetzwerke in der Produktionstechnik", GRK 124/2-96. This work was partly supported by the EU ESPRIT Long Term Research Project 20244 (ALCOM-IT).

C. Meinel and S. Tison (Eds.): STACS'99, LNCS 1563, pp. 571–580, 1999.

these problems is increased by the *online* nature of the environment. The continuous delivering of a data stream can be preempted by a customer or it can be restarted at a completely different time index. Additionally, customers can arrive to the system or leave it at any time. So, future demands of server resources are unknown. However, *scheduling* decisions (assign appropriate resources to the requests) have to be taken under this uncertainty.

It is not difficult to construct a formal model for such a server system including all dependencies. However, its analysis is a challenge. To start investigations we performed a process of abstractions and simplifications: A continuous data stream is divided into packages of roughly equal size. Under this assumption, a simple, discrete time model is used. Abstracting from additional restrictions that model a restricted communication network between memory modules and customers, the scheduling problem becomes a matching problem in a bipartite graph. The scheduler has to match requests for unit size data packages to appropriate server resources and time slots. However, these decisions does not need to take immediately when a request arrives. Instead, an online scheduler has to determine how to use the current resources.

In this work we start our studies with a very basic model for this problem and we call it *online request server matching*. It is highly idealized and therefore it is not able to represent the inherent structure of the initial problem. The ORSM model is a matching problem in a rather general bipartite graph in a new online variant. Nevertheless, the definition was selected in such a way, that an extension to more realistic models is possible. A development to a weighted matching problem is also discussed here and it is motivated by additional priorities of requests.

To investigate our online problems, we apply *competitive analysis*. Roughly speaking, it determines the worst case ratio (called *competitive ratio*) of the solution quality of an online algorithm and an optimal solution over all inputs. Nowadays, it is an established method and it counterbalances results which assume an input distribution. A comprehensive introduction to online problems and competitive analysis is the textbook by Borodin and El-Yaniv [BEY98].

Previous and Related Work. Over the last years a vast amount of publications and single results on online scheduling can be found. [Sga98] gives an encyclopedic survey. However, studies of scheduling models which allow deadlines are rare and cover very restricted cases only. None of them seems to be closely related to our model. The opposite is true for bipartite matching problems in online settings. The study of such problems was initiated by Karp, Vazirani and Vazirani in [KVV90]. Their model consists of a 'known' and an 'unknown' partition. Vertices of the unknown partition are revealed over time and an online algorithm can put at most one edge of a just revealed vertex in the online matching. The simple GREEDY algorithm is an optimal, 2-competitive, deterministic online algorithm for this problem. The key contribution of [KVV90] is the analysis of the optimal, randomized algorithm RANKING which is $\frac{e}{e-1} \approx 1.582$-competitive.

Various extensions and variants of this problem have been studied later. A comprehensive survey and further references can be found in [KP98].

An online matching problem in general graphs is the *roommates problem*
[BR93]: Guests arrive by and by at a hotel for a conference. The hotel consists
of double rooms only. Every guest has a list of acceptable roommates (it is
assumed that these lists are symmetric; they represent the adjacency lists of
vertices in an undirected graph). The manager must immediately assign a room
to every guest and wants to minimize the number of occupied rooms. In [BR93]
an optimal, deterministic, 1.5-competitive algorithm is shown. For a weighted
extension of the problem a lower bound of 3, and a 4-competitive algorithm are
given.

Results and Organization of the Material. In this work we investigate a different
online variant of the bipartite matching problem. The precise description can be
found in Sect. 2. Section 3 presents the results of the unweighted case. A lower
bound and a matching upper bound are given. It includes a 1.5-competitive
algorithm. The weighted extension of our model is studied in Sect. 4. It is shown,
that the golden ratio ($\frac{\sqrt{5}+1}{2} \approx 1.618$) is a lower bound for the competitive ratio.
A 2-competitive, deterministic online algorithm is presented as well as an outline
of its analysis. Due to space limitations, the proof of the upper bound had to be
omitted. A few concluding remarks in Sect. 5 complete this work.

2   The Model

A formal definition of the *online request server matching* problem (ORSM) and
its weighted extension (wORSM) is now presented.

A bipartite graph $G := (R \cup S, E)$ represents the underlying structure of the
problem. Both partitions R and S are totally ordered. We denote the vertices by
$r_1, r_2, r_3, \ldots$ and by $s_1, s_2, s_3, \ldots$ with $r_i \in R$, $s_i \in S$, and the indices indicating
the position within the order. We interpret this order as a discrete time model.
The vertices of partition S represent a single resource called *server*. It is available
for one unit each time step. Partition R is interpreted as a set of tasks. Such a
task has a demand of one server unit to be completed. They are called *requests*,
and every time step one of them might occur. An edge $\{r_i, s_j\}$ between a request
vertex r_i and a server vertex s_j means that request r_i can be served in time step
j. The set of edges $E \subset R \times S$ is constructed with the following restriction:

$$\{r_i, s_j\} \in E \Rightarrow i \leq j . \tag{1}$$

This means a request that occurs at time step i must not specify a possible service
time in the past. Without this restriction the modelled scheduling problem does
not make sense and no competitive online algorithm exists.

Now we have to specify how this model works online: When the system starts
partition R is completely unknown[1]. In the beginning of a time step i the request

---

[1] When taking a close look this is not the truth. Every time step i a new vertex r_i is
inserted but its set of incident edges is before unknown. For reasons of convenience
the input process is interpreted in the introduced way.

r_i is revealed as input, i.e., vertex r_i and all of its edges are inserted into the previously known part of G. If no request appears, vertex r_i is isolated. After this input, an algorithm has to decide on how to use the server in the current time step i. It can add an edge incident to s_i to the online matching M. It is worth noting that due to restriction (1) *all* edges incident to s_i are known when this decision has to be taken. The online algorithm has the objective to maximize the cardinality of matching M, i.e. to serve as many requests as it can.

We want to emphasize the difference between the model in [KVV90] and ours. Karp *et al.* study an online version of the bipartite matching problem where the input of a time step is a vertex out of the 'unknown' partition and all of its incident edges. Then, an online algorithm has to select at most one of these *just* *revealed* edges for the online matching. Contrarily, the decision on a matching edge in the ORSM problem is taken by selecting one of the edges which are incident to a vertex of the *opposite* partition. Nevertheless, the ORSM problem can be interpreted as a special case of the online bipartite matching. Then, the server partition is the 'unknown' partition and whenever a decision has to be taken, all edges incident to that vertex are known. Furthermore, all additional information provided by the ORSM model is ignored. When the focus is shifted, the ORSM problem can also be recognized as a special case of the *roommates* *problem*. The precise relationship is shown in [Rie98].

So far, the graph G is unweighted. The objective changes when a weight function $w : E \rightarrow \mathbb{R}_+$ is added. In this case, a matching with maximal total weight has to be constructed. This version is called *online request server weighted* *matching* problem or in short wORSM problem.

In the discussions and proof of this work M_{ALG} is used in order to denote an online matching which is constructed by algorithm ALG, $|M|$ to denote the total weight (cardinality respectively) of matching M, and OPT denotes an optimal solution.

3   Analysis of the ORSM Problem

This section starts with a general lower bound for the competitive ratio of the ORSM problem for deterministic online algorithms. Thereafter, the optimal 1.5-competitive algorithm LMM is presented and analysed.

3.1   The Lower Bound

By applying the standard argument of an adversary strategy, we will show the following general lower bound:

Theorem 1. *Every deterministic online algorithm* ALG *for the ORSM problem has a competitive ratio of at least* 1.5.

Proof. The adversary strategy starts with the following input structure (see Fig. 1): $E = \{\{r_1, s_2\}, \{r_1, s_3\}, \{r_2, s_2\}, \{r_2, s_4\}\}$.

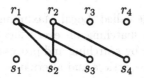

Fig. 1. Situation at time $t = 2$.

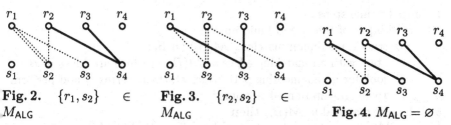

Fig. 2. $\{r_1, s_2\}$ ∈ M_{ALG}

Fig. 3. $\{r_2, s_2\}$ ∈ M_{ALG}

Fig. 4. $M_{\mathsf{ALG}} = \varnothing$

(the figures show the situations at time $t = 3$, according to different decisions of ALG and the reactions of the adversary; a double line is a matching edge; dotted lines are edges hich have been removed before $t = 3$)

An online algorithm ALG can react to this input at time $t = 2$ in three different ways:

Case 1 (Fig. 2): ALG puts edge $\{r_1, s_2\}$ to the online matching M_{ALG}. In the next step the adversary presents edge $\{r_3, s_4\}$. ALG cannot use the server vertex s_3. Therefore, $|M_{\mathsf{ALG}}| \leq 2$ whereas the optimal solution results in $|M_{\mathsf{OPT}}| = 3$.

Case 2 (Fig. 3): ALG puts edge $\{r_2, s_2\}$ to the online matching M_{ALG}. In the next step the adversary presents edge $\{r_3, s_3\}$. ALG cannot use s_4. Again $|M_{\mathsf{ALG}}| \leq 2$ and the maximum matching results in $|M_{\mathsf{OPT}}| = 3$.

Case 3 (Fig. 4): ALG decides not to match s_2. The adversary will present the input of Case 1 and $|M_{\mathsf{ALG}}| \leq 2$, $|M_{\mathsf{OPT}}| = 3$ holds. Alternatively, it is fairly obvious that ALG cannot take advantage of such a decision.

This strategy can be infinitely repeated every four time steps and this fact shows the ratio

$$\frac{|M_{\mathsf{OPT}}|}{|M_{\mathsf{ALG}}|} \geq \frac{3}{2} .$$
□

3.2 The Algorithm LMM

At a time step i the graph G representing the input of an online algorithm is known up to request vertex r_i. More precisely, the subgraph of G induced by the set $\{r_k | r_k \in R, 1 \leq k \leq i\} \cup S$ is known. Due to the irreversible decisions taken n former time steps all previous server vertices $s_1, \ldots, s_{i-1}$ and all hitherto matched request vertices cannot be rearranged anymore. It remains a vertex nduced bipartite subgraph B_i of G with the vertex set V_i:

$$V_i = \{r_k | r_k \in R, r_k \text{ not matched so far}, 1 \leq k \leq i\} \cup \{s_k | s_k \in S, i \leq k\} .$$

Our online algorithm is called 'Local Maximum Matching' (LMM) because it constructs a maximum matching on every local subgraph B_i (denoted by $\mathcal{M}(B_i)$). This can be done by searching and processing 'augmenting paths', i. e. paths with unmatched end vertices and alternately sequent matching and non-matching edges (see e. g. [Edm65] or any comprehensive text book on algorithms for an explanation of how to do so). The exact function of LMM is:

1: **loop** $\{\forall$ time steps $i\}$
2:    read input of time i and build up B_i
3:    construct a maximum matching $\mathcal{M}(B_i)$ on B_i :
        start with all matching edges of $\mathcal{M}(B_{i-1})$ which are edges in B_i ;
4:        look for an augmenting path which starts at vertex r_i and perform
        the augmentation when found[2]
5:    **if** s_i is matched in $\mathcal{M}(B_i)$ **then**
6:        add the matching edge of s_i to the online matching M_{LMM}
7:    **else if** s_i is not isolated in B_i **then** $\{$all neighbours of s_i are matched in $\mathcal{M}(B_i)\}$
8:        add an arbitrary edge $\{s_i, r\}$ of B_i to the matching M_{LMM} and delete
        the matching edge of r in $\mathcal{M}(B_i)$
9:    **end if**
10: **end loop**

Line 8 of this algorithm is essential and prefers the current server vertex s_i.

3.3 The Upper Bound

To analyse the performance of LMM three observations are needed:

1. After a request vertex r_i has been matched in B_i (in line 4 of LMM), it is in all following maximum matchings[3] up to the time step where its current matching edge is added to M_{LMM} (in line 6 or 8 of LMM).

2. If s_i is not isolated in B_i, then s_i is matched in M_{LMM} (lines 5 to 8 of LMM).

3. Let M be a matching such that no augmenting path has length less than $2\ell + 1$, then an maximum matching M_{OPT} holds $|M_{\text{OPT}}| \leq \frac{\ell+1}{\ell}|M|$. (An augmenting path of length $2\ell + 1$ has ℓ matching edges in M and $\ell + 1$ in M_{OPT}.)

Theorem 2. *The deterministic online algorithm* LMM *is 1.5-competitive.*

Proof. It will be shown that no online matching M_{LMM} can be extended by augmenting paths of length one or three. Therefore, shortest augmenting paths

---

[2] Due to the maximum cardinality of $\mathcal{M}(B_{i-1})$, every augmenting path must have r_i at one end.

[3] See the copy process of line 3 and remember that augmentations change matching edges but do not remove vertices out of the matching.

for M_{LMM} must have a length of five. This fact combined with observation 3 completes the proof.

Applying a complete distinction of cases, the non-existence of augmenting paths of length one and three in the online matching M_{LMM} is proven by contradiction.

Fig. 5. Structure of augmenting paths of length one and three.

Case 1: augmenting path of length one $\{s_i, r_a\} \notin M_{\mathsf{LMM}}$: Vertex r_a was never matched because r_a is not in M_{LMM} (reverse application of observation 1). So $\{s_i, r_a\}$ is in B_i and thus contradicts observation 2.

Case 2: augmenting path of length three and $i < j$: r_a was matched not until time j, which implies the edge $\{s_i, r_a\}$ was in B_i. This is a contradiction to observation 2.

Case 3: augmenting path of length three and $i > j$: At time j, request vertices r_a and r_b are not in M_{LMM} and so the whole path $\{\{s_i, r_a\}, \{r_a, s_j\}, \{s_j, r_b\}\}$ is in B_j. The case s_i is not in $\mathcal{M}(B_j)$ contradicts the optimality of $\mathcal{M}(B_j)$ because the path is an augmenting one (see Fig. 5). Therefore, at time j, s_i must be matched in $\mathcal{M}(B_j)$, i.e., there exists a request vertex r_c with $\{r_c, s_i\} \in \mathcal{M}(B_j)$. Later at time k ($j < k < i$), line 8 of LMM deletes the matching edge $\{r_c, s_i\}$ and adds $\{s_k, r_c\}$ to M_{LMM}.

Due to the definition of the ORSM problem, both edges $\{r_c, s_i\}$ and $\{r_c, s_k\}$ are known at time j. Now the above argument about s_i can be recursively applied to s_k and due to the finite structure of B_i, this fact contradicts the existence of an augmenting path of length three in M_{LMM}. $\qquad\square$

4   The Weighted Model

Similar to Sect. 3, a lower bound for the wORSM problem is shown first. Then the algorithm wLMM is presented as well as an outline of its analysis.

4.1   A General Lower Bound

Let $\phi := \frac{\sqrt{5}+1}{2} \approx 1.618$ be the golden ratio.

Theorem 3. *Every deterministic online algorithm ALG for the wORSM problem has a competitive ratio of at least $\phi = \frac{\sqrt{5}+1}{2}$.*

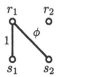

Fig. 6.   **Fig. 7.** $\{r_1, s_2\}$ ∈ **Fig. 8.** $M_{\mathsf{ALG}} = \varnothing$
Situation at time $t = 1$. M_{ALG}

(Fig. 7 and 8 show the situations at time $t = 2$ after the different decisions and the reactions of the adversary; a double line is a matching edge; dotted lines are edges which have been removed before $t = 2$)

Proof. The adversary strategy starts with input edges $\{r_1, s_1\}$ and $\{r_1, s_2\}$ and their weights are $w(\{r_1, s_1\}) = 1$ and $w(\{r_1, s_2\}) = \phi$ as you can see in Fig. 6. ALG can react to this input at time $t = 1$ in two different ways:

Case 1 (Fig. 7): ALG adds edge $\{r_1, s_1\}$ to the weighted online matching M_{ALG}. Thereafter, the adversary does not present any new edge incident to s_2. So $|M_{\mathsf{ALG}}| = 1$ and $|M_{\mathsf{OPT}}| = \phi$ holds.

Case 2 (Fig. 8): ALG does not change the online matching M_{ALG}. The adversary presents edge $\{r_2, s_2\}$ with weight $w(\{r_2, s_2\}) = \phi$. Now ALG can construct a matching with weight $|M_{\mathsf{ALG}}| \leq \phi$ only, whereas it holds $|M_{\mathsf{OPT}}| = 1 + \phi$. The ratio of these two values is also ϕ.

Every two time steps the adversary can repeat this strategy up to infinity and this fact shows the lower bound of the competitive ratio of $\phi = \frac{\sqrt{5}+1}{2}$.   ⊏

4.2   The Algorithm wLMM

The algorithm wLMM works similarly to LMM. wLMM computes a maximum weighted matching on the local bipartite graph B_i. Furthermore, the algorithm works without the special preference of vertex s_i. The way LMM does it cannot help for the weighted problem because the weights of edges can differ by very small values.[4] The problems arising from this fact will be demonstrated in the following analysis. The formal description of wLMM is:

```
1: i ⇐ 0
2: loop
3:     i ⇐ i + 1
4:     read input of time step i and build up B_i
5:     construct a maximum weighted matching M(B_i) on B_i
6:     if s_i is matched in M(B_i) then
7:         add the matching edge of s_i to the online weighted matching M_wLMM
8:     end if
9: end loop
```

---

[4] Taking advantage of this possibility, an adversary can force such kind of 'clever' online algorithm to take the same decisions as wLMM. Therefore, the weights of the input edges need to be changed by a very small ε only.

The Lower Bound of wLMM

Theorem 4. *The competitive ratio of wLMM cannot be less than 2.*

Proof. A simple input structure is applied to show the theorem. This structure (which can be repeated arbitrarily often) is the following: $w(\{r_1, s_1\}) = 1$, $w(\{r_1, s_2\}) = 1 + \varepsilon$, and $w(\{r_2, s_2\}) = 1$

The online algorithm wLMM is too greedy and matches r_1 to s_2. Then it holds $|M_{\text{wLMM}}| = 1 + \varepsilon$, whereas $|M_{\text{OPT}}| = 2$. Therefore, this lower bound of the competitive ratio exceeds any fixed real number less than 2 for a proper small value of ε. □

The Upper Bound of wLMM

Due to strict space limitations the next theorem is presented without a proof. It can be found in [Rie98].

Theorem 5. *The deterministic online algorithm wLMM is 2-competitive.*

The proof is based on a technique developed in [BR93]. It extensively uses arguments of how dynamic changes of the local bipartite graph (insertion and deletion of edges) modify the structure and property of augmenting paths. The lemmata which are describing these characteristics are not difficult but rather lengthy and a bit technical. So we had to omit them and decide to omit the whole proof because its presentation is rather senseless without the preparatory explanations. However, this proof may be of minor interest only. Theorems 3 and 5 reveal a large gap in the analysis of the wORSM problem. We have the feeling that the upper bound can be improved. Therefore, an online algorithm must prefer the current server vertex, e. g. by increasing the weights of its edges. Unfortunately, such an algorithm causes different changes in the local bipartite graph and our proof collapses completely. Hence, a different proof technique needs to be developed.

5 Concluding Remarks

Very often the following question arises: Can a randomized algorithm perform better'? However, randomized online algorithms are studied in a different model of competitive analysis (weaker adversary models) and that was not the subject of our investigations. Nevertheless, a simple extension of the lower bound construction of Theorem 1 immediately shows a $\frac{6}{5} = 1.2$ lower bound for the competitive ratio of randomized online algorithms against the oblivious adversary. The two input structures used by the adversary in Theorem 1 are chosen independently, uniformly at random to build up the input of the next four time steps. It is also not difficult to transform LMM into a randomized online algorithm. It selects one of the proper maximum cardinality matchings on B_i at random. However, we do not have an improved analysis yet. Note that a straight forward application of the RANKING algorithm [KVV90] cannot achieve an improved bound for the ORSM problem.

A different direction for further research is to study restricted model variants. It is possible to model additional concepts like parallel resources, lookahead, or individual time windows for the service of requests, by restricting the structure of the set of edges. Some of these model variants are more closely related to real world problems and a few decreased competitive ratios have been already discovered.

Acknowledgments

The author would like to thank Yossi Azar for a discussion and the comment that the ORSM problem can be regarded as a special case of the online bipartite matching problem.

References

[BEY98] Allan Borodin and Ran El-Yaniv. *Online Computation and Competitive Analysis*. Cambridge University Press, 1998.

[BR93]   Ethan Bernstein and Sridhar Rajagopalan. The roommates problem: Online matching on general graphs. Technical Report CSD-93-757, University of California, Berkeley, CS Department, 1993.

[Edm65] Jack Edmonds. Paths, trees, and flowers. *Canadian Journal on Mathematics* 7:449-467, 1965.

[KP98]   Bala Kalyanasundaram and Kirk Pruhs. On-line network optimization problems. In Amos Fiat and Gerhard J. Woeginger, editors, *Online Algorithms: The State of the Art*, LNCS 1442, pages 268-280. Springer-Verlag, Berlin-Heidelberg-New York, 1998.

[KVV90] Richard M. Karp, Umesh V. Vazirani, and Vijay V. Vazirani. An optimal algorithm for on-line bipartite matching. In *Proceedings of the 22nd Annual ACM Symposium on Theory of Computing*, pages 352-358, New York, 1990. ACM Press.

[Rie98] Marco Riedel. Online request server matching. Technical Report tr-ri-98-195, University of Paderborn, Germany, April 1998. available via: ftp://ftp.uni-paderborn.de/doc/techreports/Informatik/tr-ri-98-195.ps.Z

[Sga98] Jiří Sgall. On-line scheduling. In Amos Fiat and Gerhard J. Woeginger, editors, *Online Algorithms: The State of the Art*, LNCS 1442, pages 196-231. Springer-Verlag, Berlin-Heidelberg-New York, 1998.

Author Index

Springer
and the
environment

At Springer we firmly believe that an international science publisher has a special obligation to the environment, and our corporate policies consistently reflect this conviction.
We also expect our business partners – paper mills, printers, packaging manufacturers, etc. – to commit themselves to using materials and production processes that do not harm the environment. The paper in this book is made from low- or no-chlorine pulp and is acid free, in conformance with international standards for paper permanency.

Lecture Notes in Computer Science

For information about Vols. 1–1478
please contact your bookseller or Springer-Verlag

Vol. 1514: K. Ohta, D. Pei (Eds.), Advances in Cryptology – ASIACRYPT'98. Proceedings, 1998. XII, 436 pages. 1998.

Vol. 1515: F. Moreira de Oliveira (Ed.), Advances in Artificial Intelligence. Proceedings, 1998. X, 259 pages. 1998. (Subseries LNAI).

Vol. 1516: W. Ehrenberger (Ed.), Computer Safety, Reliability and Security. Proceedings, 1998. XVI, 392 pages. 1998.

Vol. 1517: J. Hromkovič, O. Sýkora (Eds.), Graph-Theoretic Concepts in Computer Science. Proceedings, 1998. X, 385 pages. 1998.

Vol. 1518: M. Luby, J. Rolim, M. Serna (Eds.), Randomization and Approximation Techniques in Computer Science. Proceedings, 1998. IX, 385 pages. 1998.

1519: T. Ishida (Ed.), Community Computing and Support Systems. VIII, 393 pages. 1998.

Vol. 1520: M. Maher, J.-F. Puget (Eds.), Principles and Practice of Constraint Programming - CP98. Proceedings, 1998. XI, 482 pages. 1998.

Vol. 1521: B. Rovan (Ed.), SOFSEM'98: Theory and Practice of Informatics. Proceedings, 1998. XI, 453 pages. 1998.

Vol. 1522: G. Gopalakrishnan, P. Windley (Eds.), Formal Methods in Computer-Aided Design. Proceedings, 1998. IX, 529 pages. 1998.

Vol. 1524: G.B. Orr, K.-R. Müller (Eds.), Neural Networks: Tricks of the Trade. VI, 432 pages. 1998.

Vol. 1525: D. Aucsmith (Ed.), Information Hiding. Proceedings, 1998. IX, 369 pages. 1998.

Vol. 1526: M. Broy, B. Rumpe (Eds.), Requirements Targeting Software and Systems Engineering. Proceedings, 1997. VIII, 357 pages. 1998.

Vol. 1527: P. Baumgartner, Theory Reasoning in Connection Calculi. IX, 283. 1999. (Subseries LNAI).

Vol. 1528: B. Preneel, V. Rijmen (Eds.), State of the Art in Applied Cryptography. Revised Lectures, 1997. VIII, 395 pages. 1998.

Vol. 1529: D. Farwell, L. Gerber, E. Hovy (Eds.), Machine Translation and the Information Soup. Proceedings, 1998. XIX, 532 pages. 1998. (Subseries LNAI).

Vol. 1530: V. Arvind, R. Ramanujam (Eds.), Foundations of Software Technology and Theoretical Computer Science. XII, 369 pages. 1998.

Vol. 1531: H.-Y. Lee, H. Motoda (Eds.), PRICAI'98: Topics in Artificial Intelligence. XIX, 646 pages. 1998. (Subseries LNAI).

Vol. 1096: T. Schael, Workflow Management Systems for Process Organisations. Second Edition. XII, 229 pages. 1998.

Vol. 1532: S. Arikawa, H. Motoda (Eds.), Discovery Science. Proceedings, 1998. XI, 456 pages. 1998. (Subseries LNAI).

Vol. 1533: K.-Y. Chwa, O.H. Ibarra (Eds.), Algorithms and Computation. Proceedings, 1998. XIII, 478 pages. 1998.

Vol. 1534: J.S. Sichman, R. Conte, N. Gilbert (Eds.), Multi-Agent Systems and Agent-Based Simulation. Proceedings, 1998. VIII, 237 pages. 1998. (Subseries LNAI).

Vol. 1535: S. Ossowski, Co-ordination in Artificial Agent Societies. XV; 221 pages. 1999. (Subseries LNAI).

Vol. 1536: W.-P. de Roever, H. Langmaack, A. Pnueli (Eds.), Compositionality: The Significant Difference. Proceedings, 1997. VIII, 647 pages. 1998.

Vol. 1538: J. Hsiang, A. Ohori (Eds.), Advances in Computing Science – ASIAN'98. Proceedings, 1998. X, pages. 1998.

Vol. 1539: O. Rüthing, Interacting Code Motion Transformations: Their Impact and Their Complexity. XXI, pages. 1998.

Vol. 1540: C. Beeri, P. Buneman (Eds.), Database Theory – ICDT'99. Proceedings, 1999. XI, 489 pages. 1999.

Vol. 1541: B. Kågström, J. Dongarra, E. Elmroth, J. Waśniewski (Eds.), Applied Parallel Computing. Proceedings, 1998. XIV, 586 pages. 1998.

Vol. 1542: H.I. Christensen (Ed.), Computer Vision Systems. Proceedings, 1999. XI, 554 pages. 1999.

Vol. 1543: S. Demeyer, J. Bosch (Eds.), Object-Oriented Technology ECOOP'98 Workshop Reader. 1998. X, 573 pages. 1998.

Vol. 1544: C. Zhang, D. Lukose (Eds.), Multi-Agent Systems. Proceedings, 1998. VII, 195 pages. 1998. (Subseries LNAI).

Vol. 1545: A. Birk, J. Demiris (Eds.), Learning Robots. Proceedings, 1996. IX, 188 pages. 1998. (Subseries LNAI).

Vol. 1546: B. Möller, J.V. Tucker (Eds.), Prospects for Hardware Foundations. Survey Chapters, 1998. X, pages. 1998.

Vol. 1547: S.H. Whitesides (Ed.), Graph Drawing. Proceedings 1998. XII, 468 pages. 1998.

Vol. 1548: A.M. Haeberer (Ed.), Algebraic Methodology and Software Technology. Proceedings, 1999. XI, pages. 1999.

Vol. 1550: B. Christianson, B. Crispo, W.S. Harbison, M. Roe (Eds.), Security Protocols. Proceedings, 1998. VIII, 241 pages. 1999.

Vol. 1551: G. Gupta (Ed.), Practical Aspects of Declarative Languages. Proceedings, 1999. VIII, 367 pages. 1999.

Vol. 1552: Y. Kambayashi, D.L. Lee, E.-P. Lim, M.K. Mohania, Y. Masunaga (Eds.), Advances in Database Technologies. Proceedings, 1998. XIX, 592 pages. 1999.

Vol. 1553: S.T. Andler, J. Hansson (Eds.), Active, Real-Time, and Temporal Database Systems. Proceedings 1997. VIII, 245 pages. 1998.

Vol. 1557: P. Zinterhof, M. Vajteršic, A. Uhl (Eds.), Parallel Computation. Proceedings, 1999. XV, 604 pages. 1999.

Vol. 1560: K. Imai, Y. Zheng (Eds.), Public Key Cryptography. Proceedings, 1999. IX, 327 pages. 1999.

Vol. 1563: Ch. Meinel, S. Tison (Eds.), STACS 99. Proceedings, 1999. XIV, 582 pages. 1999.

Vol. 1567: P. Antsaklis, W. Kohn, M. Lemmon, A. Nerode, S. Sastry (Eds.), Hybrid Systems V. X, 445 pages. 1999.

Vol. 1570: F. Puppe (Ed.), XPS-99: Knowledge-Based Systems. VIII, 227 pages. 1999. (Subseries LNAI).